CLINICAL EXERCISE PHYSIOLOGY

Jonathan K. Ehrman, PhD, FACSM
Henry Ford Heart and Vascular Institute

Paul M. Gordon, PhD, MPH, FACSM
West Virginia University

Paul S. Visich, PhD, MPH
Central Michigan University

Steven J. Keteyian, PhD
Henry Ford Heart and Vascular Institute

Editors

Human Kinetics

Library of Congress Cataloging-in-Publication Data

Clinical exercise physiology / Jonathan Erhman . . . [et al.], editors.
    p.   ;   cm.
  Includes bibliographical references and index.
    ISBN 0-7360-0252-9 (hard cover)
    1. Exercise therapy.
      [DNLM: 1. Exercise Therapy--Case Report.   2.
Exercise--physiology--Case Report.   WB 541 C6408 2003]   I. Erhman,
Jonathan, 1962-
    RM725.C582   2003
    615.8'2--dc21   2002155320

ISBN: 0-7360-0252-9

The Web addresses cited in this text were current as of 12/03/2002, unless otherwise noted.

Acquisitions Editor: Michael S. Bahrke, PhD; **Developmental Editor:** Judy Park; **Assistant Editor:** Lee Alexander; **Copyeditor:** Julie Anderson; **Proofreader:** Red Inc.; **Indexer:** Marie Rizzo; **Permission Manager:** Dalene Reeder; **Graphic Designer:** Robert Reuther; **Graphic Artists:** Sandra Meier and Francine Hamerski; **Photo Manager:** Kareema McLendon; **Cover Designer:** Keith Blomberg; **Photographer (cover):** Tom Roberts; **Art Manager:** Kelly Hendren; **Illustrator:** Roberto Sabas; **Printer:** Sheridan Books

Printed in the United States of America

10 9 8 7 6 5 4 3 2 1

Human Kinetics
Web site: www.HumanKinetics.com

United States: Human Kinetics
P.O. Box 5076
Champaign, IL 61825-5076
800-747-4457
e-mail: humank@hkusa.com

Canada: Human Kinetics
475 Devonshire Road Unit 100
Windsor, ON N8Y 2L5
800-465-7301 (in Canada only)
e-mail: orders@hkcanada.com

Europe: Human Kinetics
107 Bradford Road
Stanningley
Leeds LS28 6AT, United Kingdom
+44 (0) 113 255 5665
e-mail: hk@hkeurope.com

Australia: Human Kinetics
57A Price Avenue
Lower Mitcham, South Australia 5062
08 8277 1555
e-mail: liahka@senet.com.au

New Zealand: Human Kinetics
P.O. Box 105-231, Auckland Central
09-523-3462
e-mail: hkp@ihug.co.nz

CONTENTS

FOREWORD

Many Americans may be surprised at the extent and strength of the evidence linking physical activity to numerous health improvements. Most significantly, regular physical activity greatly reduces the risk of dying from coronary heart disease, the leading cause of death in the United States. Physical activity also reduces the risk of developing diabetes, hypertension, and colon cancer; enhances mental function; fosters healthy muscles, bones, and joints; and helps maintain function and preserve independence in older adults.

Although exercise professionals have known for many years that regular physical activity produces many functional and health benefits, our society is just beginning to recognize the role of regular physical activity in good health management. Endurance and resistance exercise training, when coupled with other self-help techniques such as choosing a healthy diet and managing stress, work synergistically to develop a wide range of health benefits (see *Physical Activity and Health: A Report of the Surgeon General* 1996). As health professionals' and the general public's understanding of these benefits has grown, the role of exercise testing and exercise programming in the primary and secondary prevention of disease has gained greater acceptance. Perhaps the greatest advance in recent years is that many physicians and other allied health professionals now accept and approve of the use of exercise in the prevention, diagnosis, and rehabilitation of various clinical conditions and chronic health problems.

Exercise evaluation, including assessment of muscular strength, muscular endurance, flexibility, and cardiovascular capacity, provides valuable information that enables exercise professionals to determine a client's overall functional capability, assess his or her safety for physical exertion, prescribe exercise interventions, and evaluate the impact of these interventions. Exercise evaluation also provides morbidity and mortality prognostic information from hemodynamic, electrocardiographic, cardiovascular, and pulmonary exercise responses. A key benefit of exercise testing is that it provides information that allows exercise professionals to develop safe and personalized exercise interventions for their clients. Personalized interventions result in fewer complications and more favorable outcomes than generic interventions. Some of these outcomes include a more favorable blood lipid and lipoprotein profile (reduced blood triglyceride and increased high density lipoprotein cholesterol), reductions in blood pressure (especially in persons with high blood pressure), and enhanced glucose tolerance, insulin sensitivity, and bone mineral density. Most importantly, such exercise interventions result in functional benefits such as increased ventilatory threshold and aerobic capacity.

For almost 30 years, the American College of Sports Medicine (ACSM) has published the basic principles of exercise testing and training for healthy persons and for persons with cardiopulmonary diseases (*ACSM's Guidelines for Exercise Testing and Prescription, Sixth Edition*). Until recently, there have been few definitive research publications on the impact of exercise testing and training for people with non-cardiopulmonary diseases and disabilities. Because each disease and disability has unique characteristics and within each there is a wide range of functional abilities, these conditions were previously given little attention relative to the need for exercise testing and intervention. In addition, many people who have a chronic disease or disability will at some time enter a downward spiral toward exercise intolerance. This loss of exercise capacity can result in diminished self-efficacy, greater dependence on others for daily living, depression, and a reduced capability for normal social interaction. Thus, the role of exercise intervention for improving the health, physical fitness, and rehabilitation potential for clients who are challenged by a wide range of diseases and disabilities has recently received increased attention. Given the well-established benefits of exercise for healthy individuals and those with cardiovascular disease, we can assume (and have learned) that regular exercise training has the same potential benefits for most people with other chronic diseases or disabilities.

Although the goals of a personalized exercise intervention for a person with a chronic disease or disability are similar to the goals for a healthy person, the intervention goals for these persons should emphasize reducing the sedentary lifestyle that exacerbates disease-specific consequences and reduces functional capacity. Because the health outcomes for many diseases and disabilities are difficult to anticipate, the effects of exercise programming for persons with those conditions may also be difficult to predict. Practically speaking, exercise programming for a person with a chronic condition (in conjunction with appropriate medical management) may yield improvements in functional capacity or may merely prevent further deterioration. Whatever the individual exercise goal(s), one constant objective for the chronically ill person is to optimize the overall medical management plan and help the person achieve greater independence and improved quality of life.

Recently, the ACSM created a registry of clinical physiologists (ACSM's Registry of Clinical Exercise Physiologists — RCEP) and has developed a list of the knowledge, skills, and abilities (KSAs) needed by such persons in order for them to provide safe and effective exercise testing and exercise interventions to clients with chronic diseases and disabilities. The development of this registry and its associated KSAs created a need for an evidence-based text to serve as a comprehensive resource on exercise and chronic diseases and disabilities. Consequently, I am honored to accept the invitation to introduce this book, *Clinical Exercise Physiology*. This all-inclusive textbook, written and edited by Drs. Ehrman, Gordon, Keteyian, and Visich, is divided into eight sections and contains thirty-one chapters that address the practical issues of clinical exercise physiology across a spectrum of chronic diseases and disabilities. Section topics include: *Introduction to Clinical Exercise Physiology; Endocrinology and Metabolic Disorders; Cardiovascular Diseases; Diseases of the Respiratory System; Oncology and the Immune System; Disorders of Bones and the Joints; Selected Neuromuscular Disorders;* and *Special Populations*. Two of the sections (*Introduction to Clinical Exercise Physiology* and *Special Populations*) contain 10 chapters that cover general exercise physiology-related topics. The remaining 21 chapters contain information regarding specific chronic diseases and disabilities. Each chapter follows a structured format that presents pathophysiology, clinical considerations, signs and symptoms, treatments (including adjunctive [e.g., weight loss, dietary alteration] and pharmacologic therapy), graded exercise testing, exercise prescription, special considerations, and a case study. And each chapter is written in a manner that makes often-difficult concepts easy to understand.

This book is a comprehensive text, with its content based on both science and clinical applications. The information presented will assist any qualified clinician in providing safe and effective exercise testing and programming for clients with disease and disability. Finally, this book is not only essential for preparing advanced undergraduate and graduate students to sit for the ACSM-RCEP examination, it should also serve as a model reference for all health professionals, including physicians, physician assistants, nurses, physical and occupational therapists, and rehabilitation specialists who work with chronically diseased and disabled persons.

J. Larry Durstine, PhD, FACSM, FAACVPR
Professor and Chair,
   Department of Exercise Science
Director of Clinical Exercise Programs
Department of Exercise Science
The University of South Carolina
Columbia, SC

CONTRIBUTORS

Ann L. Albright, PhD, RD
University of California, San Francisco
California Department of Health Services
Sacramento, CA

Dawn Bell, PharmD, BCPS
Director, Pharmacy Affairs
The Medicines Company
Morgantown, WV

Michael J. Berry, PhD
Department of Health and Exercise Science
Wake Forest University
Winston-Salem, NC

Steven R. Boas, MD
Children's Memorial Hospital
Chicago, IL

Kimberly A. Bonzheim, MSA, FACSM
William Beaumont Hospital
Cardiac Rehabilitation and Exercise Laboratories
Royal Oak, MI

Brian W. Carlin, MD
Assistant Professor of Medicine
Drexel University College of Medicine, Philadedelphia, PA
Allegheny General Hospital, Pittsburg, PA

James J. Collum, JD, Attorney at Law, Of Counsel
Herbert & Benson, Attorneys at Law
Canton, OH

A.S. Contractor, MD
Director, Cardiac Rehabilitation
Cumballa Hill Hospital and Heart Institute
Bombay, India

Michael J. Danduran, MS
Children's Memorial Hospital
Chicago, IL

Jonathan K. Ehrman, PhD, FASCM
Henry Ford Heart and Vascular Institute
Detroit, MI

David R. Gater, Jr., MD, PhD
Medical Director, University of Michigan Model
Spinal Cord Injury Care System
Research Associate, Ann Arbor VAMC
Assistant Professor, Physical Medicine and
Rehabilitation
University of Michigan
Ann Arbor, MI

Neil F. Gordon, MD, PhD, MPH
President and CEO, Intervent USA, Inc.
Savannah, GA

Paul M. Gordon, PhD, MPH, FACSM
School of Medicine
West Virginia University
Morgantown, WV

Gregory W. Heath, DHSc, MPH
Physical Activity and Health Branch
Division of Nutrition and Physical Activity
National Center for Chronic Disease Prevention and Health Promotion
Centers for Disease Control and Prevention
Atlanta, GA

David L. Herbert, JD, Senior Partner
Herbert & Benson, Attorneys at Law
Canton, OH

William G. Herbert, PhD, Professor
Virginia Polytechnic Institute and State University
Blacksburg, VA

Steven J. Keteyian, PhD
Henry Ford Heart and Vascular Institute
Detroit, MI

Virginia B. Kraus, MD, PhD
Division of Rheumatology, Allergy, and Clinical Immunology
Medical Director of the Duke Center for Living Arthritis Rehabilitation Program
Department of Medicine
Duke University Medical Center
Durham, NC

Charles P. Lambert, PhD
Nutrition, Metabolism, and Exercise Laboratory
Department of Geriatrics
University of Arkansas for Medical Sciences
Little Rock, AR

Richard M. Lampman, PhD
Director of Research, Department of Surgery
St. Joseph Mercy Hospital
Adjunct Associate Professor
Department of Physical Medicine and Rehabilitation
University of Michigan Medical School
Ann Arbor, MI

Charles A. Laubach, Jr., MD, FACC
Geisinger Medical Center
Penn State Geisinger Health System
Danville, PA

Nicole Y.J.M. Leenders, PhD
The Ohio State University
General Clinical Research Center
Columbus, OH

Timothy R. McConnell, PhD, FACSM, FAACVPR
Geisinger Medical Center
Penn State Geisinger Health System
Danville, PA

Peter A. McCullough, MD, MPH, FACC
Consultant Cardiologist and Chief
Divison of Nutrition and Preventative Medicine
William Beaumont Hospital
Royal Oak, MI

David L. Nichols, PhD
Assistant Research Professor
Institute for Women's Health
Texas Woman's University
Denton, TX

Patricia Painter, PhD
Adjunct Associate Professor
Department of Physiological Nursing
University of California at San Francisco
San Francisco, CA

Richard B. Parr, EdD
School of Health Sciences
Central Michigan University
Mt. Pleasant, MI

Jan Perkins, PT, MSc
School of Rehabilitation and Medical Sciences
Central Michigan University
Mt. Pleasant, MI

Farah A. Ramírez-Marrero, PhD
Associate Professor
Department of Physical Education and Recreation and HIV/AIDS
Research and Education Center
University of Puerto Rico, Río Piedras Campus
San Juan, Puerto Rico

James Raper, CFNP, DNS
School of Nursing and School of Medicine
University of Alabama at Birmingham
Birmingham, AL

Michael Saag, MD
School of Medicine
University of Alabama at Birmingham
Birmingham, AL

William Saltarelli, PhD
School of Health Sciences
Central Michigan University
Mt. Pleasant, MI

John R. Schairer, DO
Advanced Cardiovascular Health Specialists
Livonia, MI

David Seigneur, MS
Director, Cardiopulmonary Rehabilitation
Allegheny General Hospital
Pittsburgh, PA

Kimberly A. Skelding, MD
William Beaumont Hospital
Cardiac Rehabilitation and Exercise Laboratories
Royal Oak, MI

Barbara Smith, RN, PhD
School of Nursing
University of Alabama at Birmingham
Birmingham, AL

Tom Spring, MS
Henry Ford Heart and Vascular Institute
Detroit, MI

Kerry J. Stewart, EdD
Division of Cardiology
Johns Hopkins School of Medicine
Bayview Medical Center
Baltimore, MD

Paul S. Visich, PhD, MPH
School of Health Sciences
Central Michigan University
Mt. Pleasant, MI

Diane Wiggin, MS, PT
Clinical Coordinator of the Duke Center for Living
Arthritis Rehabilitation Program
Department of Physical and Occupational Therapy
Duke University Medical Center
Durham, NC

C. Mark Woodard, MS, MHA, MBA
Department of Health and Exercise Science
Wake Forest University
Winston-Salem, NC

Seth W. Wolk, MD
Vascular Surgeon
Michigan Vascular and Heart Institute
Department of Surgery
St. Joseph Mercy Hospital
Clinical Assistant Professor, Department of Surgery
University of Michigan Medical School
Ann Arbor, MI

J. Tim Zipple, MS, PT, OCS, FAAOMPT
School of Rehabilitation and Medical Sciences
Central Michigan University
Mt.Pleasant, MI

PREFACE

The concept for this book is based on the expanding field of clinical exercise physiology. Evidence of this expansion includes an increasing amount of research and publications related to clinical exercise physiology as well as certifications and registry examinations by several professional organizations such as the American College of Sports Medicine (ACSM), the American Council on Exercise (ACE), and the American Society of Exercise Physiologists (ASEP). Between the late 1960s and early 1980s, undergraduate- and graduate-level exercise science/physiology curriculums were developed in U.S. universities. A subspecialty emerged from these programs in the 1970s and has changed little since the 1980s. The disease population focus then, and even today, has been primarily on cardiac patients and those with pulmonary disease. However, the scientific literature provides numerous examples of clinical populations that can benefit from regular physical activity and exercise training. In fact, exercise can be used as both a treatment and a secondary preventive measure in many diseased populations.

The 1996 Surgeon General's Report on Physical Activity and Health provides strong evidence supporting exercise as a primary and secondary preventive tool to reduce the risk of overall mortality, colon cancer, osteoporosis, diabetes, psychological disturbances, and obesity, to name just a few. Additionally, emerging evidence suggests that exercise can protect against most types of reproductive cancer. Exercise training has also been used to treat physical or other limitations caused by diseases including, but not limited to, diabetes, obesity, multiple sclerosis, and cardiovascular disease.

Few, if any, clinical exercise physiology programs currently provide students with the breadth of information required to sit for the ACSM Clinical Exercise Physiology Registry (RCEP) examination. Additionally, ACE and ASEP have developed clinical exercise physiology examinations designed for those individuals who provide exercise-training advice to a variety of clinically diseased populations. If you plan to study and take any of these examinations, you should understand that there is no single text that provides in-depth information addressing all of the clinical populations that benefit from physical activity and exercise. This text, however, provides an up-to-date look at seven different clinical categories, each containing specific populations. These categories are endocrine and metabolic, cardiovascular, respiratory, immunological and oncological, bones and joints, neuromuscular, and specific populations. In all, 24 different specific diseases and populations are reviewed in this book.

The first seven chapters provide a foundation by reviewing general exercise physiology–related issues. Included are chapters on compliance to exercise, legal issues, pharmacology, medical history and examination, and exercise testing and prescription. These are designed to form an infrastructure on which the rest of the book is based. Often a population-specific chapter will reference one of these foundation chapters. The disease-specific chapters were designed and presented with a standardized format. This is to provide continuity to those using the text as a resource manual and a logical sequence for those using it as a textbook. Additionally, at the beginning of each chapter is a table of contents for each section within the chapter. This is designed to provide quick access to specific information contained within a chapter.

Each of the disease-specific chapters begins with an introduction to the specific disease including the definition and scope of the disease and a discussion of the pathophysiology. The second section focuses on medical and clinical considerations including signs and symptoms, diagnosis, exercise testing, and state-of-the-art treatment. The third section provides an overview of the exercise prescription designed to highlight unique disease-specific issues of the exercise prescription.

Each chapter also contains several practical application boxes that provide additional information summarizing unique chapter-specific information. Two of these practical application boxes found in each disease-specific chapter focus on the exercise prescription and on information to consider when interacting with the patient. Another includes a review of the exercise training literature discussing physiological adaptations to exercise training and how exercise can influence primary and secondary disease prevention.

Each chapter also contains actual patient cases from initial presentation and diagnosis to therapy and exercise treatment. Each case study concludes with several questions aimed at facilitating group discussion in the classroom or for the individual learner to consider when preparing for the RCEP examination.

In addition to serving as an upper level undergraduate- or graduate-level textbook for students studying clinical exercise physiology, *Clinical Exercise Physiology* is an excellent resource guide that you will want to have on your desk or office bookshelf. The features of consistent organization, case studies, discussion questions, up-to-date references, and practical application boxes are designed to provide information required for effective study for the previously mentioned clinical exercise physiology exams. In fact, the content was developed based on the KSAs (i.e., knowledge, skills, abilities) of the ACSM-RCEP examination. We hope that this text is a valuable resource both for teaching and as a reference for clinical exercise physiologists.

—Jonathan K. Ehrman, PhD, FACSM
—Paul M. Gordon, PhD, MPH, FACSM
—Paul S. Visich, PhD, MPH
—Steven J. Keteyian, PhD

ACKNOWLEDGMENTS

Thank you to my family of J's for your support. I love you all!

—Jonathan K. Ehrman

A special thanks to the distinguished contributors who worked diligently to write this text and to those who provided expert reviews. Also, to the many students and staff who provided administrative assistance. To my wonderful wife, Ina, and my children, Joshua and Natalie, thanks for putting up with me and providing much love, support, and understanding.

—Paul M. Gordon

Over time I have come to appreciate that the writing of a classroom or reference textbook is as much about me learning new information as it is about sharing what little knowledge I might have in a specific area. Having said this, I cannot begin to express how much I have enjoyed witnessing the broad application that clinical exercise physiology enjoys today – even at this early stage in its development as a subspecialty. Many thanks to my co-authors, contributing authors, and reviewers for the tireless efforts made on behalf of advancing the profession. My never-ending thanks to W. Douglas Weaver, MD, co-director, Henry Ford Heart and Vascular Institute, for his continued support of my efforts to contribute to the field. Many blessings and all my love to Lynette, Stephanie, Courtland, Jacob, and Aram for making my life . . . interesting to say the least. Finally, in loving memory of Albert Z. and Virginia Keteyian.

—Steven J. Keteyian

I have learned that completing a textbook of this quality requires a great deal of work from many individuals. I would like to first thank Human Kinetics for recognizing the importance in completing an in-depth textbook in clinical exercise physiology and their constant support to see this to fruition. Secondly, I would like to acknowledge the effort of the individual chapter authors who gave a great deal of time and effort to make sure the information was relevant to the exercise science field. Without the effort of each chapter author, this book would not be possible. Lastly, I would like to thank my family for putting up with me during the very busy times of completing this book (Diane, Matt, and Tim) and my parents that have always encouraged me to give my best effort in everything that I do.

—Paul S. Visich

We would like to thank our colleagues who reviewed original chapters of the book and provided valuable insight, including: Dr. Susan A. Bloomfield, Mr. Clinton A. Brawner, Dr. David E. Bush, Dr. Barbara N. Campaigne, Dr. Frank Cerny, Dr. Steven F. Crouse, Dr. Steven T. Devor, Dr. Philip T. Diaz, Dr. James F. Donohue, Dr. Herman J. Engels, Dr. Stephen F. Figoni, Dr. Carl Foster, Thomas Gmelich, Esq., Dr. Timothy E. Kirby, Dr. Nicole Y.J.M. Leenders, Dr. David Murdy, Dr. Mark Randall, Ms. Rosemary Snyder, Dr. Ray Squires, Dr. James Tinsdale, Dr. Michael Vredenburg, and Dr. Chrisopher J. Womack.

Jonathan K. Ehrman, PhD, FACSM

Paul M. Gordon, PhD, MPH, FACSM

Paul S. Visich, PhD, MPH

Steven J. Keteyian, PhD

Introduction

The Past, Present, and Future of Clinical Exercise Physiology

Clinical exercise physiology is a subspecialty of exercise physiology that investigates how one's ability to exercise is affected by a chronic disease or health problem and how exercise training can be used to treat such disorders. Someone who practices clinical exercise physiology (i.e., the clinical exercise physiologist; see practical application 1.1) must be knowledgeable about the broad range of exercise responses that can occur both within a disease class and across different chronic diseases. This typically relates to various organ system responses but can include behavioral, psychosocial, and spiritual issues as well. Proficiency as a clinical exercise physiologist requires a knowledge base that includes the fields of anatomy, physiology (systems and exercise), chemistry (organic and biochemistry), biology (cellular and molecular), and psychology (behavioral medicine, counseling). Typically, this requires completing an undergraduate or graduate degree in the field, performing a clinical internship, and passing a certification examination. To help students prepare for careers as clinical exercise physiologists, a

Scope of Practice

The following are several definitions of the (clinical) exercise physiologist/specialist from several organizations. It is clear that there is no consensus regarding the title of the person who works with clinical patients in an exercise/rehabilitative setting. Titles include *exercise physiologist, exercise specialist, clinical exercise specialist, certified exercise physiologist,* and *registered clinical exercise physiologist.* The American College of Sports Medicine (ACSM) provides separate definitions for both *clinical exercise physiologist* and *exercise physiologist.*

Although one may consider this strictly an issue of semantics, there may be peripheral effects such as confusion among the public with regard to which title should be sought when considering attending a clinical exercise program. Additionally, health professionals may also be confused as to the appropriate job title along with the duties of these individuals. For instance, in many institutions the job title of someone performing a technical role in a noninvasive cardiology laboratory is *cardiovascular technician* (a Department of Labor–defined occupation). However, these duties are stated in the Department of Labor and ACSM descriptions of (clinical) exercise physiologist, and often a person with a clinical exercise background is hired for these positions. It is clear that a solution must be sought. Both the ACSM and the American Society of Exercise Physiologists (ASEP) are addressing this issue separately at the time of this writing.

U.S. Department of Labor

An *exercise physiologist* is one who develops, implements, and coordinates exercise programs and administers medical tests under a physician's supervision to promote physical fitness. This person explains program and test procedures to participants, interviews participants to obtain vital statistics and medical history, and records information. The exercise physiologist records heart activity, using an electrocardiograph machine, while the participant undergoes a stress test on a treadmill, under a physician's supervision. The exercise physiologist measures oxygen consumption and lung functioning, using a spirometer, and measures the amount of body fat, using such equipment as a hydrostatic scale, skinfold calipers, and a tape measure, to assess body composition. This professional performs routine laboratory blood tests for cholesterol level and glucose tolerance or interprets test results. He or she schedules other examinations and tests, such as a physical examination, chest X-ray, and urinalysis, and records test data in the patient's chart or enters data into a computer. The exercise physiologist writes initial and follow-up exercise prescriptions for participants, following a physician's recommendations, specifying equipment, such as a treadmill, track, or bike, and demonstrates correct use of exercise equipment and exercise routines. This individual conducts individual and group aerobic, strength, and flexibility exercises and observes participants during exercise for signs of stress. The exercise physiologist teaches behavior modification classes, such as stress management, weight control, and related subjects. He or she orders material and supplies, calibrates equipment, and may supervise work activities of other staff members.

Note: This definition is from the 1991 Department of Labor's Definition of Occupational Titles. A new, searchable version of occupation titles, termed O*Net™ (http://online.onetcenter.org), does not list an occupation with the term *exercise* in its title in any form. The only location where the term *exercise physiology* is used is for the occupation of recreation and fitness studies teacher. This job description states that this person is responsible for teaching exercise physiology courses.

ACSM

The *clinical exercise physiologist* applies exercise and physical activity for those clinical and pathological situations where exercise has been shown to provide therapeutic or functional benefit. Patients for whom services are appropriate may include, but are not limited to, those with cardiovascular, pulmonary, metabolic, immunological, inflammatory, orthopedic, and neuromuscular diseases and conditions. This list will be modified as indications and procedures of application are further developed and mature. Furthermore, the clinical exercise physiologist applies exercise principles to groups such as geriatric, pediatric, or obstetric populations and to society as a whole in preventive activities. The clinical exercise physiologist performs exercise evaluation, exercise prescription,

continued

PRACTICAL APPLICATION 1.1 (continued)

exercise supervision, exercise education, and exercise outcome evaluation. The practice of the clinical exercise physiologist should be restricted to clients who are referred by and are under the continued care of a licensed physician.

Exercise physiologists are scientists who conduct controlled investigations of responses and adaptations to muscular activity using human subjects or animals within a clinical setting, a research institute, or an academic institution. Very often such a person teaches academic courses in exercise physiology, environmental physiology, or applied human physiology for students of medicine, physiology, physical education, or other health-related fields.

ASEP

Certified exercise physiologists are committed to health and fitness promotion programs, private homes and community agencies, community integration with corporate wellness and training centers, cardiac and pulmonary rehabilitation, universities, industrial settings, retail businesses, professional lifestyle managers, and research activities. Exercise physiologists work with subjects, patients, and clients in various roles including, but not limited to, education, consultation, research, administration, and management.

American Council on Exercise (ACE)

ACE does not provide a scope of practice for the *clinical exercise physiologist/specialist*. However, at the following Internet address the ACE states that the clinical exercise specialist, as certified through ACE, is an advanced personal trainer with skills to work with special populations in a variety of settings: www.acefitness.org/getcertified/certifications.cfm.

text is needed that comprehensively presents relevant exercise physiology information across the spectrum of chronic diseases. This text is the first of its kind to accomplish this.

The Past

Exercise has been used to assess and treat chronic disease for more than 30 years. However, exercise physiology can trace its roots back to the late 1800s (e.g., Fernand LaGrange's textbook *Physiology of Bodily Exercise* published in 1889) and early 1900s (e.g., the Harvard Fatigue Laboratory). At that time, much of the focus was on the physiological response to exercise in healthy and athletic populations. In the late 1930s, Sid Robinson and colleagues studied the effects of the aging process on exercise performance. In addition to the United States, other countries also contributed to the knowledge base of exercise physiology. These predominantly included the Scandinavian and other European countries in the 1950s and 1960s.

The development of modern clinical exercise physiology dates back to the 1960s, around the time that the term *aerobics* was popularized by Dr. Kenneth Cooper and when regular exercise was beginning to be considered important to maintain optimal health (5). About this time it was determined

that bed rest was extremely detrimental to physical functioning in healthy individuals (6). In addition, pioneers such as Herman Hellerstein demonstrated that bed rest was detrimental to people with heart disease as well. Previously, patients were placed on 4 to 12 weeks of complete bed rest following a myocardial infarction. This gave way to the development of inpatient and, subsequently, outpatient cardiac rehabilitation programs throughout the 1970s and 1980s and into the 1990s. These programs focused on patient ambulation soon after surgery or a myocardial infarction and on lifestyle and behavioral modification to reduce the risk of future events. An excellent, in-depth historical perspective of exercise physiology is provided by Wilmore and Costill in the preface of their text *Physiology of Sport and Exercise* (8).

The Present

The Surgeon General's report on physical activity and health, published in 1996, was a landmark report for the field of clinical exercise physiology (7). It identified numerous chronic diseases and conditions for which there is valid research evidence that a lack of exercise or physical activity places individuals at an increased risk (see figure 1.1). In addition, there are other diseases and conditions for which exercise limits disability and improves outcome.

Cancer (colon, breast, reproductive)
Cardiovascular disease
Falling
Health-related quality of life
Non-insulin-dependent diabetes
Mental health
Obesity
Osteoarthritis
Osteoporosis
Overall mortality

FIGURE 1.1   Diseases and conditions related to a lack of exercise.

Adapted from the Surgeon General's Report on Physical Activity and Health (1996).

These include multiple sclerosis, spinal cord injury, chronic obstructive pulmonary disease, and many others. For these, exercise training should be part of a comprehensive treatment plan. In addition, several population groups are at an increased risk for chronic disease development and disability as a result of physical inactivity. These include women, children, and people from minority races.

Because of the Surgeon General's report, published practice guidelines, position statements, and ongoing research conducted in the past 3 decades, cardiac rehabilitation programs have evolved and now are often sophisticated exercise training and behavioral management programs administered by a multidisciplinary team of clinical exercise physiologists, nurses, physicians, dietitians, physical therapists, psychologists, and others. In addition, many programs have expanded their focus to include many of the other chronic diseases presented in the Surgeon General's report.

The Future

To date, many of the individuals trained in the exercise sciences who have sought employment in the clinical field have been restricted to working in cardiac rehabilitation programs or academia. However, with the emerging importance of physical activity and health and an increase in both professional and public awareness of the skills of a competent clinical exercise physiologist, there are increasing employment opportunities for those with an academic background in exercise physiology. These include, but are not limited to, personal training, corporate fitness programs, medically affiliated fitness centers, professional and amateur sports consulting, weight

management programs, and noninvasive cardiology testing laboratories.

The profession of clinical exercise physiology has developed a unique body of knowledge that includes exercise prescription development and implementation of both primary and secondary prevention services. This information has been published through the years in a growing number of biomedical journals (figure 1.2). This body of knowledge has led to the emergence of evidence-based recommendations for both the general public and the chronically diseased populations with respect to physical activity and exercise. In addition, the clinical exercise physiologist is uniquely trained to identify individual lifestyle-related issues that promote poor health and to design and implement a treatment plan to modify these lifestyle factors. Thus, the profession of clinical exercise physiology and the clinical exercise physiologist fill a void in healthcare that is becoming increasingly important, especially in the United States, as the average age of the population rapidly increases, thus increasing the number of people living with chronic diseases.

ACSM's Health and Fitness Journal
American Journal of Cardiology
American Journal of Clinical Nutrition
American Journal of Physiology
American Journal of Sports Medicine
Annals of Internal Medicine
Canadian Journal of Applied Physiology
Circulation
Clinical Exercise Physiology
European Journal of Applied Physiology
European Journal of Sport Science
International Journal of Obesity
Journal of Aging and Physical Activity
Journal of Applied Physiology
Journal of Cardiopulmonary Rehabilitation
JAMA—Journal of the American Medical Association
Medicine and Science in Sports and Exercise
New England Journal of Medicine
Pediatric Exercise Science
Sports Medicine
The Physician and Sportsmedicine
Research Quarterly for Exercise and Sport

FIGURE 1.2   Selected biomedical journals contributing to the clinical exercise physiology body of knowledge.

Professional Organizations and Certifications

There are literally hundreds of exercise- and fitness-related certifications available; however, only a few exist for the purpose of certifying exercise professionals to work with people with a chronic disease. The primary professional organizations providing clinically oriented exercise physiology certifications are the American Council on Exercise (ACE), the American Society of Exercise Physiologists (ASEP), and the American College of Sports Medicine (ACSM). Table 1.1 lists the aspects of each of these organization's certifications. These are also reviewed in the next several paragraphs.

American Council on Exercise

The ACE has a long history of certifying exercise and fitness professionals, predominantly in the area of personal training (2). Nevertheless, with the foreseeable growth in the aging population and the increased prevalence of chronic diseases, ACE recently developed an advanced personal training certification to work with special populations. The ACE Clinical Exercise Specialist certification is designed to test individual competencies to work with the chronically diseased populations by using a written examination covering the following topics: screening and assessment (35%), program design (20%), and program implementation and management (45%). To sit for the examination, an individual must be at least 18 years of age, have proof of current cardiopulmonary resuscitation (CPR) certification, and have a bachelor's degree in an exercise-related field or a current ACE certification at another level. Although highly recommended, academic training in exercise physiology is not specifically required.

American Society of Exercise Physiologists

ASEP was organized to provide professional visibility for the increasing numbers of academically trained exercise physiologists (4). With that, the ASEP Board of Directors sought to develop a certification that would assess the candidate's academic and technical competence in the area of exercise physiology, with the purpose of helping college-prepared exercise physiologists access key positions in the health, fitness, rehabilitation, and research fields of professional work. Additional goals include a higher level of patient/client care,

improved consumer protection, higher standards of ethical conduct, and improved employment opportunities.

The ASEP Exercise Physiology Certification (EPC) has both a written and applied/practical examination. Requirements to sit for the examination include an academic degree in exercise physiology or a concentration in exercise science and proof of academic competency (C or 2.0 grade point average or better) in specific topic areas. In addition, candidates must be current members of ASEP and must have completed 400 hr of documented, hands-on laboratory or internship experience. To maintain the EPC candidates must accrue continuing education credits over several years.

American College of Sports Medicine

ACSM has been internationally recognized as a leader in exercise and sports medicine for several decades (3). In addition to clinical exercise physiologists, ACSM members include professionals with expertise in a variety of sports medicine fields. These professionals include physicians, physical therapists, athletic trainers, dietitians, nurses, and other allied health professionals.

For many years, ACSM has been recognized throughout the industry for its certification program. Although ACSM provides certification for the exercise professional interested in preventive exercise programming (Health Fitness Instructor), the ACSM rehabilitative track is most appropriate for the clinical exercise physiologist. For example, the ACSM Exercise Specialist® certification is designed for the professional who plans to administer and supervise an exercise program for patients with cardiovascular or pulmonary disease or the patient with metabolic syndrome. The general requirements for taking the Exercise Specialist certification examination include a minimum of a bachelor's degree in a relevant health field or the equivalent (i.e., 2-year degree plus 2 years of experience in a cardiac rehabilitation program), 600 hr of internship in a cardiopulmonary rehabilitation setting, and current basic cardiac life support certification. The test format includes both a written and practical component designed to test the candidate's competence in graded exercise testing, exercise prescription, exercise leadership, and more. Continuing education credits (80 every 4 years) must also be completed to maintain active certification.

In response to the growing body of evidence outlining the benefits of exercise for both disease

TABLE 1.1
Comparison of Certifications for Clinical Exercise Physiologists

Clinical exercise specialist (ACE)[a]	Exercise physiologist certified (ASEP)[a]	Exercise specialist (ACSM)[a]	RCEP$_{sm}$ (ACSM)[a]
REQUIREMENTS			
Degree not required but recommended. Practical experience not required but recommended.	Academic degree with major or emphasis in exercise science, exercise physiology, or physiology (type of degree not specified) with C grade or better in five of the nine content areas. Current ASEP membership. >400 hr laboratory or internship experience.	Bachelor of science degree in any health field. >600 hr of practical, clinical experience.	Masters of science degree in exercise science, exercise physiology, or physiology. >1200 hr of clinical experience, preferably in all domains.
WRITTEN			
150 questions (25 are nonscored and are used for future exam preparation). Approximate weighting: screening and assessment (35%), program design (20%), and program implementation and management (45%). Content areas: strengthening, physical conditioning, proprioceptive training, improving range of motion, neuromuscular reeducation, balancing, extremity strength, increasing muscular endurance, and increasing cardiovascular function.	200 questions. Approximate weightings and content areas: exercise physiology (cardio-respiratory/training; 36%), cardiac rehabilitation (including ECG/health fitness; 18.5%), exercise metabolism and regulation (11.5%), kinesiology (including neuromuscular; 10.5%), research (6.5%), sport biomechanics (6%), environmental exercise physiology (6%), and sport nutrition (5%).	130 questions, including two case studies (30 are nonscored and are used for future exam preparation). Approximate weightings and content areas: functional anatomy (5%), exercise physiology (18%), human development (2%), human behavior/psychology (4%), pathophysiology/risk factors (13%), health appraisal/fitness assessment (22%), ECG (17%), emergency procedures (7%), and exercise programming (11%).	230 questions, including six case studies. Approximate weightings and content areas: cardiovascular (30%), pulmonary (10%), metabolic (20%), orthopedic/musculoskeletal (20%), neuromuscular (10%), and immunological (10%).
PRACTICAL			
No practical testing.	10 content areas.	Four stations/content areas.	No practical testing.
RECERTIFICATION			
Not required.	Every 5 years. Requires 25 continuing education credits.	Every 4 years. Requires 80 continuing education credits.	Every 4 years. Requires 80 continuing education credits.

Note. ACE = American Council on Exercise; ASEP = American Society of Exercise Physiologists; RCEP = Registered Clinical Exercise Physiologist; ACSM = American College of Sports Medicine; ECG = electrocardiogram.

[a]Information gathered from the ACE (2), ASEP (4), and ACSM (3) Web sites and promotional materials.

prevention and rehabilitation across a multitude of diseases and conditions (e.g., cardiovascular, pulmonary, metabolic, musculoskeletal/orthopedic, neuromuscular, immunological, inflammatory, geriatric, obstetric, pediatric), in 1998 the ACSM developed a registry of clinical exercise physiologists. This registry is designed to credential the clinical exercise physiologist who cares for patients with these conditions. To become an ACSM Registered Clinical Exercise Physiologist (RCEP), an individual must have a minimum of a master's degree in exercise physiology, exercise science, or physiology along with 1200 hr of documented practical experience. It is recommended that practical hours be divided across various diseases and conditions to better prepare candidates for written competencies. Goals of the RCEP are to assure the consumer and employer that successful candidates are adequately prepared to work with chronically diseased patients and special populations and to improve visibility and acceptance of the clinical exercise physiologist among the public and other health professionals.

Employment Opportunities and Professional Viability

The biomedical literature continues to expand on the knowledge that physical inactivity has an inverse relationship with the development of many different chronic diseases and can play a positive role in inhibiting the progression of many diseases. Given this and the increased number of people older than 65 years in the population, the clinical exercise physiologist has increasing employment opportunities.

To date, university curriculums that prepare students for careers as clinical exercise physiologists have not been accredited by a governing body, and there is often a great deal of variance in coursework between schools. Therefore, students interested in becoming clinical exercise physiologists should seek the guidance of professionals in the field to select a program that will provide a comprehensive curriculum that is accepted by those who hire graduates. Despite the lack of a standardized accreditation program to date, both the ACSM and ASEP have begun to develop programs to accredit clinical exercise physiology/science academic programs.

Currently, most clinical exercise physiologists are employed in cardiopulmonary rehabilitation programs; therefore, it would be wise for those interested in this area to become familiar with the American Association of Cardiovascular and Pulmo-

nary Rehabilitation (AACVPR). The AACVPR is an organization of professionals who work in cardiopulmonary rehabilitation programs and is also the accrediting body for these programs. The AACVPR publishes a directory of programs in the United States and in other countries, which is an ideal resource for identifying potential sites of employment. In addition to the program directory, the AACVPR offers a Career Hotline to members and nonmembers that lists open positions and offers a location to post one's resume (4). The ACSM journal *Medicine and Science in Sports and Exercise* lists exercise physiology–related employment and educational positions each month. In addition, these positions can be accessed through their Web site (3). Finally, the Web site www.exercisejobs.com lists positions nationwide and allows the opportunity to post a resume online.

As mentioned, there are other opportunities for the clinical exercise physiologist to consider, as evidenced by the creation of the RCEP. For example, positions for those with neuromuscular rehabilitative skills can be found in some rehabilitation and medical facilities that care for patients who have had a stroke or have multiple sclerosis, muscular dystrophy, or a closed-head injury. In this setting, the clinical exercise physiologist should have a strong understanding of these disease process and natural progression and should know how to appropriately prescribe safe and effective exercise. Typically, these individuals, and those who work with other patient groups, work under the guidance of a physician.

Another area experiencing growth is the use of exercise in primary prevention. Many hospitals and corporations recognize the benefits of exercise training as a means to reduce future medical expenses and increase productivity. Often these programs are implemented in a fitness club setting. In fact, most medically based fitness facilities now exclusively hire individuals with exercise-related degrees in an effort to promote safety and provide staff who have experience with an increasingly diverse clientele, both healthy and diseased individuals.

For those seeking employment or attempting to maintain a professional certification, it would be beneficial to join one or more professional organizations. In addition to joining those organizations previously mentioned in this section, many clinical exercise physiologists join regional chapters of the ACSM or state chapters of the AACVPR or ASEP. These often have local educational programs that allow professionals to gain continuing education credits and an opportunity to network with those who hire clinical exercise physiologists.

Other professional organizations publish or present exercise-related research or information that focuses on specific disease groups. Some of these include the American Heart Association, the American Diabetes Association, the American Academy of Orthopedic Surgeons, the American Cancer Society, the Multiple Sclerosis Society, and the Arthritis Foundation. Table 1.2 provides the Web sites for these and other organizations, most of which serve as excellent information resources. Finally, clinical exercise physiologists should also stay current by reviewing current research in general exercise physiology as well as research specific to their disease area of interest. Along with the growth in professional opportunity in clinical exercise physiology, there has been a growth in the body of scientific information. Regular journal reading is an excellent way to keep current (see figure 1.2).

Conclusion

Although the profession of clinical exercise physiology is in its infancy compared with many other allied health professions, there is a wealth of information in its body of knowledge, as evidenced by this text. Both aspiring and practicing clinical exercise physiologists require up-to-date information to either prepare for or maintain certification and to continue to provide up-to-date and evidence-based

---

TABLE 1.2

Selected Internet Sites of Chronic Disease Organizations and Institutes

Category	Organizations/institutes	Internet sites
Endocrinology and metabolic disorders	American Diabetes Association National Institute of Diabetes and Digestive and Kidney Diseases National Kidney Foundation North American Association for the Study of Obesity	www.diabetes.org www.niddk.nih.gov www.kidney.org www.naaso.org
Cardiovascular diseases	American College of Cardiology American Heart Association Heart Failure Society of America National Heart, Lung, and Blood Institute Vascular Disease Foundation	www.acc.org www.americanheart.org www.hfsa.org www.nhlbi.nih.gov www.vdf.org
Respiratory diseases	American Lung Association National Heart, Lung, and Blood Institute	www.lungusa.org www.nhlbi.nih.gov
Oncology and immune diseases	American Cancer Society National Cancer Institute	www.cancer.org www.nci.nih.gov
Bone and joint diseases and disorders	American Academy of Orthopaedic Surgeons American College of Rheumatology/Association of Rheumatology Health Professionals Arthritis Foundation International Osteoporosis Foundation National Institute of Arthritis and Musculoskeletal and Skin Diseases Spondylitis Association of America	www.aaos.org www.rheumatology.org www.arthritis.org www.osteofound.org www.niams.nih.gov www.spondylitis.org
Neuromuscular disorders	National Institute of Neurological Disorders and Stroke National Multiple Sclerosis Society National Spinal Cord Injury Association	www.ninds.nih.gov www.nmss.org www.spinalcord.org
Special populations	American Geriatrics Society National Institute on Aging National Institute of Child Health and Human Development The National Women's Health Information Center	www.americangeriatrics.org www.nia.nih.gov www.nichd.nih.gov www.4woman.gov

patient care. This text provides a comprehensive and practical review for the most common chronic diseases and serves as a guide from the first client–clinician interaction through the development of a complete exercise prescription for the client.

References

1. American Association of Cardiovascular and Pulmonary Rehabilitation: www.aacvpr.org

2. American Council on Exercise: www.acefitness.org

3. American College of Sports Medicine: www.acsm.org

4. American Society of Exercise Physiologists: www.asep.org

5. Cooper, KH. *Aerobics.* 1968, New York: Evans.

6. Saltin B, et al. Response to submaximal and maximal exercise after bedrest and training. *Circulation* 38 (Suppl. 7): VI 1-78, 1968.

7. U.S. Department of Health and Human Services. *Physical Activity and Health: A Report of the Surgeon General.* 1996, Atlanta, GA: U.S. Department of Health and Human Services, Centers for Disease Control and Prevention, National Center for Chronic Disease Prevention and Health Promotion.

8. Wilmore JH and Costill DL. *Physiology of Sport and Exercise* (2nd edition). 1999, Champaign, IL: Human Kinetics.

Gregory W. Heath, DHSc, MPH
Physical Activity and Health Branch
Division of Nutrition and Physical Activity
National Center for Chronic Disease Prevention and Health Promotion
Centers for Disease Control and Prevention
Atlanta, GA

Behavioral Approaches to Physical Activity Promotion

The clinical exercise physiologist can use a number of behavioral strategies in assessing and counseling individual patients/clients about their physical activity behavior change. The behavioral strategies discussed are intended to be used in the context of supportive social and physical environments. A nonsupportive environment is considered a barrier to regular physical activity participation, and consequently, if this barrier is not altered, change is unlikely to occur. Thus, one of the clinician's goals is to identify environmental barriers with the client, include steps on how to overcome these barriers, and build supportive social and physical environments as part of the counseling strategy. Although no guarantees can be made, the literature suggests that if clinicians take a behavior-based approach to physical activity counseling, within the context of a supportive environment, they may indeed experience greater success in getting their clients moving. Therefore, information is also presented about the role of social and contextual settings in promoting health- and fitness-related levels of physical activity.

The most important task of the clinical exercise physiologist is to guide a client into a lifelong pattern of regular, safe, and effective physical activity behavior. The exercise physiologist, who understands the physiological basis for activity as well as the impact of pathology on human performance, is well positioned for such counseling. However, to be an effective counselor, the clinical exercise physiologist also needs to understand human behavior in the context of the individual client's social and physical milieu. This chapter seeks to underscore some of the important theories and models of behavior that have been shown to be important adjuncts for clinicians seeking to help people make positive changes in their physical activity behavior.

Persons who engage in regular physical activity have an increased physical working capacity (1), decreased body fat (2), increased lean body tissue (2), increased bone density (3), and lower rates of coronary heart disease (CHD) (4), diabetes mellitus, hypertension (5), and cancer (6). Increased physical activity is also associated with greater longevity (7). Regular physical activity and exercise can also assist persons in improving mood and motivational climate (8), enhancing their quality of life, improving their capacity for work and recreation, and altering their rate of decline in functional status (9).

When one is promoting planned exercise and physical activity, attention must be paid to specifically designed outcomes. Notably, the health and fitness outcomes of a well-designed exercise prescription need to be accounted for. Finally, a number of physiological, anatomical, and behavioral characteristics should also be considered to ensure a safe, effective, and enjoyable exercise experience for the participant.

Health Benefits of Physical Activity

Physical activity has been defined as any bodily movement produced by skeletal muscles that results in caloric expenditure (10). Because caloric expenditure uses energy and energy utilization enhances weight loss or weight maintenance, caloric expenditure is important in the prevention and management of obesity, CHD, and diabetes mellitus. The *Healthy People 2010* Physical Activity and Fitness Objective 22.2 (11) highlights the need for every person to engage in regular, moderate physical activity for at least 30 min per day, preferably daily. Current research suggests that engaging regularly in moderate physical activity for at least 30 min per day will help ensure that calories are expended with the conferring of specific health benefits. For example, daily physical activity equivalent to a sustained walk for 30 min per day would result in an energy expenditure of about 1050 kcal per week. Epidemiologic studies suggest that a weekly expenditure of 1000 kcal could have significant individual and public health benefits for CHD prevention, especially among those who are initially inactive.

More recently, the American College of Sports Medicine and the Centers for Disease Control and Prevention concluded that the scientific evidence clearly demonstrates that regular, moderate-intensity physical activity provides substantial health benefits (12). Therefore, after an extensive review of the physiological, epidemiological, and clinical evidence, an expert panel formulated the following recommendation:

> Every U.S. adult should accumulate 30 min or more of moderate-intensity physical activity on most, preferably all, days of the week.

This recommendation emphasizes the benefits of moderate physical activity that can be accumulated in bouts of 8 to 10 min of exercise. Intermittent activity has been shown to confer substantial benefits. Therefore, the recommended 30 min of activity can be accumulated in shorter bouts of 8 to 10 min spaced throughout the day. Although the accumulation of 30 min of moderate-intensity physical activity has been shown to confer important health benefits, this recommendation is not intended to represent the optimal amount of physical activity for health but instead a minimum standard or base on which to build to obtain more specific physical activity and exercise-related outcomes.

Specifically, selected fitness-related outcomes may be a desired result for the physical activity participant, who may seek the additional benefits of improved cardiorespiratory fitness, muscle endurance, muscle strength, flexibility, and body composition.

Fitness Benefits of Physical Activity

Regular vigorous physical activity helps achieve and maintain higher levels of cardiorespiratory fitness than moderate physical activity. There are five components of health-related fitness: cardiorespiratory fitness, muscle strength, muscle endurance, flexibility, and enhanced body composition (table 2.1). **Cardiorespiratory fitness** or aerobic capacity describes the body's ability to perform high-intensity activity for a prolonged period of time without undue physical stress or fatigue. Having higher levels

---

TABLE 2.1

Examples of Health and Fitness Benefits of Physical Activity

Health benefits	Fitness benefits
Reduction in premature mortality	Cardiorespiratory fitness—maximal oxygen uptake
Reduction in cardiovascular disease risk	Muscle strength and endurance
Reduction in colon cancer	Enhanced body composition
Reduction in non-insulin-dependent diabetes mellitus	Flexibility
Improved mental health	

---

of cardiorespiratory fitness enables people to carry out their daily occupational tasks and leisure pursuits more easily and with greater efficiency. Vigorous physical activities such as the following help to achieve and maintain cardiorespiratory fitness and can also contribute substantially to caloric expenditure:

— Very brisk walking
— Jogging/running
— Lap swimming
— Cycling
— Fast dancing
— Skating
— Rope jumping
— Soccer
— Basketball
— Volleyball

These activities may also provide additional protection against CHD over moderate forms of physical activity. Higher levels of cardiorespiratory fitness can be achieved by increasing the frequency, duration, or intensity of an activity beyond the minimum recommendation of 20 min per occasion, three occasions per week, at more than 50% of aerobic capacity (13).

Muscular strength and endurance are the ability of skeletal muscles to perform hard and/or prolonged work (14). Regular use of skeletal muscles helps to improve and maintain strength and endurance, which greatly affects the ability to perform the tasks of daily living without undue physical stress and fatigue. Examples of tasks of daily living include home maintenance and household activities such as sweeping, gardening, and raking. Engaging in

regular physical activity such as weight training or the regular lifting and carrying of heavy objects appears to maintain essential muscle strength and endurance for the efficient and effective completion of most activities of daily living throughout the life cycle (15).

Musculoskeletal flexibility describes the range of motion in a joint or sequence of joints. Joint movement throughout the full range of motion helps to improve and maintain flexibility. Those with greater total body flexibility may have a lower risk of back injury (16). Older adults with better joint flexibility may be able to drive an automobile more safely (17). Engaging regularly in stretching exercises and a variety of physical activities that require one to stoop, bend, crouch, and reach may help to maintain a level of flexibility that is compatible with quality activities of daily living.

The maintenance of an acceptable ratio of fat to lean body weight is another desired component of health-related fitness. **Excess body weight** occurs when too few calories are expended and too many consumed for individual metabolic requirements (18). The results of weight loss programs focused on dietary restrictions alone have not been encouraging. Physical activity burns calories, increases the proportion of lean to fat body mass, and raises the metabolic rate (19). Therefore, a combination of both caloric control and increased physical activity is important for attaining a healthy body weight.

Participation in Regular Physical Activity

In designing any exercise prescription, the professional needs to consider various physiological, behavioral, and psychological variables that are related to participation in physical activity (20). Two commonly identified determinants of physical activity participation include **self-efficacy** and **social support.** Self-efficacy, a construct from social cognitive theory that is most characterized by the person's confidence to exercise under a number of circumstances, appears to be positively associated with greater participation in physical activity. Social support from family and friends has consistently been shown to be associated with greater levels of physical activity participation. Incorporating some mechanism of social support within the exercise prescription appears to be an important strategy for enhancing compliance with a physical activity plan (21).

Common barriers to participation in physical activity are time constraints and injury. These barriers can be taken into account by encouraging participants to include physical activity as part of their lifestyle, thus not only engaging in planned exercise but also incorporating transportation, occupational, and household physical activity into their daily routine. Participants can also be counseled to help prevent injury. A program of low- to moderate-intensity physical activities is more likely to be adhered to than high-intensity activities during the early phases of an exercise program. Moreover, moderate activity is less likely to cause injury or undue discomfort (22). A number of physical and social **environmental factors** can also affect physical activity behavior (23). Family and friends can be role models, can provide encouragement, or can be companions during physical activity. The physical environment often presents important barriers to participation in physical activity, including a lack of bicycle trails and walking paths away from vehicular traffic, inclement weather, and unsafe neighborhoods (24). Sedentary behaviors such as excessive television viewing or computer use may also deter persons from being physically active.

Risk Assessment

Exercise prescription may be fulfilled in at least three different ways: (a) a program-based level that consists primarily of supervised exercise training (25); (b) exercise counseling and exercise prescription followed by a self-monitored exercise program (26); and (c) community-based exercise programming that is self-directed and self-monitored (27).

Within supervised exercise programs and programs offering exercise counseling and prescription, participants should complete a brief medical history and risk factor questionnaire and a preprogram

evaluation (28). More information on the medical history and risk factor questionnaire and preprogram evaluation is discussed in chapter 6.

When one is developing a community-based, self-directed program, medical clearance is left to the judgment of the individual participant. An active physical activity promotion campaign in the community seeks to educate the population regarding precautions and recommendations for moderate and vigorous physical activity (29). These messages should provide information that participants must know before beginning a regular program of moderate to vigorous physical activity. This information should include the following:

1. Awareness of preexisting medical problems (i.e., CHD, arthritis, osteoporosis, or diabetes mellitus)
2. Consultation before starting a program, with a physician or other appropriate health professional, if any of the previously mentioned problems are suspected
3. Appropriate mode of activity and tips on different types of activities
4. Principles of training intensity and general guidelines as to rating of perceived exertion and training heart rate
5. Progression of activity and principles of starting slowly and gradually increasing activity time and intensity
6. Principles of monitoring symptoms of excessive fatigue
7. Making exercise fun and enjoyable

Theories and Models of Physical Activity Promotion

Historically, the most common approach to exercise prescription taken by health professionals has been direct information. In the past, the counseling sequence often consisted of the following:

1. Exercise assessment, usually cardiorespiratory fitness measures
2. Formulation of the exercise prescription
3. Counseling the patient regarding
   — mode (usually large-muscle activity),
   — frequency (three to five sessions per week),
   — duration (20-30 min per session), and
   — intensity (assigned target heart rate based on the exercise assessment) (30).

4. Review of the exercise prescription by the health professional and participant
5. Follow-up
   — Visits (reassessments and revising of the exercise prescription)
   — Phone contact

Most of the research evaluating this traditional approach to exercise prescription has not been too favorable in terms of its results in long-term compliance and benefits (31). That is, most people who begin an exercise program drop out during the first 6 months. Why has the traditional information-sharing approach been used? Because it's easiest for the clinician, requires less time, and is prescriptive in nature. However, it is not interactive with the client.

More recently, contemporary theories and models of human behavior have been examined and developed for use in exercise counseling and interventions (32). These theories, referred to as cognitive–behavioral techniques, represent the most salient theories and models that have been used to promote the initiation of and adherence to physical activity. These approaches vary in their applicability to physical activity promotion. Some models and theories were designed primarily as guides to understanding behavior, not as guides for designing intervention protocols. Others were specifically constructed with a view toward developing cognitive–behavioral techniques for physical activity behavior initiation and maintenance. Consequently, the clinical exercise physiologist may find that the majority of the theories reviewed will assist in understanding physical activity behavior change. Nevertheless, other reviewed theories have evolved sufficiently to provide specific intervention techniques to assist in behavior change.

Health Belief Model

The **health belief model,** one of the oldest theories in health behavior research, proposes that only psychological variables influence health behaviors. This model stipulates that a person's health-related behavior depends on the person's perception of four critical areas: the severity of a potential illness, the person's susceptibility to that illness, the benefits of taking a preventive action, and the barriers to taking that action (33). The model also incorporates cues to action, such as placing a reminder to exercise in the person's appointment book. These cues are used as important elements in eliciting or maintaining patterns of behavior. The construct of self-efficacy, or a person's confidence in his or her ability to successfully perform an action (e.g., exercise), has been added to the model, perhaps allowing it to better account for habitual behaviors such as physical activity. By understanding the individual's perception of these critical areas, the clinician can coach the participant to incorporate cues to action that develop his or her self-efficacy.

Relapse Prevention Model

Another behavioral model that has been successfully incorporated into some exercise prescription protocols is the **relapse prevention model** (34). The concepts of relapse prevention have been used to help new exercisers anticipate problems with adherence. Factors that contribute to relapse include negative emotional or physiological states, limited coping skills, social pressure, interpersonal conflict, limited social support, low motivation, high-risk situations, and stress. Principles of relapse prevention include identifying high-risk situations for relapse (e.g., change in season) and developing appropriate solutions (e.g., finding a place to walk inside during the winter). Helping people distinguish between a lapse and relapse is thought to improve adherence (35).

Theory of Reasoned Action

The **theory of reasoned action** (36) states that individual performance of a given behavior is primarily determined by a person's intention to perform that behavior. This intention is determined by two major factors: the person's attitude toward the behavior, and the influence of the person's social environment or subjective norm (i.e., beliefs about what other people think the person should do, as well as the person's motivation to comply with the opinions of others). This theory has some application to the transtheoretical model (stages of change model) and is linked to precontemplation and contemplation of the behavior.

Theory of Planned Behavior

The **theory of planned behavior** (37) adds to the theory of reasoned action with the concept of perceived control over the opportunities, resources, and skills necessary to perform a behavior. As individuals develop "perceived control" over the skills and resources needed to be physically active, they begin to incorporate the behavior into their lifestyle. Ajzen's concept of perceived behavioral control is similar to Bandura's concept of self-efficacy (38)—a person's perception of his or her ability to perform the behavior. Perceived behavioral control over opportunities, resources, and skills necessary to perform a behavior is believed to be a critical aspect of behavioral change processes. Thus, as people gain an understanding of the behavior and demonstrate that they can accomplish it (e.g., resources, skills, opportunities), they become empowered to change.

Social Cognitive Theory

Social learning theory (39), later renamed **social cognitive theory** (40), proposes that behavior change is affected by environmental influences, personal factors, and attributes of the behavior itself. Each may affect or be affected by either of the other two. A central tenet of social cognitive theory is the concept of self-efficacy. A person must believe in his or her capability to perform the behavior and must perceive an incentive to do so. Additionally, a person must value the outcomes or consequences that he or she believes will occur as a result of performing a specific behavior or action. Outcomes may be classified as having immediate benefits (e.g., feeling energized following physical activity) or long-term benefits (e.g., experiencing improvements in cardiovascular health as a result of physical activity). But because these expected outcomes are filtered through a person's expectations or perceptions of being able to perform the behavior in the first place, self-efficacy is believed to be the single most important characteristic that determines a person's behavior change. Providing clear instructions, providing the opportunity for skill development or training, and modeling the desired behavior can increase self-efficacy. To be effective, role models must evoke trust, admiration, and respect from the observer; models must not, however, appear to represent a level of behavior that the observer is unable to visualize attaining.

Social Support

Often associated with health behaviors such as physical activity, social support is frequently used in behavioral and social research. Social support for physical activity can be instrumental, as in giving a nondriver a ride to an exercise class; informational, as in telling someone about a walking program in the neighborhood; emotional, as in calling to see how someone is faring with a new walking program; or appraisal, as in providing feedback and reinforcement in learning a new skill. Sources of support for physical activity include family members, friends, neighbors, coworkers, and exercise program leaders and participants.

Transtheoretical Model of Behavior Change

A model that integrates well with other theories and models of health behavior is the **transtheoretical model of behavior change** (41). This model has been adapted for use in the context of the health site and has been shown to be effective in changing physical activity behaviors (42). In this model, behavior change has been conceptualized as a five-stage process or continuum related to a person's readiness to change: precontemplation, contemplation, preparation, action, and maintenance (43). People are thought to progress through these stages at varying rates, often moving back and forth along the continuum a number of times before attaining the goal of maintenance. Therefore, the stages of change are better described as spiraling or cyclical rather than linear. In this model, people use different processes of change as they move from one stage of change to another. Efficient self-change thus depends on doing the right thing (processes) at the right time (stages). According to this theory, tailoring interventions to match a person's readiness or stage of change is essential. For example, for people who are not yet contemplating becoming more active, encouraging a step-by-step movement along the continuum of change may be more effective than encouraging them to move directly into action (44). Recently, in collaboration with the Centers for Disease Control and Prevention, investigators from the San Diego State University have developed physical activity assessment and counseling materials. The Patient-Centered Assessment & Counseling for Exercise & Nutrition (PACE) materials were developed for use by the primary care provider in the clinical setting targeting apparently clinically healthy adults (43). The materials have been evaluated for both acceptability and effectiveness in a number of different clinical settings (42). Sample materials taken from PACE have been included in practical application 2.1. These materials are intended to provide a quick look at the steps in assessing and counseling an individual for physical activity. The materials incorporate many of the principles from a number of theoretical constructs previously discussed. For further explanation of PACE materials, visit their Web site at www.paceproject.org.

Wankel et al. (45) demonstrated the effectiveness of cognitive–behavioral techniques in enhancing physical activity promotion efforts, where the use

The PACE Model

Telling patients what to do doesn't work, especially over the long term. An effective behavioral model helps to facilitate long-term changes by telling them how to change. PACE (Patient-Centered Assessment & Counseling for Exercise & Nutrition) is a comprehensive approach to physical activity and nutrition counseling that uses materials developed by a team of researchers at San Diego State University. The curriculum draws heavily on the "stages of change" model, which suggests that individuals change their habits in stages. Taking into account each person's readiness to make changes, PACE provides tailored recommendations for patients in each stage. PACE offers three different counseling protocols. Empirically derived behavioral strategies are applied in each protocol.

The development of the PACE model began in 1990, funded originally by the Centers for Disease Control and Prevention, the Association of Teachers of Preventive Medicine, and San Diego State University. The original PACE materials, first released in 1994, dealt only with physical activity. The program was originally developed to overcome barriers to physician counseling for physical activity—especially lack of time for counseling, lack of standardized counseling protocols, and lack of training in behavioral counseling. Counseling was designed to be delivered in 2 to 5 min during a general patient check-up.

The PACE materials were thoroughly tested and found to be acceptable and usable by healthcare providers and patients across the United States. Physicians also found PACE to be practical, improving their confidence in counseling patients about physical activity (51). In a controlled efficacy study of 212 sedentary adults, patients who received PACE counseling increased their minutes of weekly walking by 38.1 compared with 7.5 among the control group. Additionally, 52% of the patients who received PACE counseling adopted some physical activity compared with 12% of the control group (52).

Since these earlier studies, the current PACE materials have been revised to include the recommendations from the Surgeon General's Report on Physical Activity and Health (1) and also address nutrition behaviors such as decreasing dietary fat consumption; increasing fruit, vegetable, and fiber consumption; and balancing caloric intake and expenditure for weight control.

of increased social support and decisional strategies improved adherence to exercise classes among participants. Martin et al. (46) demonstrated through a series of studies the positive effects of personalized praise and feedback and the use of flexible goal setting among participants on exercise class adherence. Participants in the enhanced intervention group demonstrated an 80% attendance rate during the intervention compared with the control group's 50% attendance rate (46).

McAuley et al. (47) successfully emphasized strategies to increase self-efficacy and thereby increase physical activity levels among adult participants in a community-based physical activity promotion program. These successful strategies included social modeling and social persuasion to improve compliance and exercise adherence. Promoting physical activity through home-based strategies holds much promise and might prove to be cost effective (48). Through tailored mail and telephone interventions, significant levels of social support and reinforcement have been shown to enhance participants' self-

efficacy in complying with exercise prescription, thus significantly improving levels of physical activity (49).

Finally, **lifestyle-based physical activity** promotion has been demonstrated to increase the levels of moderate physical activity among adults. Lifestyle-based physical activity focuses on home/community-based participation in many forms of activity that include much of a person's daily routine (e.g., transport, home repair/maintenance, yard maintenance) (50). This approach evolved from the idea that physical activity health benefits may accrue from an accumulation of physical activity minutes over the course of the day (1). Because lack of time is a common barrier to regular physical activity, some researchers recommended promoting lifestyle changes whereby physical activity can be enjoyed throughout the day as part of one's lifestyle. Taking the stairs at work, taking a walk during lunch, and walking or biking for transportation are all effective forms of lifestyle physical activity.

Ecological Perspective

A criticism of most theories and models of behavior change is that they emphasize individual behavior change and pay little attention to sociocultural and physical environmental influences on behavior (51). Recently, interest has developed in ecological approaches to increasing participation in physical activity (52). These approaches place the creation of supportive environments on par with the development of personal skills and the reorientation of health services. Creation of supportive physical environments is as important as intrapersonal factors when behavior change is the defined outcome. Stokols (53) illustrated this concept of a health-promoting environment by describing how physical activity could be promoted by establishing environmental supports, such as bike paths, parks, and incentives to encourage walking or bicycling to work. An underlying theme of ecological perspectives is that the most effective interventions occur on multiple levels. Interventions that simultaneously influence multiple levels and multiple settings (e.g.,

schools, work sites) may be expected to lead to greater and longer lasting changes and maintenance of existing health-promoting habits. In addition, investigators have recently demonstrated that behavioral interventions primarily work by means of the mediating variables of intrapersonal and environmental factors (54). Mediating variables are those that facilitate and shape behaviors—we all have a set of intrapersonal factors (e.g., personality type, motivation, genetic predispositions) and environmental factors (e.g., social networks such as family, cultural influences, and the built/physical environment).

However, few researchers have attempted to delineate the role of these mediating factors in facilitating health behavior change. Sallis et al. (55) recently described how difficult it is to assess the effectiveness of environmental and policy interventions because of the relatively few evaluation studies available. However, based on the experience of the New South Wales (Australia) Physical Activity Task Force, a model has been proposed to help understand the steps necessary to implement these interventions (56). Figure 2.1 presents an adaptation of

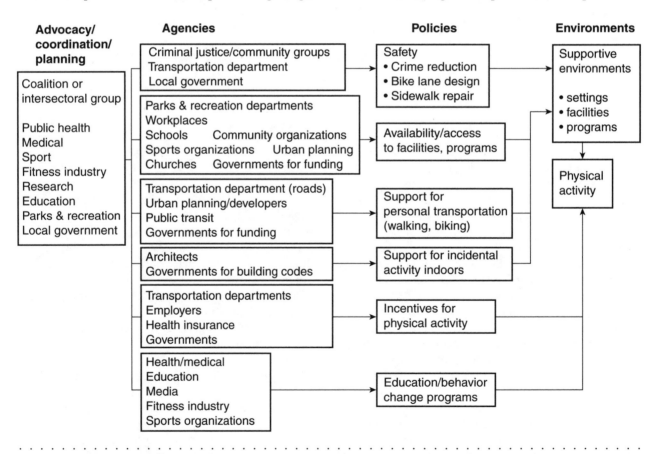

Figure 2.1 Conceptualization of the development of policy and environmental interventions to promote physical activity.

this model as prepared by Sallis (55) and outlines the necessary interaction between planning/advocacy, agencies/organizations, policies, and environments to make such interventions a reality.

Another pragmatic model that appears to have relevance for the promotion of physical activity has been proposed by McLeroy and colleagues (57). This model specifies five levels of determinants for health behavior:

1. Intrapersonal factors, including psychological and biological variables, as well as developmental history

2. Interpersonal processes and primary social groups, including family, friends, and coworkers

3. Institutional factors, including organizations such as companies, schools, health agencies, or healthcare facilities

4. Community factors, which include relationships among organizations, institutions, and social networks in a defined area

5. Public policy, which consists of laws and policies at the local, state, national, and supranational levels

Important in implementing this concept of behavioral determinants is realizing the key role of behavioral settings, which are the physical and social situations in which behavior occurs. Simply stated, human behavior such as physical activity is shaped by its surroundings—if you're in a supportive social environment with access to space/facilities you are more likely to be active. It is important to acknowledge the determinant role of selected behavioral settings: Some are designed to encourage healthy behavior (e.g., sports fields, gymnasiums, health clubs, and bicycle paths), whereas others encourage unhealthy (or less healthy) behaviors (e.g., fast food restaurants, vending machines with high-fat/high-sugar foods, movie theaters). You need to understand the environment in which your client lives. These structures (e.g., fields, gymnasiums, community centers) are part of each of our living environments—people who disregard them are less likely to be active. For the physical activity provider, a potentially important adjunct in assessing and prescribing physical activity interventions for participants is understanding the physical and social contexts in which their patients live. This information can be obtained from various sources and can be at the level of the individual or at the more general community level. When individual physical activity behavior information is coupled with

sociodemographic, physical, and social context information, physical activity interventions can be further tailored to maximize the participant's physical activity behavior change and maintenance plan. Exercise physiologists cannot alter the client's physical environment; however, they should be able to address environmental barriers and provide insights into how to overcome these barriers. In the long run, we all should be a part of changing our environments for the better.

An example of such tailoring for physical activity promotion that alters physical activity behaviors includes the work of Linenger et al. (58), where efforts to increase physical activity levels among Naval personnel were accomplished by a multifactorial environmental/policy approach to physical activity promotion. These investigators compared an "enhanced base" to a "control base." The enhancements included increasing the number of bike trails on base, acquiring new exercise equipment for the local gym, opening a women's fitness center, instituting activity clubs, and providing release time for physical activity and exercise (58). The changes were positive for those living on the enhanced base—that is, they increased their physical activity levels.

Another example, this time emphasizing an incentive-based approach to promoting physical activity, is the work of Epstein and Wing (59). Although this work was undertaken quite some time ago, the lessons are very relevant in today's inactive culture. In this study, contracts and the use of a lottery (a popular enterprise today!) were used to boost exercise attendance with the consequence of increasing participants' overall physical activity levels. Compared with a "usual care" comparison group, adherence and activity levels were significantly improved and sustained (56). However, caution in using an incentive-based approach has been urged by some researchers who believe that over the long term participants never internalize the health behavior, meaning that they are likely to stray back to sedentary habits once the incentive is removed or loses its appeal. Nevertheless, incentives have been proven to be effective in the short term.

Additional community-based environmental efforts to influence physical activity behavior have included the use of signs in public settings to increase use of stairs and walkways (60, 61). These latter studies are examples of single intervention efforts that can be carried out in concert with systematic exercise prescription efforts among individuals. Thus, the increase in stair usage as a result of a promotional campaign can help individuals meet their prescribed energy expenditure requirements.

Conclusion

Within the past decade, physical activity has emerged as a key factor in the prevention and management of chronic conditions. Although the role of exercise in health promotion has been appreciated and practiced for decades, recent findings regarding the mode, frequency, duration, and intensity of physical activity have modified exercise prescription practices. Included in these modifications has been the delineation between health and fitness outcomes relative to the physical activity prescription. Most importantly, new approaches to physical activity prescription and promotion that emphasize a behavioral approach with documented improvements in compliance have now become available to health professionals.

Behavioral science has contributed greatly to the understanding of health behaviors such as physical activity. Behavioral theories and models of health behavior have been reexamined in light of physical activity and exercise. Although more research is needed to further develop successful, well-defined applications that are easily adaptable for intervention purposes, behavioral principles and guidelines have evolved that are designed to help the health professional understand health behavior change and guide people into lifelong patterns of increased physical activity and improved exercise compliance.

New frontiers in the application of exercise prescription to specific populations as well as efforts to define the specific dose (frequency, intensity, duration) of physical activity for specific health and fitness outcomes are now being explored. As this information becomes available, it must be introduced to the participant via the most effective behavioral paradigms, such as the models discussed in this chapter. Moreover, positive changes in the participant's physical and social environments must also occur to enhance compliance with exercise prescriptions. In turn, increased levels of physical activity among all people will improve health and function.

Case Study 2

Mrs. KY is a 45-year-old Caucasian female, married with two teenage sons. She is employed as a senior manager at a large bank and reports experiencing an "above average" level of tension and stress. She presents at the referral of her primary care physician, who has observed that the client has elevated blood pressure and cholesterol levels that may be attributed to her stressful and highly sedentary job. In addition, the client admitted that she would like to lose 30 to 40 lb and improve her fitness so that she can ride her bike with her husband on a new community rail trail recently installed by her neighborhood.

The client is 5 feet 6 in. tall and weighs 196 lb with a corresponding body mass index of 31.7. She has a resting heart rate of 85 beats · min^{-1} and a resting blood pressure of 136/89 mmHg. Her total cholesterol is 198 mg · dl^{-1} untreated and her high-density lipoproteins are 34 mg · dl^{-1}. Her graded treadmill stress test reveals that she has a $\dot{V}O_2$max of 20.5 ml · kg^{-1} · min^{-1}, which is normal for an unfit woman of her age range. Her electrocardiogram was also unremarkable at rest, as well as during and following her test. In addition, she reported smoking from age 17 to 40 years. She also complains of occasional joint stiffness in her hands and ankles. The client describes herself as nonathletic and admits to never participating in an organized sport or exercise setting. She is aware of the benefits of exercise but did not feel the incentive to begin a formal program until her doctor's recommendation. The client jokes that although her workday is highly organized and structured, the rest of her life is chaotic and it is only due to the support of her husband and kids that anything gets done at home. She laments that her eating habits are atrocious and that she is so tired when she gets home from work that she only has energy to make dinner before crashing in front of the television. She presents to you to start a workout program that will help achieve her goals.

continued

Case Study 2   Discussion Questions

1. Applying the transtheoretical model, at what stage of exercise adoption is this client?

2. Based on your response to question 1, what types of interventions are most appropriate for this stage of change and why?

3. If you used the health belief model, what factors would you emphasize to achieve optimal exercise adherence?

4. How would Bandura's social cognitive theory be relevant to fostering exercise adherence for this client?

Glossary

cardiorespiratory fitness—Also known as aerobic capacity, this describes the body's ability to perform high-intensity activity for a prolonged period of time without undue physical stress or fatigue.

environmental factors—Physical and social factors that can influence participation in physical activity (e.g., vehicular traffic, inclement weather, and unsafe neighborhoods).

excess body weight—Occurs when too few calories are expended and too many consumed for individual metabolic requirements. Overweight (>25) and obese (≥30) as defined by body mass index $(kg \cdot m^{-2})$.

health belief model—Theory which proposes that only psychological variables influence health behaviors.

lifestyle-based physical activity—Home/community-based participation in many forms of activity that include much of a person's daily routine (e.g., transport, home repair/maintenance, yard maintenance).

muscular strength and endurance—The ability of skeletal muscles to perform hard and/or prolonged work.

musculoskeletal flexibility—The range of motion in a joint or sequence of joints.

physical activity—Any bodily movement produced by skeletal muscles that results in caloric expenditure.

relapse prevention model—A model used to help new exercisers anticipate problems with adherence. Factors that contribute to relapse include negative emotional or physiological states, limited coping skills, social pressure, interpersonal conflict, limited social support, low motivation, high-risk situations, and stress.

self-efficacy—A person's belief in his or her capability to perform the behavior and perceived incentive to do so.

social cognitive theory—Theory that behavior change is affected by environmental influences, personal factors, and attributes of the behavior itself.

social support—Support and encouragement that an individual receives from others to maintain behavior change.

theory of planned behavior—Theory that adds to the theory of reasoned action with the concept of perceived control over the opportunities, resources, and skills necessary to perform a behavior.

theory of reasoned action—Theory that performance of a given behavior (e.g., exercise) is primarily determined by the person's attitude toward the behavior and the influence of the person's social environment or subjective norm (i.e., beliefs about what other people think the person should do as well as the person's motivation to comply with the opinions of others).

transtheoretical model of behavior change—Model wherein behavior change is conceptualized as a five-stage process or continuum related to a person's readiness to change: precontemplation, contemplation, preparation, action, and maintenance. People are thought to progress throughout the stages.

References

1. U.S. Department of Health and Human Services. Physiologic responses and long-term adaptations to exercise. 1996. In Physical Activity and Health: A Report of the Surgeon General (pp. 61-73). Atlanta, GA: U.S. Department of Health and Human Services, Centers for Disease Control and Prevention, National Center for Chronic Disease Prevention and Health Promotion.

2. Sidney KH, Shephard RJ, Harrison JE. Endurance training and body composition of the elderly. **Am J Clin Nutr,** 30: 326-333, 1977.

3. Smith DM, Khairi MRA, Norton J, et al. Age and activity effects on rate of bone mineral loss. **J Clin Invest,** 58: 716-721, 1976.

4. Paffenbarger RS, Hyde RT, Wing AL. Physical activity as an index of heart attack risk in college alumni. **Am J Epidemiol**, 108: 161-175, 1978.

5. Tipton CH. Exercise, training, and hypertension: An update. **Exerc Sport Sci Rev,** 19: 447-505, 1991.

6. Lee IM, Paffenbarger RS, Hsieh CC. Physical activity and risk of developing colorectal cancer among college alumni. **J Natl Cancer Inst,** 83: 1324-1329, 1991.

7. Paffenbarger RS Jr, Hyde RT, Wing AL, Hsieh CC. Physical activity, all-cause mortality, and longevity of college alumni. **N Engl J Med,** 314: 605-613, 1986.

8. Ntoumanis N, Biddle SJ. A review of motivational climate in physical activity. **J Sports Sci,** 17: 643-665, 1999.

9. Shephard RJ. Exercise and aging: Extending independence in older adults. **Geriatrics,** 48: 61-64, 1993.

10. Caspersen CJ, Powell KE, Christenson GM. Physical activity, exercise, and physical fitness: Definitions and distinctions for health-related research. **Public Health Rep,** 100: 126-130, 1985.

11. U.S. Department of Health and Human Services. Healthy People 2010 (2nd ed). Understanding and Improving Health and Objectives for Improving Health. 2 vols. Washington, DC: U.S. Government Printing Office, November 2000.

12. Pate RR, Pratt M, Blair SN, et al. Physical activity and public health: A recommendation from the Centers for Disease Control and Prevention and the American College of Sports Medicine. **JAMA,** 273: 402-407, 1995.

13. American College of Sports Medicine. Position stand on the recommended quantity and quality of exercise for developing and maintaining cardiorespiratory and muscular fitness in healthy adults. **Med Sci Sports Exerc,** 22: 265-274, 1990.

14. Braith RW, Graves JE, Pollock JL, Leggett SL, Carpenter DM, Colvin AB. Comparison of 2 vs. 3 days/week of variable resistance training during 10- and 18-week programs. **Int J Sports Med,** 10: 450-454, 1989.

15. Barry HC, Eathorne SW. Exercise and aging: Issues for the practitioner. **Med Clin North Am,** 78: 357-376, 1994.

16. Cady LD, Bischoff DP, O'Connell ER, Thomas PC, Allan JH. Strength and fitness and subsequent back injuries in firefighters. **J Occup Med,** 21: 269-272, 1979.

17. West Virginia University, Department of Safety and Health Studies and Department of Sports and Exercise Studies. Physical Fitness and the Aging Driver: Phase I. AAA Foundation of Traffic Safety, Washington, DC, 1988.

18. Passmore R. The regulation of body weight in man. **Proc Nutr Soc,** 30: 122-127, 1971.

19. Wood PD, Stefanick ML, Dreon DM, Frey-Hewitt D, Garay BC, Williams PT, Superko HR, Fortmann SP, Albers JJ, Vranizan KM, Ellworth NM, Terry RB, Haskell WL. Changes in plasma lipids and lipoproteins in overweight men during weight loss through dieting as compared with exercise. **N Engl J Med,** 319: 1173-1179, 1988.

20. Sallis JF, Hovell MF. Determinants of exercise behavior. **Exerc Sports Sci Rev,** 18: 307-330, 1990.

21. King AC. Intervention strategies and determinants of physical activity and exercise behavior in adult and older adult men and women. **World Rev Nutr Diet,** 82: 148-158, 1997.

22. Pollock ML. Prescribing exercise for fitness and adherence. In: Dishman RK, ed. Exercise Adherence. Champaign, IL: Human Kinetics; pp. 259-277, 1988.

23. Sallis JF, Hovell MF, Hofstetter CR. Predictors of adoption and maintenance of vigorous physical activity in men and women. **Prev Med,** 21: 237-251, 1992.

24. Sallis JF, Hovell MF, Hofstetter CR, et al. A multivariate study of determinants of vigorous exercise in a community sample. **Prev Med,** 18: 20-34, 1989.

25. King AC, Haskell WL, Taylor CB, Kraemer HC, DeBusk RF. Group- vs. home-based exercise training in healthy older men and women. **JAMA,** 266: 1535-1542, 1991.

26. Kriska AM, Bayles C, Cauley JA, et al. Randomized exercise trial in older women: Increased activity over two years and the factors associated with compliance. **Med Sci Sports Exerc,** 18: 557-562, 1986.

27. Young DR, Haskell WL, Taylor CB, Fortmann SP. Effect of community health education on physical activity knowledge, attitudes, and behavior: The Stanford Five-City Project. **Am J Epidemiol,** 144: 264-274, 1996.

28. Hassman PR, Ceci R, Backman L. Exercise for older women: A training method and its influences on physical and cognitive performance. **Eur J Appl Physiol,** 64: 460-466, 1992.

29. King AC, Haskell WL, Young DR, Oka RK, Stefanick ML. Long-term effects of varying intensities and formats of physical activity on participation rates, fitness, and lipoproteins in men and women aged 50 to 65 years. **Circulation,** 91: 2596-2604, 1995.

30. American College of Sports Medicine. Guidelines for Exercise Testing and Prescription (4th ed). Philadelphia: Lea & Febiger, 1991.

31. Dishman RK. Increasing and maintaining exercise and physical activity. **Behav Ther,** 22: 345-378, 1991.

32. U.S. Department of Health and Human Services. Understanding and promoting physical activity. In Physical Activity and Health: A Report of the Surgeon General (pp. 209-259). Atlanta, GA: U.S. Department of Health and Human Services, Centers for Disease Control and Prevention, National Center for Chronic Disease Prevention and Health Promotion, 1996.

33. Rosenstock IM. The health belief model: Explaining health behavior through expectancies. In Health Behavior and Health Education. Theory, Research, and Practice (pp. 39-62). San Francisco: Jossey-Bass, 1990.

34. Marlatt GA, George WH. Relapse prevention and the maintenance of optimal health. In The Handbook of Health Behavior Change (Shumaker SA, Schron EB, Ockene J, eds; pp. 44-63). New York: Springer, 1990.

35. Marcus BH, Stanton AL. Evaluation of relapse prevention and reinforcement interventions to promote exercise adherence in sedentary females. **Res Q Exerc Sport,** 64: 447-452, 1993.

36. Ajzen I, Fishbein M. Understanding Attitudes and Predicting Social Behavior. Englewood Cliffs, NJ: Prentice Hall, 1980.

37. Ajzen I. Attitudes, Personality, and Behavior. Chicago: Dorsey Press, 1988.

38. Bandura A. Self-efficacy: Toward a unifying theory of behavioral change. **Psychol Rev,** 84: 191-215, 1977.

39. Bandura A. Social Learning Theory. Englewood Cliffs, NJ: Prentice Hall, 1977.

40. Bandura A. Social Foundations of Thought and Action: A Social–Cognitive Theory. Englewood Cliffs, NJ: Prentice Hall, 1986.

41. Prochaska JO, DiClemente CC. The Transtheoretical Approach: Crossing Traditional Boundaries of Change. Homewood, IL: Dorsey Press, 1984.

42. Long BJ, Calfas KJ, Patrick K, Sallis JF, Wooten WJ, Goldstein M, Marcus B, Schwenck T, Carter R, Torez T, Palinkas L, Heath G. Acceptability, usability, and practicality of physician counseling for physical activity promotion: Project PACE. **Am J Prev Med,** 12: 73-81, 1996.

43. Calfas KJ, Long BJ, Sallis JF, Wooten WJ, Pratt M, Patrick K. A controlled trial of physician counseling to promote the adoption of physical activity. **Prev Med,** 25: 225-233, 1996.

44. Marcus BH, Banspach SW, Leffebvre RC, Rossi JS, Carleton RA, Abrams DB. Using the stages of change model to increase adoption of physical activity among community participants. **Am J Health Promot,** 6: 424-429, 1992.

45. Wankel LM, Yardley JK, Graham J. The effects of motivational interventions upon the exercise adherence of high and low self-motivated adults. **Can J Appl Sport Sci,** 10: 147-156, 1985.

46. Martin JE, Dubbert PM, Katell AD, et al. The behavioral control of exercise in sedentary adults: Studies 1 through 6. **J Consult Clin Psychol,** 52: 795-811, 1984.

47. McAuley E, Courneya DS, Rudolph DL, Lox CL. Enhancing exercise adherence in middle-aged males and females. **Prev Med,** 23: 498-506, 1994.

48. Taylor CB, Miller NH, Smith PM, DeBusk RF. The effect of home-based, case-managed, multifactorial risk-reduction program on reducing psychological distress in patients with cardiovascular disease. **J Cardiopulm Rehabil,** 17: 157-162, 1997.

49. Dishman RK, Buckworth J. Increasing physical activity: A quantitative synthesis. **Med Sci Sports Exerc,** 28: 706-719, 1996.

50. Dunn AL, Marcus BH, Kampert JB, et al. Comparison of lifestyle and structured interventions to increase physical activity and cardiorespiratory fitness: A randomized trial. **JAMA,** 281: 327-334, 1999.

51. Winett R. A framework for health promotion and disease prevention programs. **Am Psychol,** 50: 341-350, 1995.

52. King AC. Community and public health approaches to the promotion of physical activity. **Med Sci Sports Exerc,** 26: 1405-1412, 1994.

53. Stokols D. Establishing and maintaining healthy environments: Toward a social ecology of health promotion. **Am Psychol,** 47: 6-22, 1992.

54. Baranowski T, Lin LS, Wetter DW, Resnicow K, Hearn MD. Theory as mediating variables: Why aren't community interventions working as desired? **Ann Epidemiol,** S7: S89-S95, 1997.

55. Sallis JF, Bauman A, Pratt M. Environmental and policy interventions to promote physical activity. **Am J Prev Med,** 15: 379-396, 1998.

56. New South Wales Physical Activity Task Force. Simply Active Every Day: A Discussion Document From the NSW Physical Activity Task Force on Proposals to Promote Physical Activity in NSW, 1997-2002. Summary Report. Sydney, Australia: New South Wales Health Department, 1997.

57. McLeroy KR, Bibeau D, Steckler A, Glanz K. An ecological perspective on health promotion programs. **Health Educ Q,** 15: 351-377, 1988.

58. Linenger JM, Chesson CV 2nd, Nice DS. Physical fitness gains following simple environmental change. **Am J Prev Med,** 7: 298-310, 1991.

59. Epstein LH, Wing RR. Aerobic exercise and weight. **Addict Behav,** 5: 371-388, 1980.

60. Blamey A, Mutrie N, Aitchison T. Health promotion by encouraged use of stairs. **Br Med J,** 311: 289-290, 1995.

61. Brownell K, Stunkard AJ, Albaum J. Evaluation and modification of exercise patterns in the natural environment. **Am J Psychiatry,** 137: 1540-1545, 1980.

David L. Herbert, JD, Senior Partner
Herbert & Benson, Attorneys at Law
Canton, OH

William G. Herbert, PhD, Professor
Virginia Polytechnic Institute and State University
Blacksburg, VA

James J. Collum, JD, Attorney at Law, Of Counsel
Herbert & Benson, Attorneys at Law
Canton, OH

Legal Considerations

The involvement of exercise physiologists in a wide variety of settings has grown dramatically over the last 25 years. Historically, most exercise physiologists were engaged in human performance–related research or academic instruction. However, many now provide professional advice and service in clinical, preventive, and recreational fitness programs located in health clubs, corporate facilities, and hospital-based complexes. Exercise physiologists provide a number of important services, including fitness assessments or screenings, exercise testing, exercise prescriptions or recommendations, exercise leadership, and exercise supervision. Exercise physiologists are also expected to initiate referrals to physicians, nutritionists or dietitians, and behavioral medicine professionals as needed.

To date in the United States, only Louisiana has opted to license clinical exercise physiologists to formally sanction their delivery of specified and defined services to consumers (41). Efforts undertaken in other states such as California and Kentucky have not yet resulted in the passage, as law, of any successful licensing enactment (31-33, 46).

Whether such efforts result in the passage of at least some relevant licensing laws in states other than Louisiana is a matter of speculation (30). Professionals within the exercise science and fitness industries have not even been able to agree on whether a need exists to adopt such statutes (52, 56). It seems likely that efforts to enact legislation that requires the licensure of exercise physiologists will not occur until the need is well documented from a consumer safety perspective. Evidence justifying licensure must first be gathered and published and then appreciated by some reasonable segment of the public and state legislators. In large part, this effort will likely focus on whether such legislation is necessary to protect the public from harm as a result of services provided by exercise physiologists. To date, this evidence does not exist (31-33, 46), which may be attributable to either a low number of adverse events or the underreporting of such events.

Additionally, healthcare provider groups such as occupational therapists, physical therapists, and physicians may well oppose licensing efforts for exercise physiologists, as a means to protect their "turf" (17, 35). Such opposition seems evident, despite the fact relevant services might be delivered more cost-effectively by exercise physiologists than by other professionals.

The lack of applicable state licensing requirements for exercise physiologists has, in some respects, "muddied the waters" for those who attempt to assess and define the practice roles for such professionals in the delivery of services to consumers. This may be particularly true from the perspective of the legal system, where questions arise relative to a variety of concerns related to service delivery. Several of the important legal questions are shown in figure 3.1 (36).

This chapter examines many of these concerns; however, it represents but a brief overview of rather broader issues. Moreover, it is not and cannot be any substitute for individual legal counseling and professional advice, where advice and counseling are needed for an actual problem or situation. If the services of a legal professional are necessary, readers are encouraged to seek such advice.

The U.S. Legal System

The U.S. system of jurisprudence can affect the delivery of services by clinical exercise physiologists in several different ways. In situations where services delivered to a particular consumer by an exercise physiologist result in injury or death, one avenue of **civil action** called the **tort** system comes into play. Within the tort system, legal claims related to personal injury and **wrongful death** are asserted, examined, and determined as either **malpractice** or **negligence** claims. Traditionally, in malpractice

- What services may exercise physiologists lawfully provide given the absence or lack of licensure in almost every state?
- What practices performed by exercise physiologists may be prohibited as a matter of law due to state unauthorized practice of medicine statutes?
- What practices performed by exercise physiologists may be prohibited as a matter of law due to unauthorized practice of other licensed healthcare services (e.g., physical therapy, nursing)?
- What potential liabilities may exercise physiologists face when their delivery of service results in harm, injury, or death attributable to alleged negligence or malpractice?
- What recognition may be given to exercise physiologists and their opinions in a wide variety of legal settings (evaluate disability or working capacity in matters involving insurance, personal injury, or workers' compensation)?

FIGURE 3.1   Legal aspects and issues to consider in clinical exercise physiology.

claims asserted against healthcare providers, the claim infers that the delivery of professional service was substandard compared with the prevailing practices performed by other professionals in like circumstances. Because clinical exercise physiologists are not licensed except in one state, actions related to **personal injury** or death arising from the provision of such services would probably be determined within the negligence rather than malpractice setting.

In tort actions, parties who are injured seek compensation for those injuries by way of an award for civil damages. The parties who assert such claims are referred to as **plaintiffs,** and the parties against whom such claims are made are referred to as **defendants.** Such cases are traditionally submitted to judges or juries for determination following the testimony of various individuals and experts who are allowed to express opinions on certain matters based on their education, background, training, and experience. Clinical exercise physiologists have been permitted to testify in a number of tort cases and express opinions as **expert witnesses** (40). Undoubtedly, such professionals are qualified to deliver expert opinions in a variety of cases. Generally, opinions like those are provided to establish the so-called **standard of care** applicable to the delivery of service and any breach from or adherence to that standard.

Aside from potential involvement as a witness or defendant in the civil tort law system, clinical exercise physiologists can also be exposed to the criminal justice system. Criminal charges related to the unauthorized practice of medicine or some other state-licensed healthcare service can be asserted against those perceived to be delivering the unlicensed service. In these situations, charges might be lodged against some unlicensed providers (e.g., clinical exercise physiologists), where services are rendered to treat, cure, mitigate, or otherwise beneficially affect a disease process, injury, or infirmity. Those who are not licensed to provide such services would be well advised to consult with their own legal advisors to determine whether the services they render or contemplate rendering are reserved for provision by others who are state-licensed healthcare providers.

Important State Statutory Definitions

The exercise physiologist working in the clinical setting typically provides services based on the recommendation and under the oversight of licensed health professionals. The material that follows is provided to help clinical exercise physiologists better appreciate the scope, definition, and interaction of their professional duties versus those of other healthcare professionals.

State Licensure of Clinical Exercise Physiologists

Aside from the state of Louisiana (see practical application 3.1), where clinical exercise physiologists are licensed to provide delineated services, no other state has licensed the delivery of such services by these professionals. As a consequence, for clinical exercise physiologists in states other than Louisiana, there are no state laws or regulations to determine exactly what services they may and may not provide to consumers and under what circumstances.

For those practicing in hospital-based or rehabilitative settings, services rendered by clinical exercise physiologists are sometimes provided in conjunction with (or under the supervision of) licensed healthcare providers such as physicians, physician assistants, physical therapists, or others. To help you better appreciate the scope of practice for clinical exercise physiologists, figure 3.2 details the core patient care services for clinical exercise physiologists

registered through the American College of Sports Medicine.

Ultimate responsibility for the delivery of service in those circumstances typically rests not with the clinical exercise physiologist but with the licensed healthcare provider. However, some state laws may provide for a rather broad delegation of authority to nonlicensed individuals and others who provide healthcare services under particular circumstances. For example, section 333.16215 of the Michigan Public Health Code (53) states the following:

> A licensee . . . may delegate to a licensed or unlicensed individual who is otherwise qualified by education, training or experience the performance of selected acts, tasks or functions where the acts, tasks or functions fall within the scope of practice of the licensee's profession and will be performed under the licensee's supervision. An act, task or function shall not be delegated under this section which, under standards of acceptable and prevailing practice, requires the level of education, skill, and judgement required of a licensee under this article.

However, some state laws clearly prohibit the rendition of certain services by individuals other than licensed physicians (see practical application 3.1) (63).

State Statutory Definitions for the Practice of Medicine

The various definitions that provide for the practice of medicine may be of particular importance to exercise physiologists conducting exercise activities in the clinical, health maintenance, or adult fitness setting. It is important for clinical exercise physiologists to understand how their scope of practice relates to the practice of medicine.

Although not universally defined, the practice of medicine is generally regarded as the diagnosis of an individual's symptoms to determine with what disease or illness he or she is afflicted, and then to determine, based on this information, what remedy or treatment should be given or prescribed to treat that disease or relieve the symptoms (36). The practice of medicine may even be broader, as defined in some state laws. For example, the practice of medicine may include the following:

- Advice given for nutritional counseling (21, 23, 25)
- Advice given to an individual to "take a walk" (61)
- Advice related to the prevention of disease or illness (60)

PRACTICAL APPLICATION 3.1

State Licensure of Clinical Exercise Physiologists

The clinical exercise physiologist legislation adopted in the state of Louisiana in 1995 (11, 38, 54) provided that clinical exercise physiologists, as licensed in that state, would be officially sanctioned to provide defined services. In this regard, that statute defined a clinical exercise physiologist as a "person who, under the direction, approval, and supervision of a licensed physician, formulates, analyzes, and implements exercise protocols and programs, administers graded exercise tests, and provides education regarding such exercise programs and tests in cardiopulmonary rehabilitation programs to individuals with deficiencies of the cardiovascular system, diabetes, lipid disorders, hypertension, cancer, chronic obstructive pulmonary disease, arthritis, renal disease, organ transplant, peripheral vascular disease, and obesity." *Exercise protocols and programs* in this regard are defined as "the intensity, duration, frequency and mode of activity to improve the cardiovascular system." The act prohibits persons other than those who are licensed from using the title "licensed clinical exercise physiologist."

Although this enactment exempts certain persons from its requirements (such as other licensed healthcare providers), it perhaps may not apply to clinical exercise physiologists who do not hold themselves out as being licensed. Additionally, it might not apply to those who provide service outside of a cardiopulmonary rehabilitation program, which is not defined in this enactment. Therefore, one might be led to believe that wellness, recreation, and preventive settings might not be covered by this Louisiana law.

This conclusion, however, is not entirely clear inasmuch as the first statutory exemption in the Louisiana law provides that the following persons and their activities are exempted from the licensing requirements:

> Any person employed or supervised by a licensed physician whose primary duty is to provide graded exercise testing within the confines of the physician's office.

Because services under these circumstances are not provided in a cardiopulmonary program, one might wonder if this exemption really goes beyond the language of the statute—meaning that the law might also apply to clinical exercise physiologists not working in rehabilitation programs. As you can see, enacting and interpreting licensure statutes are complex and confusing. Given that there is no reported case law on this subject, much of this discussion amounts to speculation, at best.

Clearly, by the terms of the revised Louisiana statute, clinical exercise physiologists are prohibited from practicing physical therapy: "Nothing in this Chapter shall be construed to allow a licensed clinical exercise physiologist to practice physical therapy . . . nor shall any licensed clinical exercise physiologist hold himself out as a physical therapist or a physical therapist assistant" (63).

— Prescription or recommendation of mineral water, vitamins, or food (21)

— Use/misuse of written wellness assessments or questionnaires (20, 24)

The key to these determinations centers on whether the prescription, advice, or recommendation is given to treat, cure, or prevent a disease or condition (21). In such circumstances, releases or **waivers** signed by participants who are deemed to be the recipients of medical services might well be invalid.

Frequently, the practice of medicine is defined by state statute, often positively by reference to the scope of practice or negatively by referring to the unauthorized practice of same. Various states have defined the practice of medicine in a number of different ways. In the state of Ohio (Rev. Code Sec. 4731.34), for example, the applicable statute provides the following:

> A person shall be regarded as practicing medicine . . . who examines or diagnoses for compensation of any kind, or prescribes, advises or recommends, administers or dispenses for compensation of any kind, direct or indirect, a drug or medicine, appliance, mold or cast, application, operation or treatment, of whatever nature, for the cure or relief of a wound, fracture or bodily injury, infirmity or disease.

- Modalities of Service
    - Exercise evaluation and interventions provided to improve function, physical fitness, and physical performance (occupational, recreational, and athletic).
    - Counseling and education to increase exercise safety and self-maintenance skills, as well as behaviors that promote changes in exercise habits aimed at reducing health risks for specified chronic diseases.
- Specific Exercise-Related Services
    - Preexercise health and fitness screening.
    - Assessment of physical activity history and symptomatology.
    - Preexercise stratification of risk for exercise-related injury.
    - Identification of demographic, health history, and physical activity history attributes that require referral for evaluation by other healthcare providers.
    - Exercise testing and evaluation, using both standardized methods and methods adapted to various chronic diseases and disabling conditions; this may include the use of simple and sophisticated technologies suitable for definitive assessments made in the laboratory and in the field. Examples include cardiorespiratory efficiency; measured blood lactate threshold; ventilatory-derived lactate threshold; peak aerobic capacity with other ventilatory parameters; muscular strength, power, endurance, and flexibility measured with technologies ranging from electrically controlled isokinetic machines to hand-held isometric dynamometers; and body composition.
    - Preparation of reports, including interpretations for use by allied health professional; includes assessment of fitness levels relative to normative databases and an explanation useful for judging health risks and physical performance limitations in occupational and recreational physical activities.
    - Physical work simulation testing and use of observational techniques to estimate activities of daily living; includes preparation of reports.
    - Physical activity prescription, individualized to meet the needs and goals of the client and the recommendations of allied healthcare professionals who may have referred the client or provided consultation to the clinical exercise physiologist; this includes risk appraisal for untoward exercise-induced events that affect physical activity planning and exercise supervision recommendations (risk stratification).
    - Exercise instruction and supervision that include teaching safe and effective participatory skills that improve components of physical fitness and are appropriate or adapted to accommodate client limitations related to a chronic disease or disabling condition; supervision includes monitoring according to safety needs and capabilities of the client and, as needed, electronic technologies or self-monitoring methods using continuous and discontinuous schedules.
    - Outcome assessment, which includes documentation of changes in performance, fitness, and disease-related risk factors that occurred as a result of exercise and work simulation training programs delivered by the registered clinical exercise physiologist; this includes reporting to the client, primary care provider, and any third parties legally authorized to receive such information.

FIGURE 3.2 Core patient care activities that are central to the scope of practice for the clinical exercise physiologist registered through the American College of Sports Medicine.

Most state statutes are broadly written and all-encompassing. Courts interpreting these statutes, which are generally enacted for the benefit of the public, apply the statutes broadly (except in criminal prosecutions) to serve the legislative purpose and the public good.

For example, in the 1975 case of Mirsa versus State Medical Board, 42 Ohio St.2d 399, the Ohio Supreme Court in examining the status of a commercial blood plasma donor bank held that because the bank's practice of determining acceptability of donors included an in-depth medical history and physical examination (i.e., blood pressure, pulse, temperature, weight, and hematocrit), they engaged in the practice of medicine. On the basis of this decision, the plasma bank was faced with terminating its operations or incorporating into its procedures the services of a licensed physician. The parallels between some exercise testing or exercise prescription procedures and the blood plasma bank and its procedures may be frightening to some in light of the Ohio Supreme Court's pronouncement.

In an earlier Ohio case, State versus Winterich (60), the defendant, a licensed chiropractor and electromechanical therapist, was criminally charged with the unauthorized practice of medicine. The state contended that she prescribed, dispensed, or recommended certain minerals, foods, and fluids to cure and relieve certain infirmities and thereby engaged in the unlawful practice of medicine. Although a jury initially found her guilty, the conviction was reversed on appeal, with the court finding that there was no proof that the foods or other items were intended for use in the cure, mitigation, or treatment of disease. However, the court noted that food items could be deemed to be drugs if given for "care, mitigation, treatment or prevention of disease."

In examining these decisions as well as others like them (61), we see that practices which include recommendations or treatments given to prevent disease, or even the prescription of exercise, may well be deemed the unauthorized practice of medicine under some circumstances. In fact, when recommending exercise regimens within a rehabilitative or preventive model, it seems that the prescription of exercise may indeed be the practice of medicine. A number of other recent cases may apply to this discussion (12, 43, 45, 47, 58, 66).

In addition, many states have enacted criminal statutes prohibiting the unauthorized practice of medicine. Often, such statutes make the unauthorized practice of medicine a **misdemeanor,** subjecting violators to a fine and a period of imprisonment generally not to exceed 1 year. Very serious criminal charges are possible if it is determined that one is engaged in the unauthorized practice of medicine and a patient death results from procedures deemed to be medical procedures. Based on such a factual finding in connection with a death, charges of **manslaughter** or even **murder** to one degree or another might be possible (51). These charges could even be enhanced by a state-advocated determination that certain exercise-related procedures must be physician supervised (49).

An example of when the unauthorized practice of medicine allegations were put forth is in the case of Gathers versus Loyola Marymount University, wherein the basketball player's estate claimed that the university, athletic director, and coach became involved in the unauthorized practice of medicine. These allegations centered on the claim that the coach contacted Mr. Gathers' physicians, whereby he allegedly suggested reducing the medication then being given to Mr. Gathers to treat a heart condition (27). The unauthorized practice of medicine as it relates to exercise or sport is a potential reality, one that persons practicing in this area must consider and avoid.

Clinical exercise physiologists and their legal counsel should evaluate all state statutes, local medical customs, and their program practices. The intention is to reduce the risks of civil as well as criminal exposure among the nonphysician exercise staff.

State Statutory Definitions for Allied Health Professionals

Most states provide practice definitions for certain allied health professionals such as physician assistants, midwives, physical therapists, emergency medical technicians, paramedics, nurses, registered nurses, and licensed practical nurses. Some state definitions could potentially affect those engaged in exercise testing and prescription or even those involved in adult fitness, preventive, or rehabilitation programs. Generally, the defined services for these allied health professionals are more restrictive than the definitions pertaining to the practice of medicine. In addition, the definitions for allied healthcare providers sometimes refer specifically to supervision by or subservience to physicians during the course of service provision. For example, the term *physician assistant* is sometimes defined as "a skilled person qualified by academic and clinical training to provide services to patients under the supervision and direction of a licensed physician or group of physicians who are responsible for his performance" (50). Other allied health professionals such as physical therapists or even nurses often have similar statutes in place.

Exercise professionals and their individual legal advisors must review these enactments to determine what statutory constraints may be in effect. Besides supervision requirements, these regulations may also mandate the development and use of an emergency plan, minimum standards for facilities and equipment, and a statutory basis for minimum personnel competencies. These regulations may even specify requirements for program records, organizational policies and procedures, and a means to review and expunge (when necessary) program records. Regulations will probably become more common in the years ahead, although to date, state involvement in this area has been minimal at best (59).

Exercise Testing and Prescription As Medical Procedures

The prescriptive and therapeutic applications of exercise have gained wide acceptance for the management of eligible patients with ischemic heart disease, intermittent claudication attributable to peripheral arterial disease, cardiac pacemakers, heart failure, diabetes mellitus, chronic obstructive pulmonary disease, obesity, arthritic disorders, renal disease, and even mood disturbances (e.g., depression). Also, many physicians, after a patient concludes treatment in a therapeutic cardiopulmonary rehabilitation program, will want that individual to continue physical activity in a medically unsupervised cardiovascular maintenance program. Clinical exercise physiologists have a clear role in these referrals. However, without state statutory provisions defining practice areas, some may feel uneasy in their practice roles. This is especially true if they are not working in close proximity to the referring physician.

There is no question that physicians, exercise professionals, and others with varying levels of competency are independently and routinely engaged in designing, writing, and issuing exercise prescriptions for both apparently healthy people and those with clinically manifest disease. No doubt, many nonphysicians are also supervising and leading activity classes for both healthy people and patients with known disease. Even if they refer to these activities as program planning, fitness counseling, or exercise consultations, they are in fact providing prescriptions and, perhaps, even medical treatment in some settings. In these respects, the possible legal ramifications of such activities are potentially devastating to those who are nonphysicians.

Practical application 3.2 presents pertinent litigation involving clinical exercise physiology and exercise testing.

The use of graded exercise testing as a medical procedure has increased substantially in recent years (15, 65). Test protocols involving electrocardiography, echocardiography, or myocardial perfusion imaging are now widely used to diagnose coronary artery disease, measure changes in disease status, stratify risk, and guide treatment plans. Although graded exercise testing initially was applied in the 1970s to stable patients cared for in the outpatient or ambulatory care setting, by the late 1980s more than 40% of patients suffering an acute myocardial infarction underwent a graded exercise test (GXT)—many as early as 1 to 2 weeks after their event. Given this, it's no surprise that many physicians now carefully and deliberately evaluate their patients as a means to minimize the likelihood of malpractice claims due to alleged omission.

In fact, a 1988 Los Angeles verdict awarded $500,000 to a plaintiff's family after a 40-year-old male died after sustaining a myocardial infarction. The decedent had sought medical clearance prior to beginning an exercise program. A prior physical examination without stress testing failed to reveal the presence of exercise-induced ischemia, and the family sought compensation from the physician, contending he was negligent. After the award, some attorneys contended that a GXT is now necessary to screen all individuals over 40 years of age to determine the presence of silent ischemia (9, 22, 26, 28, 37, 44).

The clinical exercise physiologist should consider two important issues relative to his or her role in exercise testing and prescribing exercise as a medical therapy. First, issuing prescriptions without completing necessary prerequisites such as a GXT can readily expose the prescriber to potential liability for negligence or malpractice (6). Clearly, a properly conducted GXT is almost a universally accepted prerequisite to any exercise prescription, especially among individuals 40 years of age or older with one or more important risk factors for ischemic heart disease (9, 26, 28).

Second, if the issuance of an exercise prescription by a nonphysician is legally interpreted to be the practice of medicine (based on applicable state statutes or case **precedents**), the standard of care owed by the professional who (inadvertently or otherwise) engages in the practice of medicine would be substantially increased. Specifically, he or she would be held to the standard of care normally expected of physicians, as distinguished from that expected and provided by exercise professionals. Such a determination could have two important consequences:

1. In the event of participant injury, the provider's conduct will be compared with that of a physician acting under the same circumstances. If the person's actions do not

Pertinent Litigation

Each clinical exercise physiologist should be aware that substantial litigation has taken place in cases dealing with exercise testing procedures (62), prescriptions or recommendations for exercise activity (9, 18), and even alleged deficiencies in emergency response procedures (42). We examine below two cases arising out of these areas.

Tart vs. McGann

This case involved an airplane pilot who suffered a heart attack following a medically supervised GXT. The pivotal question in the case was whether the stress test was performed in accordance with proper medical standards and whether the test should have been stopped sooner. The plaintiff's testimony indicated that in the fourth segment of the test he complained of feeling fatigued, but that the doctor encouraged him to complete the segment (62).

As in most medical malpractice cases, eminently qualified doctors testified for both the plaintiff, who claimed he had been overstressed during the procedure, and for the defendant doctor who supervised the plaintiff's test. The defendant doctor testified that the plaintiff did not look unduly stressed during the test. He supported the view that unless there was a marked showing of distress, there was no need to stop the test.

Although all protocols require stopping a stress test on signs of obvious distress, the plaintiff's expert cited the standards promulgated by the American Heart Association (which he helped write). He also testified that the plaintiff's heart rate was brought up too high and he exceeded 12 metabolic equivalents (METs), when 6 METs would have been proper based on the plaintiff's physical condition and history.

The jury received the case for determination and focused their deliberations on whether the plaintiff's facial expressions should have been sufficient to require the doctor to stop the test, even though the plaintiff made no other obvious or overt request to stop the procedure. The trial judge instructed the jury that a mere facial expression of fatigue could not be equated with a request to stop the test. The jury brought in a verdict for the defense. This case was appealed and the appellate court held that the trial judge's answer to the jury's questions about facial expression as a form of fatigue was inappropriate. Based in part on that ruling, the plaintiff was granted a new trial.

Malpractice is the failure to follow reasonably competent medical practice or the appropriate standards of care. In cases like this one involving injuries alleged to have been suffered during a claimed inappropriately prolonged stress test, expert testimony is often based on professional standards, such as those provided by the American Heart Association and the American College of Sports Medicine.

From a legal standpoint, observance of the most conservative and accepted protocols will always offer the largest umbrella of legal protection. Therefore, it is important for clinical exercise physiologists to follow monitoring procedures as published in national guidelines. This includes but is not limited to assessment of blood pressure, heart rate, heart rhythm, fatigue, perspiration, and facial expressions of strain. As well, any expressed statements of extreme or unusual fatigue or pain may provide strong evidence of a mandatory signal to end a test. Failure to properly monitor such conditions or terminate a test where appropriate might result in a claim similar to that presented in this case.

Following the Court of Appeals decision, the Tart vs. McGann case was retried by another jury. There was another defense verdict but the plaintiff received $50,000 by agreement.

Although the experts in the case quoted the standards published by the American Heart Association, it is interesting to note that a 1980 publication by the American College of Sports Medicine titled *Guidelines for Graded Exercise Testing and Exercise Prescription* (2nd ed.) clearly made a patient's facial expression of fatigue, along with other indicators, a determining factor to be considered in deciding whether to terminate a GXT. Use of this statement might have been helpful in either the first or second trial of the Tart vs. McGann case; however, it was not cited by any expert. Reference to stopping an exercise test for facial expressions was deleted from the third edition of the American College of Sports Medicine guidelines published in 1986. However, the deletion of this passage

continued

does not, in the absence of some justifying rationale, completely diminish the potential use of this standard in court for testimony addressing proper exercise stress testing procedures. Moreover, absent revocation of the second edition by reason of the publication of the third or later editions, it might be possible to cite the second edition for authority during court testimony in litigations addressing injuries after 1985. Certainly, the second edition would be available for use in incidents occurring during the period of its publication, 1980 to 1985.

Mandel vs. Canyon Ranch, Inc., et al.

The complaint in the case sought unspecified damages arising from the death of a 50-year-old Canyon Ranch guest who died at the spa while playing walleyball in 1995. The decedent had a history of coronary disease and, according to the pretrial testimony of one of the plaintiff's experts, "went to Canyon Ranch . . . was screened there for medical problems and subsequently had a cardiac arrest, a witnessed cardiac arrest and was unsuccessfully resuscitated" (34, 42).

Cardiopulmonary resuscitation and other care were provided by other guests and members of the defendants' staff; however, according to the allegations of the complaint, no staff member provided any type of cardiac resuscitation equipment or cart (e.g., a defibrillator) in the course of providing care. The plaintiffs alleged that the defendants were negligent in their screening and evaluation of the decedent, failed to appropriately handle his medical emergency, and failed to monitor and bring the appropriate medical equipment to handle the emergency. The latter included, but was not limited to, failure to bring and use a heart defibrillator unit that (the plaintiffs further alleged) is kept on the Canyon Ranch premises for this type of situation.

One of the critical issues in the case centered on whether the defendant (Canyon Ranch) was required, as part of the standard of care for health and fitness facilities, to have available (and use) a defibrillator in instances like this case. The plaintiff's expert witness, a practicing cardiologist (who also was a part owner of another health and fitness facility) took the position that the Canyon Ranch was more than an ordinary fitness facility. Instead, he believed it was more like a medical facility and, as such, owed a higher standard to those who used the facility. This meant, in his view, responding to an untoward event like that which occurred here, with a defibrillator.

One of the defendants' experts, a clinical exercise physiologist and coauthor of this chapter, testified that the defendant (Canyon Ranch) did adhere to the appropriate standard of care in dealing with the decedent's untoward event. In the end, a lay jury, presumably relying on instructions of law from the judge overseeing the case, made a defense ruling. As the decision in this case seems to indicate, the availability and use of defibrillators in health and fitness facilities (at least those that are not truly and perhaps solely based on a medical model or those that are rehabilitative in nature) for untoward events do not as of 1995 appear to be part of the standard of care owed to consumers. At the time, neither the standards of the American Heart Association, the American College of Sports Medicine, nor the International Health, Racquet and Sportsclub Association presently require the presence and use of defibrillators in health and fitness facilities. (Interestingly, the International Health, Racquet and Sportsclub Association now encourages their member clubs to have an automatic external defibrillator (AED) on site and many facilities have opted to do so.)

Despite the verdict in this case and the present state of the applicable standards, the foregoing conclusion may not always hold true. Standards of practice are developing and changing. Future editions of such standards may mandate what is not yet required. Juries in other cases may also decide these cases in different ways.

Interestingly, in the second case, both experts testified that each had previously had three instances where individuals in a health/fitness facility or cardiac rehabilitation program had suffered an event and had to be defibrillated. In each case, the individual survived.

correspond with the presumed actions of a physician, he or she will be found to have failed to exercise proper due care. Because such individuals will generally not be in a legal position to defibrillate or administer care in the event of medical emergencies (48), those professionals will almost always be found negligent if such failure proximately caused injury or death to a program participant.

2. In any case where an individual is found to have engaged in the unauthorized practice of medicine, he or she could be criminally liable under many state statutes wherein the practice of medicine without a license is determined to be a crime. Thus, such professionals could be faced with two possible suits brought by two separate and distinct entities, charging two different violations of law: one civil and one criminal.

In 1996, the American College of Sports Medicine approved a scope of practice statement for registered clinical exercise physiologists that may solve some of these practice problems. That scope of practice statement is as follows:

> The Clinical Exercise Physiologist works in the application of exercise and physical activity for those clinical and pathological situations where it has been shown to provide therapeutic or functional benefit. Patients for whom services are appropriate may include, but not be limited to those with cardiovascular, pulmonary, metabolic, immunological, inflammatory, orthopedic, and neuromuscular diseases and conditions. This list will be modified as indications and procedures of application are further developed and mature. Furthermore, the Clinical Exercise Physiologist applies exercise principles to groups such as geriatric, pediatric, or obstetric populations, and to society as a whole in preventative activities. The Clinical Exercise Physiologist performs exercise evaluation, exercise prescription, exercise supervision, exercise education, and exercise outcome evaluation. The practice of Clinical Exercise Physiologists should be restricted to clients who are referred by and are under the continued care of a licensed physician.

Many exercise professionals by training or for ethical reasons take an involved or significant role in exercise testing, prescription, and supervision, even in programs developed for patients with known disease. Until state statutes are modified to provide clear, statutory definitions for professionals engaged in exercise prescription, leadership, and supervision, and until applicable professional guidelines become uniform, practical solutions are necessary to mini-mize the legal risks for nonphysicians. The potential legal exposure of exercise professionals to claims of civil or criminal unauthorized practice of medicine is best minimized through individual legal and other professional advice. Figure 3.3 provides several general suggestions for limiting legal exposure.

Aside from adopting approaches highlighted in figure 3.3, clinical exercise physiologists must ensure that various levels of emergency support are available for untoward events. However, emergency support should be used only in ways deigned to preclude claims related to the unauthorized practice of medicine while still responding effectively to the event.

The Clinical Exercise Physiologist's Duties and Responsibilities

In addition to state definitions and statutes, professional standards and guidelines also guide the professional duties and responsibilities of clinical exercise physiologists. These duties are summarized in figure 3.4, many of which are the subject of a number of cases and significant decisions (57). Cases are presently pending that may further influence what particular responsibilities may be imposed on exercise physiologists in some circumstances (13).

Also, within the confines of principally the civil justice system, under which personal injury and wrongful death cases are determined, clinical exercise physiologists could be held accountable to a variety of professional standards and guidelines (29). These standards and guidelines, sometimes referred to as practice guidelines, professional statements, or parameters of practice, can apply to some practices carried out by clinical exercise physiologists and individuals who provide services within health and fitness facilities. These guidelines or standards dealing with the provision of service by fitness professionals, including clinical exercise physiologists, have been developed and published by a variety of professional organizations (1-5, 7, 8, 10, 29, 39). Practically, standards and guidelines statements are used by professionals to identify probable and evidence-based benchmarks of expected service delivery owed to patients and clients, which helps ensure that the delivery of service is appropriate and uniform.

In the course of litigation, these standards statements are used through expert witnesses to establish the standard of care to which providers will be held accountable in the event of patient injury or

— In programs that include patients with known disease, and where diagnosis, prescription, or treatment is the probable or actual reason for performing the procedures, a physician should be involved in a significant way. This involvement must be meaningful and real if it is to be legally effective. Examples include referring patients, receiving and acting on results provided by rehabilitation staff, and discussing findings with the rehabilitation or therapy staff.

— In programs with patients who demonstrate a high risk of suffering an adverse event or injury during exercise, a physician should be on call and physically present whenever exercise sessions are held. During initial testing of such participants, the physician should be within immediate proximity to the participant, although not necessarily watching the patient or the electrocardiogram monitor, but controlling the staff during the procedure. Thereafter, the physician's proximity of supervision in further exercise testing should be dictated by the medical interpretation of each patient's initial test result.

— Concerning apparently healthy adult participants of any age who present with exertional symptoms or medical contraindications from the referring physician, physician involvement may be limited to consultation. This consultation should focus on the development of proper policies and procedures as well as periodic supervision and continuing staff education. An appropriately trained physician or a trained and state-authorized allied healthcare provider may need to be available to properly respond to emergency test-related complications (22).

FIGURE 3.3   Means to limit legal exposure and risk among clinical exercise physiologists.

— Evaluate and screen participants before recommending or prescribing activity, especially in rehabilitative or preventive settings.

— Secure a medically mandated consent-type document when involved in procedures such as a graded exercise test.

— Develop and assist individuals with implementing a safe and evidenced-based exercise prescription that is aimed at improving health, fitness, and quality of life yet considers a patient's goals and extent of disease-related disability.

— Recognize and refer individuals who have conditions that need evaluation before commencing or continuing with various activity programs.

— Provide appropriate leadership and supervision for activity that is prescribed or recommended.

— Provide feedback to referring professionals relative to the progress of participants in activity.

— Provide appropriate, timely, and effective emergency care as needed.

FIGURE 3.4   Summary of expected behaviors for clinical exercise professionals.

death. In years past, expert witnesses have often used their own subjective and personalized opinions to establish the standard of care and then provided an opinion as to whether that standard was violated in the actual delivery of service. Many organizations sought to develop evidence-based standards of care to achieve the following:

— Achieve a consensus among professionals that practitioners could aspire to meet in their own delivery of services in accordance with known and established benchmarks of expected behavior

— Reduce cases of negligence

— Minimize the significance of individually and subjectively expressed opinions in court proceedings

— Reduce adverse findings of negligence or malpractice

— Reduce potential inconsistent verdicts and judgments that arose from the expression of individual opinions

However, despite the promulgation and publication of a variety of such statements, many professionals choose not to follow such standards because they feel that the delivery of relevant healthcare services cannot be provided by using a cookbook approach. Studies undertaken by researchers at Harvard University determined that healthcare-related standards of practice were actually being used more often than not to attack the care that was provided under some circumstances. However, those same researchers determined that when standards had been followed in the rendition of services, there was a greater likelihood that claims against healthcare providers would not be asserted or would be dismissed without final determination by a judge or jury (14, 16).

Regardless of whether clinical exercise physiologists decide to follow practice guidelines, those statements are a ready reference available to compare the actual delivery of service to what the profession as a whole has determined to be established benchmarks. As a consequence, clinical exercise physiologists should review and become aware of relevant published practice guidelines and then consider in what situations the standards might act as a "shield" to protect against negligent actions. Others may engage in the same process to determine if such standards should be used as a "sword" to attack the care provided to clients in particular cases (14).

Conclusion

The legal concerns associated with the practice of clinical exercise physiology are many and multifaceted. This chapter only scratches the surface of what is involved. Legal concerns related to clinical exercise physiologists providing service within the health and fitness industry are constantly developing and evolving. Thus, clinical exercise physiologists should consult other resources (36, 64) to stay abreast of pertinent legal issues. Doing so will help protect the public and minimize one's own legal exposure.

Case Study 3

To appreciate what can be at stake for health professionals who become involved in courtroom litigation, consider this case of an exercise physiologist working in the clinical setting (55). She wrote of her litigation experiences in the first person, and excerpts follow.

"The hospital where I work is a tertiary care facility. . . . The staff members in Cardiac Rehabilitation are required to have a Masters Degree in Exercise Physiology, American Heart Association Basic Cardiac Life Support certification, and American College of Sports Medicine certification as an Exercise Specialist. . . .

"In May of 1991, Cardiac Rehabilitation was consulted to see a 78 year old woman 4 days post aortic valve surgery. Her medical history included severe aortic stenosis, coronary artery disease (CAD), moderate-severe left ventricular dysfunction, carotid artery disease, atrial fibrillation, hypertension and non-insulin dependent diabetes. . . . She was in atrial fibrillation, but had stable hemodynamics, and was ambulating without problems in her room.

"The patient was first seen by Cardiac Rehabilitation seven days post surgery. Chart review . . . revealed no contraindications to activity. . . . Initial assessment revealed normal supine, sitting, standing and ambulating hemodynamics. She walked independently in her room, and approximately 100 feet in five minutes . . . using the hand rails for support. . . . General instructions for independent walking later that day were given. . . .

"On post-op day 8, there was documentation in the chart regarding the patient's need for an assist device while ambulating. . . . The patient ambulated approximately 200 feet with a quad cane. . . . She appeared stable. . . .

"On post-op day 9, I was informed by the patient's nurse, as well as by documentation in the chart, that the patient had fallen in her room that morning apparently hitting her head on the floor. . . . A Computed Tomography scan of the head was negative.

"On post-op day 10, I observed the patient sitting with her daughter in the hall. . . . Her mental status had noticeably changed. She did not recognize her daughter, where she was or why she was

continued

in the hospital. . . . No further formal exercise was performed. On post-op day 11, the patient was transferred to the intensive care unit and later died of a massive brain hemorrhage.

"Two years after the incident, a hospital lawyer contacted me and informed me that the family had brought suit against the hospital contending that the patient fell because she was not steady enough to walk independently. As a cardiac rehabilitation/exercise specialist assigned to the patient prior to the incident, I was called to testify and explain my participation in her care. . . .

"I met with the hospital's defense lawyer on two separate occasions. The first meeting, the lawyer inquired about my education, certification, years of employment and what specifically my job entailed. . . . two weeks before trial, the hospital's defense lawyer again questioned me. . . . He also played the role of the plaintiff's lawyer and reworded similar questions in a slightly different manner. The defense lawyer encouraged me to maintain consistency in my answers, and remain calm and assured, even under cross-examination. But to prove our case, we had to establish, by a preponderance of the evidence (i.e., more likely than not) that the patient was indeed ambulating independently prior to her fall. My documentation was crucial. . . .

"As I took the stand . . . The defense attorney was the first to question me, and I soon realized the benefits of the briefings. No surprises were encountered here. . . . Answering the plaintiffs' lawyer's questions during cross examination was like a mental chess game. . . . my answers were crucial, as was my composure and the belief that all my training and certification DID qualify me as a professional health care provider. . . . Although the only hard evidence I had to work with in the courtroom were my chart notes, the judge did allow me to supplement them with oral testimony. . . .

"Documenting the actual exercise session is written in our Policies and Procedures Manual and covered in each employee's orientation. Little did I realize how those documented notes written in the patient's chart so long ago would come back to save me . . .

"Whether you document in the patient's chart or on a separate summary sheet that is filed in your office, your documentation is a permanent record that provides powerful legal evidence. . . . Your documentation can be your best defense showing that the appropriate standard of care was provided. Or, it can be the plaintiff's best evidence, proving that substandard care was delivered.

"A lawsuit can be won or lost, by what is, or is not, documented in the records. . . .

"Later I was told that my testimony helped the hospital prove the patient's fall was not a result of negligence. The court ruled in favor of our hospital."

Case Study 3   Discussion Questions

1. Describe the important interrelationships between policy and procedure manuals that define departmental standards of care, a thorough employee orientation and training period, and the resolution of potential legal issues.

2. Clinically, why is it important to review notes made by other health professionals before you provide care to a patient? In what manner do your notes help others who provide care in the future?

3. Are you aware of any professional or scientific exercise guidelines that lawyers for a plaintiff or defendant might review to see if the level and nature of care provided were consistent with the accepted standard of care?

4. What are the key elements or notes that should be recorded or documented in a patient's record after you are finished caring for the patient?

Glossary

civil action—A type of action asserted to preserve a private right or to correct a private wrong.

defendant—Party against whom claims of injury are made.

expert witness—A person whose particular wisdom of a subject matter, obtained through his or her training, education, achievements, accomplishments, or experience, enables him or her to give opinions and form conclusions in testimony during litigation.

malpractice—A failure by a doctor, attorney, accountant, or other type of professional to render services in accordance with prevailing standards.

manslaughter—The unlawful or negligent killing of a person without malice, plan, or premeditation.

misdemeanor—A crime penalized by imprisonment for less than 1 year or a fine; not as serious as a felony.

murder—An illegal killing of another individual with malice aforethought, committed either intentionally or recklessly with a lack of interest for the importance of human life.

negligence—The unintentional failure to apply reasonable care that a careful person would exercise, or behavior that breaches specific legal principles of due care.

personal injury—An injury to one's person or reputation.

plaintiff—Injured party who seeks compensation by asserting a claim.

precedent—A formerly determined case that is used as an illustration or rule for later cases that are similar.

standard of care—A fundamental proposition of negligence law which provides that we ordinarily owe other people an obligation to exercise reasonable care in our behavior. This obligation states that we should not injure others by reckless behavior, and if a person's behavior is reckless, the duty is breached for not exercising reasonable care.

tort—Conduct by a defendant that is the consequence of a breach of some legal obligation due to the plaintiff by a defendant that proximately produces injury to the plaintiff.

waiver—Voluntary and willful relinquishment of a known benefit or right.

wrongful death—A type of claim in which a survivor of a decedent's estate brings suit for some unlawful action that caused the decedent's death.

References

1. Aerobic and Fitness Association of America. Fitness: Theory and Practice. Sherman Oaks, CA: Aerobic and Fitness Association of America, 1990.

2. American Association of Cardiovascular and Pulmonary Rehabilitation. Guidelines for Cardiac Rehabilitation and Secondary Prevention Programs, 3rd Ed. Champaign, IL: Human Kinetics, 1998.

3. American College of Cardiology. Recommendations of the American College of Cardiology on Cardiovascular Rehabilitation. J Am Coll Cardiol 7: 451-453, 1986.

4. American College of Sports Medicine. ACSM's Guidelines for Exercise Testing and Prescription, 6th Ed. Baltimore: Lippincott, Williams & Wilkins, 2000.

5. American Heart Association/American College of Sports Medicine. Rehabilitation for Cardiovascular Science, Safety and Emergency Policies at Health/Fitness Facilities. Circulation 97: 2283-2293, 1998.

6. American Heart Association. Exercise Standards. Circulation 82: 2286-2322, 1990.

7. American Heart Association. Exercise Standards: A Statement from the American Heart Association. Circulation 91: 580-615, 1995.

8. American Medical Association. Promotion of Exercise Within Medicine and Society. AMA Policy Compendium, Chicago: American Medical Association, 1990.

9. Are GXTs required for screening of all men over 40? The Exercise Standards and Malpractice Reporter 2(2): 18, 1988.

10. Balady, G.J., P.A. Ades, P. Comoss, et al. Core competencies of cardiac rehabilitation/secondary prevention. Circulation 102: 1069-1073, 2000.

11. Boulet, B. Licensure of clinical exercise physiologists in Louisiana: A retrospective look at the process. The Exercise Standards and Malpractice Reporter 9(6): 81-85, 1995.

12. California (King vs. Board of Medical Examiners, 65 Cal. App. 2d 644, 1944).

13. Chai vs. Sports and Fitness Clubs of America, Inc., Case No. 98-16053CA(05), Broward County Circuit Court (Florida, 17th Judicial Circuit) in failure to defibrillate results in new litigation. The Exercise Standards and Malpractice Reporter 13(4): 55-56, 1999.

14. Doctors' Own Guidelines Hurt Them in Court. Wall Street Journal October 19, 1994.

15. Exercise stress testing procedures to be increased? The Exercise Standards and Malpractice Reporter 4(5): 74, 1990.

16. Future of clinical guidelines could turn on 'perversion' by plaintiff's bar, doctors say. Medical Liability Monitor 20(2): 1-2, 1995.

17. Georgia is ground zero for scope-of-practice firefight. American Medical News 43(5): 1, 30-31, 2000.

18. Gonzales vs. Roth, et al., Case No. 90CV1570, Dist. Court, City and County of Denver, Colorado (1991 Defense verdict) in Herbert. New litigation: Exercise within GXT results in defense verdict. The Exercise Standards and Malpractice Reporter 5(6): 90-92, 1991.

19. Gore, J.M., R.J. Goldberg, J.S. Alpert, and J.E. Dalen. The increased use of diagnostic procedures in patients with acute myocardial infarction. A community-wide perspective. Arch Int Med 147: 1729-1732, 1987.

20. Hawkins, J. The role of written wellness assessments in exercise programming. The Exercise Standards and Malpractice Reporter 3(1): 8-11, 1989.

21. Herbert, D. Are nutritionists engaged in the unauthorized practice of medicine? The Exercise Standards and Malpractice Reporter 2(5): 76-78, 1988.

22. Herbert, D. Health screening questionnaires: An alleged failure to diagnose impending MI by non-physician results in defense verdict. The Exercise Standards and Malpractice Reporter 3(6): 92-93, 1989.

23. Herbert, D. Legal aspects of nutritional advice. Fitness Management February 1989.

24. Herbert, D. Appropriate use of wellness appraisal. Fitness Management September 1989.

25. Herbert, D. Liability associated with dieting advice and publications. The Exercise Standards and Malpractice Reporter 4(1): 9, 1990.

26. Herbert, D. New litigation: Exercise recommendation without GXT results in defense verdict. The Exercise Standards and Malpractice Reporter 5(6): 90-92, 1992.

27. Herbert, D. The death of Hank Gathers: An examination of the legal issues. The Exercise Standards and Malpractice Reporter 4(4): 57-58, 1990.

28. Herbert, D. Exercise prescription without testing results in a plaintiffs award. The Exercise Standards and Malpractice Reporter 6(1): 5, 1992.

29. Herbert, D. Standards Book for Exercise Programs. Canton, OH: PRC, 1992.

30. Herbert, D. Is license "the future" for CEPs? The Exercise Standards and Malpractice Reporter 10(1): 1, 4, 1996.

31. Herbert, D. California considers new exercise professional legislation. The Exercise Standards and Malpractice Reporter 11(3): 41-43, 1997.

32. Herbert, D. Concerns related to California Senate Bill 891 and licensure of exercise/fitness professionals. The Exercise Standards and Malpractice Reporter 12(1): 1, 4-7, 1998.

33. Herbert, D. Update in California legislative proposal for exercise professionals. The Exercise Standards and Malpractice Reporter 12(1): 8, 1998.

34. Herbert, D. Are defibrillators part of the emergency response standard of care owed to patrons by health and fitness facilities? The Exercise Standards and Malpractice Reporter 12(4): 49, 52, 1998.

35. Herbert, D. Expanding the scope of practice for fitness professionals—Is it a question of turf protection? The Exercise Standards and Malpractice Reporter 13(1): 1, 4, 1999.

36. Herbert, D., and W. Herbert. Legal Aspects of Preventative, Rehabilitative and Recreational Exercise Programs, 4th Ed. Canton, OH: PRC, 2002.

37. Herbert, W. The death of Hank Gathers: Implications to the standard of care for preparticipation exercise screening? The Exercise Standards and Malpractice Reporter 4(4): 56-57, 1990.

38. Herbert, W. Licensure of clinical exercise physiologists: Impressions concerning the new law in Louisiana. The Exercise Standards and Malpractice Reporter 9(5): 65, 68-70, 1995.

39. International Health, Racquet and Sportsclub Association. Standards for Member Clubs. Boston: International Health, Racquet and Sportsclub Association, 1997.

40. Is an exercise physiologist qualified to express expert opinions in litigation? The Exercise Standards and Malpractice Reporter 12(6): 81, 84, 1998.

41. Louisiana licenses clinical exercise physiologists. The Exercise Standards and Malpractice Reporter 9(4): 56-59, 1995.

42. Mandel vs. Canyon Ranch, Inc., et al., Superior Court of Arizona, Prima County, Case No. 312777, 1998 Defense verdict, in Herbert. Failure to defibrillate case results in a defense verdict. The Exercise Standards and Malpractice Reporter 12(5): 71-76, 1998.

43. Michigan (People vs. Bovee, 92 Mich. App. 42, (1979), Attorney General vs. Beno, 422 Mich. 293, 1985).

44. "Minimal" standard of care for non-specific arrhythmia required monitoring and exercise testing. The Exercise Standards and Malpractice Reporter 6(5): 74, 1992.

45. Mississippi (Norville vs. Mississippi State Medical Association, 364 So.2d 1084, Miss. 1978).

46. More proposals to license exercise professionals. The Exercise Standards and Malpractice Reporter 12(1): 12-14, 1998.

47. North Carolina (State vs. Baker, 229 N.C. 73, 1948).

48. New standards for CPR. The Exercise Standards and Malpractice Reporter 6(6): 93, 1992.

49. New York cardiovascular exercise training programs must be physician supervised. The Exercise Standards and Malpractice Reporter 6(3): 44-45, 1992.

50. Ohio Revised Code §4730.01.

51. Peoples vs. Burroughs, Supreme Court of California, Crim. 23151 (April 19, 1984).

52. Ribisl, P. Certification or licensure of health/fitness professionals. The Exercise Standards and Malpractice Reporter 5(2): 22-24, 1991.

53. Section 333.16215, Michigan Public Health Code.

54. Senate Bill No. 597 (Louisiana), Reprinted in Louisiana licenses clinical exercise physiologists. The Exercise Standards and Malpractice Reporter 9(4): 56-59, 1995.

55. Snader, C. Cardiac rehabilitation, legal issues and documentation: A case report. The Exercise Standards and Malpractice Reporter 8(6): 84-86, 1994.

56. Sol, N. Certification or licensure of fitness professionals: The debate begins. The Exercise Standards and Malpractice Reporter 4(5): 65-69, 1990.

57. Speigler vs. State of Arizona, Case No. CV-92-1300 (Arizona, Maricopa County Superior Court). The Exercise Standards and Malpractice Reporter 10(2): 30-31, 1996.

58. State of Georgia (Foster vs. Board of Chiropractic Examiners, 359 S.E.2d 877, 1977).

59. State regulation of exercise test administration. The Exercise Standards and Malpractice Reporter 6(6): 95, 1992.

60. State vs. Winterich, 157 Ohio St. 414 (1952).

61. Stetina vs. State Medical Licensing Board, 513 N.E. 2d (1234 Ind. App. 2 District, 1987).

62. Tart vs. McGann, 697 F 2d 75 (2nd Cir. 1982) in Edelman. The Case of Tate vs. McGann: Legal implications associated with exercise with exercise stress testing. The Exercise Standards and Malpractice Reporter 1(2): 21-25, 1987.

63. Title 37 of the Louisiana Revised Statute of 1950, 37: 3433.

64. The Exercise Standards and Malpractice Reporter, published six times each year by PRC Publishing, Inc., Canton, OH.

65. The exclusive use of GXTs to determine ischemia related disability stricken by court. The Exercise Standards and Malpractice Reporter 5(1): 9, 1991

66. Washington (State vs. Wilson, 528 P.2d 279, 1974).

Peter A. McCullough, MD, MPH, FACC
Consultant Cardiologist and Chief
Division of Nutrition and Preventative Medicine
William Beaumont Hospital
Royal Oak, MI

General Evaluation and Examination Skills

Clinical exercise physiologists work in many settings that require them to assess patients with various health problems. This chapter focuses on the elements of the clinical evaluation conducted by a physician or physician extender, as well as the measurements that the clinical exercise physiologist may need to obtain to determine if it is safe for the patient to exercise.

For the most part, the clinical evaluation of any patient usually involves two steps. First, a general interview is conducted to obtain historical and current information. This is followed by a physical assessment or examination, the extent of which may vary based on who is conducting the examination and the nature of the patient's symptoms or illness. Once these are completed, a brief numerical list is generated that summarizes the assessment, relative to both prior and current findings and diagnoses. Finally, a numerical plan is generated indicating the one, two, or three key actions that are to be taken as they relate to the care of the patient. This chapter describes in detail the general interview and physical examination components of the clinical evaluation.

Clinical Evaluation

The general interview is a key step in establishing the patient database, which is the working body of knowledge that the patient and the clinical exercise physiologist will share throughout the course of treatment. This database is primarily built from information obtained from the patient's hospital or clinical records. However, you will need to interview the patient to obtain information that is missing, as well as update data to address any changes in the patient's clinical status since his or her last clinic visit of record. With experience you will learn which information is incomplete or requires questions of the patient. Figure 4.1 lists the essential components of the general interview, most of which you will need to enter into the patient file or database.

- Demographics (age, gender, ethnicity)
- Reasons for referral
- History of present illness (HPI)
- Current medications
- Allergies
- Past medical history
- Family history
- Social history
- Physical examination
- Assessment
- Plan

FIGURE 4.1   Essentials of clinical evaluation for the new patient referred for exercise therapy.

Reprinted, by permission, from P.A. McCullough, 1999, Clinical Exercise Physiology 1(1): 33-41.

Reason for Referral

The reason for referral for exercise therapy is generally self-explanatory and may include one or more of the following: to improve exercise tolerance, improve muscle strength, increase range of motion, or provide relevant education and behavioral strategies to reduce future risk. However, the clinical exercise physiologist may need to reconcile the physician's reason for referral and the patient's understanding of the need for therapy. Differences can exist between the two. For example, consider the 48-year-old sedentary patient employed as an automotive company executive who undergoes single-vessel coronary artery angioplasty and returns to work in just a few days without physical limitations. Unfortunately, sometimes these patients perceive that they are "cured" after their coronary revascularization, and therefore no rehabilitation or lifestyle adjustments

are required. When such a patient is referred for cardiac rehabilitation, he or she must understand that coronary artery disease is a dynamic disorder that is influenced, just like his or her original problem, by lifestyle habits and medications. The clinical exercise physiologist plays an important role in enforcing long-term compliance to physical activity, hypertension and diabetes management, proper nutrition, and medical compliance—all key components of secondary prevention.

Demographics

Patient demographics such as age, sex, and ethnicity are the basic building blocks of clinical knowledge. A great deal of medical literature describes the relationship between this type of demographic information and health problems. For example, the prevalence of **congestive or chronic heart failure (CHF)** increases at an exponential rate over each decile of age beginning with age 60, with nearly 15% of the population over age 80 having some degree of heart failure (7). Age is also the most powerful risk factor for osteoarthritis, with 68% of individuals over the age of 65 having some clinical or radiographic evidence. Finally, age is an independent predictor of survival in virtually every cardiopulmonary condition, including acute myocardial infarction, stroke, peripheral arterial disease, and chronic obstructive pulmonary disease (12). Because these and other age-related diseases can influence a patient's ability to exercise, age becomes a key piece of information to consider when developing an exercise prescription.

It is also important to relate sex to outcomes such as behavioral compliance or disease management in patients with chronic diseases. For example, although the onset of cardiovascular disease is, in general, 10 years later in women than in men, the **morbidity** and **mortality** after revascularization procedures (i.e., coronary bypass or angioplasty) are higher in women (10). This sex-based difference remains an area of intense investigation, as clinical scientists strive to determine which biological or socioeconomic factors account for the poorer outcomes sometimes observed in women. Additionally, there are few data describing the positive or negative effects of exercise as an intervention for many chronic diseases in women, especially those with cardiopulmonary disease (3). When an exercise prescription is developed for women, unique compliance- and disease-related barriers and confounders need to be solved to improve exercise-related outcomes.

There is a great deal of information with respect to differences in health status between various ethnic groups. Most of these differences are attributable to socioeconomic status and access to care. There are, however, a few ethnic-related differences that are worth mentioning. For example, obesity, hypertension, renal insufficiency, and left ventricular hypertrophy are all more common in African American patients with cardiovascular disease than their age- and sex-matched Caucasian counterparts. Likewise, diabetes and insulin resistance are more common in Hispanic Americans and some tribes of Native Americans (4).

The clinical exercise physiologist should consider these issues when developing, implementing, and evaluating an exercise treatment plan. This information may influence one's decision when trying to decide which risk factors should be addressed first. Also, program expectations and outcomes may be influenced as well.

History of Present Illness

The purpose of this element is to record and convey the primary information related to the condition that led to the patient to be referred to a clinical exercise physiologist. It usually begins with the "chief complaint," which is usually one sentence that sums up the patient's comments. The body of the history of present illness is a paragraph summarizing the manifestations of the illness as they pertain to pain, mobility, nervous system dysfunction, or alterations in various other organ system functions (e.g., circulatory, pulmonary, musculoskeletal, skin, and gastrointestinal). Important elements of the illness are reviewed such as the date of onset, chronicity of symptoms, types of symptoms, exacerbating or alleviating factors, major interventions, and current disease status. Traditionally, this is a paragraph describing events in the patient's own words. A practical approach for the clinical exercise physiologist is to incorporate reported symptoms with information from the patient's medical record.

For patients with cardiovascular disease, the features of chest pain should be described (table 4.1). Such a description can help in the future application of diagnostic testing, when it comes to assessing pretest probability of underlying obstructive coronary artery disease (5). Standard classifications should be used, if possible, such as the Canadian Cardiovascular Society functional class for angina or the New York Heart Association functional class (table 4.2) (2). The Specific Activity Scale is the least commonly used classification scheme (1). For patients with pain attributable to muscular, orthopedic, or abdominal problems, the important elements of the illness such as chronicity, type of symptoms, and exacerbating or alleviating factors need to be addressed.

The last sentence of the history of present illness can serve as an abbreviated review of other organ systems. Specifically, it is important to include an abbreviated review of organ systems not directly affected by the present illness. Doing so indicates that the examiner has inquired about symptoms associated with other organ system problems (e.g., joint discomfort, gait, balance, speech). Absence of problems or "negative" responses are important because they establish the baseline condition in the patient database.

TABLE 4.1
. . . . . .
Features of Chest Pain

Type of discomfort	Quality	Radiation	Exacerbating and alleviating factors
Typical	Heaviness, pressure-like squeezing	To neck, jaw, back, left arm, less commonly the right arm	Worsened with exertion or stress, relieved with rest or nitroglycerin
Atypical	Sharp, stabbing, pricking, tingling	None	No clear pattern, can happen anytime
Non-cardiac	Discomfort clearly attributed to another cause	Not applicable	Not applicable

Reprinted, by permission, from P.A. McCullough, 1999, *Clinical Exercise Physiology* 1(1): 33-41.

TABLE 4.2
Classification of Cardiovascular Disability

Class I	Class II	Class III	Class IV
NEW YORK HEART ASSOCIATION FUNCTIONAL CLASSIFICATION			
Patients with cardiac disease but no physical limitations (e.g., undue fatigue, palpitation, dyspnea, or anginal pain)	Patients with cardiac disease resulting in slight limitation (fatigue, dyspnea, angina) of physical activity	Patients with cardiac disease with marked limitation of physical activity	Patients with cardiac disease who are unable to carry out physical activity without discomfort
	Comfortable at rest	Symptoms such as fatigue, palpitations, and dyspnea occur with less than ordinary physical activity	Symptoms may even be present at rest
			Physical activity worsens symptoms
SPECIFIC ACTIVITY SCALE			
Can perform activity requiring ≥7 METs such as doing outdoor work (shovel snow), play basketball, jog or walk 5 miles · hr^{-1}, and carry objects	Can perform activities requiring ≥5 and ≤7 METs; sexual intercourse without stopping, rake leaves, weed garden, in-line skate, walk at 4 miles · hr^{-1} on level ground	Can perform activity requiring between ≥2 and ≤5 METs; shower without stopping, make bed, clean windows, sweep garage, walk 3 miles · hr^{-1} on level ground, bowl, golf	Cannot perform activities requiring ≥2 METs
			Cannot carry out activities listed in Class III
CANADIAN CARDIOVASCULAR SOCIETY FUNCTIONAL CLASSIFICATION			
Routine daily physical activity does not produce angina	Slight limitation of ordinary activity when walking or climbing stairs rapidly, walking uphill, doing exertion after meals, in cold or windy conditions, under emotional stress, or within a few hours after awakening	Marked limitation of routine daily activities such as walking one or two blocks on level ground or climbing more than one flight of stairs in normal conditions	Unable to carry out physical activity without discomfort
Angina occurs with rapid pace or prolonged exertion	No symptoms when walking more than two blocks on level ground or climbing more than one flight of ordinary stairs at a normal pace and in normal conditions		Angina may be present at rest

Reprinted, by permission, from American Alliance of Health, Physical Education, Recreation and Dance, 1999, *Physical education for lifelong fitness: The physical best teacher's guide* (Champaign, IL: Human Kinetics), 127.

Medications and Allergies

A current medication list is an essential part of the clinical evaluation, especially for the practicing clinical exercise physiologist, because certain medications can alter physiological responses during exercise. In fact, asking about current medications is an excellent segue into obtaining relevant past medical history. Compare the medications that patients state they are taking against what you think they should be taking given the medical information they report. For example, medical therapy for patients with chronic heart failure usually includes a diuretic agent, digoxin, a β-adrenergic blocking agent (i.e., β-blocker), and a vasodilator such as an angiotensin-converting enzyme inhibitor. If during your first evaluation of a new patient with chronic heart failure you learn that the patient is not taking one of these agents, it is correct to ask if his or her doctor has prescribed it in the past. In doing so, you may learn if the patient has been found to be intolerant to an agent, or you may simply refresh his or her memory.

When describing the current medical regimen, be sure to include frequency, dose of administration, and time taken during the day. The latter may be especially important if a medication affects heart rate, because you must allow sufficient time between when the medication is taken and when the patient begins to exercise. Specifically, the medication must have time to be absorbed and exert its therapeutic effect.

A medical allergy history, with a comment on the type of reaction, is also a necessary part of the database. If a patient is unaware of any drug allergies, note this as "no known allergies (nka)" in your database. It is also helpful to note medicines that the patient does not tolerate (e.g., "Patient is intolerant to nitrates, which cause severe headache").

Medical History

This section should contain a concise, relevant list of past medical problems with attention to dates. Be sure your database of past medical illnesses is complete, because orthopedic, muscular, neurological, gastrointestinal, immunological, respiratory, and cardiovascular problems all have the potential to influence exercise response and the type, progression, duration, and intensity of exercise.

For example, for patients with coronary heart disease, it is essential to have a record of coronary anatomy, the severity of lesions, types of conduits used if bypass was performed, target vessels, and most current assessment of left ventricular function (i.e., ejection fraction). For patients with cerebral and peripheral disease, the same degree of detail is needed with respect to the arterial beds treated.

Among patients with intrinsic lung disease, an attempt should be made to clarify asthma versus chronic obstructive pulmonary disease attributable to cigarette smoking. Such information may help explain why certain medications are used when the patient is symptomatic and why others are part of a patient's long-term, chronic medical regimen. In addition, any available results from prior pulmonary function tests should be noted, including the forced expiratory volume in 1 s (FEV_1), forced vital capacity (FVC), and FEV_1/FVC ratio. Any prior pulmonary surgery should be noted.

Inquire and investigate about other organ systems. For example, if a diagnostic exercise test is ordered for a patient with intermittent claudication it may be better to use a dual-action stationary cycle in place of a treadmill. Additionally, knowing about any previous low back pain is important when developing an exercise prescription.

Family History

This element should be restricted to known, relevant heritable disorders in first-degree family members (parents, siblings, and offspring). Relevant heritable disorders include certain cancers (e.g., breast), adult-onset diabetes mellitus, familial hypercholesterolemia, sudden death, and premature coronary artery disease defined as new-onset disease before the age of 45 in men or 55 in women. An example of a heritable, although rare, pulmonary disorder is β-1 antitrypsin deficiency–mediated emphysema. While assessing family history, it is also a good time to suggest to the patient that first-degree family members be screened for pertinent risk factors and possible early disease detection, if indicated.

Social History

This section collects information about marital/significant partner status, employment status, relevant transportation, housing information, and routine and leisure activities. Because long-term compliance to healthy behaviors is influenced, in part, by conflicts with transportation, work hours, and childcare/family responsibilities, it is important to inquire about and discuss these issues. Conclude by making reasonable suggestions that are meant to improve long-term compliance.

Also, inquire about diet, nicotine or alcohol use, illegal substance abuse, and exercise. Because prior physical activity and dietary habits often predict a patient's ability to comply with behavioral change, extra attention should be given to obtaining an accurate physical activity and dietary history. Therefore, assess both prior and current habits and ask specific questions about type and frequency of the habit over a fixed time period.

For example, if a patient states that she does a lot of walking, ask how many times in the past month she walked for 20 min or more. Does she walk by herself? What time of the day does she walk? How do her current exercise habits differ from six months ago? Collecting this information is an important element of the evaluation, because it identifies those behaviors that may need to be reinforced or modified to reduce future risk. After all, many of the job duties that clinical exercise physiologists perform boil down to helping patients initiate and maintain healthy behaviors.

Assessing a patient's risk factors or risk profile not only helps the clinician quantify future risk but also helps focus future primary and secondary preventive strategies as well. However, it is not useful to list nonmodifiable risk factors (age, family history, menopausal status) for the patient with established disease (e.g., coronary artery disease).

Physical Examination

Physicians and physician extenders are taught to take a complete "head to toe" approach to the physical examination. For every part of the body examined, an orderly process of inspection, palpation, and, if applicable, **auscultation** and percussion is followed. However, the clinical exercise physiologist can take a more focused approach and concentrate on abnormal findings, given patient complaints or symptoms and information from prior examinations performed by others.

Specifically, you must develop the skills needed to determine if, on any given day, it is safe to exercise a patient who presents with symptoms that may or may not be related to a current illness. For example, a patient with a past history of dilated cardiomyopathy complains of being short of breath and having slept on three pillows last night so he could breathe comfortably. This complaint should raise a red flag for you, because it may indicate that the patient is experiencing pulmonary edema attributable to heart failure. In addition to other questions that you might ask, you should assess body weight, peripheral edema, and lung sounds before allowing the person to exercise. A telephone conversation with the patient's doctor concerning your findings, if meaningful, might also be warranted.

At no time should your examination be represented as being performed in lieu of the evaluation conducted by a professional licensed to do so (see chapter 3). However, you are responsible for ensuring patient safety. Therefore, the information and data you gather are very important and should be communicated to the referring physician and become part of the patient's permanent medical record.

Other important red flags, other than shortness of breath, that should be identified and evaluated by the clinical exercise physiologist include new-onset or change in pattern of chest pain, **syncope** (loss of consciousness) or near syncope, neurological symptoms suggestive of a transient ischemic attack (i.e., vision or speech disturbances), severe headaches, recent falling, ischemic lower extremity pain at rest (also called critical leg ischemia), or a recently discovered abdominal bruit. Any of these new findings should be discussed with the patient's physician prior to exercise. The same holds true for a resting blood pressure greater than 200/115 mmHg and physical examination findings such as pulmonary rales and active wheezing.

Baseline Examination

Identifying the relevant aspects of the baseline examination will provide a reference point for future comparison, particularly if complications occur. Key components include general state and vital signs.

General State
An initial comment about the patient's general appearance is warranted. This is a quick, subjective evaluation, usually made soon after meeting the patient. Assess whether the patient appears as follows:

— Comfortable or distressed or anxious

— Healthy or frail

— Well-developed or undernourished

Blood Pressure, Heart Rate, and Respiratory Rate
Expected competencies of the clinical exercise physiologist are the accurate determination of blood pressure in each arm, pulse (heart) rate, **body mass index (BMI),** and respiratory rate. A difference of more than 20 mmHg between the right and left brachial systolic blood pressures, taken in close time proximity, may indicate significant subclavian atherosclerosis and deserves additional evaluation by a physician. Additionally, the identification of either **hypertension** (i.e., systolic consistently >140 mmHg or diastolic consistently >90 mmHg) or **hypotension** is important.

Hypotension can be either symptomatic or asymptomatic, with the latter sometimes attributable to the blood pressure–lowering effect of medications used to treat a problem other than hypertension. For example, patients with chronic heart failure are often given medications designed to decrease peripheral vascular resistance. As a result, it is not uncommon to record a blood pressure of 90/70 mmHg in these patients.

Tachycardia (heart rate >100 beats · min^{-1}) after 15 min of sitting rest is always abnormal and indicates extremely impaired left ventricular systolic function, severely impaired pulmonary function, hyperthyroidism, anemia, volume depletion, or the rare effect of medications such as over-the-counter decongestants or appetite suppressants. **Bradycardia** (heart rate < 60 beats · min^{-1}) can be due to both abnormal and normal physiological function. In sedentary patients with a history of prior illnesses, heart rates less than 40 beats · min^{-1} may represent

an underlying problem and should be brought to the attention of their physician.

Obesity

An extremely important chronic "vital sign" is an assessment of obesity. The best measure of obesity is the BMI. The BMI is mass in kilograms divided by the height in meters squared (kg · m^{-2}). Every patient who is being evaluated for the first time should have the BMI recorded in your database. Practically, this is accomplished by weighing the patient in kilograms and then asking for self-reported height in feet and inches. An easy to use conversion chart such as the one in figure 4.2 should be accessible for the quick calculation of BMI.

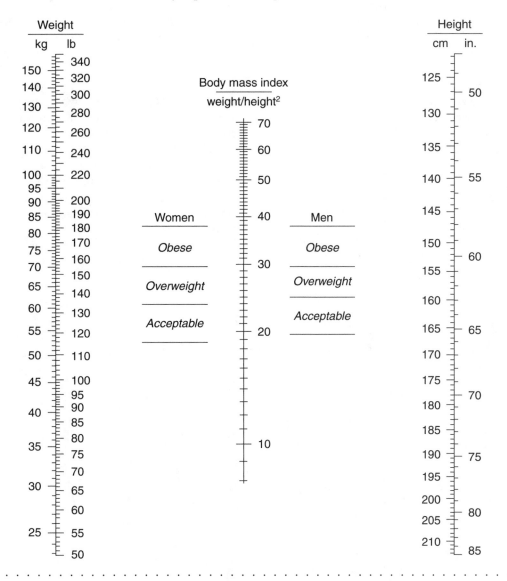

FIGURE 4.2   Body mass index (BMI) conversion chart that is used with weight and height. To use this nomogram, place a straight edge between the body weight (without clothes) in kg or lb located on the left-hand line and the height (without shoes) in cm or in. located on the right-hand line. Where the straight edge intersects the middle line indicates the BMI.

Reprinted, by permission, from G.A. Bray, 1978, "Definitions, measurements and classifications of the syndromes of obesity," *Journal of Obesity* 2: 99-112. http://www.nature.com

Obesity distribution is obtained by asking the patient to measure his or her own waist circumference with a measuring tape. In general, a waist circumference greater than 40 in. (100 cm) in a male or more than 35 in. (90 cm) in a female is indicative of central adiposity. Android or central obesity is one component of the "deadly quartet," which also includes insulin resistance, hypertension, and dyslipidemia (9). This simple measure of waist circumference is usually preferred over the waist-to-hip ratio, which involves two measures and, as a result, is more subject to measurement error.

Pulmonary System

The thorax should be inspected and deformities of the chest wall and thoracic spine noted (figure 4.3). Thoracic surgical incisions and implantable pacemaker or defibrillator sites should also be inspected and palpated. Redness or tenderness of any incision is always abnormal and often signifies a wound infection, which should prompt a physician evaluation. With the patient sitting, the lungs should be auscultated with the diaphragm of the stethoscope in both anterior and posterior positions, and breath sounds should be characterized as normal, decreased, absent, coarse, wheezing, or crackling (i.e., rales). Decreased or absent breath sounds should prompt percussion of the chest wall for dullness. If dullness is found, it signifies a pleural effusion, which is an abnormal collection of fluid in the pleural space that does not readily transmit sound. This

finding on physical examination would prompt withholding exercise and notification of the patient's physician. Coarse breath sounds can signify pulmonary congestion or chronic bronchitis. Crackles or rales can be caused by atelectasis (inadequate alveolar expansion after thoracic surgery), pulmonary edema attributable to congestive heart failure, or intrinsic lung disease such as pulmonary fibrosis.

Cardiovascular System

With the patient supine, the cardiac examination should start with inspection of the anterior chest wall. A cardiac pulse that can be visualized on the chest wall is often abnormal and represents a left ventricular hyperdynamic state. Standing at the patient's right side, palpate the heart with the right hand while the patient lies comfortably. Effort should be made to characterize where you feel or palpate the cardiac **point of maximal cardiac impulse (PMI)** or cardiac apex, as shown in figure 4.4. A normal PMI is the size of a dime and is located in the fourth or fifth intercostal space at the midclavicular line. The PMI should be characterized as normal, diffuse (enlarged), hyperdynamic, or sustained. The location of the PMI should be identified as normal or laterally displaced. A diffuse and laterally displaced PMI indicates left ventricular systolic impairment, often with enlargement of that chamber. If two cardiac impulses are felt, this often indicates a right and left ventricular heave sugges-

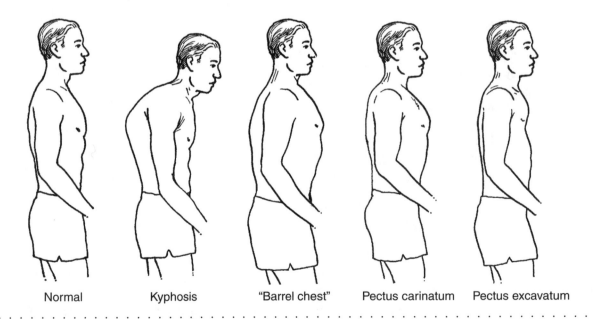

Normal   Kyphosis   "Barrel chest"   Pectus carinatum   Pectus excavatum

FIGURE 4.3   Common chest configurations.

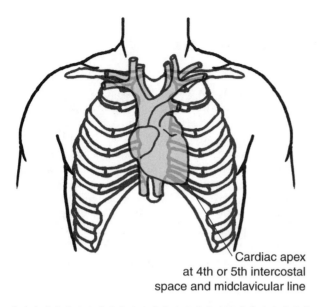

Cardiac apex
at 4th or 5th intercostal
space and midclavicular line

FIGURE 4.4 Surface topography of the heart.

tive of biventricular impairment in patients with heart failure.

An essential component in the examination of the heart is auscultation. An introduction to the assessment of heart sounds is reviewed in practical application 4.1. Practice listening to heart and lung sounds in apparently healthy people and patients known to have cardiac and pulmonary disorders that are confirmed in the medical record.

The cardiovascular physical examination should also include some evaluation of the status of peripheral vascular circulation. Extremities that are well perfused with blood are warm and dry. Poorly perfused extremities are often cold and clammy. Measuring and grading the characteristics of the arterial pulse in a region assesses adequacy of blood flow in arteries. Arterial pulses are graded as follows: 2/2 = normal, 1/2 = palpable but decreased, and 0/2 = absent or nonpalpable. Using a stethoscope, the clinical exercise physiologist can also listen for **bruits**, which are high-velocity "swooshing" sounds created as blood becomes turbulent when it flows past a narrowing or through a tortuous artery. Volume should be assessed, and bruits should be characterized as soft or loud. Bruits detected in the carotid arteries that were not previously mentioned in the medical record should be brought to the attention of a physician, as they may indicate severe carotid atherosclerosis. A bruit heard in the abdomen during diastole (diastolic bruit), near the level of the umbilicus, may indicate stenosis of the renal artery and should also be brought to the attention of a physician. Bruits in the common femoral arteries are

suggestive of peripheral arterial disease but by themselves do not call for immediate physician evaluation.

Peripheral **edema** (e.g., swelling of the lower legs, ankles, or feet) is a cardinal sign of congestive chronic heart failure. Because of elevated left ventricular end-diastolic pressure and consequently the backward cascade of increased pressure to the left atrium, pulmonary capillaries, pulmonary artery, and right-sided cardiac chambers, increased hydrostatic forces move extracellular fluid from within the blood vessels into the tissue spaces of the lower extremities. Edema is graded on a 1 to 3 scale, with 1 being mild, 2 being moderate, and 3 being severe. Additionally, "pitting edema" can be present, which is easily identified by pressing a thumb into an edematous area (e.g., distal anterior tibia) and observing that an indentation remains. A patient with 3+ pitting edema of the lower legs and ankles obviously has a great deal of fluid that has left the vascular compartment and moved into the surrounding tissue.

All edema, however, is not due to congestive heart failure. Minor edema is a side effect of many medications such as slow-channel calcium entry blockers (e.g., nifedipine). In addition, chronic venous incompetence associated with prior vascular surgery, obesity, or lymphatic obstruction can cause edema in the setting of normal cardiac hemodynamics and heart function. A practical point for the clinical exercise physiologist is that an increase in edema or body mass (>1.5 kg) over a 2- or 3-day period is often the first sign of worsening congestive heart failure and warrants a call to a physician.

Musculoskeletal System

Approximately one person in seven suffers from some sort of musculoskeletal disorder. The history of present illness or past medical history should note the major areas of discomfort and self-reported limitation of motion. In addition, prior major orthopedic surgeries, such as a hip or knee joint replacement, should be noted.

The approach to the musculoskeletal physical examination should be grounded in observation. For example, observe the patient as he or she gets up from a chair and walks into a rehabilitation area, gets onto an examination table, or handles personal belongings. Observe gait and characterize it as normal (narrow-based, steady, deliberate), antalgic (limping because of pain), slow, hemiplegic (attributable to weakness or paralysis), shuffling (parkinsonian), wide-based (cerebellar ataxia or loss of position information), foot drop, or slapping (sensory ataxia or loss of position information) (figure 4.7).

Auscultation of the Human Heart

Auscultation or listening to the sounds made by the heart is but one part of a comprehensive cardiac examination. Initiate a habit of auscultating the heart in a systematic fashion that is the same for every patient you evaluate. Establishing a uniform approach will more quickly familiarize you with normal heart sounds as well as help to identify abnormal sounds. To aid your concentration, try to auscultate the heart in a quiet room. Approach the patient from the right side and do all you can to minimize anxiety. This includes attempting to warm your stethoscope before using it and communicating with the patient as you progress through this part of the physical examination. And remember, maintaining patient modesty is always a top priority.

Auscultation is usually done with the patient lying on his or her back; however, before an exercise test, auscultation can be performed in the sitting position. As mentioned, when auscultating the heart for cardiac sounds, do so systematically. Begin by placing the diaphragm of the stethoscope firmly on the chest wall in the lower left parasternal region. First, characterize the rhythm as regular, occasionally irregular, or irregularly irregular. An irregularly irregular rhythm is usually attributable to atrial fibrillation. The diaphragm of the stethoscope is best used to hear high-pitched sounds, whereas the bell portion is used for low-pitched sounds.

Next, move the stethoscope to the point on the chest where the first heart sound (S_1, the sounds of mitral and tricuspid valves closing) is best characterized. For most people, this location is found at the apex of the heart at the left midclavicular line and near the fourth and fifth intercostal spaces. You will hear two sounds, the first sound (S_1) being the louder and more distinct sound of the two. Then, move your stethoscope upward and to the right side of the sternal border at the second intercostal space. This location is generally the best location to characterize the second heart sound (S_2, the sounds of aortic and pulmonic valves closing), and it is here that the second sound (S_2) is louder and more pronounced.

Soft heart sounds occur with low cardiac output states, obesity, and significant pulmonary disease (e.g., diseased or hyperinflated lung tissue between the chest wall and the heart). Loud heart tones occur in thin people and in hyperdynamic states such as pregnancy. The second heart sound normally splits with inspiration as right ventricular ejection is delayed with the increased volume it receives from the augmented venous return of inspiration. This delays or splits the pulmonic component of S_2 (sometime referred to as P_2) from the aortic component of S_2 (referred to as A_2, figure 4.5).

With practice and as you begin to care for patients with various cardiac problems, you will become exposed to and appreciate third (S_3) and fourth (S_4) heart sounds. S_3 and S_4 are low-pitched sounds that are both heard during diastole and best appreciated by using the bell portion of the stethoscope. The presence of either of these two heart sounds is most often associated with a heart problem and should be brought to the attention of a physician. S_3 is best heard at the apex and occurs right after S_2. S_4 is also well heard at the apex and occurs just before S_1. An S_3 commonly indicates severe left ventricular systolic impairment with volume overload and dilation. An S_4 commonly indicates chronic stiffness or poor compliance of the left ventricle, usually attributable to long-term hypertension. As you can see, learning and identifying heart sounds will require a great deal of practice.

The clinical exercise physiologist should be able not only to appreciate systole and diastole but, with advanced training, to listen for murmurs in the mitral, tricuspid, pulmonic, and aortic areas (figure 4.6) as well. Murmurs are characterized by the timing in the cardiac cycle (systolic, diastolic, or both), location where best heard, radiation, duration (short or long), intensity, pitch (low, high), quality (musical, rumbling, blowing), and change with respiration (9). A central concept to keep in mind while listening with the stethoscope is that the sounds heard are attributable to changes in blood velocity and the movement of cardiac valve leaflets, both of which are driven by pressure gradients and result in flow. Systolic murmurs are more common and are often characterized as ejection-type (e.g., diamond-shaped or holosystolic). Diastolic murmurs are distinctly less common.

Again, exercise physiologists working in the clinical setting must acquire the basic skills needed to auscultate the heart.

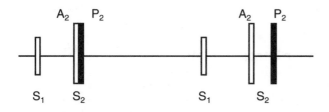

FIGURE 4.5 Normal physiological splitting of the second heart sound due to augmented venous return to the right heart during inspiration.

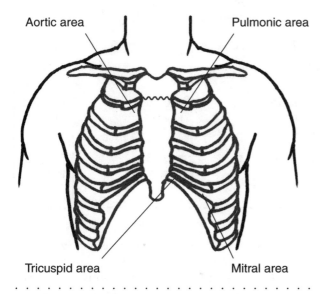

FIGURE 4.6 Auscultatory areas of the heart.

An antalgic gait is a limp, which reflects unilateral pain and compensation for that pain. A slow gait is often a tip-off for back disease, hip arthritis, or underlying neurological problems. Hemiplegic, shuffling, wide-based, foot-drop, and slapping gaits all represent compensation for underlying neurologic disease such as a spinal cord injury, cerebellar dysfunction (e.g., attributable to alcohol), or midbrain dysfunction (e.g., Parkinson's disease). These "neurological gaits" are all unsteady and leave the patient prone to falling. Thus, the clinical exercise physiologist must pay special attention to the safety of exercise, modifying the exercise prescription as necessary.

The core of the musculoskeletal physical examination is an assessment of range of motion of the moveable joints. Careful notation of a restricted range of motion (table 4.3), for example, a frozen shoulder caused by adhesive capsulitis (i.e., a thickening of the shoulder joint capsule), should be noted in the clinical record, because it may influence test and rehabilitation performance. Important terminology for describing limb motion is given in table 4.4.

Palpation of the major joints (shoulders, elbows, wrists, hips, knees, and ankles) can be performed to note thickening of the joint capsules, swelling or effusion, and tenderness of ligaments or tendons. Redness, warmth, swelling, and fever are all signs of active inflammation and require evaluation by a physician before proceeding with exercise testing or

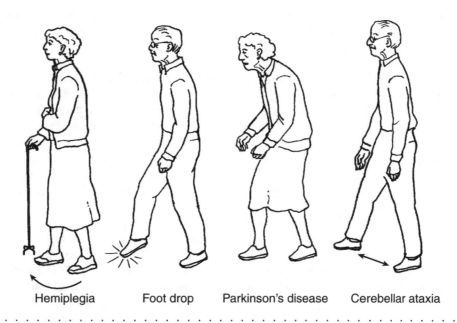

FIGURE 4.7 Description of common gait abnormalities.

Reprinted from *Physical Diagnosis: History and Examination*, Swartz, 452, © 1999, with permission from Elsevier Science.

TABLE 4.3
Typical Ranges of Motion for Specific Joints

	SPINE		UPPER BODY			LOWER BODY			
	Cervical	Thoracic and lumbar spine	Shoulder	Elbow	Wrist	Hip	Hip	Knee	Ankle
Flexion	45°	75°	180°	150°	80°	90° with knee extended	120° with knee flexed	135°	50°
Extension	55°	30°	50°	180°	70°	30° with knee extended	—	0°-10°	15°
Lateral bending	40°	35°	Abduction to 180°	Pronation to 80°	Radial motion to 20°	40°	Abduction to 45°	—	Inversion to 20°
Rotation	70°	30°	Adduction to 50°	Supination to 80°	Ulnar motion to 55°	45°	Adduction to 30°	—	Eversion to 10°

therapy. If these signs are found in conjunction with a prosthetic joint, such as a total knee replacement, they may indicate an infection and require immediate contact with the patient's surgeon. Inspection and palpation of the cervical, thoracic, lumbar, and sacral spine areas are important features of the complete examination given the involvement of the spine in virtually every form of exercise therapy.

In addition to joint health, muscle strength should be examined and graded on a scale of 0 to 5, with 0 indicating flaccid paralysis and 5 indicating sufficient power to overcome the resistance of the examiner. Muscle stiffness and soreness (not related to exercise) should be noted because they are often the sign of a chronic underlying inflammatory condition that requires medical evaluation.

Low Back Pain

Because low back pain (see also chapter 26) is one of the most common physical complaints in human medicine, it is worth mentioning that the etiology of this problem can range from a mild muscle strain to a ruptured abdominal aortic aneurysm. The clinical exercise physiologist must have a rational approach to this problem and tailor aspects of care based on etiology, severity, and prognosis. In the young stable patient with no evidence of neurological compromise (e.g., radiating pain, numbness), a physician evaluation is likely not necessary prior to exercise testing or therapy. In the geriatric patient (i.e., >65 years old), however, new-onset low back pain that has not been evaluated by a physician deserves referral to the primary care physician or a back specialist before exercise testing or training. This is because one must consider a spontaneous compression fracture attributable to osteoporosis.

Arthritis

A final musculoskeletal condition worth mentioning, one that is especially common among the elderly, is arthritis (see also chapter 24). **Rheumatoid arthritis (RA)** is a chronic inflammatory condition manifested by early morning stiffness (>1 hr) with improvement through the rest of the day. RA begins insidiously with fatigue, anorexia, generalized weakness, and vague musculoskeletal symptoms until the appearance of synovitis (inflammation of joint lining) becomes apparent. It has a predilection for the small joints in the hand, especially the metacarpophalangeal joints (knuckles). RA results in pain, inflammation, thickening of the joint capsule, lateral deviation of the fingers, and significant disability. RA can involve large joints and the spine, especially the cervical vertebra. RA is usually apparent to the clinical exercise physiologist by history, physical examination, and the presence of disease-modifying

TABLE 4.4	
Terminology for Describing Joint Motion	
Motion	**Definition**
Flexion	Motion away from zero position
Extension	Motion toward zero position
Dorsiflexion	Movement toward dorsal surface
Plantar (or palmar) flexion	Movement toward plantar (or palmar) surface
Adduction	Movement toward midline
Abduction	Movement away from midline
Inversion	Turning of plantar foot surface inward
Eversion	Rotation of plantar foot surface outward
Internal rotation	Rotation of anterior limb surface inward
External rotation	Turning of anterior limb surface outward
Pronation	Rotation of palm downward
Supination	Rotation of the palm upward

Reprinted from *Physical Diagnosis: History and Examination*, 261, © 1999, with permission from Elsevier Science.

drugs such as D-penicillamine, gold, antimalarials, and sulfasalazine.

It is generally safe to for patients with RA to exercise unless neck pain or lancinating pains into the shoulders or arms are reported. These symptoms are indicative of cervical spine involvement and require physician attention. In some cases, bracing or surgery to prevent cervical spine subluxation and paralysis is needed.

Osteoarthritis

Osteoarthritis (OA) is the most common skeletal disorder in adults. It can result from longstanding wear and tear on large and small joints including shoulders, elbows, wrists, hands, hips, knees, and back. The correlation between the pathological severity of OA and patient symptoms is poor. Patients with OA may show signs of synovitis and secondary muscle spasm.

Exercise rehabilitation may improve the symptoms of OA in one location (e.g., the back), only to worsen pain in another location (e.g., hips and knees). The clinical exercise physiologist should take a pragmatic approach and work in a coordinated fashion with physical and occupational therapy colleagues to find activities that minimize joint loading at involved sites.

Nervous System

Like the examination of the musculoskeletal system, the neurological examination is mainly carried out by observation. In general, a comment should be made in the patient file regarding level of understanding, orientation, and cognition. Obvious disabilities of speech, balance, and muscle tremor and disabilities of the eyes, ears, mouth, face, and swallowing should be cross-referenced for confirmation with the patient's medical record.

A detailed neurological examination is not typically performed by the vast majority of practicing clinical exercise physiologists, and describing the conduct for such an examination is beyond the scope of this chapter. However, a notation regarding gross **hemiparesis** or complete paralysis of a limb should be made. As mentioned previously (figure 4.7), "neurological gaits" should be identified as well. For obvious reasons, patients with any of these problems may require an exercise prescription that is modified accordingly.

Although the clinical exercise physiologist may work with the spinal cord–injured patient (see chapter 27) or patients with multiple sclerosis (see chapter 28), the single most common neurological problem that most exercise physiologists encounter will be patients suffering a previous cerebrovascular event or stroke. Stroke is defined as an ischemic insult to the brain resulting in neurological deficits lasting for more than 24 hr. The etiology of stroke includes local arterial thrombosis, cardiac and carotid thromboembolism, and intracranial hemorrhage. Risk factors for stroke include the following:

— Prior history of stroke
— Atrial fibrillation
— Left ventricular dysfunction
— Aneurysms
— Carotid artery stenoses
— Uncontrolled hypertension

If the clinical exercise physiologist determines that any of these risk factors are new, the physician should be notified before the patient starts exercise. Common abnormalities after a stroke are a drooping of one side of the face, drooling, and garbled speech (dysarthria).

Metabolic and Other Organ Systems

Signs and symptoms of abnormalities of metabolism and of other organ systems of the body (e.g., endocrine, immune, and hematological systems) are generally less common and less obvious on physical examination. Therefore, they are typically beyond the scope of practice for most clinical exercise physiologists. However, the clinical exercise physiologist should still consider some important points in these areas.

Diabetes

Prior to starting an exercise program, patients with diabetes should have a foot examination to look for blisters, cuts, and abrasions. Any new signs of infection should be reported to a doctor. Also, diabetes is associated with a higher probability of silent ischemia, multivessel coronary disease, and decreased left ventricular function. Therefore, the clinical exercise physiologist should determine if symptomatic versus asymptomatic ischemia is present, based on previous graded exercise tests, functional studies, and angiographic information.

Evaluation of the diabetic must include notation of the doses of diabetic medications and the presence of any diabetic complications such as peripheral or autonomic **neuropathies.** The clinical exercise physiologist should be aware of the signs and symptoms of hypoglycemia, such as sudden weakness, pallor, sweatiness, and confusion, and should be prepared to treat with orange juice (with sugar added) or a carbohydrate source.

Obesity

We previously described how obesity is identified and quantified via BMI. Obesity (BMI > 30) is the most rapidly growing metabolic disorder in the United States. However, identifying during examination exactly where excess fat is stored is important, because the site of excess fat deposition is related to overall mortality rate and increased risk of various diseases. In general, women tend to store fat in the lower half of the body around the hip and thigh area (gynoid obesity), whereas men tend to store fat on the upper part of the body around the abdomen (android obesity). The latter of the two types is associated with the greater risk for developing heart disease, diabetes, and hypertension.

Given the unlikely occurrence of rapid shifts in the human U.S. gene pool over the past 30 to 50 years, lack of physical activity and absolute increases in caloric intake are therefore responsible for the secular increase in mean population body weight. However, it is likely that patients with obesity have a "reporting bias." During your evaluation, be aware that this often occurs in some obese patients, who commonly self-report a caloric intake that is less than needed to even maintain a body weight that has not changed or increased over the past several months or year.

Anemia

A history of anemia should be noted in the general evaluation. Anemia is generally defined as a hemoglobin less than $12 \text{ g} \cdot \text{dl}^{-1}$ in women and less than $13 \text{ g} \cdot \text{dl}^{-1}$ in men. Complaints of new-onset excessive fatigue or weakness, excessive paleness, excessive vaginal bleeding, or gastrointestinal bleeding should be reported to the physician prior to starting exercise therapy. In addition, many patients with cancer, renal disease, and human immunodeficiency virus (HIV) suffer from chronic anemia that is attributable to either the disease itself or some of the drugs used to treat the disease. Anemia is commonly observed in patients age 80 years and older and patients who undergo cardiac or pulmonary surgery. For example, after coronary artery bypass surgery, it is not uncommon for a patient's hemoglobin to fall to $10 \text{ g} \cdot \text{dl}^{-1}$. Less common, but still occurring, is the incidence of anemia among highly active, usually younger female athletes with poor dietary habits.

Inquire about and document any history of easy bruising or excessive gum or nasal bleeding. Skin pallor, especially pallor of the conjunctiva (inner eyelid) and the oral mucosa, is an examination finding supportive of anemia. Unusual bruising on the trunk and extremities should be noted as well. Again, if these are new, a physician should be notified before the patient begins a particularly strenuous exercise program.

Human Immunodeficiency Virus

You will be alerted to HIV infection by the clinical history and multiple drugs used to treat HIV (see chapter 23). If tuberculosis has been identified in a patient who is HIV-positive, records should indicate a completed treatment course with medical therapy and the patient should be on maintenance therapy before admission to exercise sessions with other patients.

Principles of exercise therapy should include careful attention to changes in exercise tolerance and dyspnea. Patients with fever and marked changes in general strength and in dyspnea on exertion should be referred to a physician for evaluation of opportunistic infection.

Widespread use of antiretroviral therapy has brought out an unexpected side effect profile including dyslipidemia, abnormal adipose deposition on the back of the neck ("buffalo hump"), and early atherosclerosis (11). Be on the lookout for both the buffalo hump on the posterior neck and abnormal lipoprotein values in patients with HIV.

Cancer

It is increasingly common for the clinical exercise physiologist to encounter patients with different malignancies. Because cancer can involve more than one organ or organ system, the clinical evaluation should identify the organ systems or tissues involved, the date of first discovery, and the treatment received. It is also helpful to assess the patient's understanding of the cancer status: cured, in remission, or active. In general, most cancers with no objective evidence of recurrence 5 years after treatment are considered cured. Metastases to bone, often manifested as pain in the bone area, are a red flag for the clinical exercise physiologist and should prompt postponement of exercise therapy until discussed with the patient's physician. For patients in remission or with active neoplastic processes, anemia with fatigue and decreased appetite is common.

Conclusion

This chapter provides a platform for the clinical evaluation of the new patient by the clinical exercise physiologist. Special emphasis is given to those day-to-day interview and examination skills that you may need to help you decide if it is safe for a patient to exercise. As mentioned at the beginning of this chapter and as shown in the accompanying case study, the clinical evaluation typically concludes with both an assessment list and a plan list. The former summarizes past and current primary findings and diagnoses, whereas the latter lists the one, two, or three key actions to be taken as they relate to the care of the patient.

Finally, we have emphasized that the information you gather through clinical evaluation, when viewed in conjunction with physician examination findings and existing medical record information, provides a point of reference should complications occur during exercise treatment. If the evaluation is broadened to include a comprehensive assessment of future health risk and consideration of disease, your clinical evaluation also becomes a useful decision-making tool and guide for patient education.

Case Study 4

Ms. WY

Medical Record Number: 123-45-678

Date: October 17, 2000

DOB: September 12, 1959

Demographics and Reason for Referral

A 41-year-old Caucasian female seen for a graded exercise test prior to starting a cardiorespiratory and resistance training exercise program.

> History of present illness: An 8-year history of HIV disease with no opportunistic infections. She now complains of chronic fatigue and lack of energy. She has shortness of breath on moderate exertion such as carrying a full laundry basket up one flight of stairs. Patient denies any chest pain or discomfort at rest or with exertion.
>
> Medications: efavirenz, zidovudine, didanosine.
>
> Past medical history: HIV disease for 8 years, no complications.
>
> Family history: negative for premature coronary artery disease.
>
> Social history: unemployed, unmarried, lives alone.
>
> Lifestyle history: unrestricted diet, sedentary with very rare physical exertion.

continued

Risk factors: nonsmoker; fasting lipoprotein profile from medical record of March 12, 2000 (total cholesterol = 266 mg · dl^{-1}, low-density lipoprotein cholesterol = 185 mg · dl^{-1}, high-density lipoprotein cholesterol = 47 mg · dl^{-1}, triglycerides = 150 mg · dl^{-1}); no diabetes in the medical record.

Fasting blood glucose (taken in office this morning): 88 mg · dl^{-1}.

Physical Examination

General appearance: appears chronically ill, pale.

Vital signs: blood pressure = 110/50 mmHg in both arms, pulse = 108 beats · min^{-1}, respiration = 22 breaths · min^{-1}, mass = 66 kg, height = 5 ft 4 in. (64 in.), BMI = 25, waist circumference = 30 in.

Head and neck: no oral lesions, pale conjunctival (inner eyelid) and oral mucosal membranes, "buffalo hump" noted.

Cardiovascular: no carotid bruits, no jugular venous distension, clear lungs, normal point of maximal impulse, regular rate and rhythm, normal S$_1$ and S$_2$, 2/6 short early peaking systolic ejection murmur best heard in the pulmonic area without radiation, no extra heart sounds, no edema, normal peripheral vascular exam.

Skin: no rashes.

Musculoskeletal: normal gait, joints are normal, good range of motion throughout.

Neurologic: normal cranial nerves, normal motor exam, sensory exam reveals decrease in light touch sensation from toes to the knees bilaterally.

Resting electrocardiogram before exercise test: sinus rhythm, rate = 110 beats · min^{-1}, PR interval = 0.14 s, QRS duration = 0.09 s, no Q waves and no ST or T-wave abnormalities, computer interpretation is normal electrocardiogram.

Assessment: (a) HIV-positive with associated exercise intolerance and fatigue; (b) likely medical therapy–induced hyperlipidemia and deposition of fat (buffalo hump) in posterior neck region.

Plan: (a) Complete graded exercise test; (b) initiate exercise training program aimed at increasing muscle mass, muscle endurance, and aerobic capacity; (c) initiate counseling about low-fat diet and refer to registered dietitian for specific meal planning and counseling.

Case Study 4   Discussion Questions

1. What is your assessment of the functional class? Explain your answer.

2. Are there any red flags on the clinical evaluation that require contacting, at least by telephone, her physician or another supervising physician prior to exercise testing?

3. What is the likely cause of the resting tachycardia?

4. Do any preventive cardiology issues need to be addressed in the near future?

5. Are there special risks to physical injury with exercise?

Glossary

auscultation—Listening to the sounds made by various body structures as a diagnostic method.

body mass index (BMI)—A unit of anthropometric measure related to health status. Presented as mass/height^2 (kg · m^{-2}).

bradycardia—Heart rate less than 60 beats · min^{-1}.

bruit—An acquired sound of venous or arterial origin that is caused by turbulent blood flow; heard by auscultation.

congestive or chronic heart failure (CHF)—The symptom complex associated with shortness of breath, edema, and exercise intolerance attributable to abnormal left ventricular function.

edema—A condition in which body tissue contains an excessive amount of fluid.

hemiparesis—Paralysis affecting only one side of the body.

hypotension—Abnormally low blood pressure; typically associated with symptoms.

hypertension—Abnormally high blood pressure (consistently >140/90 mmHg).

morbidity—A diseased state.

mortality—Death.

neuropathy—A disease involving the cranial nerves or the peripheral or autonomic nervous system.

osteoarthritis (OA)—Erosion of articular or joint cartilage that leads to pain and loss of function.

point of maximal cardiac impulse (PMI)—Point identified by palpation and inspection of the chest wall during physical examination as the most prominent location for cardiac impulse.

rheumatoid arthritis (RA)—Inflammation of the joints attributable to autoimmune attack that leads to pain and loss of function.

syncope—Loss of consciousness caused by diminished cerebral blood flow.

tachycardia—Heart rate greater than 100 beats · min⁻¹.

References

1. Braunwald, E. The history. *Heart Disease: A Textbook of Cardiovascular Medicine* (4th ed.). Philadelphia: Saunders, 1992.

2. Campeau, L. Grading of angina pectoris. *Circulation* 54: 522-523, 1976.

3. Carhart, R.L., and P.A. Ades. Gender differences in cardiac rehabilitation. *Cardiol Clin* 16: 37-43, 1998.

4. Centers for Disease Control and Prevention. Prevalence of diagnosed diabetes among American Indians/Alaskan Natives—United States. *Morb Mortal Wkly Rep* 47: 901-904, 1998.

5. Diamond, G.A., and J.S. Forrester. Analysis of probability as an aid in the clinical diagnosis of coronary-artery disease. *N Engl J Med* 300: 1350-1358, 1979.

6. Fauci, A. *Harrison's Principles of Internal Medicine* (4th ed.) [online]. Available: www.harrisononline.com [1999].

7. Kannel, W.B., P.A. Wolf, E.J. Benjamin, and D. Levy. Prevalence, incidence, prognosis, and predisposing conditions for atrial fibrillation: Population-based estimates. *Am J Cardiol* 82: 2N-9N, 1998.

8. Kaplan, N.M. The deadly quartet. Upper-body obesity, glucose intolerance, hypertriglyceridemia, and hypertension. *Arch Int Med* 149: 1514-1520, 1989.

9. Kloner, R.A. *The Guide to Cardiology*. Greenwich, CT: Le Jacq Communications, 1995.

10. Lloyd-Jones, D.M., M.G. Larson, A. Beiser, and D. Levy. Lifetime risk of developing coronary heart disease. *Lancet* 353: 89-92, 1999.

11. Périard, D., A. Telenti, P. Sudre, J. Cheseaux, P. Halfon, M.J. Reymond, S.M. Marcovina, et al. Atherogenic dyslipidemia in HIV-infected individuals treated with protease inhibitors. *Circulation* 100: 700-705, 1999.

12. Smith, D.W. Changing causes of death of elderly people in the United States, 1950-1990. *Gerontology* 44: 331-335, 1998.

13. Swartz, M.H. *Textbook of Physical Diagnosis: History and Examination*. Philadelphia: Saunders, 1998.

Dawn Bell, PharmD, BCPS
Director, Pharmacy Affairs
The Medicines Company
Morgantown, WV

Pharmacotherapy

The study of a drug's action on the body is referred to as *pharmacology*, whereas the art and science of individualizing drug therapy based on patient-specific characteristics are referred to as *therapeutics* or *pharmacotherapy*. This chapter is meant to outline basic information about the pharmacology and therapeutics of commonly used drug therapies. It is not meant to serve as a comprehensive reference, but instead, should provide general information in an easy to use format, as much of the textual information is summarized in tabular form. This chapter is organized into four general categories based on systems (metabolic, endocrine, and cardiovascular) or situations (drugs of abuse) that the clinical exercise physiologist may commonly encounter when working with patients. The effects of the agents on exercise capacity, if any, are noted in the text. Agents banned by either the United States Olympic Committee (USOC) (1) or the National Collegiate Athletic Association (NCAA) (2) are noted as well.

Metabolic Agents

Several different patient populations may be prescribed various metabolic agents for treatment (table 5.1). Cancer and renal patients may be anemic as a result of treatments and may need agents to

---

TABLE 5.1
· · · · · ·
Metabolic Agents and Their Exercise Effects

Medications	Heart rate	Blood pressure	ECG	Exercise capacity
Iron supplementation	↔	↔	↔	May increase in those with iron deficiency but no effect with normal iron levels
Erythropoietin	↔	↑	↔	May increase in those with anemia; questionable effect in persons with normal hemoglobin
NSAIDs and selective COX-2 inhibitors	↔	↔	↔	Do not affect exercise directly but by reducing pain allow the individual to perform

Note. ECG = electrocardiogram; NSAIDs = nonsteroidal anti-inflammatory drugs; COX-2 = cyclooxygenase-2; ↔ = no change; ↑ = increase.

---

improve oxygen-carrying capacity (3). Arthritic patients are likely candidates for nonsteroidal anti-inflammatory therapy (4). Insulin and hypoglycemic agents are necessary for controlling blood glucose levels among diabetics. The clinician should thoroughly understand these forms of treatments.

Iron Supplementation

Anemia is a decrease in the red blood cells, which carry hemoglobin, a substance that transports oxygen. If the concentration of hemoglobin is reduced, the oxygen-transporting capacity of the body is impaired. Persons with anemia will have a decreased exercise tolerance and will be more prone to fatigue. The primary causes of anemia include reduced red blood cell and hemoglobin production, **hemolysis** of red blood cells, and loss of blood. Although an inadequate dietary intake of several nutrients may reduce the production of red blood cells and hemoglobin, the most common cause of anemia throughout the world is iron deficiency (3).

Sports anemia is a term often used to describe a low hemoglobin condition that is relatively common at the beginning of rigorous training or when training requirements increase. It is characterized by exhaustion and fatigue. After adaptation to training, sports anemia seems to subside. The severity and exact causes of this condition have not yet been determined. Possible explanations for this condition are inadequate dietary iron intakes by athletes or the use of protein for tasks other than red blood cell production during the early training stages. As a result, athletes may have higher daily iron requirements than average (5). Iron requirements are determined by physiological losses and the needs deter-

mined by growth. For example, an adult male needs 13 μg $\cdot$ kg^{-1} $\cdot$ day^{-1} (about 1 mg), whereas a menstruating female requires about 21 μg $\cdot$ kg^{-1} $\cdot$ day^{-1} (about 1.4 mg). Pregnant women require 80 μg $\cdot$ kg^{-1} $\cdot$ day^{-1} in the last two trimesters, and infants have similar requirements because of rapid growth (3).

Iron-deficiency anemia is treated with orally administered iron supplements. Ferrous sulfate, the least expensive of iron preparations, is the treatment of choice. Iron supplementation carries several side effects (3):

— Heartburn
— Nausea
— Upper gastrointestinal discomfort
— Constipation
— Diarrhea

Iron poisoning in adults is rarely fatal, but children between 12 and 24 months are susceptible to death from acute iron toxicity. Multivitamins and prenatal vitamins should be kept out of the reach of children for these reasons, and a child-resistant cap should be used if children are present in the household.

Iron has no effect on exercise capacity in persons with normal serum iron levels. However, in athletes who are iron deficient, supplementation can increase hemoglobin levels with subsequent increases in oxygen-carrying capacity and performance enhancement (6) (figure 5.1).

Erythropoietin

Erythropoietin (EPO) is a naturally occurring hormone that is secreted by the kidneys when hemoglobin concentrations are low and stimulates the

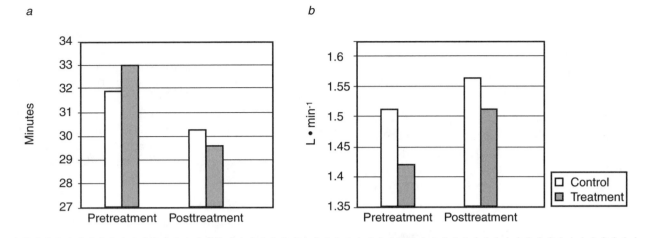

FIGURE 5.1   Changes in *(a)* 15-km time trial and *(b)* oxygen consumption following iron supplementation in iron-depleted nonanemic women. *Note.* In *(a)*, a group pre/post difference was observed ($p < .05$) and a difference between the iron-supplemented group and control group was observed posttreatment ($p < .05$). In *(b)*, a pre/post increase was observed in both groups ($p < .05$).

Reprinted, by permission, from P.S. Kinton et al., 2000, "Iron supplementation improves endurance after training in iron-depleted, nonanemic women," *Journal of Applied Physiology* 88: 1103-1111.

bone marrow stem cells to make red blood cells (3). Although iron is required as a cofactor for red blood cell production, the bone marrow is essential for providing the cells. A synthetic form of the hormone is available and used clinically in persons who have lost the ability to make the hormone on their own (patients with cancer, human immunodeficiency virus, or kidney failure) (3). EPO is generally well tolerated, but side effects can include the following (7):

— Hypertension

— Seizures

— Thrombosis

EPO has been used by endurance athletes to enhance performance by artificially increasing the oxygen-carrying capacity of the blood. However, by increasing production of red blood cells, EPO increases the viscosity of the blood, which can lead to thrombosis. Use of EPO to increase hemoglobin is different from "blood doping." In doping, an athlete typically receives a blood transfusion (his or her own blood previously donated) prior to competition to increase hemoglobin concentrations (8). Both EPO and blood doping are banned by the USOC and the NCAA (1, 2).

Nonsteroidal Anti-Inflammatory Agents

Common nonsteroidal anti-inflammatories (NSAIDs) include ibuprofen, naproxen sodium, ketorolac, sulindac, and etodolac (many are available without a prescription). These agents have both anti-inflammatory and fever-reducing activities. Although the exact mechanism of action is unknown, the main mechanism of action appears to be inhibition of cyclooxygenase (COX) activity (both COX-1 and COX-2) and prostaglandin synthesis (production of certain types of prostaglandins is important for inflammation) (9). They are also effective analgesics and can be used to relieve mild to moderate pain. Chronic use of these agents can erode the gastric mucosa with subsequent gastrointestinal bleeding (10). In addition, renal damage also has been reported with repeated use of higher than recommended daily doses of NSAIDs (11). Patients may also experience some gastric upset with short-term use, but this can be minimized by taking these agents with food or milk. Age appears to increase the risk of adverse drug reactions, so patients over the age of 65 years should be cautioned to not exceed the recommended daily dose. NSAIDs do not affect exercise capacity directly but may facilitate continued exercise by decreasing pain and inflammation resulting from exercise-induced injuries.

Selective COX-2 Inhibitors

Celecoxib (Celebrex®) and rofecoxib (Vioxx®) are both NSAIDs that selectively inhibit COX-2. Selective inhibition of COX-2 decreases inflammatory prostaglandins without a subsequent decrease in other prostaglandins that protect the gastric mucosa. For this reason, it is believed that COX-2 inhibitors may be less likely to cause gastric ulceration. Common side effects include these:

— Abdominal pain (12)

— Diarrhea

— Dyspepsia

— Flatulence

— Nausea

However, these occur less often than they do with the other NSAIDs. Like other NSAIDs, COX-2 inhibitors do not affect exercise capacity directly but may facilitate continued exercise by decreasing pain and inflammation resulting from exercise-induced injuries.

Insulin and Hypoglycemic Agents

Several different types of medications are used to treat high blood glucose levels found in the diabetic patient. Patients with type 1 diabetes will primarily be on supplemental or replacement insulin, which promotes glucose uptake. Patients with type 2 diabetes may be treated with oral hypoglycemic agents (sulfonylureas, biguanides, and thiazolidinediones), which lower blood glucose by aiding insulin production in the pancreas or allowing the cells to use insulin more effectively. In severe cases, patients with type 2 diabetes will be prescribed oral hypoglycemic agents in conjunction with insulin.

Insulin

Insulin, secreted by the β-cells of the pancreas, is the principal hormone required for proper glucose use in normal metabolic processes. Currently, most patients take a recombinant human form of insulin that is formulated in different preparations to increase its duration of action. Although beef and pork insulin are still available, the recombinant human form is preferred because the results are more predictable and there is no risk of antibody production.

Sulfonylureas

These agents are used to treat type 2 diabetes mellitus, also known as non-insulin-dependent diabetes mellitus. They bind to the plasma membrane of functional β-cells in the pancreas, leading to a calcium influx and insulin release from secretory granules. This process is also stimulated by glucose and other insulin-releasing fuels but appears to occur without elevated glucose concentrations when sulfonylureas are present. Common agents in this class include chlorpropamide, tolbutamide, glipizide, glyburide, and glimepiride. Patients should be advised to monitor their blood glucose regularly and to eat at regular intervals to avoid wide fluctuations in blood glucose.

Biguanides

Metformin is an oral hypoglycemic agent not related to the sulfonylureas. It decreases hepatic glucose production, decreases intestinal absorption of glucose, and improves insulin sensitivity. Insulin secretion remains unchanged but plasma blood glucose decreases. Patients with impaired renal function (serum creatinine >1.5 mg · dl^{-1}) should not take metformin because it can lead to metabolic acidosis and death. Other side effects include nausea and gastrointestinal upset.

Thiazolidinediones

Thiazolidinediones are antidiabetic agents that increase insulin sensitivity through a variety of mechanisms. They also decrease hepatic gluconeogenesis and increase insulin-dependent muscle glucose uptake. Agents in this class include pioglitazone and rosiglitazone. In early 2000, the agent troglitazone was removed from the market by the Food and Drug Administration because of several reports of hepatic toxicity with resulting liver failure. Other agents in this class have the propensity to cause liver toxicity and therefore should be closely monitored. Other adverse effects include fluid retention and edema. In addition, **hypoglycemia** can occur when thiazolidinediones are used in combination with insulin or sulfonylureas. Important drug interactions include oral contraceptives, with subsequent loss of contraceptive effectiveness, and decreased concentrations of cyclosporine, an immunosuppressant drug. None of these agents affect exercise capacity.

Hormones and Steroids

Pharmaceutical agents in this section include anabolic steroids and human growth hormone (table 5.2). These agents are of particular importance to the exercise physiologist. Although they are often used in treating various illnesses, they are often abused, particularly by athletes and bodybuilders because of their reputed performance-enhancing effects. The purpose of this section is to alert the clinician to the myriad dangers associated with abuse of these substances.

Anabolic Steroids

Anabolic steroids are a family of hormones that includes the natural male hormone, testosterone, together with a large number of synthetic compounds (13). All drugs in this family possess both anabolic and **androgenic** effects. Although various steroids differ in the relative amounts of anabolic and an-

TABLE 5.2
Hormones and Steroids and Their Exercise Effects

Medications	Heart rate	Blood pressure	ECG	Exercise capacity
Anabolic steroids	↔	May increase	↔	Performance enhancing
Androgenic steroids	↔	May increase	↔	Performance enhancing
Human growth hormone	↔	↔	↔	Performance enhancing

Note. ECG = electrocardiogram; ↔ = no change.

drogenic effects they produce, none even approaches being purely anabolic or purely androgenic. Examples of anabolic steroids include nandrolone, testosterone enanthate, and ethlestrenol. The anabolic effects include increased protein synthesis, decreased nitrogen excretion, and consequent gains in muscle size and strength (14). The androgenic effects are the "masculinizing" effects of the drugs, such as growth of male hair patterns and male sexual characteristics. Anabolic steroids promote body tissue-building processes and reverse catabolic, or tissue-depleting, processes. They are used clinically for various illnesses including certain types of anemia (to stimulate **erythropoiesis**), burns, and breast cancer (to control metastasis). These products are controlled substances because of their abuse potential (15-17).

Anabolic steroids are taken as pills or are injected. Steroid abusers may take hundreds of times more than the medically recommended dose. Users often combine several different types of steroids to boost their effectiveness—a method called stacking. In another method, called cycling, users take steroids for 6 to 12 weeks or more, stop for several weeks, and then start again.

Athletes, as well as some coaches, trainers, and physicians, report significant increases in muscle mass, strength, and endurance from steroid use. In acknowledgment of these effects, the International Olympic Committee has placed 20 anabolic steroids and related compounds on its list of banned drugs (1). However, no well-controlled studies have documented that the drugs improve agility, skill, cardiovascular capacity, or overall athletic performance.

Use of agents in amounts of up to 40 times the recommended therapeutic dose has been reported. An abuse or addiction syndrome occurs with the chronic use of these drugs (18). Long-term use can lead to a preoccupation with drug use, difficulty

stopping despite side effects, and drug craving (19). A type of withdrawal syndrome may be noted as well when drug concentrations fluctuate, with symptoms that are similar to those seen with alcohol, cocaine, and narcotic withdrawal. To detect steroid use or abuse, be aware of physical, psychological, and behavioral changes.

Males who take large doses of anabolic steroids typically experience changes in sexual characteristics. Although derived from a male sex hormone, the drug can trigger a mechanism in the body that can actually shut down the healthy functioning of the male reproductive system. Some possible side effects are as follows (20):

— Shrinking of the testicles

— Reduced sperm count

— Impotence

— Baldness

— Difficulty or pain in urinating

— Development of breasts

— Enlarged prostate

Females may experience "masculinization" as well as other problems (20):

— Growth of facial hair

— Changes in or cessation of the menstrual cycle

— Enlargement of the clitoris

— Deepened voice

— Breast reduction

For both sexes, chronic use of anabolic steroids may cause the following side effects (20):

— Acne

— Jaundice

— Trembling
— Swelling of feet or ankles
— Bad breath
— Reduction in high-density lipoprotein cholesterol
— High blood pressure
— Liver damage and cancers
— Aching joints
— Increased chance of injury to tendons, ligaments, and muscles

Psychological effects of anabolic steroid abuse are important to note. Wide mood swings ranging from periods of violent, even homicidal, episodes known as "roid rages" to bouts of depression when the drugs are stopped have been shown to occur. In addition, anabolic steroid users may suffer from paranoid jealousy, extreme irritability, delusions, and impaired judgment stemming from feelings of invincibility (20).

When used in combination with exercise training and a high-protein diet, anabolic steroids can increase the size and strength of muscles, improve endurance, and decrease recovery time between workouts. All anabolic steroids are banned by the USOC and the NCAA (1, 2).

Human Growth Hormone

Human growth hormone (GH), which consists of 191 amino acids linked in a specific sequence, is secreted from the anterior pituitary gland (21). Once secreted by the pituitary gland, circulating levels of GH stimulate production of insulinlike growth factor-1 (IGF-1) from the liver. Most of the positive effects of GH are mediated by the IGF-1 system. Both growth hormone (GH) and IGF-1 are anabolic, non-androgenic compounds that decrease body fat and increase lean muscle mass. Selected populations have lower than normal circulating GH: the aging population, those with hypothalamic–pituitary disorders, and patients with acute diseases associated with enhanced tissue breakdown, including AIDS, malnutrition, postoperative wounds, infections, bony fractures, and burns. Administration of synthetic GH increases circulating levels of IGF-1, skin thickness, and bone mineral content and reduces fat mass (22).

GH supplementation is touted as a veritable "fountain of youth" by many. Among athletes, doping with GH has become an increasing problem during the last 10 years despite the lack of evidence of its benefits (1). GH has a reputation among GH users of being fairly effective; however, the few controlled studies that have been performed in athletes with normal GH levels showed no significant performance-enhancing effects of GH.

The most common side effect of excess GH is a condition known as acromegaly (21). When GH is given after growth is completed, thickening of the bones and skull can occur, leading to bony changes that alter the patient's facial features: The brow and lower jaw protrude, the nasal bone enlarges, and spacing of the teeth increases (21). Other symptoms of acromegaly include thick, coarse, oily skin; enlarged lips, nose, and tongue; deepening of the voice attributable to enlarged sinuses and vocal cords; snoring attributable to upper airway obstruction; excessive sweating and skin odor; fatigue and weakness; headaches; impaired vision; abnormalities of the menstrual cycle and sometimes breast discharge in women; and impotence in men (21). There may be enlargement of body organs, including the liver, spleen, kidneys, and heart.

Cardiovascular Agents

Cardiovascular disease (CVD) is the number one killer of men and women in developed countries (23). Patients with CVD often receive multiple medical therapies that can affect exercise capacity. The clinical exercise physiologist is likely to encounter patients with CVD as they enter cardiac rehabilitation programs or other lifestyle modification programs designed for weight management and CVD risk factor reduction (table 5.3).

Diuretics

Thiazide diuretics are mostly used for the treatment of hypertension but may be used in patients who have mild edema from congestive heart failure or vascular insufficiency. The mechanism for their antihypertensive effects is not well defined but may be attributable, in part, to altered sodium handling by the kidney. Decreased blood volume is not likely to contribute to the antihypertensive effects because a decrease in blood pressure is noticed at doses far below those required to elicit net fluid losses (24). Thiazide diuretics have the following effects:

— Increase the urinary excretion of sodium and chloride in equivalent amounts
— Inhibit reabsorption of sodium and chloride in the ascending limb of the loop of Henle and the distal tubules
— Increase potassium and bicarbonate excretion
— Decrease calcium and uric acid excretion (24)

TABLE 5.3
Cardiovascular Agents and Their Exercise Effects

Treatment choice	Medications	Heart rate	Blood pressure	ECG	Exercise capacity
Diuretic agents	Hydrochlorothiazide Furosemide spironolactone	↔ (R and E)	↔ or ↓ (R and E)	↔ or PVCs (R). May cause PVCs and false-positive test results if hypokalemia occurs. May cause PVCs if hypomagnesemia occurs (E)	↔, except possibly in patients with CHF
Antianginal agents	Nitrates	↑ (R) ↑ or ↔ (E)	↓ (R) ↓ or ↔ (E)	↑ HR (R) ↑ or ↔ HR (E) ↓ ischemia (E)	↑ in patients with angina; ↔ in patients without angina; ↑ or ↔ in patients with CHF
Antiarrhythmic agents[a]	Quinidine Disopyramide	↑ or ↔ (R and E)	↓ or ↔ (R) ↔ (E)	↑ or ↔ HR (R). May prolong QRS complex and Q-T interval (R). Quinidine may result in false-negative test results (E)	↔
	Procainamide	↔ (R and E)	↔ (R and E)	May prolong QRS complex and Q-T interval (R). May result in false-positive test results (E)	↔
	Phenytoin Tocainide Mexiletine	↔ (R and E)	↔ (R and E)	↔ (R and E)	↔
	Flecainide Moricizine	↔ (R and E)	↔ (R and E)	May prolong QRS complex and Q-T interval (R) ↔ (E)	↔
	Propafenone	↓ (R) ↓ or ↔ (E)	↔ (R and E)	↓ HR (R) ↓ or ↔ HR (E). Prolongs P-R interval	↔
	Amiodarone Bretylium Sotalol Ibutilide Dofetilide	↓ (R and E)	↔ (R and E)	↓ HR (R) ↔ (E) Prolong Q-T interval and P-R interval to a lesser extent	↔

CALCIUM CHANNEL–BLOCKING AGENTS

Treatment choice	Medications	Heart rate	Blood pressure	ECG	Exercise capacity
Antihypertensive agents	Verapamil Diltiazem	↓	↓	Prolong P-R interval, may cause heart block	No effect
	Amlodipine Isradipine Nifedipine Felodipine	No effect	↓	No effect	No effect

continued

65

TABLE 5.3 (CONTINUED)

Treatment choice	Medications	Heart rate	Blood pressure	ECG	Exercise capacity
	β-ADRENERGIC BLOCKING AGENTS				
	Propranolol Metoprolol Timolol	↓	↓	Prolong P-R interval, may cause heart block	↓
	OTHER ANTIADRENERGIC AGENTS				
	Prazosin Terazosin Doxazosin	No effect	↓	No effect	No effect
	Carvedilol Labetalol	↓	↓	Prolong P-R interval, may cause heart block	May decrease
	CENTRALLY ACTING AGENTS				
	Methyldopa cloindine	May decrease	↓	May prolong P-R interval	May decrease
	VASODILATORS				
	Hydralazine Minoxidil	No effect	No effect	No effect	No effect
	ANGIOTENSIN-CONVERTING ENZYME INHIBITORS				
	Captopril Enalapril Lisinopril	No effect	No effect	No effect	No effect but may increase in patients with heart failure
	ANGIOTENSIN RECEPTOR BLOCKERS				
	Losartan Valsartan Irbesartan	No effect	No effect	No effect	No effect
Cardiac glycoside agents	Digoxin	↓ (R)	No effect	May prolong P-R inter-valor cause heart block or various arrhythmias at toxic doses	No effect
Antilipemic agents	Lovastatin Simvastatin Fluvastatin Atorvastatin Cholestyramine Gemfibrozil	No effect	No effect	No effect	No effect
Anticoagulant agents	Warfarin	↔ (R and E)	↔ (R and E)	↔ (R and E)	↔

Note. ª = All antiarrhythmic agents may cause new or worsened arrhythmias (proarrhythmic effect); ECG = electrocardiogram; R = rest; E = exercise; PVC = premature ventricular contraction; CHF = congestive heart failure; HR = heart rate; ↔ = no change; ↓ = decrease; ↑ = increase.

The thiazide diuretics include bendroflumethiazide, benzthiazide, chlorothiazide, chlorthalidone, hydrochlorothiazide, indapamide, methyclothiazide, metolazone, polythiazide, quinethazone, and trichlormethiazide. These agents are generally well tolerated, but some side effects occur:

— Exacerbations of diabetes or gout in certain individuals

— Phototoxicity (patients should be advised to use sunscreen regularly)

— Electrolyte depletion, which may manifest as muscle cramping or weakness

Loop diuretics inhibit the reabsorption of sodium and chloride in the loop of Henle. These agents are extremely potent diuretics and can cause profound diuresis with subsequent electrolyte depletion (25). Potassium supplementation is often required. The loop diuretics include bumetanide (Bumex®), ethacrynic acid (Edecrin®), furosemide (Lasix®), and torsemide (Demedex®). The side effects of loop diuretics are the same as listed previously, with the increased potential to cause electrolyte disturbances.

Potassium-sparing diuretics interfere with sodium reabsorption in the distal tubule, thus inhibiting water and sodium reabsorption without affecting potassium (24). Because potassium is filtered at the glomerulus and then absorbed parallel to sodium throughout the proximal tubule and the ascending limb of the loop of Henle, only minor amounts reach the distal tubule. Potassium-sparing diuretics include amiloride, spironolactone, and triamterene. **Hyperkalemia** is the most serious side effect of the potassium-sparing diuretics, and concomitant potassium supplementation should be avoided (24). Coadministration of angiotensin-converting enzyme (ACE) inhibitors can precipitate hyperkalemia with these drugs. In addition, spironolactone (Aldactone) may sometimes cause gynecomastia (enlargement of the breasts). Diuretics do not affect exercise response but are banned by both the USOC and the NCAA because their use can mask the presence of banned anabolic steroids (1, 2). The side effects of potassium-sparing diuretics are the same as listed previously, with the addition of gynecomastia and hyperkalemia.

Antianginal Agents

The symptoms of angina are caused by a mismatch between oxygen supply (decreased because of blockages in the coronary arteries) and oxygen demand (26). Patients with chronic stable angina will have enough blood flow at rest to meet metabolic needs because of internal processes that cause the coronary arteries to dilate. However, with exercise the internal mechanisms are not sufficient to further dilate the arteries, and a mismatch between myocardial oxygen supply and demand occurs, causing anginal symptoms. The only way to increase oxygen supply is through surgery, but medication can help to decrease oxygen demand and thereby decrease anginal symptoms (26).

There are four types of antianginal agents: long-acting nitrate formulations, β-adrenergic antagonists, and calcium channel antagonists for the prevention of chronic stable angina, and fast-acting nitrates used to relieve acute angina (26). β-blockers reduce the resting heart rate and therefore reduce the demand for oxygen (27). Calcium channel antagonists prevent the blood vessels from constricting and thus prevent coronary artery spasm (28). They also dilate arteries in the peripheral vascular system to decrease the force that the heart has to pump against (decreasing oxygen demand). Certain calcium antagonists, such as verapamil and diltiazem, also slow the heart rate (29). β-antagonists and calcium antagonists are covered in detail in the section on antihypertensive agents (pp. 68-69).

Nitrates, such as nitroglycerin, cause dilation of the blood vessels. There are short-acting and long-acting nitrates. Nitroglycerin is available as a tablet (sublingual) or an oral spray. A tablet of nitroglycerin placed under the tongue or inhalation of the oral spray usually relieves an episode of angina in 1 to 3 min; the effect of these short-acting nitrates lasts 30 min. Anyone with chronic stable angina must carry nitroglycerin tablets or spray at all times (26).

Long-acting nitrates are available as tablets, skin patches, or paste. Tablets are taken one to four times daily. Nitro paste and skin patches, in which the drug is absorbed through the skin over many hours, are also effective. Side effects of nitrates include the following (26):

— Headache

— Low blood pressure

— Dizziness

— Fainting

All antianginals improve exercise capacity in the patient with chronic stable angina by decreasing anginal symptoms with exertion. They function as antianginal agents by lowering heart rate and blood pressure to decrease the work of the heart (decrease myocardial oxygen demand) by reducing preload on the heart.

Antiarrhythmic Agents

Cardiac arrhythmias (irregular heartbeat) can range from life threatening to asymptomatic (30). Drug treatment of cardiac arrhythmias is sometimes required to either prevent fatal arrhythmias or decrease symptoms or other adverse effects that could be caused by the arrhythmia. Optimal therapy of cardiac arrhythmias requires accurate diagnosis and assessment of risk, optimal management of any precipitating factors, and careful selection of antiarrhythmic drug therapy if required. Antiarrhythmic agents work by blocking calcium, sodium, and potassium channels in the heart's electrical conduction system (30). The most serious adverse effect of antiarrhythmic agents is proarrhythmia, or the precipitation of a life-threatening arrhythmia from the antiarrhythmic drug itself. For that reason, judicious use of antiarrhythmic drugs is prudent. All antiarrhythmic agents can decrease exercise tolerance by decreasing contractility of the heart, which lowers cardiac output. In addition, exercise can sometimes precipitate an arrhythmia in susceptible individuals.

Antihypertensive Agents

Several classes of cardiovascular drugs are used to treat high blood pressure (31). They work by different mechanisms, and it is unclear if any agent is more effective at preventing the clinical sequelae of hypertension. All are effective at lowering blood pressure.

Calcium Channel–Blocking Agents

In addition to the calcium channel blockers (CCB) verapamil and diltiazem, which are also class IV antiarrhythmic agents, several other CCBs can be used to treat hypertension (32). These agents are collectively known as the dihydropyridine CCBs. Whereas the nondihydropyridine CCBs verapamil and diltiazem lower heart rate by inhibiting calcium influx in conduction tissue, the dihydropyridine CCBs cause vasodilation by inhibiting calcium influx in vascular smooth muscle (29). In fact, a common effect of short-acting dihydropyridines (nifedipine, isradipine) is that they may actually increase heart rate through a reflex mechanism resulting from rapid vasodilation (29). These agents are generally well tolerated but may cause lower extremity edema in some patients. Sustained-release products or long-acting agents such as amlodipine (Norvasc®) are preferred over short-acting agents (31). Long-acting agents aid compliance by offering once daily dosing and also cause less activation of the sympathetic nervous system because onset of vaso-

dilation is more gradual. Dihydropyridine CCBs exhibit antiplatelet effects and can cause bruising. They also lower blood pressure and systemic vascular resistance. CCBs can increase exercise tolerance in patients with chronic stable angina by decreasing angina on exertion.

β-Adrenergic Blocking Agents

β-adrenergic blocking agents compete with catecholamines for available β-receptor sites. Propranolol, nadolol, timolol, penbutolol, carteolol, sotalol, and pindolol inhibit both the β_1-receptor (cardiac muscle) and β_2-receptors (bronchial and vascular smooth muscle) (33). Acebutolol, atenolol, betaxolol, bisoprolol, esmolol, and metoprolol are specific to the β_1-receptors at usual doses (33). At higher doses, specificity weakens. Nonspecific agents can precipitate wheezing in patients with reactive airway disease. In addition, unopposed β-adrenergic activity in the peripheral vasculature can cause vasoconstriction and may impair blood flow in patients with peripheral vascular disease (33). Metoprolol and bisoprolol decrease mortality rates in patients with New York Heart Association class II or III congestive heart failure (34). Patients with diabetes mellitus receiving insulin or oral hypoglycemic agents should be cautioned that β-blockers blunt the tachycardia associated with hypoglycemia (33). β-blockers can also cause sexual dysfunction. All β-blockers depress myocardial contractility, decrease heart rate and blood pressure, and can decrease exercise tolerance (33). In addition, β-blockers suppress tremors and can be abused by athletes in shooting sports. For this reason, they are banned by the NCAA for riflery/shooting only (2) and by the USOC in the biathlon, bobsled, luge, ski jumping, freestyle skiing, archery, diving, equestrian, fencing, gymnastics, modern pentathlon, riflery/shooting, sailing, and synchronized swimming (1).

Other Antiadrenergic Agents

Prazosin, terazosin, and doxazosin selectively block postsynaptic α_1-adrenergic receptors (35). These drugs dilate both arterioles and veins. Orthostatic or postural hypotension can occur and is more common when these agents are initiated (35). Otherwise, α_1-adrenergic blockers are generally well tolerated. α-blockers do not effect exercise tolerance.

Labetalol and carvedilol are combination α/β-blockers. They inhibit β_1- and β_2-receptors as well as peripheral α_1-adrenergic receptors. Carvedilol has been shown to decrease mortality rates in patients with New York Heart Association class II or III chronic heart failure. It can cause dizziness and worsening of chronic heart failure when first started, and therefore it should be started at the lowest dose possible and titrated upward over several weeks. Both of these agents can decrease exercise tolerance because of their β-adrenergic-blocking effects (34).

Centrally Acting Agents

Methyldopa, methyldopate, clonidine, reserpine, guanfacine, and guanabenz lower blood pressure by stimulating central inhibitory α-adrenergic receptors. Reduction in sympathetic outflow from the central nervous system decreases peripheral vascular resistance, renal vascular resistance, heart rate, and blood pressure (36). Sedation is a common side effect with these agents. Tolerance to centrally acting agents may occur and requires the addition of diuretics to the antihypertensive regimen to maintain efficacy. Although methyldopa is used less frequently than most antihypertensives, it is the drug of choice for pregnant patients with hypertension because of extensive experience in managing pregnant hypertensive women with this agent (37). Clonidine (Catapres®) is sometimes also used to aid in withdrawal from narcotic and nicotine addiction. These agents do not affect exercise response (36).

Vasodilators

Hydralazine and minoxidil exert a peripheral vasodilating effect through a direct relaxation of vascular smooth muscle. This results in a decrease in peripheral vascular resistance and a subsequent increase in cardiac output. Minoxidil can cause serious adverse effects including **pericardial effusion,** which is a collection of fluid in the lining of the heart that can lead to cardiac failure. Minoxidil can also result in excessive growth of hair on the palms of the hands and face, which has led to the development of topical formulations to treat male pattern baldness. These agents do not affect exercise response (38).

Angiotensin-Converting Enzyme Inhibitors

ACE inhibitors prevent the formation of the potent vasoconstrictor angiotensin II (39). Reduction in circulating angiotensin II concentrations results in decreased peripheral vascular resistance with a subsequent increase in cardiac output and no change in heart rate. ACE inhibitors are the antihypertensive of choice in patients with diabetes mellitus (31). In addition, ACE inhibitors are first-line treatment for chronic heart failure (40). ACE inhibitors are generally well tolerated, but a small percentage of patients may develop a cough. In addition, angioedema, a life-threatening allergic reaction, can rarely occur with all ACE inhibitors (39). ACE inhibitors

increase exercise duration in patients with congestive heart failure but otherwise have no effect on exercise capacity.

Angiotensin Receptor Blockers

Angiotensin receptor blockers inhibit the binding of angiotensin II to angiotensin II type 1 receptors, which block the vasoconstriction and aldosterone-secreting effects of angiotensin II. Available agents include losartan (Cozaar®), valsartan (Diovan®), candesartan (Atacand®), and irbesartan (Avapro®) (41). These agents do not affect exercise response.

Cardiac Glycoside Agents

The cardiac (or digitalis) glycosides include digitoxin and digoxin (42). Cardiac glycosides have a direct action on cardiac muscle that increases force and velocity of cardiac muscle contractions (inotropic actions). These inotropic effects result from inhibition of the sodium–potassium–adenosine-triphosphatase pump in the sarcolemmal membrane that leads to increases in intracellular calcium (42). In addition, cardiac glycosides increase vagal tone to the atrioventricular node, which decreases heart rate and prolongs the P-R interval on the electrocardiogram (42). Serum drug concentrations of digoxin are often used to adjust dosing and check for toxicity.

Cardiac glycosides are used in chronic heart failure and for heart rate control in atrial fibrillation or flutter (42). Digoxin toxicity may manifest as severe gastrointestinal complaints, visual disturbances, heart block, or any number of ventricular arrhythmias (42). Although digoxin slows the heart rate and increases the P-R interval on the electrocardiogram, these effects are obliterated by exercise when vagal tone to the atrioventricular node is overcome by the sympathetic activation produced by exercise (42). Digoxin has no effect on exercise capacity.

Antilipemic Agents

Cholesterol is produced by the liver and can also enter the bloodstream via dietary absorption. Although cholesterol is needed for normal cell function, excess amounts in the bloodstream increase the risk of atherosclerotic diseases (43). Two of the primary agents used for lowering cholesterol are bile acid sequestrants and statins. Bile acid sequestrants interfere with cholesterol synthesis by binding bile acids in the intestine to form an insoluble complex that is excreted in the feces (44). This results in removal of bile acids from the body, preventing their absorption.

Because cholesterol is the major precursor of bile acids, serum concentrations of cholesterol are lowered as more cholesterol is channeled into the manufacture of bile acids. This results in a reduction in low-density lipoprotein (LDL) cholesterol. Bile acid sequestrants include cholestyramine and colestipol (44). They interfere with the absorption of many drugs, and should not be taken with other medications. In general they are poorly tolerated, causing flatulence, bloating, and cramping. They are useful in smaller doses as an adjunctive therapy for patients receiving other cholesterol-lowering medications who need additional cholesterol lowering (44).

The "statin" agents competitively inhibit 3-hydroxy-3-methyl-glutaryl-coenzyme A (HMG-CoA) reductase, the enzyme that catalyzes the conversion of HMG-CoA to mevalonate (44). This conversion is an early, rate-limiting step in cholesterol synthesis (43). The statins potently decrease low-density lipoprotein cholesterol and modestly increase high-density lipoprotein cholesterol and triglycerides (44). Common side effects include bloating and abdominal cramping. Liver function tests can be elevated, but hepatic failure is extremely rare. **Rhabdo-myolysis** with renal dysfunction or acute renal failure secondary to **myoglobinuria** has occurred. Patients should be advised to report myalgias or muscle weakness as soon as possible. In addition, muscle soreness without other complications has been reported with atorvastatin and fluvastatin (43). Risk of systemic muscle damage and breakdown is increased when these agents are taken with cyclosporine, erythromycin, gemfibrozil, fibric acid derivatives, or lipid-lowering doses of niacin (44). None of the lipid-lowering agents have an effect on exercise capacity. Additional information regarding cholesterol treatments can be found in chapter 10.

Anticoagulant Agents

Warfarin is used to prevent clot formation and subsequent stroke that can result from atrial fibrillation or prosthetic heart valves or to prevent recurrence of venous clots that cause deep venous thrombosis or pulmonary embolism (45). Warfarin interferes with the hepatic synthesis of vitamin K-dependent clotting factors, which results in depletion of factors VII, IX, X, and II (prothrombin) (46).

PRACTICAL APPLICATION 5.1

Client–Clinician Interaction

It is critical that the clinical exercise physiologist obtain all medications and their dosage from the patient during the medical evaluation. The mechanism of action for each medication should be clearly understood, and their effect on cardiorespiratory parameters (heart rate, blood pressure, electrocardiogram) should be noted before testing or prescribing exercise. Bear in mind that some medications will influence the patient's exercise capacity. For instance, a β-blocker can improve the exercise capacity in a patient with ischemic disease but hampers the exercise capacity of a healthy individual. In some cases, medications can be added or removed while the patient is exercising in your program. If there are no graded exercise test data with the new medications, it may be necessary to use several precautions until appropriate exercise responses can be established. These may include using electrocardiogram monitoring for signs of ischemia or decreased cardiac output; administering prophylactic nitroglycerin to high-risk patients with coronary artery disease before beginning exercise; and avoiding exercise in the presence of angina, dyspnea, or extreme fatigue. Another critical issue that a clinician often confronts is the timing between administering the medication and the exercise session. Taking a medication just before exercise (i.e., breakfast and morning exercise) may not be optimal because the elapsed time may be insufficient to allow for the medication to take full effect. This is particularly true with longer acting medications that influence heart rate or blood pressure.

On occasion, the clinician may become aware of drug abuse by a client. Whether drug abuse is intentional or not (e.g., the client becomes overly reliant on pain killers), it is important to counsel the patient about drug abuse and refer the patient to his or her primary care provider for appropriate treatment. It is also important to provide positive encouragement and support to help the patient receive the appropriate treatment. As such, it is critical to maintain professional conduct by using the utmost sensitivity and discretion in handling this matter and maintaining patient confidentiality.

Because these factors are involved in stimulating coagulation, inhibition results in decreased propensity to form a blood clot. Warfarin has a narrow therapeutic index and extreme interindividual variability. For these reasons, it must be intensely monitored to ensure that the level of anticoagulation remains within acceptable parameters (45). Monitoring of warfarin therapy is achieved with the international normalized ratio, a globally standardized measurement of anticoagulation (46). Common side effects of warfarin include bleeding, and patients should be counseled about signs and symptoms of acute blood loss (46). In addition, there are many significant drug interactions with warfarin, and patients should not stop or start any new drugs without notifying their healthcare provider. Patients should be advised to avoid activity that may lead to trauma, such as rock climbing, contact sports (basketball and football), and motor sports. Otherwise, warfarin has no effect on exercise capacity.

Drugs of Abuse

Abuse of prescription, legal, and illegal drugs is a major problem facing society today (47) (table 5.4). Commonly abused legal substances include nicotine, alcohol, and caffeine. Commonly abused prescription drugs include central nervous system stimulants such as amphetamines and diet pills and central nervous system depressants such as benzodiazepines, barbiturates, and narcotic analgesics (opiates). Illegal drugs of abuse include stimulants such as cocaine, depressants like heroin, and other agents such as LSD and ecstasy (47).

The essential feature of substance dependence is the continued use of the substance despite adverse substance-related problems. Repeated use of the drug is often associated with the development of tolerance, withdrawal, and compulsive use, but it is possible to meet criteria for dependence in the absence of physical dependence (47).

TABLE 5.4

Drugs of Abuse and Narcotic Analgesics and Their Exercise Effects

Medications	Heart rate	Blood pressure	ECG	Exercise capacity
Nicotine	↑ or ↔ (R and E)	↑ (R and E)	↑ or ↔ HR May provoke ischemia, arrhythmias (R and E)	↔, except ↓ or ↔ in patients with angina
Alcohol	↔ (R and E)	Chronic use may have role in ↑ BP (R and E)	May provoke arrhythmias (R and E)	↔
Caffeine	Variable effects depending on previous use		May provoke arrhythmias	Variable effects on exercise capacity
CNS STIMULANTS				
Amphetamines Diet pills Cocaine	↑ (R and E)	↑ (R and E)	↑ or ↔	No effect
CNS DEPRESSANTS				
Benzodiazepines Barbiturates Opiates	May decrease HR by controlling anxiety. Withdrawal may precipitate anxiety and increased HR	No effect	No effect	No effect

Note. ECG = electrocardiogram; R = rest; E = exercise; HR = heart rate; BP = blood pressure; CNS = central nervous system; ↔ = no change; ↑ = increase; ↓ = decrease.

Conclusion

The widespread use of pharmacological agents in our society necessitates that the clinician be familiar with the impact of these agents on exercise and exercise capacity. Whether involved in training the elite athlete or in cardiac rehabilitation, the exercise physiologist is likely to encounter patients taking various pharmacological agents. This chapter is not meant to be a comprehensive overview of all therapeutic agents available but rather targets the specific types of therapies likely to be encountered by exercise physiologists in daily practice.

Case Study 5

Medical History and Diagnosis

Mr. MT is a 46-year-old white male with a family history of cardiovascular disease, high blood pressure (150/94), and hypercholesterolemia (total cholesterol = 284, high-density lipoprotein = 35). He is referred to you for an exercise prescription as part of a comprehensive lifestyle modification program to deal with his cardiovascular risk factors. Because of his extremely high risk of cardiac disease, his primary care physician has started him on the following medications:

— Metoprolol 100 mg per day (a β-blocker)
— Atorvastatin 10 mg per day (a cholesterol-lowering agent)
— Niacin 1000 mg per day (a cholesterol-lowering agent)

Exercise Prescription

The goal is to determine an appropriate exercise regimen given this patient's current medical therapy. Given the lack of a preliminary exercise test and the potential influence of his medical therapy (β-blocker) on heart rate responses, the rating of perceived exertion (RPE) is used to determine the appropriate exercise intensity. In addition, the subject's target heart rate is determined by plotting the heart rate responses against a graded workload.

Given this assessment, the following exercise prescription is determined:

— Frequency = 5 days per week
— Intensity = 78 to 95 beats $\cdot$ min^{-1}
— RPE = 11 to 15
— Duration = 30 min
— Mode = Aerobic

At a scheduled follow-up visit, the patient complains of increased sluggishness throughout the day and an increased fatigue when performing the prescribed exercise.

Case Study 5 Discussion Questions

1. What are some of the reasons this patient could be experiencing these symptoms?
2. What changes in the patient's current exercise and medication regimen could be made to decrease these symptoms and increase his quality of life?
3. Given the change in medical therapy, how might you update the exercise prescription?

Glossary

androgenic—Masculinizing effects; stimulation of male sex characteristics and male hair characteristics.

anemia—A decrease in the red blood cells that carry hemoglobin, resulting in a reduced oxygen-carrying capacity.

erythropoiesis—Stimulation of red blood cell production.

hemolysis—Alteration or destruction of red blood cells.

hyperkalemia—Excess concentrations of potassium in the bloodstream.

hypoglycemia—Abnormally small concentration of glucose in the circulating blood.

myoglobinuria—Excretion of myoglobin in the urine resulting from muscle degeneration.

pericardial effusion—Increased amounts of fluid within the pericardial sac usually attributable to inflammation.

rhabdomyolysis—Acute, potentially fatal disease of the skeletal muscle that entails destruction of the muscle as evidenced by myoglobinuria.

References

1. USADA Guide to Prohibited Classes of Substances and Prohibited Methods of Doping. United States Anti-Doping Agency. 2nd ed. Colorado Springs, USADA, July 2001.

2. National Collegiate Athletic Association Banned Substances List [Online]. Available: www.ncaa.org/sports_sciences/drugtesting/banned_list.html [May 26, 2001].

3. Sproat TT. Anemias. In Dipiro JT, ed. Pharmacotherapy: A Pathophysiologic Approach, 4th ed. Stamford, CT, Appleton & Lange, 1999, pp. 1531-1548.

4. Hochberg MC. Clinical features and treatment. In Klippel JH, ed. Primer on the Rheumatic Diseases, 11th ed. Atlanta, GA, Arthritis Foundation, 1997, pp. 218-221.

5. Balaban EP. Sports anemia. Clin Sports Med 1992; 11(2): 313-325.

6. Nielsen P, Nachtigall D. Iron supplementation in athletes. Current recommendations. Sports Med 1998; 26(4):207-216.

7. Macdougall IC. An overview of the efficacy and safety of novel erythropoiesis stimulating protein (NESP). Nephrol Dial Transplant 2001; 16(Suppl 3): 14-21.

8. Ekblom BT. Blood boosting and sport. Baillieres Best Pract Res Clin Endocrinol Metab 2000; 14(1): 89-98.

9. Simon LS. Biologic effects of nonsteroidal anti-inflammatory drugs. Curr Opin Rheumatol 1997; 9: 178-182.

10. Allison MC, Howatson AG, Torrance CJ, et al. Gastro-intestinal damage associated with the use of nonsteroidal anti-inflammatory drugs. N Engl J Med 1992; 327: 749-754.

11. Bennett WM, Henrich WL, Stoff JS. The renal effects of nonsteroidal anti-inflammatory drugs: Summary and recommendations. Am J Kidney Dis 1996; 28(Suppl 1): S56-S62.

12. Lane JM. Anti-inflammatory medications: Selective COX-2 inhibitors. J Am Acad Orthop Surg 2002; 10(2): 75-78.

13. Parssinen M, Seppala T. Steroid use and long-term health risks in former athletes. Sports Med 2002; 32(2): 83-94.

14. Giorgi A, Weatherby RP, Murphy PW. Muscular strength, body composition and health responses to the use of testosterone enanthate: A double blind study. J Sci Med Sport 1999; 2(4): 341-355.

15. Demling RH, DeSanti L. The rate of restoration of body weight after burn injury, using the anabolic agent oxandrolone, is not age dependent. Burns 2001; 27(1): 46-51.

16. Gascon A, Belvis JJ, Berisa F, et al. Nandrolone decanoate is a good alternative for the treatment of anemia in elderly male patients on hemodialysis. Geriatr Nephrol Urol 1999; 9(2): 67-72.

17. Rose C, Kamby C, Mouridsen HT, et al. Combined endocrine treatment of elderly postmenopausal patients with metastatic breast cancer. A randomized trial of tamoxifen vs. tamoxifen + aminoglutethimide and hydrocortisone and tamoxifen + fluoxymesterone in women above 65 years of age. Breast Cancer Res Treat 2000; 61(2): 103-110.

18. Bond AJ, Choi PY, Pope HG. Assessment of attentional bias and mood in users and non-users of anabolic-androgenic steroids. Drug Alcohol Depend 1995; 37(3): 241-245.

19. Pope HG Jr, Kouri EM, Hudson JI. Effects of supraphysiologic doses of testosterone on mood and aggression in normal men: A randomized controlled trial. Arch Gen Psychiatry 2000; 57(2): 133-140.

20. NIDA Research Report—Steroid Abuse and Addiction: NIH Publication No. 00-3721. Bethesda, MD, National Institutes of Health, 2000.

21. Heck AM, Calis KA, Yanovski JA. Pituitary gland disorders. In Dipiro JT, ed. Pharmacotherapy: A Pathophysiologic Approach, 4th ed. Stamford, CT, Appleton & Lange, 1999, pp. 1281-1297.

22. Demling RH. Comparison of the anabolic effects and complications of human growth hormone and the testosterone analog, oxandrolone, after severe burn injury. Burns 1999; 25(3): 215-221.

23. American Heart Association. Unpublished data. Dallas, American Heart Association, 2001.

24. Valvo E, D'Angelo G. Diuretics in hypertension. Kidney Int 1997; 59(Suppl): S36-S38.

25. Dormans TP, van Meyel JJM, Gerlag PGG, et al. Diuretic efficacy of high dose furosemide in severe heart failure: Bolus injection versus continuous infusion. J Am Coll Cardiol 1996; 28: 376-382.

26. Talbert RL. Ischemic heart disease. In Dipiro JT, ed. Pharmacotherapy: A Pathophysiologic Approach, 4th ed. Stamford, CT, Appleton & Lange, 1999, pp. 182-210.

27. Goldstein S. Beta-blocking drugs and coronary heart disease. Cardiovasc Drugs Ther 1997; 11: 219-225.

28. Opie LH. Calcium channel antagonists in the treatment of coronary artery disease: Fundamental pharmacological properties relevant to clinical use. Prog Cardiovasc Dis 1996; 38: 273-290.

29. Ferrari R. Major differences among the three classes of calcium antagonists. Eur Heart J 1997; 18(Suppl A): A56-A70.

30. Bauman JL, Schoen MD. Arrhythmias. In Dipiro JT, ed. Pharmacotherapy: A Pathophysiologic Approach, 4th ed. Stamford, CT, Appleton & Lange, 1999, pp. 232-264.

31. The Sixth Report of the National Committee on Detection, Evaluation, and Treatment of High Blood Pressure (JNC-VI). Arch Intern Med 1997; 157: 2413-2446.

32. Kaplan NM. Calcium entry blockers in the treatment of hypertension. JAMA 1989; 262: 817-823.

33. Nadelmann J, Frishman WH. Clinical use of beta-adrenoreceptor blockade in systemic hypertension. Drugs 1990; 39: 862-876.

34. Heidenreich PA, Lee TT, Massie BM. Effect of beta-blockade on mortality in patients with heart failure: A meta-analysis of randomized clinical trials. J Am Coll Cardiol 1997; 30: 27-34.

35. Perry HM. Central and peripheral sympatholytics. In Izzlo JL, Black HR eds. Hypertension Primer. Dallas, American Heart Association, 1993, pp. 306-308.

36. Perry HM. Central and peripheral sympatholytics. In Izzlo JL, Black HR eds. Hypertension Primer. Dallas, American Heart Association, 1993, pp. 306-308.

37. Henriksen T. Hypertension in pregnancy: Use of anti-hypertensive drugs. Acta Obstet Gynecol Scand 1997; 76: 96-106.

38. Venkata C, Ram S, Featherston WE. Vasodilators. In Izzlo JL, Black HR eds. Hypertension Primer. Dallas, American Heart Association, 1993, pp. 314-316.

39. Ferrario CM. Importance of the rennin-angiotensin-aldosterone system (RAS) in the physiology and pathology of hypertension. Drugs 1990; 39(Suppl 2): 1-8.

40. American College of Cardiology/American Heart Association Task Force on Practice Guidelines. Guidelines for the evaluation and management of heart failure. Circulation 1995; 92: 2764-84.

41. Messerli FH, Weber MA, Brunner HR. Angiotensin II receptor inhibition. A new therapeutic principle. Arch Intern Med 1996; 156: 1957-1965.

42. Kelly RA, Smith TW. Pharmacologic treatment of heart failure. In Hardman JG, Limbird LE, Molinoff PB, et al. eds. Goodman and Gilman's Pharmacological Basis of Therapeutics, 9th ed. New York, McGraw Hill, 1996, pp. 875-898.

43. Talbert RL. Hyperlipidemia. In Dipiro JT, ed. Pharmacotherapy: A Pathophysiologic Approach, 4th ed. Stamford, CT, Appleton & Lange, 1999, pp. 350-373.

44. Dailey JH, Gray DR, Bradberry JC, et al. Lipid-modifying drugs. In McKenney JM, Hawkins D eds. Handbook on the Management of Lipid Disorders, 2nd ed. St. Louis, National Pharmacy Cardiovascular Council, 2001, pp. 124-166.

45. Ansell JE, Buttaro ML, Thomas OV, et al. Consensus guidelines for coordinated outpatient oral anticoagulation therapy management. Ann Pharmacother 1997; 31: 604-615.

46. Hirsh J, Dalen JE, Deykin D, Piller L. Oral anticoagulants: Mechanism of action, clinical effectiveness, and optimal therapeutic range. Chest 1995; 108(Suppl): 231S-246S.

47. Doering PL. Substance-related disorders: Overview of depressants, stimulants and hallucinogens. In Dipiro JT, ed. Pharmacotherapy: A Pathophysiologic Approach, 4th ed. Stamford, CT, Appleton & Lange, 1999, pp. 1083-1103.

48. Doering PL. Substance-related disorders: Alcohol, nicotine, and caffeine. In Dipiro JT, ed. Pharmacotherapy: A Pathophysiologic Approach, 4th ed. Stamford, CT, Appleton & Lange, 1999, pp. 1104-1117.

Paul S. Visich, PhD, MPH
School of Health Sciences
Central Michigan University
Mt. Pleasant, MI

Graded Exercise Testing

Graded exercise testing was first used in approximately 1846, when Edward Smith began to evaluate the response of different physiological parameters (heart rate [HR], respiratory rate, inspired air) during exertion (5). Since that time, graded exercise testing has become a valuable tool to evaluate cardiorespiratory function in many different settings. Even with the addition of the **radionuclide agents** and pharmacological agents available today, graded exercise testing remains the first choice in evaluating cardiovascular function in many situations. Graded exercise testing in a clinical perspective consists of continuous electrocardiogram (ECG) monitoring (12 leads), patient observation, and periodic blood pressure (BP) assessment before, during progressively increasing exercise intensity, and after exercise. The exercise test is generally completed on a treadmill or bicycle ergometer. The termination point of exercise depends on the objectives of the test and is generally the point of maximal effort or a predetermined submaximal point.

This chapter provides a general overview of graded exercise testing that is commonly used in today's clinical setting. Other chapters in this textbook address unique features of exercise testing that are population specific. This chapter focuses on the information that a clinical exercise physiologist, along with other allied health professionals (i.e., nurses, physician assistants, and physical therapists) should acquire to perform a graded exercise test (GXT) in physician-supervised and nonsupervised settings.

Personnel

Before 1980, graded exercise testing in a clinical setting was primarily (90%) supervised by a cardiologist (25). Since that time, GXTs have been performed by many healthcare professionals (internists, family practitioners, physician assistants, exercise physiologists/exercise specialists, nurses, and physical therapists). The use of healthcare professionals and not cardiologists is based on a better understanding of graded exercise testing over the years, cost-containment initiatives, time constraints on physicians, and more sophisticated ECG analysis (computerized exercise ST-segment interpretation). The American Heart Association guidelines for exercise testing laboratories state that the paramedical personnel listed previously, when appropriately trained and possessing specific performance skills (e.g., American College of Sports Medicine [ACSM] certification), can safely supervise clinical GXTs (18). In addition, paramedical personnel should be trained in basic life support, and advanced cardiac life support is highly suggested. Suggested skills that the paramedical professional needs prior to supervising a GXT are listed in figure 6.1 and are presented throughout this chapter.

During graded exercise testing in patients with high-risk medical conditions, such as heart failure or high-grade dysrhythmias, it has been suggested that there be direct physician supervision. Otherwise, it is acceptable to have a supervising physician in the immediate area and readily available to respond to emergencies and questionable interpretations. Four studies found that the average morbidity and mortality rates during a GXT with physician supervision (>85% of the tests directly supervised by a physician) were 3.6 and 0.44 per 10,000 tests, respectively (2, 19, 20, 25). In three studies involving nonphysician supervision, average morbidity and mortality rates of 2.4 and 0.77 per 10,000 tests were observed, respectively (8, 12, 13). These data suggest that there are no differences in graded exercise testing-related morbidity and mortality rates when direct physician and paramedical staff supervision are compared. In addition, when a symptom-limited GXT is performed on a high-risk population (**left ventricular dysfunction** with **ejection fraction** <35%), nonphysician supervision has been observed

to be safe, when a physician is immediately available (24).

Indications

Although the overall risk of death during a GXT is small (zero to five sudden cardiac deaths per 10,000 tests), before a GXT is performed it is important to understand the purpose or reason for completing the test and, most importantly, to be sure that the benefits outweigh the risks (10). When a patient is referred for a GXT, it is imperative that the physician clearly state the indication or purpose of the test (e.g., evaluate chest pain, rule out coronary artery disease, assess function after myocardial infarction). In addition, the physician should inform the GXT supervisor if medications should be continued or discontinued before testing. If there is ever a question concerning the purpose for completing the GXT, the referring physician should be contacted for clarity.

Evaluating potential or existing cardiac problems is the most common reason that graded exercise testing with 12-lead ECG is performed. With respect to cardiac abnormalities, indications for graded exercise testing are separated into two classes (table 6.1). Class I represents conditions in which there is agreement that a GXT is justified. In class II, a GXT is frequently used, but there is a difference in opinion with respect to the value of performing the test (22).

In most cases, a GXT is the first test of choice because of its noninvasive nature and lower cost. The diagnostic value of determining the presence of coronary artery disease (CAD) is greatest when evaluating individuals with an intermediate to high probability of CAD (multiple risk factors, symptomatic, or both). Conversely, there appears to be little value in improving patient outcomes when exercise testing young (<40 years of age) and apparently healthy (no risk factors) people. Previous literature has observed a low yield for graded exercise testing to detect CAD in asymptomatic individuals. In 1998, it was calculated that $39,623 was spent in graded exercise testing to identify one individual who would benefit from surgical revascularization (17).

More recently, as part of the triage of patients with chest pain, a GXT is now used as a screening tool prior to hospitalization. If the test is interpreted as positive, the results may be sufficient to admit the patient for further diagnostic testing. Therefore, using a GXT as part of the triage protocol can help reduce length of stay and hospital costs.

- Indications for exercise testing for different populations.
- Absolute and relative contraindications for exercise testing.
- Ability to properly communicate with the client to effectively complete a medical history and informed consent.
- Ability to explain to the client the purpose of completing the graded exercise test (GXT), procedures of the test, and the responsibilities of the client during the test.
- Competence in cardiopulmonary resuscitation; certified by American Heart Association Basic Cardiac Life Support and preferably Advanced Cardiac Life Support.
- Knowledge of specificity, sensitivity, and predictive value of a positive and negative test and diagnostic accuracy of exercise testing in different patient populations.
- Understanding of causes that produce false-positive and false-negative test results.
- Knowledge of appropriate mode of activity and exercise protocol (e.g., Bruce, Naughton, Balke-Ware) for each individual, based on his or her medical history.
- Knowledge of normal and abnormal hemodynamic responses to graded exercise (blood pressure and heart rate response) in different age groups and with various cardiovascular conditions.
- Knowledge of normal and abnormal acute cardiovascular physiological responses to exercise for different age groups and cardiovascular conditions.
- Knowledge of metabolic data collected during a GXT and knowledge of how to properly interpret the data (e.g., maximal oxygen uptake, metabolic equivalent) for different medical conditions.
- Knowledge of 12-lead electrocardiography and changes in the electrocardiogram that may result from exercise, especially ischemia, arrhythmias, and conduction abnormalities.
- Knowledge of proper lead placement and skin preparation for 12-lead electrocardiogram.
- Knowledge and skills for accurately taking blood pressure under resting and exercise conditions.
- Knowledge of how to properly run and troubleshoot the medical equipment used for graded exercise testing (treadmill, electrocardiogram, bicycle ergometer, metabolic cart).
- Ability to communicate with the client to assess signs and symptoms of cardiovascular disease before, during, and after the GXT.
- Knowledge of how to assess signs and symptoms by using appropriate scales (e.g., chest pain, shortness of breath, rating of perceived exertion).
- Knowledge of the appropriate time to end the GXT and absolute and relative indications for test termination.
- Knowledge of knowing when to ask for physician support, when the physician is not directly involved in the GXT.
- Ability to appropriately communicate results of the GXT to the supervising physician.

FIGURE 6.1 Skills and knowledge needed to perform graded exercise testing for the paramedical professional.

Reprinted, by permission, from "American Heart Association clinical competence statement on stress testing: A report of the American College of Cardiology," *Circulation* 102: 1726-1738.

In individuals with documented CAD, graded exercise testing is used to evaluate the severity of the disease. This information can be beneficial in regard to evaluating the need for further intervention, such as coronary revascularization. In individuals who have an abnormal response such as ST depression 1 mm or greater at low work levels (e.g., unable to complete stage 1 of the Bruce protocol), a higher mortality rate (5%) is observed compared with those who were able to complete III stages without any ST changes (<1% mortality rate) (28).

In respect to minimizing healthcare costs, decreasing the length of stay in the hospital is looked on favorably. It has been observed that graded exercise testing can safely be performed 3 days after an uncomplicated myocardial infarction, which improves hospital efficiency and aids in risk stratification of the patient (8). If the GXT is interpreted as normal (nonischemic, normal hemodynamic response with exertion, and no significant dysrhythmias, and the patient achieves an acceptable **metabolic equivalent** [MET] level), this information helps the physician

TABLE 6.1

Common Reasons for Exercise Testing

Class I: Conditions in which there is agreement that graded exercise testing is justified

To aid in the diagnosis of coronary artery disease (CAD) in patients with symptoms that are atypical for myocardial ischemia

To assess the prognosis and severity of disease after an uncomplicated myocardial infarction

To evaluate the effect of surgical therapy (coronary artery revascularization or angioplasty)

To evaluate exercise-induced cardiac arrhythmias

To evaluate functional capacity of selected patients (e.g., congenital heart disease) and aid in formulating an exercise prescription

To evaluate patients with rate-responsive pacemakers

To evaluate drug therapy in special populations (e.g., hypertension, ischemia, dysrhythmias)

Class II: Conditions for which exercise testing is frequently performed but in which there is a divergence of opinion with respect to its value and appropriateness

To evaluate asymptomatic male patients over the age of 40 with special occupations (pilots, air traffic controllers, firefighters, police officers, critical process operators, bus or truck drivers, and railroad engineers)

To evaluate asymptomatic males over the age of 40 with two or more risk factors for CAD

To evaluate sedentary male patients over the age of 40 who plan to enter a vigorous exercise program

To assist in the diagnosis of CAD in women with a history of typical or atypical angina pectoris

To assist in the diagnosis of CAD in patients who are taking digitalis or who have complete right bundle branch block

To evaluate the functional capacity and response to therapy with cardiovascular drugs in patients with CAD or heart failure

To evaluate patients with variant angina

To follow up serially (at 1-year intervals or longer) patients with known CAD

To evaluate patients with a class 1 indication who have baseline electrocardiographic changes or coexisting medical problems that limit the value of the test (in some of these patients, exercise testing may still yield clinically useful information such as duration of exercise, blood pressure response, and production of chest discomfort)

To evaluate patients who have sustained a complicated myocardial infarction but who have subsequently "stabilized" (before discharge or early after discharge)

To evaluate on a routine, yearly basis patients who remain asymptomatic after a revascularization procedure

To evaluate the functional capacity of selected patients with valvular heart disease

To evaluate the blood pressure response of patients being treated for systemic arterial hypertension who wish to engage in vigorous dynamic or static exercise

To evaluate selected children and adolescents with valvular or congenital heart disease

Reprinted with permission from the American College of Cardiology Foundation Journal American College of Cardiology, 1986, 8: 725-738.

in deciding to discharge the patient. However, an abnormal response most likely will delay the discharge process and lead to additional diagnostic testing.

Following coronary revascularization, a GXT is commonly used to assess the outcome of the procedure and help determine a safe and effective exercise prescription. To evaluate the effectiveness of medical therapy, a GXT is used as an objective measure to evaluate changes in the onset of dysrhythmias, chest pain, and hypertension. In individuals with documented CAD, a GXT can be used yearly to evaluate disease progression. An increase in chest pain, greater ST depression, or ischemia that occurs at a lower rate–pressure product can suggest disease progression.

When one first notices a change in a GXT that would suggest disease progression, as listed previously, it is important to consider other factors that promote unfavorable changes without necessarily increasing disease progression. Factors that would have a negative impact include the following:

— Substantial weight gain
— Significant decrease or stoppage of aerobic exercise training
— A decrease or discontinuation in medications that negatively influence aerobic capacity

When an individual has a substantial weight gain, this causes the cardiovascular system to work harder, per given workload, which can enhance the onset of chest pain, ST depression, and ischemia. If one stops aerobic exercise or decreases activity substantially to the point of a significant decline in **functional capacity (FC),** this could cause abnormal signs or symptoms prematurely, if underlying CAD is present. Last, alterations in one's medical regimen could favorably or unfavorably alter the onset of abnormal signs or symptoms. **β-blockers,** calcium channel blockers, and nitrates are common medications given to cardiac patients, because they decrease myocardial oxygen consumption (per given workload) by decreasing HR, contractility, and vascular resistance. Ultimately, these medications delay the onset of ischemia and increase one's work tolerance in the presence of ischemic heart disease. Decreasing the dosage or frequency of these drugs can enhance abnormal signs or symptoms at an earlier workload. Therefore, it is important to keep accurate records of the patient's drug regimen, especially before a GXT.

Based on the severity of disease, a GXT can be a useful tool to **stratify** patients. In individuals at high risk (i.e., abnormal GXT response at low workloads; <5 METs), further medical intervention may be immediately required. However, if an abnormal ECG response doesn't occur until the individual reaches 10 METs, the prognosis is much more favorable, and additional medical or surgical intervention may not be needed.

Graded exercise testing can also be used for the primary purpose of evaluating FC. There are predicted normal values for individuals based on one's age, sex, and activity level. If values are below expected, one's **functional aerobic impairment (FAI)** can be determined and offers information that guides the safe return to work, activities of daily living, and participation in certain hobbies (4).

$$\%FAI = (predicted\ \dot{V}O_2 - observed\ \dot{V}O_2) / (predicted\ \dot{V}O_2) \times 100$$

Predicted $\dot{V}O_2$ (ml · kg^{-1} · min^{-1}) can be determined by using the following equations:

Active men: 69.7 – (0.612 · age in years)
Sedentary men: 57.8 – (0.445 · age in years)
Active women: 42.9 – (0.312 · age in years)
Sedentary women: 42.3 – (0.356 · age in years)

Note: When using these equations, a person is considered active when they exert themselves physically to the point of sweating at least once a week. FAI scores can be described as follows:

— 27-40% = mild
— 41-54% = moderate
— 55-68% = marked
— >68% = extreme

Negative scores are indicative of an above average fitness level, whereas a positive score is indicative of a physical impairment. In respect to completing an exercise prescription, knowing one's FC can be very beneficial in determining a training range. After FC, magnitude of ST changes, and angina response during the GXT are determined, a patient's prognosis can be estimated by using the **Duke nomogram** (14), shown in figure 6.2. In addition, it was observed recently that HR recovery after exercise helps predict mortality rate. It appears that HR recovery at 2 min best predicts mortality rate, and that a HR decrease less than 22 beats · min^{-1} produced a **hazard ratio** of 2.6, irrespective of β-blocker use. However, HR recovery was not able to discriminate those with coronary disease, and the authors suggest that the decrease in HR recovery should supplement other predictors of mortality rate, such as the Duke nomogram (23).

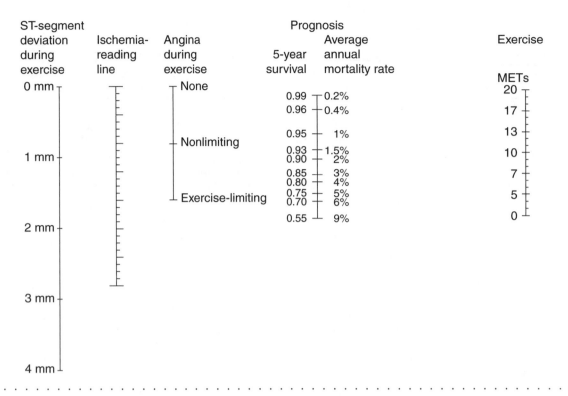

FIGURE 6.2 Duke nomogram. Step one; mark the largest level of ST-segment deviation. Step two; mark the observed degree of angina. Step three; connect the marks from steps one and two with a straight line and mark where this line crosses the ischemia line. Step four; mark the total METs achieved during the testing. Step five, connect the marks from steps three and four with a straight line. Where this line crosses the prognosis line indicates the five-year survival rate and annual mortality rate for the subject.

Reprinted, by permission, from D.B. Mark, et al., 1991, "Prognostic value of a treadmill exercise score in outpatients with suspected coronary artery disease," *New England Journal of Medicine* 325: 849-853.

To accurately determine FC, indirect measurement of oxygen consumption by use of a metabolic cart is needed; however, this is not financially practical in many institutions. A less accurate alternative is the use of metabolic prediction equations. Functional capacity can be predicted from final grade (percentage) and speed (meters per minute) of a treadmill, maximal workload (watts or kilopond-meters per minute) from arm or leg ergometer, and fastest stepping rate and height in stair stepping. The standard ACSM metabolic equations (1) are commonly used; however, they rely on achieving a steady state, which does not occur at maximal effort and leads to an overestimation of one's FC (figure 6.3). Other metabolic prediction equations have been developed for predicting FC but are specific to the Bruce GXT protocol, use of hand rail support, ramping on a treadmill, and watt increment per minute on a bicycle ergometer (table 6.2). Because numerous maximal GXT protocols are used in the field, there is a need for additional equations in predicting FC.

Just before the patient is discharged from the hospital, a GXT may be used. This test is commonly a submaximal (following myocardial infarction) effort to assess future risk and verify that it is safe for the patient to resume activities of daily living following discharge. In addition, this GXT can be used to determine a safe exercise prescription for home-based and initial cardiac rehabilitation programs.

The need for a GXT before starting an exercise program depends on age, sex, number of risk factors, signs or symptoms related to cardiopulmonary disease, and documented cardiopulmonary disease. The ACSM provides specific recommendations detailing when a physical examination and GXT should be completed and the need for physician supervision prior to starting an exercise program (1).

More recently, these guidelines were challenged relative to older people (>75 years) who are starting an exercise program. It has been suggested that the current ACSM guidelines encourage a large number of older individuals to have a GXT before starting an exercise program, because of the commonly observed risk factors in this population and non-specific ascertainment of signs and symptoms related to cardiopulmonary disease (11). It has been suggested that these guidelines create an excessive

---

Walking

$ml \cdot kg^{-1} \cdot min^{-1} = (0.1 \cdot S) + (1.8 \cdot S \cdot G) + 3.5$
Most appropriate between 50 and 100 m $\cdot$ min^{-1} or 1.9-3.7 mph
OK to use above this range (up to 5 mph), if the patient is truly walking
S = speed in m $\cdot$ min^{-1}, G = percent grade expressed as a fraction

Running

$ml \cdot kg^{-1} \cdot min^{-1} = S \cdot 0.2 + (0.9 \cdot S \cdot G) + 3.5$
Most accurate for speeds >134 m $\cdot$ min^{-1} or 5 mph or down to 80 m $\cdot$ min^{-1} or 3 mph, if truly jogging
S = speed in m $\cdot$ min^{-1}, G = percent grade expressed as a fraction

Leg ergometer

$ml \cdot kg^{-1} \cdot min^{-1} = (10.8 \cdot W \cdot M^{-1}) + 3.5$
Most appropriate for power outputs between 50 and 200 W
W = power in watts, M = body mass in kg

Arm ergometer

$ml \cdot kg^{-1} \cdot min^{-1} = (18 \cdot W \cdot M^{-1}) + 3.5$
Most appropriate for power outputs between 25 and 125 W
W = power in watts, M = body mass in kg

Stepping

$ml \cdot kg^{-1} \cdot min^{-1} = (0.2 \cdot F) + (1.33 \cdot 1.8 \cdot H \cdot F) + 3.5$
Most appropriate for stepping rates between 12 and 30 steps $\cdot$ min^{-1}, and step height is in meters (1 in. = 0.0254 m)
F = stepping frequency per minute, H= step height in meters

. . . .

FIGURE 6.3   Equations to predict energy expenditure (assuming steady state is achieved).

cost burden on the healthcare system and deter an already inactive population from exercising (suggested guidelines prior to exercising for this population are presented in practical application 6.1).

Contraindications

Although the risk of death during a GXT is very small (1 out of every 2500 tests), there are situations where GXT is not suggested because the risks of completing the test outweigh the benefits (9). Current guidelines separate **absolute** versus **relative contraindications for test termination** (figure 6.4). Absolute contraindications represent potential serious medical consequences if a GXT is completed, whereas relative contraindications suggest that there is a potential medical concern and that further inquiry (benefits vs. risk) should be considered before the GXT is conducted (18).

In addition to contraindications of GXT, there are conditions in which GXT is not recommended for diagnostic purposes because of the decreased ability to identify CAD (i.e., ST-T changes). Resting ECG abnormalities that limit the diagnostic abilities of a GXT are as follows:

— Complete left bundle branch block
— Preexcitation syndrome (Wolff-Parkinson-White)
— ST elevation or depression exceeding 1 mm (17)

The latter is often attributable to left ventricular hypertrophy, **digoxin** therapy, or prior myocardial infarction. If a client presents one of these ECG abnormalities at rest or is taking digoxin, it is prudent to inform the supervising physician and referring physician to make sure they are aware of the condition prior to starting the GXT. To increase sensitivity

<table>
<tr><td colspan="2" align="center">TABLE 6.2
· · · · ·
Estimating Functional Capacity at Maximal Effort</td></tr>
<tr><td colspan="2">Standard treadmill test</td></tr>
<tr><td>Using the Bruce Protocol without holding the handrail</td><td>$\dot{V}O_2max\ (ml \cdot kg^{-1} \cdot min^{-1}) = 14.8 - 1.379\ (time\ in\ min) + 0.451\ (time^2) - 0.012\ (time^3)$</td></tr>
<tr><td>Using the Bruce Protocol while holding the handrail</td><td>$\dot{V}O_2max\ (ml \cdot kg^{-1} \cdot min^{-1}) = 2.282\ (time\ in\ min) + 8.545$</td></tr>
<tr><td colspan="2">Ramping treadmill protocols</td></tr>
<tr><td>Individualized ramp protocol</td><td>$\dot{V}O_2max\ (ml \cdot kg^{-1} \cdot min^{-1}) = 0.72x + 3.67$
x = predicted $\dot{V}O_2$ based on peak speed/grade using ACSM walking equation</td></tr>
<tr><td>Standardized Bruce ramping protocol</td><td>$\dot{V}O_2max\ (ml \cdot kg^{-1} \cdot min^{-1}) = 3.9\ (time\ in\ min) - 7.0$</td></tr>
<tr><td colspan="2">Cycle ergometry (based on the final power completed in a 15-W $\cdot$ min^{-1} protocol)</td></tr>
<tr><td>Males</td><td>$\dot{V}O_2max\ (ml \cdot min^{-1}) = 10.51\ (power\ in\ W) + 6.35\ (body\ mass\ in\ kg) - 10.49\ (age\ in\ years) + 519.3$</td></tr>
<tr><td>Females</td><td>$\dot{V}O_2max\ (ml \cdot min^{-1}) = 9.39\ (power\ in\ W) + 7.7\ (body\ mass\ in\ kg) - 5.88\ (age\ in\ years) + 136.7$</td></tr>
</table>

Note. ACSM = American College of Sports Medicine.

Adapted, by permission, from American College of Sports Medicine, 2000, *ACSM's guidelines for exercise testing and prescription*, 6th ed. (Philadelphia, PA: Lippincott, Williams, and Wilkins).

to detect **exertional ischemia,** a GXT with **nuclear perfusion** or **echocardiography** is a better diagnostic procedure to consider when an individual has these resting abnormalities (see practical application 6.2 for more about sensitivity, specificity, and predictive value). If a GXT is being performed for reasons other than evaluating exertional myocardial ischemia (i.e., functional capacity or efficacy of medical therapy to control dysrhythmias or hypertension), the GXT may still be appropriate.

Graded Exercise Testing Protocols

Numerous protocols to perform graded exercise testing have been developed (table 6.3). The supervisor of the GXT must use a protocol that enables the client to achieve maximal effort without premature fatigue. It is recommended that the client exercise for a minimum of 6 min but not exceed 12 min. It is the responsibility of the GXT supervisor along with the supervising physician to choose the most appropriate protocol. When reviewing the medical history of a client, it is important to evaluate exercise/activity habits to determine an appropriate exercise protocol. In situations where a client has exercised for 12 min and perceives the work as fairly easy, it may be appropriate to increase the elevation or speed beyond the normal increment to avoid excessive time on the treadmill or bicycle ergometer.

Ideally, each clinical site should have specific protocols for testing clients with a wide variety of functional capacities and should avoid changing protocols if possible from one test to the next on the same client (unless their exercise time is < 6 min or > 12 min). Such consistency means comparisons can more easily be made, such as defining the workload that corresponds to the onset of chest pain or ST changes.

Protocol Considerations

For diagnostic purposes, the Bruce treadmill protocol is the most widely used test because of its long-

PRACTICAL APPLICATION 6.1

Recommendations for Starting an Exercise Program in Apparently Healthy Sedentary Individuals Over Age 75[a]

1. Prior to starting an exercise program, the individual should have a complete physical exam by his or her physician to determine if there are any contraindications to exercise. Contraindications include the following:
   - Myocardial infarction within past 6 months
   - Angina symptoms
   - Physical signs/symptoms of coronary heart failure (i.e., shortness of breath, bilateral rales, pedal edema)
   - Resting BP exceeding a systolic BP of 200 mmHg or diastolic BP of 110 mmHg
   - Abnormal symptoms[b] of cardiac reserve (angina and/or unusual shortness of breath)
   - New Q waves, ST depression, and T-wave inversion on ECG
   - Overt cardiovascular disease (needs to be risk stratified, and beyond the scope of these recommendations)

2. Start exercise at a low intensity (i.e., gait training, balance exercises, t'ai chi, self-paced walking, and lower extremity resistance training using elastic tubing or ankle weights).

3. In the elderly, instruction should be given on proper exercise techniques, and depending on the exercise, there should be supervision on at least one occasion to ensure safety and successful completion. If BP becomes abnormal—decrease in systolic BP greater than 20 mm Hg, increase in systolic BP greater than 250 mmHg or diastolic BP greater than 120 mmHg—or exercise HR is greater than 90% of one's predicted HRmax, then consider less intense exercise.

4. Intensity should be gradually increased, based on improvements in exercise capacity.

5. Exercise sessions should include an appropriate warm-up and cool-down.

6. Their physician should be informed of any unusual symptoms associated with exertion (chest pain, dizziness, unusual shortness of breath).

[a]These general guidelines are recommended based on anecdotal evidence from 30 years of experience with older populations and refer to individuals who plan to start moderate-intensity exercise programs.
[b]It is suggested that cardiac reserve be evaluated in the physician's office by one of the following: having the individual (1) get up and down from the examination table, (2) walk for 15 m, (3) climb one flight of stairs, and (4) cycle in the air for 1 min while sitting or lying on an examination table.

Modified from Gill, T., L. DiPietro, and H. Krumholz. Role of exercise stress testing and safety monitoring for older populations starting an exercise program. JAMA, 284:3, July, 19, 2000.

standing history, training familiarity, frequent citation in the medical literature, and normative data for many different populations. In individuals who are not frail, do not have an extremely low functional capacity, and are free from orthopedic problems that are irritated by walking uphill, the Bruce protocol is acceptable for diagnostic purposes. In addition, the Bruce protocol has been shown to be a feasible and a safe form of exercise testing within 3 days of an acute myocardial infarction in the majority of patients without resting angina, uncontrolled heart failure, dysrhythmias, and intravenous **nitroglycerine** use (26).

However, there are negative consequences of using the Bruce protocol; these include large increments in workload that make it difficult to determine one's functional capacity (2- to 3-MET increments), 3-min stages that for some are too fast to walk and too slow to run, and a large vertical component, which can lead to premature leg fatigue. When one is trying to determine if the Bruce protocol is appropriate, it is helpful to ask the client if he or she can walk several flights of stairs without stopping. Because of these concerns, other treadmill protocols should be considered. Commonly used protocols with smaller MET increments between

Absolute contraindications

1. Acute myocardial infarction (within 2 days) or other acute cardiac event
2. Significant change on the electrocardiogram suggesting ischemia, myocardial infarction, or other acute cardiac event
3. Unstable angina not stabilized by medical therapy
4. Uncontrolled cardiac dysrhythmias causing symptoms or hemodynamic compromise
5. Symptomatic severe aortic stenosis
6. Uncontrolled symptomatic heart failure
7. Acute pulmonary embolus or pulmonary infarction
8. Acute myocarditis or pericarditis
9. Suspected or known dissecting ventricular or aortic aneurysm
10. Acute infections (influenza, rhinovirus)

Relative contraindications

1. Left main coronary stenosis
2. Moderate stenotic valvular heart disease
3. Severe arterial hypertension (systolic blood pressure >200 mmHg and/or diastolic blood pressure >110 mmHg)
4. Tachycardic or bradycardic dysrhythmias
5. Hypertrophic cardiomyopathy and other forms of outflow tract obstruction
6. Mental or physical impairment that leads to inability to exercise adequately or is exacerbated with exercise
7. High-degree atrioventricular block
8. Ventricular aneurysm
9. Chronic infectious disease (AIDS, mononucleosis, hepatitis)
10. Electrolyte abnormalities (hypokalemia, hypomagnesemia)
11. Uncontrolled metabolic diseases (diabetes, myxedema, thyrotoxicosis)

It may be acceptable to perform a graded exercise test on an individual with a relative contraindication, if the benefits outweigh the risks.

FIGURE 6.4   Absolute versus relative contraindications to graded exercise testing.
Modified from (1, 9).

stages include the Balke-Ware and Naughton protocols. These protocols advance in 1- and 2-min stages with one or fewer MET increments. These protocols are commonly used with the elderly and individuals who are deconditioned because of a chronic medical problem such as heart disease. One of the more common problems with these protocols is that the supervisor underestimates the client's functional capacity and the individual exercises for longer than 12 min. More recently, **ramping treadmill protocols** have been introduced along with the more automated ECG/treadmill or bicycle ergometer systems that are now available. The advantage of ramping tests is that there are no large incremental changes, ramp rates can be adjusted, and one's functional capacity can more accurately be determined. Companies that include a ramping option allow conversion of standardized protocols (i.e., Bruce) to a ramping format, where treadmill speed and grade can change in small time increments (i.e., 6-15 s). When the Bruce protocol was compared between the standard 3-min stage versus ramping in 15-s intervals, similar hemodynamic changes were observed, but the ramping test produced a significantly greater duration in time and a higher peak MET level. In addition, subjects perceived the ramping protocol to be significantly easier than the standard Bruce protocol (29).

Sensitivity, Specificity, and Predictive Value

Graded exercise testing is compared with coronary angiography (the gold standard) to determine the accuracy of detecting or not detecting CAD. *Sensitivity* and *specificity* are terms used frequently to describe the likelihood of detecting or not detecting CAD. To determine sensitivity and specificity, four variables need to be determined: (a) true positive, those with CAD who have a positive GXT, (b) true negative, those without CAD who don't have a positive GXT, (c) false positive, those without CAD who have a positive GXT, and (d) false negative, those with CAD who have a negative GXT response.

Sensitivity determines the percentage of people who have a positive GXT who have significant CAD or how reliable the GXT is in identifying the presence of CAD. A positive GXT is indicated when the ECG corresponds to ischemia with exertion (>1 mm ST depression). Significant CAD is generally considered greater than 70% narrowing in at least one coronary artery.

Sensitivity and specificity are calculated by using a 2 × 2 contingency table:

Test result	Disease present	Disease absent
Positive	True positive (TP)	False positive (FP)
Negative	False negative (FN)	True negative (TN)

Sensitivity = [TP/(TP + FN)] × 100

Specificity determines the percentage of people who have a negative GXT result who do not have significant CAD, or how reliable the GXT is in identifying the absence of CAD.

Specificity = [TN/(TN + FP)] × 100

In addition to determining sensitivity and specificity, you can calculate the predictive value of a GXT to determine how accurately a test result can predict the presence or absence of CAD or the percentage of those identified correctly. Bayes' theorem states that the predictive value of a GXT is dependent on the prevalence of disease in the population being studied. The greater the prevalence of CAD, the higher is the predictive value of a positive test and lower the predictive value of a negative test. The predictive value of a positive or negative test can be determined as follows:

Predictive value of a positive test = [TP/(TP + FP)] × 100

Predictive value of a negative test = [TN/(TN + FN)] × 100

Equipment Considerations

If a treadmill cannot be used, the bicycle ergometer is the usual second choice. However, when comparing the treadmill to bicycle ergometer, an individual will achieve a 5% to 20% lower functional capacity on the bicycle ergometer, based on one's leg strength and conditioning level (1). However, bicycle ergometer testing is helpful for diagnostic reasons and to assess functional capacity in clients with weight-bearing and gait problems. In electronically braked bicycles, ramping protocols are commonly used, where there is a slow increase in workload during each minute. A typical bicycle ramping protocol consists of an increase between 15 and 25 $W \cdot min^{-1}$.

In more standardized bicycle ergometers, a work rate increase of 150 $kpm \cdot min^{-1}$ or 25 W every 2- to 3-min stage is commonly used for the general population. However, in frail clients or individuals with a low functional capacity, smaller work rate increments should be considered. In contrast, in heavier clients, the standard incremental increase of 25 W may be insufficient to reach maximal effort within 12 min. With respect to cadence, 50 to 60 $rev \cdot min^{-1}$ are suggested with standard bicycle ergometers. In the electronically braked bicycles, cadence is

TABLE 6.3

Commonly Used Treadmill/Bicycle Protocols

Protocol	Stage	Time (min)	Speed (mph)	Grade (%)	$\dot{V}O_2$ (ml · kg^{-1} · min^{-1})	METs
Bruce[a]	1	3	1.7	0	8.1	2.3
	2	3	1.7	5	12.2	3.5
	3	3	1.7	10	16.3	4.6
	4	3	2.5	12	24.7	7.0
	5	3	3.4	14	35.6	10.2
	6	3	4.2	16	47.2	13.5
	7	3	5	18	52.0	14.9
	8	3	5.5	20	59.5	17.0

Protocol	Stage	Time (min)	Speed (mph)	Grade (%)	$\dot{V}O_2$ (ml · kg^{-1} · min^{-1})	METs
Naughton	1	2	1	0.0	8.9	2.5
	2	2	2	0.0	14.2	4.1
	3	2	2	3.5	15.9	4.5
	4	2	2	7.0	17.6	5.0
	5	2	2	10.5	19.3	5.5
	6	2	2	14.0	21.0	6.0
	7	2	2	17.5	22.7	6.5

Protocol	Stage	Time (min)	Speed (mph)	Grade (%)	$\dot{V}O_2$ (ml · kg^{-1} · min^{-1})	METs
Balke-Ware	1	1	3.3	2.0	15.5	4.4
	2	1	3.3	3.0	17.1	4.9
	3	1	3.3	4.0	18.7	5.3
	4	1	3.3	5.0	20.3	5.8
	5	1	3.3	6.0	21.9	6.3
	6	1	3.3	7.0	23.5	6.7
	7	1	3.3	8.0	25.1	7.2

Protocol	Stage	Time (min)	Speed (mph)	Grade (%)	$\dot{V}O_2$ (ml · kg^{-1} · min^{-1})	METs
Ellestead	1	3	1.7	10	16.3	4.6
	2	2	3	10	26.0	7.4
	3	2	4	10	33.5	9.6
	4	3	5	10	41.0	11.7
	5	2	5	15	53.1	15.2
	6	3	6	15	63.0	18.0

Protocol	Stage	Time (min)	Speed (mph)	Grade (%)	$\dot{V}O_2$ (ml · kg^{-1} · min^{-1})	METs
Substandard Balke	1	2	2.0	0.0	8.9	2.5
	2	2	2.0	2.5	11.3	3.2
	3	2	2.0	5.0	13.7	3.9
	4	2	2.0	7.5	16.1	4.6
	5	2	2.0	10.0	18.5	5.3
	6	2	2.0	12.5	20.9	6.0
	7	2	2.0	15.0	23.3	6.7

Protocol	Stage	Time (min)	Speed (mph)	Grade (%)	$\dot{V}O_2$ (ml · kg^{-1} · min^{-1})	METs
Standard Balke	1	2	3.0	2.5	15.2	4.3
	2	2	3.0	5.0	18.8	5.4
	3	2	3.0	7.5	22.4	6.4
	4	2	3.0	10.0	26.0	7.4
	5	2	3.0	12.5	29.6	8.5
	6	2	3.0	15.0	33.2	9.5
	7	2	3.0	17.5	36.9	10.5

Protocol	Stage	Time (min)	Speed (mph)	Grade (%)	$\dot{V}O_2$ (ml · kg^{-1} · min^{-1})	METs
Superstandard Balke	1	2	3.4	2.0	15.9	4.5
	2	2	3.4	4.0	19.2	5.5
	3	2	3.4	6.0	22.5	6.4
	4	2	3.4	8.0	25.7	7.4
	5	2	3.4	10.0	29.0	8.3
	6	2	3.4	12	32.3	9.2
	7	2	3.4	14	35.6	10.2

Protocol	Stage	Time (min)	Cadence (rev · min^{-1})	W (kpm · min^{-1})	$\dot{V}O_2$ (ml · kg^{-1} · min^{-1})[b]	METs
Standard bicycle test	1	2 or 3	50	150: 25	10.9	3.1
	2	2 or 3	50	300: 50	14.7	4.2
	3	2 or 3	50	450: 75	18.6	5.3
	4	2 or 3	50	600: 100	22.4	6.4
	5	2 or 3	50	750: 125	26.3	7.5
	6	2 or 3	50	900: 150	30.1	8.6

Note. When trying to determine the appropriate protocol, one needs to consider the client's age, current functional capacity, activity level, and medical history/disease status. MET = metabolic equivalent.

[a]$\dot{V}O_2$ is calculated while the client is walking through stage 6. If the client is running at stage 6, the $\dot{V}O_2$ would be 42.4 ml · kg^{-1} · min^{-1} and 12.1 METs. The conventional Bruce protocol begins at 1.7 mph/10% grade; modified versions start at 1.7 mph/0% or 1.7mph/5%. [b]$\dot{V}O_2$ and METs determined for a 70-kg individual. Values need to be adjusted based on body weight.

not a major concern, because resistance automatically changes, based on cadence, to maintain a specific work rate (e.g., 100 W).

Among heavier and more active clients where you expect that the client will need to accomplish a high workload to reach maximum effort, increasing work rate by 300 kpm · min^{-1} or 50 W every 2 or 3 min may be more appropriate. Once the client achieves 75% of his or her predicted maximal HR or a **rating of perceived exertion (RPE)** of 13, the workload would increase by 150 kpm · min^{-1} every stage or 25 W, so that the ending workload could be more clearly defined. This protocol may more accurately determine the client's functional capacity.

In the past, arm ergometry testing was used for diagnostic purposes in individuals with severe orthopedic problems, **peripheral vascular disease,** lower extremity amputation, and neurological conditions that inhibited the ability to exercise on a treadmill or bicycle ergometer. However, in a clinical setting, the arm ergometer has now been almost completely replaced with pharmacological stress testing. Clients for whom an arm ergometry GXT may still be beneficial include those with symptoms of myocardial ischemia during dynamic upper body activity or those who suffered a myocardial infarction and plan to return to an occupation that requires upper body activity. Because of a smaller muscular mass in the arms versus the legs, a typical arm protocol involves 10- to 15-W increments for every 2- to 3-min stage at a cranking rate of 50 to 60 rev · min^{-1}.

In individuals who are symptomatic while doing a combination of isometric upper body work with aerobic activity (walking while carrying an object), or repetitive lifting, exercise testing can be modified in an effort to reproduce similar symptoms (see practical application 6.3).

Graded Exercise Testing Procedures

Prior to the start of the GXT, a number of procedures must be completed. A medical history/physical assessment needs to be performed prior to testing. The client needs to be given instructions about what he or she is required to do, along with reviewing and signing the informed consent. The client needs to be properly prepped for a 12-lead ECG, followed by recording resting measurements (ECG and BP). If resting measurements are acceptable, the GXT can be started. During the GXT, the client needs to be monitored closely for any abnormalities and the tester needs to be responsible for discontinuing the test at the appropriate time.

Patient Instructions

It is important that the patient receive clear instructions on how to prepare for the GXT along with an appropriate explanation of the test. A lack of proper instruction can delay the test, increase patient anxiety, and potentially increase the health risk to the patient. Instructions for patient preparation should include information on clothing; footwear; food consumption; avoidance of alcohol, cigarettes, caffeine, and over-the-counter medicines that could influence the GXT results; and whether to take one's prescribed drug regimen. A prescribed medication may be discontinued because certain agents can inhibit the ability to observe ischemic responses during exercise (e.g., β-blockers blunt HR and BP, which can inhibit the occurrence of ST-segment changes associated with ischemia). It is important to note that if a physician wants the patient to discontinue medications prior to GXT, the patient should receive these instructions from the physician. The process of weaning an individual off medications can vary, and one needs to be concerned with any rebound effect. This is generally not the responsibility of the paramedical professional, unless instructed by the physician.

Medical Evaluation

Prior to any testing, it is important that the test supervisor understand the individual's medical history. A medical history will help determine if there are any contraindications to the GXT, will aid in choosing an appropriate testing protocol, and will help identify areas that may require more supervision during the test (e.g., history of hypertension, previous knee injury). A comprehensive medical history should evaluate height and weight; assess risk factors and any recent signs or symptoms of cardiovascular disease; document previous cardiovascular disease; determine other chronic diseases that may influence the outcome (e.g., arthritis, gout); and obtain information about current medications (including dose, frequency), drug allergies, recent hospitalizations or illnesses, and exercise/work history. A plethora of medical history forms are available, and you should feel comfortable with the form you are using and that appropriate questions are asked, based on the subject's sex and age.

In addition to medical history, a brief physical examination of heart and lung sounds and peripheral edema is most always required. Specific skills

PRACTICAL APPLICATION 6.3

Graded Exercise With Upper Body Resistance

Depending on the type of activity that is producing symptoms of CVD, a standard GXT may need to be modified to reproduce symptoms in a clinical setting. Upper body work can be reproduced in a laboratory setting with the following protocols.

Static Upper Body Work With Aerobic Exercise

— Test: discontinuous exercise

— Modality: treadmill and variable weights, starting at 10 lb

— Exercise: 2- to 3-min stages

— Rest: 1 to 3 min

— Upper body resistance: incremental increase of 10 to 20 lb per stage, depending on the client's strength and symptoms with arms holding the weight in a flexed position

— Fixed walking speed: 2 to 3 mph

— Monitoring: HR, BP, ECG, and symptoms (ideally at the end of each exercise stage or immediately into recovery)

A discontinuous exercise test is commonly used, where the client rests 1 to 3 min after every 2 to 3 min of walking between 2 and 3 mph. Typical protocol would consist of walking without a load, followed by a 20-lb load at stage 2, with each following stage increasing the resistance by 10 lb.

Repetitive Lifting

— Test: discontinuous exercise

— Modality: variable weights starting at 20 to 30 lb

— Exercise: 5- to 6-min stages lifting a weight from the floor to a bench that is self-paced

— Rest: 1 to 3 min

— Upper body resistance: start at 20 to 30 lb and increase by 10 lb as tolerated

— Monitoring: HR, BP, ECG, and symptoms (ideally at the end of each exercise stage or immediately into recovery)

This type of test would most likely be used with a client who is symptomatic or possibly hypertensive with upper body repetitive lifting. Discontinuous exercise would be used where the client lifts a weight (i.e., box) repeatedly for 5 to 6 min. The weight is increased after each rest period.

---

are described in chapter 4. Last, all clients should complete an informed consent. An informed consent properly informs clients of what they will be asked to do, complications associated with completing the test (risks and discomforts), test supervisor responsibilities, and client responsibilities. Informed consent provides both an opportunity to ask questions about the procedure and freedom to refuse participation. If the client agrees to participate, he or she must sign the form, and clients under 18 years of age must have a parent or guardian sign the consent. Examples of both consent and medical history forms are available in Pina et al. (18). It is important to recognize that the client may be very anxious prior to completing the GXT, and the test

supervisor should answer all questions in a professional and caring manner. Also, if the test supervisor is prepared and confident, the client will be more at ease and potentially more willing to give his or her best effort.

ECG Preparation

The clarity of the ECG tracing during exercise is of utmost importance, especially for diagnostic purposes. Therefore, the client needs to be properly prepped. In males, hair from all electrode positions should be removed with a razor blade or battery-operated shaver. For a good connection it is important to make sure the entire electrode is hair free.

After the site is free of hair, alcohol-saturated wipes should be used to clean the skin surface and remove any oils. To help improve electrode adherence and remove the superficial layer of skin, light abrasion is used on the skin surface with very fine sandpaper, commercially made abrasion tape, or an abrasion pad. Achieving optimal skin preparation without excessive skin irritation requires practice.

When the electrode is placed, the conducting gel should be placed directly at the designated site. Ideally, the subject should be in a standing position when electrodes are placed on the chest. Lying in a supine position or sitting is not ideal, because of possible shifts in the electrode placement when one stands, which potentially could lead to increased artifact. The electrode should be placed on the skin surface gently, the edges sealed, and then pressure applied to the center to make sure the conducting gel makes good contact with the skin surface. During exercise, electrodes need to adhere well to the skin surface, especially when the client begins to sweat. Conductance through the skin surface is critical for achieving a clear ECG recording. Silver–silver chloride electrodes are ideal because they offer the lowest offset voltage and are the most dependable in regard to minimizing **motion artifact** (18). The conduction solution found in the center of the electrode is either a saturated sponge or an electrolyte solution. After electrodes are in place and cables are attached, each electrode placement site should be assessed for excessive resistance. On the newer ECG stress systems there is a built-in assessment to determine if the electrode sites have excessive resistance that would require re-prepping that specific site. Another good test is to lightly tap on the electrode and observe the ECG. If the ECG produces a great deal of artifact, it is important to check the electrode and lead wire for potential interference. Adequate electrode preparation can decrease the chances of motion artifact, whereas muscle artifact cannot be reduced unless upper body skeletal muscle activity is decreased. Because of skin movement while the patient walks on the treadmill, it is important to keep electrode and wire movement to a minimum, especially in large breasted and obese individuals. To help decrease movement, using some sort of wrap or mesh tubing over the electrodes and lead wires may be beneficial.

Resting Measurements

After the patient has rested for a several minutes, 12-lead ECG and BP should be recorded in both the supine and standing positions. Evaluating both con-

ditions allows the test supervisor to assess the effects of body position on hemodynamics. Hyperventilation is sometimes completed prior to testing to induce ST-T wave changes not related to ischemia. Changes in T waves versus ST depression is more commonly observed (18). This measurement is not universally accepted because of the small incidence of producing an abnormal ECG response. Generally, the consensus is that hyperventilation for the purpose of determining false-positive tests has not been sufficiently validated (6). It may be more prudent to have the client hyperventilate after the GXT if there are no signs or symptoms indicative of CAD but exercise-induced ST-T wave changes are observed. The resting ECG should be compared with a prior resting ECG for any changes. If there are any significant differences that may contraindicate starting the GXT (see section on contraindications), the paramedical supervisor should contact the supervising physician or the referring physician with the updated information, so that a decision can be made with regard to completing the GXT.

Test Procedures

After the tester prepares the patient, chooses the appropriate protocol, and reviews the resting ECG for any abnormalities that may inhibit the evaluation, the GXT can start once the patient is instructed on how to use the related mode of testing (treadmill, bicycle ergometer). In elderly individuals asked to walk on a treadmill for the first time, it is important to take the time to make sure they feel comfortable walking on the unit with little or no rail support assistance. Clients who are not comfortable may simply stop prematurely because of apprehension. In addition to the supervisor of the GXT (physician or paramedical professional), a technician is needed to monitor the patient's BP and potential signs and symptoms of exercise intolerance. ECG, HR, RPE, and BP should be recorded at the end of each stage. Approximately 30 to 45 s should be given to accurately measure BP. Manual auscultation of BP is the most common measurement; however, automated units are becoming more prevalent. Automated units are expensive, and there is still some question as to their accuracy during high-intensity exercise. If automated units are being considered, it is important to first validate the model you are considering against standard auscultation through a range of BPs.

As the exercise test progresses from one stage to the next, it is imperative that testing staff observe the client and communicate constantly. Before the test begins, the patient must understand that it is his or her responsibility to communicate at the onset of any discomforts. At minimum, at the end of each stage of testing staff should ask clients how they are feeling and if they are having any discomforts. At each stage the supervisor should record any discomforts the client has. This includes the absence of any stated discomforts, because doing so implies that the question was asked. During the GXT, the ECG must be continually monitored. A 12-lead ECG should be printed and reviewed at the end of each stage, because it is possible that ECG changes can take place in any of the lead combinations. Between stages, the supervisor should monitor leads that reflect the different walls of the myocardium.

If gas exchange is analyzed during the GXT, communication becomes more difficult because of the use of a mouthpiece or mask. It is important to determine specific hand signals so that symptoms (e.g., chest pain, shortness of breath) along with RPE can be accurately determined. Ways in which to communicate during a GXT without talking include having the client point to a chart, having the client show a thumbs up or down to continue or stop, and asking yes and no questions that the client can answer with a hand signal or a head nod. The first time gas exchange is measured for a client, there can be a lot of apprehension because of the fear that one's breathing and speaking abilities are inhibited. The test supervisor must properly instruct the client on the use of the gas exchange apparatus and allow the client to wear the gas exchange apparatus while resting, so that he or she is comfortable, prior to starting the test.

Exercise Data

The use of RPE with exercise testing is now common and provides a monitor of how hard the client perceives his or her work and how much longer he or she will be able to continue. One's RPE is generally recorded the last 15 s of each stage. Two RPE scales are commonly used: the 6- to 20-point category scale and the 0- to 10-point category-ratio scale (table 6.4).

To improve the accuracy of these scales, the following instructions (13) should be given to the client:

"During the exercise test we want you to pay close attention to how hard you feel the exercise work rate is. This feeling should reflect your total amount of exertion and fatigue, combining all sensations and feelings of physical stress, effort, and fatigue. Don't concern yourself with any one factor such as leg pain, shortness of breath, or exercise intensity, but try to concentrate on your total, inner feeling of exertion. Try not to underestimate or overestimate your feelings of exertion; be as accurate as you can."

Other scales that are helpful in evaluating clients' specific symptoms include angina, **dyspnea,** and peripheral vascular disease scales (table 6.5). The

---

TABLE 6.4

Category and Category-Ratio Scales for Ratings of Perceived Exertion (RPE)

Category scale	Category-ratio scale
6 No exertion at all	0 Nothing at all "No P"
7 Extremely light	0.3
8	0.5 Extremely weak Just noticeable
9 Very light	1 Very weak
10	1.5
11 Light	2 Weak Light
12	2.5
13 Somewhat hard	3 Moderate
14	4
15 Hard (heavy)	5 Strong Heavy
16	6
17 Very hard	7 Very strong
18	8
19 Extremely hard	9
20 Maximal exertion	10 Extremely strong "Max P"
	11
	* Absolute maximum Highest possible

TABLE 6.5

Chest Pain, Dyspnea, and Claudication Scales

Chest pain scale

1+	Light, barely noticeable
2+	Moderate, bothersome
3+	Moderately severe, very uncomfortable
4+	Most severe or intense pain ever experienced

Dyspnea scale

+1	Mild, noticeable to patient but not observer
+2	Mild, some difficulty, noticeable to observer
+3	Moderate difficulty but patient can continue
+4	Severe difficulty, patient cannot continue

Claudication scale

1+	Definite discomfort or pain but only of initial or modest levels (established but minimal)
2+	Moderate discomfort or pain from which the patient's attention can be diverted by a number of common stimuli (e.g., conversation, interesting TV show)
3+	Intense pain from which the patient's attention cannot be diverted except by catastrophic events (e.g., fire, explosion)
4+	Excruciating and unbearable pain

---

sion of dyspnea. The peripheral vascular disease scale is beneficial when evaluating clients with documented or questionable **claudication,** based on the client's medical history.

In respect to prescribing exercise, the information gathered from the scales mentioned here during the GXT can be beneficial. By comparing a client's RPE with a prescribed training HR range, you can determine if the intensity level is achievable and sustainable by the client. Be aware, however, that differences may exist when mode of testing differs from mode of training. It is of interest to note that cardiac clients perceive exercise to be more difficult during exercise training compared with exercise testing (3). In addition to the use of RPE, comparing prescribed training HR ranges to the onset of angina, dyspnea, and claudication symptoms can be very beneficial in determining a safe and feasible training range. Finally, although these scales can be helpful, they need to be interpreted and considered in conjunction with other medical findings.

As clients approach their maximum effort, they may become anxious and want to stop the test, so it is important that the supervisor and technician are prepared to record peak values (ECG, BP, RPE). Following is a suggested order of events at maximal effort:

1. Record BP 30 to 45 s before the test stops.
2. Record ECG 15 s before the test stops.
3. Record symptoms and RPE immediately before test termination.
4. Reduce workload to a comfortable level.
5. Remove gas exchange instrument after peak values are measured, if applicable.

Submaximal GXTs are often terminated when a certain MET level, or a percentage of maximal HR (i.e., 85% of predicted maximum HR or 20-30 beats · min^{-1} above resting HR), is achieved. If HR is affected by medical therapy such as a β-blocking agent, an RPE of 13 may be used to terminate a submaximal test. Submaximal tests are commonly used in patients being discharged from a hospital, to make sure future risk of a cardiovascular event is sufficiently low and it is safe to return to activities of daily living. Symptom-limited or maximal GXTs are commonly used for diagnostic purposes and assessment of functional capacity. *Symptom-limited* refers to a test being terminated because of the onset of symptoms that put the client at an increased risk for further medical problems (see figure 6.4). *Maximal* implies that the test will terminate when an individual reaches his or her maximal level of exertion, with-

angina scale is beneficial because it evaluates the intensity of chest pain and helps determine if the test should be terminated based on standard criteria (see figure 6.5). In addition, it is important to document the onset of angina with the corresponding MET level and **rate pressure product (RPP).** The progression of angina can be evaluated with the scale, can be beneficial when comparing previous test results (i.e., whether the intensity of angina is occurring at a lower or higher workload), and may provide information about the progression of CAD. The dyspnea scale is commonly used when one is testing patients with pulmonary disease and serves a similar role as that used with clients who have angina, in regard to rating the intensity and progres-

Absolute indications

1. Decrease in systolic blood pressure of >10 mmHg from baseline blood pressure despite an increase in workload, when accompanied by other evidence of ischemia

2. Moderate to severe angina (2-3+)

3. Increasing nervous system symptoms (e.g., ataxia, dizziness, or near syncope)

4. Signs of poor perfusion (cyanosis or pallor)

5. Technical difficulties monitoring the electrocardiogram or systolic blood pressure

6. Subject's desire to stop

7. Sustained ventricular tachycardia

8. ST elevation (>1.0 mm) in leads without diagnostic Q waves (other than V_1 or aVR)

Relative indications

1. Decrease in systolic blood pressure of >10 mmHg from baseline blood pressure despite an increase in workload, in the absence of other evidence of ischemia

2. ST or QRS changes such as excessive ST depression (>2 mm horizontal or down-sloping ST-segment depression) or marked axis shift

3. Arrhythmias other than sustained ventricular tachycardia, including multifocal PVCs, triplets of PVCs, supraventricular tachycardia, heart block, or bradyarrhythmias

4. Fatigue, shortness of breath, wheezing, leg cramps, or claudication

5. Development of bundle branch block or intraventricular conduction delay that cannot be distinguished from ventricular tachycardia

6. Increasing chest pain

7. Hypertensive response[a]

Note. PVC = premature ventricular contractions.

[a]Systolic blood pressure of >250 mmHg and/or a diastolic blood pressure of >115 mmHg.

FIGURE 6.5 Indications for termination of graded exercise test.

Adapted from ACC/AHA Guidelines for the Perioperative Cardiovascular Evaluation for Noncardiac Surgery. J Am Coll Cardiol 1996; 27: 910-48. Copyright 1996 by the American College of Cardiology and American Heart Association, Inc.

out being limited by any abnormal signs or symptoms as listed in figure 6.5. When a client gives a maximal effort, the test is normally terminated because of volitional or voluntary fatigue. Criteria that are used to determine whether one has reached maximal effort include these:

1. A plateau in $\dot{V}O_2$ (<2.1 ml · kg^{-1} · min^{-1} increase)

2. A respiratory exchange ratio value greater than 1.1

3. Venous blood lactate exceeding 8 to 10 mM (7)

These criteria rely on having a metabolic cart to measure oxygen consumption and a lactate analyzer, which is not feasible at many clinical facilities. Therefore, other criteria are commonly used in conjunction with voluntary fatigue; these include an RPE greater than 17, a plateau in HR despite an increasing workload or attainment of 85% of predicted maximal HR in non-β-blocked clients, and attainment of 240 rate pressure product.

When to Stop a GXT

The supervisor of the GXT must be aware of and understand potential complications that may occur and know when to terminate the GXT. This knowledge is especially crucial if a physician is not immediately present. Knowledge of the normal and abnormal physiological responses that take place with GXT is imperative. Reasons for test termination are presented as absolute and relative contraindications (figure 6.5). Absolute contraindications are considered high-risk indications (except a subject's desire to stop and technical difficulties) with the potential to result in serious complications if the test is continued. Therefore the test should be terminated immediately. If the physician is not immediately present, he or she should be contacted and informed of the outcome as soon as possible.

The client may be anxious and may wonder why the test was terminated prematurely. Try to keep the client calm relative to what you observed that led to stopping the test. Refer specific questions to his or her physician. Following is a reasonable response to why the test was terminated early: "An abnormality occurred during the GXT, which caused us to stop the test. This abnormality needs to be evaluated by your physician. When you review the test with your physician he or she will give you more specific information pertaining to the test outcome."

With respect to relative contraindications, these findings might represent reasons for test termination, but generally the test is not stopped unless other abnormal signs or symptoms occur simultaneously. The reason for performing the GXT may also influence relative reasons for test termination. For example, when evaluating a patient for suspected CAD, you observe left bundle branch block with exertion, which precludes interpretation of ST changes and is cause for test termination. It is imperative that each facility have policies and procedures that describe absolute and relative indications for test termination. Doing so eliminates any confusion about why a test was terminated. In addition, and potentially most important, if a serious complication does arise during the GXT, policies and procedures are followed, and the test is appropriately terminated, there will be little opportunity for legal issues to arise. If you find yourself questioning whether a test should be stopped based on what you observe, always lean toward the conservative side and consider safety of the patient as the highest priority (i.e., the test can always be repeated if necessary).

Exercise ECG Analysis

Within this section, analysis of the 12-lead ECG is presented, which is based on recognizing normal and abnormal responses during a GXT. This section should not replace an in-depth study of ECG analysis but serves as a review for GXT-specific issues.

When one is reviewing the resting ECG prior to the GXT, it is important to review ECG findings along with the client's medical history to see if there are any discrepancies (e.g., the client states no previ-

ous myocardial infarction, but the current ECG shows significant Q waves throughout the lateral leads). It is possible that since the client's last physician visit (at which time the GXT was ordered), cardiac status may have changed. In addition, it is important to compare the client's current resting 12-lead ECG with a previous ECG, especially when an abnormality is detected (e.g., client's current resting ECG rhythm shows atrial fibrillation; is this dysrhythmia new or old?). When resting differences are detected on the current ECG and this appears to be a new finding, it is important that the supervising or referring physician be informed about this prior to the test. When determining the safety of performing the GXT, refer to figure 6.4 for contraindications for GXT.

Resting ECG Abnormalities

The most common reason to complete a GXT with 12-lead ECG is to assess potential CAD. As a result, it is important to accurately interpret ST changes. Specific ECG abnormalities that prevent using ST changes to assess ischemia are left bundle branch block, right bundle branch block (unable to interpret ST changes in anterior leads, V1-V3, but lateral and inferior leads are interpretable), and preexcitation syndrome (figures 6.6, 6.7, and 6.8).

With respect to ST depression on the resting ECG attributable to left ventricular hypertrophy or digoxin use, there appears to be a difference of opinion about the usefulness of ST changes during exercise. If there is less than 1 mm of resting ST depression, it is generally accepted that GXT represents a reasonable test to assess exercise-induced ischemia; however, specificity is reduced. See Practical Application 6.2 for more information. In conditions in which a client has greater than 1 mm ST depression at rest, GXT with 12-lead ECG alone offers very little diagnostic information. Use of cardiac imaging via echocardiography or radionuclide testing is preferred (8).

Exercise ECG Changes

During the GXT, it is important that one individual constantly observe the ECG recording to identify the onset of any ECG change. The evaluation of the ST segment is of great importance because of its ability to suggest the onset of ischemia. In addition, the onset of dysrhythmias should be identified, especially those that are related to absolute and relative contraindications to test termination (figure 6.5). Although the onset and progression of ST depres-

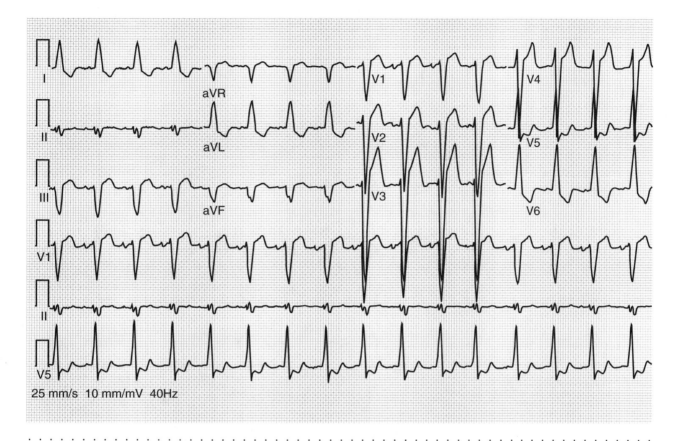

FIGURE 6.6 Electrocardiogram showing left bundle branch block.

sion is in most cases subtle, dysrhythmias can occur suddenly and can either be brief, intermittent, or sustained.

ST-Segment Changes

When exercise testing is used as a diagnostic tool for CAD, the most common abnormality that needs to be constantly monitored is ST-segment changes. All 12 leads should be monitored; however, V5 is the most diagnostic, whereas the inferior leads (II, III, and aVF) are associated with a high incidence of false-positive findings (i.e., ST depression occurs, but it truly is not related to ischemia). V5 represents the most diagnostic lead because when true ischemia occurs with exertion, it is most commonly observed in this lead. When the test supervisor recognizes ST-segment changes, he or she should note when (time and work rate) it begins, the **morphology,** the magnitude, and any additional changes that take place at each stage. Most ECG stress systems provide summaries of ST segments throughout the testing period, which is very beneficial but needs to be verified. In addition, it is important to document any symptoms associated with the ST changes (e.g., chest pain, shortness of breath, dizziness).

ST-Segment Depression Of the potential ST-segment changes, ST-segment depression is the most frequent response and is suggestive of **subendocardial ischemia.** One or more millimeters of horizontal or down-sloping ST-segment depression that occurs 0.08 s past the J point is recognized as a positive test for myocardial ischemia. When ST-segment changes of this type occur along with typical angina symptoms, the likelihood of CAD is very high (approximately 95%). In addition, the greater the ST depression, the more leads with ST depression, and the more time it takes for the ST depression to resolve in recovery, the more likely that significant CAD is present. In addition, in some cases ST-segment depression is only observed in recovery, and should be treated as an abnormal response. Additionally, J-point depression with up-sloping ST segments that are greater than 1.5 mm depressed at 0.08 s past the J-point are suggestive of exercise-induced ischemia. However, if the rate of up-sloping ST depression is gradual (< 1 mV $\cdot$ s^{-1}) there is an increased probability of CAD (25), although the rate of false-positive tests will be higher. As the rate of false-positive tests increases, there is a decrease in specificity and an increase in sensitivity

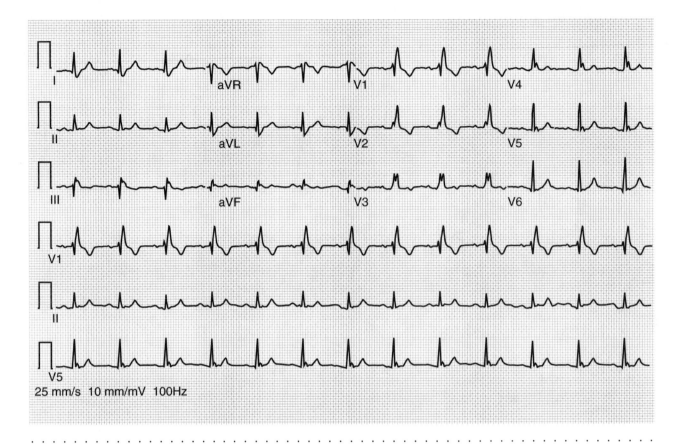

FIGURE 6.7   Electrocardiogram showing right bundle branch block.

(i.e., the test is more likely to suggest that CAD is present when it truly is not, but is also more likely to detect people who truly have CAD). The negative consequence to decreased specificity is that this will increase client anxiety because of the need for additional diagnostic testing (i.e., radionuclide GXT).

ST-Segment Elevation   ST-segment elevation observed on a resting ECG attributable to early repolarization is not abnormal in healthy individuals; however, this should be documented on the GXT report. With exertion, this type of ST elevation normally returns to the isoelectric line. ST-segment elevation with exertion (assuming the resting ECG is normal) may suggest **transmural ischemia** or a coronary artery spasm. ST-segment elevation is associated with serious arrhythmias and is an absolute reason for stopping the test (see figure 6.5). When Q waves are present on the resting ECG from a previous infarction, ST elevation with exertion may suggest a left ventricular **aneurysm** or wall motion abnormality which is a potential reason for discontinuing the GXT and referral for further testing. ST-segment elevation can localize the ischemic area

and the arteries involved, whereas this is not always the case with ST-segment depression (27).

T-Wave Changes

When exercise is started in healthy individuals initially, there is a gradual decrease in T-wave amplitude, followed by an increase in T-wave amplitude at maximal exercise. These T-wave changes are believed to be associated with an increase in stroke volume. It has been postulated that an increase in serum potassium following exercise is responsible for increased T waves.

In the past, T-wave inversion with exertion was suggested to reflect an ischemic response. However, it is now believed that flattening or inversion of T waves is not associated with ischemia. T-wave inversion is common in the presence of left ventricular hypertrophy. Inverted T waves present on the ECG at rest that normalize with exertion have also been suggested to reflect ischemia. However, normalization of T waves is commonly observed in subjects with and without exercise-induced ischemia. Normalization of T waves has also been shown to be present in ischemic responses associated with coronary spasms. This finding has the

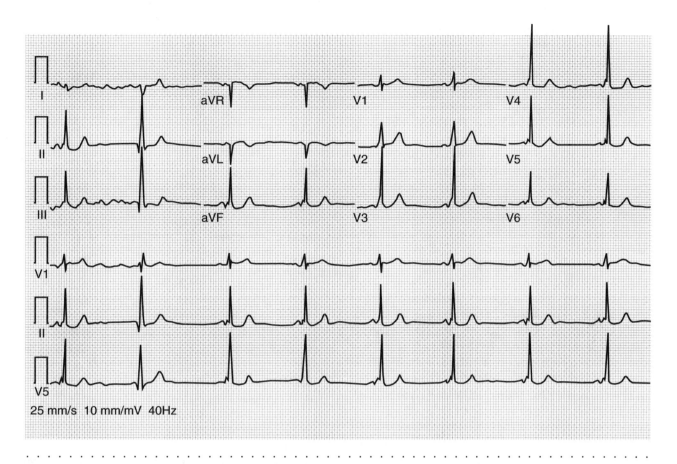

FIGURE 6.8 Electrocardiogram showing Wolfe-Parkinson-White preexcitation abnormality.

greatest significance under resting conditions. Overall, T-wave changes with exertion are not specific to exercise-induced ischemia but should be correlated with other signs and symptoms. It also should be noted that T-wave changes frequently follow ST changes and can be difficult to isolate with exertion.

R-Wave Changes

The normal response of R waves during exercise is an increase in amplitude to a submaximal HR; then as one approaches maximum, R-wave amplitude decreases below resting values. Previously, it has been suggested that increased R-wave amplitude at peak exercise is suggestive of CAD. However, it is important to recognize that if the effort was submaximal, R-wave amplitude may not have had the opportunity to decrease. If an individual was able to give a maximal effort and R waves increased in amplitude with no other signs or symptoms of CAD, the predictive power of R-wave changes alone and CAD is poor. In general, it appears that abnormal R-wave changes provide no substantial additional information, unless combined with other abnormal signs or symptoms.

Dysrhythmias

In relation to ST changes with exertion, dysrhythmias are of equal importance and potentially more life threatening, based on the suddenness in which dysrhythmias may appear. Though a GXT is most commonly used to diagnosis potential CAD, this test is also commonly used to evaluate symptoms (e.g., near syncope) attributable to dysrhythmias. In addition, GXT may be used to evaluate the effectiveness of medical therapy in controlling dysrhythmias. It is of utmost importance that the supervisor of the GXT have a strong knowledge of dysrhythmia detection and be able to respond appropriately and quickly. When dysrhythmias appear during the GXT, the onset of the dysrhythmias and any signs or symptoms associated with the dysrhythmias should be documented, along with any other ECG changes (e.g., ST depression). Of the potential dysrhythmias, there are primarily three major areas that the supervisor should be cognitive of, each requiring that the test be terminated prematurely:

1. Supraventricular dysrhythmias that compromise cardiac function

2. Ventricular dysrhythmias that have the potential to progress to life-threatening dysrhythmias

3. The onset of high-grade conduction abnormalities

Supraventricular Dysrhythmias   Isolated premature atrial and junctional contractions and short runs of supraventricular tachycardia are generally considered benign and provide little diagnostic value relative to CAD. Supraventricular tachycardia may simply be related to an increased adrenergic response that occurs with physical exertion, although sustained supraventricular tachycardia is generally considered an absolute indication for test termination, if the client begins to experience symptoms that may be related to the fast rate of supraventricular tachycardia (e.g., dizziness, syncope). The onset of supraventricular tachycardia is most noticeable at low levels of exertion, where a large change in HR can be easily observed. If the dysrhythmias persist following test termination, the supervising physician should be informed immediately if he or she is not in the testing room. Additionally, exercise-induced atrial flutter or fibrillation is generally considered a reason for terminating the GXT. As previously mentioned, the testing supervisor must be knowledgeable of the policy and procedures for test termination. By following stated guidelines, you will more likely avoid legal allegations if a problem should arise during the GXT.

Ventricular Dysrhythmias   In general, ventricular dysrhythmias are considered to be of greater importance than supraventricular dysrhythmias, because they can more easily progress to more life-threatening dysrhythmias. However, occasional (fewer than six per minute) ventricular premature beats are benign and do not warrant stopping the test. It has been estimated that approximately 20% to 30% of the healthy population and 50% to 60% of the cardiac population experience ventricular premature beats under resting conditions. Generally, ventricular premature beats at rest that disappear with exertion are considered benign, but when ventricular premature beats increase in frequency with exertion, there is more reason to suspect an underlying cardiac problem (i.e., cardiomyopathy, CAD).

Ventricular premature beats can be either unifocal or multiform in nature. Unifocal beats originate from the same place within the ventricle, whereas multifocal beats suggest that the abnormality is originating from multiple places in the ventricles. Multiform ventricular premature beats are considered of greater significance, because of a greater risk to progress to more dangerous ventricular dysrhythmias. The time in which the ventricular premature beat occurs during the cardiac cycle is important. Ventricular premature beats that occur on the T wave are referred to as the R on T phenomena and, though rare, are more likely to induce ventricular tachycardia and are a reason for test termination, especially in the presence of ischemia.

Ventricular couplets or pairs should be noted but generally are not considered an absolute reason for test termination. If there is an increase in the rate of ventricular couplets or if they become multiform, this is considered a relative contraindication for test termination. When ventricular premature beats occur as a triplet (ventricular tachycardia) and the rate is less than 150 beats $\cdot$ min^{-1}, this too is considered a relative contraindication. If the ventricular tachycardia reaches 150 to 250 beats $\cdot$ min^{-1}, this is an absolute contraindication for test termination, because of the increased likelihood of progressing to a faster rate and eventually ventricular flutter and ventricular fibrillation. The test supervisor must recognize ventricular tachycardia and record the onset. Generally, ventricular tachycardia is easily recognized by the wide QRS complex; however, the onset of left bundle branch block can resemble ventricular tachycardia. If an individual converts to bundle branch block, P waves still proceed the QRS complex, whereas this is not true in ventricular tachycardia. However, because heart rate increases during exercise, it is often difficult to identify P waves. If the tester has difficulty trying to distinguish between these two dysrhythmias, the test should be terminated. The test supervisor should always lean toward safety and not be afraid to terminate the test (the worst-case scenario is that the test would need to be repeated at a later date).

Emergency Procedures

Emergency procedures during GXT must be developed before testing begins. Procedures must be written, and these procedures should be reviewed and practiced regularly. There can be a great deal of difference in emergency procedures, based on the location of the testing site (in or outside of a hospital) and immediate personnel. All paramedical professionals should have American Heart Association basic cardiac life support (and ideally advanced cardiac life support) (23). A crash cart with emergency equipment and drugs should be in close proximity to the testing room and preferably in the room. The American Heart Association has given guidance for recommended emergency drugs (figure 6.9) and emergency equipment (figure 6.10).

Required drugs

- Atropine
- Isoproterenol
- Lidocaine
- Bretylium
- Adenosine
- Sublingual nitroglycerin
- Epinephrine
- Procainamide
- Verapamil
- Dopamine
- Dobutamine
- Intravenous fluids
    - Normal saline (0.9%)
    - D5W

Optional (depending on facility)

- Loop diuretic (intravenous)
- Morphine sulfate
- Sodium bicarbonate
- Theophylline for laboratories

FIGURE 6.9   Emergency drugs and solutions.

From Emergency Cardiac Care Committee and Subcommittees, American Heart Association. Guidelines for cardiopulmonary resuscitation and emergency cardiac care. *JAMA.* 268:2171-2302, 1992.

1. Defibrillator (portable)
2. Oxygen tank (portable, if possible, for transport)
3. Nasal cannula, ventimask, nonbreathing mask, oxygen mask
4. Airways (oral)
5. Bag-value-mask hand respirator (Ambu bag)
6. Syringes and needles
7. Intravenous tubing, solutions, and stand
8. Adhesive tape
9. Suction apparatus and supplies (gloves, tubing)[a]

[a]May be immediately available when a team arrives in certain centers.

FIGURE 6.10   Emergency equipment.

Pina, I.L., Balady, G.J., Hanson, P. etal. Guidelines for clinical exercise testing laboratories. A statement for healthcare professionals from the committee on exercise and cardiac rehabilitation, American Heart Association. *Circulation* 91(3):912-921, 1995.

Conclusion

The GXT continues to be a useful tool and is generally the first diagnostic tool to assess the presence of significant coronary artery disease with or without nuclear perfusion imaging. Previous supervision of the GXT was primarily completed by a cardiologist. However, as the knowledge of testing has increased in regard to identifying clients who are at increased risk for further complications, as ECG assessment technology has advanced, and because of cost-containment issues and time constraints of cardiologists, other healthcare professionals have been shown to perform graded exercise testing safely. Though the GXT is primarily used for diagnostic purposes in a clinical setting, there are a number of other indications (i.e., prehospital admission, evaluation of disease severity, pharmacological intervention, predischarge evaluation, assessment of functional capacity, evaluation of revascularization, assessment of disease progression,

and prerequisite to starting an exercise program in select patients).

Prior to completing the GXT, it is imperative that the supervisor of the test properly evaluate the client's medical history, especially one's signs and symptoms of cardiopulmonary disease and cardiovascular disease risk factors. This medical history information is needed to determine the safety of performing the test. It is important that the supervisor recognize the reason for completing the test and confirm that the benefits outweigh the risks. After determining that the GXT is safe, the supervisor is responsible for determining what type of test is appropriate in regard to mode and protocol. In most cases the mode and protocol are predetermined by the referring physician, but there may be situations in which the type of test has not been determined or the proposed test is not appropriate. The supervisor of the test needs to communicate with the referring physician, especially if there are any discrepancies.

If the GXT is going to provide important clinical information, the client needs to be properly informed of what is expected of him or her (i.e., good effort, communicate symptoms). In addition, the client needs to be properly prepped so that the ECG recordings can be clearly interpreted. Following resting measurements, at the onset of starting the exercise test, the client needs to be monitored closely and accurately in regard to the onset of symptoms and ECG and BP changes. Symptoms can be accurately evaluated by using a number of different tables (RPE, dyspnea, angina, and claudication). One of the most important things that the supervisor needs to oversee is the safety of the client during the GXT. Therefore, the supervisor needs to know and understand the absolute and relative contraindications to test termination. Along with knowing contraindications of test termination, the supervisor needs to understand ECG analysis and recognize and respond to potential ECG abnormalities that can occur with exertion. Last, the supervisor needs to be properly trained and certified in responding to medical emergencies that can happen during a GXT.

Glossary

absolute contraindications for test termination—Conditions that occur during a GXT that require the test to be terminated.

aneurysm—Dilation of an artery that is connected with the lumen of the artery or cardiac chamber. Usually occurs because of a congenital or acquired (e.g., myocardial infarction) weakness in the wall of the artery or chamber. Different forms of aneurysms include true, dissecting, and false.

β-blocker—Medication used to block β-receptors in the myocardium, which decreases myocardial work, by decreasing HR and myocardial contractility.

claudication—Limping, lameness, and pain that occur in individuals who have an ischemia response in the muscles of the legs, which is brought on with physical activity (e.g., walking). A scale can be used to determine the severity of claudication.

digoxin—A cardioactive steroid glycoside used to increase myocardial contractility.

Duke nomogram—Five-step tool to estimate one's prognosis (5-year survival or average annual mortality rate) following the completion of a maximum GXT.

dyspnea—Shortness of breath that is perceived by an individual at rest or with exertion. A scale can be used to determine the severity of dyspnea.

echocardiography—Use of ultrasound images to evaluate the heart and great vessels.

ejection fraction (EF)—Percentage of blood that is ejected from the left ventricle per beat (normal 55-60%), EF = [(EDV − ESV)/EDV] × 100, where EDV = end diastolic volume and ESV = end systolic volume.

exertional ischemia—Myocardial ischemic response produced by exerting oneself physically.

functional aerobic impairment (FAI)—Percentage of an individual's observed functional capacity that is below that expected for the individual's sex, age, and conditioning level. %FAI = (predicted $\dot{V}O_2$ − observed $\dot{V}O_2$)/(predicted $\dot{V}O_2$) × 100.

functional capacity—One's maximum level of oxygen consumption. Can be measured at one's maximal effort with the use of a metabolic cart or predicted based on the maximum workload achieved.

graded exercise testing—Gradual increase in exercise workload to a predetermined point or until volitional fatigue, unless symptoms occur prior to this point. Generally completed on a treadmill or bicycle ergometer.

hazard ratio—Multiplicative measure of association. Exposure to a certain risk factor or certain characteristic is associated with a fixed instantaneous risk, compared with the hazard in the unexposed.

left ventricular dysfunction—Abnormal function of the left ventricle (i.e., poor wall motion).

metabolic equivalent (MET)—Multiples of one's resting oxygen consumption (3.5 ml of oxygen per kilogram of body weight per minute). Example: 5 METs are equivalent to five times one's resting metabolic rate.

morphology—Configuration or shape (e.g., shape of the ST segment: down-sloping, up-sloping, or horizontal).

motion artifact—Incidental activity that is picked up on one's ECG during body movement.

nitroglycerine—Medication that is used to promote vasodilation in patients with angina pectoris.

nuclear perfusion—Radioactive isotope that has the ability to perfuse through tissue, so that select organs can be imaged.

peripheral vascular disease—Disease of the vascular system that can be found in the periphery (i.e., commonly observed in the legs, which leads to claudication with physical exertion).

radionuclide agent—Isotope (natural or artificial) that exhibits radioactivity. Used in nuclear cardiology medicine to image the myocardium for potential ischemia.

ramping treadmill protocol—Continuous gradual increase in workload (speed and grade) over a select period of time.

rate pressure product (RPP)—Indirect indication of how hard the heart is working. RPP = [(systolic BP × HR)/100].

rating of perceived exertion (RPE)—One's perception of how hard he or she is working physically. There are currently two commonly used scales to assess RPE.

relative contraindications for test termination—Conditions that occur during a GXT that require strong clinical judgment concerning the safety of continuing the exercise test.

stratify—The process of separating individuals or samples into subcategories based on variables of interest (e.g., sex, age, number of risk factors, symptoms).

subendocardial ischemia—Myocardial ischemic response beneath the endocardium.

transmural ischemia—Myocardial ischemic response that occurs throughout the myocardial wall.

References

1. *ACSM's Guidelines for Exercise Testing and Prescription*, 6th edition. Lippincott, Williams & Wilkins, Philadelphia, 2000.

2. Atterhog, J.H., B. Jonsson, and R. Samuelsson. Exercise testing: A prospective study of complication rates. *Am Heart J* 98: 572-579, 1979.

3. Brubaker, P.H., W.J. Rejeski, H.D. Law, et al. Cardiac patients' perception of work intensity during graded exercise testing: Do they generalize to field testing? *J Cardiopulm Rehabil* 14: 127-133, 1994.

4. Bruce, R.A., F. Kusumi, and D. Hosmer. Maximal oxygen intake and nomographic assessment of functional aerobic impairment in cardiovascular disease. *Am Heart J* 85:546-562, 1973.

5. Chapman, C.B. Edward Smith (? 1818-1874), physiologist, human ecologist, reformer. *J Hist Med Allied Sci* 22: 1-26, 1967.

6. Detrano, R., R. Gianrossi, and V. Froelicher. The diagnostic accuracy of the exercise electrocardiogram: A meta-analysis of 33 years of research. *Prog Cardiovasc Dis* 32(3): 173-206, 1989.

7. Foss, M.L., and S.J. Keteyian (Eds). *Fox's Physiological Basis for Exercise and Sport,* 6th edition. McGraw-Hill, Boston, 1998.

8. Franklin, B.A., R. Dressendorfer, K. Bonzbeim, et al. Safety of exercise testing by non-physician health care providers: Eighteen year experience [abstract]. *Circulation* 92(Suppl I): I-37, 1995.

9. Gibbons, R.J., G.J. Balady, J.W. Beasley, et al. ACC/AHA Guidelines for exercise testing. *J Am Coll Cardiol* 30(1): 260-315, 1997.

10. Gill, T., L. DiPietro, and H. Krumholz. Role of exercise stress testing and safety monitoring for older populations starting an exercise program. *JAMA* 284: 3, 2000.

11. Gordon, N.F., and H.W. Kohl. Exercise testing and sudden cardiac death. *J Cardiopulm Rehabil* 13: 381-386, 1993.

12. Kinight, J.A., C.A. Laubach, R.J. Butcher, et al. Supervision of clinical exercise testing by exercise physiologist. *Am J Cardiol* 75: 390-391, 1995.

13. Lem, V., J. Krivokapich, and J.S. Child. A nurse-supervised exercise stress testing laboratory. *Heart Lung* 14: 280-284, 1985.

14. Mark, D.B., L. Shaw, F.E. Harrell, et al. Prognostic value of a treadmill exercise score in outpatients with suspected coronary artery disease. *N Engl J Med* 325: 849-853, 1991.

15. Morgan, W., and G. Borg. Perception of effort in the prescription of physical activity. In Nelson T (Ed.), *Mental Health and Emotional Aspects of Sports*. Chicago: American Medical Association, 1976, pp. 126-129.

16. Nostratian, F., and V.F. Froelicher. ST elevation during exercise testing: a review. *Am J Cardiol* 63: 986-988, 1989.

17. Pilote, L., F. Pashkow, J. Thomas, et al. Clinical yield and cost of exercise treadmill testing to screen for coronary artery disease in asymptomatic adults. *Am J Cardiol* 81: 219-224, 1998.

18. Pina, I.L., G.J. Balady, P. Hanson, et al. Guidelines for clinical exercise testing laboratories. A statement for healthcare professionals from the committee on exercise and cardiac rehabilitation, American Heart Association. *Circulation* 91(3): 912-921, 1995.

19. Rochmis, P., and H. Blackburn. Exercise tests: A survey of procedures, safety, and litigation experience in approximately 170,000 tests. *JAMA* 217: 1061-1066, 1971.

20. Rodgers, G.P., J.Z. Ayanian, G. Balady, et al. American college of cardiology/American heart association clinical competence statement on stress testing. A report of the American college of cardiology/American heart association/American college of physicians-American society of internal medicine task force on clinical competence. *Circulation* 102:1726-1738, 2000.

21. Scherer, D., and M. Kaltenbach. Frequency of life-threatening complications associated with exercise testing. *Dtsch Med Wochenschr* 33: 1161-1165, 1979.

22. Schlant, R.C., G.C. Friesinger, and J.J. Leonard. Clinical competence in exercise testing. A statement for physicians from the ACP/ACC/AHA task force on clinical privileges in cardiology. *Circulation* 28(5): 1884-1888, 1990.

23. Senaratne, M.P.J., et al. Feasibility and safety of early exercise testing using the Bruce protocol after acute myocardial infarction. *J Am Coll Cardiol* 35: 1212-1220, 2000.

24. Shetler, K., R. Marcus, V.F. Froelicher, et al. Heart rate recovery: Validation and methodologic issues. *J Am Coll Cardiol* 38: 1980-1987, 2001

25. Squires, R.W., T.G. Allison, B.D. Johnson, and G.T. Gau. Non-physician supervision of cardiopulmonary exercise testing in chronic heart failure: Safety and results of a preliminary investigation. *J Cardiopulm Rehabil* 19: 249-253, 1999.

26. Stuart, R.J., Jr., and M.H. Ellestad. National survey of exercise stress testing facilities. *Chest* 17: 94-97, 1980.

27. Stuart R.J., and M.H. Ellestad. Upsloping S-T segments in exercise stress testing: Six year follow-up study of 438 patients and correlation with 248 angiograms. *Am J Cardiol* 37: 19-22, 1976.

28. Weiner, D.A., T.J. Ryan, C.H. McCabe, et al. Prognostic importance of a clinical profile and exercise test in medically treated patients with coronary artery disease. *J Am Coll Cardiol* 3(3): 772-779, 1984.

29. Will, P.M., and J.D. Walter. Exercise testing: Improving performance with a ramped Bruce protocol. *Am Heart J* 138: 1033-1037, 1999.

Jonathan K. Ehrman, PhD, FASCM
Henry Ford Heart and Vascular Institute
Detroit, MI

General Exercise Prescription Development

Optimal adaptation to exercise training requires the development of, and adherence to, an exercise prescription. An exercise prescription is a specific guide provided to an individual for the performance of an exercise training program. Despite the term *prescription* within, the development of an exercise prescription does not necessarily require approval of a physician. However, in some situations, especially with clinical patients, a physician's signature may be required (e.g., possible requirement for Medicare reimbursement for cardiac rehabilitation) or desired (i.e., to limit liability). Individuals practicing clinical exercise physiology must possess the knowledge, skills, and practical experience to combine both "art" and science when developing an exercise prescription (see practical application 7.1). The primary purposes of the exercise prescription are to provide a reliable and valid guide for optimal health and physical fitness improvements and to provide a safe environment in which to achieve these improvements.

The exercise prescription can be general or very specific with regard to desired benefits and chronic disease prevention and management. Specificity of the exercise prescription is a burgeoning area of research, and the American College of Sports Medicine (ACSM) has published several position statements

PRACTICAL APPLICATION 7.1

The Art of Exercise Prescription

Definition: The art of exercise prescription is the successful integration of exercise science with behavioral techniques that result in long-term program compliance and attainment of the individual's goals (2).

Unlike chemistry or physics, physiology and psychology are not exact sciences. This means that we cannot always predict physiological or psychological responses because numerous factors can affect these. These include, but are not limited to, age, physical and environmental condition, sex, previous experiences, and diet. When developing an exercise prescription, we should follow the basic guidelines that are provided in this chapter. By doing so, we can elicit a response that is desired both during a single exercise training session and over the course of an extended training period. When we consider effectors of the exercise response, not all individuals respond as expected. This is especially true for individuals with chronic disease. For example, people with coronary artery disease often require modification of exercise intensity because of myocardial ischemia. Those who are extremely deconditioned may not be able to handle a typical intensity level for their age. And others may have a variable training response based on their type and severity of disease.

ACSM's Guidelines for Exercise Testing and Prescription (2) lists several reasons for altering an exercise prescription in selected individuals:

— Variance in objective (physiological) and subjective (perceptual) responses to an exercise training bout

— Variance in the amount and rate of exercise training responses

— Differences in goals between individuals

— Variance in behavioral changes relative to the exercise prescription

Each of these should be considered for each individual during both the initial development and subsequent review of an individual's exercise prescription. A modified exercise prescription should not be considered adequate until follow-up is performed. As a general rule, a person's exercise prescription should be reevaluated weekly until it appears that the parameters of the exercise prescription are adequate to improve health-related behaviors and selected physiological indexes while maintaining safety.

---

on special populations including healthy individuals, those with coronary artery disease, those with osteoporosis, those with hypertension, those with diabetes, and older individuals (2, 4-7). For instance, we may make very different recommendations for someone wishing to lose a great amount of weight versus someone who wants to increase his or her pain-free walking distance. The exercise prescription can also be specific to any of the following five health-related components of physical fitness:

1. Cardiorespiratory endurance: Often termed *aerobic fitness,* this is determined by the combined ability of the cardiorespiratory system to supply oxygen to active skeletal muscles during prolonged submaximal exercise and the ability of the skeletal muscles to perform aerobic metabolism.

2. Skeletal muscle strength: The peak ability of the skeletal muscles to produce force. Force may be

developed via **isometric, isotonic,** or **isokinetic** contraction.

3. Skeletal muscle endurance: The ability of the skeletal muscles to produce a submaximal force for an extended period of time.

4. Flexibility: The ability of a joint to move through its full, capable **range of motion.**

5. Body composition: The relative percentage of fat or nonfat mass of the body weight. Body fat is most important when relating body composition to health. Chapter 9 provides details regarding body composition assessment and exercise prescription for fat or weight loss, and therefore this topic is not covered in this chapter.

Each of these components of physical fitness is related to at least one aspect of health, and each can be improved by regular, specific exercise training. For instance, the relationship between cardiorespiratory endurance and the risk and recurrence of coronary artery disease is well documented (10, 45, 63). Another example is the relationship between poor body composition and the risk and incidence of adult-onset diabetes, hypertension, and coronary artery disease (63). Importantly, each component of the health-related physical fitness model can be positively influenced by exercise training, which will likely reduce the risk of a primary or secondary chronic disease. Thus, the importance of regular exercise and physical activity for overall health is well established. The general benefits of exercise training are presented in table 7.1 (63).

Several principles of exercise prescription must be considered during its development. These include the following:

1. Specificity of training
2. Progressive overload
3. Reversibility

One must also consider several aspects of a person's psychological state that are specific to beginning and adhering to an exercise training program. When these are considered, the likelihood that regular exercise will be successfully incorporated into a person's lifestyle is increased. These issues are presented in detail in chapter 2.

This chapter focuses on the general development of individualized exercise prescriptions. The purpose is to provide the basic principles of exercise prescription development for all components of health-related physical fitness. Parts II through VII of this text provide specific guidelines for exercise prescription development with respect to the specific clinical situation.

The Exercise Training Sequence

A comprehensive training program should include flexibility, resistance, and cardiorespiratory (aerobic) exercises. The order of the exercise training routine is important for both safety and effectiveness. However, information on this topic is lacking. It is generally recommended that flexibility training take place following a warm-up period or following an aerobic or resistance-training routine. Muscular injury and soreness are less likely to occur and range of motion may be enhanced to a greater degree if the flexibility training takes place in this manner.

In a clinical population, if an aerobic and resistance-training bout take place on the same day, it may be best to first perform the activity that is the primary focus of the day's training. In this manner, if one tires from the initial exercise mode and the second is ended prematurely, the primary mode will have been completed. On the other hand, performing resistance training following aerobic training may be beneficial because the skeletal muscles will

---

TABLE 7.1

Clinical Benefits of Exercise Training

Physiological/health and disease	Psychological/quality of life
Improved cardiorespiratory and musculoskeletal fitness	Improved sleep patterns
Improved metabolic, endocrine, and immune function	Reduced depression and anxiety
Reduced overall mortality rate	Improved health behavior
Reduced cardiovascular disease risk	Improved mood levels
Reduced cancer risk (colon, breast)	Overall improved health-related quality of life
Reduced risk of osteoporosis and osteoarthritis	Reduced risk of falling
Reduced risk of non-insulin-dependent diabetes mellitus	
Reduced risk of obesity	

Adapted from the Surgeon General's report on physical activity and health (63).

be more compliant and arterial beds near maximal dilation from aerobic exercise.

Goal Setting

A comprehensive exercise prescription should consider goals of each individual. Common goals of a person beginning an exercise training program include these:

— Feel better
— Look better
— Lose weight
— Prepare for competition
— Improve general health and reduce primary or secondary risk of disease
— Reduce the negative effects of a chronic disease or condition on the ability to perform physical activity

People with specific diseases often have goals that relate directly to reversing or reducing the progression of their disease and its side effects. For instance, a person with multiple sclerosis may wish to improve flexibility and strength, whereas someone with coronary artery disease may want to stop the progression of atherosclerosis. Because of these and other goals, a clinical exercise professional must have a comprehensive understanding of how to alter a general exercise prescription to provide the best chance of success to achieve a desired goal. Within this text (parts II-VIII), specific exercise prescription development considerations to achieve desired goals are addressed. Additionally, chapter 2 reviews methods to improve compliance that can be used to help individuals best achieve their goals.

Principles of Exercise Prescription

To gain the optimal benefits of exercise training, no matter the area of emphasis (i.e., cardiorespiratory, strength, endurance, body composition, range of motion), people need to follow several basic guidelines. The following sections review these concepts, which are specificity, progressive overload, training frequency, intensity and duration, and reversibility.

Specificity of Training

The clinical exercise professional must consider the principle of specificity of training and how this af-

fects the training response. The principle of specificity of training states that the body will adapt in specific ways to specific types of exercise. This is true for changes in physiological function in response to an acute bout of exercise (short-term exercise) or a chronic series of exercise (long-term exercise). In more simplistic terms, specificity of training states that what you do for exercise relates directly to the improvements that occur. Investigations have demonstrated this to be true for numerous specific physiological adaptations.

The selection of an exercise training mode is important with respect to desired adaptations, because specific training mode adaptations do not carry over completely to other modes. For example, let's compare running and cycling. In one study, experienced runners were able to achieve a 14% higher peak oxygen uptake (i.e., $\dot{V}O_2$peak) during treadmill running than during cycle ergometry (64). A range of 5% to 15% has been reported by several studies that compared the $\dot{V}O_2$peak difference in normal, healthy individuals tested using a treadmill and cycle ergometer (28, 39, 40, 42, 66). In each of these investigations, the $\dot{V}O_2$peak was greater for the treadmill. Although part of this difference is related to a smaller total muscle mass used during cycling, much of it is related to specific training adaptations. To further illustrate this point, note that trained cyclists are able to achieve a similar $\dot{V}O_2$peak on the cycle compared with a treadmill. This is because of their cycle-specific training. Changes in $\dot{V}O_2$peak following a defined training period are also mode specific. For instance, McArdle et al. (38) reported that during swimming, $\dot{V}O_2$peak values were 19% lower than during running, and following run training the $\dot{V}O_2$peak improved during treadmill running but not for swimming. These specific adaptations are the result of local physiological changes within the skeletal muscles that are used during training. However, there can be some crossover of training response to a nontraining mode of exercise. For example, McArdle et al. (38) reported that heart rate was reduced during a standardized-speed, submaximal running bout following a 10-week interval run training program. Heart rate was also reduced to a similar degree in these same subjects during submaximal swimming. This and other crossover effects are likely the result of a combination of central cardiac improvements (e.g., increased stroke volume and cardiac output) and involvement of specifically trained skeletal muscles that are then used in the alternative exercise mode (e.g., train the legs by running and then use legs for kicking while swimming), which is the basis for the **cross-training** concept. Similarly, this exemplifies the carryover effect from mode-specific training to other exercise modes. Similar results occur in clinical patients who train using a specific mode.

Based on the specificity concept, a general recommendation for exercise prescription development is that a training regimen should use modes of exercise that will provide adaptations as close as possible to those desired (table 7.2). For instance, a factory worker who primarily uses upper body movements during the workday should focus on upper body work during an exercise rehabilitation program for an injury or cardiovascular problem. This will improve the likelihood that this individual will return to his or her occupation. On the other hand, general improvements (e.g., cardiorespiratory endur-ance, body composition) can be achieved by any aerobic mode of exercise as long as the caloric expenditure of the training regimen is adequate (37, 44, 49). Table 7.3 provides questions that can be asked of an individual for whom an exercise professional is developing an exercise prescription. This will help to ensure that the exercise prescription is specific to that individual.

Progressive Overload

The progressive overload principle refers to the relationship noted between the "dose" of exercise and the benefits gained. There appears to be a physical activity or exercise threshold at which physiological benefits and disease and mortality benefits are

---

TABLE 7.2

Selected Modes/Types of Exercise Training and General Skeletal Muscle Adaptations

I. CARDIOVASCULAR: GENERAL CARDIORESPIRATORY ENDURANCE AND IMPROVED BODY COMPOSITION (AEROBIC)

Mode	Adaptation	Comment
Walking	Leg endurance	Easy to perform for most
Running	Leg endurance	Increased injury risk over walking
Cycling	Leg endurance	Good for those with walking difficulty
Rowing	Arm and leg endurance	Often difficult to perform
Stair stepping	Leg endurance	Seated is better option for many clinical patients
Swimming	Arm and leg endurance	Poor technique will require a lot of exertion to perform; easy on joints
Rope skipping	Primarily leg endurance	Likely difficult for most clinical patients to perform
Elliptical training	Leg and arm endurance	Easy on joints
Cross-country skiing	Leg and arm endurance	Stationary machines require good coordination

II. POWER: SUBMAXIMAL AND PEAK EXERCISE POWER INCREASES (ANAEROBIC)

Mode	Adaptation	Comment
Sprint running	Leg power	Any of the sprinting modes are potentially dangerous for most clinical populations
Sprint cycling	Leg power	
Swim sprinting	Arm and leg power	
Resistance training	Muscle-specific power	Excellent for most clinical patients but requires proper instruction

---

TABLE 7.3

Questions to Ask an Individual When Developing an Exercise Prescription

Specificity	Mode	Frequency	Intensity	Time
What are your specific goals when performing exercise?	What types of exercise or activity do you like the best?	Do you know how many days per week of exercise or physical activity are required for you to reach your goals?	Do your goals include optimal improvement of your fitness level?	How much time per day do you have to perform an exercise routine?
Do you want to walk farther?	What types of exercise do you like the least?	How often during a week do you have 30-60 min of continuous free time?	Or are your goals primarily related to your health?	What is the best time for you to exercise?
Do you want to do more activities of daily living?				Can you get up early or take 30-40 min at lunch time?
Do you want to perform something you currently cannot?				
Do you want to feel better?				
Is there something else?				

---

gained and even lost. For instance, in the Harvard Alumni study, a dose response was reported for all-cause mortality from a caloric expenditure of less than 500 kcal · week^{-1} to 3000 to 3499 kcal · week^{-1} (45). However, the all-cause mortality rate rose slightly when more than 3500 kcal was expended per week. Based on these findings optimal health-related threshold for caloric expenditure likely lies between 1000 and 2000 kcal · week^{-1}.

Overload refers to the increase in total work performed above and beyond that normally performed on a day-to-day basis. Let's look at walking to illustrate this point. Walking is an activity that all physically able people perform daily to move around in the home, at school, or at work. When a person performs walking as part of an exercise training regimen, the pace and duration should be above those typically encountered on a daily basis to gain fitness benefits. Progressive overload is the gradual increase in the amount of work performed in response to the continual adaptation of the body to the work. This would relate to walking more often, farther, or at a faster pace.

The general adaptation syndrome was developed by Dr. Hans Selye (55) and is used to explain the process of physiological adaptation from overload. The three stages of general adaptation syndrome are alarm reaction, resistance development, and exhaustion. Initially, an exercise training program or increased physical activity level brings about a stress to the body that may cause short-term fatigue or pain (i.e., alarm reaction). To maintain homeostasis, the body has the ability to change, or adapt, to an overload stimulus to reduce the risk of fatigue, pain, or injury (i.e., resistance development). This adaptation results in less effort required to perform a standardized task. If the stress placed on the body becomes intolerable (e.g., too great of an overload or too little recovery between exercise bouts), there is an increased risk of injury and illness and the body loses its ability to adapt (i.e., exhaustion).

In the walking example given previously, the individual will, in time, adapt to the pace and duration of the walk, and either the pace or duration, or both, will need to be increased (i.e., overload) for further beneficial adaptation to occur. The overload is applied by using the FIT principle of overload. FIT is an acronym for frequency, intensity, and time (i.e., duration) of the exercise that is performed.

Frequency

Frequency is the number of times that an exercise routine or physical activity is performed. This commonly refers to the number of times per week exercise is performed, but it can also refer to the number of times per day or per session that an activity is performed.

Intensity and Duration

Intensity and duration are presented together because they are of equal importance when determining the total volume or load of exercise or physical activity. Investigators have evaluated the relationship between intensity and duration, and this has led to our current knowledge of this important relationship.

The intensity of exercise or physical activity refers to either the objectively measured work or the subjectively determined level of effort performed by an individual. Typical objective measures of work that are important to the clinical exercise professional include oxygen consumption ($\dot{V}O_2$, or **metabolic equivalent**), caloric expenditure **(kilocalories [kcal] or joules [J])**, and power output (kilograms per minute [kg · min^{-1}] or watts [W]). The anaerobic or lactate threshold may also be used to determine exercise intensity; however, it is impractical to use as a guide during exercise training. The subjective level of effort can be evaluated by either a verbalized statement from a person performing exercise (e.g., "I'm tired" or "This is easy") or by using a standardized scale (e.g., Borg rating of perceived exertion). Chapter 6 presents the rating of perceived exertion scale in detail.

Duration (or time in the FIT acronym) refers to the amount of time that is spent performing exercise or physical activity. During exercise training, the duration of training is typically accumulated without interruption or with very short rest periods in order to gain fitness benefits. To the contrary, the Centers for Disease Control and Prevention (CDC)/ACSM statement on physical activity states that every U.S. adult should accumulate 30 min or more of physical activity each day of the week, suggesting that one can gain health-related benefits from the accumulation of duration throughout a day (47).

Reversibility

The reversibility principle refers to the loss of exercise training adaptations because of inactivity. Positive adaptations accrue at a rate specific to the overloaded physiological processes. Typically, most untrained individuals can expect a 10% to 30% improvement in $\dot{V}O_2$peak and work capacity following an 8- to 12-week period of training (7). Alterations in other physiological variables, such as body weight and blood pressure, may take a variable amount of time. As a general rule, less fit individuals can expect to achieve gains at a faster rate and to a greater absolute degree than more fit people.

Maintaining improvements in fitness is an important issue. It is generally believed that total training volume can be reduced to a level that still allows for the maintenance of a desired physical fitness level. If a minimal training volume is not maintained, a loss of training effects will occur. This reversal of fitness is often termed *deconditioning* or *detraining*. Reversibility of fitness gains can occur in recreational and competitive athletes as well as in individuals who exercise train for general health and fitness. Importantly, reversibility also occurs in clinical populations.

Cardiorespiratory Endurance

Cardiovascular conditioning to improve aerobic endurance requires that individuals perform modes of training that use large muscle groups and are continuous and repetitive or rhythmic (3). The traditional forms of aerobic exercise involve use of either the legs or arms. Common leg exercise modes include walking, running, cycling, skating, stair stepping, rope skipping, or group aerobics (e.g., dance, step, tae-bo, spinning, water, seated). Exercise using strictly the arms is limited but includes upper body crank ergometry, dual-action cycles using only the arms, and wheelchair ambulation. Several other popular modes of exercise use both the arms and legs: rowing, swimming, and some types of stationary equipment (e.g., Airdyne-type cycles, cross-country skiing, elliptical trainers, seated dual-action steppers). When an exercise professional is providing advice to an individual regarding the selection of an exercise mode, one must consider the client's goals and his or her disease-specific or orthopedic limitations.

Specificity of Training

General benefits gained from aerobic exercise training appear to be independent of any specific type or mode of training. For instance, the Harvard Alumni study reported that men who were physically active and had a high weekly caloric expenditure had a lower incidence of all-cause mortality than those who were less active (45). Total caloric expenditure was estimated from a variety of modes, including walking, stair climbing, sports activity, and yard work. The type of physical activity appeared to be independent of the benefits gained. Further interpretation of the Harvard Alumni database demonstrates that all-cause mortality rate is improved in those who perform higher intensity exercises than those who perform less vigorous activity (36). These reports suggest that the specific type of physical activity is less important for general mortality benefits than the amount and intensity of the activity.

Progressive Overload

To gain benefits from cardiorespiratory training, one must follow the principle of progressive overload (FIT principle).

Frequency

Recommendations for aerobic activity frequency vary between sources. The ACSM position stand titled "The Recommended Quantity and Quality of Exercise for Developing and Maintaining Cardiorespiratory and Muscular Fitness, and Flexibility in Healthy Adults" states a desired frequency of 3 to 5 days per week of cardiorespiratory training (7). A slightly different recommendation was published in a joint recommendation paper from the CDC and the ACSM (47). This statement recommends "physical activity on most, preferably all, days of the week." The difference in these recommendations is twofold. First, the ACSM position stand recommends exercise to optimally improve one's cardiorespiratory fitness level, whereas the CDC/ACSM recommendation focuses on improving overall health. Cardiorespiratory fitness and health benefits related to the performance of regular exercise training or physical activities, respectively, are not mutually exclusive. The difference in these recommendations appears to lie in the desired effects of exercise or physical activity. The second difference in these recommendations is that the CDC/ACSM joint statement recommends a reduced intensity of activity. Therefore, to expend a sufficient amount of calories per week (e.g., 2000 as recommended by Paffenbarger), the individual must increase number of days (i.e., >3-5) that physical activity is performed (45). One may also perform physical activity more than once per day.

Investigations of the effect of different numbers of training days per week have, in general, demon-

strated a minimal threshold of 3 days per week for optimal $\dot{V}O_2$peak improvements and that training less than 2 days per week does not result in significant $\dot{V}O_2$peak changes (23). Let's look at a study that held total work volume constant while varying the number of training days per week in a group of healthy individuals (57). A group who exercise trained 1 day per week did not improve, whereas other groups who trained 2 or 4 days per week raised their $\dot{V}O_2$peak. However, the 4 day per week group improved $\dot{V}O_2$peak to a greater degree than the 2 day per week group. This suggests that although 2 days of training per week is sufficient to raise $\dot{V}O_2$peak, it requires more days per week to maximally increase aerobic capacity. Several other studies have demonstrated similar increases in $\dot{V}O_2$peak when 2 and 4 days per week were compared (8, 48).

Aerobic training 5 days per week has only a minimal effect on the improvement of $\dot{V}O_2$peak compared with fewer days. This was demonstrated in a study in which 20 college-aged men trained for 5 weeks (31). The training consisted of 10 min of treadmill running on either 1, 2, 3, or 5 days per week. There was no difference for improvement in $\dot{V}O_2$peak between either the 2 or 3 day per week groups and the 5 day per week group. The authors concluded that exercise training 5 days per week was as effective as 2 or 3 days for improving $\dot{V}O_2$peak. This suggests that it is possible that an increased time commitment to exercise training of more than 3 days per week may not be an efficient use of time for the person with little time to spare. In fact, in a study that held total exercise volume (i.e., intensity × duration × frequency) constant, no difference in $\dot{V}O_2$peak was reported for those who exercised 3 versus 5 days per week (57).

To conclude this topic, the recommendation of 3 to 5 days per week of aerobic exercise put forth in a position stand by the ACSM is based on studies that show optimal improvements in $\dot{V}O_2$peak and other variables within this period. In the general population, exercise training fewer than 2 days per week appears to be less effective than 3 to 5 days per week, and exercise training 5 days per week does not provide further improvement in $\dot{V}O_2$peak. However, exercising more than 5 days per week may play a positive role by increasing total caloric expenditure, optimizing health improvements, and reducing all-cause mortality rates.

Intensity and Duration

Intensity and duration of training are interdependent with respect to the overall training load. Generally, the higher the intensity the shorter the duration of an exercise training bout and vice versa. For instance, the caloric expenditure is similar for 30 min of walking at 4 mph and 60 min of walking at 2 mph. The decision of which intensity and duration one should select for an individual exercise program should be made with the consideration of several factors. These include the current cardiovascular conditioning level of the individual, the existence of underlying chronic diseases such as coronary artery disease or obesity, the possibility of injury or death, and the individual's goals (58, 60).

An important consideration for determining the appropriate individual exercise training intensity is that the primary metabolic pathway used must be specific to the desired benefits. In those individuals who desire to improve cardiovascular endurance, an overload must occur of the aerobic metabolic pathway. To achieve this, most individuals must exercise at an intensity between 50% and 85% of their $\dot{V}O_2$peak in order for the cardiorespiratory system to adapt and for aerobic capacity to increase (33). In some clinical populations with extreme deconditioning (e.g., those with heart failure, the elderly, obese people), improvements in $\dot{V}O_2$peak may be noted at intensities as low as 40% of their maximal ability (2). This is because the person's fitness level prior to initiating a conditioning program affects the intensity at which cardiorespiratory benefits begin to accrue (16, 54). Generally, the lower a person's initial fitness level, the lower the required intensity level to produce adaptations.

The upper level for healthy individuals to improve $\dot{V}O_2$peak is typically set at 85%, as recommended by the ACSM (7). This upper level is guided by the principle of diminishing returns in that little fitness is gained at intensity levels greater than 85% of peak. Additionally, the risks of orthopedic injury or adverse cardiovascular events are increased at training intensities above 85% of peak $\dot{V}O_2$. Training at a high intensity may also be difficult for healthy and clinically diseased individuals who have an anaerobic/lactate threshold that is less than 85% of their $\dot{V}O_2$peak. However, some may be able to train at intensities up to 90% of $\dot{V}O_2$peak, but this is typically limited to those who wish to perform in competitive athletic events. Practical application 7.2 and 7.3 provide additional information about determining appropriate heart rate and prescribing exercise by using $\dot{V}O_2$ reserve.

ACSM's position stand (7) recommends 20 to 60 min of continuous or intermittent (no less than 10 min per bout) aerobic activity when performed, whereas the CDC/ACSM statement suggests that accumulating a minimum of 30 min of exercise per day is sufficient (7, 47). These differing recommendations reflect different goal-related focuses. Accumulating activity throughout a day is designed to affect health by reducing the primary and secondary risk of cardiovascular disease and other inactivity-related chronic diseases. Achieving 20 to 60 min of continuous aerobic exercise reflects an attempt to improve the parameters of physical fitness. Both regimens help to reduce the primary and secondary risks of chronic disease. However, there is evidence that exercise training is more effective than accumulating physical activity at positively influencing a variety of cardiovascular disease risk factors such as plasma cholesterol and hypertension in a healthy population (41).

Reversibility

Several studies have investigated the physiological effects of detraining before or after a period of conditioning. The classic study of Saltin et al. (54) reported on the effects of bed rest over a 3-week period in five subjects. All subjects had reductions in cardiac output resulting from a reduced stroke volume, and this was related to reductions in total heart volume. This trend was reversed when training was implemented. Additionally, heart rate at a standardized submaximal $\dot{V}O_2$ rose during bed rest and decreased with training.

Because this study followed bed rest with training, the authors could not answer questions about the time course of deconditioning (54). This issue was addressed in a separate study of seven subjects who exercise trained for 10 to 12 months, followed by 3 months of detraining (15). The detraining consisted of no exercise but did allow light daily activity. $\dot{V}O_2$peak decreased from 0 to 12 days of detraining by 7% and by another 7% from 21 to 56 days. The early reduction in $\dot{V}O_2$peak was the result of a near-equal reduction in stroke volume. The later reduction was the result of a decline in arteriovenous oxygen difference. Thus, it appears that $\dot{V}O_2$peak reduction from deconditioning occurs in a biphasic manner and that there are separate mechanisms for this decline. It is interesting to note that declines in $\dot{V}O_2$peak occurred in both well- and less well-trained subjects and in sedentary subjects, emphasizing that all persons are prone to reductions in fitness attributable to becoming sedentary. Interestingly, the reduction in $\dot{V}O_2$ was greater in the complete bed rest study than the study allowing limited daily activity (15, 54). This suggests that a minimal amount of activity can be effective at maintaining fitness levels or can attenuate the effects of deconditioning. Also,

Determining the Appropriate Heart Rate Range

It is impossible for an individual to guide his or her exercise intensity by using $\dot{V}O_2$. However, between about 50% and 90% of $\dot{V}O_2$peak, the $\dot{V}O_2$ relationship with heart rate (HR) is linear (see figure). This allows HR to be an excellent guide of intensity because it relates directly to $\dot{V}O_2$ in the effective $\dot{V}O_2$ intensity range (i.e., ~50-90%).

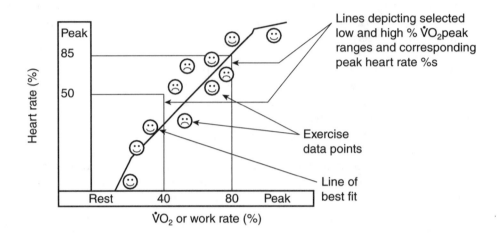

Several methods exist to determine an exercise training HR range (THRR). By using these methods, one can find the HR range that an individual can use to safely and effectively enhance aerobic capacity. The following demonstrates how to develop a HR rate–based exercise prescription.

HRmax

When the **true maximal heart rate** (HRmax) is unknown, you may estimate the maximal HR by using one of the following equations:

      Equation 1: 220 – age = estimated HRmax

      Equation 2: 210 – (0.5 × age) = estimated HRmax

Note: Equation 2 produces a higher value than the former for adults (i.e., >20 years).

For example, when predicting HRmax in a 32-year-old person by using equation 1, the yield is 220 – 32 = 188. Using equation 2 yields 210 – (0.5 × 32) = 194.

The disadvantage of estimating HRmax is that the standard deviation (*SD*) is ±10 - 12 beats · min^{-1}. What this means is that 68% of the population will be within a 20 beats · min^{-1} range around the predicted HRmax value (i.e., 10 beats above and below the predicted HRmax = +10 beats · min^{-1}), 27% will be within a 20 to 40 beats · min^{-1} range, and 5% will be within a 40 to 60 beats · min^{-1} range or greater. In the example using equation 1, 1 *SD* would result in the predicted HRmax ranging from 178 to 198. The take-home message is that estimating HRmax may produce a significant miscalculation of peak HR.

Another disadvantage of estimating HRmax is the inaccuracy of equations 1 and 2 in people on β-blocker medications. Equation 3 is a newly developed equation for use in these persons (12).

      Equation 3: 162 – (0.7 × age) = estimated HRmax

It is preferable to use the true HRmax to calculate the THRR. However, there may be problems with using the peak HR determined from a maximal exercise test. The HR achieved at the end of an exercise stress test may be inaccurate in certain instances:

— The subject did not attain a true peak exercise level during the exercise stress evaluation (e.g., stopped because of intermittent claudication, arrhythmias, poor effort).

continued

— The subject did not take a prescribed chronotropic medication (e.g., β-blocker) prior to the exercise stress evaluation.

— The subject had a change in his or her chronotropic medication (type or dose) since the exercise evaluation.

Each of these must be considered when one is using the peak HR from an exercise evaluation to develop a THRR.

THRR

Three methods can be used to determine the THRR.

— Percentage HR (%HR) method

Equation 4: HRmax × desired %HR = THR

For example, when HRmax (estimated or actual) = 161 and desired %HR range is 60% to 90%, equation 4 yields the following:

161 × 0.6 = 97 (lower end)

161 × 0.9 = 145 (upper end)

Note: Because the relationship between $\dot{V}O_2$ and HR is not a straight line, the percentage peak exercise HR will not match the exact same percentage for $\dot{V}O_2$peak within the training range. At lower intensity levels, the percentage peak HR is approximately 10% higher than the %$\dot{V}O_2$peak (i.e., 60% of peak HR is equivalent to about 50% of $\dot{V}O_2$peak). This difference in percentage is reduced at higher intensity levels (i.e., 90% of peak HR is roughly equivalent to 85% of $\dot{V}O_2$peak).

— Heart Rate Reserve method (or Karvonen method)

Equation 5: Step I. HRmax – HRrest = HR reserve (HRR)

Step II. (HRR × desired $\dot{V}O_2$ percentages) + HRrest = THRR

Example: given HRmax (estimated or actual) = 170, HRrest = 68, and desired $\dot{V}O_2$ percentages are 50% and 85%.

Step I. 170 – 68 = 102 (HRR)

Step II. 102 × 0.5 + 68 = 119 (lower end)

102 × 0.85 + 68 = 155 (upper end)

Note: Recently, several investigators have noted that a specific %HRR is equivalent to the same percentage of $\dot{V}O_2$ reserve. $\dot{V}O_2$ reserve is defined as the difference between resting and peak exercise $\dot{V}O_2$ ($\dot{V}O_2$max – $\dot{V}O_2$ rest). See practical application 7.3 for a complete explanation of this relationship.

— Direct method

You can also use a graphing method to plot the absolute HR from an exercise test versus the measured $\dot{V}O_2$ or work rate. Then a line of best fit is drawn and used to determine the HRs at a desired lower and upper %$\dot{V}O_2$peak or work rate level. An example is provided in *Line of identity between percentage of heart rate reserve (%HRR) and percentage of oxygen uptake reserve (%$\dot{V}O_2$R)* figure.

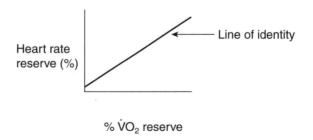

continued

The direct method can be used to determine either the desired THRR or a desired work rate setting (e.g., metabolic equivalent, watts) on a piece of exercise equipment. The data collected to use for this method (HRs and work rates or $\dot{V}O_2$) often come from submaximal testing. If submaximal data are used, peak HR (see equation 1, 2, or 3) must be estimated. The $\dot{V}O_2$ peak or peak work rate is then determined by extrapolation. As stated previously, there is potential error in estimating peak HR. This in turn may result in an incorrect estimation of $\dot{V}O_2$ peak or peak work rate. Also, if you are calculating a target work rate instead of a THRR, you should understand that an estimation of $\dot{V}O_2$ peak from a submaximal or maximal exercise test may result in overestimation. This often happens when patients use handrail support. There is also error associated with established prediction equations. When possible, a maximal exercise test should be performed for best accuracy, and the true submaximal and peak HRs and $\dot{V}O_2$ should be used to determine specific intensity ranges by using the direct method.

Despite the potential errors associated with these techniques, each will provide useful information from which to guide exercise training intensity for optimal safety and improvement purposes.

the gravitational stress of the upright position provides a natural cardiovascular stress versus a prone body position and likely reduces the negative effects of deconditioning.

Reductions in fitness attributable to aging are enhanced when activity or exercise levels are very low or nonexistent. Also, those who actively exercise trained in their formative years but discontinued training after high school or college are not protected against the risk imposed by a sedentary lifestyle (45, 46). This includes overall mortality rate, cardiac-specific mortality rate, and myocardial infarction.

The take home points of the reversibility concept are (a) cardiorespiratory conditioning can be maintained with a reduced level of exercise training; (b) the more sedentary one becomes, the greater the loss of fitness; (c) all individuals, no matter what their conditioning level or disease status, are prone to deconditioning; and (d) a sedentary lifestyle results in an additive effect on the loss of fitness that naturally occurs with aging.

Skeletal Muscle Strength and Endurance

Resistance training improves muscular strength and power and also reduces levels of muscular fatigue. The definition of muscular strength is the maximum ability to develop force or tension by a muscle. The definition of muscular power is the maximal ability to apply a force or tension at a given velocity. The definition of muscular endurance is the reduction in either muscular strength or power over a number

of times of muscular contraction. Fatigue is the inability to sustain a desired level of muscular strength or power. The general benefits of a resistance training program are listed in Figure 7.1.

For general skeletal muscle conditioning, the focus should be on the primary muscle groups. Table 7.4 lists several of the standard lifting techniques that can be performed using a barbell/dumbbell set and minimal other equipment. In addition, these can be performed on most commercially available Universal-type resistance equipment. Most of these can also be performed without tra-

---

- Improved muscular strength and power
- Improved muscular endurance
- Improvements in cardiorespiratory fitness
- Reduced effort for activities of daily living as well as leisure and vocational activities
- Improved flexibility
- Reduced skeletal muscle fatigue
- Elevated skeletal density and improved connective tissue integrity
- Reduced risk of falling
- Improved body composition
- Possible reduction of blood pressure values
- Improved glucose tolerance
- Possible improvement in blood lipid profile

FIGURE 7.1   Benefits of resistance training.

PRACTICAL APPLICATION 7.3

Exercise Prescription Using $\dot{V}O_2$ Reserve

It is now well accepted that the HR reserve (HRR, see equation 1) method of determining a THRR (see equation 2) is a valid method to use when prescribing exercise intensity in people with cardiac disease, other chronic diseases, and disabilities.

Equation 1: HRR = peak exercise HR – upright, resting HR

Equation 2: THRR = HRR (% intensity range) + upright, resting HR

Prevailing dogma assumes that a selected %HRR is equivalent to the same percentage of $\dot{V}O_2$peak. This assumption is presumably based on Karvonen's work from a 1957 publication (33). However, Karvonen did not test this or claim this assumption.

$\dot{V}O_2$ reserve ($\dot{V}O_2$ R) is the difference between peak and resting $\dot{V}O_2$ (see equation 3). Recent investigations by Swain et al. (61, 62) demonstrated that a better relationship exists between %HRR and %$\dot{V}O_2$ R (see equation 4) than between %HRR and %$\dot{V}O_2$peak.

Equation 3: $\dot{V}O_2$ R = peak exercise $\dot{V}O_2$ – upright, resting $\dot{V}O_2$

Note: May substitute $3.5 \text{ ml} \cdot \text{min}^{-1} \cdot \text{kg}^{-1}$ for upright, resting $\dot{V}O_2$

Equation 4: %$\dot{V}O_2$ R = (exercise $\dot{V}O_2$ – resting $\dot{V}O_2$) / $\dot{V}O_2$ R $\times$ 100

The highest value of HRR (i.e., 100% HRR) occurs at 100% of peak $\dot{V}O_2$. However, at rest the %HRR does not equal the same %$\dot{V}O_2$peak. This is because 0% of HRR (at rest) is equal to a percentage (not 0%) of $\dot{V}O_2$peak. Remember, resting $\dot{V}O_2$ is never at 0 (unless you're dead!) but is at an average of $3.5 \text{ ml} \cdot \text{min}^{-1} \cdot \text{kg}^{-1}$. So if $\dot{V}O_2$peak = 14.0, then the average resting $\dot{V}O_2$ represents 25% of the peak $\dot{V}O_2$ value.

At rest, the %$\dot{V}O_2$ R = 0% (see equation 4). Thus, at rest 0% HRR equals 0% $\dot{V}O_2$ R, and at peak 100% HRR equals 100% $\dot{V}O_2$ R. The question then becomes: Is the %HRR equal to the %$\dot{V}O_2$ R at any given submaximal work rate? If so, then the relationship between %HRR and %$\dot{V}O_2$ R will be similar to the line of identity (slope = 1, intercept = 0). Several recent investigations have answered this question.

Swain et al. (61) evaluated 63 healthy adults during incremental exercise on a cycle ergometer. The authors reported a similar relationship to the line of identity between %HRR and %$\dot{V}O_2$ R. Also, they reported that the relationship between %HRR and %$\dot{V}O_2$peak was different than the line of identity. They also reported similar findings during treadmill testing (61). Brawner et al. (11) evaluated 65 post–myocardial infarction patients during incremental treadmill exercise testing and found no difference between the slope and intercept of the relationship between %HRR and %$\dot{V}O_2$ R and the line of identity. This was true both for patients taking and those not taking β-blockers. Brawner et al. (11) also reported similar findings in a group of heart failure patients.

Although a better relationship exists between HRR and $\dot{V}O_2$ R (vs. HRR and absolute $\dot{V}O_2$), there should be no alteration in the method for determining the THRR using the HRR method based on these findings. We simply now know that the %HRR has closer to a 1:1 relationship with %$\dot{V}O_2$ R than with %$\dot{V}O_2$peak. This means that 50% HRR = 50% $\dot{V}O_2$ R and 75% HRR = 75% $\dot{V}O_2$ R. As recommended by the ACSM (2), the range used for intensity with the HRR method should remain 40% or 50% to 85% of peak $\dot{V}O_2$ R for cardiac patients and most other populations.

ditional resistance equipment using items such as elastic bands, water resistance, and even an individual's own body weight. There are numerous variations of the standard resistance maneuvers listed in table 7.4. Most of these are designed to isolate specific skeletal muscles within a group. These are most often used by individuals who are rehabilitating an injury, performing a specific athletic event, or bodybuilding. Figure 7.2 demonstrates the common resistance training maneuvers and their variations.

Proper technique is important during resistance training because it will reduce the risk of injury and increase the effectiveness of an exercise. General recommendations include the following:

— Lift throughout the range of motion unless otherwise specified.

TABLE 7.4
Selected Resistance Training Exercises and General Skeletal Muscle Adaptations

Resistance maneuver	Area of enhanced submaximal endurance and peak power/strength
Bench press	Chest, arm endurance, power, strength
Military press	Shoulder endurance, power, strength
Arm biceps curl	Biceps brachii endurance, power, strength
Arm triceps curl	Triceps brachii endurance, power, strength
Curl-up/sit-up	Abdominal strength
Leg extension/squats	Quadriceps endurance, power, strength
Leg curl	Hamstrings endurance, power, strength
Toe/calf raises	Calf (soleus, gastrocnemius) endurance, power, strength

— Breath out (exhale) during the lifting phase and in (inhale) during the recovery phase.

— Do not arch the back.

— Do not recover the weight passively by allowing weights to "crash" down before beginning the next lift (i.e., always control the recovery phase of the lift).

In certain clinical populations the following may be prudent:

— Ensure that a fully equipped crash cart and advanced cardiac life support–trained staff is available.

— Monitor blood pressure before and after a resistance training session and periodically during a session.

— Ensure that a clinical exercise professional familiar with resistance training in the specific clinical population conducted the initial patient orientation and regular reevaluations of lifting technique.

— Constantly reiterate signs and symptoms of exercise intolerance that may occur during resistance training.

— Instruct participants to always train with a partner.

It is important to order resistance exercises so that large muscle groups are worked first and smaller groups thereafter (26). This allows for optimal training of the primary locomotive and postural muscles.

If smaller muscle groups are trained first (i.e., those associated with fine movement), the larger muscle groups may become fatigued earlier when they are used.

Resistance training, like other types of training, can be very general or very specialized. Specialized training programs and specific lifts are abundant. Fleck and Kraemer (21) provided a review of specific specialized resistance training programs that includes explanations and examples of circuit programs, the triangle program, and different systems such as the single-set, heavy to light and light to heavy, super-set, and split-routine. Isotonic training using eccentric contractions and performed at the end of a fatiguing concentric set is also often used as an additional resistance training approach. The purported benefits of these specific exercises and resistance training programs vary from building muscle bulk and definition (such as needed by a bodybuilder) to the development of greater strength and power. However, these specific responses have little relation to improvement of skeletal muscle fitness for healthy and chronically diseased individuals. These populations will improve sufficiently with a standard resistance training program as suggested by the ACSM (2).

Circuit programs have become a popular method to incorporate a cardiovascular stimulus during a resistance training program (24, 43). These "circuits" are a series of resistance and aerobic training stations that are performed for a certain time period (e.g., 20 s to 1 min). An individual moves from station to station after a timed rest period. However,

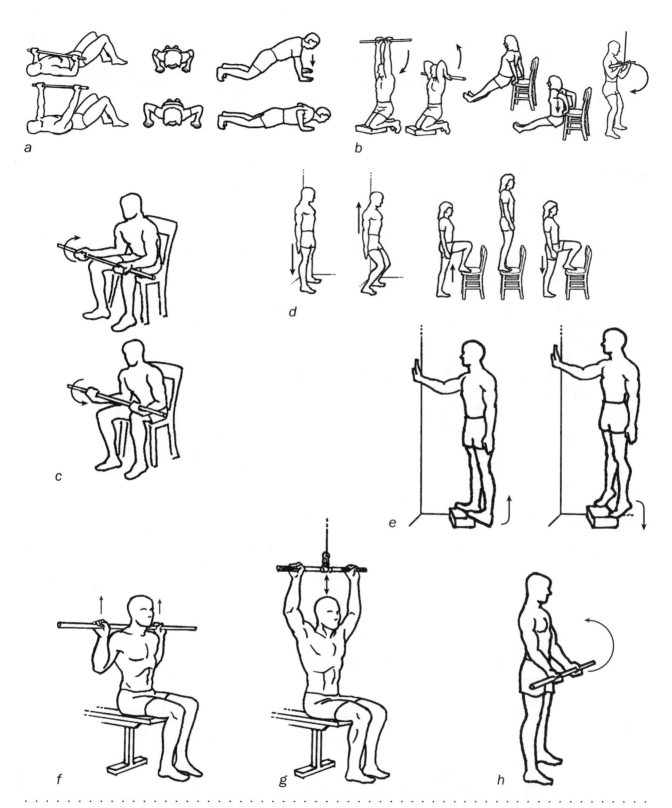

FIGURE 7.2 Examples of specific resistance training exercises *(a)* bench press: keep knees bent, do not arch back, hands slightly wider than shoulders, consider alternative exercise; *(b)* elbow extensor: do not arch back, tighten buttocks or wear a weight belt to support back, consider alternative triceps exercise; *(c)* double wrist curl: support forearms on thighs; *(d)* half squat: support back on wall, place feet away from wall so that knees do not move past the ankles, slide down wall slightly and back up, consider alternative exercise; *(e)* toe raises: can perform on stairs, use hands for balance; *(f)* shoulder press: keep back straight, hands wider than shoulders; *(g)* pull downs: keep back straight, hands wider than shoulders; *(h)* arm curls: do not arch back, consider variations supporting elbows.

although impressive strength gains have been reported, only modest cardiorespiratory benefits result, or even none at all (23, 30). Another investigation noted that similar strength gains were observed in subjects who specifically performed resistance training compared with those who also added endurance training (9).

Specificity of Training

Resistance training can be performed to provide general benefits or specific adaptations. For specificity, the lifting routine should closely resemble the desired movements in which gains in muscular fitness are desired. However, because the components of skeletal muscle fitness are related, any one type of resistance training program will provide benefits in each area of muscular fitness (i.e., strength, power, endurance). Subject evaluation should focus on the person's desired benefits. For example, individual goals might include improvements related to easier lifting for daily activities or for reduced fatigue when performing a leisure activity such as golf or gardening. Thus, the resistance training program for most individuals should be designed in a general fashion to promote overall muscular fitness improvement. Static (i.e., isometric) and dynamic (i.e., isotonic and isokinetic) resistance exercises have unique advantages and limitations (table 7.5).

Progressive Overload

The ACSM recommends the performance of resistance exercise training for 8 to 12 repetitions per set to maximally improve both skeletal muscle strength and endurance (7). In general, reducing repetitions (i.e., <8), because of increased resistance or load, leads to improvements in strength via fiber hypertrophy. Increased repetitions (i.e., >12) with a lower resistance will improve skeletal muscle endurance more so than strength. However, as long as the total volume of work during training for skeletal muscle strength or endurance is above that typically encountered during daily activities, regardless of the load or number of repetitions per set, improvement will likely occur. The greater the overload, typically the greater the improvement. However, as with endurance training, an overload can lead to **maladaptation,** which increases the risk of overtraining effects and skeletal muscle injury.

Frequency

Each type of resistance routine has specific guidelines with respect to intensity, frequency, and duration. Most studies have reported optimal gains when subjects perform resistance training from 1 to 3 days per week (17, 22, 27). The ACSM recommends performing a general or circuit resistance training program, designed to develop general muscular fitness, on at least 2 or 3 days per week (7). There is little evidence that substantial gains can be realized from performing resistance exercise on more than 3 days per week.

Intensity and Duration

The intensity of resistance training is also important in determining the load or overload placed on the skeletal muscle system to produce adaptation.

TABLE 7.5

Advantages and Limitations of Resistance Training

Type of exercise	Advantages	Limitations
Isometric	Little or no equipment required Useful for rehabilitation of injuries	Limited ROM effects Takes longer time to get full ROM effect
Isotonic	Large ROM effects Transfers to daily/athletic activities Highly motivating	Requires equipment Can be costly
Isokinetic	Specific to desired ROM and speed Possibly best method for improvement	Extremely expensive equipment

Note. ROM = range of motion.

Lifting a resistance that represents a very low percentage of a skeletal muscle group's ability, for 8 to 12 repetitions, will likely not place a sufficient overload on the system to generate an adaptive stimulus. For maximal strength and endurance improvement, the resistance should be at an individual's 8- to 12-**repetition maximum** (RM) so that the individual is at or near maximal exertion at the end of the repetitions (however, this may not be appropriate for some clinical conditions—see specific recommendations in chapters 8-31). The RM can be determined using either a direct or indirect method, as described in practical application 7.4.

The total training load placed on the skeletal muscle system is a combination of the number of repetitions performed per set and the number of sets per resistance exercise. The ACSM recommends between one and three sets per exercise (7). Several well-conducted studies indicate little benefit of performing resistance training for more than one set per resistance exercise (20). An additional benefit of performing only one set per exercise is a reduced total time commitment for resistance training. For instance, if an individual performs two or three sets for each muscle group exercise, he or she may spend over 60 min doing resistance training. Pollock (50) reported that the dropout rate from resistance training is higher if a training session requires more than 60 min. If more than one set is performed per exercise, it may be prudent to keep the between-set period to a minimum to reduce the total exercise time. Generally, no more than 1 min of rest is required between sets.

Reversibility

Maximal adaptations in skeletal muscle strength and endurance can be expected within 8 to 12 weeks (29, 50). The anticipated mean improvement in strength is about 25% to 30% (21). As with aerobic training, reductions in total resistance training volume without complete cessation of training allow for the maintenance of much of the gained resistance training effects. For instance, Graves et al. (27) reduced resistance training intensity by one third to one half and demonstrated maintenance of

---

PRACTICAL APPLICATION 7.4

Performing the Direct and Indirect 1RM

Direct

The direct 1RM method is performed by beginning with three or four repetitions at an estimated 40% to 60% of the estimated 1RM. After a short rest (2-3 min), this is repeated with 60% to 80% of the estimated 1RM. With recovery periods of no more than 5 to 10 min, the individual then lifts an estimated 95% of the 1RM, one time only, and continues the process by adding small load increments until the weight cannot be lifted. The weight of the last successful lift is the 1RM value. Note: The direct 1RM test is not considered to increase the risk of an untoward cardiac or blood pressure response in cardiac patients (13, 25, 59).

Indirect

A modified version of the direct 1RM assessment allows for the prediction of the 1RM by using resistances between 2RM and 20RM (1). When one is using the indirect 1RM method, it is assumed that for each time a weight is lifted, the %RM is reduced by approximately 2.5%. Thus, a load lifted only once represents 100% of the 1RM, whereas the lift of a weight five times, without the ability to lift it a sixth time, represents about 90% of the 1RM (i.e., a reduction of 2.5% for each successful lift beyond 1, so that $4 \times 2.5\% = 10\%$ and $100\% - 10\% = 90\%$). From this, the 1RM can be estimated from the following:

$$\text{weight lifted} / [1.0 - (\text{number of lifts} \times 0.025)] = 1\text{ RM}$$

The indirect 1RM method is believed to reduce the risk of an abnormal response such as orthopedic injury, hypertension, arrhythmia, and left ventricular dysfunction in some clinical populations. Other methods for determining the proper training resistance intensity include beginning with the lowest possible weight and starting the resistance at 30% to 50% of the 1, 2, or 3RM resistance (65).

strength gains for up to 12 weeks. Complete loss of training adaptations will occur after several weeks to months of inactivity.

Flexibility Training

Flexibility is the ability to move a joint throughout a full capable range of motion (ROM). Poor flexibility is associated with the following:

— An increased risk of lower back pain (acute and chronic)

— An increased risk of muscle and/or tendon injury (34, 35, 68)

— A reduced ability to perform activities of daily living and tasks requiring high-intensity muscular exertion (67, 68)

— Reduced postural stability, which increases the risk of falling (32)

The determination of joint ROM is important to establish a baseline prior to initiating a training program. This information can be used to determine the relative ROM of a joint compared with normative data and to uncover imbalances. This will provide a focus for a training program. Several devices are used to assess ROM. These include a goniometer, which is a protractor device; the Leighton flexometer, which is strapped to a limb and reveals the ROM in degrees as the limb moves around its joint; and the sit-and-reach box, which assesses the ability to forward flex the torso while in a seated position.

The sit-and-reach test assesses the limitation in torso forward flexion ROM attributable to poor lower back, hamstring, and calf flexibility. Poor ROM from this assessment is related to poor performance, injury, and low back pain. An excellent review of the methodology of the sit-and-reach test is provided by Adams (1). Other commonly assessed joints include the shoulders, ankles, and knee. These are best assessed using the aforementioned goniometer or flexometer. In some clinical disease states, flexibility may be reduced as the course of the disease progresses (e.g., multiple sclerosis, osteoporosis, obesity). These populations would benefit from regular ROM assessment and an exercise training program designed to enhance flexibility.

A well rounded flexibility program will focus on all of the major joints of the body:

— Neck

— Shoulders

— Upper trunk

— Lower trunk and back

— Hips

— Knees

— Ankles

An important consideration with respect to the mode of stretching used in a routine to enhance flexibility is the goal of the individual. An athletic goal commonly requires a different type of flexibility training than does a goal of reducing chronic pain (e.g., lower back) or allowing for independent living. The following are brief descriptions of the three primary stretching modes of flexibility training.

1. Static: A stretch of the muscles surrounding a joint that is held without movement for a period of time (e.g., 10-30 s) and may be repeated several times.

2. Ballistic: A method of rapidly moving a muscle to stretch and relaxation very quickly for several repetitions.

3. Proprioceptive neuromuscular facilitation (PNF): A method whereby a muscle is isometrically contracted, relaxed, and subsequently stretched. The theory is that the contraction activates the **muscle spindle receptors** or **Golgi tendon organs,** which results in a reflex relaxation (i.e., inhibition of contraction) in either the **agonist** or **antagonist** muscle. There are two types of PNF stretching (14, 18).

   — Contract–relax is when a muscle is contracted and then relaxed and passively stretched. Enhanced relaxation is theoretically produced via the muscle spindle reflex.

   — Contract–relax with agonist contraction occurs initially in the same manner as contract–relax, but during the static stretch the opposing muscle is contracted. This theoretically induces more relaxation in the stretched muscle via a reflex of the Golgi tendon organs.

Static and ballistic types of stretching are simple to perform and require only basic instruction. PNF stretching is somewhat complex and may require close supervision by a clinical exercise professional, especially when individuals are initially learning the technique. Any of the stretching techniques can be performed alone or with a partner. Additionally, devices such as towels, balls, ropes, resistance equipment, stairs, walls, and many other items can be used to assist or enhance a stretch. Once again, in a clinical population learning a stretching routine, using these items may initially require close supervision by a clinical exercise professional.

Static stretching is typically believed to be the safest method for enhancing the range of motion of a joint. Both ballistic and PNF stretching may increase the risk of delayed-onset muscle soreness and muscle fiber injury. PNF is possibly the most effective of the methods of stretching at improving joint ROM (7, 19, 53). A review of the physiological adaptations that occur as a result of a flexibility training program is beyond the scope of this chapter. The ACSM provides a comprehensive review of these adaptations (7).

Specificity of Training

Flexibility of a joint and the increase in range of motion from a flexibility training program are specific to a joint. This means that good flexibility in one joint does not necessarily mean good flexibility in other joints. The flexibility of a joint depends on the joint structure, the surrounding muscles and tendons, and the use of that joint for activities. A joint used during physical activity, especially if it requires a good ROM, will typically demonstrate good flexibility.

Progressive Overload

Flexibility routines should be performed as often as possible. As ROM increases, an individual should enhance the stretch to a comfortable degree. This will allow for optimal increases in ROM.

Frequency

It is generally recommended that a stretching routine using any of the previously described methods be performed a minimum 2 to 3 days per week (7). However, as stated previously, daily stretching is advised for optimal improvement in ROM.

Intensity and Duration

These routines should provide a full-body stretch allowing for increased focus on specific joints, especially the lower back and thighs (2). As previously stated, static stretches should be held for 10 to 30 s. PNF stretches should also be held for 10 to 30 s following a 6-s contraction period. These recommendations are for a general population. Each stretch should also be performed for three to five repetitions and to a point of only mild discomfort or "feeling of stretch."

The chapters of this text focusing on clinical populations provide specific modifications of these recommendations that reflect enhanced safety and effectiveness of the stretching routine. Table 7.6 and figure 7.3 provide many of the common maneuvers

---

TABLE 7.6

Selected Flexibility Training Exercises and Subsequent General Skeletal Muscle Adaptations

Flexibility maneuver	Affected joints
Sit and reach	Lower back, hamstring, calf region
Seated toe reach	Calf region
Supine trunk twist	Lower spine
Neck flexion, extension, rotation	Neck
Arm rotations	Shoulder girdle
Knees to chest hug	Lower spine
Back flexion, extension	Upper and middle back
Ankle to buttocks	Quadriceps
Overhead bent-arm elbow push	Triceps

continued

FIGURE 7.3   Examples of specific range of motion exercises, *(a)* thighs, *(b)* lower back, *(c)* abdominals, *(d)* arms, shoulders, and neck, *(e)* hamstrings, calves, and ankles.

used in a comprehensive flexibility training program.

Reversibility

Few data exist on the rate of loss of ROM. Many factors are involved in the rate of loss of ROM: injury, specific individual physiology, degree of overall in-

activity, and posture. The reintroduction of a flexibility training routine should result in rapid improvements in ROM.

Conclusion

Any type of training routine, whether it is cardiorespiratory conditioning, resistance training, or

FIGURE 7.3 *(continued)*

ROM training, should follow the FIT principle to ensure an optimal rate of improvement and safety during training. This chapter provides the general principles of exercise training. When one is working with specific clinical populations, modifications of these general principles may be necessary to ensure a safe and effective training routine. Please refer to chapters 8 through 31 for clinical condition and population-specific exercise training guidelines.

Medical History

Ms. WB was a high school and college cross country runner. She is Caucasian, 40 years old, and the mother of two children, ages 8 and 6. Her children are in school and she works full-time. She wishes to begin a regular exercise training routine. She has an interest in improving all five areas of her health-related physical fitness. She does not smoke and has no known serious medical condition. Her body mass index is 31.4 and she has a goal of 50 lb weight loss (currently 64 in. tall and 180 lb). She has not run regularly in 15 years and only walks occasionally.

Diagnosis

Ms. WB is evaluated by her physician and cleared for participation in an exercise program. She is apparently healthy. With her doctor's permission she decides to take advantage of her local hospital's performance assessment program to establish a baseline fitness assessment from which to gain specific exercise training recommendations.

Exercise Test Results

A symptom-limited exercise test is performed and the following reported:

Protocol: ramp running
Rest HR: 76
Rest blood pressure: 136/88
Rest electrocardiogram: normal
Peak HR: 187
Peak blood pressure: 210/90
Exercise electrocardiogram: rare premature atrial contractions and premature ventricular contractions; 0.5 mm J-point depression with quickly up-sloping ST segments at peak exercise.
$\dot{V}O_2$peak: 32.7 ml $\cdot$ kg^{-1} $\cdot$ min^{-1}
$\dot{V}O_2$ at anaerobic threshold: 25.6 ml $\cdot$ kg^{-1} $\cdot$ min^{-1}
Peak exercise pace: 8.2 mph
Pace at anaerobic threshold: 6.3 mph

Other procedures

Body fat: 32%
Bench press 1-repetition maximum (1RM): 40 lb
Sit and reach: lowest quintile (poor)
Leg press 1RM: 150 lb

Case Study 7 Discussion Questions

1. What would be an appropriate comprehensive exercise prescription for Ms. WB?

continued

2. Ms. SJ is 56 years old and healthy. She wants to begin her own exercise program. What are your general recommendations and concerns?

3. Mr. BW performed a stress test on a treadmill with the following results: stage 4 of Bruce protocol (4.2 mph, 16% grade), HRmax = 190, maximum blood pressure = 200/70, $\dot{V}O_2$peak = 43.2 ml · min^{-1} · kg^{-1}, no symptoms, stopped because of leg fatigue. His resting HR and blood pressure were normal. What specific intensity recommendations can you make for him with regard to beginning a jogging program? Use each of the HR based methods discussed in this chapter (see feature box 7.2).

Glossary

agonist—Denoting a muscle in a state of contraction, with reference to its opposing muscle, or antagonist.

antagonist—Something opposing or resisting the action of another; certain structures, agents, diseases, or physiological processes that tend to neutralize or impede the action or effect of others.

cross-training—The concept of training in another mode that allows for the development of physiology that will have a carryover effect to another mode; for example, resistance training is often performed to develop sport-specific strength.

Golgi tendon organ—A proprioceptive sensory nerve ending embedded among the fibers of a tendon, often near the musculotendinous junction; it is compressed and activated by any increase of the tendon's tension, caused either by active contraction or passive stretch of the corresponding muscle.

isokinetic—Denoting the condition when muscle fibers shorten at a constant speed in such a manner that the tension developed may be maximal over the full range of joint motion.

isometric—Denoting the condition when the ends of a contracting muscle are held fixed so that contraction produces increased tension at a constant overall length.

isotonic—Denoting the condition in which muscle fibers shorten with varying tension as the result of a constant load.

Joule (J)—A unit of energy; the heat generated, or energy expended, by an ampere flowing through an ohm for 1 s; equal to 107 erg and to a Newton-meter. It is an approved multiple of the SI fundamental unit of energy, the erg, and is intended to replace the calorie (4.184 J).

kilocalorie (kcal)—A unit of heat content or energy. The amount of heat necessary to raise 1 g of water from 14.5° C to 15.5° C, times 1000.

maladaptation—Adaptation to a progressive stimulus (e.g., exercise) that results in an overload to the system to the degree that performance is reduced and the risk of injury is increased.

metabolic equivalent—The equivalent of 3.5 mL · kg^{-1} · min^{-1} of oxygen consumption.

muscle spindle receptors—A fusiform end organ in skeletal muscle in which afferent and a few efferent nerve fibers terminate; this sensory end organ is particularly sensitive to passive stretch of the muscle in which it is enclosed.

range of motion—The total degrees of movement that a joint can move through.

repetition maximum (RM)—The number of times that a weight can be lifted; 1RM is the maximal amount of weight that can be lifted one time only.

true maximal heart rate—The highest heart rate as measured at, or near, the end of an exercise test; as opposed to peak heart rate estimated by the equation 220 – age.

References

1. Adams GM. *Exercise Physiology: Laboratory Manual*. 3rd ed. Boston, WCB McGraw-Hill, 1998.

2. American College of Sports Medicine. *ACSM's Guidelines for Exercise Testing and Prescription*. 6th ed. Baltimore, Lippincott Williams & Wilkins, 2000.

3. American College of Sports Medicine. Exercise for patients with coronary artery disease. *Med Sci Sports Exerc* 26(3): i-v, 1994.

4. American College of Sports Medicine. Position stand. Exercise and physical activity for older adults. *Med Sci Sports Exerc* 30(6): 992-1008, 1998.

5. American College of Sports Medicine. Position stand. Exercise and type 2 diabetes. *Med Sci Sports Exerc* 32: 1345-1362, 2000.

6. American College of Sports Medicine. Position stand. Physical activity, physical fitness, and hypertension. *Med Sci Sports Exerc* 25(10): i-x, 1993.

7. American College of Sports Medicine. Position stand. The recommended quantity and quality of exercise for developing and maintaining cardiorespiratory and muscular fitness, and flexibility in healthy adults. *Med Sci Sports Exerc* 30(6): 975-991, 1998.

8. Bartels R, Billings CE, Fox EL, Mathews DK, O'Brien R, Tanz D, Webb W. AAHPER Con:123, 1968.

9. Bell GJ, Petersen SR, Wessel J, Bagnall K, Quinnery HA. Physiological adaptations to concurrent endurance

training and low velocity resistance training. *Int J Sports Med* 12: 384-390, 1991.

10. Blair SN, Kohl HW III, Paffenbarger RS Jr, Clark DG, Cooper KH, Gibbons LW. Physical fitness and all-cause mortality: A prospective study of healthy men and women. *JAMA* 262: 2395-2401, 1989.

11. Brawner CA, Keteyian SJ, Ehrman JK. The relationship of heart rate reserve to $\dot{V}O_2$ reserve in patients with heart disease. *Med Sci Sports Exerc* 34(3): 418-422, 2002.

12. Brawner CA, Keteyian SJ, Ehrman JK. Predicting maximum heart rate in patients with heart disease: Influence of beta-adrenergic blockade therapy. *Med Sci Sports Exerc* 34(5): s269, 2002.

13. Butler RM, Belerwalters WH, Rodger FJ. The cardiovascular response to circuit weight training in patients with cardiac disease. *J Cardiac Rehabil* 7: 402-409, 1987.

14. Cornelius WL, Craft-Hamm K. Proprioceptive neuromuscular facilitation flexibility techniques: Acute effects on arterial blood pressure. *Physician Sportsmed* 16(4): 152-161, 1988.

15. Coyle EF, Martin WH, Sinacore DR, Joyner MJ, Hagberg JM, Holloszy JO. Time course of loss of adaptations after stopping prolonged intense endurance training. *J Appl Physiol* 57(6): 1857-1864, 1984.

16. Crews TR, Roberts JA. Effects of interaction of frequency and intensity of training. *Res Q* 47: 48-55, 1976.

17. DeMichele PD, Pollock ML, Graves JE, et al. Isometric torso rotation strength: Effect of training frequency on its development. *Arch Physiol Med Rehabil* 78: 64-69, 1997.

18. Etnyre BR, Abraham LD. Antagonist muscle activity during stretching: A paradox re-assessed. *Med Sci Sports Exerc* 20: 285-289, 1988.

19. Etnyre BR, Lee JA. Chronic and acute flexibility of men and women using three different stretching techniques. *Res Q Exerc Sport* 59: 222-228, 1988.

20. Feigenbaum MS, Pollock ML. Strength training: Rationale for current guidelines for adult fitness programs. *Physician Sportsmed* 25: 44-64, 1997.

21. Fleck SJ, Kraemer WJ. Designing Resistance Training Programs. 2nd ed. Champaign IL, Human Kinetics, 1997.

22. Frontera WR, Meredith CN, O'Reilly KP, Evans WJ. Strength conditioning in older men: Skeletal muscle hypertrophy and improved function. *J Appl Physiol* 68: 1038-1044, 1988.

23. Gettman LR, Pollock ML, Durstine JL, Ward A, Ayres J, Linnerud AC. Physiological responses of men to 1, 3, and 5 days per week training program. *Res Q* 47: 638-646, 1976.

24. Gettman LR, Pollock ML. Circuit weight training: A critical review of its physiological benefits. *Physician Sportsmed* 9: 44-60, 1981.

25. Ghilarducci LE, Holly RG, Amsterdam EA. Effects of high resistance training in coronary artery disease. *Am J Cardiol* 64: 866-870, 1989.

26. Graves JE, Pollock ML, Jones AE, et al. Specificity of limited range of motion variable resistance training. *Med Sci Sports Exerc* 21: 84-89, 1989.

27. Graves JE, Pollock ML, Leggett SH, Braith RW, Carpenter DM, Bishop LE. Effect of reduced training frequency on muscular strength. *Int J Sports Med* 9: 316-319, 1988.

28. Hermansen L, Saltin B. Oxygen uptake during maximal treadmill and bicycle exercise. *J Appl Physiol* 26: 31-37, 1969.

29. Hickson RC, Rosenkoetter MA, Brown MM. Strength training effects on aerobic power and short-term endurance. *Med Sci Sports Exerc* 12: 336-339, 1980.

30. Hurley BF, Seals DR, Ehsani AA, Cartier LJ, Dalsky GP, Hagberg JM, Holloszy JO. Effects of high-intensity strength training on cardiovascular function. *Med Sci Sports Exerc* 16(5): 483-488, 1984.

31. Jackson J, Sharkey B, Johnston L. Cardiorespiratory adaptations to training at specified frequencies. *Res Q* 39: 295-300, 1968.

32. Judge JO, Lindsey C, Underwood M, Winsemius D. Balance improvements in older women: Effects of exercise training. *Phys Ther* 73: 254-265, 1993.

33. Karvonen M, Kentala K, Mustala O. The effects of training heart rate: A longitudinal study. *Ann Med Exp Biol Fenn* 35: 307-315, 1957.

34. Kibler WB, Goldberg C, Chandler TJ. Functional biomechanical deficits in running athletes with plantar fasciitis. *Am J Sports Med* 266: 185-196, 1991.

35. Leach RE, James S, Wasilewski S. Achilles tendinitis. *Am J Sports Med* 9: 93-98, 1981.

36. Lee IM, Hsieh C, Paffenbarger RS. Exercise intensity and longevity in men. *JAMA* 273: 1179-1184, 1995.

37. Lieber DC, Liever RL, Adams WC. Effects of run-training and swim-training at similar absolute intensities on treadmill $\dot{V}O_2$ max. *Med Sci Sports Exerc* 21: 655-661, 1989.

38. McArdle WD, Magel JR, Delio DJ, Toner M, Chase JM. Specificity of run training on VO_2 max and heart rate changes during running and swimming. *Med Sci Sports Exerc* 10: 16-20, 1978.

39. McArdle WD, Magel JR. Physical work capacity and maximum oxygen uptake in treadmill and bicycle exercise. *Med Sci Sports* 2: 118-123, 1970.

40. McKay GA, Banister EW. A comparison of maximum oxygen uptake determination by bicycle ergometry of various pedaling frequencies and by treadmill running at various speeds. *Eur J Appl Physiol* 35: 191-200, 1976.

41. McMurray RG, Ainsworth BE, Harrell JS, Griggs TR, Williams OD. Is physical activity or aerobic power more influential on reducing cardiovascular disease risk factors? *Med Sci Sports Exerc* 30(10): 1521-1529, 1998.

42. Miyamura M, Kitamura K, Yamads A, Matsui H. Cardiorespiratory responses to maximal treadmill and bicycle exercise in trained and untrained subjects. *J Sports Med Phys Fitness* 18:25-32, 1978.

43. Morgan RE, Adamson GT. *Circuit weight training*. London, G. Bell and Sons, 1961.

44. Olree HD, Corbin DB, Penrod J, Smith C. *Methods of Achieving and Maintaining Physical Fitness for Prolonged Space Flight. Final progress report to NASA*. Grant no. NGR-04-002-004, 1969.

45. Paffenbarger RS, Hyde RT, Wing AL, Hsieh CC. Physical activity, all-cause mortality, and longevity of college alumni. *N Engl J Med* 314: 605-613, 1986.

46. Paffenbarger RS, Hyde RT, Wing AL, Lee IM, Jung DL, Kampert JB. The association of changes in physical-activity level and other lifestyle characteristics with mortality among men. *N Engl J Med* 328: 538-545, 1993.

47. Pate RR, Pratt M, Blair SN, et al. Physical activity and public health: A recommendation from the Centers for Disease Control and Prevention and the American College of Sports Medicine. *JAMA* 273: 402-407, 1995.

48. Pollock ML, Cureton TK, Greninger L. Effects of frequency of training on working capacity, cardiovascular function, and body composition of adult men. *Med Sci Sports* 1: 70-74, 1969.

49. Pollock ML, Dimmick LJ, Miller HS, Kendrick Z, Linnerud AC. Effects of mode of training on cardiovascular function and body composition of middle-aged men. *Med Sci Sports* 7: 139-145, 1975.

50. Pollock ML, Leggett SH, Graves JE, Jones A, Fulton M, Cirulli J. Effect of resistance training on lumbar extension strength. *Am J Sports Med* 17: 624-629, 1989.

51. Pollock ML, Miller H, Janeway R, Linnerud AC, Robertson B, Valentino R. Effects of walking on body composition and cardiovascular function of middle-aged men. *J Appl Physiol* 30: 126-130, 1971.

52. Prochaska JO, Diclemente CC, Norcross JC. In search of how people change. *Am Psychol* 47: 1102-1114, 1992.

53. Sady SP, Wortman M, Blanke D. Flexibility training: Ballistic, static or proprioceptive neuromuscular facilitation? *Arch Phys Med Rehabil* 63: 261-263, 1982.

54. Saltin B, Blomqvist G, Mitchell JH, Johnson RL, Wildenthal K, Chapman CB. Response to exercise after bed rest and after training. *Circulation* 38(Suppl. VII): VII-1–VII-55, 1968.

55. Seyle H. *The Stress of Life*. New York, McGraw-Hill, 1976.

56. Sharkey BJ. Intensity and duration of training and the development of cardiorespiratory endurance. *Med Sci Sports* 2: 197-202, 1970.

57. Sidney KH, et al. *Training: Scientific basis and application*. Taylor AW (Ed.). Springfield, IL, Charles C Thomas, 1972.

58. Siscovick DS, Weiss NS, Fletcher RH, Lasky T. The incidence of primary cardiac arrest during vigorous exercise. *N Engl J Med* 311: 874-877, 1984.

59. Squires RW, Muri AJ, Anderson LJ, et al. Weight training during phase II (early outpatient) cardiac rehabilitation: Heart rate and blood pressure responses. *J Cardiac Rehabil* 11: 360-364, 1991.

60. Stanish WD. Overuse injuries in athletes: A perspective. *Med Sci Sports Exerc* 16: 1, 1984.

61. Swain D, Leutholtz B, King M, Haas L, Branch J. Relationship between % heart rate reserve and % $\dot{V}O_2$ reserve in treadmill exercise. *Med Sci Sports Exerc* 30: 318-321, 1998.

62. Swain D, Leutholtz B. Heart rate reserve is equivalent to %$\dot{V}O_2$ reserve, not to %$\dot{V}O_2$max. *Med Sci Sports Exerc* 29: 410-414, 1997.

63. U.S. Department of Health and Human Services. *Physical Activity and Health: A report of the Surgeon General*. Atlanta, GA, U.S. Department of Health and Human Services, Centers for Disease Control and Prevention, National Center for Chronic Disease Prevention and Health Promotion, 1996.

64. Verstappen FTJ, Huppertz RM, Snoeckx LHEH. Effect of training specificity on maximal treadmill and bicycle ergometer exercise. *Int J Sports Med* 3:43-46, 1982.

65. Welsch MA, Pollock ML, Brechue WF, Graves JE. Using the exercise test to develop the exercise prescription in health and disease. *Primary Care* 21(3): 589-609, 1994.

66. Wilmore J, Davis JA, O'Brien RS, Vodak PA, Wolder G, Amsterdam EA. Physiological alterations consequent to 20-week conditioning programs of bicycling, tennis, and jogging. *Med Sci Sports Exerc* 12:1-8 1980.

67. Wilson GJ, Elliott BC, Wood GA. Stretch shorten cycle performance enhancement through flexibility training. *Med Sci Sports Exerc* 24: 116-123, 1992.

68. Worrell TW. Factors associated with hamstring injuries: An approach to treatment and preventative measures. *Sports Med* 17: 335-345, 1994.

Ann L. Albright, PhD, RD
University of California, San Francisco
California Department of Health Services
Sacramento, CA

Diabetes

Diabetes mellitus (diabetes) is a group of metabolic diseases characterized by an inability to sufficiently produce or properly use insulin, resulting in **hyperglycemia** (10). Insulin, a hormone produced by the β-cells of the **pancreas,** is needed by muscle, fat, and the liver to utilize glucose. The hyperglycemia resulting from diabetes places individuals with this disease at risk for developing **microvascular diseases** including retinopathy and nephropathy, **macrovascular disease,** and various neuropathies (both autonomic and peripheral).

Scope

Approximately 17 million people in the United States have diabetes. One-third of these are undiagnosed, in large part because symptoms may develop gradually and it can be years before severe symptoms appear (34). However, even before symptom development, these individuals are at increased risk for developing complications (14, 46, 69, 82, 106).

The prevalence rate for diagnosed diabetes increased from 0.37% in 1935 to 3.0% in 1994 (26, 60). Currently, 6% of the United States population has diabetes, with 1 million new cases of diabetes each year (27). This is estimated to double in the next 15 to 20 years and is a worldwide problem. The Centers for Disease Control and Prevention consider diabetes to be at epidemic proportions in the United States. The reasons for this are likely threefold:

1. An increasingly sedentary lifestyle
2. The increase in high-risk ethnic populations in the United States
3. Aging of the population

Diabetes is currently the seventh leading cause of death in the United States (27). African Americans, Hispanics/Latinos, American Indians/Alaskan Natives/Native Hawaiians/other Pacific Islanders, and Asians have higher rates diabetes than the non-Hispanic white population (27). Death rates of middle-aged people with diabetes are twice those of people without diabetes. More than 15% of those who develop diabetes under the age of 30 will die by the age of 40, which is 20 times the normal rate of death (89). As serious as these mortality statistics are, it is important to note that they underestimate the impact of diabetes. Because people with diabetes usually die from the complications of this disease, diabetes is underreported as the underlying or contributing cause of death and is listed on only half of the death certificates of those with the disease (15, 25).

The economic impact of diabetes is staggering. The estimated direct (cost of medical treatment and services) and indirect (cost of time lost from work) costs of diabetes are estimated to be between $98 and $105 billion per year (29, 34). A large portion of the economic burden of diabetes is attributable to the long-term complications, especially heart and end-stage renal disease.

Physiology and Pathophysiology

There are various forms of diabetes, and this affects the options for treatment. What all forms share is the risk for developing complications. This section reviews the types of diabetes and associated complications.

Diabetes Categories

The American Diabetes Association Expert Committee on the Diagnosis and Classification of Diabetes recognizes four categories of diabetes (figure 8.1).

Type 1 Diabetes

Type 1 diabetes is caused by β-cell destruction resulting in an absolute deficiency of insulin (figure 8.1). Consequently, insulin must be supplied by regular

---

Type 1 diabetes[a]

β-cell destruction, usually leading to absolute insulin deficiency

Immune mediated

Idiopathic

Type 2 diabetes[a]

May range from predominant insulin resistance with relative insulin deficiency to predominant secretory defect with insulin resistance

Other specific types

Genetic defects of β-cell function

Genetic defects in insulin action

Diseases of the exocrine pancreas

Endocrinopathies

Drug or chemical induced

Infections

Uncommon forms of immune-mediated diabetes

Other genetic syndromes sometimes associated with diabetes

Gestational diabetes mellitus

Resulting from pregnancy

[a]Patients with any form of diabetes may require insulin treatment at some stage of their disease. Such use of insulin does not, of itself, classify the patient.

FIGURE 8.1   Etiologic classificatio of diabetes mellitus.

Reprinted, by permission, from American Diabetes Association, 1999, "Report of the expert committee on the diagnosis and classification of diabetes mellitus," *Diabetes Care* 26:S7.

---

injections or an insulin pump. Those with type 1 diabetes are prone to develop ketoacidosis when marked hyperglycemia occurs. Approximately 5% to 10% of those with diabetes have type 1 (10).

There are two subgroups within type 1diabetes: immune-mediated and **idiopathic.** Type 1 immune-mediated diabetes was formerly known as juvenile-onset or insulin-dependent diabetes. This form of the disease usually occurs before the age of 30 with most cases occurring in childhood or adolescence, and symptoms develop abruptly. Type 1 immune-mediated diabetes is considered an **autoimmune** disease in which the immune system attacks the body's own tissues. Type 1 idiopathic diabetes is a new subgroup and represents only a small number of people with β-cell destruction. It is now estimated that 10% to 20% of Caucasians developing diabetes in adulthood may have immune-mediated β-cell destruction in which the disease develops over several months to years (10, 119). These patients have variable insulin deficiency and only intermittently require insulin treatment.

Type 2 Diabetes

Type 2 diabetes was formerly called adult-onset or non-insulin-dependent diabetes. This is the most common form of the disease and affects approximately 90% to 95% of all those with diabetes (10). Its onset usually occurs after 40 years of age; however, type 2 diabetes is seen at increasing frequency in adolescents (28).

The pathophysiology of type 2 diabetes is complex and multifactorial. Insulin resistance of the peripheral tissues and defective insulin secretion are common features. With insulin resistance, the body cannot effectively use insulin in the muscle or liver even though sufficient insulin is being produced early in the course of the disease (31, 84, 90, 104). Type 2 diabetes is progressive, and, over time, the pancreas cannot increase insulin secretion enough to compensate for the insulin resistance, and hyperglycemia occurs (figure 8.2). The treatment options are **medical nutrition therapy** and exercise and, if medication is needed, oral agents or insulin. Ketoacidosis seldom occurs.

There is a clear genetic influence for type 2 diabetes. The risk in children of parents with type 2 diabetes is about two times greater than normal (64). Along with the genetic influences, there are other risk factors. Obesity significantly contributes to insulin resistance, and the majority (80%) of people with type 2 diabetes are overweight or obese at disease onset (24, 67). An abdominal distribution of body fat (e.g., male belt size >40 in. and female >35 in.) is associated with type 2 diabetes (61). The risk of developing type 2 diabetes also increases with age, lack of physical activity, history of **gestational diabetes,** and presence of hypertension or dyslipidemia (49, 118). The combination of hypertension, dyslipidemia, obesity, and diabetes is often termed the "metabolic syndrome."

Other Specific Types

The third category of diabetes is termed "other specific types" and accounts for only about 1% to 2% of all diagnosed cases of diabetes (10). In these cases, certain diseases, injuries, infections, medications, or genetic syndromes cause the diabetes. This form may or may not require insulin treatment.

Gestational Diabetes

Gestational diabetes occurs during 2% to 5% of pregnancies (10). It is usually diagnosed during pregnancy by an oral glucose tolerance test, often performed routinely during the second trimester.

Stages / Types	Normoglycemia	Hyperglycemia			
	Normal glucose regulation	Impaired glucose tolerance or impaired fasting glucose	Diabetes mellitus		
			Non insulin requiring	Insulin requiring for control	Insulin requiring for survival
Type 1[a]	◄———————————————————————————————————————►				
Type 2	◄—————————————————————————————————►				
Other specific types[b]	◄—————————————————————————————————►				
Gestational diabetes[b]	◄—————————————————————————————————►				

FIGURE 8.2 Disorders of glycemia.

[a]Even after presenting in ketoacidosis, these patients can briefly return to normal glycemia without requiring continuous therapy (i.e., "honeymoon" remission). [b]In rare instances, patients in these categories may require insulin for survival.

Reprinted, by permission, from American Diabetes Association, 1999, "Report of the expert committee on the diagnosis and classification of diabetes mellitus," *Diabetes Care* 26: S8.

Risk factors for the development of this form of diabetes include family history of gestational diabetes, previous delivery of a large birth-weight (>9 lb) baby, and obesity. Although glucose tolerance usually returns to normal after delivery, approximately 50% of the women who develop gestational diabetes will go on to develop type 2 diabetes within a 10-year period and should receive frequent long-term follow-up (8, 79).

Complications of Diabetes

Associated complications of diabetes are categorized as acute and chronic. This section reviews these complications.

Acute Complications

The acute complications of diabetes are hyperglycemia (high blood sugar) and **hypoglycemia** (low blood sugar). Each of these acute complications must be quickly identified to ensure proper treatment and reduce the risk of serious consequences.

Hyperglycemia The manifestations of hyperglycemia are as follows:

1. Diabetes out of control
2. Diabetic ketoacidosis
3. Hyperosmolar nonketotic syndrome (30)

"Diabetes out of control" is a term used to describe frequent blood glucose levels that are above the patient's **glycemic goals** (see following discussion and table 8.1). This causes the kidneys to excrete glucose and water, which causes increased urine production and dehydration. Symptoms of high blood glucose levels and dehydration are headache, weakness, and fatigue (30). The best treatment for a patient with "diabetes out of control" includes drinking plenty of non-carbohydrate-containing beverages, regular self-monitoring of blood glucose, and, when instructed by a healthcare professional, an increase in diabetes medication. Frequent high blood glucose levels damage target organs or tissues, which increases the risk of chronic complications.

Diabetic ketoacidosis occurs in patients whose diabetes is in poor control and in whom the amount of **effective insulin** is very low or absent. This is much more likely to occur in those with type 1 diabetes. **Ketones** form because without insulin, the

---

TABLE 8.1

Suggested Treatment Goals for Blood Glucose, Blood Pressure, and Lipids

Blood glucose–related biochemical indexes	Normal	Goal	Additional action suggested
Average preprandial glucose (mg · dl^{-1})[a]	<110	80-120	<80 >140
Average bedtime glucose (mg · dl^{-1})[a]	<120	100-140	<100 >160
Hemoglobin A$_{1c}$ (%)	<6	<7	>8

Blood lipid–related risk	LDL cholesterol (mg · dl^{-1})	HDL cholesterol[b] (mg · dl^{-1})	Triglyceride (mg · dl^{-1})
High	≥130	<35	≥400
Borderline	100-129	35-45	200-399
Low	<100	>45	<200

Note. Data are given in milligrams per deciliter. The primary goal of therapy for adults should be to decrease blood pressure to <130/80 mmHg. The values shown in this table are by necessity generalized to the entire population of individuals with diabetes. Patients with comorbid diseases, the very young and older adults, and others with unusual conditions or circumstances may warrant different treatment goals. These values are for nonpregnant adults. Hemoglobin A$_{1c}$ is referenced to a nondiabetic range of 4.0-6.0% (mean 5.0%, *SD* 0.5%).

[a]Measurement of capillary blood glucose. [b]For women, HDL cholesterol values should be increased by 10 mg · dl^{-1}.

Adapted, by permission, from American Diabetes Association, 1999, "Report of the expert committee on the diagnosis and classification of diabetes mellitus," *Diabetes Care* 26: S5-S20.

body cannot use glucose effectively and a high amount of fat metabolism occurs to provide necessary energy. A by-product of fat metabolism in the absence of adequate carbohydrate is ketone body formation by the liver, causing an increased risk of coma and death. Other symptoms include abdominal pain, nausea, vomiting, rapid or deep breathing, and sweet or fruity smelling breath. Exercise is contraindicated in anyone experiencing diabetic ketoacidosis.

Hyperglycemic hyperosmolar nonketotic syndrome occurs in patients with type 2 diabetes when hyperglycemia is profound and prolonged. This is most likely to happen during periods of illness or stress, in the elderly, or in those who are undiagnosed (30). The syndrome results in severe dehydration attributable to rising blood glucose levels resulting in excessive urination. Extreme dehydration eventually leads to decreased **mentation** and possible coma. Exercise is contraindicated during periods of hyperglycemic hyperosmolar nonketotic syndrome.

Hypoglycemia   Hypoglycemia (also called insulin shock and insulin reaction) is a potential side effect of diabetes treatment and usually occurs when blood glucose levels drop below 60 to 70 mg · dl^{-1}. This may occur in the presence of the following factors:

— Too much insulin or selected antidiabetic oral agents

— Too little carbohydrate intake

— Missed meals

— Exercise that is excessive or not planned appropriately (30)

Hypoglycemia can occur either during exercise or several hours later. The postexercise late-onset hypoglycemia generally occurs following moderate- to high-intensity exercise greater than 30 min duration. It results from increased insulin sensitivity, ongoing glucose utilization, and physiological replacement of glycogen stores (3). It is important that patients are instructed to monitor blood glucose before and periodically after exercise to assess glucose response. This is also recommended in clinical exercise programs, such as cardiac rehabilitation, especially in patients new to exercise. Specifics for pre- and postexercise blood glucose assessment are provided later in this chapter.

There are two categories of symptoms of hypoglycemia: autonomic and **neuroglycopenic.** As blood glucose decreases, **glucagon,** epinephrine, growth hormone, and cortisol are released to help increase circulating glucose. Autonomic symptoms such as shakiness, weakness, sweating, nervousness,

anxiety, tingling of the mouth and fingers, and hunger result from epinephrine release. As the blood glucose delivery to the brain decreases, neuroglycopenic symptoms such as headache, visual disturbances, mental dullness, confusion, amnesia, seizures, or coma may occur (30).

Some people with diabetes lose their ability to sense hypoglycemic symptoms (termed hypoglycemia unawareness). By instituting tight control of blood glucose, the threshold may be lowered so that symptoms do not occur until blood glucose drops quite low. Intensity of control may need to be slightly reduced to alleviate hypoglycemia unawareness. To the contrary, those patients who have been in poor control may sense low blood glucose symptoms at levels with much higher values than 60 to 70 mg · dl^{-1}.

Treatment of hypoglycemia consists of testing the blood glucose to confirm the hypoglycemia, and if the person is conscious, consumption of approximately 15 g of carbohydrate (e.g., glucose, sucrose, or lactose) that does not contain fat. Commercial products (glucose tablets) are available that allow a precise amount of carbohydrate to be eaten. Other sources include 1 cup nonfat milk, 2 tablespoons of raisins, one half can of regular soda, one half cup of orange juice, six or seven Life Savers, or one tablespoon of sugar, honey, or corn syrup. The person with diabetes should wait about 15-20 min to allow the symptoms to resolve. If necessary, another 15 g of carbohydrate should be consumed (43). If the patient becomes unconscious due to hypoglycemia, an injection of glucagon should be administered. If glucagon is not available, 911 should be called immediately.

Chronic Complications

Diabetes is the leading cause of adult-onset blindness, nontraumatic lower limb amputation, and end-stage renal failure (51). In addition, those with diabetes are at two to four times the normal risk of heart disease and stroke (27). The hyperglycemia of diabetes is considered of primary importance in the development of the chronic complications along with hypertension and hyperlipidemia. Tight blood glucose control can reduce the risk of developing diabetic complications in patients with either type 1 or 2 diabetes (32, 105). The clinical exercise professional who is involved in the exercise training of people with diabetes must obtain information about the presence and stage of complications. This information should be used when developing an exercise prescription and behavior modification plan designed to help those with diabetes reduce their risk of developing or amplifying the complications of the disease. Cardiac rehabilitation programs may be suitable for those high-risk patients wishing to incorporate an exercise program into their lifestyle.

The chronic complications are more clearly described when considered in three different categories:

1. Macrovascular (large vessel or atherosclerotic) disease, which includes coronary artery disease with or without angina, myocardial infarction, cerebrovascular accident, and peripheral arterial disease

2. Microvascular (small vessel) disease, which includes diabetic retinopathy (eye disease) and diabetic nephropathy (kidney disease)

3. Neuropathy that involves both the peripheral and autonomic nervous systems (30)

Macrovascular Disease   Diabetes is a risk factor for macrovascular disease. The vessels to the heart, brain, and lower extremities can be affected. Figure 8.3 illustrates the relationship of the insulin resistance syndrome (including hyperglycemia) to coronary artery disease. Blockage of the blood vessels in the legs results in peripheral artery disease, intermittent claudication, and exercise intolerance (7). Reduction and control of vascular risk factors are especially important in those with diabetes. The methods used for this are similar to those used for coronary heart disease. The myocardial infarction

chapter (chapter 12) in this text reviews the vascular risk factor control methods in detail. The symptoms of peripheral arterial disease can be improved with exercise training, as reviewed in chapter 17. It is important to note that the National Cholesterol Education Program (NCEP) Adult Treatment Panel III (ATP III) guidelines recommend that lipid and other risk factor treatment for those with diabetes be to the level of those with coronary artery disease (see chapter 10).

Microvascular Disease   Microvascular disease causes retinopathy and nephropathy, which cause abnormal function and damage to the small vessels of the eyes and kidneys, respectively. The ultimate result of retinopathy can be blindness, whereas end-stage renal failure is the most serious complication of nephropathy (see chapter 11). Prevention or appropriate management requires periodic (often yearly) dilated eye examinations and renal function tests, along with optimal blood glucose and blood pressure control. You must give careful attention to the stage of complications when prescribing exercise for those with microvascular involvement; this is discussed in detail in the exercise prescription section of this chapter.

Peripheral and Autonomic Neuropathy   Both peripheral and autonomic neuropathy have implications for exercise. **Peripheral neuropathy** typically affects the legs before the hands. Patients initially

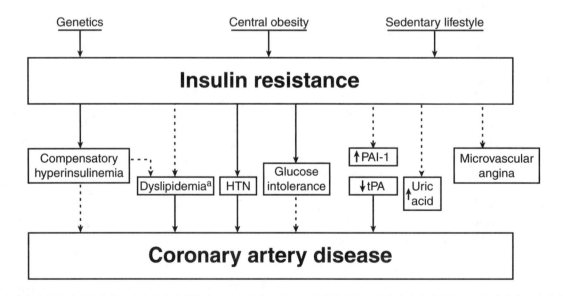

FIGURE 8.3   Components of Insulin Resistance Syndrome.

→Indicates probable or established causal relationship. ⇢Indicates association present but causal relationship not determined. ªDyslipidemia includes increased triglyceride levels, decreased HDL cholesterol levels, small dense LDL particles, and large postprandial triglyceride-rich lipoprotein particles. HTN, hypertension; PAI-1, plasminogen activator inhibitor, type 1; tPA, tissue plasminogen activator.

Reprinted, by permission, from M.B. Davidson, 1998, *Diabetes Mellitus Diagnosis and Treatment*, 4th ed. (Philadelphia: Saunders), 267-298.

experience sensory symptoms (paresthesia, burning sensations, and hyperesthesia) and loss of tendon reflexes. As the complication progresses, the feet become numb and the patients are at high risk for foot injuries because they have difficulty realizing when they are injured. Muscle weakness and atrophy can also occur. This causes foot deformities, resulting in areas that receive increased pressure from shoe wear or foot strike, placing them at risk for injury (30). The large number of lower limb amputations from diabetes is the result of loss of sensation that places the patient at risk for injury and also from diminished circulation attributable to peripheral artery disease. This impairs healing and can lead to severe reductions in blood flow, potential gangrene, and amputation. Persons with diabetes must be given instruction on how to examine their feet and practice good foot care. This is especially important when someone with peripheral neuropathy begins an exercise program, because increased walking and cycle pedaling increase the risk of foot injury.

Diabetic autonomic neuropathy may occur in any system of the body (e.g., cardiovascular, respiratory, neuroendocrine, gastrointestinal). Many of these systems are integral to the ability to perform exercise (1). **Cardiovascular autonomic neuropathy** is manifested by a high resting heart rate, an attenuated exercise heart rate response, an abnormal blood pressure, and redistribution of blood flow response during exercise.

Clinical Considerations

In the clinical setting, laboratory tests and examinations are used to diagnose diabetes or for ongoing monitoring. The following sections review these.

Signs and Symptoms

The symptoms of diabetes include excessive thirst (**polydipsia**), frequent urination (**polyuria**), unexplained weight loss, infections and cuts that are slow to heal, blurry vision, and fatigue. Many who develop type 1 diabetes have some or all of these symptoms, but those with type 2 diabetes may remain asymptomatic. The fact that one third of the population with diabetes do not know they have the disease underscores the lack of symptoms experienced by many.

Diagnostic and Laboratory Evaluations

The American Diabetes Association recommends that all people over the age of 45 years be tested for

diabetes, which should be repeated at 3-year intervals (10). Testing should occur earlier or more frequently if someone has the following risk factors:

1. Is overweight or obese (>120% desirable body weight or a body mass index >27 kg $\cdot$ m^{-2})
2. Has a first-degree relative with diabetes
3. Is a member of a high-risk ethnic population
4. Delivered a baby weighing more than 9 lb or was diagnosed with gestational diabetes
5. Is hypertensive (>140/90 mmHg)
6. Has a high-density cholesterol less than or equal to 35 mg $\cdot$ dl^{-1} or a triglyceride of greater than or equal to 250 mg $\cdot$ dl^{-1}
7. Had an impaired fasting glucose or glucose tolerance test

Three criteria are used to diagnose diabetes (figure 8.4). In the absence of hyperglycemia with acute metabolic decompensation (i.e., ketoacidosis), these criteria should be confirmed on a subsequent day. The goal of lowering the value considered diagnostic for diabetes (i.e., 126 mg $\cdot$ dl^{-1}) is to treat those earlier who are at risk for microvascular complications.

Those found to meet the criteria for the diagnosis of diabetes should be told they have diabetes and

---

1. Symptoms of diabetes plus casual plasma glucose concentration ≥200 mg $\cdot$ dl^{-1} (11.1 mmol $\cdot$ L^{-1}). Casual is defined as any time of day without regard to time since last meal. The classic symptoms of diabetes include polyuria, polydipsia, and unexplained weight loss.

   or

2. Fasting plasma glucose ≥126 mg $\cdot$ dl^{-1} (7.0 mmol $\cdot$ L^{-1}). Fasting is defined as no caloric intake for at least 8 hr.

   or

3. Two-hour plasma glucose ≥200 mg $\cdot$ dl^{-1} (11.1 mmol $\cdot$ L^{-1}) during oral glucose tolerance test. The test should be performed as described by the World Health Organization (2), using a glucose load containing the equivalent of 75 g of anhydrous glucose dissolved in water.

FIGURE 8.4   Criteria for the diagnosis of diabetes mellitus.

Reprinted, by permission, from American Diabetes Association, 1999, "Report of the expert committee on the diagnosis and classification of diabetes mellitus," *Diabetes Care* 26:S12.

not "borderline" diabetes, because the patient may get the impression that the disease is not serious. The method of therapy used to treat diabetes should not be mistaken to determine the seriousness of the disease. Regardless of treatment, diabetes is a serious disease that requires diligent self-care and appropriate medical intervention. A fasting blood glucose ranging from 110 to 125 mg · dl^{-1} is considered a risk factor for developing diabetes and is termed impaired fasting glucose. When an oral glucose tolerance test is used, a 2-hr postload glucose between 140 mg · dl^{-1} and 199 mg · dl^{-1} is termed impaired glucose tolerance. Impaired fasting glucose and impaired glucose tolerance are now called "prediabetes." This group of patients should receive instruction and encouragement to lower their risk of developing diabetes, including beginning an exercise training program.

Treatment

There is currently no cure for diabetes. It must be managed with a program of exercise, medical nutrition therapy, self-blood glucose monitoring, diabetes self-management education, and, when needed, medication. The patient and his or her healthcare team must work together to develop a program to achieve individual treatment goals. Few diseases require the same level of ongoing daily patient involvement as required by diabetes. Because so much patient involvement is required in diabetes, it is extremely important that patients receive infor-

mation and training on disease management. Other members of the healthcare team may include the patient's primary care physician and/or an endocrinologist, a nurse, a diabetes educator, a registered dietitian, a clinical exercise professional, a behavioral/psychosocial counselor, and a pharmacist. In many instances, these healthcare professionals work together in a diabetes education program. The American Diabetes Association has developed standards for diabetes education programs (13).

The patient must understand and be involved in developing appropriate treatment goals. These are developed with consideration of the patient's desires, abilities, willingness, cultural background, and comorbidities. Suggested treatment goals for blood glucose, blood pressure, and plasma lipids from the American Diabetes Association are provided in table 8.1. There are **evidence-based** care guidelines such as regular hemoglobin A_{1C} testing, dilated eye exam, foot exam, blood pressure monitoring, lipid panel, renal function tests, smoking cessation counseling, flu/pneumococcal immunizations, and diabetes education (see figure 8.5 for example guidelines) that should be followed to help insure appropriate care. The patient should be educated about the purpose and importance of the medical tests and feedback on his or her results.

Medical nutrition therapy is essential to the management of diabetes and is often the most challenging aspect of therapy. Nutrition recommendations were developed by the American Diabetes Association (11). These guidelines promote individually

---

Physical and emotional assessment

— Blood pressure, weight (for children, add height; plot on growth chart)

Every visit. Blood pressure target goal < 130/80 mmHg (children: < 90th percentile age standard). Children: normal weight for height (see standard growth charts).

— Foot exam (for adults)

Thorough visual inspection every "diabetes visit"; pedal pulses, neurological exam annually.

— Dilated eye exams

Type 1 (insulin-dependent diabetes mellitus): 5 years post-diagnosis, then every year by a trained expert. Type 2 (non-insulin-dependent diabetes mellitus): shortly after diagnosis, then every year by a trained expert.

— Depression

Probe for emotional/physical factors linked to depression annually; treat aggressively with counseling, medication, and referral.

continued

FIGURE 8.5   Basic guidelines for diabetes care.

These materials have been produced through the collaborative efforts of the Diabetes Coalition of California and the California Diabetes Control Program©. These guidelines are consistent with ADA Clinical Practice Recommendations.

Lab exam

— HbA1c

Quarterly, if treatment changes or is not meeting goals; one to two times per year if stable. Target goal < 7.0% or < 1.0% above lab norms; children, modify if necessary to prevent significant hypoglycemia.

— Microalbuminuria (albumin/creatine ratio)

Type 1: five years post-diagnosis, then every year
Type 2: begin at diagnosis, then every year.

— Blood lipids (for adults)

On initial visit, then annually for adults. Target goals: cholesterol, triglycerides (mg · dl^{-1}) < 200; LDL < 130 unless CHD, then < 100; HDL > 45 for men, > 55 for women.

Self-management

— Management principles and complications

Initially, then annually: assess knowledge of diabetes, medications, self-monitoring, acute/chronic complications, and problem-solving skills. Each visit: screen for problems with barriers to self-care; help patient identify achievable self-care goals. Children: appropriate for developmental stage.

— Self glucose monitoring

Type 1: typically test four times a day. Type 2 and others: as needed to meet treatment goals.

— Medical nutrition therapy

Initial: assess needs, assist patient in setting nutritional goals. Follow-up: assess progress toward goals, identify problem areas; assessment should be done by a trained expert.

— Physical activity

Assess patient initially and in follow-up visits: prescribe physical activity based on patient's needs and condition.

— Weight management

Initially and in follow-up visits: must be individualized for patient.

Interventions

— Preconception counseling and management

Consult with high-risk perinatal programs where available (e.g., "Sweet Success" Regional Perinatal Programs of California). Adolescents: special counseling advisable, beginning with puberty.

— Pregnancy management

Consult with high-risk perinatal program where available.

— Aspirin therapy

81-325 mg · day^{-1} in adults as primary and secondary prevention of CHD, unless contraindicated.

— Smoking cessation

Initially, then annually: screen, advise, and assist.

— Vaccinations

Influenza and pneumococcal, per CDC recommendations.

— Dental exams

At least twice yearly.

Note. LDL = low-density lipoprotein; HDL = high-density lipoprotein; CHD = coronary heart disease; CDC = Centers for Disease Control and Prevention.

FIGURE 8.5 *(continued)*

developed dietary plans based on metabolic, nutrition, and lifestyle requirements in place of a calculated caloric prescription. This is because a single diet cannot adequately treat all types of diabetes or individuals.

Consideration must be given to each macronutrient (i.e., protein, fat, carbohydrate) when developing a nutrition plan for the person with diabetes. Protein intake should be approximately 10% to 20% of daily caloric intake, as no evidence exists that a lower or higher intake is of value.

Based on the risk of atherosclerosis, fat intake should be limited, with less than 10% from saturated fats and up to 10% from polyunsaturated fats. Cholesterol intake should be limited to 300 mg daily. Carbohydrate and monosaturated fat make up the remaining calories and need to be individualized based on glucose, lipid, and weight goals. The most common nutritional assumption about diabetes is that simple sugars should be avoided and replaced with starches. There is little evidence to support this assumption (17, 74). Priority should first be given to the total amount of carbohydrate consumed rather than the source, because all carbohydrates can raise blood glucose. Nutritional value must also be considered.

Self-monitoring of blood glucose is also an important part of managing diabetes. All people with diabetes should test their blood glucose regardless of whether they use insulin. There is no standard frequency for self-monitoring, but it should be performed frequently enough to help the patient meet treatment goals. Increased frequency of testing is often required when initiating an exercise program to assess blood glucose prior to and following exercise and allow for safe exercise participation. Patients must be given guidance on how to use the information to make exercise, food, and medication adjustments.

For those requiring medication (insulin and/or oral agents), it is necessary to understand how their medications work with food and exercise to ensure the greatest success and safety. The clinical exercise professional must also understand diabetes medications in order to safely prescribe exercise and provide guidance on exercise training to patients with diabetes. Refer to chapter 5 for specific information about diabetes medications.

Graded Exercise Testing

Most people with diabetes can benefit from participating in regular exercise. Participation in exercise is not without risk, however, and each individual should be assessed for safety (table 8.2). Priority

must be given to minimizing the potential adverse effects of exercise through appropriate screening, program design, monitoring, and patient education (7). Exercise testing may be viewed as a barrier or unnecessary for some. Discretion must be used to determine the need for exercise testing. The clinical exercise professional must be prepared to provide input to the physician to assist in this decision-making process. Practical application 8.1 provides information about client–clinician interaction.

Medical Evaluation

The American College of Sports Medicine recommends that those with known metabolic disease

---

TABLE 8.2
Potential Adverse Effects of Exercise in Patients With Diabetes
Cardiovascular
Cardiac dysfunction and arrhythmias attributable to ischemic heart disease (often silent)
Excessive increments in blood pressure during exercise
Postexercise orthostatic hypotension
Microvascular
Retinal hemorrhage
Increased proteinuria
Acceleration of microvascular lesions
Metabolic
Worsening of hyperglycemia and ketosis
Hypoglycemia in patients on insulin or insulin secretagogue oral therapy
Musculoskeletal and traumatic
Foot ulcers (especially in presence of neuropathy)
Orthopedic injury related to neuropathy
Accelerated degenerative joint disease
Eye injuries and retinal hemorrhage

Reprinted, by permission, from American Diabetes Association, 1993, "Exercise and NIDDM (Technical Review)," *Diabetes Care* 16: 54-58.

Client–Clinician Interaction

The interaction between the client and the clinician at the time of exercise evaluation, and especially during ongoing exercise training visits, is an important consideration. Living with diabetes poses many challenges and fears for the patient and his or her family. The exercise professional must be aware of the psychosocial components of living with a chronic disease and strategies that can be used to help the patient maintain participation in exercise. Some general considerations include the following:

— Treat the patient as an individual who is much more than his or her diagnosis. Be cautious about referring to the patient as a "diabetic," because this terminology labels the patient by the disease.

— Remember that it usually takes a great deal of effort and discipline to live with diabetes. Acknowledge that diabetes is challenging and listen to the patient's particular challenges.

In general, do not use terms like *noncompliance* when discussing an exercise program. Inherent in the definition of noncompliance is the concept that an individual is not following rules or regulations enforced by someone else. This concept is incongruent with self-management and patient empowerment, which consider the patient to be the key member of the healthcare team. It is not the healthcare professional's role to make decisions for the patient. Instead, the clinical exercise professional should equip the patient with information so the patient can make his or her own decisions. The following are some strategies for exercise maintenance:

— Ask the patient to consider the following questions: How easily can I engage in my activity of choice where I live? How suitable is the activity in terms of my physical attributes and lifestyle (1, 77)?

— Have the patient identify exercise benefits he or she finds personally motivating.

— Be sure that exercise goals are not too vague, ambitious, or distant (3).

— Establish a routine to help exercise become more habitual.

— Have the patient identify any social support systems he or she may have.

— Provide positive feedback to the patient.

---

have a medical exam and clinical exercise test prior to participation in moderate (40-60% $\dot{V}O_2max$) and vigorous intensity (>60% $\dot{V}O_2max$) exercise training (5). When reviewing the medical history of a patient with diabetes, it is important to consider the following:

1. The presence or absence of acute and chronic complications and, if chronic complications exist, the stage of complications

2. Laboratory values for hemoglobin A_{1C}, plasma glucose, lipids, and **proteinuria**

3. Blood pressure

4. Self-monitoring blood glucose results

5. Body weight and body mass index

6. Medication use and timing

7. Exercise history

8. Nutrition plan, particularly timing, amount, and type of most recent food intake

9. Other non-diabetes-related health issues

Contraindications

Contraindications for exercise testing are listed in chapter 6. A physician should be consulted prior to exercise testing if any absolute contraindication exists. Exercise testing may need to be postponed until it can be safely conducted.

Recommendations and Anticipated Responses

Age, duration of diabetes, and presence of complications should be considered before a patient begins an exercise test. In most cases, standard methods should be used (7) and physician supervision is recommended (5). Chapter 6 provides details on protocol selection.

Because of the high risk for cardiovascular disease in those with diabetes, low-level treadmill protocols or cycle or arm ergometer modes are recommended for the following groups:

1. Type 1 diabetes and over 30 years old or had diabetes longer than 15 years
2. Type 2 diabetes and over 35 years old
3. Type 1 or type 2 diabetes plus one or more other coronary artery disease risk factors
4. Suspected or known coronary artery disease
5. Any microvascular or neurological diabetic complications (7, 12)
6. Peripheral arterial disease or peripheral neuropathy

Exercise Prescription

Exercise is a vital component of diabetes management. It is considered a method of treatment for type 2 diabetes because it can improve insulin resistance. Although exercise is not considered a method of treating type 1 diabetes because of the absolute requirement for insulin, it is a very important part of a healthy lifestyle for these individuals.

Special Exercise Considerations

When one is developing an exercise prescription for persons with diabetes, it is important to consider the topic of fitness versus the health benefits of exercise. Methods to enhance maximal oxygen uptake are often extrapolated to the exercise prescription for disease prevention and management (7). However, changes in health status do not necessarily parallel increases in maximal oxygen uptake. In fact, evidence strongly suggests that regular participation in light- to moderate-intensity exercise may help prevent diseases such as coronary artery disease, hypertension, and type 2 diabetes but will not have an optimal effect on maximal oxygen uptake (4, 36, 50). Therefore, when frequency and duration are sufficient, exercise can be performed at an intensity below the threshold for an increase in maximal oxygen uptake and still be beneficial to health (47).

Exercise must be prescribed with careful consideration given to risks and benefits. The consequences of disuse combined with the complications of diabetes are likely to lead to more disability than the complications alone (1). Exercises that can be readily maintained at a constant intensity, and in which there is little interindividual variation in energy

---

PRACTICAL APPLICATION 8.2

Literature Review

Exercise has long been recognized as an important component of diabetes care (71). Benefits for diabetes are seen with both acute and chronic exercise. Acute bouts of exercise can improve blood glucose, particularly in those with type 2 diabetes (23, 72). The response of blood glucose to exercise is related to preexercise blood glucose level as well as the duration and intensity of exercise. Several studies in type 2 diabetes have demonstrated a reduction in blood glucose levels that is sustained into the postexercise period following mild to moderate exercise (55, 65, 81, 103). The reduction in blood glucose is attributed to an attenuation of hepatic glucose production with muscle glucose utilization increasing normally (22, 62, 81). The impact of acute exercise on blood glucose levels in those with type 1 diabetes and lean type 2 patients is more variable and unpredictable.

Physiological Adaptations

A rise in blood glucose with exercise can be seen in patients who are very insulin deficient (usually type 1) and with short-term, high-intensity exercise (19, 54, 62). Also, an elevation in blood glucose has been shown in type 2 diabetes, but this was with short-term, high-intensity exercise (82).

Most of the benefits of exercise for those with diabetes come from regular, long-term exercise. These benefits can include improvements in metabolic control (glucose control and insulin resistance), hypertension, lipids, body composition/weight loss, and psychological well-being.

Epidemiological evidence is accruing that strongly supports the role of exercise in the primary prevention or delay of type 2 diabetes (85). Early studies showing an increase in type 2 diabetes in societies that had abandoned traditional active lifestyles suggest a relationship between physical activity and diabetes (112). Several cross-sectional studies have shown that blood glucose and

continued

insulin values after an oral glucose tolerance test were significantly higher in less active, compared with more active, individuals (35, 40, 73, 87, 91, 110). Prospective studies of several different groups have also demonstrated that a sedentary lifestyle may play a role in the development of type 2 diabetes (45, 50, 75, 76, 88, 96). The strongest data in support of exercise in the prevention of type 2 diabetes come from a 6-year clinical trial in which subjects with impaired glucose tolerance were randomized into one of four groups: exercise only, diet only, diet plus exercise, or control group. The exercise group was encouraged to increase their daily physical activity to a level that was comparable to a 20-min brisk walk. The incidence of diabetes in the exercise intervention groups was significantly lower than in the control group (85). A randomized, multicenter clinical trial of type 2 diabetes prevention in those with impaired glucose tolerance at numerous sites around the United States found a 58% reduction in the incidence of type 2 diabetes with lifestyle intervention that had goals of 7% weight loss and 150 min of physical activity per week (33).

Like acute exercise, exercise training can improve blood glucose. Exercise training has been shown to improve glucose control as measured by hemoglobin A_{1C} and/or glucose tolerance, primarily in those with type 2 diabetes (52, 70, 92, 103). These studies used training programs ranging from 6 weeks to 12 months, and improved glucose tolerance was seen in as little as 7 consecutive days of training in subjects with early type 2 diabetes (94). Improvements in blood glucose deteriorate within 72 hr of the last bout of exercise, emphasizing the need for consistent exercise (99).

Following exercise training, insulin-mediated glucose disposal is improved. Insulin sensitivity of both skeletal muscle and adipose tissue can improve with or without a change in body composition (53, 66, 78). There are several mechanisms by which exercise may improve insulin sensitivity, including changes in body composition, muscle mass, capillary density, and glucose transporters in muscle (GLUT 4) (2). The effect of exercise on insulin action is lost within a few days, once again emphasizing the importance of consistent exercise participation (7).

The data supporting a positive effect of exercise on blood pressure come primarily from studies done in nondiabetic subjects. Two studies showed a reduction in blood pressure with exercise training in those with type 2 diabetes (69, 98). However, additional research is needed to better understand the impact of exercise training on blood pressure in diabetes. This information is necessary because essential hypertension is present in more than 60% of those with type 2 diabetes (9).

The information regarding the effect of exercise on lipids in diabetes shows primarily positive results, with most data available on type 2 diabetes. An increased aerobic capacity in people with type 2 diabetes is related to a less atherogenic lipid profile. Improvements have been demonstrated in triglycerides, total cholesterol, and the ratio between high-density lipoprotein cholesterol and total cholesterol following physical training in patients with type 2 diabetes (16, 18, 94, 95). Most of these studies included dietary modifications, so it is difficult to single out the effect of exercise.

Weight loss is often a therapeutic goal for those with type 2 diabetes, because 80% of those with type 2 are obese. Moderate weight loss has been shown to improve glucose control (111, 113) and decrease insulin resistance. Medical nutrition therapy and exercise combined are more effective than either alone in achieving moderate weight loss (93, 114, 115). Visceral or abdominal body fat is negatively associated with insulin sensitivity so that increased abdominal body fat decreases peripheral insulin sensitivity. This body fat is a significant source of free fatty acids and may be preferentially oxidized over glucose, contributing to hyperglycemia (20, 86). Exercise results in preferential mobilization of visceral body fat, likely contributing to the metabolic improvements (83, 116). Very importantly, exercise is one of the strongest predictors of success of long-term weight control (63). This is an extremely important feature of exercise because there is such high regain of weight loss.

Psychological Benefits

Psychological benefits of regular exercise have been demonstrated in those without diabetes including reduced stress, reductions in depression, and improved self-esteem (59, 101, 102). It is believed that these benefits are equally applicable to those with diabetes, and because of the burden of diabetes, it is especially important to try to use exercise to maximize psychological well-being in these patients (107).

expenditure, are preferred for those with complications where more precise control of intensity is needed (7).

Macrovascular Disease

Macrovascular disease is often a complication affecting patients with diabetes. These primarily include coronary artery disease and peripheral artery disease. Chapters 12 and 17 review specifics regarding preexercise evaluation and exercise prescription for coronary and peripheral artery disease. These should be incorporated for the patient with diabetes and coronary and peripheral artery disease.

Peripheral Neuropathy

The major consideration in patients with peripheral neuropathy is the loss of protective sensation in the feet and legs that can lead to musculoskeletal injury and infection. Non-weight-bearing activities are recommended to minimize the risk of injury (table 8.3). Proper footwear and examination of the feet are especially important for these patients. The clinical exercise professional should reinforce instruction

TABLE 8.3

Exercises for Patients With Diabetes and Loss of Protective Sensation

Contraindicated exercise	Recommended exercise
Treadmill	Swimming
Prolonged walking	Bicycling
Jogging	Rowing
Step exercises	Chair exercises
	Arm exercises
	Other non-weight-bearing exercise

Reprinted, by permission, from American Diabetes Association, 1999, "Physical activity, exercise, and diabetes," *Diabetes Care* 26: S51.

PRACTICAL APPLICATION 8.3

Prevention and Treatment of Abnormal Blood Glucose Before and After Exercise

Preexercise Hypoglycemia

Blood glucose levels should be monitored before an exercise session to determine if it is safe to begin exercising. The preexercise assessment of blood glucose and carbohydrate consumption is conducted to prevent exercise-induced hypoglycemia. Consideration must be given to the intensity and duration of exercise and whether the patient is managed with medication (insulin or oral agents). The following guidelines can be used to determine if additional carbohydrate intake is necessary (44).

— If blood glucose is less than 100 mg · dl^{-1} and the exercise will be low intensity and short duration (e.g., bike riding or walking for <30 min), 10 to 15 g of carbohydrate should be consumed. If blood glucose is greater than 100 mg · dl^{-1}, no extra carbohydrate is needed.

— If blood glucose is less than 100 mg · dl^{-1} and exercise is moderate intensity, moderate duration (e.g., jogging for 30-60 min), 30 to 45 g of carbohydrate should be consumed. If blood glucose is 100 to 180 mg · dl^{-1}, then 15 g of carbohydrate is needed.

— If blood glucose is less than 100 mg · dl^{-1} and exercise is of moderate intensity, long duration (e.g., ≥1 hr of bicycling), then 45 g of carbohydrate should be consumed. If blood glucose is 100 to 180 mg · dl^{-1}, then 35 to 45 g of carbohydrate is needed.

It is important to remember that these are guidelines and should be modified in certain cases. For instance, someone trying to lose weight might benefit from a medication adjustment rather than increased food intake.

continued

Preexercise Hyperglycemia

If the preexercise blood glucose is greater than 250 mg · dl⁻¹, urine should be checked for ketones. If ketones are present (moderate to high) or blood glucose is greater than 300 mg · dl⁻¹, irrespective of whether ketones are present, exercise should usually be postponed until glucose control is improved. The blood glucose values given previously are guidelines and actions should be verified with the patient's doctor.

Patients who use medication as part of diabetes treatment should be assessed to determine if the timing and dosage of medication will allow exercise to have a positive effect on blood glucose. For example, a patient who uses insulin who had a blood glucose of 270 mg · dl⁻¹, had no ketones, and had taken regular insulin within 30 min will see a reduction in blood glucose from both the insulin and exercise. If this patient has not just administered fast-acting insulin and the previous insulin injection has run its duration, the patient is underinsulinized and additional insulin is needed to help reduce the blood glucose before he or she exercises. In this case, exercise would likely increase blood glucose level. In all cases, adding additional medication must be cleared by a physician.

Those with type 2 diabetes who are appropriately managed by diet and exercise alone will usually experience a reduction in blood glucose with low to moderate exercise. Timing of exercise after meals can often help many patients with type 2 diabetes reduce postprandial hyperglycemia. Blood glucose should be monitored after an exercise session to determine the patient's response to exercise.

Postexercise Hypoglycemia

It is more likely for a patient to experience hypoglycemia (usually <70 mg · dl⁻¹) after exercise than during exercise because of the replacement of muscle glycogen, which uses blood glucose (113). Periodic monitoring of blood glucose is necessary in the hours following exercise to determine if blood glucose is dropping. More frequent monitoring is especially important when initiating exercise. If the patient is hypoglycemic, he or she needs to take appropriate steps to treat this medical emergency as previously presented.

Postexercise Hyperglycemia

In poorly controlled diabetes, insulin levels are often too low, resulting in an increase in counterregulatory hormones with exercise. This causes production of glucose by the liver and enhanced free fatty acid release by adipose tissue, and muscle uptake of glucose is reduced. The result is an increased blood glucose level during and after exercise. High-intensity exercise can also result in hyperglycemia. In this case, the intensity and duration of exercise should be reduced as needed.

---

given to the patient on self-foot exams, learn how to recognize related injuries, and encourage the patient to have his or her feet examined regularly.

Autonomic Neuropathy

Cardiovascular autonomic neuropathy can affect the patient with diabetes and is manifested by abnormal heart rate, blood pressure, and redistribution of blood flow. Patients with cardiovascular autonomic neuropathy have a higher resting and lower maximal exercise heart rate than those without this condition (57). Thus, estimating peak heart rate in this population may lead to an overestimation of the training heart rate range if heart rate–based methods are used (see chapter 7). Early warning signs of ischemia may be absent in these patients. The risk

of exercise hypotension and sudden death also is elevated (39, 58). An active cool-down is very important to reduce the possibility of a postexercise hypotensive response. Exercise in this patient population should focus on lower intensity activities where mild changes in heart rate and blood pressure can be accommodated (48). Because of difficulty with thermoregulation, these patients should be advised to stay hydrated and not to exercise in hot or cold environments (12).

Retinopathy

The exercise recommendations for those with diabetic retinopathy are contingent on the stage of the complication and should focus on limiting systolic blood pressure and jarring activities. Table 8.4

TABLE 8.4
Considerations for Activity Limitation in Those With Diabetic Retinopathy

Level of DR	Acceptable activities	Discouraged activities	Ocular reevaluation
No DR	Dictated by medical status	Dictated by medical status	12 months
Mild NPDR	Dictated by medical status	Dictated by medical status	6-12 months
Moderate NPDR	Dictated by medical status	Activities that dramatically elevate blood pressure (e.g., power lifting, heavy lifting, Valsalva maneuver)	4-6 months
Severe NPDR	Dictated by medical status	Activities that substantially increase systolic blood pressure (e.g., Valsalva maneuver, active jarring, boxing, heavy competitive sports)	2-4 months (may require laser surgery)
PDR	Low-impact, cardiovascular conditioning (e.g., swimming, walking, low-impact aerobics, stationary cycling, endurance exercises)	Strenuous activities (e.g., Valsalva maneuver, pounding or jarring, weight lifting, jogging, high-impact aerobics, racket sports, strenuous trumpet playing)	1-2 months (may require laser surgery)

DR = diabetic retinopathy; NPDR = nonproliferative diabetic retinopathy; PDR = proliferative DR.

Reprinted, by permission, from American Diabetes Association, 1999, "Physical activity, exercise, and diabetes," *Diabetes Care* 26: S51.

provides information on selection of appropriate activities based on severity of retinopathy. For optimal improvement and safety, exercise should be conducted in a supervised environment when retinopathy is significant (68, 108, 117).

Nephropathy

Elevated blood pressure is related to the onset and progression of diabetic nephropathy. There is no clear reason to limit low- to moderate-intensity activity, but strenuous exercises should likely be discouraged in those with diabetic nephropathy because of the elevation in blood pressure (12). Patients on renal dialysis or who have received a kidney transplant can also benefit from exercise (6). See chapter 11 for details regarding renal failure.

Exercise Recommendations

Endurance, resistance, and range of motion exercise training are all appropriate modes for most patients with diabetes. For those patients who are trying to lose weight (especially those with type 2 diabetes), it is strongly recommended that they expend a minimum cumulative total of 1000 kcal per week in aerobic activity and participate in a well-rounded resistance training program (21, 41). Patient interests, goals of therapy, type of diabetes along with medication use (if appropriate), and presence and severity of complications must be carefully evaluated in the development of the exercise prescription. The following exercise prescription recommendations are guidelines, and individual patient circumstances must always determine the specific prescription (7).

Mode

Personal interest and the desired goals of the exercise program should drive the type of physical activity that is selected. Caloric expenditure is often a key goal for those with diabetes. Walking is the most commonly performed mode of activity (38). It is a convenient, low-impact activity that can be safely and effectively used to maximize caloric expenditure. Non-weight-bearing modes should be used if necessary (e.g., if the patient has peripheral arterial disease or peripheral neuropathy). For a given level of energy expenditure, the health-related benefits of exercise appear to be independent of the mode.

Frequency

The frequency of exercise should be 3 to 7 days per week. Exercise duration, intensity, weight loss goals,

PRACTICAL APPLICATION 8.4

Summary of Exercise Recommendations for Diabetes Screening

Type: aerobic

- Mode: walking, cycling, swimming
- Frequency: 3 to 5 times/week, or most days of the week
- Duration: 20 to 60 min
- Intensity: 50% to 75% of maximal aerobic capacity
- Energy expenditure: modulate type, frequency, duration, and intensity to attain an energy expenditure of 700 to 2000 kcal · week^{-1}
- Timing: time participation so that it does not coincide with peak insulin action

Type: resistance

- Mode: free weights, machines, elastic bands
- Frequency: at least two times per week, but never on consecutive days
- Duration: 10-15 repetitions per set, one to two sets per type of specific resistance exercise
- Intensity: ~60% of 1 repetition maximum

Type: range of motion

- Mode: static stretching
- Frequency: postaerobic exercise
- Duration: 10 to 30 s per exercise of each major muscle group

Special Considerations

- Search for vascular and neurological complications, including silent ischemia, that might alter exercise prescription or contraindicate exercise.
- Consider an exercise electrocardiogram in patients with known or suspected coronary artery disease, those who are greater than 30 years of age with type 1, those who have had type 1 for more than 15 years, or those who are older than 35 years with type 2.
- Instruct patients to warm-up and cool-down.
- Carefully select exercise type and intensity.
- Promote patient education.
- Encourage patients to wear proper footwear.
- Teach patients to avoid exercise in extreme heat or cold.
- Encourage patients to inspect feet daily and after exercise.
- Teach patients to avoid exercise when metabolic control is poor.
- Instruct patients to maintain adequate hydration.
- Instruct patients to monitor blood glucose if taking insulin or oral hypoglycemic agents and follow guidelines to prevent hypoglycemia.

Adapted from American Diabetes Association 2002. *Handbook of Exercise in Diabetes*. Eds. N. Ruderman, J. Devlin, S. Schneider, pp. 269-288. Alexandria: American Diabetes Association, Inc.

and personal interests determine the specific frequency. Additionally, the blood glucose improvements with exercise in those with diabetes are seen for greater than 12 but less than 72 hr (109). These data indicate that exercise done on 3 nonconsecutive days each week, and ideally 5 or more days per week, is recommended. Those who take insulin and have difficulty balancing caloric needs with insulin dosage may prefer to exercise daily. This will result in less daily adjustment of insulin dosage and caloric intake than if exercise is performed every other day or sporadically, and it will reduce the likelihood of a hypoglycemic or hyperglycemic response. Also, patients who are trying to lose weight will maximize caloric expenditure by participating in daily physical activity (7).

Intensity

Programs of moderate intensity are preferable for most people with diabetes, because there is a reduced cardiovascular risk and chance for musculoskeletal injury and greater likelihood of maintaining the exercise program. However, some low fit individuals may increase $\dot{V}O_2$peak at an intensity level as low as 40% of $\dot{V}O_2$peak. It is generally recommended that exercise be prescribed at an intensity of 60% to 80% of maximal heart rate, 50% to 75% of $\dot{V}O_2$max, or a rating of perceived exertion of 12 to 13 (7). Chapter 7 provides specifics for determining and calculating proper exercise intensity.

Duration and Rate of Progression

Exercise duration and the rate of progression can be at the standard levels for a chronic diseased population. Chapter 7 reviews the specifics of determining appropriate duration and progression of exercise.

Resistance Training

There is evidence to support the inclusion of resistance training in a patient's program. In nondiabetic subjects, resistance training improves glucose tolerance and insulin sensitivity (56, 80, 97). The limited data available for diabetes suggest that resistance training is safe and effective (37, 42). All patients should be screened for contraindications, specifically retinopathy and nephropathy, prior to beginning resistance training. Proper instruction and monitoring are also needed.

A recommended resistance training program consists of a minimum of 8 to 10 exercises involving major muscle groups performed with a minimum of one set of 10 to 15 repetitions to near fatigue. Resistance training exercises should be done at least 2 days per week (7). Modifications such as lowering the intensity of lifting, preventing exercise to the point of exhaustion, and eliminating the amount of sustained gripping or isometric contractions should be considered to ensure safety. Chapter 7 provides specific information regarding general resistance training that should be considered for these patients.

Timing

Exercise should be performed at the time of day most convenient for the participant. The time of day that exercise is performed should be given careful consideration by those taking insulin because of the risk of hypoglycemia. Exercise should not be performed when insulin action is peaking. Because exercise acts like insulin in that it promotes peripheral glucose uptake, the combination of exercise and peak insulin action increases the risk of hypoglycemia. Because of this, and the need to replace muscle glycogen, it is more likely for hypoglycemia to occur after exercise than during exercise (100). Exercising late in the evening when insulin and oral medications that increase insulin production may peak is not recommended because of the possible occurrence of hypoglycemia when sleeping.

Conclusion

Living with a chronic illness poses special issues. Diabetes management requires ongoing dedication, and the patient must cope with complications if they develop and, at the very least, must deal with the threat of their development. Exercise training should be an essential component of the treatment plan for patients with diabetes because it improves blood glucose control, lipid levels, blood pressure, and body weight; reduces stress; and has the potential to reduce the burden of this metabolic disease.

Medical History

Medication: Glucophage taken two times per day, Captopril (for blood pressure control and protection of kidneys), and Lipitor for control of hyperlipidemia.

Laboratory values: Last Hb A_{1C} = 8.8% (normal 4-6%); cholesterol 200 mg · dl^{-1}, low-density lipoprotein cholesterol 130 mg · dl^{-1}; high-density lipoprotein cholesterol 35 mg · dl^{-1}; triglycerides 160 mg · dl^{-1}; microproteinuria.

Physical exam: blood pressure 130/80 mmHg; resting heart rate 70 beats · min^{-1}; height 5 ft 11 in.; weight 230 lb with 27% body fat (skinfold).

Complications history: Acute—periodic episodes of diabetes out of control but has never experienced hyperosmolar nonketotic syndrome or diabetic ketoacidosis. Chronic—two-vessel bypass surgery 5 years ago, moderate peripheral neuropathy, and early (stage 3) diabetic nephropathy.

Diagnosis

Mr. SR is 63 years old and was diagnosed with type 2 diabetes 5 years ago.

Exercise Test Result

No abnormal electrocardiogram changes, maximum blood pressure 180/83 mmHg; maximum heart rate 150 beats · min^{-1}; $\dot{V}O_2$max 25.5 ml · kg^{-1} · min^{-1}; random blood glucose before test 180 mg · dl^{-1}.

Development of Exercise Prescription

The goals of the exercise program, mutually agreed upon by the patient and the clinical exercise professional, are to lose weight and improve body composition, improve blood glucose levels, and reduce risk for another cardiac event. When asked about interests and hobbies, the patient indicates that he enjoys traveling, wine tasting, playing with his dog, and classic movies.

Participation in a supervised exercise program and frequent contact with an exercise professional is advised. A warm-up and cool-down of static stretches and low-intensity aerobic activity are prescribed.

Mode: Stationary cycling or water exercise. Low- to non-weight-bearing activities are selected because of the peripheral neuropathy. Walking his dog and walking while traveling are discussed to help him take care of his feet and do these activities as safely as possible.

Frequency: 3 to 5 days per week with a goal of increasing to daily.

Intensity: This patient is taught to monitor heart rate and to use the rating of perceived exertion (RPE) scale. The intensity is prescribed at 60% of maximum heart rate (150 beats · min^{-1}) or 90 beats · min^{-1}. An RPE rating of 12 to 13 on a 6- to 20-point Borg scale is advised.

Duration: An initial duration of 15-30 min is suggested and should be eventually increased to 60 min per session to facilitate weight loss.

Rate of progression: Attention is first given to frequency of exercise. Once he has reached 5 days per week or more, duration will be increased.

Other information: Mr. SR is instructed to increase his blood glucose monitoring frequency to assess the impact of exercise on his blood glucose control.

continued

Case Study 8   *(continued)*

> ### Case Study 8   Discussion Questions
>
> 1. Is this patient likely to have problems with hypoglycemia during or following exercise? Why or why not?
> 2. What are potential risks of exercise for this patient? What cautions should he be given?
> 3. What strategies can be given to help the patient stay motivated and maintain a regular exercise program?
> 4. What general nutrition suggestions might be helpful for this patient?
> 5. What other healthcare team members should this patient work with as he begins his exercise program? Why?
> 6. What else should be added to his program to help him attain his goals?

Glossary

autoimmune—Referring to cells and/or antibodies arising from and directed against the individual's own tissues, as in autoimmune disease.

cardiovascular autonomic neuropathy—Neural damage to the autonomic nerves of the cardiovascular system, which can result in a high resting and low peak exercise heart rate and severe orthostatic hypotension.

diabetic ketoacidosis—A type of metabolic acidosis caused by accumulation of ketone bodies in diabetes mellitus.

effective insulin—Insulin available for use by body tissues.

evidence-based—Using the best available clinical research to guide treatment.

gestational diabetes—Carbohydrate intolerance of variable severity with onset or first recognition during pregnancy.

glucagon—A hormone produced by the pancreas that stimulates the liver to release glucose causing an increase in blood glucose levels, thus opposing the action of insulin.

glycemic goals—A goal range for blood glucose concentration.

hyperglycemia—An abnormally high concentration of glucose in the circulating blood, seen especially in people with diabetes mellitus.

hypoglycemia—Symptoms resulting from low blood glucose (normal glucose range 60-100 mg · dl^{-1}, or 3.3-5.6 mmol · L^{-1}) that are either autonomic or neuroglycopenic.

idiopathic—Denoting a disease of unknown cause.

ketones—A substance with the carbonyl group linking two carbon atoms.

macrovascular disease—Atherosclerosis affecting large vessels such as the aorta, femoral artery, and carotid artery.

medical nutrition therapy—The use of nutrition as a treatment for a clinical condition or disease.

mentation—The process of reasoning and thinking.

microvascular disease—Atherosclerosis affecting small blood vessels such as those of the kidney, eye, heart, and brain.

neuroglycopenic hypoglycemia—Symptoms of hypoglycemia that include feelings of dizziness, confusion, tiredness, difficulty speaking, headache, and inability to concentrate.

pancreas—A gland lying behind the stomach that secretes pancreatic enzymes into the duodenum and insulin, glucagon, and somatostatin into the bloodstream.

peripheral neuropathy—Damage to the nerves of the legs or arms resulting in a loss of sensation (e.g., touch, temperature).

polydipsia—Excessive thirst that is relatively prolonged.

polyuria—Excessive excretion of urine.

postprandial—One to two hours following a meal.

proteinuria—The presence of abnormal amounts of protein in the urine.

References

1. Albright, A.L. Exercise precautions and recommendations for patients with autonomic neuropathy. *Diabetes Spectrum* 11: 231-237, 1998.

2. Albright, A., Franz, M., Hornsby, G., Kriska, A., Marrero, D., Ullrich, I., Verity, L. Exercise and type 2 diabetes. *Med Sci Sports Exerc* 32: 1345-1362, 2000.

3. American Association of Diabetes Educators. *A Core Curriculum for Diabetes Education.* 2nd Edition. Eds. J. Schwarz, K. Moline, M. Urban. Chicago: American Association of Diabetes Educators, p. 203, 1993.

4. American College of Sports Medicine. Physical activity, physical fitness, and hypertension (position stand). *Med Sci Sports Exerc* 25: i-x, 1993.

5. American College of Sports Medicine. *ACSM's Guidelines for Exercise Testing and Prescription.* 6th Edition. Eds. B. Franklin, M. Whaley, E Howley. . Baltimore: Williams & Wilkins, 2000.

6. American College of Sports Medicine. *ACSM's Exercise Management for Persons With Chronic Diseases and Disabilities.* Champaign, IL: Human Kinetics, 1997.

7. American Diabetes Association. *The Health Professional's Guide to Diabetes and Exercise.* Eds. N. Ruderman, J. Devlin. Alexandria, VA: American Diabetes Association, pp. 71-158, 1995.

8. American Diabetes Association. Economic consequences of diabetes mellitus in the U.S. in 1997. *Diabetes Care* 21: 296-309, 1998.

9. American Diabetes Association. *Medical Management of Type 2 Diabetes.* 4th Edition. Alexandria, VA: American Diabetes Association, 1998.

10. American Diabetes Association. Report of the Expert Committee on the Diagnosis and Classification of Diabetes Mellitus. *Diabetes Care* 26: S5-S20, 2003.

11. American Diabetes Association. Evidence-Based Nutrition Principles and Recommendations for the Treatment and Prevention of Diabetes and Related Complications. *Diabetes Care* 25: S50-S60, 2002.

12 American Diabetes Association. Physical activity, exercise and diabetes. *Diabetes Care* 26: S73-S77, 2003.

13. American Diabetes Association. National standards for diabetes self-management education. *Diabetes Care* 25: S140-S147, 2002.

14. Andersson, D.K.G., Svaardsudd, K. Long-term glycemic control related to mortality in type II diabetes. *Diabetes Care* 18: 1534-1543, 1995.

15. Andressen, E.M., Lee, J.A., Pecoraro, R.E., Koepsell, T.D., Hallstrom, A.P., Siscovick, D.S. Under reporting of diabetes on death certificates, King County, Washington. *Am J Public Health* 83: 1021-1024, 1993.

16. Bandura, A. *Social Foundations of Thought and Action.* Englewood Cliffs, NJ: Prentice Hall, 1986.

17. Bantle, J.P., Swanson, J.E., Thomas, W., Laine, D.C. Metabolic effects of dietary sucrose in type II diabetic subjects. *Diabetes Care* 19: 1249-1256, 1996.

18. Barnard, R.J., Lattimore L., Holly, R.G., Cherny, S., Pritikin, N. Response of non-insulin-dependent diabetic patients to an intensive program of diet and exercise. *Diabetes Care* 5: 370-374, 1982.

19. Berger, M., Berchtold, P., Cuppers, H.J., et al. Metabolic and hormonal effects of muscular exercise in juvenile type diabetics. *Diabetologia* 13: 355-365, 1977.

20. Björntorp, P. Portal adipose tissue as a generator of risk factors for cardiovascular disease and diabetes. *Arteriosclerosis* 10: 493-496, 1990.

21. Blair, S.N., Kohl, H.W., Paffenbarger, R.S., Clark, D.G., Cooper, K.H., Gibbons, L.W. Physical fitness and all-cause mortality. A prospective study of healthy men and women. *JAMA* 262: 2395-2401, 1989.

22. Blake, G.A., Levin, S.R., Koyal, S.N. Exercise induced hypertension in normotensive patients with NIDDM. *Diabetes Care* 13: 799-801, 1990.

23. Bogardus, C., Ravussin, E., Robins, D.C., Wolfe, R.R., Horton, E.S., Sims, E.A.H. Effects of physical training with diet therapy on carbohydrate metabolism in patients with glucose intolerance and non-insulin-dependent diabetes mellitus. *Diabetes* 33: 311-318, 1984.

24. Bogardus, C., Lillioja, S., Mott, D.M., Hollenbeck, C., Reaven, G. Relationship between degree of obesity and in vivo insulin action in man. *Am J Physiol* 248: E286-E291, 1985.

25. Brosseau, J.D. Occurrence of diabetes among decedents in North Dakota. *Diabetes Care* 10: 542-543, 1987.

26. Centers for Disease Control and Prevention. Trends in the prevalence and incidence of self-reported diabetes mellitus—United States, 1980-1994. *Morb Mortal Wkly Rep* 46:1027-1028, 1997.

27. Centers for Disease Control and Prevention. *National Diabetes Fact Sheet: National Estimates and General Information on Diabetes in the United States.* Revised edition. Atlanta: U.S. Department of Health and Human Services, Centers for Disease Control and Prevention, 1998.

28. Centers for Disease Control and Prevention. Special focus: Diabetes. *Chronic Dis Notes and Reports* 12: 1-28, 1999.

29. Cowie, C.C., Eberhardt, M.S. Sociodemographic characteristics of persons with diabetes. In *Diabetes in America* (NIH publication 95-1468). 2nd Edition. Eds. M.I. Harris, C.C. Cowie, M.P. Stern, et al. Bethesda, MD: National Institutes of Health, National Institute of Diabetes and Digestive and Kidney Diseases, pp. 85-101, 1995.

30. Davidson, M.B. *Diabetes Mellitus Diagnosis and Treatment.* 4th Edition. Philadelphia: Saunders, 1998.

31. DeFronzo, R., Deibert, D., Hendler, R., Felig, P. Insulin sensitivity and insulin binding to monocytes in maturity-onset diabetes. *J Clin Invest* 63: 939-946, 1979.

32. Diabetes Control and Complications Trial Research Group. The effect of intensive treatment of diabetes on the development and progression of long-term complications in insulin-dependent diabetes mellitus. *New Engl J Med* 329: 977-986, 1993.

33. Diabetes Prevention Program Research Group. The Diabetes Prevention Program. Design and methods for a clinical trial in the prevention of type 2 diabetes. *Diabetes Care* 22: 623-634, 1999.

34. Diabetes Research Working Group. *Conquering Diabetes: A Strategic Plan for the 21st Century* (NIH publication 99-4398). Bethesda, MD: National Institutes of Health, 1999.

35. Dowse, G.K., Zimmet, P.Z., Gareeboo, H., Alberti, K.G.M.M., Tuomilehto, J., Finch, C.F., Chitson, P., Tulsidas, H. Abdominal obesity and physical inactivity are risk factors for NIDDM and impaired glucose tolerance in Indian, Creole, and Chinese Mauritians. *Diabetes Care* 14: 271-282, 1991.

36. Duncan, J.J., Gordon, N.F., Scott, C.B. Women walking for health and fitness: How much is enough? *JAMA* 266: 3295-3299, 1991.

37. Eriksson, J., Taimela, S., Eriksson, K., Parvianen, S., Peltonen, J., Kujala, U. Resistance training in the treatment of non-insulin-dependent diabetes mellitus. *Int J Sports Med* 18: 242-246, 1997.

38. Estacio, R.O., Regensteiner, J.G., Wolfel, E.E., Jeffers, B., Dickenson, M., Schrier, R.W. The association between

diabetic complications and exercise capacity in NIDDM patients. *Diabetes Care* 21: 291-295, 1998.

39. Ewing, D.J., Boland, O., Neilson, J.M., Cho, C.G., Clarke, B.F. Autonomic neuropathy, QT interval lengthening, and unexpected deaths in diabetic autonomic neuropathy. *J Clin Endocrinol Metab* 54: 751-754, 1991.

40. Feskens, E.J., Loeber, J.G., Kromhout, D. Diet and physical activity as determinants of hyperinsulinemia: The Zutphen elderly study. *Am J Epidemiol* 140: 350-360, 1994.

41. Fletcher, G.F., Blair, S.N., Blumenthal, J., Caspersen, C., Chaitman, B., Epstein, S. Statement on exercise: Benefits and recommendations for physical activity programs for all Americans. *Circulation* 86(1): 340-344, 1992.

42. Fluckey, J.D., Hickey, M.S., Brambrink J.K., Hart, K.K., Alexander, K., Craig, B.W. Effects of resistance exercise on glucose tolerance in normal and glucose-intolerant subjects. *J Appl Physiol* 77: 1087-1092, 1994.

43. Franz, M.J., Etzwiler, D.D., Joynes, J.O., Hollander, P.M. *Learning to Live Well With Diabetes*. Minneapolis: DCI, 1991.

44. Franz, M. Medical nutrition therapy. In *Diabetes Mellitus—Diagnosis and Treatment*. Philadelphia: Saunders, pp. 45-79, 1998.

45. Frisch, R.E., Wyshak, G., Albright, T.E., Albright, N.L., Schiff, I. Lower prevalence of diabetes in female former college athletes compared with nonathletes. *Diabetes* 35: 1101-1105, 1986.

46. Fujimoto, W.Y., Leonetti, D.L., Kinyoun, J.L., Shuman, W.P., Stolov, W.C., Wahl, P.W. Prevalence of complications among second-generation Japanese-American men with diabetes, impaired glucose tolerance or normal glucose tolerance. *Diabetes* 36: 730-739, 1987.

47. Gordon, N.D., Kohl, H.W., Blair, S.N. Lifestyle exercise: A new strategy to promote physical activity for adults. *J Cardiopulm Rehabil* 13: 161-163, 1993.

48. Graham, C., Lasko-Mccarthey, P. Exercise options for persons with diabetic complications. *Diabetes Educator* 16: 212-220, 1990.

49. Harris, M.I., Couric, C.C., Reiber, G., Boyko, E., Stern, M., Bennett P., eds. *Diabetes in America* (NIH publication 95-1468). 2nd Edition. Washington, DC: U.S. Government Printing Office, 1995.

50. Helmrish, S.P., Ragland, D.R., Leung, R.W., Paffenbarger, R.W. Physical activity and reduced occurrence of non-insulin-dependent diabetes mellitus. *New Engl J Med* 325: 147-152, 1991.

51. Herman, W.H., Eastman, R.C., Songer, T.J., Dasbach, E.J. The cost-effectiveness of intensive therapy for diabetes mellitus. *Endocrinol Metab Clin North Am* 26: 679-695, 1997.

52. Holloszy, J.O., Schultz, J., Kusnierkiewicz, J., Hagberg, J.M., Ehsani, A.A. Effects of exercise on glucose tolerance and insulin resistance: Brief review and some preliminary results. *Acta Med Scand* 711(Suppl.): 55-65, 1987.

53. Horton, E.S. Exercise and physical training: Effects on insulin sensitivity and glucose metabolism. *Diabetes Metab Rev* 2:1-17, 1986.

54. Horton, E.S. Role and management of exercise in diabetes mellitus. *Diabetes Care* 11: 201-211, 1988.

55. Hubinger, A., Franzen, A., Gries, A. Hormonal and metabolic response to physical exercise in hyperinsulinemic and non-hyperinsulinemic type 2 diabetics. *Diabetes Res* 4: 57-61, 1987.

56. Hurley, B.F., Seals, D.R., Ehsani, A.A., Carter, L.J., Dalsky, G.P., Hagberg, J.M., Holloszy, J.O. Effects of high-intensity strength training on cardiovascular function. *Med Sci Sports Exerc* 16: 483-488, 1984.

57. Kahn, J.K., Zola, B., Juni, J., Vinik, A. Decreased exercise heart rate and blood pressure response in diabetic subjects with cardiac autonomic neuropathy. *Diabetes Care* 9: 389-394, 1986.

58. Kahn, J.K., Sisson, J.C., and Vinik, A.I. Prediction of sudden cardiac death in diabetic autonomic neuropathy. *J Nucl Med* 29: 1605-1606, 1988.

59. Kelley, S., Seraganian, P. Physical fitness level and autonomic reactivity to psychosocial stress. *J Psychosom Res* 28: 279-287, 1984.

60. Kenny, S.J., Aubert, R.E., Geiss, L.S. Prevalence and incidence of non-insulin-dependent diabetes. In *Diabetes in America* (NIH publication 95-1468). 2nd Edition. Eds. M.I. Harris, C.C. Cowie, M.P. Stern, et al.. Bethesda, MD: National Institutes of Health, National Institute of Diabetes and Digestive and Kidney Diseases, pp. 46-67, 1995.

61. Kissebah, A.H., Vydelingum, N., Murray, R., Evans, D.F., Hartz, A.J., Kalkhoff, R.K., Adams, P.W. Relationship of body fat distribution to metabolic complications of obesity. *J Clin Endocrinol Metab* 54: 254-260, 1982.

62. Kjaer, M., Hollenbeck, C.B., Frey-Hewitt, B., Galbo, H., Haskell, W., Reaven. G.M. Glucoregulation and hormonal responses to maximal exercise in non-insulin-dependent diabetes. *J Appl Physiol* 68: 2067-2074, 1990.

63. Klem, M.L., Wing, R.R., Mcguire, M.T., Seagle, H.M., Hill, J.O. A descriptive study of individuals successful at long-term maintenance of substantial weight loss. *Am J Clin Nutr* 66: 239-246, 1997.

64. Kobberling, J., Tillil, H. Empiric risk figures for first degree relatives of non-insulin-dependent diabetics. In *The Genetics of Diabetes Mellitus* (Serono Symposium no. 47). Eds. J. Kobberling, R. Tattersall. New York: Academic Press, pp. 201-209, 1982.

65. Koivisto, V.A., Defronzo, R.A. Exercise in the treatment of type 2 diabetes. *Acta Endocrinol* 262: 107-111, 1984.

66. Koivisto, V.A., Yki-Jarvinen, H., Defronzo, R.A. Physical training and insulin sensitivity. *Diabetes Metab Rev* 1: 445-481, 1986.

67. Kolterman, O.G., Gray, R.S., Griffin, J., Burstein, P., Insel, J., Scarlett, J.A., Olefsky, J.M. Receptor and postreceptor defects contribute to the insulin resistance in noninsulin-dependent diabetes mellitus. *J Clin Invest* 68: 957-969, 1981.

68. Krentz, A.J., Ferner, R.E., Bailey, C.J. Comparative tolerability profiles of oral antidiabetic agents. *Drug Safety* 11: 223-241, 1994.

69. Krotkiewski, M., Lonnroth, P., Mandroukas, K., Wroblewski, Z., Rebuffe-Scrive, M., Holm, G., et al.. The effects of physical training on insulin secretion and effectiveness and on glucose metabolism in obesity and type 2 (non-insulin-dependent) diabetes mellitus. *Diabetologia* 28: 881-890, 1985.

70. Lampman, R.M., Schteingart, D.E. Effects of exercise training on glucose control, lipid metabolism, and insulin sensitivity in hypertriglyceridemia and non-insulin dependent diabetes mellitus. *Med Sci Sports Exerc* 23: 703-712, 1991.

71. Lawrence R.H. The effects of exercise on insulin action in diabetes. *Better Med J* 1: 648-652, 1926.

72. Laws, A., Reaven, G.M. Physical activity, glucose tolerance, and diabetes in older adults. *Ann Behav Med* 13:125-131, 1991.

73. Lindgärde, F., Saltin, B. Daily physical activity, work capacity and glucose tolerance in lean and obese normoglycaemic middle-aged men. *Diabetologia* 20:134-138, 1981.

74. Loghmani, E., Rickard, K., Washburne, L., Vandagriff, J., Fineberg, N., Golden, M. Glycemic response to sucrose-containing mixed meals in diets of children with insulin-dependent diabetes mellitus. *J Pediatr* 119: 531-537, 1991.

75. Manson, J.E., Rimm, E.B., Stampfer, M.J., Colditz, G.A., Willett, W.C., Krolewski, A.S., Rosner, B., Hennekens, C.H., Speizer, F.E. Physical activity and incidence of non-insulin-dependent diabetes mellitus in women. *Lancet* 338: 774-778, 1991.

76. Manson, J.E., Nathan, D.M., Krolewski, A.S., Stampfer, M.J., Willett, W.C., Hennekens, C.H. A prospective study of exercise and incidence of diabetes among US male physicians. *JAMA* 268: 63-67, 1992.

77. Marrero, D.G., Sizemore, J.M. Motivating patients with diabetes to exercise. In *Practical Psychology for Diabetes Physicians: How to Deal With Key Behavioral Issues Faced by Health Care Teams*. Eds. B.J. Anderson, R.R. Ruben. Alexandria, VA: American Diabetes Association, 1996.

78. Mayer-Davis, E.J., D'agostino, R., Karta, A.J., Haffner, S.M., Rewers, M.J., Saad, M., Bergman, R.N. Intensity and amount of physical activity in relation to insulin sensitivity. *JAMA* 279: 669-674, 1998.

79. Metzget, B.E., Cho, N.H., Roston, S.M., and Radvany, R.. Prepregnancy weight and antepartum insulin secretion predict glucose tolerance five years after gestational diabetes mellitus. *Diabetes Care* 16: 1598-1605, 1993.

80. Miller, W.J., Sherman, W.M., Ivy, J.L. Effect of strength training on glucose tolerance and post-glucose insulin response. *Med Sci Sports Exerc* 16: 539-543, 1984.

81. Minuk, H.L., Vranic, M., Marliss, E.B., Hanna, A.K., Albisser, A.M., Zinman, B. Glucoregulatory and metabolic response to exercise in obese non-insulin-dependent diabetes. *Am J Physiol* 240: E458-E464, 1981.

82. Moss, S.E., Klein, R., Klein, B.E.K., Meuer, M.S. The association of glycemia and cause-specific mortality in a diabetic population. *Arch Intern Med* 154: 2473-2479, 1984.

83. Mourier, A., Gautier, J-F., Dekerviler, E., Biagard, A.X., Villette, J-M., Garnier, J.P., Duvallet, A., Guezennec, C.Y., Cathelineau, G. Mobilization of visceral adipose tissue related to the improvement in insulin sensitivity in response to physical training in NIDDM. *Diabetes Care* 20: 385-392, 1997.

84. Olefsky, J.M., Kolterman, O.G., Scarlett, J.A. Insulin action and resistance in obesity and noninsulin-dependent type II diabetes mellitus. *Am J Physiol* 243: E15-E30, 1982.

85. Pan, X-P., Li, G-W., Hu, Y-H., Wang, J., Yang, W., Hu, Z-X., Lin, J., Xiao, J-Z., Cao, H-B., Liu, P., Jiang, X-G., Jiang, Y-Y., Wang, J-P., Zheng, H., Zhang, H., Bennet, P.H., Howard, B.V. Effects of diet and exercise in preventing NIDDM in people with impaired glucose tolerance. *Diabetes Care* 20: 537-544, 1997.

86. Paternostro-Bayles, M., Wing, R.R., Robertson, R.J. Effect of life-style activity of varying duration on glycemic control in type 2 diabetic women. *Diabetes Care* 12: 34-37, 1989.

87. Pereira, M., Kriska, A., Joswiak, M., Dowse, G., Collins, V., Zimmet, P., Gareeboo, H., Chitson, P., Hemraj, F., Purran, A., Fareed, D. Physical inactivity and glucose intolerance in the multi-ethnic island of Mauritius. *Med Sci Sports Exerc* 27: 1626-1634, 1995.

88. Perry, I., Wannamethee, S., Walker, M., et al. Prospective study of risk factors for development of non-insulin-dependent diabetes in middle aged British men. *Br Med J* 310: 560-564, 1995.

89. Portuese, E., Orchard, T. Mortality in insulin-dependent diabetes. In *Diabetes in America* (NIH publication 95-1468). 2nd Edition. Eds. M.I. Harris, C.C. Cowie, M.P. Stern, et al. Bethesda, MD: National Institutes of Health, National Institute of Diabetes and Digestive and Kidney Diseases, pp. 221-232, 1995.

90. Reaven, G.M., Bernstein, R., Davis, B., Olefsky, J.M. Nonketotic diabetes mellitus: Insulin deficiency or insulin resistance? *Am J Med* 60: 80-88, 1976.

91. Regensteiner, J.G., Shetterly, S.M., Mayer, E.J., Eckel, R.H., Haskell, W.L., Baxter, J., Hamman, R.F. Relationship between habitual physical activity and insulin area among individuals with impaired glucose tolerance. *Diabetes Care* 18: 490-497, 1995.

92. Reitman, J.S., Vasquez, B., Klimes, I., Nagulusparan, M. Improvement of glucose homeostasis after exercise training in non-insulin-dependent diabetes. *Diabetes Care* 7: 434-441, 1984.

93. Rice, B., Janssen, I., Hudson, R., Ross, R. Effects of aerobic exercise and/or diet on glucose tolerance and plasma levels in obese men. *Diabetes Care* 22: 684-691, 1999.

94. Rogers, M.A., Yamamoto, C., King, D.S., Hagberg, J.M., Ehsani, A.A., Holloszy, J.O. Improvement in glucose tolerance after 1 wk of exercise in patients with mild NIDDM. *Diabetes Care* 11: 613-618, 1988.

95. Ruderman, N.B., Ganda, O.P., Johansen, K. The effect of physical training on glucose tolerance and plasma lipids in maturity-onset diabetes. *Diabetes* 28(Suppl. 1): 89-92, 1979.

96. Ruderman, N., Chisholm, D., Pi-Sunyer, X., Schneider, S. The metabolically obese, normal weight individual revisited. *Diabetes* 47:699-713, 1998.

97. Ryan, A.S., Pratley, R.E., Goldberg, A.P., Elahi, D. Resistive training increases insulin action in postmenopausal women. *J Gerontol Biol Sci Med* 51: M199-M205, 1996.

98. Schneider, S.H., Khachadurian, A.K., Amorosa, L.F., Clemow, L., Ruderman, N.B. Ten-year experience with an exercise-based outpatient lifestyle modification program in the treatment of diabetes mellitus. *Diabetes Care* 15(Suppl. 4): 1800-1810, 1992.

99. Schneider, S.H., Amorosa, L.F., Khachadurian, A.K., Ruderman, N.B. Studies on the mechanism of improved glucose control during regular exercise in type 2 (non-insulin-dependent) diabetes. *Diabetologia* 26: 325-360, 1984.

100. Sherman, W.M., Ferrara, C., Schneider, B. Nutritional strategies to optimize athletic performance. In *The Health Professional's Guide to Diabetes and Exercise*. Eds. N. Ruderman, J. Devlin. Alexandria, VA: American Diabetes Association, 71-158, 1995.

101. Sonstroem, R.J., Morgan, W.P. Exercise and self-esteem: Rationale and model. *Med Sci Sports Exerc* 21: 329-337, 1989.

102. Sothmann, M.S., Horn, T.S., Hart, B.A., Gustafson, A.B. Comparison of discrete cardiovascular fitness groups on plasma catecholamine and selected behavioral responses to psychological stress. *Psychophysiology* 24: 47-54, 1987.

103. Trovati, M., Carta, Q., Cavalot, F., Vitali, S., Banaudi, C., Lucchina, P.G., et al. Influence of physical training on blood glucose control, glucose tolerance, insulin secretion, and insulin action in non-insulin-dependent diabetic patients. *Diabetes Care* 7: 416-420, 1984.

104. Turner, R.C., Holman, R.R., Matthews, D., Hockaday, T.D.R., Peto, J. Insulin deficiency and insulin resistance interaction in diabetes: Estimation of their interaction in diabetes: Estimation of their relative contribution by feedback analysis from basal plasma insulin and glucose concentrations. *Metabolism* 28: 1086-1096, 1979.

105. UKPDS Group. UK Prospective Diabetes Study 33: Intensive blood-glucose control with sulphonylureas or insulin compared with conventional treatment and risk of complications in patients with type 2 diabetes. *Lancet* 352: 837-853, 1998.

106. Uusitupaa, M.I.J., Niskanen, L.K., Siitonen, O., Voutilainen, E., Pyorala, K. Ten year cardiovascular mortality in relation to risk factors and abnormalities in lipoprotein composition in type 2 (non-insulin-dependent) diabetic and non-diabetic subjects. *Diabetologia* 18: 1534-1543, 1993.

107. Vasterling, J.J., Sementilli, M.E., Burish, T.G. The role of aerobic exercise in reducing stress in diabetic patients. *The Diabetes Educator* 14(3): 197-201, 1988.

108. Vinik, A.I. Neuropathy. In *The Health Professional's Guide to Diabetes and Exercise*. Eds. N. Ruderman, J. Devlin. Alexandria, VA: American Diabetes Association, pp. 183-197, 1995.

109. Vranic, M., Wasserman, D. Exercise, fitness, and diabetes. In *Exercise, Fitness and Health*. Eds. C. Bouchard, R.J. Shephard, T. Stephens, J. Sutton, B. McPherson. Champaign, IL: Human Kinetics, pp. 467-490, 1990.

110. Wang, J.T., Ho, L.T., Tang, K.T., Wang, L.M., Chen, Y.D.I., Reaven, G.M. Effect of habitual physical activity on age-related glucose intolerance. *J Am Geriatr Soc* 37: 203-209, 1989.

111. Watts, N.B., Spanheimer, R.G., Digirolamo, A., Gebhart, S.S.P., Musey, V.C., Siddiq, K., Phillips, L.S. Prediction of glucose response to weight loss in patients with non-insulin-dependent diabetes mellitus. *Arch Intern Med* 150: 803-806, 1990.

112. West, K.M. *Epidemiology of Diabetes and Its Vascular Lesions*. New York: Elsevier, 1978.

113. Wing, R.R., Koeske, R., Epstein, L.H., Nowalk, M.P., Gooding, W., Becker, D. Long-term effects of modest weight loss in type II diabetic patients. *Arch Intern Med* 147: 1749-1753, 1987.

114. Wing, R.R. Behavioral strategies for weight reduction in obese type 2 diabetic patients. *Diabetes Care* 12: 139-144, 1989.

115. Wing, R.R., Epstein, L.H., Nowalk, M.P., Koeske R., Hagg, S. Behavior change, weight loss, and physiological improvements in type II diabetic patients. *J Consult Clin Psychol* 53: 111-122, 1985.

116. Yki-Jarvinen, H. Glucose toxicity. *Endocrinol Rev* 13: 415-431, 1992.

117. Zamboni, M., Armellini, F., Turcato, E., Todesco, T., Bissoli, L., Bergamo-Andreis, I.A., Bosello, O. Effect of weight loss on regional body fat distribution in premenopausal women. *Am J Clin Nutr* 58: 29-34, 1993.

118. Zimmet, P.Z. Kelly West Lecture 1991: Challenges in diabetes epidemiology: From west to the rest. *Diabetes Care* 15: 232-252, 1992.

119. Zimmet, P.Z., Tuomi, T., Mackay, R., Rowley, M.J., Knowles, W., Cohen, M., Lang, D.A. Latent autoimmune diabetes mellitus in adults (LADA): The role of antibodies to glutamic acid decarboxylase in diagnosis and prediction of insulin dependency. *Diabetic Med* 11: 299-303, 1994.

120. American Diabetes Association. *Handbook of Exercise in Diabetes*. Eds. N. Ruderman, J. Devlin, and S. Schneider. Alexandria, VA: American Diabetes Association, Inc., pp. 268-288, 2002.

Richard B. Parr, EdD
School of Health Sciences
Central Michigan University
Mt. Pleasant, MI

Obesity

Obesity is a chronic disease for which short-term intervention has limited effectiveness and long-term success is rare and primarily limited to surgical treatment. Clinicians should approach obesity as a disease in itself as well as managing its comorbidities. Exercise has an important contribution in the management of obesity as a vital adjunct to diet, drugs, and surgical treatment. An exercise prescription for weight loss requires manipulating intensity and duration to benefit caloric expenditure and may not provide cardiovascular improvements in those patients with comorbidities that require associated fitness benefits.

The term **obesity** refers to being overfat, and several methods of assessing fatness (skinfolds, bioelectrical impedance, underwater weighing, and dual energy X-ray absorptiometry) are available. However, using these expensive and time-consuming techniques may be difficult, and patient cooperation may preclude their use. A more practical approach for the clinical setting is to use **body mass index (BMI).** BMI is recommended by the National Institutes of Health (31) to classify **overweight** and obesity and to estimate relative risk of disease. BMI indicates overweight for height but does not discriminate between fat mass and lean tissue. The BMI does, however, significantly correlate with total body fat and therefore is an acceptable measure of overweight and obesity in the clinical setting. It is calculated as weight in kilograms divided by height in meters squared. BMI can also be estimated using pounds and inches.

$$BMI = kg/m^2 \text{ or } BMI = lb/(in.)^2 \times 703$$

The classification of overweight and obesity by BMI is based on the 1998 *Clinical Guidelines on the Identification, Evaluation, and Treatment of Overweight and Obesity in Adults* (see table 9.1). Table 9.2 shows, for comparison purposes, the percentage fat and percentage overweight that are roughly equated to the BMI classifications. Table 9.2 also shows the estimated percentage of the U.S population within each category. This information is helpful for classifying patients, determining realistic patient goals, and prescribing exercise.

Overweight is defined as having a BMI 25 to 29.9, and those with BMI greater than 30 are considered obese. Medically significant obesity refers to adults who have gained 30 lb or more as adults and increased their waist circumference more than 6 in. (11). The National Institutes of Health guidelines (31) define this population as adults with a BMI greater than 30 kg · m^{-2} or those with greater than 25 kg · m^{-2} with a family history of obesity or an obesity-related comorbidity.

Although BMI is recommended to evaluate obesity, patients often feel more comfortable with a table of recommended weight rather than the more obscure BMI. The *Dietary Guidelines for Americans* (5th ed) recommends "healthy weights" for Americans (14). These guidelines (table 9.3) provide a range of weight where the higher weights apply to men and women with more muscle and bone.

Scope

Obesity is a public health problem that affects 34 million Americans, who spend $33 billion each year in efforts to control their weight. In recent years, several professional guidelines have been written that have focused on obesity:

— The 5th edition of the *Dietary Guidelines for Americans* (2000) (14)

— The Surgeon General's Report of Physical Activity and Health (1996) (39)

— The National Institutes of Health *Clinical Guidelines on the Identification, Evaluation, and Treatment of Overweight and Obesity in Adults* (1998) (31)

TABLE 9.1

Classification of Overweight and Obesity by Body Mass Index (BMI)

	Obesity class	BMI (kg · m^{-2})
Underweight		<18.5
Normal		18.5-24.9
Overweight		25.0-29.9
Obesity	I	30.0-34.9
	II	35.0-39.9
	III	≥40.0

TABLE 9.2

Overweight/Obesity Continuum

	Underweight	Normal	Overweight	Mildly obese (Class I)	Moderately obese (Class II)	Severely obese (Class III)
Weight[a]		0-10%	11-20%	21-40%	41-100%	>100%
BMI[b]	<18.5	18.5-24	25-29	30-34	35-39	≥40
% Fat[c]	<20	20-25	26-31	32-37	38-45	>45
% Pop[d]	25%	20%	32%	22%	1%	<1%

[a]Percentage over standard height/weight tables. [b]Body mass index (kg · m^{-2}) adapted from National Institutes of Health Clinical Guidelines on Obesity, 1998. [c]Calculated body fat expressed as percentage of total weight. [d]Estimated percentage of population with each category, from the National Health and Nutrition Examination Survey III.

TABLE 9.3
Healthy Weight Ranges
for Men and Women

Height (ft, in.)	Weight (lb)
4, 10	91-119
4, 11	94-124
5, 0	97-128
5, 1	101-132
5, 2	104-137
5, 3	107-141
5, 4	111-146
5, 5	114-150
5, 6	118-155
5, 7	121-160
5, 8	125-164
5, 9	129-169
5, 10	132-174
5, 11	136-179
6, 0	140-184
6, 1	144-189
6, 2	148-195
6, 3	152-200
6, 4	156-205
6, 5	160-211
6, 6	164-216

— The American College of Sports Medicine/ Centers for Disease Control and Prevention report on *Physical Activity in the Treatment of Obesity and Its Comorbidities* (33)

— The American Heart Association's medical/scientific statement on obesity and heart disease (1997) (3)

— The *Healthy People 2010: National Health Promotion and Disease Prevention Objectives* (38)

Currently, 33% of adult Americans are considered obese (23). This represents a 32% increase in obesity from 1980. Obesity often begins in childhood and early adolescence; however, 70% of all obesity begins in adulthood. Approximately 20% to 25% of children are obese, and many of them will carry their obesity into adulthood. The risks of childhood obesity persisting into adult obesity depend on the severity of obesity, age of onset, and parental obesity (43). After age 25, it is estimated that the average person gains 1.5 lb of fat each year. This weight gain is attributed to a decrease in physical activity and increase in food intake, including large portion sizes and high fat content.

The 34 million obese adults in America represent an accumulated economic cost of $56.3 billion annually. A prevalence-based approach in relationship to the cost of several comorbidities is summarized in table 9.4 (13). The prevalence of these comorbidities indicates the seriousness of obesity. The social and economic consequences of obesity have been evaluated, and it was found that obese individuals were less likely to be married, had lower household incomes, had a higher rate of poverty, and had less formal education (19). These consequences of obesity were greater for women than men.

TABLE 9.4
Economic Costs of Obesity

Comorbidity	Prevalence (%)	Estimated cost (billions of dollars)
Diabetes (non-insulin-dependent)	57.0%	11.3
Cardiovascular disease	19.0%	22.2
Gall bladder	30.0%	2.4
Hypertension	20.0%	1.5
Cancer (breast, colon)	2.5%	1.9
Musculoskeletal	50.0%	17.0
Total		56.3

Adapted from G.A. Colditz, 1992, "Economic cost of obesity," *American Journal of Clinical Nutrition* 55: 5035-5075.

Physiology and Pathophysiology

Obesity results from a positive energy balance. Caloric intake has increased by 200 kcal · day^{-1} in the last decade (16), and physical activity has diminished significantly (16, 30, 39). Current research focuses on control mechanisms that affect the satiety and hunger centers of the brain. The relationship of **leptin** and appetite has contributed significantly to our understanding of the causes of obesity. Leptin is secreted by fat cells, and its role is to control body fat by affecting the satiety center of the hypothalamus. Although not completely understood, leptin appears to be the only hormone secreted from fat cells that acts on the brain to regulate appetite (10). Another control mechanism appears to be the **uncoupling protein-2**, which has been identified in muscle and fat tissue and when activated increases thermogenesis. Other control mechanisms are found in the form of the brain's transmitter chemicals. Increased serotonin production reduces appetite whereas an increased release of dopamine is associated with increased appetite. Several genes have been identified that regulate many of these control mechanisms (16).

It is often difficult to assess the importance of environment and heredity to obesity. In studies of identical twins where heredity plays a more significant role, the heritability is 50% to 90%; adoption studies show 10% to 50% heritability (12). Genes that play a role in obesity are likely susceptibility genes that interact with the environment to influence appetite and metabolism. A dysfunctioning gene may increase the risk of obesity rather than act solely as the causative agent. More than 20 genes have been implicated in obesity, and current research is focusing on three specific genes (16). The ob gene regulates the production of leptin, which increases satiety and energy expenditure. The B$_3$ adrenoreceptor gene is located mainly on adipose tissue and is thought to regulate resting metabolism and lipolysis. Finally, the D$_2$ dopamine receptor gene regulates dopamine production.

The number of fat cells predisposes people to developing obesity. Fat cellularity is influenced by heredity, which sets the limits to cell number and content, and environment, which sets the functional level. At 1 year of age, weight gain is associated with increased size of fat cells but not an indication of obesity in adulthood. Obesity at ages 4 to 11 years is associated with increased number of fat cells, which become a lifelong risk. Adult obesity is usually associated with increased fat cell size, but when these cells reach a finite capacity there will be an increased cell number (37).

Clinical Considerations

The obese patient presents with a unique set of problems that often stem from years of unhealthy eating and inactivity patterns. The location of fat deposits affects the medical complications of obesity and, at the same time, presents challenges in the treatment of this disease. A comprehensive history of both obesity and exercise patterns can guide the clinician to more effective treatment modalities.

Signs and Symptoms

Weight that is 20% above desirable carries an increased risk to health (30); however, the patterns of fat distribution also effect risk. Central or android obesity (upper body) carries a higher risk for diabetes and coronary heart disease than lower body obesity. The enzyme lipoprotein lipase, which regulates the storage of fats as triglyceride, is more active in abdominal obesity and therefore increases fat storage. Upper body obesity is measured by the **waist-to-hip ratio,** and there is an increased health risk in women when the waist-to-hip ratio exceeds 0.8 and in men when this ratio exceeds 0.9 (37). A more important factor in fat distribution is to distinguish abdominal **visceral fat** from subcutaneous fat. Visceral fat lies deep within the body cavities and is associated with a higher risk than subcutaneous fat because of the metabolic characteristics, which include insulin resistance and glucose intolerance. Visceral fat is measured by magnetic resonance imaging, which is expensive and unavailable to most practitioners. A more practical measurement of visceral fat has been described that uses sagittal diameters (36). This technique requires the patient to lie on his or her back, and the sagittal diameter is obtained by measuring the distance from the examination table to a horizontal level placed over the abdomen at the site of the iliac crest. This technique has promise, and it is currently the best practical predictor of visceral fat. The assessment of visceral fat provides additional information regarding health risk and should be used in counseling patients about realistic weight loss. Realistic weight loss should be based on total weight loss and the resulting redistribution of fats.

Diagnostic and Laboratory Evaluations

The diagnoses of most diseases are based on their etiology or pathology; however, with obesity, etiology and pathology are not well defined and are poorly understood. The diagnosis for obesity is based on the degree of overweight or overfat. In respect to obesity, medical and clinical considerations include the following:

1. Medical history
2. Assessment of the degree of obesity
3. Exercise and obesity history
4. Assessment of obesity-related comorbidities
5. Contraindications to exercise
6. Exercise testing
7. Treatment modalities

An obesity history can help the clinician determine compliance to various modes of intervention and may help the patient understand the disease. The purpose of the obesity history is to access the social, psychological, and developmental aspects of the patient's obesity. Figure 9.1 outlines the important components of an obesity history. Family history supplies background information regarding the heritability of obesity and the family environment. In general, if both parents are obese, there is an 80% chance that the child will be obese. If neither parent is obese, there is only a 10% chance that the child will be obese, and if one parent is obese, there is a 50% chance the child will be obese. Age of onset of obesity is important in understanding the potential difficulty in developing behavioral changes. Pound years of overweight is an attempt to combine the degree of overweight with length of time being over-

weight. Comorbidities are related to pound/years overweight, and the success rate at intervention decreases as pound/years overweight increases. Most patients can identify a medical weight as the weight when medical complications become apparent. It may be the weight at which blood pressure or blood lipids become abnormal. The lowest weight maintained for 2 years helps the practitioner develop realistic weight loss goals. Weight cycling can be determined from the loss/regain assessment and helps explain the frustration the patient has with managing obesity. The assessment of social support, coping skills, and control over eating based on previous weight loss attempts provides information about behavioral techniques that may have been useful in the past and helps to establish appropriate behavioral intervention strategies.

The exercise history (figure 9.2) is used to assess the exercise characteristics of the patient and the present level of physical activity. The exercise history helps the clinician understand the patient and his or her experience with physical activity. Types of activities may include competitive sports, recreational activities, gardening, occupational, or specific activities. This helps the patient to realize that all forms of activity contribute to weight loss. Time of day is important in planning a program and recognizes that adherence is improved when the time to exercise is compatible with the patient's lifestyle. Planned activities are those that are scheduled into the day as specific time to exercise. Incidental activities occur within the day that add exercise to the person's lifestyle; they include taking the stairs, park and walk, and increasing activities into daily routines. The social component should be recognized as an important contributor to exercise

— Family history
— Age of onset
— Pound/years overweight
— Landmark medical weight
— Lowest weight maintained 2 years
— Loss/regain history
— Social support
— Coping skills
— Control eating

FIGURE 9.1   Obesity history.

— Types of activities
— Time of day
— Planned vs. incidental
— Social component
— Clothing/shoes
— Perception of intensity/duration
— Frequency/consistency
— Typical interruptions
— Long-term patterns
— Seasonal patterns

FIGURE 9.2   Exercise history.

adherence. Advice on clothing for hot and humid conditions as well as cold and windy days also increases adherence rates. Shoe selection helps prevent soreness and injury. An assessment of exercise intensity, duration, and frequency helps to establish a starting point for an exercise program. The patient's typical interruptions and patterns of activity can be addressed with behavioral intervention and appropriate preplanning.

The term **metabolic syndrome** is commonly used when the following abnormalities are associated with obesity: glucose intolerance, insulin resistance, hyperinsulinism, elevated plasma lipids, and hypertension. These metabolic disorders increase the risk of type 2 diabetes (non-insulin-dependent diabetes) (9), coronary heart disease (7), and stroke (1, 35). Between 80% and 90% of all patients with type 2 diabetes are obese, mainly as a result of insulin resistance. Moderate to severe comorbidities of obesity include increased risk for gallstones, osteoarthritis, sleep apnea, and depression. Cancers of the colon, rectum, and prostate are a greater risk for obese men, whereas obese women have increased risk for cancer of the gallbladder, breast (postmenopausal), cervix, ovaries, and endometrium. Clinical exercise physiologists should recognize that obese patients often have health problems that are discomforting and impair lifestyles. Dermatitis, impaired agility, and heat intolerance must be considered when a patient is counseled about exercise. The risk of these comorbidities can be attributed to the degree and duration of overweight, age, and regional fat distribution (34).

The benefits of a modest weight loss of 10% for patients with these obesity-related comorbidities is well documented (9, 18, 20). Differences in methodology, treatment modalities, patient selection, rate of weight loss, and length of maintained lost weight influence the degree of benefits attained (18, 21).

A J-shaped relationship between BMI and overall mortality rate has been observed in women (26). When women who had never smoked were examined separately, no increase in risk of mortality was found in the leanest women. The relationship of severe obesity (**class III obesity**) and increased mortality rate is clear and unchallenged. Mortality rate related to moderate (**class II**) and mild (**class I**) obesity is more controversial. It has been suggested that BMI is not associated with increased mortality rate when accompanied by moderate to high levels of physical fitness (8, 25). Low levels of cardiorespiratory fitness are a strong independent predictor of cardiovascular disease and all-cause mortality in all classes of obesity (42). Also, active and fit obese patients are at lower death rates compared with sedentary, unfit individuals with normal BMIs.

Treatment

The initial goal in the treatment of obesity is to lose 10% of body weight within 4 to 6 months and reevaluate for further weight loss or for maintenance strategies (see practical application 9.1). Substantial health benefits have been achieved with as little as 10% weight loss (9, 18). The recommended rate of weight loss is an average of 1 to 2 lb per week over the intervention phase (31). During the first 1 to 3 weeks, patients may lose 2 to 5 lb, which can be accounted for from water loss attributable to a greater metabolic contribution from glycogen stores. Three grams of water is stored with each gram of glycogen. The goal is to lose weight while maintaining lean body mass. This can be accomplished by using a **low-calorie diet** (LCD) that provides adequate high-quality protein and exercise. Typically, weight loss with diet alone results in 75% loss of fat and 25% loss of lean tissue (9). A major goal of weight loss programs is to preserve lean body mass, and this is accomplished by appropriate exercise and inclusion of adequate high-quality protein. A comprehensive program for weight loss for class I obesity should include diet, exercise, and behavioral treatment modalities. Drug intervention and surgery may be indicated as adjunct therapy for class II and III obesity (table 9.5).

PRACTICAL APPLICATION 9.1

Treatment Goals

The primary goal of obesity treatment is to maintain a loss of 10% to 15% of body weight. Appropriate weight loss is 1 to 2 lb per week; however, greater weight loss during the first 2 or 3 weeks is common because of increased water loss. Early weight loss may include glycogen stores as well as fat, and each gram of stored glycogen includes 3 g of water. On average, weight loss includes three fourths fat weight and one fourth lean body mass. The ultimate goal is to lose weight while preserving lean body mass. This can be accomplished by designing a program that includes adequate high-quality protein and appropriate exercise.

TABLE 9.5

Suggested Modes of Intervention

	BMI	Intervention
Overweight	25.0-29.9	Diet/exercise
Class I obesity	30.0-34.9	Diet/exercise (possibly drugs)
Class II obesity	35.0-39.9	Drugs (possibly surgery)
Class III obesity	≥40	Surgery

Diet

Weight loss diets are generally categorized as LCDs (800-1500 kcal) or **very low calorie diets** (VLCDs; 800 kcal or less) (31). VLCDs were popular in the early 1990s but were criticized for failure to produce long-term weight loss (40, 41). Recently, however, VLCDs have regained popularity in the clinical setting and have become more effective in the long term. VLCDs improve compliance and provide the physician an opportunity to adjust medications, monitor laboratory data, and counsel on health-related issues. Partial meal replacements have shown promise especially in the maintenance phase, as this technique allows patients to "treat themselves" by using meal replacements as they monitor their own weight and notice weight gain. VLCDs result in more rapid weight loss during the treatment phase; however, a greater proportion of lean body mass is lost and weight is regained more quickly. When exercise is combined with a VLCD, there is little improvement in weight loss during the treatment phase, because the calorie deficit from the VLCD has a greater impact than the relatively little calorie expenditure from exercise. Exercise, however, provides the benefits of improved cardiovascular function, endurance, and self-esteem even when additional weight loss is not evident. Those who continue to exercise realize the weight loss benefits of exercise during the maintenance phase.

LCDs are recommended for weight loss in class I and class II obesity because they can contribute enough calories to provide adequate nutrients, they are easier to adhere to, and they result in less loss of lean tissue, especially when exercise is used as an adjunct. LCDs require patient education for food preference, food preparation, shopping, eating out, portion size, and understanding food labels. The addition of exercise to an LCD improves weight loss during treatment as well as during maintenance

because the additional calorie deficit from exercise adds to the calorie deficit from the LCD.

The macronutrient composition should follow the recommendations of the *Dietary Guidelines for Americans* (5th ed.) (14) with approximately 30% of calories from fat, 58% of calories from carbohydrates, and 12% of calories from protein. As the diet becomes more hypocaloric, it must be adjusted to provide adequate protein ($0.8\text{-}1.0\,\text{g}\cdot\text{kg}^{-1}$) based on body weight. A 1200-kcal diet will require 15% or more of its calories from protein, depending on the person's body weight. The Food Pyramid (14) recommends 1600 to 2800 kcal per day depending on whether one is following the low or high end of the number of servings. These recommendations are designed for healthy Americans to maintain their weight. Adaptations can be made to provide a 1200-kcal, low-calorie diet. The Dietary Exchange List developed by the American Dietetic Association and the American Diabetes Association is also a popular approach to weight loss and is highly recommended because of its "exchange list" approach that provides information on protein, carbohydrate, fat, and calorie content of the various exchanges (15). Popular (fad) diets have been used with successful weight loss outcomes because they are hypocaloric in nature, but weight regain tends to occur rapidly. Patients should be advised of the potential complications of popular/fad diets and their high rates of recidivism (see practical application 9.2).

The optimal macronutrient composition of a weight loss diet has not been established. High-carbohydrate, low-fat diets or high-fat, low-carbohydrate diets have not been shown to be any

PRACTICAL APPLICATION 9.2

Popular Diets

Popular and fad diets have gained the support of those seeking weight loss because of testimonials from many who have experienced early success. Popular diets, however, frequently do not comply to guidelines for evaluating weight loss programs. These guidelines include (a) a balance of fat, carbohydrates, and protein; (b) sources of key nutrients; (c) long-term eating patterns; (d) inclusion of exercise; and (e) weight loss of 1 to 2 lb per week. Commonly, early success in weight loss is followed by quick weight regain.

more beneficial than the standard recommendation in the *Dietary Guidelines for Americans* (14). When researchers compared diets of different carbohydrate composition, there were no differences in weight loss between a 25% carbohydrate diet (45% fat, 30% protein) and a 45% carbohydrate diet (26% fat, 29% protein) (17). A similar study compared diets composed of 25%, 45%, and 75% carbohydrates and also found no significant differences in weight loss (4).

Behavior Therapy

Behavior therapy should be used in conjunction with dietary, exercise, pharmacological, and surgical interventions. The goal is to make patients aware of their eating and exercise habits and to restructure their environment to minimize eating cues and maximize activity cues. Behavioral strategies include teaching a variety of techniques to help the patient make positive changes. These include self-monitoring, stimulus control, contingency management, problem solving, cognitive restructuring, stress management, and social support.

Self-monitoring requires the patient to observe and document his or her weight, dietary behaviors, and exercise patterns. The act of observation and writing down the behaviors is a constant reminder of eating and activity patterns. Periodic assessment of the records sensitizes the patient to inappropriate behaviors and highlights positive changes in lifestyle. Self-monitoring is the mainstay of behavior modification and the most helpful technique for behavior change.

Stimulus control uses procedures that control external cues including sight or smell of food, time of day, and advertisements. Lean people tend to respond to internal cues such as feelings of hunger. Techniques used to help eliminate these external cues include limiting the time and place of eating, limiting exposure to high-calorie foods, and avoiding high-risk situations.

Contingency management uses reward systems to reinforce positive behaviors. Initially, rewards should come from the therapist, but patients should learn principles of rewarding positive behaviors and use self-rewards for motivation. Rewards for short-term behavior changes of 1 to 4 weeks tend to be most useful. Rewards can be given for weight loss as well as changing exercise and eating behaviors. Typical rewards include money, treating oneself to a movie, making a long-distance telephone call, or purchasing clothing.

Problem-solving techniques help the patient identify specific problems associated with eating or activity and design strategies to promote healthier alternatives. The patient learns to identify problems, choose a solution from several alternatives, and evaluate the outcome. For instance, a patient may identify the movie theater as a high-risk situation because of its association with snack foods. An alternative may be to eat before going to the movies or plan to chew sugarless gum. An outcome evaluation would reflect the behavior chosen by the patient.

Cognitive restructuring is used to identify and modify negative thoughts and attitudes related to diet and exercise. The purpose is to alter feelings, beliefs, and perceptions related to weight loss. Visual imagery is useful, whereby the patient visualizes a problem situation and seeing him- or herself responding positively in an otherwise negative situation. This technique is especially effective when clients attend parties where food is plentiful or when eating away from home.

Stress management is a technique which recognizes that many people eat in response to stress. Techniques to manage stress include imagery, relaxation, and coping skills, and they can often diffuse situations that lead to overeating or to a sedentary lifestyle. Stress management can be used when eating and exercise lifestyles change as a result of overcommitment, poor planning, or time management problems.

Social support from friends, coworkers, spouses, and family is essential for weight loss. Patients should learn when and how to use support persons and understand the potential of abusing the opportunity for support. Refusing to go with friends to a restaurant that may be high risk for you is inappropriate; however, asking for support in choosing a low-calorie meal is suitable.

Pharmacology

Weight loss drugs approved by the Food and Drug Administration (FDA) are useful for selected patients with Class II and III obesity. Pharmacological intervention should be used in combination with diet and exercise. Clinical guidelines (30) suggest patient selection for drug therapy to include a BMI greater than 30 kg · m^{-2} or a BMI greater than 27 kg · m^{-2} with existing obesity-related comorbidities or risk factors. Currently two FDA weight loss drugs have been approved for long-term use in the treatment of obesity. Sibutramine (Meridia), which was accepted by the FDA in 1997, is a norepinephrine and serotonin uptake inhibitor that reduces appetite as well as increasing metabolism rate. Patients lose 7% to 8% of initial weight in 1 year compared with a 1% to 2% loss in a placebo group (24). Orlistat (Xenical) was accepted by the FDA in 1999 and inhibits pancreatic lipase and decreases fat absorption. Orlistat prevents the absorption of about 30% of dietary fat, which may account for 200 kcal per day assuming the patient follows a low-calorie, fat-restricted diet. Orlistat results in 8% to 10% weight loss in one year (6). Phentermine is accepted by the FDA for short-term use (3 months or less) and increases norepinephrine levels in the brain that reduce hunger. Phentermine's popularity has declined substantially since the withdrawal of fenfluramine in 1997. The combination of fenfluramine and phentermine (Fen-phen) was extremely popular until fenfluramine was taken off the market because of its association with valvular heart disease (6).

Surgery

Gastrointestinal surgery can result in substantial long-term weight loss and is recommended for selected patients with a BMI greater than 40 kg · m^{-2} or BMI greater than 35 kg · m^{-2} who have morbid conditions and acceptable surgical risk (30). Intestinal bypass (jejunoileal bypass) surgeries were performed in the 1950s; however, gastric surgeries are often preferred today because of fewer side effects and less complicated procedures. **Gastric bypass** (gastric stapling) were performed in the mid 1960s and **vertical banded gastroplasty** became popular in the 1980s. More recently, vertical banded gastroplasty has been successfully performed with laparoscopic techniques. In general, surgery results in 30% weight loss the first year with an additional 10% loss the second year. Weight loss generally plateaus at 24 to 36 months.

Exercise

Exercise is an important and favorable adjunct to dietary, pharmacological, and surgical treatment of obesity. Exercise alone without calorie restriction has not shown long-term weight loss success; however, the improvements in cardiovascular function can be beneficial to related comorbidities. Exercise improves self-esteem, which increases adherence to both diet and physical activity. The approach to physical activity is to develop a planned and monitored program of activities based on the patient's needs. Additionally, educational sessions should focus on restructuring the environment to increase incidental activities of daily living. Exercise should begin slowly with walking to avoid soreness and injury and should increase intensity as functional capacity improves.

Graded Exercise Testing

Graded exercise testing is commonly used in obese individuals to evaluate numerous chronic diseases associated with obesity. Depending on the degree of obesity, the graded exercise test may need to be modified to evaluate the specific chronic disease.

Medical Evaluation

Obese patients often exhibit the "metabolic syndrome," a cluster of metabolic characteristics that include hypertension, hyperlipidemia, insulin resistance, and hyperinsulinemia. These metabolic disorders can lead to type 2 diabetes (non-insulin-dependent diabetes), cardiovascular disease, and stroke.

Contraindications

Obesity itself is not a contraindication to graded exercise testing. However, the comorbidities associated with obesity may present obstacles to graded exercise testing. Specific contraindications to exercise testing are discussed in chapter 6.

Recommendations and Anticipated Responses

The purpose of a graded exercise test for obese patients is to improve the diagnosis for coronary heart disease, determine functional capacity, and derive an individual training intensity. Stress testing of obese patients is often difficult because they frequently have low functional capacity, orthopedic problems that limit total work, and fat distribution that affects balance.

Selecting an appropriate test protocol that reaches maximal limits in 8 to 12 min can optimize exercise testing. Exercise protocols that increase workloads at 1 metabolic equivalent each minute or those that use the Ramp protocol are most appropriate for obese patients because they allow time to adjust to the cardiac, respiratory, and metabolic demands of each workload. A nomogram for estimating exercise capacity (29) is used to predict the end point for the Ramp protocol. The Ramp protocol uses computerized workload increments that produce small but frequent changes that reach the estimated end point in approximately 10 min (27, 28). Some treadmills have a weight capacity of 350 lb, and therefore alternate methods of exercise testing may have to be used for extremely heavy patients. The treadmill is most often the first choice for testing because patients generally will be walking in an exercise program and it is a mode of activity that is less threatening to the patient. Cycle ergometers may be used; however, many obese patients are unfamiliar with the activity and may experience leg fatigue that limits total work capacity.

Other special considerations for the obese patient include the use of an appropriate size blood pressure cuff. Often a leg cuff is used and taped in position to prevent slipping. The patient should be advised to remove the hands from the treadmill bar because the vibrations from the treadmill often interfere with blood pressure measurements. Electrode placement on the obese patient may present additional problems by increasing artifact. Use of elastic bandages may reduce artifact, or placement of limb leads on the back, especially for large-breasted women, is recommended. A belt extension for holding the electrocardiogram pack is often required.

Exercise Prescription

Treatment of the obese patient requires both dietary and exercise intervention using principles of behavior change to increase long-term adherence. The American College of Sports Medicine guidelines recommend that an exercise program focus on physical activities that promote an energy expenditure of 300 to 500 kcal each day or 1000 to 2000 kcal per week (2). These recommendations may be inadequate, and a greater caloric expenditure may be necessary to maintain weight loss. The National Weight Control Registry, which lists people who successfully maintained an average weight loss of approximately 30 lb for 6 years, showed that these people expended 2800 kcal per week. This goal is appropriate; however, some Class II patients and most Class III patients will have to gradually progress to these higher levels of energy expenditure. Counseling about physical activity helps practitioners develop realistic goals and motivates the patient to take ownership of his or her exercise program (32).

Intensity and duration have to be manipulated such that the intensity is low enough to allow suitable duration to expend the recommended caloric energy. For many obese patients, the intensity will not be great enough to improve cardiovascular fitness; however, the initial focus should be on weight loss and therefore caloric expenditure. As the exercise progresses and the patient improves, higher intensity activities should be encouraged. Patients should be encouraged to progressively increase the duration from 20 min per day to 60 min, 7 days a week. An exercise program for obese patients should include both the supervised and nonsupervised

phases with adaptations in modes, intensity, duration, and frequency to provide adequate calorie expenditure while preventing soreness and injury. Patients with existing comorbidities should preferably participate in a supervised exercise program 3 to 5 days per week with a prescribed intensity and duration to treat their comorbidities. Patients should be physically active 1 hr each day including the days of supervised exercise; therefore, they may have to supplement the time with walking to accumulate 60 min. The remaining days of the week (2-4 days) can be nonsupervised with self-reported exercise to accumulate 1 hr of physical activity each day.

Special Exercise Considerations

Obese patients often lack motivation to become physically active. Barriers to exercise affect obese patients more than their nonobese counterparts. Special considerations in the treatment of obesity with exercise can increase patient adherence.

Non-Weight-Bearing Activities

In addition to aquatic activities, other non-weight-bearing options are tolerated well by obese patients. Stationary cycling, recumbent cycling, and rowing add variety to exercise choices and reduce joint pain. Obese patients often lack balance, and these activities may be difficult. The practitioner should adapt these modes of exercise by providing larger seats and more stable equipment.

Recreational Activities

Bowling, tennis, volleyball, croquet, biking, hiking, and golfing are examples of recreational activities that promote calorie expenditure and should be encouraged. Patients should be educated to estimate the total activity time during recreation and not count the inactive portion of time. Two hours of bowling may only represent 30 min of actual physical activity.

Planned Versus Incidental

Planned activity provides 60 min of exercise that should be required each day. Patients need to manage their time and plan when they will exercise and which activities they will do, remembering that walking at any pace is a minimal requirement and that several bouts of activity can be used to accumulate the required time. Incidental exercise is the increased lifestyle activity that takes place throughout the day at home, at work, and during recreation. Taking the stairs rather than an elevator or escalator, parking and walking, or using public transportation can improve activity levels. The contribution of incidental exercise to total caloric expenditure is significant (7) and may be as beneficial as the planned exercise.

Supervised Versus Nonsupervised

Patients with existing comorbidities should exercise 3 to 5 days a week in a supervised exercise program at an intensity appropriate for anticipated cardiovascular and metabolic improvements. The purpose of the supervised program is to monitor patients' responses to exercise, provide a safe environment, and give feedback to enhance motivation and adherence. If the patient has not accumulated 60 min of exercise, he or she should include low-level intensity work to accumulate 60 min. Patients should supplement the 3 to 5 days of supervised exercise with additional days of nonsupervised activity to ensure exercise each day of the week. Nonsupervised exercise can be done at home and should emphasize walking and other low-intensity activities.

Exercise Recommendations

Modes of activity should be selected that are compatible with the patient's interest, comfort zone, and lifestyle. Obese patients tend to resist new experiences with exercise and remain committed to activities that may be boring but with which they feel comfortable. These patients need to be encouraged to try alternate forms of physical activity to increase variety in their exercise program. Non-weight-bearing activities reduce the problems of injury and are recommended for patients who develop soreness during the initial stages of exercise. All forms of physical activity should be recognized as exercise because they contribute to calorie expenditure. Therefore, recreational activities, active hobbies, and daily activities should be encouraged and monitored for documentation and motivation.

Cardiovascular Training

Cardiovascular training is suggested to be the most important form of exercise for obese individuals. This form of exercise is preferred because it promotes the greatest caloric expenditure and ultimately has the greatest impact on weight loss.

Walking   There are few disadvantages to walking; it is an activity that most patients have experienced, and they report less soreness and fewer injuries when walking. Walking is available to most patients and does not require special facilities. A minimum amount of attention is necessary, which makes socializing easy and convenient. Patients should include walking in a prescribed program and should be encouraged to walk in parks, around their neighborhood, on walking trails, and at shopping malls and to increase walking in all daily activities.

Jogging   Jogging can be an appropriate activity for some patients with Class I obesity because it may be tolerated well and it has the advantage of burning more calories than walking. Additionally, jogging can provide greater cardiorespiratory endurance benefits. However, the exercise specialist must consider the increased risk of skeletal muscle soreness and injury and the patient's ability to sustain jogging for the recommended duration.

Water Activities   Water activities are recommended as a non-weight-bearing exercise alternative for patients with joint soreness. Patients who experience heat intolerance with other activities are often able to perform aquatic exercise. Because of the buoyancy of the body in water, the caloric expenditure is reduced. Patients often will not consider water activities because of the effort necessary to get into and out of the pool and because of their sensitivity to their appearance in a bathing suit. Devoting pool time to classes strictly for obese patients can diminish these concerns.

Resistance Training

Resistance exercise can preserve lean body mass during weight loss, and it adds variety in the selection of activities. The strength equipment may be threatening to some patients, as it requires some mobility to maneuver about the weights. The exercise prescription for obese patients should include resistance that is 30% to 40% of peak strength, 15 to 20 repetitions in 30 s, 8 to 12 stations, with approximately 20 to 30 min per session. Resistance exercises can be used 2 to 3 days a week. The primary benefit of the prescribed resistance program is to improve muscle endurance and, secondarily, to increase muscle strength.

Range of Motion

Obese patients may have a reduced range of motion as a result of increased fat mass surrounding joints of the body, in conjunction with a lack of stretching. As a result, these patients often respond slowly to changes in body position and have poor balance.

Intensity

The intensity of exercise must be adjusted so that the patient can endure 1 hr of activity each day. Exercise designed to treat obesity-related comorbidities should take place in a supervised program with intensity based on desired outcomes. Supervised programs may provide an intensity of 70% to 85% of maximal heart rate to improve cardiovascular function. The duration is generally less than 60 min; therefore, the patient should include low-intensity activities (walking) to accumulate 1 hr of exercise. Nonsupervised home activities should be at an intensity of only 50% to 60% of maximal heart rate, and the emphasis should be on duration.

Duration

Recommendations for increased physical activity should include prescribed exercise as well as an emphasis on increasing lifestyle physical activity, such as park and walk and increased use of stairs. Obese patients should begin with 20 min each day and progressively increase the duration of exercise to accumulate 60 min each day of physical activity. Some previously sedentary Class II and III patients may need to start with 5 min of walking at a slow pace and progress to 20 min each day for 2 weeks, 40 min each day for 2 weeks, and finally 60 min. This progress is intended to increase compliance to the duration of each session as well as to daily exercise. An accumulation of time over several sessions in a day is equally as beneficial as one continuous work bout for calorie expenditure.

Frequency

It is clear that behavioral changes in activity must be consistent and long lasting in order for the pa-

tient to lose weight and maintain weight loss over extended time. Physical activity must become part of a lifestyle that values its contribution to weight management. Therefore, it is important to motivate patients to be physically active, with a minimum of leisure walking all days of the week for 60 min each day. All obese patients can attain this goal when they understand that all activity counts and an accumulation of activity with several short bouts throughout the day is acceptable.

Conclusion

Management of the obese patient requires an understanding of the causes of obesity and a comprehensive history specific to each patient. The obese patient should be treated not only for the co-morbidities associated with obesity but also for the obesity itself. Dietary restriction, with an emphasis on reducing fat intake and choosing appropriate portion sizes, is the hallmark of dietary intervention. Physical activity is an important adjunct to be used in combination with diet, medication, and surgery. Exercise alone has little effect on weight loss; however, it diminishes a number of risk factors that lead to morbidity and mortality in obese patients. Finally, exercise improves self-esteem, which in turn improves adherence to both diet and exercise. These important and lasting benefits promote weight loss and improve maintenance in the long term.

Case Study 9

Medical History and Diagnosis

Mr. CJ is a 52-year-old white male security guard who is screened for health risks at the Clinic's Wellness Center; at 36% fat, he is told he should lose "significant weight." Additionally, because of his obesity, his employer believes he is unable to perform duties, which may include running up stairs or chasing trespassers. He has a 60 pack per year history of smoking but quit 5 years ago. He has progressively gained weight over the past 10 years. In the past year there was a 30-lb weight gain. Hypertension diagnosed 5 years ago is being treated with Tenormin.

The patient admits to a sedentary lifestyle, being "active" only when forced to by job requirements or family events. He was a high school football player and recently has thought about diet and exercise to get back to his competitive weight, but he can't seem to get motivated. When asked why he isn't more physically active, he complains of time constraints, and, because of being sensitive to body size, he "is not about to join a health club."

continued

Clinical Data

Age (years) = 52

Sex = M

Height (in.) = 70

Weight (lb) = 230

BMI (kg · m^{-2}) = 32.8

Body fat (%) = 36

Blood pressure (mm Hg) = 142/90

Total cholesterol (mg · dl^{-1}) = 235

High-density lipoprotein (mg · dl^{-1}) = 47

Low density lipoprotein (mg · dl^{-1}) = 151

Triglycerides (mg · dl^{-1}) = 187

Glucose (mg · dl^{-1}) = 126

Case Study 9   Discussion Questions

1. Based on the National Institutes of Health *Clinical Guidelines on the Identification, Evaluation, and Treatment of Overweight and Obesity in Adults*, what is this subject's obesity classification?

2. How much weight loss (in pounds) is necessary to get this subject into the "overweight" category?

3. What would the corresponding BMI be at this weight?

4. What suggestions would you have for supervised versus unsupervised exercise?

5. What suggestions would you provide for weight-bearing and non-weight-bearing modes of activity?

6. Design an exercise program over a 12-week period that includes the following variables: (a) energy expenditure = 1000 kcal per week, (b) three modes of aerobic activity, (c) appropriate intensity, frequency, and duration. Based on the mode of activity, frequency, intensity, and duration, determine the weekly energy expenditure and weight loss attributed to physical activity.

Glossary

behavior therapy—Strategies, based on learning principles such as reinforcement, that provide tools for overcoming barriers to compliance with dietary and physical activity.

body mass index (BMI)—Relative weight for height. It is calculated as weight (kg)/height squared (m)^2 or weight (lb)/height squared (in.)^2 × 703.

Class I obesity—A BMI of 30.0 to 34.9.

Class II obesity—A BMI of 35.0 to 39.9.

Class III obesity—A BMI equal to or greater than 40.0.

exercise prescription—A recommendation for a program of physical activity that includes the type, intensity, frequency, and duration of activity and the amount of total work.

gastric bypass—A surgical procedure that combines the creation of a small stomach pouch to restrict food intake and the construction of bypasses of the duodenum of the small intestine to cause food malabsorption.

gastroplasty—A surgical procedure that limits the amount of food the stomach can hold by closing off part of the stomach. Food intake is restricted by creating a small pouch at the top of the stomach where food enters from the esophagus and a lower outlet that is small, causing a delay of stomach emptying and feeling of fullness.

leptin—A protein messenger from adipose tissue to the satiety center in the hypothalamus and it regulates appetite.

low calorie diet (LCD)—A hypocaloric diet of 800 to 1500 kcal per day.

metabolic syndrome—A cluster of metabolic characteristics that include insulin resistance, hyperinsulinism, hypertension, and hyperlipidemia. The metabolic syndrome often leads to non-insulin-dependent diabetes, cardiovascular disease, and stroke.

obesity—A condition of having an abnormally high proportion of body fat. Defined as having a body mass index greater than 30.

overweight—A body mass index of 25.0 to 29.9.

uncoupling protein 2—A messenger protein found in muscle and adipose tissue that, when activated, can increase thermogenesis (metabolism).

vertical banded gastroplasty—A surgical procedure on the stomach that involves constructing a small pouch in the stomach that empties through a narrow opening into the distal stomach and duodenum.

very low calorie diet (VLCD)—A hypocaloric diet of 600 to 800 kcal/day.

visceral fat—One of three compartments of abdominal fat. The other two compartments are retroperitoneal and subcutaneous fat.

waist-to-hip ratio—The ratio of a person's waist circumference to hip circumference. For men a ratio of 0.90 or less is considered safe, and for women 0.80 or less.

References

1. Abott RD, Behrens GR, Shart DS. Body mass index and thromboembolic stroke in nonsmoking men in older middle age: The Honolulu Heart Program. Stroke 25: 2370-2376, 1994.

2. ACSM's Guidelines for Exercise Testing and Prescription (5th ed.). American College of Sports Medicine. Williams & Wilkins, Baltimore, 1995.

3. AHA Medical/Scientific Statement on Obesity and Heart Disease. Circulation 96: 3248-3250, 1997.

4. Alford BB, Blankenship AC, Hagen RD. The effects of variations in carbohydrate, protein, and fat content of diet upon weight loss, blood values and nutrient intake of adult obese men. J Am Diet Assoc 90: 534-540, 1990.

5. Anderson RE, Wadden TA, Bartlett SJ, Zemel B, Verde TJ, Franckowiak SC. Effects of lifestyle activity vs. structured aerobic exercise in obese women. JAMA 281: 335-340, 1999.

6. Anderson RE, Wadden TA. Treating the obese patient: Suggestions for primary care practice. Arch Fam Med 8: 156-167, 1999.

7. Anderson RE, Wadden TA, Bartlett ST, Vogt RA, Weinstock RS. Relation of weight loss in serum lipids and lipoproteins in obese women. Am J Clin Nutr 62: 350-357, 1995.

8. Barlow CE, Kohl HW, Gibbon LW, Blair SN. Physical fitness, mortality, and obesity. Int J Obes 19: 541-544, 1995.

9. Basello O, Armellini F, Zamboni M, Fitchat M. The benefits of modest weight loss on Type II diabetes. Int J Obes 12: 510-513, 1997.

10. Behme MT. Leptin: Product of the obese gene. Nutr Today 31: 138-141, 1996.

11. Blackburn GL, Dwyer J, Flanders WD, Hill JO, Keller CH, Pi-Sunyer FX, St Jear ST, Willett WC. Report of the American Institute of Nutrition (AIN) Steering Committee on Health Weight. J Nutr 124: 2240-2243, 1994.

12. Bray GA. Progress in understanding the genetics of obesity. J Nutr 127: S940-S942, 1997.

13. Colditz GA. Economic costs of obesity. Am J Clin Nutr 55: 5035-5075, 1992.

14. Department of Agriculture, Department of Health and Human Services. Nutrition and Your Health: Dietary Guidelines for Americans (5th ed.). Government Printing Office, Washington, DC, 2000.

15. Exchange List in Weight Management. American Dietetic Association, Chicago, 1989.

16. Foryet JP, Carlos Post WS. Diet, genetics, and obesity. Food Technol 51: 70-73, 1997.

17. Golay A, Eigenheer G, Morel Y, Kujowski P, Lehman T, deTonnac N. Weight-loss with low or high carbohydrate diet? Int J Obes 20: 1067-1072, 1996.

18. Goldstein DJ. Beneficial health effects of modest weight loss. Int J Obes 16: 397-415, 1992.

19. Gortmaker SL, Must A, Perrin JM, Sobel AM, Dietz WH. Social and economic consequences of overweight in adolescence and young adulthood. N Engl J Med 329: 1008-1012, 1993.

20. Kanders BS, Peterson FJ, Lavin PT, Norton DE, Istfan NW, Blackburn GL. Long term health effects associated with significant weight loss: A study of dose-response effect. In: Blackburn GL, Kanders BS, eds. Obesity Pathophysiology, Psychology and Treatment. Chapman and Hall, New York, 1994, pp. 167-185.

21. Kaplan RM, Atkins CJ. Selective attrition causes overestimates of treatment effects in studies of weight loss. Addict Behav 12: 297-302, 1987.

22. Klem ML, Wing RR, McGuire MT, Seagle HM, Hill JO. A descriptive study of individuals successful at long-term maintenance of substantial weight loss. Am J Clin Nutr 66: 239-246, 1997.

23. Kuczmanski RJ, Flegal KM, Campbell SM, Johnson CL. Increasing prevalence of overweight among US adults. The National Health and Nutrition Examination Survey, 1960 to 1991. JAMA 272: 205-211, 1994.

24. Lean MJ. Sibutramine: A review of clinical efficacy. Int J Obes 21: 530-536, 1997.

25. Lee CD, Jackson AS, Blair SN. US weight guidelines: Is it also important to consider cardiorespiratory fitness? Int J Obes 22: 52-57, 1998.

26. Manson JE, Willett WC, Stampfer MJ, Colditz GA, Hunter DJ, Hankinson SE, Hennekens LH, Speizer FE. Body weight and mortality among women. N Engl J Med 333: 677-685, 1995.

27. Myers J, Buchanan N, Smith D, Neutel J, Bowes E, Walsh D, Froelicher V. Individualized ramp treadmill observations one new protocol. Chest 101: 2365-2415, 1992.

28. Myers J, Buchanan N, Walsh D, Kraemer M, McAuley P, Hamilton-Wessler M, Froelicher V. Comparison of ramp versus standard exercise protocols. J Am Coll Cardiol 17: 1334-1342, 1991.

29. Myers J, Do D, Herbert W, Ribisl P, Froelicher F. A nomogram to predict exercise capacity from a specific activity questionnaire and clinical data. Am J Cardiol 73: 591-596, 1994.

30. Nation Institutes of Health Consensus Development Conference Statement. Ann Intern Med 103: 1073-1077, 1995.

31. National Institutes of Health. Clinical Guidelines on the Identification, Evaluation, and Treatment of Overweight and Obesity in Adults. National Institutes of Health, Bethesda, MD, 1998.

32. Parr RB, Capozzi L. Counseling patients about physical activity. J Am Acad Phys Assist 10: 45-49, 1997.

33. Physical Activity in the Treatment of Obesity and Its Co-Morbidities. American College of Sports Medicine/Centers for Disease Control and Prevention. In Press.

34. Rexrode KM, Carey VJ, Henmakens CH, Walters EE, Colditz GA, Stampfer MJ, Willett WC, Manson JE. Abdominal adiposity and coronary heart disease in women. JAMA 280: 1843-1848, 1998.

35. Rexrode KM, Henneken CH, Willett WC, Colditz GA, Stampfer MJ, Rick-Edwards JW, Speizer FE, Manson JE. A prospective study of body mass index, weight change, and risk of stroke in women. JAMA 277: 1539-1545, 1997.

36. Sjostrom L. Impact of body weight, body composition, and adipose tissue distribution on morbidity and mortality. In: Stunkard AJ, Wadden TA, eds. Obesity: Theory and Therapy. Raven Press, New York, 1993, pp. 13-41.

37. Skelton NK, Skelton WP. Medical implication of obesity. Postgrad Med 92: 151-152, 1992.

38. U.S. Department of Health and Human Services. Healthy People 2010: National Health Promotion and Disease Prevention Objectives (DHHS publication PHS 91-50212). U.S. Department of Health and Human Services, Washington, DC, 2000.

39. U.S. Department of Health and Human Services. Physical Activity and Health: A Report of the Surgeon General. U.S. Department of Health and Human Services, Center for Disease Control and Prevention, National Center for Chronic Disease Prevention and Health Promotion, Atlanta, GA, 1996.

40. Very low-calorie diets: National Task Force on the Prevention and Treatment of Obesity. National Institutes of Health. JAMA 270: 967-974, 1993.

41. Wadden TA, Van Itallie TB, Blackburn GA. Responsible and irresponsible use of very-low-calorie diet in the treatment of obesity. JAMA 263: 83-85, 1990.

42. Wei W, Kamport JB, Barlow CE, Nichaman MZ, Gibbons LW, Paffenbarger RS, Blair SN. Relationship between low cardiorespiratory fitness and mortality in normal-weight, overweight, and obese men. JAMA 282: 1547-1553, 1999.

43. Whitaker RC, Wright JA, Pepe MS, Seidel KD, Dietz WH. Predicting obesity in young adulthood from childhood and parental obesity. N Engl J Med 337: 869-873, 1997.

Paul M. Gordon, PhD, MPH, FASCM
School of Medicine
West Virginia University
Morgantown, WV

Hyperlipidemia and Dyslipidemia

A linear association between elevated blood cholesterol levels and coronary artery disease and atherosclerotic disease in general is well established. A variety of abnormal lipid patterns have been identified that may contribute to the development of atherosclerotic diseases. This chapter discusses the primary lipid abnormalities seen in the general population and their medical and clinical relevance. Furthermore, the potential for exercise to improve lipid profiles is reviewed, and specific recommendations are provided for prescribing exercise in the hyperlipidemic patient.

Unfortunately, the term *hyperlipidemia* has been incorrectly used by much of the medical community to mean any form of lipid abnormality. More appropriately, hyperlipidemia is any condition that elevates fasting blood triglyceride or cholesterol levels. However, when genetic, environmental, and pathological factors combine to abnormally alter blood lipid and lipoprotein concentrations, the condition is better termed dyslipidemia. Although severe forms of dyslipidemia are linked to genetic defects in cholesterol metabolism, less severe forms may result secondarily either because of other diseases (e.g., diabetes) or as a result of combining one's specific genetic pattern with various environmental exposures such as diet, exercise, body composition, and smoking. Secondary dyslipidemia may result from diabetes mellitus, hypothyroidism, renal insufficiency, nephrotic kidney disease, or biliary obstruction.

Scope

The prevalence of high blood cholesterol in the United States has been on the decline. The Third National Health and Nutrition Examination Survey (NHANES III; 1988-1994) demonstrates that the public's intake of saturated fat and total fat has declined. NHANES III also shows that blood cholesterol levels have dropped. Since 1978, average total cholesterol levels among U.S. adults have fallen from 213 to 203 mg · dl^{-1}; the prevalence of cholesterol of 240 mg · dl^{-1} or higher has declined from 26% to 19%; and the prevalence of high blood cholesterol requiring medical advice and dietary therapy has dropped from 36% to 29%. These findings appear to mirror the decline in coronary heart disease (CHD) mortality rates. The decline in blood cholesterol levels is likely related to several factors. Cholesterol awareness among the public has improved. The latest Cholesterol Awareness Survey of physicians and the public shows that from 1983 to 1995, the percentage of the public who had ever had their blood cholesterol checked increased from 35% to 75%. This means that approximately 70 to 80 million Americans who in 1983 were unaware of their blood cholesterol level have taken action to learn where they stand. Moreover, physicians now report initiating diet and drug treatment at much lower cholesterol levels than in 1983. The targeted efforts to treat lower blood cholesterol levels are probably attributable to the increased availability of effective medications that have relatively minor side effects and the evidence suggesting that their use appears to lower CHD mortality rates.

Physiology and Pathophysiology

Lipids are not water-soluble and consequently are transported through the plasma compartment as a lipoprotein. A lipoprotein is spherical in shape, made up of a lipid core that contains triglyceride, free and esterified cholesterol, and phospholipid and surrounded by an apoprotein to provide water solubility (figure 10.1). Lipoproteins differ slightly by their content of apoprotein, cholesterol (free and esterified), and triglyceride, resulting in differing density levels as determined by ultracentrifugation and electrophoretic mobility.

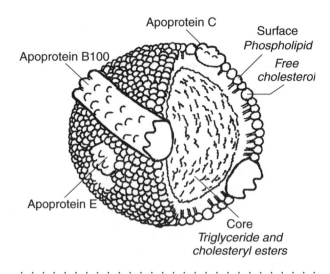

FIGURE 10.1   General structure of lipoproteins (a schematic representation of very low density lipoprotein).

Reprinted, from *William's textbook of endocrinology*, 9th ed. Wilson, © 1998, with permission from Elsevier Science.

The primary lipoprotein classifications (see table 10.1) include chylomicron, very low-density lipoprotein (VLDL), low-density lipoprotein (LDL), and high-density lipoprotein (HDL). The chylomicron is synthesized from the intestinal absorption of dietary fat and is responsible for the transport of dietary fat to extrahepatic tissue. Chylomicrons are catabolized in the liver via the apolipoprotein (apo) E receptor-mediated pathway. VLDL is synthesized in the liver and is the primary mechanism for transporting endogenous triglycerides. The VLDL is converted into an intermediate density lipoprotein (IDL) following the removal of triglyceride at the extrahepatic tissue. The IDL, which is higher in cholesteryl ester content, is rapidly converted into LDL. LDL is the primary carrier of cholesterol in the bloodstream and is usually the most prevalent lipoprotein form in humans. Elevated levels of LDL are linearly associated with an increase in coronary artery disease (CAD) risk (4). Lipoprotein (a), or Lp(a), is a subclassification of LDL and is also highly **atherogenic.** Lp(a) contains apo A, a protein that is linked to apo B and has a similar structure to plasminogen. Consequently, Lp(a) may inhibit normal **fibrinolytic** activity, increasing the likelihood of the development of a thrombus (28). HDLs have proven to be cardioprotective, in part through a reverse cholesterol transport mechanism. Free cholesterol from circulating lipoproteins and possibly extrahepatic tissue is taken up and transported via circulating HDL to the liver, where it is catabolized. The exchange of lipids and biotransformation of lipoproteins can occur in the intravascular space and are

TABLE 10.1

Lipoprotein Classifications and Corresponding Forms of Dyslipidemia

Lipoprotein class	Major lipids	Apolipoproteins	Density (g · ml⁻¹)	Diameter (Å)	Conditions where elevated
Chylomicron	Dietary triglyceride	Major: A-IV, B-48, B-100, H Minor: C-I, C-II, C-III, A-I, A-II, E	<0.95	800-5000	LPL deficiency, apo C-II deficiency, apo E-2
VLDL	Endogenous triglycerides	Major: B-100, C-III, E Minor: A-I, A-II, B-48, C-II, D, G	0.96-1.006	300-800	Familial combined hyperlipidemia Familial hyper-triglyceridemia Apo C-II deficiency Type III hyper-lipoproteinemia
IDL		Major: B-100, C-III, apo A Minor: C-I, C-II, E	1.006-1.019	250-350	Type III hyper-lipoproteinemia
LDL		Major: B-100, apo A Minor: C-I, C-II	1.055-1.120	180-280	Familial hyper-cholesterolemia Familial combined hypercholestero-lemia
HDL		Major: A-I, A-II, F Minor: A-IV, C-I, C-II, C-III, E	1.063-1.210	50-120	Hyper-α-lipo-proteinemia, CETP deficiency
HDL₂			1.063-1.125	90-120	
HDL₃			1.125-1.210	50-90	

Note. VLDL = very low density lipoprotein; IDL = intermediate density lipoprotein; LDL = low-density lipoprotein; HDL = high-density lipoprotein; CETP = cholesteryl ester transfer protein.

also catalyzed by another enzyme, cholesteryl ester transfer protein (CETP).

The association between elevated plasma cholesterol and increased incidence of CAD is well established. A clear and direct correlation has been observed in individuals who have plasma cholesterol levels of 200 mg · dl⁻¹ or higher. An even stronger association has been observed between elevated LDL and **atherosclerotic** diseases. Even within the LDL subclass, the smaller and denser LDL is more strongly linked to the atherogenic process (4). Furthermore, in addition to LDL, it appears that any other lipid containing apo B can be atherogenic. Thus, patients with elevated VLDL containing apo B are at risk for premature development of atherosclerosis.

The atherosclerotic process appears to occur mainly in large to medium-sized elastic and muscular arteries of the heart, brain, or extremities. Numerous studies have suggested that the first step in atherogenesis occurs in response to an injury causing endothelial dysfunction in the arterial wall. Several possible factors that can create endothelial dysfunction have been observed:

1. Elevated and modified LDL
2. Free radicals caused by cigarette smoking
3. Hypertension
4. Diabetes mellitus
5. Elevated plasma homocysteine levels
6. Genetic alterations
7. Infectious organisms such as herpes viruses and *C. pneumoniae* (31)

As a result, the endothelial wall is rendered more permeable and develops an adhesiveness to platelets, leukocytes, and other procoagulant factors. Highly susceptible areas to progressive atherogenesis include places in the arterial tree where bifurcations, curvatures, and branching occur, resulting in alterations in blood flow and increased turbulence (31). It appears that turbulent blood flow against the endothelium induces a release of procoagulant factors such as vascular adhesion molecule-1, leaving these sites prone to progressive lesion development.

The initiation of lesion development appears to begin with the attachment of circulating monocytes, T-cells, and other atherogenic lipoproteins (e.g., adhesive molecules, oxidized LDL) to the endothelial surface. These particles modify the endothelial surface, which begins a cascade of events resulting in the proliferation of plaque within the vascular wall (9). In addition, circulating LDL enters this space, where it undergoes oxidation. Nearby macrophages take up the oxidized LDL and begin to take the appearance of a foam cell. As lipid continues to accumulate in the foam cells, smooth muscle cell proliferation occurs through the release of growth factors. At this point the first visible signs, fatty streaks, begin to appear. The smooth muscle cells along with monocytes begin to migrate to the intimal space within the artery wall. As this cycle of interactions continues, the fatty streaks mature into a fibrous lesion, which begins to extend into the lumen of the artery. The foam cells continue to accumulate LDL, which begins to crystallize. At this point the foam cells can rupture, spilling their cholesterol contents, which leads to further proliferation of lipid deposition throughout the lesion. Smooth muscle cells, accompanied by collagen synthesis, continue the progression of the lesion. Advanced stages of lesion development include calcification, which hardens the plaque.

A high level of HDL cholesterol has been shown to lower the risk of CHD development (21, 26). Circulating HDL is involved in a reverse cholesterol transport process, where cholesterol from circulating lipoproteins and possibly lesions is gathered up by the HDL in the intravascular space and transported to the liver for catabolism. Consequently, low plasma HDL (<40 mg · dl⁻¹) is considered a risk factor for CHD.

Clinical Considerations

Several dyslipidemia and hyperlipidemia conditions can present themselves clinically. Although many of the lipid abnormalities can be genetically influenced

and most conditions can place the individual at risk for developing premature CHD, some may actually lower risk (e.g., hyper-α-lipoproteinemia).

Signs and Symptoms

The dyslipidemia condition can be classified based on which lipid or lipoprotein is abnormal. Although it is helpful to classify conditions for treatment purposes, some of the following classifications may only be identified through very extensive testing that may not be available in every clinical setting. Nevertheless, the following are dyslipidemia conditions.

Hypercholesterolemia

Blood cholesterol levels greater than 240 mg · dl⁻¹ are classified as hypercholesterolemia. This level of total cholesterol or higher has been shown to result in a very high risk of coronary heart disease and also corresponds to the 80th percentile of the adult U.S. population. Severe forms of hypercholesterolemia conditions include the following:

— Familial hypercholesterolemia: Characterized by a hyper-β-lipoproteinemia, this genetic condition results from defective LDL receptors or a lack of LDL receptors, which reduce LDL clearance rates. Patients with this disease present with an elevated total cholesterol and LDL levels often over 260 mg · dl⁻¹, and these levels may reach has high as 500 mg · dl⁻¹. Xanthomas and atheromas may be present as a result of excessive lipid deposition. These patients are at increased risk for premature atherosclerosis. Familial hypercholesterolemia occurs in approximately 1 in 500 people in the United States.

— Familial polygenic hypercholesterolemia: A more common genetic condition, which manifests as high total cholesterol and LDL cholesterol levels (>220 mg · dl⁻¹) and occurs in approximately 1 in 100 individuals. Premature atherosclerotic disease is commonly observed in this population but to a lesser extent than in the familial hypercholesterolemia patient.

— Familial combined hypercholesterolemia: Characterized by elevated total cholesterol and triglyceride levels and is seen in multiple individuals within the same family. This condition is common in the United States, found in approximately 1% of the general population. Patients affected with this condition are at increased risk for premature atherosclerotic disease.

Hypertriglyceridemia

Elevated fasting triglyceride levels may be classified as hypertriglyceridemia. Triglycerides may be either consumed in the diet (chylomicron) or synthesized (VLDL) in the liver. In addition to familial combined hypercholesterolemia, additional forms of hypertriglyceridemia may result from the following:

— Apoprotein E2 genotype: There are three primary apo E isoforms (E2, E3, and E4). However, the apo E2 genotype does not allow efficient binding to the apo E receptor, which is needed for triglyceride hydrolysis. Consequently, patients having the apo E2 genotype are unable to efficiently hydrolyze triglycerides in the liver. The condition results in an increase in both chylomicron and VLDL levels.

— Familial lipoprotein lipase (LPL) deficiency: Caused by either a deficiency in apo C-II, which is responsible for LPL activation, or an insufficient production of LPL, this condition limits the clearance of chylomicron and VLDL. In addition to elevated triglyceride levels, patients typically have low values of HDL and LDL.

— Familial hypertriglyceridemia: A genetic condition that causes abnormally high levels of endogenously produced triglycerides. A very high VLDL concentration is present in the patient's lipid profile. Cholesterol levels will also rise in proportion to the level of triglycerides resulting in an elevated CAD risk.

Hyperlipoproteinemia

Hyperlipoproteinemia is a generic term used to imply an elevated lipoprotein class. Virtually all dyslipoproteinemias result from a defect in the stage of lipoprotein development, transportation, or destruction. A hyperlipoproteinemia may result for many reasons including a familial trait, as a secondary characteristic of another disease (e.g., thyroid disease, diabetes), or possibly as a result of excess body weight, lack of exercise, and poor nutrition.

— Hyper-β-lipoproteinemia: A condition characterized by excess LDL levels usually as a result of too much β-band protein (apo B). Hyper-β-lipoproteinemia may result from excess production but is most likely the result of a reduced binding capacity at the liver for LDL hydrolysis. As an example, individuals with a familial polygenic hypercholesterolemia have a hyper-β-lipoproteinemia pattern and are at increased risk for the development of atherosclerosis.

— Hyper-α-lipoproteinemia: A rare condition characterized by an increased concentration of HDL. The etiology of this genetic disease is likely linked to both increased synthesis of the apo A (α-band) and a reduction in HDL clearance. Nevertheless, this dyslipidemia is apparently beneficial to health, because elevated HDL is inversely related to atherosclerotic diseases.

Hypolipoproteinemia

Hypolipoproteinemia is a generic term used to identify a low lipoprotein level. Individuals suffering from hypolipoproteinemia, as in a hyperlipoproteinemia, may have developed the condition from a familial trait, as a secondary characteristic of another disease (e.g., thyroid disease, diabetes), or possibly as a result of excess body weight, lack of exercise, and poor nutrition. Specific forms of hypolipoproteinemia include the following:

— A-β-lipoproteinemia: A rare genetic disorder characterized by absence of β-band lipoprotein (LDL) in the plasma. The condition is caused by a defect in apo B synthesis. It is important to note that no chylomicrons or VLDL is formed.

— Familial hypo-β-lipoproteinemia: A genetic condition that manifests a lower than normal LDL concentration (10-50%), but chylomicrons are developed. This condition probably stems from a lack of synthesis of apo B-100, which is associated with the LDL.

— Hypo-α-lipoproteinemia: A term used for low levels of HDL (<40 mg · dl^{-1}) and can result from a lack of synthesis of apo A or insufficient LPL activity. Low HDL levels may also result from behavioral influences such as excess body weight or smoking.

Diagnostic and Laboratory Evaluations

General dyslipidemia conditions can be identified through a complete blood lipid profile obtained via forearm venipuncture following a 12-hr fast (water only). Most clinical laboratories will provide a measure of total cholesterol, LDL, HDL and triglycerides. More advanced testing may be needed to identify a specific dyslipidemia classification such as those mentioned earlier. Enzymatic measurements, lipoprotein subfractions, and apoprotein analyses are usually only obtained when the patient is under the care of a lipid specialist.

General guidelines for the evaluation of blood lipids and lipoproteins have been established. The National Cholesterol Education Program (NCEP) (18) first published a systematic approach to the management of high cholesterol in 1988. The program establishes goals for both individual management and the nation as a whole. The NCEP Adult Treatment Panel III lipoprotein (47) classifications are listed in table 10.2. In general, the program calls

TABLE 10.2

National Cholesterol Education Program Adult Treatment Panel III Classification

LDL cholesterol (mg · dl^{-1})	Classification
<100	Optimal
100-129	Near optimal
130-159	Borderline high
160-189	High
≥190	Very high

Total cholesterol (mg · dl^{-1})	Classification
<200	Desirable
200-239	Borderline high
≥240	High

HDL cholesterol (mg · dl^{-1})	Classification
<40	Low
≥60	High

Note. LDL = low-density lipoprotein; HDL = high-density lipoprotein.

for all adults 20 years and older to have their total cholesterol measured at least once every 5 years.

Treatment

The intensity of an individual's treatment depends on the patient's overall risk status. Typically, persons at higher risk for CHD should be given more aggressive treatment. Patients with high blood cholesterol may be categorized into three general levels of risk:

1. CHD and CHD risk equivalent: persons having CHD, other atherosclerotic disease (e.g., peripheral vascular disease or carotid artery disease), or diabetes
2. Multiple risk factors: patients with no known CHD but at high risk because of high blood cholesterol in combination with other CHD risk factors
3. 0-1 risk factors: patients with no known CHD who have no other CHD risk factors

Over the past two decades, several pharmacological agents have been introduced that can effectively control many hyperlipidemia or dyslipidemia con-

ditions. However, in accordance with national cholesterol guidelines, before choosing to use a medication physicians will frequently allow patients to attempt to control influencing behavioral factors such as weight loss, dietary alterations, and exercise. Several treatments are reviewed next.

Diet Therapy

The Adult Treatment Panel III clinical guidelines developed by NCEP recommend a first-line intervention to improve cholesterol levels prior to administering medications through therapeutic lifestyle changes (TLC) that include diet and exercise. The recommended nutrient composition of the TLC diet includes reducing total fat intake to 25% to 35% of total calories. Allowing for this, saturated fat should be less than 7%, polyunsaturated fat 10% or less, and monounsaturated fats less than 20% of total calories. If LDL levels have not reached targeted goals after 6 weeks, increases in dietary fiber and plant stanols/sterols are recommended. As part of the TLC approach, weight reduction may be recommended to achieve optimal weight goals. Success in adhering to the nutrient recommendations may require the use of a qualified nutritional professional or registered dietician.

Patients switching from a standard Western diet (35-40% fat calories, with 15-20% saturated fat) to the TLC-recommended diet can expect an average decrease in their total blood cholesterol of 5-10% (10, 30). However, this form of dieting has poor adherence rates. Nevertheless, aggressive dietary changes are an important component in controlling the patient's total cholesterol level and should not be overlooked. Allison et al. (1) found that 35% of cardiac patients were able to reduce their total cholesterol levels to NCEP-recommended levels using diet therapy, thus eliminating the need for costly medications.

The effect that various types of dietary fat have on serum total cholesterol levels has been reviewed extensively. The recommendation to restrict saturated fats in the diet in particular stems from several studies that have identified a clear and direct association between level of dietary saturated fat and total cholesterol (19). It appears that high amounts of saturated fat intake decrease LDL clearance rates and thus prolong the exposure of LDL within the circulation. In a similar manner, a high amount of dietary cholesterol also reduces LDL clearance from the circulation and increases LDL synthesis. A reduced LDL catabolism and subsequent longer cycle within the circulation may promote the formation of the smaller, dense LDL subfraction, which has greater atherogenic properties (4). **Polyunsaturated fats** appear to have little effect on plasma cholesterol levels. However, high polyunsaturate diets may reduce plasma HDL cholesterol levels. Moreover, the long-term consequences of this type of diet are unknown. Consequently, the American Heart Association currently recommends that polyunsaturates account for no more than 10% of total calories. In addition, polyunsaturated fats are often hydrogenated to harden their appearance (e.g., margarine). This process alters the chemical structure of the double bonds in the fatty acid from a *cis* to a *trans* configuration. *Trans* fatty acids appear to act in a similar manner as saturated fats, raising plasma cholesterol (29) and LDL and lowering HDL levels (2).

Recently, research has begun to indicate that carbohydrates should not be used as a substitute for saturated fats. A high-carbohydrate diet has been shown to reduce HDL levels and raise triglycerides. Rather, saturated and trans fatty acids should be replaced with **monounsaturated fats.** Monounsaturated fats, such as oleic acid, have a single double bond configuration and appear to lower plasma cholesterol while maintaining and possibly elevating HDL levels (2).

Omega-3 fatty acids, which are long-chain fatty acids that contain a double bond in the n-3 position, have been shown to lower triglycerides; however, there are few data to support their effectiveness at preventing the development of cardiovascular disease (CVD). Nevertheless, cross-sectional data from populations whose diets are rich in omega-3 fats show that these people have lower CVD rates (25).

Plasma lipids and lipoproteins may also be modified by other dietary sources. Diets high in soluble fiber have been shown to sequester bile and consequently reduce LDL levels by approximately 5% to 10% (24). However, HDL levels may also be reduced. Other dietary components such as garlic (40) and walnuts (32) may help to lower plasma cholesterol levels. But additional research is needed to support these preliminary findings.

In general, patients should be encouraged to increase the intake of fruits, vegetables, and high-fiber foods. Moreover, saturated and trans fatty acids should be minimized and replaced with monounsaturated and omega-3 fatty acids.

Pharmacological Management

If diet therapy is insufficient to restore appropriate LDL cholesterol levels, drug therapy may be initiated. Pharmacological treatments should be considered as an adjunct to diet therapy and not as a replacement. Usually a 6-month diet therapy trial is sufficient to determine its effectiveness. However,

individuals at higher risk, such as those with existing CAD, may be given a shorter diet-alone trial before resorting to drug therapy.

The clinician and patient should carefully weigh the benefits versus the costs (health insurance, adverse effects) of drug therapy. Moreover, any improvement in cholesterol attributed to the medication may be reversed if drug therapy is discontinued. Table 10.3 lists the various medications that are used to treat dyslipidemia. The drugs used may be classified into the following categories: bile acid sequestrants, statins, nicotinic acid, fibric acids, prubocol, and estrogen replacement therapy.

— Bile acid sequestrants: A powdery resin that is usually consumed in a liquid format, the bile acid sequestrants are safe and have few adverse effects. The resins bind bile acids for excretion, resulting in a greater oxidation of cholesterol to replace lost bile. The resins have been shown to be effective at reducing moderately elevated LDL cholesterol levels (3). However, many patients may not comply well with bile acid therapy because of gastric irritability and lack of palatability. Although efforts have been made to improve palatability, the development of more effective medications has lowered the frequency of usage of bile acid resins. Moreover, since the resins bind negatively charged molecules in the intestine, they can interfere with the absorption of nutrients and other medications. Consequently, their administration should be timed so that it does not conflict with other medications.

— Statins: Currently the most effective and thus most commonly prescribed lipid-lowering medications, the statins inhibit HMG CoA reductase, the rate-limiting enzyme in cholesterol synthesis. As a consequence, an up-regulation of cellular LDL receptors occurs to clear the plasma compartment of circulating LDL. Statins have been shown to reduce cholesterol levels by 15% to 30% with a 20% to 40% decrease in LDL (27). Although these medications appear to be the most effective for therapeutic lipid management, little is known about the long-term safety of this class of drugs. Few side effects have been reported, but they include hepatotoxicity and myopathy (<1%). Frequent liver enzyme tests are recommended for patients, particularly when the drug is newly administered.

— Nicotinic acid: A B vitamin that at high doses has lipid-regulating properties, nicotinic acid has several side effects that limit its use among many patients. These side effects include minor effects such as skin irritation and flushing to the more serious such as liver toxicity and hepatic failure.

TABLE 10.3
Effect of Common Medications on Lipid and Lipoprotein Profiles

Medication	Triglycerides	VLDL	LDL	HDL	Comments
β-blockers		Increase		Decrease	Those with intrinsic sympatho-mimetic activity have little effect
Thiazides		Increase	Increase		Dose dependent
Oral contraceptives			Increase/decrease	Increase/decrease	Depending on dose of progestogen and estrogen used
Hormone replacement therapy			Decrease	Increase	Depending on dosage
Corticosteroids, cyclosporine	Increase	Increase	Increase		
Dilantin				Increase	
Alcohol	Increase[a]	Increase[a]		Increase[b]	[a]Alcohol intake >3% of total calories [b]Not more than 12 oz. per day

Note. VLDL = very low density lipoprotein; LDL = low-density lipoprotein; HDL = high-density lipoprotein.

Patients are slowly introduced to the medication through a staggered increase in dosage to reduce the likelihood of side effects. Nevertheless, frequent monitoring of the patient is important, particularly during the early stages of nicotinic acid administration. A therapeutic dose is effective at lowering total cholesterol and triglycerides as well as increasing HDL. Improvements in HDL appear to be attributable to a decrease in HDL clearance rates (5).

— Fibric acids: Fibric acids are used predominantly as a triglyceride-lowering medication. Although their mechanism of action is not completely known, they increase LPL activity, which results in triglyceride hydrolysis. In addition to the triglyceride-lowering effects, the action of LPL can also raise plasma HDL levels. A 10% to 15% increase in HDL has been observed following gemfibrozil administration (14). Fibric acids have also been shown to alter the LDL subfraction class by decreasing the concentration of the atherogenic small, dense LDL particle. Fibric acids are safe with little side effects, although risk of myopathy is increased when fibric acids are used in combination with a statin (14).

— Probucol: Probucol inhibits LDL oxidation, which may reduce the risk of atherogenesis (6). Although limited data exist in humans, a decreased rate of progression of atherosclerotic lesions has

been observed following probucol administration in rats. Probucol has largely been limited to patients with familial hyperlipidemia as an antioxidant, although concern has been raised over the reported decrease in HDL following administration. Probucol administration can prolong the QT interval of the electrocardiogram.

— Estrogen replacement therapy: The risk of CAD development among postmenopausal women approaches that of men. Significant adjustments in lipid profiles can be observed between pre- and postmenopausal women. LDL values tend to increase and HDL values decrease following menopause, which creates a more atherogenic lipid profile. Exogenous replacement of estrogen has been shown to lower LDL and raise HDL; however, limited data are available to support a reduced risk of CAD (22). Nevertheless, estrogen replacement therapy is commonly provided to postmenopausal patients because of its link to slowing osteoporosis. Estrogen administration may not be appropriate for some patients because of its link to breast and uterine cancers.

Prevention

Since hyperlipidemia is an independent risk factor for atherosclerotic diseases including cardiac, peripheral, and cerebral vascular disease, steps should be taken to control the lipid profile. As a method of prevention, individuals should be educated and encouraged to follow the NCEP guidelines for cholesterol testing. Healthcare providers have not been consistent in adhering to NCEP guidelines with their patients (15), and consequently, more attention to disease prevention guidelines is warranted. Among families where premature disease is suspect, efforts should be made to screen children early (>2 years) to therapeutically intervene and prevent the aggressive development of disease. In addition, individuals need to understand the potentially modulating effects that other risk factors place on blood lipids. For example, efforts should be made to prevent weight gain given that added weight has a negative influence on the lipid profile (high LDL and low HDL). In addition to weight maintenance, body composition, glucose tolerance, and smoking may also adversely affect the lipid profile. Thus, efforts to minimize other risk factors are critical for preventing hyperlipidemia. Although exercise has a direct, positive influence on the lipid profile, it can also indirectly improve lipids through weight loss, reduced adiposity, and improved glucose tolerance.

Based on the general risk classifications mentioned previously, cholesterol management guidelines have been developed to target LDL cholesterol levels. The guidelines are listed in table 10.4. Emphasis is placed on LDL cholesterol levels to help guide the appropriate treatment recommendations. As such, a full fasting lipoprotein analysis should be conducted at the point of screening. The current recommendations for cholesterol management

using a full lipoprotein profile emphasize the level of LDL cholesterol, because it is the principal constituent behind the atherosclerotic process.

Graded Exercise Testing

The American Heart Association and the American College of Sports Medicine have developed exercise test procedures and protocols that can guide the clinician (see chapter 6). Although no specific testing modifications are needed for this population, a cautious approach to the evaluation process is prudent given the probabilities for underlying disease. Clinicians should be very familiar with the contraindications for exercise and exercise testing before treating this population, and certain individuals may need a medical evaluation.

Medical Evaluation

A patient entering the exercise program with no history of CHD may not even know his or her lipid and lipoprotein risk classification, even though much effort has been made recently to improve awareness and management of CHD risk factors among the public. As such, patient education should be provided and a complete lipoprotein profile should be periodically obtained in accordance with current NCEP screening guidelines. Although a medical evaluation is not required for the hyperlipidemic or dyslipidemic patient to begin exercise, the clinician should be careful to probe for any signs or symptoms of underlying disease, particularly among individuals who present with multiple CHD risk factors. A complete medical evaluation is usually

---

TABLE 10.4

National Cholesterol Education Program: Clinical Guidelines for Cholesterol Management

Risk category	LDL goal	LDL level at which to initiate therapeutic lifestyle changes	LDL level at which to consider drug therapy
CHD or CHD equivalent	<100 mg · dl^{-1}	≥100 mg · dl^{-1}	≥130 mg · dl^{-1} (100-129 mg · dl^{-1}: drug optional)
Multiple risk factors[a]	<130 mg · dl^{-1}	≥130 mg · dl^{-1}	≥130 mg · dl^{-1} with 10-20% risk[b] ≥160 mg · dl^{-1} with <10% risk[b]
0-1 risk factors[a]	<160 mg · dl^{-1}	≥160 mg · dl^{-1}	≥190 mg · dl^{-1} (160-189 mg · dl^{-1}: drug optional)

Note. LDL = low-density lipoprotein; CHD = coronary heart disease.

[a]Risk factors include cigarette smoking, hypertension, low HDL, family history of premature CHD (CHD in males <55 years; females <65 years), age (men ≥45 years; women ≥55 years). [b]Risk of developing CHD over a 10-year period as determined by Framingham Scoring Procedures.

provided to patients with known CHD and will likely include a current lipoprotein profile.

Contraindications

The American College of Sports Medicine has developed contraindications for exercise, which are reported in chapter 6. By itself, the atherogenic blood lipid profile is not a contraindication. However, patients with dyslipidemia may present with existing disease or anomalies that may preclude them from exercise. Consequently, a thorough review of the patient's medical history and additional physical assessments should be conducted to rule out any established risk to exercise training.

Recommendations and Anticipated Responses

Individuals with elevated plasma cholesterol and, in particular, LDL cholesterol are considered at risk for CAD; consequently, care must be taken to ensure that there are no signs or symptoms of underlying disease. This is particularly important among middle-aged patients with a familial hyperlipidemia that may accelerate the risk for premature CAD. If signs or symptoms are present, or if the patient has known disease, the exercise test should follow the protocol established for patients with known disease. In contrast, patients who have no known disease may undergo normal exercise testing protocols for patients considered at risk. The exercise evaluation is an effective tool for ruling out underlying disease and determining the patient's fitness level, and it can be used to guide the formulation of the exercise prescription.

Exercise Prescription

In general, exercise along with diet is recommended as part of the Adult Treatment Panel (ATP) III therapeutic lifestyle changes for improving the lipid profile. Special considerations for developing the exercise prescription along with currently supported exercise recommendations are discussed next. Practical application 10.1 discusses blood lipid responses to exercise training.

Special Exercise Considerations

The mechanisms behind the alterations in blood lipid profiles following exercise training are not fully understood, although many of the improvements may be attributed to enzymatic changes. LPL, which is responsible for hydrolyzing triglyceride-rich particles such as VLDL and chylomicrons, increases following both acute exercise (17) and training (33).

The elevated LPL is directly related to the lowering of plasma triglyceride levels. Following lipid hydrolysis, the remnant particles may be transformed into nascent HDL or accepted by circulating HDL for reverse cholesterol transport. In addition to having an increase in LPL activity, exercise-trained individuals have an elevated lecithin cholesterol acyltransferase (20), which is responsible for maturation of HDL through cholesterol esterification. Another enzyme, CETP, apparently decreases following exercise training (34) and may also play a role in the maturation of HDL. CETP catalyzes the removal of esterified cholesterol from HDL to larger triglyceride-rich particles such as VLDL. Individuals with CETP deficiency typically have very high HDL levels. Nevertheless, the role of CETP following exercise training has not been fully elucidated. CETP may also help lower the amount of circulating LDL_3 that is linked to atherogenesis (12). Hepatic triglyceride lipase generally decreases following exercise training (20). The decline in hepatic triglyceride lipase activity prolongs the circulating half-life of HDL because of a decrease in HDL catabolism by the liver. The increased metabolic half-life of the HDL likely aids in the increased HDL_2 subfraction observed following exercise training.

Individuals with familial dyslipidemia may not experience the same enzymatic responses, nor may they expect the same alterations to their lipid profile as healthy individuals following exercise training. Exercise alone has not been shown to be very effective at increasing HDL levels among subjects with hypo-α-lipoproteinemia or low HDL. Moreover, LPL-deficient individuals with familial hypertriglyceridemia may not experience the same level of improvement in their lipid profile as a healthy individual. Nevertheless, exercise training among individuals with a familial dyslipidemia is very important in reducing other risk factors (e.g., body composition, glucose tolerance, hypertension) that can influence the lipid profile as well as independently affect CVD mortality and morbidity rates.

The goals for exercise prescription may vary slightly depending on the form of dyslipidemia. Individuals with congenital hyperlipidemia need to adhere to prescribed medications and use exercise to target other risk factors such as weight loss and reduced adiposity, because the direct impact of exercise on blood lipids may be limited in this population. Furthermore, excess body weight and adiposity should be reduced as a primary objective of any hyperlipidemia patient because body mass index is directly associated with total cholesterol levels. Thus, an exercise prescription that targets weight loss can profoundly affect the lipid profile. Reductions in

Blood Lipid Responses to Exercise Training

Exercise has become a recommended and valuable therapeutic treatment for improving the blood lipid profile. However, caution must be used when interpreting its therapeutic value, at least with regard to the blood lipid profile. The majority of exercise studies have been conducted in subjects with normal lipid status, and consequently few data exist regarding the effects of exercise training among individuals with various dyslipidemia conditions. Thus, despite the commonly held notion that exercise may be improve the lipid profile in those with abnormal lipids, this may simply not be the case. For instance, some individuals appear to have intractable lipid profiles. One hypothesis behind the lack of a favorable blood lipid response to exercise training may be related to one's apo E genotype. Individuals with a specific apo E genotype may not be able to adequately hydrolyze triglycerides and may have limited improvements in HDL. It is important to keep in mind, however, that physical activity is even recommended among patients whose lipid profile appears to be resistant to exercise in order to minimize other risk factors (e.g., obesity and glucose intolerance).

In many individuals, exercise has proven to have a powerful modulating effect on many components of the lipid profile. Cross-sectional studies have long identified a clear association between physical activity or fitness levels and improved lipid profile (12). They indicate that aerobically trained athletes have lipid profiles that are lower in triglycerides and higher in HDL than sedentary counterparts. Furthermore, few if any improvements in total cholesterol and LDL have been observed. In contrast, prospective studies, while showing blood lipid improvements, have not established a strong relationship between physical activity and blood lipid levels. Although some of this discrepancy can be related to the lack of sufficient length of the training study (<6 months), other potential reasons remain. For example, a selection bias may have existed in many cross-sectional studies implying that highly fit individuals or elite athletes may have other characteristics, some of which may be genetic, that give rise to the large differences observed. Furthermore, the larger increases reported from cross-sectional studies may also be related to the volume of exercise training, changes in body composition, failure to control for recent exercise, dietary intake, or weight loss.

Although many of the improvements in the lipid profile can require several months, benefits may accrue from a single exercise bout (8). Both a decrease in plasma triglycerides and an increase in HDL cholesterol have been observed 18 to 48 hr following exercise (12). Furthermore, the transient changes observed following an acute exercise bout still occur after 24 weeks of training, suggesting that exercise training and a single session of exercise exert distinct and interactive effects (8).

Exercise training has been shown to lower plasma triglycerides (38), especially when baseline levels are elevated (12, 44). Although acute decreases have been observed following a single prolonged (>60 min) exercise bout (33, 37) and also following shorter exercise bouts (8), habitually active individuals have low triglycerides even when they have not recently exercised (12). Furthermore, a greater reduction in triglyceride levels is more likely following exercise training programs where weight loss has occurred (44).

Perhaps the most prominent effect that exercise training has on the lipid profile is on HDL cholesterol. Exercise training studies indicate that HDL may increase following exercise training by 5% to 15% (12). However, little to no increases in HDL may occur following exercise training among individuals with initially low levels (46) or unless weight loss occurs (12). The increase in HDL following training appears to be specific to the HDL_2 subfraction, which may provide stronger atherogenic protection.

Alterations in total and LDL cholesterol as a result of exercise training have been less convincing. Because total cholesterol is made up of all cholesterol-carrying lipoproteins, changes within these various components can prevent any observable change in total cholesterol. However, because LDL and HDL are usually the principle carriers of cholesterol, the ratio of total cholesterol to HDL cholesterol (total/HDL) may provide a stronger indication of relative movements within the lipid profile and an indication of atherogenic risk. When the total/HDL ratio is calculated, studies have indicated that exercise training can lower this ratio if weight loss has occurred (36). Furthermore, the decrease in this ratio cannot be solely attributed to an increase in HDL. Although exercise training has not proven to be very effective at lowering LDL cholesterol unless weight loss or decreases in adiposity (36, 44) occur, changes in LDL subfraction have been observed. Exercise-trained individuals have smaller amounts of the small dense LDL subfraction (LDL_3) (39, 45), which is considered more atherogenic.

total cholesterol, LDL, and triglycerides along with increases in HDL have been observed following weight loss (36).

The exercise prescription should first address any chronic disease (i.e., CAD, diabetes) issues. Consequently, if a patient presents with CAD, the immediate exercise prescription goals may not allow an optimal training level for lipid control. Greater strides may take place in controlling lipid levels once the subject has achieved sufficient fitness levels and attended to other immediate needs of the chronically diseased patient. Furthermore, among individuals with metabolic disorders where lipid abnormalities often develop, such as diabetes, efforts need to be placed on weight control and a regimented diet to control glucose levels. In doing so, lipid levels, at least in part, may be controlled.

Among individuals who are otherwise healthy, the exercise prescription may be written to more specifically address hyperlipidemia. Although limited data are available to tailor the exercise prescription toward optimal improvements in blood lipids, the preliminary guidelines presented here are based on previous investigations, and further research is needed to specifically tailor the exercise prescription to modify blood lipids. Although findings of several investigations have supported the benefits of exercise training on the lipid profile, the majority of studies were performed in subjects with a normal lipid status. Very few data exist regarding the benefits of exercise training among the dyslipidemia populations.

It is also important to recognize that it may take several months of exercise training to provide any significant or lasting improvements to the blood lipid profile. Consequently, added encouragement along with an evaluation of the patient's behavioral and environmental milieu may be needed to help the patient stay on course. Further encouragement should be taken from several recent studies that have identified favorable improvements in the blood lipid profile after a single exercise session (8, 12, 16, 17). Moreover, these improvements have been identified in both trained and untrained individuals and among those with elevated total cholesterol. These improvements appear to be only transient, lasting only 48 to 72 hr after exercise, which should encourage patients to exercise every other day.

Exercise Recommendations

The minimum amount of exercise training that is required to improve blood lipids is not fully known. However, the results of previous studies indicate that the quantity of exercise is likely the most important

factor. A dose-response relationship has been identified between the volume of physical activity and HDL cholesterol (42). Furthermore, this relationship has been identified in both men (43) and women (41). Although the specific combination of exercise frequency and duration for lipid improvements has not been fully elucidated, a minimum of 1000 kcal per week has been recommended (35). Individuals who are able to achieve higher amounts of weekly activity will likely benefit more. Practical application 10.2 details exercise prescription for the dyslipidemia patient.

Mode

Aerobic exercise has proven to be effective at improving the blood lipid profile. Both reductions in triglycerides and increases in HDL have been observed following endurance exercise training. In contrast, resistance training may not be as effective. Cross-sectional data indicate that elite resistance-trained athletes have similar blood lipid profiles as their sedentary counterparts. Resistance training studies that have identified improvements in blood lipids have either used a circuit training protocol (which closely resembles aerobic exercise) or involved substantial reductions in body weight. An enhanced fat utilization during aerobic activity may be the metabolic stimulus that is needed to improve blood lipid levels (12). Another important consideration is that weight loss, which is considered an important goal for improving the lipid profile, may be better achieved through aerobic exercise training, where a higher caloric expenditure may be maintained. Nevertheless, resistance training should be considered an important adjunct to any successful fitness program.

Frequency

The frequency of exercise needed to improve the lipid profile also needs further examination. Studies evaluating the effects of a single exercise session on HDL cholesterol indicate that elevations in HDL can occur 18 to 24 hr after exercise and remain elevated for 48 to 72 hr (8, 12, 16, 17, 38). Furthermore, over a 24-week training period, these acute increases in HDL continued in response to the acute exercise session (8). This information gives rise to speculation that the increase in HDL following exercise training is attributed, in part, to the transient increase observed following the acute exercise session. Furthermore, the transient rise in HDL in response to a single exercise session may have important day-to-day health benefits. Consequently, exercise sessions should not occur any less than three or four times per week, and no more than one day should separate exercise sessions.

Exercise Prescription for the Dyslipidemia Patient

Training	Mode	Frequency	Intensity	Duration	Special considerations
Aerobic	Treadmill Walking Cycling Combined arm and leg exercise Rowing Stepper Combination of preceding	3-4 days per week, alternating days	60-85% of $\dot{V}O_2$max	At least 30 min. May be intermittent (e.g., 3 × 10 min) or continuous. Improvements are dose-response.	Favorable changes, particularly in HDL, may occur following a single exercise bout and persist for up to 48-72 hr.
Resistance	Elastic bands Hand weights Free weights Multistation machines All major muscle groups	Minimum of 2 days per week	Select a weight where the last repetitions feel somewhat or moderately hard without inducing significant straining (bearing down and breath holding).	8-12 repetitions	Follow general resistance training guidelines to maintain strength.
Warm-up and cool-down exercises	Range of motion and flexibility exercises	Daily	Static stretching	5-10 min	Exercises should emphasize major muscle groups, especially lower back and posterior leg muscles.

Intensity

The cross-sectional differences in HDL cholesterol levels between endurance-trained athletes and sedentary counterparts led many to believe that the lipid improvements were related to the intensity of exercise. Although endurance-trained athletes typically engage in vigorous exercise, more recent studies using sedentary individuals have been able to observe blood lipid improvements following moderate physical activity (50-60% maximum heart rate) (11). Consequently, a moderate intensity of exercise is sufficient to develop and sustain an improved lipid profile and will likely ensure a better exercise adherence rate. However, among healthy individuals, maintaining progressive and optimal improvements in HDL may require an eventual increase in the training intensity. An acute increase in HDL cholesterol following a single exercise session in moderately trained males appears to be dependent, in part, on the training intensity (17). The potential risks and benefits to raising the patient's exercise training level need to be carefully reviewed beforehand, and a prudent plan for progression should be developed.

Duration

The exercise bout should last at least 30 min with an optimal goal of 45 to 60 min. However, improvements in blood lipids are evident even when exercise time is broken up throughout the day (13). Because weight loss is also an effective modulator of blood lipids, extending the caloric expenditure rate above 1000 calories per week will likely help with both goals.

Conclusion

Exercise is a powerful modulator for improving the lipid profile among individuals who have normal and mildly elevated values. These improvements are likely attributable to enzymatic changes that primarily promote increases in HDL and reductions in triglyceride values. Less is known, however, about the benefits of exercise among those with familial dyslipidemias either as a primary or secondary adjunct. Nevertheless, because of the general overall benefits of exercise, even those with severe abnormalities should be encouraged to adopt a habitually active lifestyle.

Case Study 10

Medical History and Diagnosis

Mr. HS, a 45-year-old white businessman, has been referred to an exercise program for CVD risk reduction following a recent medical examination. He has a family history of heart disease and stroke. His father had a nonfatal myocardial infarction at 52 and his grandmother a fatal stroke at 69. He is 5 ft 8 in. and weighs 200 lb, and his medical history is unremarkable. He currently does not exercise regularly, although he jogged regularly when he was in college. He smokes half a pack of cigarettes a day and self-reports a moderate amount of stress on the job. A fasting blood lipid profile was obtained during his medical examination and revealed the following:

Total cholesterol = 224 mg $\cdot$ dl^{-1}

HDL = 30 mg $\cdot$ dl^{-1}

LDL = 166 mg $\cdot$ dl^{-1}

Triglycerides = 300 mg $\cdot$ dl^{-1}

He has a resting heart rate of 70 beats $\cdot$ min^{-1} and resting blood pressure of 130/90.

Exercise Prescription

Initial Exercise Program:

Frequency = 5 days per week

Intensity = 125 to 136 beats $\cdot$ min^{-1} (50-60% heart rate reserve)

Rating of perceived exertion = 11 to 13

Duration = 20 to 30 min

Mode = aerobic

Exercise Progression (over first 6-12 weeks):

Frequency = 5 days per week

Intensity = progress toward the upper end of intensity range as patient tolerates

Duration = gradually increase to 45 min

Exercise Progression (after 3 months):

Frequency = 3 to 4 days per week, alternating

Intensity = 140 to 160 beats $\cdot$ min^{-1} (60-85% heart rate reserve)

Rating of perceived exertion = 12 to 15

Duration = 45 to 60 min

The subject should be educated on the signs and symptoms of exercise intolerance. Supplemental weight training should be introduced using modest amounts of weight (12-15 reps) and basic exercises for general conditioning purposes. Subject's lipid profile would likely benefit from weight

continued

loss and by eliminating smoking. Patient should also be placed on a Step 2 diet to control saturated fat intake. Total fat intake should be 25-35% of total calories. In addition, saturated fat should be less than 7%, polyunsaturated fat should be lest than or equal to 10%, and monounsaturated fat should be less than 20% of total calories. Following 3 to 6 months, a repeat blood lipid profile should be conducted to reassess the patient's status. Following reassessment, the patient may need to be placed on drug therapy; however, allowing a 3- to 6-month follow-up will determine the impact of a behavioral approach.

Case Study 10   Discussion Questions

1. What objectives should be addressed in hopes of improving the patient's lipid profile?

2. To a patient who has low HDL cholesterol (<35 mg $\cdot$ dl^{-1}), what specific recommendations would you make?

3. What is the importance of an exercise program for patients who have a familial hyperlipidemia?

Glossary

atherogenic—Having the capacity to initiate, increase, or accelerate the process of the atherosclerosis.

atherosclerotic—Relating to or characterized by atherosclerosis.

fibrinolytic—Causing fibrinolysis, which is the breakdown of fibrin, a blood-coagulating protein.

monounsaturated fats—Dietary fatty acid that contains one double bond along the main carbon chain.

omega-3 fatty acids—Long-chain polyunsaturated fatty acids that contain a double bond in the n-3 position.

polyunsaturated fats—Dietary fatty acids that contain two or more double bonds along the main carbon chain.

References

1. Allison, T.G., et al., *Achieving National Cholesterol Education Program goals for low-density lipoprotein cholesterol in cardiac patients: Importance of diet, exercise, weight control, and drug therapy.* Mayo Clin Proc, 1999. 74(5): p. 466-73.

2. Ascherio, A., et al., *Trans fatty acids and coronary heart disease.* N Engl J Med, 1999. 340(25): p. 1994-8.

3. Ast, M., W.H. Frishman, *Bile acid sequestrants.* J Clin Pharmacol, 1990. 30: p. 99-106.

4. Ballantyne, C.M., *Low-density lipoproteins and risk for coronary artery disease.* Am J Cardiol, 1998. 82(9A): p. 3Q-12Q.

5. Belalcazar, L.M., C.M. Ballantyne, *Defining specific goals of therapy in treating dyslipidemia in the patient with low high-density lipoprotein cholesterol.* Prog Cardiovasc Dis, 1998. 41(2): p. 151-74.

6. Carew, T., D.C. Schenke, D. Steinberg, *Antiatherogenic effect of probucol unrelated to its hypercholesterolemic effect: Evidence that antioxidants in vivo can selectively inhibit low density lipoprotein degradation in macrophage-rich fatty streaks and slow the progression of atherosclerosis in the Watanabe heritable hyperlipidemic rabbit.* Proc Natl Acad Sci USA, 1987. 84: p. 5928-31.

7. Cobb, M.M., H.S. Teitelbaum, J.L. Breslow, *Lovastatin efficacy in reducing low-density lipoprotein cholesterol on high vs low fat diets.* JAMA, 1991. 265: p. 997-1001.

8. Crouse, S.F., et al., *Effects of training and a single session of exercise on lipids and apolipoproteins in hypercholesterolemic men.* J Appl Physiol, 1997. 83(6): p. 2019-28.

9. Cybulsky, M.I., M.A. Gimbrone, *Endothelial expression of a mononuclear leukocyte adhesion molecule during atherogenesis.* Science, 1991. 251: p. 788-91.

10. Denke, M.A., S.M. Grundy, *Individual responses to a cholesterol-lowering diet in 50 men with moderate hypercholesterolemia.* Arch Intern Med, 1994. 154: p. 317-24.

11. Duncan, J.J., N.F. Gordon, C.B. Scott, *Women walking for health and fitness.* JAMA, 1991. 266: p. 3295-9.

12. Durstine, J.L., W.L. Haskell, *Effects of exercise training on plasma lipids and lipoproteins.* Exerc Sports Sci Rev, 1994. 22: p. 477-521.

13. Ebisu, T., *Splitting the distance of endurance running: On cardiovascular endurance and blood lipids.* Jpn J Physical Ed, 1995. 30(1): p. 37-43.

14. Farmer, J.A., A.M. Gotto, *Dyslipidemia and other risk factors for coronary artery disease,* in *Heart Disease: A Textbook of Cardiovascular Medicine,* E. Braunwald, Editor. 1997, Saunders: Orlando, FL. p. 1126-60.

15. Frolkis, J.P., S.J. Zyzanski, J. Schwartz, P. Suhan, *Physician noncompliance with the 1993 National Cholesterol*

Education Program (NCEP-ATPII) guidelines. Circulation, 1998. 98: p. 851-5.

16. Gordon, P.M., et al., *Comparison of exercise and normal variability on HDL cholesterol concentrations and lipolytic activity.* Int J Sports Med, 1996. 17(5): p. 332-7.

17. Gordon, P.M., F.L. Goss, V. Warty, B. Denys, P. Visich, R. Robertson, et al., *The acute effects of exercise at different intensities on HDL-C, HDL subfractions and lipid enzymes.* Med Sci Sports Exerc, 1994. 26(6): p. 671-7.

18. Grundy, S., D. Bilheimer, A. Chait, L. Clark, M. Denke, R. Havel, W. Hazzard, S. Hulley, D. Hunninghake, R. Kreisberg, P. Kris-Etherton, J. McKenney, M. Newman, E. Schaefer, B. Sobel, C. Somelofski, M. Weinstein, *National Cholesterol Education Program: Second report of the Expert Panel on Detection, Evaluation, and Treatment of High Blood Cholesterol in Adults.* Circulation, 1994. 89: p. 1329.

19. Grundy, S.E., M.A. Denke, *Dietary influences on serum lipids and lipoproteins.* J Lipid Res, 1990. 31: p. 1149-72.

20. Gupta, A.K., E.A. Ross, J.N. Myers, M.L. Kashyap, *Increased reverse cholesterol transport in athletes.* Metabolism, 1993. 42(6): p. 684-90.

21. Harper, C.R., T.A. Jacobson, *New perspectives on the management of low levels of high-density lipoprotein cholesterol.* Arch Intern Med, 1999. 159(10): p. 1049-57.

22. Herrington, D.M., et al., *Effects of estrogen replacement on the progression of coronary-artery atherosclerosis.* N Engl J Med, 2000. 343(8): p. 522-9.

23. Hunninghake, D.B., E.A. Stein, C.A. Dujovne, *The efficacy of intensive dietary therapy alone or combined with lovastatin in outpatients with hypercholesterolemia.* N Engl J Med, 1993. 328: p. 1213-9.

24. Jenkins, D.J., T.M. Wolever, A.V. Rao, et al., *Effect on blood lipids of very high intakes of fiber in diets low in saturated fat and cholesterol.* N Engl J Med, 1993. 329: p. 21-6.

25. Kromhout, D., E.B. Bosscheiter, C. de Lezenne Coulander, *The inverse relation between fish consumption and 20-year mortality from coronary heart disease.* N Engl J Med, 1985. 312: p. 1205-9.

26. Kwiterovich, P.O., Jr., *The antiatherogenic role of high-density lipoprotein cholesterol.* Am J Cardiol, 1998. 82(9A): p. 13Q-21Q.

27. Mahley, R.W., K.H. Weisgraber, R.V. Farese, *Disorders of lipid metabolism,* in *Williams Textbook of Endocrinology,* Larsen, P.R., H.M. Kronenberg, S. Melmed, K.S. Polonsky, Editors. 1998, Saunders: Orlando, FL. p. 1099-1154.

28. Marcovina, S.M., M.L. Koschinsky, *Lipoprotein(a) as a risk factor for coronary artery disease.* Am J Cardiol, 1998. 82(12A): p. 57U-86U.

29. Mensink, R.P., M.B. Katan, *Effect of dietary trans fatty acids on high density and low density lipoprotein cholesterol levels in healthy subjects.* N Engl J Med, 1990. 323: p. 439-45.

30. Ornish, D., S.E. Brown, L.W. Schwerwitz, et al, *Can lifestyle changes reverse coronary heart disease? The Lifestyle Heart Trial.* Lancet, 1990. 336: p. 129-33.

31. Ross, R., *Atherosclerosis—An inflammatory disease.* N Engl J Med, 1999. 340(2): p. 115-26.

32. Sabate, J., G.E. Fraser, K. Burke, et al., *Effects of walnuts on serum lipid levels and blood pressure in normal men.* N Engl J Med, 1993. 328: p. 603-7.

33. Seip, R.L., C.F. Semenkovich, *Skeletal muscle lipoprotein lipase: Molecular regulation and physiological effects in relation to exercise.* Exerc Sports Sci Rev, 1998. 26: p. 191-218.

34. Seip, R.L., P. Moulin, T. Cocke, A. Tall, W. Kohrt, et al., *Exercise training decreases plasma cholesteryl ester transfer protein.* Arterioscler Thromb, 1993. 13: p. 1359-67.

35. U.S. Department of Health and Human Services, *Physical Activity and Health: A Report of the Surgeon General.* 1996, U.S. Department of Health and Human Services, Centers for Disease Control and Prevention, National Center for Chronic Disease Prevention and Health Promotion: Atlanta, GA.

36. Stefanick, M.L., et al., *Effects of diet and exercise in men and postmenopausal women with low levels of HDL cholesterol and high levels of LDL cholesterol.* N Engl J Med, 1998. 339(1): p. 12-20.

37. Thompson, P.D., E. Cullinane, L.O. Henderson, P.N. Herbert, *Acute effects of prolonged exercise on serum lipids.* Metabolism, 1980. 29(7): p. 662-5.

38. Thompson, P.D., E.M. Cullinane, S.P. Sady, M.M. Flynn, D.N. Bernier, et al., *Modest changes in high-density lipoprotein concentrations and metabolism with prolonged exercise training.* Circulation, 1988. 78(1): p. 25-34.

39. Vasankari, T.J., et al., *Reduced oxidized LDL levels after a 10-month exercise program.* Med Sci Sports Exerc, 1998. 30(10): p. 1496-501.

40. Warshafsky, S., R.S. Kramer, S.L. Sivak, *Effect of garlic on total serum cholesterol: A meta-analysis.* Ann Intern Med, 1993. 119: p. 599-605.

41. Williams, P.T., *High-density lipoprotein cholesterol and other risk factors for coronary heart disease in female runners.* New Engl J Med, 1996. 334(20): p. 1298-1325.

42. Williams, P.T., et al., *Variability of plasma HDL subclass concentrations in men and women over time.* Arterioscler Thromb Vasc Biol, 1997. 17(4): p. 702-6.

43. Williams, P.T., P.D. Woods, W.L. Haskell, M.A. Vranizan, *The effects of running mileage and duration on plasma lipoprotein levels.* JAMA, 1982. 247(19): p. 2674-9.

44. Wood, P.D., M.L. Stefanick, P.T. Williams, W.L. Haskell, *The effects of plasma lipoproteins of a prudent weight-reducing diet, with or without exercise, in overweight men and women.* N Engl J Med, 1991. 325(7): p. 461-6.

45. Ziogas, G.G., T.R. Thomas, W.S. Harris, *Exercise training, postprandial hypertriglyceridemia, and LDL subfraction distribution.* Med Sci Sports Exerc, 1997. 29(8): p. 986-91.

46. Zmuda, J.M., et al., *Exercise training has little effect on HDL levels and metabolism in men with initially low HDL cholesterol.* Atherosclerosis, 1998. 137(1): p. 215-21.

47. National Heart, Lung, Blood Institute, *Third Report of the National Cholesterol Education Program Expert Panel on the Detection, Evaluation and Treatment of High Blood Cholesterol in Adults.* National Heart, Lung, Blood Institute of the National Institutes of Health Report No. 01-3670, pp. 1-40, Washington, DC, May 2001.

Patricia Painter, PhD
Adjunct Associate Professor
Department of Physiological Nursing
University of California at San Francisco
San Francisco, CA

End-Stage Renal Disease

Chronic renal failure results from structural renal damage and progressively diminished renal function. Once initiated, the disease progresses to end-stage renal disease (ESRD), requiring some form of **renal replacement therapy** such as dialysis or transplantation.

Scope

There are approximately 308,000 patients with kidney failure in the United States (1). Although ESRD affects people of every age, race, and walk of life, 42% of the ESRD population are over the age of 60 years, and several studies have indicated lower educational and income levels among dialysis patients than the general population (2).

In the United States, prior to 1972, access to dialysis treatment was limited, and selection of patients for treatment was made by committees of medical professionals, clergy, and lay people. Essentially, they decided who would receive the lifesaving therapy of dialysis. In 1972, Congress passed landmark legislation that extended Medicare coverage to patients with ESRD. This legislation hinged on the expectation of successful vocational rehabilitation of these patients (an expectation that has not been realized). Renal replacement therapy is expensive, with the estimated cost of dialysis being $51,000 per patient per year; kidney transplant costs less over time ($18,000 per year) (1). Although the cost of renal replacement therapy has remained relatively constant since 1972, the population of patients with ESRD is increasing annually (100 individuals are diagnosed with ESRD each day); thus, the cost to the Medicare program is substantial for the number of patients involved. Additionally, ESRD patients are qualified for disability payments, another cost to government programs (2). Access to and payment for treatment as well as preferences of treatment vary in other countries.

Although the overall outcomes and well-being of patients with renal failure have significantly improved because of advances in technology and pharmacology that improve these patients' potential for rehabilitation, it is generally acknowledged that rehabilitation has not been addressed nationally in this patient group in a sustained, consistent, and integrated fashion (2). Low levels of physical functioning contribute significantly to the low levels of rehabilitation, thus indicating the need for physical rehabilitation as a part of the routine medical therapy of these patients.

Physiology and Pathophysiology

Damage to the kidney can result from long-standing diabetes mellitus, hypertension, autoimmune diseases (e.g., lupus), **glomerulonephritis**, **pyelonephritis**, some inherited diseases (i.e., **polycystic kidney disease**, Alport's syndrome), and congenital abnormalities. The damaged kidney initially responds with higher filtration and excretion rates per nephron, which masks symptoms until only 10% to 15% of renal function remains. Progressive renal failure results in the loss of both excretory and regulatory functions, resulting in uremic syndrome. **Uremia** is characterized by fatigue, nausea, malaise, anorexia, and subtle neurological symptoms. Patients present with these symptoms and often with peripheral edema, pulmonary edema, or congestive heart failure. Diagnosis is made from elevated serum **creatinine,** blood urea nitrogen, and reduced **glomerular filtration rate** (3).

The loss of the excretory function of the kidney results in the buildup of "toxins" in the blood, any of which can negatively affect enzyme activities and inhibit systems such as the sodium pump, resulting in altered active transport across cell membranes and altered membrane potentials. The loss of regulatory

function of the kidneys results in the inability to regulate extracellular volume and electrolyte concentrations, which adversely affects cardiovascular and cellular functions. Most patients are volume overloaded, resulting in hypertension and often congestive heart failure. Other malfunctions in regulation include impaired generation of ammonia and hydrogen ion excretion, resulting in metabolic acidosis and decreased production of erythropoietin, which is the primary cause of the anemia of ESRD (4, 5).

Normal substances may be excessively produced or inappropriately regulated in response to renal failure. **Parathyroid hormone** may be the most important of these. Parathyroid hormone is produced in excess secondary to hyperphosphatemia, reduced conversion of vitamin D to its most active forms and malabsorption and impaired release of calcium ions from bone. The attempt to maintain adequate circulating calcium ion concentrations in the face of hypocalcemia results in **hyperparathyroidism** and renal osteodystrophy (4, 5).

Several metabolic abnormalities are associated with uremia, including insulin resistance and hyperglycemia. Hyperlipidemia is characterized in patients treated with dialysis by hypertriglyceridemia with normal (or low) total cholesterol concentrations. Several interventions associated with the dialysis treatment or **immunosuppression** therapy (following transplant) can contribute to these metabolic abnormalities (4, 5).

Clinical Considerations

Renal failure produces specific signs and symptoms, and the diagnosis is strongly related to the evaluation of serum creatinine and blood urea nitrogen. The treatment of renal failure depends on creatinine clearance. Dietary adjustment for protein, sodium, and fluid intake plays a very important role in the initial management of renal failure. If dietary intervention is not successful, renal replacement therapy is required. Currently there are three alternatives for renal replacement therapy: hemodialysis, peritoneal dialysis, or transplantation. Although transplantation is the preferred method, patients need to be free of other life-threatening illnesses to be considered for transplantation. Hemodialysis is the most common therapy for renal failure, although it requires significant time throughout the week at a renal center. The third alternative for renal therapy replacement, peritoneal dialysis, is the least used method in United States but is more frequently used in other countries.

Signs and Symptoms

Deterioration in renal function results in an overall decline in physical well-being. Signs include anemia, fluid build-up in tissues, loss of bone minerals, and hypertension. Patients experience fatigue, shortness of breath, loss of appetite, restlessness, change in urination patterns, and overall malaise.

Diagnostic and Laboratory Evaluations

Diagnosis of renal failure is typically made by determination of levels of serum creatinine and blood urea nitrogen through a blood test. Renal biopsy can be done to determine the etiology of disease, and a renal scan or intravenous pyelogram can be performed to rule out obstruction or congenital abnormalities that may contribute to increased creatinine levels in the blood.

Treatment

Treatment of chronic renal failure consists of medical management until the **creatinine clearance** is less than 5 ml · min^{-1}, at which time more aggressive therapy is required. Management is directed at minimizing the consequences of accumulated nitrogenous waste products normally excreted by the kidneys. Dietary measures play a primary role in the initial management, with very low protein diets being prescribed to decrease the symptoms of uremia and possibly to delay the progression of the disease. In addition to protein restriction, dietary sodium and fluid restrictions are critical (4), because the fluid regulation mechanisms of the kidney are deteriorating in function. Any excess fluid taken remains in the system and, with progressing deterioration in renal function, will ultimately result in peripheral edema, congestive heart failure, and pulmonary congestion.

Progressive deterioration of renal function will ultimately require the initiation of some form of renal replacement therapy for maintenance of life. Treatment options include hemodialysis (performed in center or at home), **peritoneal dialysis,** or transplantation. The decision to initiate dialysis is determined by many factors, including cardiovascular status, electrolyte levels (specifically potassium), chronic fluid overload, severe and irreversible **oliguria** or **anuria,** significant uremic symptoms, and excessively abnormal laboratory values (usually creatinine >8-12 mg · dl^{-1}, blood urea nitrogen >100-120 mg · dl^{-1}) and creatinine clearance (<5 ml · min^{-1}). Renal replacement therapy does not correct all signs and symptoms of uremia and often presents the patient with other concerns and side effects that must be dealt with. Table 11.1 lists laboratory values for healthy patients versus those undergoing dialysis.

Hemodialysis

Hemodialysis is the most common form of renal replacement therapy in the United States, with approximately 63% of all patients treated in a center or at home. In other countries, some patients prefer more home-based treatments such as peritoneal dialysis (discussed later). Hemodialysis is a process of ultrafiltration (fluid removal) and clearance of toxic solutes from the blood. It necessitates vascular access by way of an arteriovenous connection that uses either a prosthetic conduit or native vessels. Two needles are placed in the fistula; one directs blood out of the body to the artificial kidney (dialyzer), and the other directs blood back into the body. The dialyzer has a semipermeable membrane that separates the blood from a dialysis solution, which creates an osmotic and concentration gradient to clear substances from the blood. Factors such as the characteristics of the membrane, transmembrane pressures, blood flow, and **dialysate** flow rate all determine removal of substances from the blood. Manipulation of the blood flow rate, dialysate flow rate, dialysate concentrations, and time of the treatment can be used to remove more or less substances and fluids (4).

The duration of the dialysis treatment is determined by the degree of residual renal function, body size, dietary intake, and clinical status. A typical dialysis prescription is 3 to 4 hr three times per week. Complications of the dialysis treatment include hypotension, cramping, problems with bleeding, and fatigue. Significant fluid shifts can occur between treatments if the patient is not careful with dietary and fluid restrictions. Table 11.2 and figure 11.1 list complications of dialysis.

Peritoneal Dialysis

Approximately 8% of patients in the United States are treated with peritoneal dialysis. Other countries tend to have a higher percentage of patients treated with this form of dialysis. This form of therapy is accomplished by introducing a dialysis fluid into the peritoneal cavity via a permanent catheter placed in the lower abdominal wall. The peritoneal membranes are effective for ultrafiltration of fluids and clearance of toxic substances in the blood of uremic individuals. The dialysis fluid is of a given osmotic and concentration to provide gradients to remove fluid and substances. The fluid is introduced either

TABLE 11.1

Normal Laboratory Values Compared With Typical Values for Dialysis Patients

Laboratory value	Normal range	Typical range for dialysis patients[a]
Hemoglobin (g · dl^{-1})	12.0-16.0	[b]
Hematocrit (%)	37.0-47.0	[b]
Sodium (mEq · L^{-1})	136.0-145.0	135.0-142.0
Potassium (mEq · L^{-1})	3.5-5.3	4.0-6.0
Chloride (mEq · L^{-1})	95.0-110.0	95.0-100.0
HCO$_3$ (mEq · L^{-1})	22.0-26.0	18.6-23.4
Albumin	3.7-5.2	2.7-3.3
Calcium	9.0-10.6	9.0-12.0
Phosphorous	2.5-4.7	2.5-6.0
BUN (mg · dl^{-1})	5.0-25.0	60.0-110.0
Creatinine (mg · dl^{-1})	0.5-1.4	12.0-15.0
PH	7.35-7.45	7.38-7.39
Creatinine clearance (ml · min^{-1})	85.0-150.0	0 (or minimal residual)
Glomerular filtration rate (ml · min^{-1})	90.0-125.0	0 (or minimal)

BUN = blood urea nitrogen.

[a]Assuming well-dialyzed, stable patient. [b]Hematocrit and hemoglobin levels depend on the level of erythropoietin treatment. With no treatment, hematocrit may be as low as 17%. With treatment, it can be normalized; however, most patients are treated to a range of 33% to 37%.

by a machine (cycler), which cycles fluid in and out over a 8- to 12-hr period at night, or manually by 2- to 2.5-L bags that are attached to tubing and emptied by gravity into and out of the peritoneum. The latter process is known as continuous ambulatory peritoneal dialysis and allows the patient to dialyze continuously throughout the day. Continuous ambulatory peritoneal dialysis requires exchange of fluid every 4 hr using sterile technique (4). Table 11.3 lists complications associated with peritoneal dialysis.

Peritoneal dialysis may be chosen by patients so they can experience more freedom and less dependency on a center for use of a machine. It allows patients to travel and dialyze on their own schedule. Patients may also be placed on peritoneal dialysis because of cardiac instability, since it does not involve the major fluid shifts experienced with hemodialysis. Peritoneal dialysis may also be preferable for diabetic patients, because they can inject insulin into their dialysate and achieve better glucose control.

Complications of peritoneal dialysis include problems with the catheter or catheter site, infection, hernias, low back pain, and obesity. Hypertriglyceridemia is a problem caused by the exposure and absorption of glucose from the dialysate. Patients may absorb as many as 1200 kcal from the dialysate per day, contributing to the development of obesity and hypertriglyceridemia (4).

Renal Transplantation

Transplantation of kidneys is the preferred treatment of ESRD. In the United States, 12,000 kidney transplants are performed each year (28% of ESRD patients). The source of the kidneys available for transplant can be a living related or unrelated individual or a cadaver. Because of the shortage of organs available for transplantation and improvements in immunosuppression medications, living nonrelated transplants are becoming more frequent. Patients considered for transplant are generally healthier and younger than the general dialysis population, although there are no age limits to transplantation. Patients with severe cardiac, cerebrovascular, or pulmonary disease and neoplasia are not considered candidates. Table 11.4 lists long-term complications of transplantation.

Following transplantation, patients are placed on immunosuppression medication, which includes combinations of glucocorticosteroids (prednisone), cyclosporine derivative, and monoclonal antibody

TABLE 11.2

Complications Associated With Hemodialysis Treatment

Complication	Pathophysiology
Hypotension	Decreased plasma volume with slow refilling Impaired vasoactive or cardiac responses Vasodilation Autonomic dysfunction
Cramping	Contraction of intravascular volume Reduced muscle perfusion
Anaphylactic reactions	Reaction to dialysis membrane (usually at first use)
Pyrogen or infection-induced fever	Bacterial contamination of water system Systemic infection (often at the access site)
Cardiopulmonary arrest	Dialysis line disconnection Air embolism Aberrant dialysate composition Anaphylactic membrane reaction Electrolyte abnormalities Intrinsic cardiac disease
Itching	Unknown etiology
"Restless legs"	Unknown etiology

Reprinted, by permission, from K.L. Johansen, 1999, "Physical functioning and exercise capacity in patients on dialysis," *Advances in Renal Replacement Therapy* 6(2): 142.

Metabolic abnormalities
— Metabolic acidosis
— Hyperlipidemia (type 4)
  — Increased triglycerides
  — Increased very low density lipo-protein cholesterol
  — Decreased high-density lipo-protein cholesterol
  — Normal total cholesterol
— Hyperglycemia
— Other hormonal dysfunction
Malnutrition
Cardiovascular disease
— Hypertension
— Ischemic heart disease
— Congestive heart failure
— Pericarditis
Uremic osteodystrophy (secondary hyperparathyroid-ism)
Peripheral neuropathy
Amyloidosis
Severe physical deconditioning
Frequent hospitalizations
Continuation of progressive complications in diabetic patients

FIGURE 11.1 Long-term complications of dialysis.

therapy. New immunosuppression medications are constantly being developed, allowing for minimization of side effects by altering therapies or combinations of therapies. Patients may experience **rejection** early (acute) or later (chronic), which is detected by elevation of creatinine. Rejection is treated immediately with increased dosing of immunosuppression (mostly prednisone), with a tapering back to maintenance dose. Patients must remain on immunosuppression for the lifetime of the transplanted organ. Nationwide 1-year graft survival is 85% and patient survival is 90%. Five-year rates are 67% graft survival and 85% patient survival. Causes of graft loss include chronic rejection (25%), cardiovascular deaths (20.3%), infectious deaths (8.7%), acute rejection (10.2%), technical complications

TABLE 11.3

Complications Associated With the Peritoneal Dialysis Treatment

Complication	Comments
Infections	Possible at exit site or along catheter—"tunnel infection" May be result of a break in sterile procedures during exchange
Peritonitis	Most frequent complication of peritoneal dialysis
Hypotension	Excessive ultrafiltration and sodium removal
Hernia, leaks	Associated with the increased intra-abdominal pressure

(4.7%), and other deaths (10.2%). Short-term transplant survival has been improved with new immunosuppression medications, leaving the major challenges of long-term survival of graft and patients to be investigated. Loss of kidney results in the need to return to dialysis (6).

TABLE 11.4

Long-Term Complications Associated With Transplantation

Complication	Comments
Rejection	Can be acute or chronic, in most cases treated with increased immunosuppression dosages.
Cardiovascular disease	Most frequent cause of death posttransplant. All known risk factors are prevalent, plus immunosuppression medications may exacerbate risk.
Infections	Immunosuppression may increase infection risk.
Musculoskeletal disorders	Glucocorticoid therapy (prednisone) reduces bone density and causes muscle protein breakdown.
Obesity	Very prevalent, often associated with prednisone therapy; however, more likely attributable to calorie intake/expenditure imbalance (i.e., lifestyle issues).

Complications of kidney transplantation are primarily related to immunosuppression therapy and include infection, hyperlipidemia, hypertension, obesity, steroid-induced diabetes, and osteonecrosis. The incidence of atherosclerotic cardiovascular disease is four times higher in kidney transplant recipients than the general population, and cardiovascular risk factors are prevalent in the majority of patients.

Graded Exercise Testing

Most patients on dialysis are severely limited in exercise capacity, primarily by leg fatigue. Peak oxygen uptake is reported to be only 60% to 70% of normal age expected levels, and it can be as low as 39% (7, 18) (figure 11.2). The degree to which exercise is limited in these patients is difficult to determine because of the complex nature of uremia, which effects nearly every organ system of the body. It is almost certain that the reduced exercise capacity is a multifactorial problem that is influenced by anemia, muscle blood flow, muscle oxidative capacity, myocardial function, and the individual's activity levels. Muscle function may be affected by nutritional status, dialysis adequacy, hyperparathyroidism, and other clinical variables (19).

Most studies that have measured oxygen uptake have only included the healthiest patients; thus, the majority of patients have an even lower exercise capacity, and most may be unable to perform exercise testing. Information obtained from exercise testing in this patient group is not diagnostically useful, because most patients stop exercise because of leg fatigue and do not achieve age-predicted maximal heart rates. Although many patients have abnormal

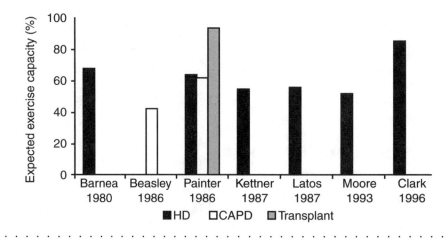

FIGURE 11.2   Exercise capacity in patients with end stage renal disease. HD = hemodialysis; CAPD = continuous ambulatory peritoneal dialysis.

Reprinted, by permission, from K.L. Johansen, 1999, "Physical functioning and exercise capacity in patients on dialysis," *Advances in Renal Replacement Therapy* 6(2): 142.

left ventricular function (32), most patients have conditions that make interpretation of stress electrocardiogram difficult, including left ventricular hypertrophy (LVH) with strain patterns, electrolyte abnormalities, and digoxin effects on the electrocardiogram. Thus, stress testing is not necessarily recommended before initiation of exercise training, and requiring stress testing may prevent some patients from becoming more physically active (20). Because exercise capacity is so low, most patients will not train at levels that are much above the energy requirements of their daily activities; thus, risk associated with such training is minimal. Heart rate is not recommended for determining training intensity because of the effects of antihypertensive medications and fluid shifts on heart rates. Thus, exercise testing is not needed to develop a training heart rate prescription. Practical application 11.1 describes client–clinician interaction for patients with end-stage renal disease.

Physical performance may best be tested in dialysis patients by using tests such as stair climbing, 6-min walk test, sit-to-stand-to-sit test, or gait speed testing. These have been standardized and used in many studies of elderly individuals and have been shown to predict outcomes such as hospitalization, discharge to nursing home, and mortality rate (21). A walking–stair-climbing test was validated in hemodialysis patients by Mercer et al. (33). In dialysis patients, these tests are effective in demonstrating improvements from exercise counseling interventions (22). Additionally, self-reported physical functioning scales such as those on the SF-36 Health Status Questionnaire are highly predictive of outcomes in dialysis patients, specifically hospitalization and death (23). Exercise training improves scores on these self-reported scales in hemodialysis patients (22).

Exercise capacity is similarly low in peritoneal dialysis patients (8, 13, 34). Following successful renal transplant, exercise capacity increases significantly, to near sedentary normal predicted values (8, 13, 24). Renal transplant recipients who were active and who participated in the 1996 U.S. Transplant Games had exercise capacity that averaged 115% of normal age-predicted values (25). Exercise

Client–Clinician Interaction

Exercise professionals can best serve the needs of dialysis patients if they understand the patients' treatment regimens and the setting in which they receive treatment. The majority of hemodialysis patients typically receive treatments 3 times per week for 3 to 4 hr. They are transported either by friends and family or by medical transportation services, so they have minimal flexibility.

The staffing at outpatient dialysis clinics typically consists of a charge nurse and patient care technicians who are not medically trained except for administration of dialysis. The schedules for dialysis are quite "tight," and the opportunity for patient education is minimal in terms of both staff and time. Therefore, it seems reasonable to assume that asking patients to attend exercise class at a supervised program may be a significant barrier to exercise participation. Thus, every effort should be made to implement home exercise programs or programs at the dialysis clinic to optimize adherence.

For a program to be successful, the exercise professional should first interact with and educate the dialysis staff, which takes the support of the administration in helping to coordinate in-service training. The dialysis staff is very close to the patients and can be very influential in their participation (or nonparticipation) in exercise. This support is critical to the efforts of the patient and the exercise personnel. Educational and motivational programs at the dialysis clinic not only increase staff and patients' awareness of the importance of exercise but also change the environment of the clinic from one of illness to one of wellness.

Following kidney transplant, many patients are afraid to exert themselves vigorously. That fear comes from lack of information provided from the transplant service, weakness despite a significant improvement in overall health, fears/concerns on the part of the patient's family, and lack of experience with exercise (because of health concerns prior to transplant). Exercise is not routinely addressed following kidney transplant—either at the time of transplant or in the routine follow-up care in clinic. Thus, again the exercise professional should do all possible to educate the transplant team about the importance of exercise and, as much as possible, become a part of the routine care team so the exercise counseling is incorporated into post-transplant care.

testing with standard protocols is more appropriate for transplant recipients who are able to push themselves in their training programs above their daily levels of activity. Exercise heart rate responses are normalized after transplant. The major abnormality noted in transplant recipients is excessive blood pressure response to exercise.

Exercise Prescription

The exercise prescription for patients on dialysis should include flexibility and range of motion, strengthening, and cardiovascular exercises. Weight

management considerations may be needed for many transplant recipients. For the dialysis patient, the key to prescription is understanding the multiple barriers to exercise that may exist. These include general feelings of malaise, time requirements of treatment, lack of encouragement and information provided by nephrology healthcare workers, fear, and accustomization or adaptation of lifestyles to low levels of functioning. Thus, any prescription should start slowly and progress gradually to prevent discouragement and additional feelings of fatigue or muscle soreness (29). Practical application 11.2 reviews the literature about exercise training and end-stage renal disease.

PRACTICAL APPLICATION 11.2

Literature Review

Several studies have reported the effects of cardiovascular exercise training in dialysis patients. The type and duration of these studies were variable (range 8 weeks to 12 months), but all studies compared exercise tests before and after training in the same individuals. Most studies were performed prior to the availability of recombinant human erythropoietin, most included small patient numbers, and very few provided adequate control or comparison data.

Most studies showed an improvement in $\dot{V}O_2$max after exercise training, with an average increase of 16.4% across all studies (range 0-52%) (9, 11, 12, 17, 18, 22, 35) (figure 11.2). Although the improvement was similar to that seen in healthy individuals, training did not normalize $\dot{V}O_2$max, and the posttraining values remain well below age-predicted values. In addition to the improvements in oxygen uptake, there are reports of improvements in hematocrit (before the availability of erythropoietin), blood pressure control, and lipid profiles. Many of these improvements have not been duplicated in other studies, and most studies had too few subjects to be conclusive (19). None of these studies included measures of quality of life or self-reported functioning. Impressive improvements have been observed in muscle fiber size and improvement in atrophy following a program of cardiovascular exercise training plus sports activity and strengthening exercises (34). Also, positive changes in cardiac function following physical training have been observed (32, 36).

One recent study showed significant improvements in gait speed and the sit-to-stand-to-sit test following exercise counseling interventions for independent home exercise and in-center cycling. This study also reported significant improvements in the physical scales of the SF-36 Health Status Questionnaire (22).

Most exercise training studies incorporated exercise on nondialysis days, three times per week. The studies reported major problems in adherence to exercise that required patients to attend on their nondialysis days at a supervised center. The dialysis treatment provides an ideal opportunity to capture a "captive audience." A high rate of adherence to bicycle exercise training during the dialysis treatment was observed (12). This programming was well tolerated within the first hour of the treatment and did not interfere with the dialysis treatment. $\dot{V}O_2$peak changes were comparable to those seen in other cardiovascular exercise studies performed on nondialysis days.

One exercise training study that included patients who were treated with peritoneal dialysis showed significant improvements in $\dot{V}O_2$peak (16.2%) and physical functioning dimensions of quality of life (34). Two studies reported significant improvements in exercise capacity in renal transplant recipients (27, 28). In a randomized controlled trial of independent home exercise over the first year of transplant, we found significant improvements in $\dot{V}O_2$peak with home exercise (increase in $\dot{V}O_2$peak to 30 ml $\cdot$ kg^{-1} $\cdot$ min^{-1} at 1 year) compared with those in the usual care group (24 ml $\cdot$ kg^{-1} $\cdot$ min^{-1} at 1 year; unpublished data). The values for the usual care group were not much different than values reported for high-functioning dialysis patients.

Special Exercise Considerations

When patients are diagnosed with end-stage renal failure, most are never given information on exercise and physical activity. If they ask, typically they are told to "take it easy" or "just don't overdo it." This poses questions and doubt in the minds of patients and their families, who will be very protective. They do not know how much is "too much" activity, and because they do not feel well and are fatigued, they opt for no activity. This inactive lifestyle is often reinforced by the dialysis staff who see the patients regularly for their treatments. It is not surprising, therefore, that many patients are skeptical about becoming physically active. Patients have little interaction and thus receive little information from others than their dialysis providers. Thus, it is very important for any exercise professional to take the time to learn about dialysis and transplant to understand what the patient must deal with daily or three times per week. This could entail watching a patient being put on the dialysis machine, and visiting with a few patients during their treatment. Patient support groups are also a good source of information. Patients will talk freely about their experiences, many of which will help the exercise professional better motivate patients and understand more about patient responses to major changes in lifestyle such as initiating exercise. The exercise professional should reach out to dialysis staff about how exercise might benefit their patients and assure them that the programs initiated will be safe and will not interfere with the treatments (31). This education should also include ideas of how the dialysis staff can encourage patients to be physically active. This additional encouragement and reinforcement can greatly facilitate patient efforts in rehabilitation. Likewise, lack of support and understanding on the part of the dialysis staff can sabotage the efforts to increase patients' activity.

Most exercise professionals practice in their own laboratory and depend on referrals of patients to their exercise facility. There are many patients who may be unable to participate in exercise at the designated facility but for whom counseling on home independent exercise can be extremely beneficial. Thus, exercise professionals are encouraged to reach out to other healthcare providers (other than cardiologists and pulmonologists), educate them on the services they can offer their patients, and discuss the benefits of exercise for their patient groups. Although there is growing interest in the nephrology community in improving physical functioning of patients, most nephrologists and kidney transplant staff are not familiar with how exercise may benefit their patients or how to evaluate physical functioning or prescribe exercise. Thus, a trained professional who knows about the problems associated with dialysis and transplant may be a welcome addition to the patient care team.

Exercise Recommendations

The timing of exercise in relation to the dialysis treatment should be considered. Hemodialysis patients can exercise anytime, although they may feel best on their nondialysis days and may be able to tolerate higher intensity or duration. Most feel extremely fatigued following their dialysis treatment, and there may be problems with hypotension following the treatment when physical activity-induced vasodilation occurs. Immediately before the dialysis treatment, some patients may have excessive fluid in their systems, because they are unable to rid their bodies of fluid taken in between treatment. Thus, they may not tolerate as much exercise prior to dialysis, because of an increased volume overload on the left ventricle; increased blood pressure at rest (and during exercise), which may increase ventricular preload and afterload; and, in extreme cases, pulmonary congestion. Exercise should be deferred if the patient is experiencing shortness of breath related to the excess fluid status. There are no specific guidelines as to the upper limit of fluid weight gain that would contraindicate exercise, although the guidelines for blood pressure as established by American College of Sports Medicine (37) should be followed. Practical applications 11.3 and 11.4 discuss exercise prescription for patients undergoing dialysis and those with transplants.

The ideal mode of exercise and the ideal time for the patient to exercise, in terms of adherence and convenience, may be recumbent stationary cycling during the hemodialysis treatment. Although this form of exercise does not interfere with the dialysis treatment, most facilities are unwilling to have cycles in the clinic for their patients. Thus, independent home exercise may be the best approach for exercise for these patients.

Cycling during dialysis is best tolerated during the first 1 to 1.5 hr of the treatment because after this time the patient has a greater risk of being hypotensive when in the chair, making it difficult to cycle. This response is caused by the continuous removal of fluid throughout the treatment, which decreases cardiac output, stroke volume, and mean arterial pressure at rest (38). Therefore, after 2 hr of dialysis, cardiovascular decompensation may preclude exercise (38).

Exercise Prescription for ESRD Treated With Dialysis

Cardiovascular Exercise

The clinician should use the following guidelines in prescribing cardiovascular exercise for dialysis patients:

— Mode: walking, cycling, swimming, low-level aerobics, stepping
— Frequency: 4 to 5 days per week
— Intensity: RPE of 12 to 15 (on 6- to 20-point scale)
— Duration: work up to ≥30 min of continuous exercise
— Progression: start with intervals of intermittent exercise and gradually increase the work intervals until continuous exercise is tolerated

Strengthening Exercise

Following are guidelines for prescribing strengthening exercise for dialysis patients:

— Mode: Theraband, isometric, very low level hand/ankle weights, body weight resistance
— Frequency: 2 to 3 days per week
— Sets: three sets of exercises for major muscle groups
— Repetitions: 12 to 15 repetitions of each exercise
— Progression: start with one set of 12 repetitions with 1- to 2-lb weights; increase gradually

Special Considerations

Several additional considerations should be recognized when one is working with dialysis patients:

— Patients will have very low fitness levels.
— Timing of exercise sessions should be coordinated with dialysis sessions.
— Patients will experience frequent hospitalizations and setbacks.
— Gradual progression is critical.
— Heart rate prescriptions are typically invalid—use of RPE is recommended.
— Maximal exercise testing is typically not tolerated well by the majority of patients and is not diagnostically useful for those with coronary artery disease because of peripheral muscle fatigue.
— Performance-based testing is more feasible and useful.
— One-repetition maximum testing for strength is not recommended because of secondary hyperparathyroidism-related bone/joint problems.
— Prevalence of orthopedic problems will be significant.
— Motivation of patients is often a challenge.
— Every attempt should be made to educate dialysis staff about the benefits of exercise so they can also help motivate patients to participate.

For patients treated with continuous ambulatory peritoneal dialysis, the exercise may be best tolerated at a time when the abdomen is drained of fluid, which allows for greater diaphragmatic excursion and less pressure against the catheter during exertion, reducing the risk of hernias or leaks around the catheter site (30). Patients may choose to exercise in the middle of a dialysis exchange—after draining fluid and before introducing the new dialysis fluid. This requires capping off the catheter for exercise, a technique that must be discussed with the dialysis nurse.

Exercise Prescription for ESRD Treated With Transplantation

Cardiovascular Exercise

The clinician should use the following guidelines in prescribing cardiovascular exercise for transplant patients:

— Mode: walking, jogging, cycling, swimming, aerobics, stepping, sports
— Frequency: 4 to 5 days per week
— Intensity: 65% to 80% of peak heart rate; RPE of 12 to 15 (on 6- to 20-point scale)
— Duration: work up to ≥30 min of continuous exercise (longer duration for weight management
— Progression: start with intervals of intermittent exercise and gradually increase the work intervals until continuous exercise is tolerated

Strengthening Exercise

Following are guidelines for prescribing strengthening exercise for transplant patients:

— Mode: Theraband, weight machines, hand weights, free weights
— Frequency: three times per week
— Sets: three sets of exercises for major muscle groups
— Repetitions: 12 to 15 repetitions of each exercise
— Progression: start with 1 set of 12 repetitions with low weights and increase gradually

Special Considerations

Several additional considerations should be recognized when one is working with transplant patients:

— Patients are initially weak, so gradual progression is recommended.
— Patients may experience a lot of orthopedic and musculoskeletal discomfort with strenuous exercise.
— Weight management often becomes an issue following transplant.
— Patients and their families are often fearful of "overexertion"; thus, gradual progression should be stressed.
— Prednisone may delay adaptations to resistance training.
— Exercise should be decreased in intensity and duration during episodes of rejection, not curtailed completely.
— Patients may experience frequent hospitalizations during the first year posttransplant.
— Because patients are immunosuppressed, every effort must be made to avoid infectious situations (e.g., strict sterilization procedures for testing and training equipment).

Mode

There is no restriction on the type of activity that can be prescribed for dialysis or transplant patients. Range of motion and strengthening exercises are critical for most patients because of their history of long periods of inactivity and resulting stiffness and weakness. Because many patients have weak muscles and joint discomfort, non-weight-bearing cardiovascular activity may be best tolerated. As for anyone else, if jarring activity causes joint discomfort, then a change in mode of exercise is indicated. The access site for the hemodialysis may be in the arm or upper leg. This should not inhibit activity at all, although many patients are told not to use the arm with their fistula in it. This restriction is typically given by the vascular surgeon at the time of placement and pertains only to the time of healing (i.e., 6-8 weeks). The only precaution for the fistula is to avoid any activity that would close off the flow of blood (e.g., having weights lying directly over the top of the vessels). Although patients should be quite protective of their access site, use of the extremity will increase flow through it and actually help develop muscles around the access site, which should make placement of needles easier.

Full sit-ups and activities that involve full flexion at the hip should be avoided for patients with a peritoneal catheter. Abdominal strengthening can be accomplished by using isometric contractions and crunches. Swimming may be a challenge for those with peritoneal catheters because of the possibility of infection. Patients must be advised to cover the catheter with some protective tape and clean around the catheter exit site after swimming. Freshwater lake swimming is not recommended, whereas swimming in chlorinated pools and in the ocean involves less risk of infection.

Although transplant recipients are often told not to participate in vigorous activities, the main concern is any contact sport that may involve direct hit to the area of the transplanted kidney (e.g., football). Vigorous activities and noncontact sports are well tolerated by transplant recipients who have worked to build adequate muscle strength and cardiovascular endurance through a comprehensive general conditioning program.

Frequency

Range of motion exercises should be encouraged daily. Hemodialysis patients will feel especially stiff after their dialysis session because of 3 to 4 hr of sitting as well as removal of fluid and often cramping. Stretching during the dialysis treatment and after may relieve this stiffness. Muscle strengthening should be done 3 days per week. Cardiovascular exercise should be prescribed for at least 3 days per week, although a prescription of 4 to 6 days per week may be most beneficial.

Intensity

Cardiovascular exercise intensity should be prescribed by using a rating of perceived exertion (RPE), because heart rates are highly variable in dialysis patients because of fluid shifts and vascular adaptations to fluid loss during the dialysis treatment. Many patients initially may only tolerate several minutes of very low-level exercise, which means that warm-up and cool-down intensities are irrelevant. These individuals should just be encouraged to gradually increase duration at whatever level they

can tolerate. Once they achieve 20 min of continuous exercise, then warm-up, conditioning, and cool-down phases can be incorporated, with an RPE of 9 to 10 for warm-up and cool-down and 12 to 15 for the conditioning time (on a 6- to 20-point scale).

Duration and Progression

It is critical to start patients slowly and progress gradually. In practice, this means that the patient should determine the duration of activity that he or she can comfortably tolerate during the initial sessions. This duration will be the starting duration of activity. If only 2 to 3 min of exercise is tolerated, the prescription may be for several intervals of 2 to 3 min, with a gradual decrease in the rest times to progress the patient to continuous activity. A progression in duration of 2 to 3 min per session or per week is recommended, depending on individual tolerance. Very weak patients may need to start with a strengthening program of low weights and high repetitions and range of motion exercise before initiating any cardiovascular activity. The progression should gradually work up to 20 to 30 (or more) min of continuous activity at an RPE of 12 to 15 (on a 20-point scale).

When a patient begins cycling during dialysis, the initial session is usually limited to 10 min, even if the patient is able to tolerate a longer duration. This precaution is to assure the dialysis staff and the patient that the cycling does not have any adverse effects on the dialysis treatment. The patient can then progress duration in subsequent sessions according to tolerance, as described previously. RPE is used for intensity prescription during the dialysis treatment also, because removal of fluid from the beginning to the end of dialysis can cause resting and exercise heart rates (standard submaximal level) to vary by 15 to 20 beats (38).

Conclusion

Exercise prescription for patients with renal failure depends on their treatment and must be individualized according to patients' limitations. The prescription should include the type of exercise (cardiovascular, range of motion, strengthening), frequency of exercise, timing of exercise in relation to treatment, duration, intensity (prescribed primarily based on RPE), and progression. The progression should be very gradual in those who are extremely debilitated. The starting levels and progression must be according to tolerance, because fluctuations in well-being, clinical status, and overall ability frequently change with changes in medical status. Often hospitalization or a medical event (e.g., clotting of the fistula or placement of a new fistula) will set a patient back in the progression of his or her program, requiring frequent evaluation of the prescription. The goal is for patients to become more active in general and, if possible, for them to work toward a regular program of 4 to 6 days per week of cardiovascular exercise, 30 min or more per session at an intensity of 12 to 14 on the RPE scale. Strengthening exercise should be recommended three times per week.

Case Study 11

Medical History and Diagnosis

Mrs. HN is a 68-year-old Hispanic female with known ESRD. She has been on hemodialysis for 28 months, and her treatment prescription is for 3 days per week with 3-hr treatment sessions each day. She has a graft in her right upper arm as her access site. She presents with the complaints of lack of energy, weakness, and decreased endurance. Her nephrologist refers her to the staff exercise physiologist. The exercise physiologist reviews Mrs. HN's chart to find that her ESRD is secondary to long-term non-insulin-dependent diabetes (15 years). Mrs. HN has also developed severe peripheral neuropathy as a result of her diabetes.

Exercise Test Results

The exercise physiologist then conducts a battery of physical function tests to assess Mrs. HN's physical ability. These tests consist of the sit-to-stand test, the 6-min walk, and the 20-foot gait speed test at both a comfortable and a fast pace. The results of these tests are as follows: sit-to-stand test, 33.01 s, which is 28% of normal age-predicted values (39); 6-min walk, 350 ft; 20-ft normal gait speed, 55.01 cm $\cdot$ s^{-1}, which is 42% of normal age normal values (40). Her self-reported physical function scale on the SF-36 Health Status Questionnaire is 55 (average age value 84). During her walking tests, Mrs. HN exhibits poor balance and endurance as a result of her peripheral neuropathy and general weakness, respectively. Physical activity questionnaires are administered to assess her current activity as well as degree of difficulty of those activities.

Exercise Prescription

With the assessment complete, an exercise prescription is developed. Mrs. HN is first counseled on exercise as it relates to her diabetes and glycemic control. Written information is also provided. The exercise prescription continues only when it is certain that Mrs. HN fully understands the balance between exercise and glycemic control. Because of her poor balance and endurance, a stationary bicycle is the preferred mode for cardiovascular exercise. She is asked to begin with a frequency of 3 to 4 days per week, on nondialysis days, because she generally feels better on those days. The duration of the exercise is 10 min with two bouts each exercise day, totaling 20 min of exercise each exercise day. The initial prescribed intensity should be light to moderate, or enjoyable. She is asked to gradually progress each week with a goal of 30 min of continuous cycling 3 to 4 days per week, minimum. Her initial exercise prescription also includes various flexibility exercises for upper and lower body as well as for the back. She is asked to do these exercises daily. Strengthening exercises are also prescribed. Again, both upper and lower body exercises that use the major muscle groups are encouraged. These exercises are prescribed for 3 days per week on nonconsecutive days. She will initially perform the exercises without resistance weight and gradually progress to performing them

continued

with weight. Her initial prescription consists of one set of 10 repetitions of each exercise. The exercise physiologist reviews Mrs. HN's progress weekly at her dialysis treatments. Progression and exercise participation are noted in the patient's chart. At the end of 8 weeks of exercise, Mrs. HN's physical functioning is again assessed with the battery of physical function tests and the activity questionnaires.

Case Study 11   Discussion Questions

1. Why is performance-based testing preferred for this patient?

2. What causes Mrs. HN to have poor balance during walking? How does this affect her exercise prescription?

3. Why is no heart rate prescription given to Mrs. HN for her cycling program?

4. How would the cycling program differ if she chose to cycle during the dialysis treatment instead of at home?

5. Is there anything in her history which suggests that cycling during dialysis might be advantageous for Mrs. HN?

Glossary

anuria—Suppression or arrest of urinary output, resulting from impairment of renal function or from obstruction in the urinary tract.

creatinine—End product of creatine metabolism excreted in the urine at a constant rate—a blood marker of renal function.

creatinine clearance—An index of the glomerular filtration rate, calculated by multiplying the concentration of creatinine in a timed volume of excreted urine by the milliliters of urine produced per minute and dividing the product by the plasma creatinine value.

dialysate—The fluid that passes through the membrane in dialysis and contains the substances of greater diffusibility in solution.

glomerular filtration rate—The rate of filtrate formation at the glomerulus.

glomerulonephritis—An acute, subacute, or chronic, usually bilateral, diffuse inflammatory kidney disease primarily affecting the glomeruli.

hyperparathyroidism—A state produced by the increased function of the parathyroid glands; results in dysregulation of calcium.

immunosuppression—Suppression of immune responses produced primarily by any of a variety of immunosuppressive agents.

oliguria—A diminution in the quantity of urine excreted: specifically, less than 400 ml in a 24-hr period.

parathyroid hormone—A peptide hormone formed by the parathyroid glands; it raises the serum calcium when administered parenterally by causing bone resorption.

peritoneal dialysis—Dialysis that is performed by introducing fluid into the peritoneal cavity. Dialysis fluid can be cycled through the peritoneal cavity via a machine over a 10- to 12-hr period daily (intermittent peritoneal dialysis) or exchanged every 4 hr with the fluid staying in the peritoneal cavity between exchanges (continuous ambulatory peritoneal dialysis). The fluid is introduced through a catheter (tube) that is placed in the abdomen.

polycystic kidney disease—Hereditary bilateral cysts distributed throughout the renal parenchyma resulting in markedly enlarged kidneys and progressive renal failure.

pyelonephritis—The disease process from the immediate and late effects of bacterial and other infections of the parenchyma and the pelvis of the kidney.

rejection—Immune response to foreign tissue (transplanted organ).

renal replacement therapy—Medical technologies that serve as substitutes for renal function include hemodialysis, peritoneal dialysis, and transplantation. Without this therapy, the patient with no renal function would die.

uremia—A complex biochemical abnormality occurring in kidney failure: characterized by azotemia, chronic acidosis, anemia, and a variety of systemic and neurologic symptoms and signs.

References

1. United States Renal Data System. *USRDS 1999 Annual Data Report.* Bethesda, MD: National Institutes of Health, NIDDK; April 1999.

2. Oberly E. *Renal Rehabilitation: Bridging the Barriers.* Madison, WI: Medical Education Institute; 1994.

3. Brenner BM, Stein JH. *Chronic Renal Failure*. New York: Churchill Livingstone; 1981.

4. Nissenson AR, Fine RN. *Dialysis Therapy*. 2nd ed. Philadelphia: Hanley & Belfus; 1993.

5. Brenner P. Quality of life: A phenomenological perspective on explanation, prediction and understanding in nursing science. *Adv Nurs Sci* 8: 14, 1985.

6. United Network for Organ Sharing. *UNOS Update 1996*. Washington, DC: Organ Procurement and Transplantation Network and Scientific Registry for Organ Transplantation; 1996.

7. Barnea N, Drory Y, Iaina A, et al. Exercise tolerance in patients on chronic hemodialysis. *Isr J Med Sci* 16: 17-21, 1980.

8. Beasley RW, Smith A, Neale J. Exercise capacity in chronic renal failure patients managed by continuous ambulatory peritoneal dialysis. *Aust N Z J Med* 16: 5-10, 1986.

9. Goldberg AP, Geltman EM, Hagberg JM, Delmez JA, Haynes ME, Harter HR. Therapeutic benefits of exercise training for hemodialysis patients. *Kidney Int* S16: S303-S309, 1983.

10. Moore GE, Brinker KR, Stray-Gundersen J, Mitchell JH. Determinants of VO_2peak in patients with end-stage renal disease: On and off dialysis. *Med Sci Sports Exerc* 25: 18-23, 1993.

11. Moore GE, Parsons DB, Painter PL, Stray-Gundersen J, Mitchell J. Uremic myopathy limits aerobic capacity in hemodialysis patients. *Am J Kid Dis* 22: 277-287, 1993.

12. Painter PL, Nelson-Worel JN, Hill MM, et al. Effects of exercise training during hemodialysis. *Nephron* 43: 87-92, 1986.

13. Painter PL, Messer-Rehak D, Hanson P, Zimmerman S, Glass NR. Exercise capacity in hemodialysis, CAPD and renal transplant patients. *Nephron* 42: 47-51, 1986.

14. Painter P. Exercise in end stage renal disease. *Exerc Sport Sci Rev* 16: 305-339, 1988.

15. Painter P. The importance of exercise training in rehabilitation of patients with end stage renal disease. *Am J Kidney Dis* 24(Suppl 1): S2-S9, 1994.

16. Ross DL, Grabeau GM, Smith S, et al. Efficacy of exercise for end-stage renal disease patients immediately following high-efficiency hemodialysis: A pilot study. *Am J Nephrol* 9: 376-383, 1989.

17. Shalom R, Blumenthal JA, Williams RS. Feasibility and benefits of exercise training in patients on maintenance dialysis. *Kidney Int* 25: 958-963, 1984.

18. Zabetakis PM, Gleim GW, Pasternak FL, Saraniti A, Nicholas JA, Michelis MF. Long-duration submaximal exercise conditioning in hemodialysis patients. *Clinical Nephrology* 18: 17-22, 1982.

19. Johansen KL. Physical functioning and exercise capacity in patients on dialysis. *Adv Ren Replace Ther* 6: 141-148, 1999.

20. Copley JB, Lindberg JS. The risks of exercise. *Adv Ren Replace Ther* 6: 165-172, 1999.

21. Painter PL, Stewart AL, Carey S. Physical functioning: Definitions, measurement, and expectations. *Adv Ren Replace Ther* 6: 110-124, 1999.

22. Painter PL, Carlson L, Carey S, Paul SM, Myll J. Physical functioning and health related quality of life changes with exercise training in hemodialysis patients. *Am J Kidney Dis* 3: 482-492, 2000.

23. DeOreo PB. Hemodialysis patient-assessed functional health status predicts continued survival, hospitalization and dialysis-attendance compliance. *Am J Kidney Dis* 30: 204-212, 1997.

24. Painter P, Hanson P, Messer-Rehak D, Zimmerman SW, Glass NR. Exercise tolerance changes following renal transplantation. *Am J Kidney Dis* 10: 452-456, 1987.

25. Painter PL, Luetkemeier MJ, Dibble S, et al. Health related fitness and quality of life in organ transplant recipients. *Transplantation* 64: 1795-1800, 1997.

26. Painter PL, Stewart AL, Carey S. Physical functioning: Definitions, measurement, and expectations. *Adv Ren Replace Ther* 6: 110-123, 1999.

27. Miller TD, Squires RW, Gau GT, Frohnert PP, Sterioff S. Graded exercise testing and training after renal transplantation: A preliminary study. *Mayo Clin Proc* 62: 773-777, 1987.

28. Kempeneers G, Myburgh KH, Wiggins T, Adams B, van Zyl-Smith R, Noakes TD. Skeletal muscle factors limiting exercise tolerance of renal transplant patients: Effects of a graded exercise training program. *Am J Kidney Dis* 14: 57-65, 1990.

29. Painter P, Blagg C, Moore GE. *Exercise for the Dialysis Patient: A Comprehensive Program*. Madison, WI: Medical Education Institute; 1995.

30. Carey S, Painter P. An exercise program for CAPD patients. *Nephrol News Issues* June: 15-18, 1997.

31. Carlson L, Carey S. Staff responsibility to exercise. *Adv Ren Replace Ther* 4: 172-181, 1999.

32. Deligianis A, Kouidid E, Tassoulas E, Gigis P, Tourkantonis A, Coats A. Cardiac response to physical training in hemodialysis patients: An echocardiographic study at rest and during exercise. *Int J Cardiol* 70: 253-266, 1999.

33. Mercer T, Naish PF, Gleeson NP, Wilcock JE, Crawford C. Development of a walking test for the assessment of functional capacity in non-anemic maintenance dialysis patients. *Nephrol Dial Transplant* 13(8): 2023-2026, 1998.

34. Lo C, Li L, Lo WK. Benefits of exercise training in patients on continuous ambulatory peritoneal dialysis. *Am J Kidney Dis* 32(6): 1011-1018, 1998.

35. Kouidi E, Albani M, Natsis K, Magalopoulos A, Gigis P, Guiba-Tziampiri O, Tourkantonis A, Deligiannis A. The effects of exercise training on muscle atrophy in hemodialysis patients. *Nephrol Dial Transplant* 13: 685-699, 1998.

36. Deligiannis A, Kouidi E, Tourkantonis A. Effects of physical training on heart rate variability in patients on hemodialysis. *Am J Cardiol* 84: 197-202, 1999.

200 PART II   *Endocrinology and Metabolic Disorders*

37. American College of Sports Medicine. *Guidelines for Exercise Testing and Prescription.* 5th ed. Philadelphia: Williams & Wilkins; 1995.

38. Moore GE, Painter PL, Brinker KR, Stray-Gundersen J, Mitchell JH. Cardiovascular response to submaximal stationary cycling during hemodialysis. *Am J Kidney Dis* 31: 631-637, 1998.

39. Csuka M, McCarty DJ. Simple method for measurement of lower extremity muscle strength. *Am J Med* 78: 77-81, 1985.

40. Bohannon RW. Comfortable and maximum walking speed of adults aged 20-79 years: Reference values and determinants. *Age Ageing* 26: 15-19, 1997.

41. Ware J. *SF-36 health survey: Manual and interpretation guide.* Boston: The Health Institute; 1993.

42. Ware JE, Kosinski M, Keller SD. *SF-36 physical and mental health summary scales: A user's manual.* 2nd ed. Boston: The Health Institute; 1994.

Jonathan K. Ehrman, PhD, FACSM
*Henry Ford Heart and Vascular Institute
Detroit, MI*

Myocardial Infarction

A myocardial infarction (MI) is the formation of infarct tissue within the layers of the heart. An MI occurs when there is a substantial decrease or complete disruption of blood flow through a coronary artery so that the "downstream" tissue is deprived of oxygen for an extended period of time. The infarct region is composed of necrotic tissue. Typically, an MI is caused by **thrombus** formation in a coronary artery that has a degree of established coronary artery disease or **atherosclerosis**. The term **acute MI** refers to the sudden occurrence of **ischemia** leading to myocardial damage and subsequent infarction (3).

Scope

MI is one of the most prevalent events resulting in hospitalization in the United States and other Western countries, with approximately 1.1 million MIs annually in the United States. The American Heart Association (AHA) estimates that approximately 45% of those with an MI will die within a year of their MI (10). About half of these never reach a hospital emergency department, and many die suddenly, within 1 hr of the onset of symptoms. This occurs for a variety of reasons, including the following:

— Failure to recognize symptoms

— Failure to act on the perception of symptoms

— Failure of timely transport to an acute care facility (48, 74)

MI incidence rates differ between males and females. In 1995, the AHA estimated that about 1.5 times more males experienced an MI than did females (10). The majority occur in persons older than age 65. The annual rate of first MI is similar in black and white people until age 45 for females and 65 for males. From there on, black people have a higher annual rate of first MI (10).

MI results in a major expenditure for diagnosis and treatment. The estimated individual cost of a fatal acute MI is $17,532 per event and a nonfatal acute MI $15,540 per event in the first year following the event (101). This amounts to approximately $5.54 billion annually for new MI cases and thus is a large burden on the U.S. healthcare system.

Physiology and Pathophysiology

The process of coronary artery disease development is the antecedent eventually resulting in MI. Atherosclerotic **plaque** development occurs in most of the major arteries in the body (figure 12.1). The genesis of atherosclerosis is likely at a very early age. Changes in the **arterial intimal layer,** such as pale yellow streaks, have been noted to occur as early as age 1 in children throughout the world. However, progression to an **atheroma** occurs significantly more often and at younger ages in individuals living in developed, affluent countries. Investigations have reported finding coronary artery atherosclerotic lesions in 45% and 77%, respectively, of young men (age 18-35) killed in the Korean and Vietnam wars (60). Severe disease was noted in 5% to 6% of these victims (118). A recent investigation confirmed these findings in 111 victims of noncardiac trauma (60).

The formation of atherosclerosis has been well documented, and its specifics are beyond the scope of this chapter (57, 88). It is hypothesized that progression of atherosclerosis depends on the existence and interaction of several factors:

1. Hemodynamic forces (e.g., elevated blood pressure, turbulent blood flow)

2. Levels of plasma atherogenic lipoproteins (e.g., low-density lipoprotein, Lipoprotein a (LPa), high-density lipoprotein)

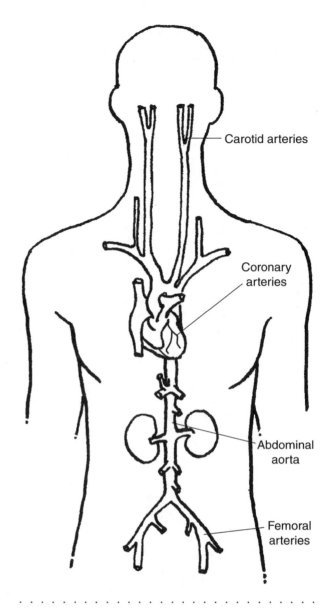

FIGURE 12.1   The arterial tree and the common locations of atheroma development.

3. Disruption of the intimal layer by free radical formation (i.e., oxidation), smoking (i.e., chemical injury), or high blood pressure (i.e., mechanical stress injury)

Intimal disruption is based on the "reaction to injury" hypothesis, in which the endothelial cells lose their ability to function normally and may lose their squamous shape. Other hypotheses of atherosclerotic development exist (e.g., monoclonal hypothesis and lysosomal theory). Factors that likely play a direct or indirect role in atherosclerotic development include diabetes (i.e., metabolic injury), physical inactivity, obesity, blood homocysteine concentration,

C-reactive protein, mental stress, the aging process, genetic makeup, and sex-related factors. There are likely other contributing factors that are unknown at this time.

The progression of an atheroma to a fibrous plaque occurs over a period of years to decades. This process includes increased **platelet aggregation** and localized **growth factor** release resulting in growth of smooth muscle cells in the medial arterial layer that absorb lipid and become **foam cells.** Within this cellular mass, collagen and eventually a fibrous cap form. Over time, insoluble calcium salts are deposited within the atheroma.

Atheromas are irregularly distributed along the arterial tree. Within the coronary artery system, discrete lesions develop. These are most often in the area of **bifurcation** within the main coronary artery branches (i.e., left main, left anterior descending, circumflex, right coronary), and it is common for a single artery to develop multiple discrete lesions along its length.

In addition to lumen narrowing, the vasodilatory process is also impaired in the region of an atheroma. Conditions that favor thrombosis development often occur when a complex atherosclerotic lesion fissures, ruptures, or ulcerates. This process is most likely to occur in smaller lesions (i.e., <50% narrowing) that have not yet formed a fibrous cap (i.e., unstable lesions). This is in contrast to previous beliefs that thrombosis development most likely occurred in atheromas significantly narrowing the arterial lumen (e.g., >75% narrowing). However, these lesions are typically stable and less likely to rupture.

The amount and severity of myocardial damage resulting from thrombotic occlusion depends on the location of the occlusion, whether blood flow is completely or partially disrupted, the duration of the blood flow disruption, the amount of collateral blood vessel circulation, and the state of myocardial oxygen demand (figure 12.2). The more severe and prolonged the blood flow limitation the more likely infarction will occur, and this results in cellular death or necrosis. These cells lose wall integrity, resulting in the release of several metabolic enzymes used in the diagnosis of MI. This results in initial thinning of the myocardial wall, which is susceptible to rupture (i.e., aneurysm). A dense fibrous network of connective tissue replaces these cells over time, resulting in permanent scarring that improves the integrity of the infarct tissue. This area of the myocardium can no longer contribute to contractile process and becomes either **akinetic** (i.e., without movement) or **dyskinetic** (i.e., moving in the opposite direction, or bulging). Four to six weeks after MI,

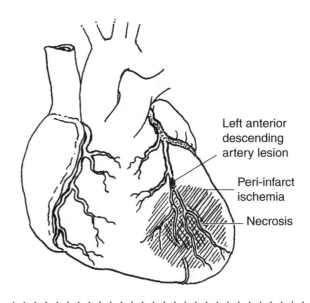

FIGURE 12.2   Necrotic tissue and surrounding ischemia in myocardial tissue following arterial occlusion and infarct.

the integrity of the myocardial wall in the infarcted tissue will have reached its strongest point. Other potential hazards during the MI recovery period include arrhythmias, **cardiogenic shock**, myocardial rupture, and thrombosis.

Clinical Considerations

Often, the initial symptoms of coronary artery disease are exposed during a bout of physical exertion because of a mismatch of myocardial oxygen demand and supply. Patients experiencing an MI can be identified by specific signs and symptoms exhibited during a physical examination. Most often these are determined in the emergency room. However, silent MIs are often discovered at the physician's office during a routine examination when an electrocardiogram (ECG) reveals Q waves. Treatment for the post-MI patient includes medications, lifestyle modification, and exercise.

Signs and Symptoms

Pain is the most common complaint in the patient experiencing an MI. Patients often describe their pain as a squeezing or burning sensation or a heavy or crushing discomfort. This symptom is the result of severe myocardial ischemia and is termed **angina pectoris** (figure 12.3). If the patient has had previous episodes of angina, the discomfort felt during the MI is typically similar in nature (i.e., location, description) but more intense.

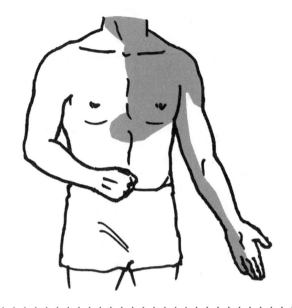

FIGURE 12.3   The common locations of anginal/ischemic discomfort (shaded area).

Other possible symptoms are associated with an MI:

— Unusual shortness of breath (i.e., dyspnea)
— Profound weakness
— Profuse sweating (i.e., diaphoresis)
— Loss of consciousness (i.e., syncope)
— Confusion

Often, during an evolving MI, patients become anxious and restless. Many, especially males, deny symptoms and do not seek immediate attention. This may occur, in part, because discomfort experienced before an MI is usually not considered painful, and therefore most do not think it is related to their heart. The AHA states that the average waiting time from the onset of symptoms to seeking medical assistance is about 2 hr (11).

About a quarter of patients with an evolving anterior MI are tachycardic and hypertensive, likely the result of an elevation in sympathetic activity in response to a reduction in cardiac output. Elevated sympathetic activity increases heart rate and contractility to counteract a reduction in cardiac output. About 50% of patients with an evolving inferior MI have an elevated parasympathetic nervous tone resulting in **bradycardia** and hypotension. Chapter 4 provides specifics regarding the cardiovascular physical examination.

Diagnostic and Laboratory Evaluations

When MI is suspected during the physical examination, the next step in the **differential diagnosis** in-

volves diagnostic and laboratory testing. These tests are divided into four groups:

1. Laboratory indicators of tissue damage
2. The ECG
3. Serum enzymes
4. Cardiac imaging

Blood analysis can be performed to determine if tissue damage has occurred to the myocardium. **Polymorphonuclear leukocytosis** and **erythrocyte sedimentation** rates both are indicative of myocardial damage. The time course of polymorphonuclear leukocytosis is from immediately after the onset of pain until 3 to 7 days after resolution of the MI. The erythrocyte sedimentation rate also increases after an MI and remains elevated for 1 to 2 weeks.

Electrocardiographic changes associated with an MI are categorized as follows:

1. Acute
2. Evolving/resolving
3. Chronic

Patients presenting with an acute **transmural** MI have ECG changes that include hyperpolarization of the T waves and ST-segment elevation (figure 12.4a). Patients experiencing a **subendocardial** MI typically demonstrate ST-segment depression. A subendocardial MI can be differentiated from transient ischemia if the ST-segment depression persists despite a reduction in myocardial oxygen demand (e.g., stopping exercise or physical activity) and blood pressure (i.e., reduced afterload). Myocardial oxygen demand is estimated by the **double product** (heart rate and systolic blood pressure). During the **evolving MI** and **resolving MI** phases, Q waves develop. Also, ST-segment changes return toward baseline and the T waves may invert. The ECG during the **chronic MI** phase is characterized by fully developed Q waves and either inverted or resolved T waves (figure 12.4b). This is typically the ECG observed when a patient enters a supervised cardiac rehabilitation program. A portion of people will not demonstrate Q waves following an MI. These are termed non-Q-wave MIs. Note that the ECG is more variable and less specific than the other methods of MI diagnosis, including plasma enzyme elevation and cardiac imaging.

Enzymatic changes are highly sensitive for evidence of myocardial injury and are often used in emergency rooms to rule out an MI. Enzymes used for MI diagnosis include creatinine kinase and

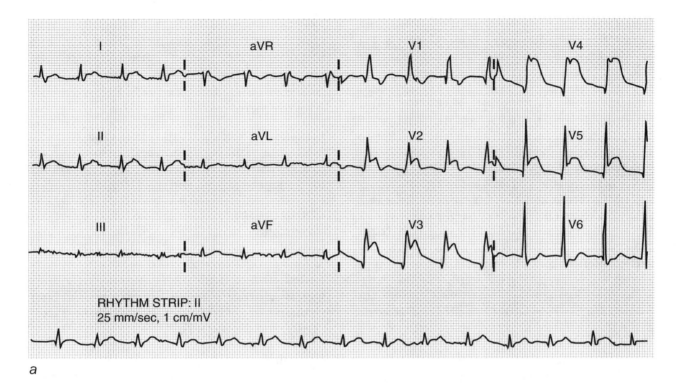

a

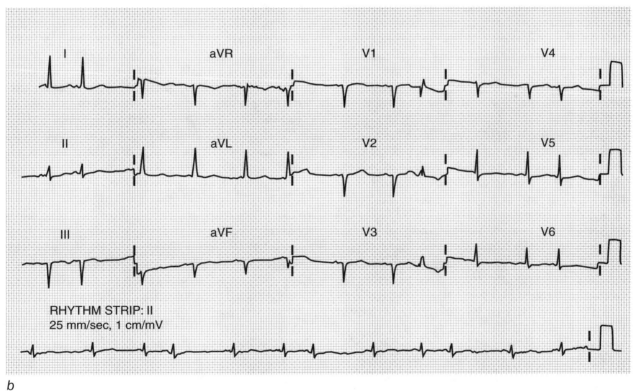

b

FIGURE 12.4 Electrocardiograms depicting myocardial infarction (a) elevated ST segments in leads V2 through V5, (b) Q-waves in leads III, aVf (old infarct), and V1 through V5 (evolving infarct) with flattened T waves.

lactate dehydrogenase. Both creatinine kinase and lactate dehydrogenase are metabolic enzymes that are released when the myocardial cell walls lose their integrity. Creatinine kinase will rise within 8 to 24 hr and remain elevated for 48 to 72 hr. Peak creatinine kinase levels occur at 12 to 24 hr after initial injury. The normal concentration of creatinine kinase is 10 to 13 units $\cdot$ L^{-1}. Lactate dehydrogenase begins to increase later, at about 24 to 48 hr, and remains elevated for 7 to 14 days.

Cardiac imaging provides information regarding **wall motion** and blood perfusion. Since wall motion abnormalities are apparent in nearly all MIs, their detection is useful for diagnosis. Two-dimensional echocardiography is used to visualize the left ventricular walls and assess the amount of movement. Note that the echocardiographic analysis cannot distinguish between wall motion abnormalities caused by an old, healed MI and an acute MI. Radionuclide ventriculography, often termed multiple gated analyses (MUGA), uses technitium-99m stannous pyrophosphate, which tags to red blood cells. By imaging tagged red blood cells, healthcare workers can assess wall motion and **ejection fraction.** Both two-dimensional echocardiography and MUGA can be performed during exercise or pharmacological (e.g., with dobutamine or diprydimole) stress evaluation. Other procedures, such as computed tomography and magnetic resonance imaging, may also be used to assess ventricular wall motion.

Radionuclide imaging procedures are used to assess either blood flow defects or wall motion. Myocardial perfusion is assessed with isotopes such as **thallium 201** or **technetium-99m Sestamibi**. Once injected, these radioisotopes are taken up by the myocardium proportionally to blood flow. On the X-ray image, areas of limited isotope uptake are darkened and reflect areas in which blood flow is reduced or ceased. Reimaging several hours later will reveal areas of permanent tissue damage by the continued appearance of a previously darkened area. This procedure can be performed during exercise or pharmacological stress.

An algorithm developed jointly by the American College of Cardiology (ACC) and AHA provides strategies to follow when evaluating a post-MI patient (102). The strategy chosen is based on whether the patient is considered at high risk of a future event. High-risk patients typically should have a cardiac catheterization performed before hospital discharge to determine if revascularization is warranted. It is recommended that stable, lower risk post-MI patients who have not had a cardiac catheterization undergo a submaximal exercise evaluation before hospital discharge or a symptom-limited evaluation within 3 weeks after discharge (102). Further diagnostic evaluation is then performed if the exercise evaluation is considered abnormal.

Treatment

In the setting of acute MI, the primary focus of treatment is on pain relief and **reperfusion.** Chronic treatment focuses on revascularization procedures, medical management, and risk factor reduction.

Acute Treatment

Treatment of ischemic or anginal pain associated with MI includes sublingual nitroglycerin and morphine. Nitroglycerine relieves pain by reducing myocardial oxygen demand (reduced venous return) and increasing oxygen supply (coronary artery dilation). Common side effects include hypotension, **tachycardia,** and dizziness. Nitroglycerin should not be administered if the systolic blood pressure is below 100 mmHg. Morphine is very effective at pain reduction via its narcotic effects. A variety of thrombolytic agents (e.g., tissue plasminogen activator, streptokinase) are available to lyse clots responsible for coronary artery occlusion. The goal is to reperfuse the affected tissue to salvage myocardium that otherwise would infarct. Supplemental oxygen is used to increase oxygen supply by increasing arterial oxygen pressure. A flow rate of 4 to 5 L $\cdot$ min^{-1} is commonly administered via nasal cannula.

Percutaneous coronary intervention (PCI) and coronary artery bypass surgery are also used as primary treatment strategies for acute MI. The Primary Angioplasty in Myocardial Infarction Trial reported significantly reduced short- and long-term mortality rates in acute MI patients treated with percutaneous transluminal coronary angioplasty versus those treated with thrombolytic agents (13). This was demonstrated in both younger and older patients. Some suggest that revascularization procedures only be used after MI to prolong life or relieve unacceptable symptoms despite optimal medical management (11). Chapter 13 is devoted to revascularization.

Admittance to a coronary care unit should occur when the patient is clinically stable. Bed rest is typically used in the initial stages of an acute MI to reduce myocardial oxygen demand and the risk of a large infarct size. However, by the second to fourth day, patients should be regularly sitting up several times a day and beginning to ambulate. This will increase orthostatic and hemodynamic stress and reduce the amount of negative consequences (e.g., orthostatic hypotension, plasma volume loss) after an MI.

Chronic Treatment

Long-term medical therapy should include β-blockade, angiotensin-converting enzyme inhibition, aspirin, and lipid management. β-blockers have been extensively investigated in placebo-controlled trials in post-MI patients (14, 38, 55, 114) and unequivocally demonstrate reduction of all-cause mortality and sudden death rate. There is also evidence that both fatal and nonfatal reinfarction rates are reduced (84, 107). The use of β-blocker therapy should begin as soon as possible (102).

Angiotensin converting enzyme inhibitors have proven effective at reducing the risk of heart failure development and lowering mortality risk in post-MI patients with reduced ejection fractions (e.g., <40%) (2, 41, 68). Calcium channel blockers (i.e., calcium antagonists) have not been shown experimentally to reduce mortality rate and may in fact increase mortality rate in some post-MI patients (113). However, calcium channel blockers may be useful for controlling ischemia by reducing systolic blood pressure, a result of its afterload-reduced mechanism of reduced vasoconstriction. Calcium channel blockers should only be used in patients who cannot tolerate β-blockade.

An aspirin should be administered in the emergency room setting, and the patient should take a daily dose of 160 to 325 mg per day thereafter. Coumadin (i.e., warfarin) anticoagulation therapy is effective in patients at risk of developing an atrial or ventricular thrombus or who cannot use aspirin. Lipid management is reviewed in chapter 10.

Primary Prevention

Primary prevention of coronary artery disease is important to reduce the risk of a first MI. There is ample epidemiological evidence that an inactive lifestyle plays a leading role in the development of coronary artery disease (15, 89). The "at-risk" individual is one who gets little regular exercise, is male, is older (i.e., >45 years for males and 55 for females), has other cardiovascular risk factors, and is symptomatic (i.e., angina-like chest discomfort, dyspnea, exertional fatigue, dizziness). In addition to standard risk factor modification, regular exercise training and/or increased daily physical activity levels are useful for reducing the risk of coronary artery disease and MI.

The risk of an acute MI is slightly elevated during an acute exercise bout (121). There are reports that each year approximately 75,000 people in the United States suffer an MI soon after performing vigorous exercise (77). However, this is fewer than 10 per 100,000 persons (115, 117). The mechanism of action of MI during exercise may be related to the **prothrombotic** effect of acute exercise. Therefore, the individual with existing coronary artery disease is at a higher risk of thrombus development during exercise than a person without coronary disease. Aspirin therapy is commonly recommended for males over 40 years of age to reduce the risk of coronary thrombosis. Regular exercise training may also improve **thrombolytic** effects during nonexercise periods.

Secondary Prevention

Secondary measures, in addition to medical therapy, are effective at preventing MI in those with established coronary artery disease or previous MI (106). The scientific evidence in favor of secondary prevention is strong and points toward the following:

1. Improved overall survival
2. Enhanced quality of life
3. Reductions in revascularization procedures
4. Reductions in subsequent MI

A comprehensive secondary preventive approach focuses on reducing the risk factors of coronary artery disease in conjunction with optimized medical therapy. Table 12.1 lists the risk factors and generalized recommendations for risk factor alteration as outlined by the ACC's Secondary Prevention Panel in an AHA Consensus Panel Statement (80).

Secondary prevention in the post-MI patient should include a referral to an outpatient cardiovascular rehabilitation program (CRP) (102). A well-designed CRP provides the post-MI patient with a risk factor assessment, goal setting, education, and exercise therapy designed to decrease the risk of a future cardiac event.

Three meta-analyses reported that individuals with a previous MI are not protected from the recurrence of a nonfatal MI following cardiac rehabilitation (73, 85, 86). However, others have demonstrated a reduction in nonfatal reinfarctions with physical activity (34, 109). Although the evidence is equivocal with respect to the exercise effects on nonfatal reinfarction, the benefits of exercise for reducing the risk of cardiac (22-24%) and all-cause (20-25%) death and improving quality of life make exercise training an important part of the therapeutic regimen for the post-MI patient (73, 85, 86). Compared with other treatment strategies, cardiac rehabilitation services are considered highly cost-effective for post-MI patients (88). Interestingly, the reduction in all-cause mortality rate as a result of CRP is similar to the reduction provided by β-blockade therapy in post-MI patients.

TABLE 12.1

Secondary Prevention for Patients With Coronary Artery Disease

Action Item	Recommendation
Smoking	Complete cessation
Lipid management	Low-density lipoprotein <100 mg · dl^{-1}, high-density lipoprotein >35 mg · dl^{-1}, triglycerides <200 mg · dl^{-1}
Physical activity	30 min of aerobic activity 3-4 times per week
Blood pressure	<140/90 mmHg
Weight control	<120% of ideal weight for height
Antiplatelet/anticoagulants	80-325 mg/day of aspirin; Coumadin if not able to take aspirin
Angiotensin-converting enzyme inhibition	Begin early after myocardial infarction, especially in high-risk patients (e.g., Killip II+)
β-blockers	Begin on day 5-28 after myocardial infarction in high-risk patients
Estrogens	Consider in all postmenopausal women

Adapted from Smith et al., 1995 (106).

Smoking cessation should focus on the complete discontinuation of all types of smoking. Smoking can cause coronary artery spasm, adversely affect β-antagonist drug effects, and increase mortality rate in post-MI patients. Importantly, the rates of reinfarction and mortality are significantly reduced following smoking cessation (56, 67, 108, 119). Continued behavioral management is important because of the high rate of relapse of smokers who quit (up to 50% within 1 year). Typically, the willingness to discontinue cigarette smoking is greatest during and immediately after the acute MI. However, this may be an extremely difficult period for the patient, whose focus on behavioral modification required for lifestyle changes is commonly diverted. Finally, most patients who are successful at quitting smoking do so only after several attempts.

Lipid management should be initiated in post-MI patients presenting with a low-density lipoprotein (LDL) level greater than 100 mg · dL^{-1} or total cholesterol level greater than 200 mg · dL^{-1}, as outlined by the National Cholesterol Education Program (80). This is important, as investigations have demonstrated that both LDL and total cholesterol levels are predictive of future MI in patients with coronary artery disease and are related to the pro-

gression of atherosclerosis (24, 80, 93, 100, 103, 122). Post-MI patients who exercise train demonstrate a reduction in LDL cholesterol and triglycerides and an elevation of high-density lipoprotein (HDL) cholesterol (116). Chapter 10 provides a comprehensive review of methods for treating lipid disorders.

Blood pressure management is important, and general recommendations are that values remain below 140/90 mmHg to reduce cardiovascular risk (24). Hypertension is an independent predictor of morbidity and mortality rates in post-MI patients (63), and aggressive blood pressure control can significantly reduce cardiac-related death in these patients (66). Chapter 16 reviews treatment strategies for hypertension.

Obesity was recently added to the list of primary risk factors for the development of coronary artery disease (10). When body mass index criteria are used, a value greater than or equal to 30 is considered a higher than normal risk for cardiovascular disease (22). Chapter 9 reviews obesity and weight loss methods.

Physical inactivity is generally believed to be related to an increased risk of future MI in patients who have had a previous MI, although there is evidence to the contrary (59). Despite this, it is well known that physical activity is useful in treating several risk factors related to reinfarction including hypertension, obesity, smoking, and diabetes (90). Patients who perform regular physical activity are also more likely to be successful at modifying their behavior as it relates to cardiovascular risk factors.

Graded Exercise Testing

Graded exercise testing is used in the post-MI patient to evaluate prognosis and functional status. The clinical exercise physiologist must consider the status of a patient before testing and be knowledgeable of the exercise response in these patients. Practical application 12.1 reviews the process of interaction between the clinical exercise professional and the post-MI patient.

Medical Evaluation

The AHA, ACC, and American College of Sports Medicine (ACSM) have published separate and collaborative guidelines for exercise testing that include lists of absolute and relative contraindications for exercise testing (10, 42, 46). These are reviewed in chapter 6. The determination of whether to perform an exercise evaluation lies with whether the specific risks of the evaluation, based on an individual's status, outweigh the potential benefits. To determine this, one must use appropriate clinical judgment when evaluating the patient's medical history and physical assessment.

Contraindications

In patients who have recently experienced an acute MI, the following recommendations are made with

---

PRACTICAL APPLICATION 12.1

Client–Clinician Interaction

The post-MI patient, like most chronic diseased patients, will come into contact with healthcare system personnel on numerous occasions. The patient referred to a cardiac rehabilitation program will, depending on length of stay in the program, make contact with the staff on 10 to 36 occasions over a 1- to 3-month period. This is increased to hundreds of contacts over a period of years in those who continue in a phase 3 or 4 cardiac rehabilitation program. Therefore, the potential to affect behavioral change may be greater for cardiac rehabilitation clinicians than the patient's physician because of the greater total amount of contact hours.

Cardiac rehabilitation staff must develop skills to use when approaching the post-MI patient to discuss behavior change. The early post-MI patient is often overwhelmed dealing with issues related to the MI. These issues include fear of death or disability, depression, family support, and medication side effects. The cardiac rehabilitation professional must develop the ability to assess patient concerns. This will increase the potential of approaching the post-MI patient at the right time, and in the right setting, to best deal with issues that are both important and within the patient's ability to handle. It is common for a patient to have to make many changes to optimally reduce his or her risk of a mortal or morbid event. This is likely to be an overwhelming task. The exercise professional must be able to address these issues with the post-MI patient in a manner that motivates the patient to make behavioral changes.

The initial patient contact is very important. A relationship must first be established before the clinician discusses behavioral change. The exercise professional is more likely to be successful if he or she has an approach that is friendly, courteous, patient, kind, and empathetic. With this approach, a patient is more likely to be honest and open. In selected patients, it may require several informal discussions before serious issues can be discussed. Some patients may have difficulty discussing issues with certain exercise professionals. This may include members of the opposite sex and those with large age discrepancies (i.e., a young exercise professional and an older patient).

Motivation to make behavioral changes will vary among post-MI patients. Some patients will be in a state of denial when entering a cardiac rehabilitation program, whereas others may have already committed themselves to change. Others will be somewhere between these extremes on the readiness to change continuum. Sometimes an exercise professional may have to search to find what motivates an individual to make a change. Some examples include remaining healthy and active to play with grandchildren, improving fitness levels to return to full employment and leisure activities, and the will to remain alive. Individuals will be motivated differently, and it is up to the exercise professional to cultivate an individual's motivational cues.

respect to the minimal amount of time that should elapse before a symptom-limited exercise test is performed: AHA, 3 to 5 days (42); ACC/AHA, 2 days (46); ACSM: no mention of days (10) but states that "testing after a recent complicated MI is contraindicated." Given the array of possible scenarios with which each individual patient may present, it is best to assess each patient individually before making a decision about when to perform an exercise test.

The AHA, ACC, and ACSM also review absolute and relative contraindications to exercise (10, 42, 46). These include the presence of unstable angina, arrhythmias, hemodynamic dysfunction, cardiomyopathy or heart failure, severe coronary artery disease, and severe hypertension. Each of these contraindicators should be assessed during a medical history, physical examination, or standard cardiac diagnostic test because they are common in the post-MI patient. Several, depending on severity, may be considered an absolute or relative contraindication. For instance, an arrhythmia may be relatively benign, such as infrequent premature atrial contractions, or potentially serious, such as a history of malignant arrhythmia (e.g., intermittent ventricular tachycardia) that increases the risk of severe hemodynamic compromise or sudden death.

Guidelines developed separately by the AHA and the American Association of Cardiovascular and Pulmonary Rehabilitation (AACVPR) address risk stratification for individuals who will perform an exercise test or who want to become more physically active (8, 39). Both are reviewed in *ACSM's Guidelines for Exercise Testing and Prescription* (9).

Recommendations and Anticipated Responses

The primary purpose of performing an exercise evaluation soon after an MI is to allow for the detection of exertional myocardial ischemia. Submaximal testing is demonstrated to be safe in stable patients (50). This includes those without angina or heart failure symptoms and those with a stable ECG for 48 to 72 hr before the exercise evaluation. If the patient is stable, most submaximal, predischarge exercise tests can be performed within 3 to 5 days after an uncomplicated MI (102). A treadmill or cycle ergometer is most commonly used, and the protocol must be low level, with increments of 1 to 2 metabolic equivalents (METs) every 2 to 3 min. Examples of commonly used low-level treadmill protocols are the modified Naughton, the modified Bruce, the Balke, and the Stanford (10). Blood pressure, ECG, and symptom monitoring should be performed throughout the exercise test. Common end-point criteria used during a submaximal exercise evaluation include:

— 120-130 beats · min^{-1} heart rate
— 70% of predicted maximal heart rate
— 5 metabolic equivalent work rate
— Mild angina or dyspnea
— 2 mm ST-segment depression
— Hypotension
— Three or more consecutive PVCs

The test should be discontinued when any one of these end points occurs.

Symptom-limited exercise evaluations may also be performed before hospital discharge. These are typically performed in stable patients at least 5 days after the acute phase of the MI (102). The end point for this type of evaluation is commonly leg fatigue. A symptom-limited evaluation is different from a submaximal test because there is no predetermined end point based on heart rate or work rate. There are concerns with respect to the prognostic value of ST-segment displacement in early post-MI patients, who are typically deconditioned, because there is evidence that this may lead to an unnecessary cardiac catheterization (102). For example, an investigation of 236 post-MI patients who performed an exercise evaluation 3 weeks after an acute MI reported that exercise-related parameters, such as ST-segment displacement and exercise duration, were poor predictors of cardiac prognosis (1). Therefore, these predischarge tests are often used only for functional assessment and hemodynamic evaluation purposes.

Posthospitalization exercise testing is similar to the symptom-limited predischarge test with respect to test end points. However, the protocol used is commonly higher level. Stage increments in work rate range from 1 to 3 METs. The most common treadmill protocols used are the modified or standard Bruce, the Ellestad, and the Naughton (10). During symptom-limited testing, the ECG and blood pressure are assessed prior to each stage increment throughout the evaluation. Gas exchange assessment is useful in accurately quantifying exercise capacity and is important in the post-MI patient who develops heart failure (see chapter 15). Rating of perceived exertion (RPE) provides an evaluation of the patient's subjective level of exertion and can be used later to guide exercise training. The posthospitalization test is better for predicting prognosis versus the predischarge test. Indicators of prognosis including functional capacity, ST-segment changes, and blood pressure response can be used to determine prognosis and guide medical management (102). For instance, patients without indications of ischemia and who achieve at least 5 METs are candidates for medical management. Those who become ischemic, show decreased systolic blood pressure, or cannot achieve 3 to 4 METs should be referred for cardiac catheterization, because their short-term prognosis, without intervention, is poor.

In addition to standard ECG exercise testing, other procedures can be used that may enhance the **sensitivity** of the test. These are typically used when the resting ECG demonstrates ST-wave abnormalities, as often seen in post-MI patients, especially those with ventricular conduction abnormalities associated with bundle branch block, left ventricular hypertrophy, and digitalis therapy. Myocardial perfusion imaging (i.e., thallium, sestamibi) provides information regarding both infarct size and peri-infarct ischemia (i.e., ischemia in the tissue surrounding infarcted myocardium). There is evidence that infarct size is related to prognosis (69). Exercise echocardiographic evaluation improves both the sensitivity and **specificity** of standard exercise testing and is performed either immediately following exercise on a treadmill or cycle ergometer or following infusion of a drug (e.g., dobutamine or adenosine). A negative echocardiographic evaluation indicates a reduced risk of future MI and overall death (65).

The response of the MI patient to exercise testing will depend on factors such as the course of the post-MI period, the amount of myocardial damage, the degree of left ventricular dysfunction, the pre-MI level of exercise or physical activity, and medications. It is common for the post-MI patient to be deconditioned. However, the ability of the skeletal muscles to extract oxygen and thus produce a large arteriovenous oxygen difference ($\dot{V}O_2$ = cardiac output × arteriovenous oxygen difference) is not limited to a great extent in post-MI patients compared with healthy normal individuals (102). Therefore, the poor exercise ability of the post-MI patient is related to myocardial function. For example, a group of male patients evaluated at 3 weeks post-MI had a mean $\dot{V}O_2$peak of 20.5 ml · kg^{-1} · min^{-1} compared with 37.5 ml · kg^{-1} · min^{-1} in a group of healthy male subjects (53). It is likely that a reduced peak exercise cardiac output, related to a comparatively low peak heart rate (137 ± 19 vs. 170 ± 13 beats · min^{-1}, respectively) and possible left ventricular dysfunction, were partly responsible for this difference. Similar $\dot{V}O_2$peak (19.4 ml · kg^{-1} · min^{-1}) and heart rate values (144 ± 22 beats · min^{-1}) were reported in another study of post-MI patients (79). In this study, peak cardiac output was 12.0 L · min^{-1}. By using this cardiac output value and the peak heart rate, researchers determined that peak stroke volume was 83 ml · beat^{-1}. These cardiac output and stroke volume values are lower than reported values in normally active individuals (22 L · min^{-1} and 112 ml · beat^{-1}, respectively). Thus, the reduced $\dot{V}O_2$peak noted in post-MI patients is directly related to depressed cardiac function (i.e., decreased cardiac output).

Exercise Prescription

The exercise prescription for the post-MI patient should follow standard procedures as presented in chapter 7. Specific modifications and considerations should be made when indicated. Open communication is an effective tool used to assess and educate the post-MI patient. Practical application 12.2 provides a brief description of the randomized, controlled trials of exercise training in post-MI patients.

Special Exercise Considerations

The early post-MI exercise prescription should focus on providing an **orthostatic** workload to counteract the negative effects of bed rest. This is achieved by having the patient spend time in an upright position, either seated or standing. Stable patients should be progressed to this as early as possible. Orthostatic stress loads the cardiovascular system by activating the sympathetic nervous system. The result is an increase in heart rate and blood pressure (afterload), which increases myocardial oxygen consumption. Patients with residual ischemia or other acute MI complications must be carefully and slowly progressed to lengthening bouts of upright posture. Ambulation should begin on the second or third day after the patient becomes clinically stable. An exercise professional (e.g., clinical exercise physiologist, exercise specialist, physical therapist) should meet with the patient to discuss the importance of early ambulation with regard to preparation for the patient's return home. During the hospitalization course, the patient should be progressed from walking several feet to several hundred feet two to four times a day.

Although the incidence is low, myocardial rupture related to exercise has occurred (18, 50, 58, 61). At this time there is not enough information to make recommendations regarding patients at risk. However, these reports suggest that post-MI patients are most vulnerable to cardiac rupture in the initial period following the acute phase of an inferior MI. There are no reports in the literature of myocardial rupture during exercise training. Mineo et al. performed exercise training, without complication, in two patients who suffered previous myocardial rupture without complication (76). The authors

Literature Review

Post-MI patients improve their physical work capacity by exercise training. In randomized controlled trials, post-MI patients participating in cardiac rehabilitation demonstrated greater improvement in functional capacity than non-exercise-trained post-MI patients (23, 28-30, 44, 45, 49, 56, 69, 70, 75, 82, 87, 99, 110, 112). Contrary to these findings, some have reported no change in functional capacity following an exercise training program (47, 51, 62, 95, 104).

Most investigations have reported a 10% to 30% improvement in $\dot{V}O_2$peak following 8 or more weeks of exercise training (see table). The mechanism for improvement in physical work capacity is related to changes in stroke volume, cardiac output, maximal oxygen consumption, minute ventilation, exercise time, work rate, and lactate threshold (4, 78). Improvements in stroke volume and cardiac output noted in post-MI patients following exercise training do not account for the total percentage increase in $\dot{V}O_2$peak (35, 36). Therefore, increased peripheral oxygen consumption (i.e., widened arteriovenous oxygen difference) accounts for the remainder of the $\dot{V}O_2$peak improvement.

Echocardiographic findings of an increased ejection fraction and reduced end-systolic volume following exercise training suggest improvement in left ventricular contractile function in post-MI patients (36). Others have corroborated these findings (37). These suggest that it is possible to improve left ventricular hemodynamic function following an MI. However, improvement may require progression to a high-intensity (>80% $\dot{V}O_2$peak) exercise training program.

Resistance training in post-MI patients improves skeletal muscle strength. For example, one study enrolled 57 post-MI patients, split into three groups of differing resistance training intensity, who performed 20 reps at 20% of 1-repetition maximum (1RM), 10 reps at 40% of 1RM, and 7 reps at 60% of 1RM. Both upper and lower body isotonic resistance exercise was performed on 3 days of the week over a period of 10 weeks (27). Each group improved their 1RM values, but there was no difference in this improvement between the groups.

There may be additional benefits for the post-MI patient who performs both resistance and aerobic exercise training. In a study of post-MI patients who resistance and cycle ergometer trained, or only cycle trained, only the combined training group increased $\dot{V}O_2$peak, whereas both groups increased arm and leg strength (111). However, the combined training group had a significantly greater increase in strength measures. Fragnoli-Munn et al. (127) compared a younger and older group of post-MI patients who participated in a cardiac rehabilitation program that combined aerobic and resistive training. Both groups improved leg extension and bench press strength to a similar degree. Although the older group had lower initial absolute strength, their improvement was similar to the younger group. Interestingly, females in the older group tended to have greater strength improvements than men.

Only a few studies have reported a lack of improvement of functional capacity in post-MI patients after exercise training. However, other than differences in total training volume, no specific indicators for a lack of positive adaptations have been identified.

Author	Methods	Outcome
Wilhelmsen et al., 1975 (120)	Randomized post-MI; exercise trained (n = 158), control (n = 157); 4 years follow-up	Significantly increased PWC in trained but no difference in re-MI or mortality rates.
Plavsic et al., 1976 (95)	Post-MI; 53 total patients in exercise-trained and control groups; 6 months	No change in PWC or mortality rate.
De Busk et al., 1979 (28)	Randomized post-MI; cardiac rehab (n = 28), home exercise (n = 12), control (n = 30); 8 weeks	All groups significantly increased PWC, most in cardiac rehab group.
Kallio et al., 1979 (62)	Randomized post-MI; cardiac rehab (n = 188), control (n = 187); 3-year follow-up	Reduced mortality rate in cardiac rehab group via less sudden deaths; no difference in re-MI.

continued

Author	Methods	Outcome
Nolewajka et al., 1979 (83)	Randomized post-MI; exercise trained (n = 10), control (n = 10); 7 months	No change in collateralization, myocardial perfusion; trained subjects had significantly higher anaerobic threshold and lower submaximal HR; both groups had disease progression.
Mayou et al., 1981 (71)	Post-MI; exercise trained 12 weeks (n = 33), control (n = 35), advice (n = 34); 18 month follow-up	Significantly increased PWC in trained but better social outcome in advice group.
Shaw et al., 1981 (103)	Randomized post-MI; exercise trained (n = 323), control (n = 328); 3-year follow-up	Possible reduction in mortality rate; no change in morbidity or re-MI rates.
Carson et al., 1982 (23)	Randomized post-MI; cardiac rehab 12 weeks (n = 151), control (n = 152); 3-year follow-up	Significantly increased PWC in rehab group; no change in mortality rate.
Sivarajan et al., 1982 (105)	Randomized early post-MI; exercise trained (n = 88), trained with teaching (n = 86), control (n = 84); 12 weeks training	Significantly improved functional aerobic impairment, rest HR, and submax exercise SBP/DBP; no effect on morbidity and mortality rates.
Bengtsson, 1983 (12)	Randomized post-MI; matched groups; cardiac rehab (n = 81), control (n = 90); 1-year follow-up	No difference for improvement of PWC, return to work rate, or psychological status.
Rechnitzer et al., 1983 (98)	Multiple-center randomized post-MI; high-intensity (n = 379) and light intensity/control (n = 354) groups; 4-year follow-up	No difference in MI recurrence rate.
Roman et al., 1983 (99)	Randomized post-MI; matched groups; cardiac rehab (n = 93), control (n = 100); 9-year follow-up	No difference in re-MI rate, strokes, ischemia, and arrhythmias; reduced cardiac deaths and angina.
Stern et al., 1983 (110)	Randomized post-MI; exercise trained 12 weeks (n = 42), counseling (n = 35), control (n = 29); 1-year follow-up	Significantly improved PWC and independence and decreased fatigue/anxiety in trained group; improved psychosocial factors in counseling; no mortality effect.
Hung et al., 1984 (56)	Randomized post-MI; exercise trained ? weeks (n = 23), control (n = 30)	Significantly greater increased PWC in exercise-trained subjects; no effect on myocardial perfusion or EF%.
DeBusk et al., 1985 (29)	Post–acute MI; randomized to home-based (n = 33) or group (n = 30) exercise training or control (n = 37) for 23 weeks	Significantly increased PWC for exercise-trained groups; no difference for adherence to training or peak work capacity between trained groups.
De Busk et al., 1985 (29)	Randomized post-MI; cardiac rehab (n = 66), home exercise trained (n = 61)	Similar adherence and PWC improvement.
Marra et al., 1985 (70)	Randomized post-MI; cardiac rehab (n = 84), control (n = 83); 4- to 5-year follow-up	Significantly improved PWC, rate pressure product, symptomatology, and triglyceride levels in rehab group.
Ehsani et al., 1986 (36)	12 months high-intensity exercise training (n = 25) or control (n = 14); 85% had previous MI; nonrandomized	Significantly improved $\dot{V}O_2$peak, EF%, SBP, and RPP and reduced ESV in trained group.

continued

Author	Methods	Outcome
Grodzinski et al., 1987 (49)	Randomized post-MI; exercise trained 4 weeks (n = 53), control (n = 46)	Significantly increased PWC in trained; no change in EF% at rest but increased after submaximal exercise; inferior MI improved more than anterior MI for EF%.
Blumenthal et al., 1988 (16)	Randomized recent MI; moderate- to high-intensity exercise trained for 12 weeks (n = 36), low intensity (n = 34)	No difference in PWC improvement, psychosocial factors between groups; depressed cohort had less depression.
Blumenthal et al., 1988 (17)	Randomized post-MI; high-intensity (n = 23) and low-intensity (n = 23) exercise trained groups; 12 weeks	No difference in improvements of PWC, HDL cholesterol, resting HR and BP, and exercise RPP.
Hamalainen et al., 1989 (50)	Randomized, post–acute MI to cardiac rehab (n = 188) or control (n = 187); 10-year follow-up	Significantly lower sudden and cardiac mortality rates for cardiac rehab group; no difference for nonfatal re-MI rate or PWC.
Newton et al., 1991 (81)	Randomized post-MI; cardiac rehab 10 weeks (n = 12), control (n = 10)	Significantly improved psychological status in rehab group but no difference in treadmill time to fatigue.
Oldridge et al., 1991 (87)	Post–acute MI; randomized to 8 weeks of exercise training with behavioral counseling (n = 99) or usual care (n = 102)	Improved quality of life scores, exercise tolerance in trained at 8 weeks but not 12 months; no difference in return to work status.
Giannuzzi et al., 1992 (125)	Post–anterior MI; randomized to physical training (n = 25) or control (n = 24) for 6 months	Significantly improved PWC and lactate threshold.
Goble et al., 1991 (47)	Post–acute MI; randomized to high- and low-intensity exercise training groups (n = 308)	No difference in PWC at 11 weeks and 1 year.
Worcester et al., 1993 (123)	Post–acute MI; randomized at 5 weeks to a high- or low-intensity exercise training group (n = 224 males)	No difference in quality of life scores between groups and over time.
Giannuzzi et al., 1993 (126)	Multicenter randomized to 6 months exercise training (n = 490) or control (n = 46); post–anterior MI	Significantly improved PWC in trained group; no effect on spontaneous LV function deterioration.
DeBusk et al., 1994 (30)	Post–acute MI; case managed, home-based cardiac rehab with coronary risk modification (n =293); usual care (n = 292)	Case managed significantly better for smoking cessation, LDL reduction, and functional capacity improvement.
Pavia et al., 1995 (92)	Recent MI; exercise trained 12 weeks with β-blockade (n = 14), without β-blockade (n = 13)	No difference in improvement of $\dot{V}O_2$peak and ventilatory threshold between groups.
Adachi et al., 1996 (4)	Post-MI; randomized to low (n = 11) or high (n = 10) intensity exercise training group or control (n = 8)	Significantly improved peak $\dot{V}O_2$ and rest SV in high group only; significantly improved peak work rate in exercise trained vs. control groups.
Daub et al., 1996 (27)	Randomized post-MI; low- (n = 14), moderate- (n = 13), and high- (n = 15) intensity resistance trained (with aerobic), control (n = 15); 12 weeks	Similar strength gains in all groups versus control; fewer untoward events during resistance than aerobic exercise training.

continued

Author	Methods	Outcome
Ehsani et al., 1997 (37)	Post-MI; 12 months high-intensity exercise training; group <45% EF (n = 8) and >50% (n = 12)	Significantly improved exercise EF%, LVESV, and SBP in >50% group only; significantly improved $\dot{V}O_2$peak, but more so in >50% group; no change in peak cardiac output.
Doi et al., 1997 (32)	Post-MI; exercise trained (n = 20), control (n = 14); trained with LV monitoring (n = 10); length of training not specified	Significantly increased PWC, anaerobic threshold, resting stroke volume, and peak exercise cardiac output in trained.
Malfatto et al., 1998 (69)	Post–anterior MI; n = 53; cardiac rehab group (n = 9); β-blocker with cardiac rehab (n = 20); β-blocker only (n = 14)	Significantly increased exercise duration and parasympathetic activity; reduced sympathetic activity in cardiac rehab groups.
Fragnoli-Munn et al., 1998 (127)	Nonrandomized recent post-MI; combined aerobic and resistance training; younger (n = 25) and older (n = 19); 12 weeks	Significant improvement in upper and lower body strength in both groups; older females improved more than males; positive body composition changes.
Stewart et al., 1998 (111)	Randomized post-MI; combined resistance and cycle trained (n = 12) and cycle only (n = 11); 10 weeks	$\dot{V}O_2$peak improved in combined group only; upper and lower strength gains in both groups but significantly more in combined; safety demonstrated.
Dorn et al., 1999 (33)	19-year follow-up of 651 male participants in the National Exercise and Heart Disease Project; subjects aerobically trained for 8 weeks	Nonsignificant improvement in mortality risk which diminished over time, suggesting only possible short-term protection.
Myers et al., 1999 (78)	Post-MI, EF<40%; randomized to exercise training (n = 12) or control (n = 13) groups; 8 weeks, 60-80% HRR	Significantly increased $\dot{V}O_2$peak and cardiac output (secondary to increased arteriovenous oxygen difference) and ventilation in trained group.

Note. MI = myocardial infarction; PWC = peak work capacity; HR = heart rate; SBP = systolic blood pressure; DBP = diastolic blood pressure; EF = ejection fraction; RPP = rate pressure product; ESV = end systolic volume; HDL = high-density lipoprotein; BP = blood pressure; LV = left ventricular; LDL = low-density lipoprotein; SV = stroke volume; LVESV = left ventricular end systolic volume; HRR = heart rate reserve.

There is some evidence that post-MI patients with lower, versus higher, ejection fraction are likely to demonstrate less improvement in hemodynamic function following exercise training. Ehsani et al. (37) found that improvement in $\dot{V}O_2$peak was less in a group with a lower ejection fraction than a group with a normal ejection fraction (25% vs. 34%, respectively) after 12 months of exercise training. Results from this study implied that patients with compromised cardiac function (i.e., low ejection fraction) secondary to MI have the ability to improve functional capacity by means of an increased arteriovenous oxygen difference. Those with higher ejection fractions improve their $\dot{V}O_2$peak by the combination of hemodynamic (i.e., oxygen delivery) and peripheral (i.e., oxygen use) improvement.

Similar changes in global and regional left ventricular dilation have been reported for both a control and exercise-trained group of post-MI patients (45, 54). Myers et al. (79) reported results after 8 weeks of exercise training in post-MI patients with reduced ejection fraction and found no change in ejection fraction, end diastolic volume, and myocardial wall thickness. These studies suggest that aerobic exercise training does not have an influence, positive or negative, on ventricular dilation.

The safety of resistance training in the stable post-MI patient has recently been reaffirmed, as cardiovascular complications were more frequent during aerobic exercise training than resistance training (27). In support, Stewart et al. (111) reported no complications in a group of post-MI patients who performed a circuit program of 10 to 15 repetitions at 40% of their 1RM.

suggested that these patients undergo low-level exercise training (<4-6 METs) with gradual progression.

Exercise training patients with exertional ischemia is controversial. There are two theories of the effects of exertional ischemia:

1. Consistent exertional ischemia causes myocardial stunning, which can result in the loss of cardiovascular function attributable to myocardial cell hibernation.

2. Ischemia during regular exertion results in preconditioning, which is associated with the development of collateral circulation and reduced risk of MI and improved survival post MI (19).

Investigations have shown evidence of myocardial stunning following exercise and also have shown that the effects, demonstrated by depressed ejection fraction and elevated end-diastolic and end-systolic volumes, last for at least 1 hr after exercise is discontinued and angina has subsided; the stunning appears to be reversible (6, 91). To the contrary, Foster et al. (126) concluded that mild ischemia might be tolerated during exercise without causing left ventricular dysfunction. Little evidence exists that collateral development is enhanced in post-MI or ischemic patients resulting from exercise-induced ischemia (25). Kay et al. (64) reported that patients with coronary artery occlusion and identified collateral vessels had less ST-segment depression following a prolonged warm-up period versus with a shorter warm-up. Because the issue of myocardial stunning versus collateral development/recruitment during exercise training remains controversial, it is prudent to remain on the conservative side when training post-MI patients with inducible ischemia. As little ischemia as possible should be allowed with the assumption that repeated bouts of ischemia can be harmful.

Exercise Recommendations

Post-MI patients should be referred to the outpatient phases of CRP (i.e., phases 2, 3, and 4). The post-MI patient is progressed through cardiac rehabilitation with the following general goals:

— Cardiac risk factor behavior modification
— Enhanced self-image
— Improved functional capacity
— Return to a normal lifestyle
— Reduced morbidity and mortality risk

There is evidence that home-based aerobic exercise is beneficial in the post-MI population (28, 29).

However, little is known about the long-term effects and clinical outcomes of home-based exercise training. When determining the type of setting that a patient will exercise within, one must consider the drawbacks of each setting and the goals of the patient. For instance, a group-based cardiac rehabilitation setting is less convenient. However, in an at-home setting, a patient does not interact daily with staff and other patients. Regardless of which is selected, standard exercise prescription guidelines should guide the exercise training. Practical application 12.3 provides exercise training recommendations for post-MI patients.

Cardiovascular Training

The development of an exercise prescription in the post-MI patient should consider the following:

— Current functional capacity
— Exercise response: ischemia, blood pressure, heart rate, arrhythmias, RPE, work rate
— Exertional symptoms: angina, dyspnea, claudication, dizziness
— Individual goals

Following hospital discharge, the post-MI patient should begin to gradually increase his or her amount of physical activity and exercise at home. Typical recommendations focus on walking but may also include other forms of aerobic type exercise. During this time of the recovery process, it is common to advise patients to increase their heart rate by no more than 20 to 30 beats above rest. However, many patients lack the ability to monitor heart rate. Because of this, it may be prudent for patients to use the Borg RPE scale (9-11 points) and to closely monitor signs and symptoms of exercise intolerance while performing exercise. Signs and symptoms of exercise intolerance in postmyocardial infarction patients include:

— Angina
— Nausea
— ST-segment depression
— Dyspnea
— Hypotension
— Cyanosis
— Dizziness
— Pallor skin color
— Confusion
— Cold/clammy skin
— Palpitations

PRACTICAL APPLICATION 12.3

Exercise Training Recommendations

Aerobic and resistance training may be best initiated in a monitored setting (e.g., cardiac rehabilitation); consider comorbid conditions such as hypertension, diabetes, chronic obstructive pulmonary disease, and peripheral arterial disease when developing and implementing an exercise program.

Type	Mode(s)	Intensity	Frequency and duration	Special considerations
Aerobic	Walking Cycling (leg ergometry or outdoors) Stepping (seated or upright) Elliptical trainer Swimming or other water activities Upper body ergometry	40-85% of HRR or $\dot{V}O_2$ peak reserve Use RPE (9-14) or "talk test" as adjunct to or in place of HR 20-30 beats above rest if no exercise evaluation available	4-7 times per week 30-60 min per session	Adjust intensity for ischemia or hemodynamic instability if necessary. Goals are to increase endurance and reduce secondary event risk and mortality risk. Supplement with lower intensity physical activity on nonexercising days. Consider timing of exercise test and training time of day in β-blocked patients, because HR response may be altered.
Resistance	Universal-type machines Dumbbells or other free weights	40-60% of direct or indirect 1RM or weight lifted comfortably 8-15 times	2-3 times per week 1 or 2 sets per lift Working through the major muscle groups (20-30 min)	Avoid Valsalva maneuver. Goals are to increase skeletal muscle strength and endurance. Circuit training may be used by selected patients.
Range of motion	Passive	Work throughout range of each upper and lower body major joint	Daily for 5-10 min	Avoid ballistic stretching. Avoid breath hold. Goals are to improve flexibility and reduce injury risk.

Note. HRR = heart rate reserve; RPE = rating of perceived exertion; HR = heart rate.

After several weeks, the post-MI patient should progress to an increased intensity and duration of exercise training and physical activity.

Another common rule used to guide intensity is the "talk test." Individuals who can talk comfortably (i.e., without dyspnea) during exercise are typically working at an intensity below 85% of their maximal heart rate (~80% $\dot{V}O_2$ peak) on treadmills and dual-action cycles (20, 26). However, some individuals may be able to exercise above 85% of maximal heart rate and continue to talk comfortably. In fact, up to 50% of healthy, sedentary subjects, while walking on an outdoor track, can exercise above 85% $\dot{V}O_2$ peak and continue to talk comfortably. Because there is individual variation in RPE and comfortable talking, and because no research exists on this topic in MI patients, an objective variable should be used (e.g., heart rate or work rate), when available, to guide exercise intensity in post-MI patients immediately after hospitalization.

Depending on the particular program philosophy, a post-MI patient may begin and even complete a 4- to 12-week phase 2 cardiac rehabilitation program without having had a maximal exercise evaluation. In these programs, exercise intensity should be based on the patient's subjective ratings or RPE and symptoms, as previously discussed. At this point, patients can exercise to an RPE between 11 and 14. Some phase 2 programs require that a symptom-limited exercise evaluation be performed either before program entry or soon after program initiation. When this is available, the exercise intensity can be guided by using a heart rate–based method with adjustment for ischemia or other signs and symptoms (as discussed in chapter 7).

It is also common to electrocardiographically monitor the post-MI patient, via telemetry, for the duration of phase 2 participation. In fact, for those with Medicare insurance, this is a requirement for reimbursement. However, there is now a trend to move away from continuous ECG monitoring, especially in low-risk and non-exercise-induced ischemia patients. Investigations have demonstrated that ECG monitoring during phase 2 cardiac rehabilitation resulted in few significant findings and very little physician referral or other intervention (72).

The ACSM position stand "Exercise for Patients With Coronary Artery Disease" addresses the issue of exercise prescription in these patients, which includes those with a previous MI (8). It is recommended that in supervised programs, post-MI patients perform aerobic modes of exercise "at a moderate, comfortable intensity, generally 40-85% of maximal functional capacity (VO_2max)" (p. iii). This corresponds to 40% to 85% of heart rate reserve and 55% to 90% of maximal exercise heart rate. Recently, it was reported that the heart rate reserve method of exercise prescription is related to a corresponding percentage of VO_2 reserve in post-MI patients (21). This does not alter the use of the current heart rate reserve method but allows a better understanding of the relationship between heart rate reserve and VO_2. VO_2 reserve should be used if one is developing an exercise prescription using work rate or METs to guide intensity (see chapter 7).

Reduced exercise intensity may be prescribed in selected patients. Severely deconditioned patients and those limited by cardiac complications should be prescribed exercise at lower intensity levels. In deconditioned and low-fit persons, positive adaptations can occur at exercise intensities as low as 40% of heart rate reserve (9). A goal is to slowly increase exercise intensity to 60% to 85% heart rate reserve as the training effect occurs.

After the initial 4 to 12 weeks of exercise training, much of the potential physiological adaptation will have taken place. However, lifelong exercise training and increased physical activity should still be encouraged. Training at this point can be carried out in a supervised setting such as a phase 3 or 4 cardiac rehabilitation program or in a nonsupervised setting such as at home or a public fitness facility. Patients who are candidates for a nonsupervised setting include those who have had an uncomplicated post-MI course; who remain asymptomatic; who demonstrate no indications of myocardial ischemia, severe arrhythmias, or signs of

exercise intolerance; who have an exercise tolerance of greater than 5 METs; and who are self-motivated to continue exercise training.

It is common to have patients attend cardiac rehabilitation minimally 3 days per week and to exercise for a minimum of 30 min per session (9). The upper end of the frequency range may be increased to 6 or 7 days per week and the duration increased to 40 to 60 min in selected patients. However, the increased risk of musculoskeletal injury and cardiovascular events should be considered (96).

Each exercise training session should begin with a warm-up period that consists of active aerobic exercise at a reduced intensity level compared with the actual training session. This may reduce the risk of arrhythmia and symptomatic ischemia during exercise training. Active cool-down invokes the skeletal muscle pump of the lower legs, which assists the recovery hemodynamic response and allows for a safer and faster return to preexercise cardiovascular status. There is some evidence that an improper warm-up can increase the risk of ischemia and arrhythmias with subsequent left ventricular dysfunction and skeletal muscle injury (43). Insufficient cool-down can result in lower limb venous pooling leading to hypotension, dizziness, syncope, arrhythmias, and myocardial ischemia.

Resistance Training

General resistance training guidelines and modalities are presented in chapter 7. The ACSM and AACVPR have published guidelines that provide indications for resistance training in cardiac diseased populations (7, 9). These include beginning a minimum of 4 to 6 weeks after an acute MI, participation in a supervised exercise program during resistance training, a diastolic blood pressure less than 105 mmHg, a peak work capacity of at least 5 METs, and stabilized heart failure symptoms and arrhythmias.

The intensity level should be set at a resistance that can be comfortably handled for 8 to 15 repetitions. Some investigations have used a 40% to 60% of 1-repetition maximum range for the initial intensity range (27, 31, 42, 111). The direct or indirect 1-repetition maximum method may be used (see chapter 7) (5, 31).

Resistance training increases skeletal muscle strength and endurance. Lighter resistance training (i.e., <20 lb) can begin early in the rehabilitative process (e.g., latter stages of phase 2 cardiac rehabilitation) and progress to greater intensities after 4 to 12 weeks of training. If a circuit type program is used, it is recommended that cardiac patients begin with one circuit per session and gradually progress to two or three circuits per session. Focus should be on the major muscle groups (see chapter 7). Resistance training should not be performed daily or on consecutive days in post-MI patients, because most are deconditioned and prone to excessive fatigue, injury, and noncompliance when the exercise load becomes too great (97).

Few data exist to guide therapy in post-MI patients using other resistance devices including elastic bands, water, and light dumbbells, or hand weights. However, because of the general level of deconditioning, it is likely that post-MI patients would gain strength and endurance with the initiation of any resistance-training program.

To reduce the risk of noncompliance during resistance training, the following are suggested:

— Begin with light resistance to reduce the risk of skeletal muscle soreness.
— Progress the training load slowly and within the patient's capability.
— Keep training sessions to 15 to 45 min.

A clinical exercise professional guiding post-MI patients' resistance training programs should regularly assess each patient individually for indications of maladaptations and progress. As with aerobic exercise, the resistance training session should be discontinued if any indicator of exercise intolerance occurs.

A potential danger of resistance training in the post-MI patient is a markedly elevated systolic or diastolic blood pressure. Systolic blood pressure increases myocardial oxygen demand, and large increases may produce ischemia. However, the rate pressure product during resistance training does not typically exceed values at peak aerobic exercise until the load reaches at least 80% of the 1RM. Conversely, elevated diastolic pressure enhances coronary blood perfusion and may satisfy elevations in myocardial oxygen demand by increasing the oxygen availability to the myocardial tissue.

Overall, resistance training in the post-MI patient appears to be safe and effective. Avoidance of resistance training might be prudent in unstable patients, and patients with impaired left ventricular function should be monitored closely.

Range of Motion

While they are hospitalized, all patients should begin a range of motion program designed to keep the joints and skeletal muscles from losing compliance and to maintain flexibility, which is often lost during prolonged bed rest. Range of motion exercise can be started passively by an exercise professional while the patient is in bed or sitting in a chair. Gradually the patient should begin an active range of motion program. During the hospitalization period, the exercise professional should educate the patient about the importance of flexibility. Specifically, an improved range of motion will allow for greater ease of movement and will reduce the risk of skeletal muscle injury.

Breath holding or a **Valsalva** maneuver during resistance or range of motion exercise can reduce venous return because of an elevated intrathoracic pressure. This results in an initial elevation of mean arterial pressure that increases afterload and myocardial work. Within a few seconds, there is a secondary reduction in cardiac output, caused by the reduced venous return, and a subsequent decrease in mean arterial pressure. This combination of events can be problematic for the post-MI patient with residual ischemia or reduced left ventricular function. Normal breathing during exercise reduces these risks. See chapter 7 for specifics of proper range of motion exercise.

Conclusion

Exercise training is an accepted and important part of the rehabilitative process for patients who have had an MI. This conclusion is based on a number of investigations that have been performed to date.

Case Study 12

Medical History

Mr. RG is a 48-year-old white male who previously had an unremarkable medical history. He is sedentary and has been for most of his adult life. He works as a foreman in an auto factory. He has smoked two packs of cigarettes a day for 30 years and his body mass index is 28.4. His total cholesterol

continued

= 274 mg · dl^{-1}, LDL = 145 mg · dl^{-1}, and HDL = 24 mg · dl^{-1}. He does not know his typical blood pressure value.

Diagnosis

Mr. RG presented to the emergency department with substernal and left arm discomfort and diaphoresis. This woke him approximately 1 hr before he usually arose and worsened on the drive to work. He saw the nurse at the factory and was then transported by ambulance to the hospital. Sublingual nitroglycerin and aspirin were given on the way to the hospital to treat his symptoms. The patient was immediately given morphine on arrival at the hospital. His blood pressure was 180/110 mmHg. An ECG demonstrated ST-segment elevation and tall T waves in leads V1 through V6 and II, III, and aVf. Blood was drawn and stat analyzed for creatine phosphokinase and lactate dehydrogenase. These values were both elevated above resting values. From this information, the diagnosis of acute myocardial infarction was made. The plan was to evaluate for thrombolytic versus primary angioplasty therapy.

Mr. RG underwent percutaneous coronary intervention (PCI) with stent placement of his right coronary artery and left coronary artery 2 hr after arrival at the hospital. The circumflex artery also had an 80% blockage that was anatomically difficult and thus was not revascularized. A subsequent in-hospital dobutamine echocardiography examination revealed a lateral wall motion abnormality at the highest dose of dobutamine.

Exercise Test Results

Following stabilization and release from the hospital, Mr. RG was referred for cardiac rehabilitation. The patient began phase II at 2 weeks post-MI and was subsequently scheduled for an exercise test to evaluate his functional status, prognosis, and ischemic status and to develop an exercise prescription.

He walked on a treadmill using a modified Bruce protocol. His medications were metoprolol, zocor, aspirin, and captopril. His ECG was interpreted as anterior and inferior MIs with repolarization abnormality. The following are the results:

Protocol: Bruce
Time: 10:46 (stage 4 of standard Bruce protocol)
Resting heart rate = 88
Resting blood pressure = 148/94 mmHg
Peak heart rate = 136
Peak blood pressure = 226/100 mmHg
$\dot{V}O_2$peak = 21.4 ml · min^{-1} · kg^{-1}
Symptoms: angina at peak exercise (+3/4); shortness of breath; right calf "cramping"
ECG: ST-segment depression = 1 mm horizontal at heart rate = 115 and 2.5 mm horizontal at peak exercise in leads V5, V6, I, and aVL

Development of Exercise Prescription

Mr. RG wishes to progress to jogging and has a goal of running a marathon. He is a candidate for coronary artery bypass surgery but wishes to medically manage his condition. He has retired from

continued

his employment and appears to be willing to make the necessary changes to control his risk of a future event.

· ·

Case Study 12   Discussion Questions

1. What would be general recommendations for a home-based exercise program immediately following hospitalization?

2. What is an appropriate exercise prescription for Mr. RG for the initial stages of phase II cardiac rehabilitation?

3. Based on your knowledge of Mr. RG's medical history, what might be his greatest challenges? How would you handle this in a cardiac rehabilitation setting?

4. Assuming Mr. RG improves as expected during the phase 2 program, what might you recommend for him to progress to jogging and subsequently running a marathon?

5. What are Mr. RG's risks during exercise?

6. What is the significance of Mr. RG's symptoms during his exercise evaluation?

7. How would you approach Mr. RG for behavior modification of his coronary artery disease risk factors?

Glossary

acute MI—The initial stages of an evolving MI.

akinetic—Loss of movement of a left ventricular wall during the normal cardiac cycle.

angina pectoris—Constricting chest pain, often radiating to the left shoulder or arm, back, or neck and jaw regions, caused by ischemia of the heart muscle.

arterial intimal layer—The innermost layer of an artery, composed of endothelial cells.

atheroma—A mass of plaque of degenerated, thickened arterial intima occurring in atherosclerosis.

atherosclerosis—An extremely common form of arteriosclerosis in which deposits of yellowish plaques (atheroma) containing cholesterol, lipid material, and lipophages are formed within the intima and inner media of large and medium-sized arteries.

bifurcation—Point at which an artery branches to form two arteries.

bradycardia—A heart rate of less than 60 beats · min^{-1}.

cardiogenic shock—Lack of cardiac and systemic oxygen supply resulting from a decline in cardiac output secondary to serious heart disease; typically follows an MI.

chronic MI—The latest phase of an MI when the heart is stable.

differential diagnosis—The determination of which of two or more diseases with similar symptoms is the one from which the patient is suffering, by a systematic comparison and contrasting of the clinical findings.

double product—The value obtained by multiplying the heart rate and systolic blood pressure; an estimate of myocardial oxygen demand that is reproducible, such as at the ischemic threshold; used to determine confidence of results of a diagnostic exercise evaluation (i.e., value should be >24,000 for highest predictive confidence).

dyskinetic—Denoting an outward or bulging movement of the myocardium during systole; often associated with aneurysm.

ejection fraction—The fraction of the blood contained in the ventricle at the end of diastole that is expelled during its contraction, that is, the stroke volume divided by end-diastolic volume, normally 0.55 (by electrocardiogram) or greater; with the onset of congestive heart failure, the ejection fraction decreases, sometimes to 0.10 or even less in severe cases.

erythrocyte sedimentation—The sinking of red blood cells in a volume of drawn blood.

evolving MI—Period of time after the acute onset of an MI when the myocardial tissue is transforming from ischemic to necrotic tissue.

foam cell—Smooth muscle cells that take up intimal lipid which accumulates in the cytoplasm and develops a bubbly appearance when observed microscopically.

growth factor—A category of hormones responsible for stimulating the process of tissue growth.

ischemia—Deficiency of blood flow, attributable to functional constriction or actual obstruction of a blood vessel.

orthostatic—Relating to upright or erect posture.

plaque—A yellow area or swelling on the intimal surface of an artery, produced by the atherosclerotic process of lipid deposition.

platelet aggregation—The congregation of platelets, which are disk-shaped fragments found in the peripheral blood and involved in the clotting process.

polymorphonuclear leukocytosis—An elevation in neutrophilic leukocyte (white blood cell) count.

prothrombotic—Condition or agent that increases the risk of formation or presence of a thrombus.

radionuclide imaging—A type of cardiac imaging that can detect ischemia and wall motion; uses an injected radioisotope (i.e., thallium 201 or technitium-99m sestamibi) that is scanned using X-ray.

reperfusion—The process of reinstituting blood flow to an area of tissue previously deprived of normal blood flow.

resolving MI—The phase of an MI in which necrotic tissue is forming a scar.

sensitivity—The proportion of affected individuals who give a positive test result for the disease that the test is intended to reveal.

specificity—The proportion of individuals with negative test results for the disease that the test is intended to reveal.

subendocardial—Referring to the endocardial surface of the heart.

tachycardia—Referring to a heart rate greater than 100 beats $\cdot$ min^{-1}.

technetium 99m sestamibi—A radioisotope, introduced into the bloodstream via a catheter, that tags red blood cells and when imaged using a gamma camera can provide a measure of ventricular volume, ejection fraction, and regional ventricular wall motion at rest and during exercise. Used to depict myocardial ischemia.

thallium 201—A white metallic substance with radioactivity, introduced into the bloodstream via a catheter, that is perfused into the myocardium; used in conjunction with stress testing (exercise and pharmacological) to image the myocardium to detect transient ischemia and tissue necrosis.

thrombolytic—Agents that degrade fibrin clots by activating plasminogen, a naturally occurring modulator of hemostatic and thrombotic processes.

thrombus—An aggregation of blood factors, primarily platelets and fibrin with entrapment of cellular elements, which frequently causes vascular obstruction at the point of formation.

transmural—Referring to effects on all tissue layers of the heart.

valsalva—The forced exhalation against a closed glottis that results in an elevation in thoracic pressure which reduces venous return to the heart.

wall motion—Relating to movement of the left ventricular segments of the heart; used to describe normal or abnormal movement during contraction and to calculate ejection fraction during two-dimensional echocardiography or some types of nuclear imaging.

References

1. Abboud L, Hir J, Eisen I, Cohen A, Markiewicz W. The current value of exercise testing soon after acute myocardial infarction. Isr J Med Sci 1992; 28: 694-699.

2. Acute Infarction Ramipril Efficacy (AIRE) Study Investigators. Effect of ramipril on mortality and morbidity of survivors of AMI with clinical evidence of heart failure. Lancet 1993; 342: 821-828.

3. Acute myocardial infarction (chapter 202). Harrison's Principles of Internal Medicine (13th edition). 1994; McGraw-Hill, New York.

4. Adachi H, Koike A, Obayashi T, Umezawa S, Aonuma K, Inada M, Korenaga M, Niwa A, Marumo F, Hiroe M. Does appropriate endurance exercise training improve cardiac function in patients with prior myocardial infarction? Eur Heart J 1996; 17: 1511-1521.

5. Adams GM. Exercise Physiology Laboratory Manual (3rd ed.). 1998; WCB McGraw-Hill, Boston.

6. Ambrosio G, Betocchi S, Pace L, Losi MA, Perrone P, Soricelli A, et al. Prolonged impairment of regional contractile function after resolution of exercise-induced angina. Evidence of myocardial stunning in patients with coronary artery disease. Circulation 1996; 94: 2455-2464.

7. American Association of Cardiovascular and Pulmonary Rehabilitation. Guidelines for Cardiac Rehabilitation and Secondary Prevention Programs (3rd ed.). 1999; Human Kinetics, Champaign, IL.

8. American College of Sports Medicine. Exercise for patients with coronary artery disease: Position stand. Med Sci Sports Exerc 1994; 26(3): i-v.

9. American College of Sports Medicine. ACSM's Guidelines for Exercise Testing and Prescription. (6th ed.). 2000; Lippincott, Williams & Wilkins, Philadelphia.

10. American Heart Association. 2002 Heart and Stroke Statistical Update. 2002; American Heart Association, Dallas.

11. Aronow WS. Management of older persons after myocardial infarction. J Am Geriatr Soc 1998; 46(11): 1459-1468.

12. Bengtsson K. Rehabilitation after myocardial infarction. Scand J Rehabil Med 1983; 15(1): 1-9.

13. Berger AK, Schulman KA, Gersh BJ, Pirzada S, Breall JA, Johnson AE, Every NR. Primary coronary angioplasty vs thrombolysis for the management of acute myocardial infarction in elderly patients. JAMA 1999; 282(4): 341-348.

14. Beta-Blocker Heart Attack Trial Research Group. A randomized trial of propranolol in patients with acute myocardial infarction. I. Mortality results. JAMA 1982; 247: 1707-1714.

15. Blair SN, Kohl HW III, Paffenbarger RS Jr, Clark DG, Cooper KH, Gibbons LW. Physical fitness and all-cause mortality: A prospective study of healthy men and women. JAMA 1989; 262: 2395-2401.

16. Blumenthal JA, Emery CF, Rejeski WJ. The effects of exercise training on psychosocial functioning after myocardial infarction. J Cardiopulm Rehabil 1988; 8: 183-193.

17. Blumenthal JA, Rejeski WJ, Walsh-Riddle M, Emery CF, et al. Comparison of high- and low-intensity exercise training early after acute myocardial infarction. Am J Cardiol 1988; 61: 26-30.

18. Bodi B, Monmeneu JV, Marin F. Acute cardiac rupture complicating pre-discharge exercise testing. A case report with complete echocardiographic follow-up. Int J Cardiol 1999; 68(3): 333-335.

19. Bolli R. Basic and clinical aspects of myocardial stunning. Prog Cardiovasc Dis 1998; 40:477-516.

20. Brawner CA, Keteyian SJ, Czaplicki T. A method of guiding exercise intensity: The talk test. Med Sci Sports Exerc 1995; 27(5, Suppl.): S241.

21. Brawner CA, Keteyian SJ, Ehrman JK. The relationship of heart rate reserve to VO2 reserve in patients with heart disease. Med Sci Sports Exerc 2002; 34(3): A18-A22.

22. Bray GA. Pathophysiology of obesity. Am J Clin Nutr 1992; 55: 488S-494S.

23. Carson P, Phillips R, Lloyd M, Tucker H, Neophytou M, Buch NJ, Gelson A, Lawton A, Simpson T. Exercise after myocardial infarction: A controlled trial. J R Coll Physicians Lond 1982; 16: 147-151.

24. Castelli WP. Categorical issues in therapy for coronary heart disease. Cardiology in Practice 1985; Jan/Feb: 267-273.

25. Conner JF, LaCamera F, Swanick EJ, Oldham MJ, Holzaepfel W, Lyczkowskyj O. Effects of exercise on coronary collateralization-angiographic studies of six patients in a supervised exercise program. Med Sci Sports Exerc 1976; 8: 145-151.

26. Czaplicki T, Keteyian SJ, Brawner CA, Weingarten MA. Guiding exercise training intensity on a treadmill and dual-action bike using the talk test. Med Sci Sports Exerc 1997; 29(5, Suppl.): S70.

27. Daub WD, Knapik GP, Black WR. Strength training early after myocardial infarction. J Cardiopulm Rehabil 1996; 16: 100-108.

28. DeBusk RF, Houston N, Haskell W, Fry F, Parker M. Exercise training soon after myocardial infarction. Am J Cardiol 1979; 44: 1223-1229.

29. DeBusk RF, Haskell WL, Miller NH, Berra K, Taylor CB, Berger WE III, Lew H. Medically directed at-home rehabilitation soon after uncomplicated acute myocardial infarction: A new model for patient care. Am J Cardiol 1985; 55: 251-257.

30. DeBusk RF, Houston Miller N, Superko HR, Dennis CA, Thomas RJ, Lew HT, Berger WE III, Heller RS, Rompf J, Gee D, et al. A case-management system for coronary risk factor modification after acute myocardial infarction. Ann Intern Med 1994; 120: 721-729.

31. DeGroot DW, Quinn TJ, Kertzer R, Vroman NB, Olney WB. Lactic acid accumulation in cardiac patients performing circuit weight training: Implications for exercise prescription. Arch Phys Med Rehabil 1998; 79: 838-841.

32. Doi M, Itoh H, Niwa A, Taniguchi K, Hiore M, Marumo F. Effect of training program based on anaerobic threshold in the early phase after acute myocardial infarction. Cardiologia 1997; 42(10): 1077-1082.

33. Dorn J, Naughton J, Imamura D, Trevisan M. Results of a multicenter randomized clinical trial of exercise and long-term survival in myocardial infarction patients: The national exercise and heart disease project (NEHDP). Circulation 1999; 100: 1764-1769.

34. Dugmore LD, Tipson RJ, Phillips MH, Flint EJ, Stentiford NH, Bone MF, Littler WA. Changes in cardiorespiratory fitness, psychological well-being, quality of life, and vocational status following a 12 month cardiac exercise rehabilitation programme. Heart 1999; 81: 359-366.

35. Ehsani AA, Heath GW, Hagberg JM, Sobel BE, Holloszy JO. Effects of 12 months of intense exercise training on ischemic ST-segment depression in patients with coronary artery disease. Circulation 1981; 64(6): 1116-1124.

36. Ehsani AA, Biello DR, Schultz J, Sobel BE, Holloszy JO. Improvement of left ventricular contractile function by exercise training in patients with coronary artery disease. Circulation 1986; 74(2): 350-358.

37. Ehsani AA, Miller TR, Miller TA, Ballard EA, Schechtman KB. Comparison of adaptations to a 12-month exercise program and late outcome in patients with healed myocardial infarction and ejection fraction <45% and >50%. Am J Cardiol 1997; 79: 1258-1260.

38. Lapin ES, Murray JA, Bruce RA, and Winterschield L. Changes in maximal exercise performance in evaluation of saphenous vein bypass surgery. Circulation 1973; 47: 1164-1173.

39. Leon MB, Almagor Y, Erbel R, Teirstein PS, Perez J, and Schatz RA. Subacute thrombotic events after coronary stent placement: Clinical spectrum and predictive factors. Circulation 1989; 80: II-174.

40. Lesperance J, Bourassa MG, Saltiel J, Campeau L, and Grondin CM. Angiographic changes in aortocoronary vein grafts. Lack of progression beyond the first year. Circulation 1973; 48: 633.

41. Levine MJ, Leonard BM, Burke JA, Nash ID, Safian RD, Diver DJ, and Baim DS. Clinical and angiographic results of balloon-expandable intracoronary stents in right coronary artery stenoses. J Am Coll Cardiol 1990; 16: 332-339.

42. McConnell TR, Klinger TA, Gardner JK, Laubach CA Jr, Herman CE, and Hauck CA. Cardiac rehabilitation without exercise tests for post-myocardial infarction and post-bypass surgery patients. J Cardiopulm Rehabil 1998; 18: 458-463.

43. Mehta J, Mehta P, and Horalek C. The significance of platelet-vessel wall prostaglandin equilibrium during exercise-induced stress. Am Heart J 1983; 105: 895-904.

44. Milani RV, Lavie CJ, and Cassidy MM. Effects of cardiac rehabilitation and exercise training programs on

depression in patients after major coronary events. Am Heart J 1996; 132: 726-732.

45. Morris GC, Reul GJ, and Howell JF. Follow-up results of distal coronary artery bypass for ischemic heart disease. Am J Cardiol 1972; 29: 180.

46. Gibbons RJ, Balady GJ, Beasley JW, et al. ACC/AHA guidelines for exercise testing: A report of the American College of Cardiology/American Heart Association task force on practice guidelines (committee on exercise testing). J Am Coll Cardiol 1997; 30(1): 260-315.

47. Goble AJ, Hare DL, Macdonald PS, Oliver RG, Reid MA, Worcester MC. Effect of early programmes of high and low intensity exercise on physical performance after transmural acute myocardial infarction. Br Med J 1991; 65: 126-131.

48. Gordon NF, Hoh HW. Exercise testing and sudden cardiac death. J Cardiopulm Rehabil 1993; 13: 381-386.

49. Grodzinski E, Jette M, Blumchen G, Borer JS. Effects of a four-week training program on left ventricular function as assessed by radionuclide ventriculography. J Cardiopulm Rehabil 1987; 7: 518-524.

50. Hamalainen H, Luurila OJ, Kallio V, Arstila M, Hakkila J. Long-term reduction in sudden deaths after a multifactorial intervention programme in patients with myocardial infarction: 10-year results of a controlled investigation. Eur Heart J 1989; 10: 55-62.

51. Hamm LF, Crow RS, Stull GA, Hannan P. Safety and characteristics of exercise testing early after acute myocardial infarction. Am J Cardiol 1989; 63: 1193-1197.

52. Hanson P, Nagle F. Isometric exercise: Cardiovascular responses in normal and cardiac populations. Cardiol Clin 1987; 5(2): 157-170.

53. Haskell WL, Savin W, Oldridge N, DeBusk R. Factors influencing estimated oxygen uptake during exercise testing soon after myocardial infarction. Am J Cardiol 1982; 50: 299-304.

54. Heldal M, Rootwelt K, Sire S, Dale J. Short-term physical training reduces left ventricular dilatation during exercise soon after myocardial infarction. Scand Cardiovasc J 2000; 34(3): 254-260.

55. Hjalmarson A, Elmfeldt D, Herlitz J, et al. Effect on mortality of metoprolol in acute myocardial infarction. A double-blind randomised trial. Lancet 1981; 8251: 823-827.

56. Hung J, Gordon EP, Houston N, Haskell WL, Goris ML, DeBusk RF. Changes in rest and exercise myocardial perfusion and left ventricular function 3 to 26 weeks after clinically uncomplicated acute myocardial infarction: Effects of exercise training. Am J Cardiol 1984; 54: 943-950.

57. Isselbacher KJ, Braunwald E, Wilson JD, Martin JB, Fauci AS, Kasper DL. Harrison's Principles of Internal Medicine (13th ed.). 1994; McGraw-Hill, St. Louis.

58. Joao I, Cotrim C, Duarte JA, do Rosario L, Freire G, Pereira H, Oliveira LM, Catarino C, Carrageta M. Cardiac rupture during exercise stress echocardiography: A case report. J Am Soc Echocardiogr 2000; 13(8): 785-787.

59. Johansson S, Rosengren A, Tsipogianni A, Ulvenstam G, Wiklund I, Wilhelmsen L. Physical inactivity as a risk factor for primary and secondary coronary events. Eur Heart J 1998; 9(Suppl L): 8-19.

60. Joseph A, Ackerman D, Talley JK, Johnstone J, Kupersmith J. Manifestations of coronary atherosclerosis in young trauma victims—An autopsy study. J Am Coll Cardiol 1993; 22(2): 459-467.

61. Juneau M, Colles P, Theroux P, de Guise P, Pelletier G, Lam J. Symptom-limited versus low level exercise testing before hospital discharge after myocardial infarction. J Am Coll Cardiol 1992; 20(4): 927-933.

62. Kallio V, Hamalainen H, Hakkila J, Luurila OJ. Reduction in sudden deaths by a multifactorial intervention programme after acute myocardial infarction. Lancet 1979; 2(8152): 1091-1094.

63. Kannel WB, Sorlie BP, Castelli WP, McGee D. Blood pressure and survival after myocardial infarction: The Framingham study. Am J Cardiol 1980; 45: 326-330.

64. Kay IP, Kittelson J, Stewart RAH. Collateral recruitment and "warm-up" after first exercise in ischemic heart disease. Am Heart J 2000; 140: 121-125.

65. Krivokapich J, Child JS, Gerber RS, Lem V, Moser D. Prognostic usefulness of positive or negative exercise stress echocardiography for predicting coronary events in ensuing twelve months. Am J Cardiol 1993; 71: 646-651.

66. Langford HG, Stamler GJ, Wasserther-Smollers S, Preneas RJ. All-cause mortality in the Hypertensive Detection and Follow-up Program. Prog Cardiovasc Dis 1986; 29: 29-54.

67. Lichtlen P, Nikutta P, Jost S, et al. and the INTACT study group. Anatomical progression of coronary artery disease in humans as seen by prospective, repeated, quantitated coronary angiography: Relation to clinical events and risk factors. Circulation 1992; 86: 828-838.

68. Mahmarian JJ, Moye LA, Chinoy DA, Sequeira RF, Habib GB, Hanry WJ, et al. Transdermal nitroglycerin patch therapy improves left ventricular function and prevents remodeling after acute myocardial infarction: Results of a multicenter prospective randomized, double-blind, placebo-controlled trial. Circulation 1998; 97(20): 2017-2024.

69. Malfatto G, Facchini M, Sala L, Branzi G, Bragato R, Leonetti G. Effects of cardiac rehabilitation and beta-blocker therapy on heart rate variability after first acute myocardial infarction. Am J Cardiol 1998; 81: 834-840.

70. Marra S, Paolillo V, Spadaccini F, Angeleno PF. Long-term follow-up after a controlled randomized post-myocardial infarction rehabilitation programme: Effects on morbidity and mortality. Eur Heart J 1985; 6: 656-663.

71. Mayou R, Sleight P, MacMahon D, Florencio MJ. Early rehabilitation after myocardial infarction. Lancet 1981; 2(8260-61): 1399-1401.

72. Mellett PA, Keteyian SJ, Davenport MJ, Fedel FJ, Stein PD. Existing criteria for electrocardiographic monitoring during cardiac rehabilitation versus observed events. Med Sci Sports Exerc 1994; 26: S47.

73. McAlister FA, Lawson FM, Teo KK, Armstrong PW. Randomised trials of secondary prevention programmes in coronary heart disease: Systematic review. Br Med J 2001; Br J Med: 957-962.

74. McNeilly RH, Pemberton J. Duration of last attack in 998 fatal cases of coronary artery disease and its relation to possible cardiac resuscitation. Br Med J 1968; 3: 139-142.

75. Miller NH, Haskell WL, Berra K, DeBusk RF. Home versus group exercise training for increasing functional capacity after myocardial infarction. Circulation 1984; 70(4): 645-649.

76. Mineo K, Takizawa A, Shimamoto M, Yamazaki F, Kimura A, Chino N, Izumi S. Graded exercise in three cases of heart rupture after acute myocardial infarction. Am J Phys Med Rehabil 1995; 74: 453-457.

77. Mittleman MA, Maclure M, Tofler GH, et al. Triggering of acute myocardial infarction by heavy physical exertion: Protection against triggering by regular exertion. N Engl J Med 1993; 329: 1677-1683.

78. Myers J, Dziekan G, Goebbels U, Dubach P. Influence of high-intensity exercise training on the ventilatory response to exercise in patients with reduced ventricular function. Med Sci Sports Exerc 1999; 31(7): 929-937.

79. Myers J, Goebbels U, Dzeikan G, Froelicher V, Bremerich J, Mueller P, Buser P, Dubach P. Exercise training and myocardial remodeling in patients with reduced ventricular function: One-year follow-up with magnetic resonance imaging. Am Heart J 2000; 139(2 Pt 1):252-261.

80. National Cholesterol Education Program. Expert Panel on Detection, Evaluation, and Treatment of High Blood Cholesterol in Adults. Executive Summary of the Third Report of the National Cholesterol Education Program (NCEP) Expert Panel on Detection, Evaluation, and Treatment of High Blood Cholesterol in Adults (Adult Treatment Panel III). JAMA 2001; 285: 2486-2497.

81. Newton M, Mutrie N, McCarthur JD. The effects of exercise in a coronary rehabilitation programme. Scott Med J 1991; 36: 38-41.

82. Nowak TJ, Handford AG. Essentials of Pathophysiology: Concepts and Applications for Health Care Professionals (2nd ed.). 1999; WCB McGraw-Hill, St. Louis.

83. Nolewajka AJ, Kostuk WJ, Rechnitzer PA, Cunningham DA. Exercise and human collateralization: An angiographic and scintigraphic assessment. Circulation 1979; 60(1): 114-121.

84. Norwegian Multicenter Study Group. Timolol-induced reduction in patients surviving acute myocardial infarction. N Engl J Med 1981; 304: 801-807.

85. O'Connor GT, Buring JE, Yusuf S, Goldhaber SZ, Olmstead EM, Paffenbarger RS, Hennekens CH. An overview of randomized trials of rehabilitation with exercise after myocardial infarction. Circulation 1989; 80(2): 234-244.

86. Oldridge NB, Guyatt GH, Fischer ME, Rimm AA. Cardiac rehabilitation after myocardial infarction: Combined experience of randomized clinical trials. JAMA 1988; 260: 945-950.

87. Oldridge NB, Guyatt G, Jones N, Crowe J, Singer J, Feeny D, McKelvie R, Runions J, Streiner D, Torrance G. Effects on quality of life with comprehensive rehabilitation after acute myocardial infarction. Am J Cardiol 1991; 67: 1084-1089.

88. Oldridge NB. Cardiac rehabilitation and risk factor management after myocardial infarction. Clinical and economic evaluation. Wien Klin Wochenschr 1997; 109(Suppl 2): 6-16.

89. Paffenbarger RS, Hyde RT, Wing AL, Hsieh CC. Physical activity, all-cause mortality, and longevity of college alumni. N Engl J Med 1986; 314: 605-613.

90. Paffenbarger RS, Hyde RT. Exercise adherence, coronary heart disease, and longevity. In Exercise Adherence: Its Impact on Public Health, (ed. R. Dishman). 1988; Human Kinetics, Champaign, IL, pp. 41-73.

91. Paul AK, Hasegawa S, Yoshioka J, Tsujimura E, Yamaguchi H, Tokita N, Maruyama A, Xiuli M, Nishimura T. Exercise-induced stunning continues for at least one hour: Evaluation with quantitative gated single-photon emission tomography. Eur J Nucl Med 1999; 26: 410-415.

92. Pavia L, Orlando G, Myers J, Maestri M, Rusconi C. The effect of beta-blockade therapy on the response to exercise training in postmyocardial infarction patients. Clin Cardiol 1995; 18: 716-720.

93. Pedersen TR for the Scandinavian Simvastatin Survival Study group. Randomised trial of cholesterol lowering in 4444 patients with coronary heart disease: The Scandinavian Simvastatin Survival Study (4S). Lancet 1994; 344: 1383-1389.

94. Pekkanen J, Linn S, Heiss G, et al. Ten-year mortality from cardiovascular disease in relation to level among men with and without preexisting cardiovascular disease. N Engl J Med 1990; 322: 1700-1707.

95. Plavsic C, Turkulin K, Perman Z, Steinel S, Oreskovic A, Dimnik R, Martic P, Fischer F, Puharic M, Bruketa I, Martic M, Stojanovic D, Ljubetic L. The results of exercise therapy in coronary prone individuals and coronary patients. G Ital Cardiol 1976; 6: 422-432.

96. Pollock ML, Gettman LR, Milesis CA, Bah MD, Durstine JL, Johnson RB. Effects of frequency and duration of training on attrition and incidence of injury. Med Sci Sports 1977; 9: 31-36.

97. Pollock ML. How much exercise is enough? Physician Sportsmed 1978; 8: 50-64.

98. Rechnitzer PA, Cunningham DA, Andrew GM, Buck CW, et al. Relation of exercise to the recurrence rate of myocardial infarction in men: Ontario exercise-heart collaborative study. Am J Cardiol 1983; 51: 65-69.

99. Roman O, Gutierrez M, Luksic I, Chavez E, Camuzzi AL, Villalon E, Klenner C, Cumsille F. Cardiac rehabilitation after acute myocardial infarction: 9-year controlled follow-up study. Cardiology 1983; 70: 223-231.

100. Rossouw JE, Lewis B, Rifkind BM. The value of lowering cholesterol after myocardial infarction. N Engl J Med 1990; 323: 1112-1119.

101. Russell MW, Huse DM, Drowns S, Hamel EC, Hartz SC. Direct medical costs of coronary artery disease in the United States. Am J Cardiol 1998; 81(9): 1110-1115.

102. Ryan TJ, Anderson JL, Antman EM, Braniff BA, Brooks NH, Califf RM, et al. ACC/AHA guidelines for the management of patients with acute myocardial infarction. J Am Coll Cardiol 1996; 28(5): 1328-1428.

103. Shaw LW for the project staff. Effects of a prescribed supervised exercise program on mortality and cardiovascular morbidity in patients after a myocardial infarction: The national exercise and heart disease project. Am J Cardiol 1981; 48: 39-46.

104. Sivarajan ES, Bruce RA, Almes MJ, Green B, Belanger L, Lindskog BD, Newton KM, Mansfield LW. In-hospital exercise after myocardial infarction does not improve treadmill performance. N Engl J Med 1981; 307: 357-362.

105. Sivarajan ES, Bruce RA, Lindskog BD, Almes MJ, et al. Treadmill test responses to an early exercise program after myocardial infarction: A randomized study. Circulation 1982; 65(7): 1420-1428.

106. Smith SC, Blair SN, Criqui MH, Fletcher GF, Fuster V, Gersh BJ, et al. Preventing heart attack and death in patients with coronary disease. J Am Coll Cardiol 1995; 26(1): 292-294.

107. Soriano JB, Hoes AW, Meems L, Grobbee DE. Increased survival with beta-blockers: Importance of ancillary properties. Prog Cardiovasc Dis 1997; 39(5): 445-456.

108. Sparrow D, Dawber TR, Colton T. The influence of cigarette smoking on prognosis after a first myocardial infarction. J Chronic Dis 1978; 31: 425-432.

109. Steffen-Batey L, Nichaman MZ, Goff DC, Frankowski RF, Hanis CL, Ramsey DJ, Labarthe DR. Change in level of physical activity and risk of all-cause mortality or reinfarction. The Corpus Christi Heart Project. Circulation 2000; 102: 2204-2209.

110. Stern MJ, Gorman PA, Kaslow P. The group counseling vs exercise therapy: A controlled intervention with subjects following myocardial infarction. Arch Intern Med 1983; 143: 1719-1725.

111. Stewart KJ, McFarland LD, Weinhofer JJ, Cottrell E, Brown CS, Shapiro EP. Safety and efficacy of weight training soon after acute myocardial infarction. J Cardiopulm Rehabil 1998; 18: 37-44.

112. Taylor CB, Houston-Miller N, Ahn DK, Haskell W, DeBusk RF. The effects of exercise training programs on psychosocial improvement in uncomplicated postmyocardial infarction patients. J Psychosom Res 1986; 30: 581-587.

113. The Israeli Sprint Study Group. Secondary Prevention Reinfarction Israeli Nifedipine Trial (SPRINT). A randomized intervention trial of nifedipine in patients with acute myocardial infarction. Eur Heart J 1988; 9(4): 354-364.

114. The MIAMI Trial Research Group. Metoprolol in Acute Myocardial Infarction (MIAMI). A randomised placebo-controlled international trial. Eur Heart J 1985; 6: 199-226.

115. Thompson PD. The cardiovascular complications of vigorous physical activity. Arch Intern Med 1996; 156: 2297-2302.

116. Tran ZV, Brammell HL. Effects of exercise training on serum lipid and lipoprotein levels in post-MI patients. A meta-analysis. J Cardiopulmonary Rehabil 1989; 9: 250-255.

117. Van Camp SP. Exercise-related sudden death: Risks and causes. Physician Sportsmed 1988; 16(5): 97-112.

118. Virmani R, Robinowitz M, Geer JC, Breslin PP, Beyer JC, McAllister HA. Coronary artery atherosclerosis revisited in Korean war combat casualties. Arch Pathol Lab Med 1987; 111(10): 972-976.

119. Waters D, Lesperance J, Galdstone P, et al. for the CCAIT study group. Effects of cigarette smoking on the angiographic evolution of coronary atherosclerosis—A Canadian coronary atherosclerosis intervention substudy. Circulation 1996; 94: 614-621.

120. Wilhelmsen L, Sanne H, Elmfeldt D, Grimby G, et al. A controlled trial of physical training after myocardial infarction: Effects on risk factors, nonfatal reinfarction, and death. Prev Med 1975; 4: 491-508.

121. Willich SN, Lewis M, Lowel H, Arntz HR, Shubert F, Schroder R. Physical exertion as a trigger of acute myocardial infarction. N Engl J Med; 329: 1684-1690, 1993.

122. Wong ND, Wilson PWF, Kannel WB. Serum cholesterol as a prognostic factor after myocardial infarction: The Framingham Study. Ann Intern Med 1991; 115: 687-693.

123. Worcester MC, Hare DL, Oliver RG, Reid MA, Goble AJ. Early programmes of high and low intensity exercise and quality of life after acute myocardial infarction. Br Med J 1993: 307: 1244-1247.

124. Giannuzzi P, Temporelli PL, Tavazzi L, Corra U, Gattone M, Imparato A, Giordano A, Schweiger C, Sala L, Malinverni C, et al. EAMI-exercise training in anterior myocardial infarction: An ongoing multicenter randomized study: Preliminary results on left ventricular function and remodeling. Chest 1992; 101(5 Suppl): 315S-321S.

125. Giannuzzi P, Tavazzi L, Temporelli PL, Corra U, Imparato A, Gattone M, Giordano A, Sala L, Schweigher C, Malinverni C. Long-term physical training and left ventricular remodeling after anterior myocardial infarction: Results of the exercise in anterior myocardial infarction (EAMI) trial. J Am Coll Cardiol 1993; 22: 1821-1829.

126. Foster C, Gal RA, Murphy P, Port SC, Schmidt DH. Left ventricular function during exercise testing and training. Med Sci Sports Exerc 1997; 29: 297-305.

127. Fragnoli-Munn K, Savage PD, Ades PA. Combined resistive aerobic training in older patients with coronary artery disease early after myocardial infarction. J Cardiopulm Rehabil 1998, 8: 416-420.

Timothy R. McConnell, PhD, FACSM, FAACVPR
Geisinger Medical Center
Penn State Geisinger Health System
Danville, PA

Charles A. Laubach, Jr, MD, FACC
Geisinger Medical Center
Penn State Geisinger Health System
Danville, PA

Revascularization of the Heart

When a person has coronary artery disease, many clinical procedures may be elected to restore blood flow throughout the myocardium. Over the past decade, referral patterns to cardiac rehabilitation for patients having coronary artery **revascularization** have evolved beyond patients who have undergone surgical revascularization—coronary artery bypass surgery. Today, because of advances in coronary invasive technology, cardiac rehabilitation programs must also prepare for a growing number of individuals who have experienced percutaneous transluminal coronary **angioplasty** alone and in combination with **stent** therapy, which involves the placement of a mesh tube along the artery wall to prevent reocclusion (36). Even though minor convalescent differences exist among the different percutaneous transluminal coronary revascularization procedures, the standards of practice for cardiac rehabilitation are similar; thus, this chapter focuses on those patients who have undergone coronary artery bypass surgery and percutaneous transluminal coronary angioplasty with or without stent therapy.

Coronary Artery Bypass Surgery

Coronary artery bypass surgery (CABS) involves revascularization by using a venous graft from an arm or leg or an arterial graft (both free—not intact such as a section—and from a regional intact native vessel, e.g., **internal mammary, gastroepiploic**) to provide blood flow to the myocardium beyond the site of the occluded or nearly occluded area in a coronary artery. Although CABS has traditionally involved a **sternotomy** and the use of a **heart and lung bypass,** technical advances now permit a growing number of procedures to be performed:

1. Procedures performed through small port incisions (or "mini-surgery") using microscopic procedures
2. Procedures performed with robotic technology
3. Surgery performed on the beating heart without the use of cardiopulmonary bypass

Subsequent to these technical advances, postoperative morbidity has significantly decreased and the postsurgical hospital stay for CABS patients without complications is now less than 5 days.

As a result of the evolution of revascularization procedures, the role for CABS has changed, now being reserved for the following patients:

1. Patients who are post–percutaneous transluminal coronary angioplasty (PTCA) with restenosis
2. Patients who are no longer candidates for angioplasty but still have target vessels offering preservation of left ventricular systolic function
3. Those with multivessel disease not amenable to angioplasty
4. Those with technically difficult vessel lesions, for example, on the curve of a vessel or in a **distal** location not readily amenable to angioplasty

Successful CABS results in an increased functional capacity, improved cycle ergometer or treadmill performance, variable improvements in left ventricular function, increased maximal heart rate, increased rate pressure product, and a reduction in ST-segment depression (18, 38, 59). Combined with medical therapy, CABS may more effectively relieve significant residual exercise-induced symptomatic or silent myocardial **ischemia** (57). Thus, the symptom relief, improved functional capacity, and improved **quality of life** may be the most practical and important patient benefits of CABS.

Percutaneous Transluminal Coronary Angioplasty

PTCA is a well-established, safe, and effective revascularization procedure for patients with symptoms attributable to coronary artery disease. It may use one or more techniques alone or in combination to open the vessel:

1. Balloon dilation is most used (figure 13.1).
2. Directional **atherectomy** may be used to debulk large **eccentric lesions** (figure 13.2).
3. Rotational atherectomy may be applied to central bulky lesions (figure 13.2).
4. Laser may be used to debulk eccentric or central lesions; however, the risk of vessel wall perforation or dissection may be greater.

The complications of angioplasty are acute vessel closure (rebound vasoconstriction) or chronic restenosis, thrombotic distal **embolism,** myocardial infarction (MI), arrhythmias, and bleeding (62).

Stent Therapy

To reduce the risk of acute closure and **restenosis** of coronary arteries after PTCA (27), several models of intracoronary stents have been advocated. Stent therapy is frequently used in conjunction with one of the previously described techniques to preserve the patency of the vessel. Stents are stainless steel mesh tube bridges that are advanced on the end of a balloon catheter, passed across the **culprit lesion,** and expanded. The stent, serving as a bridge to emergency CABS or as a permanent intravascular prosthesis, compresses the lesion resulting in an open vessel. Upon removal of the balloon catheter, the stent permanently remains in the coronary artery and is eventually covered with endothelium, becoming part of the luminal wall structure.

Clinical Considerations

The success rate of a revascularization procedure may be predicted, in part, based on the patient's disease status. The utility of each procedure and risk for reocclusion are discussed next.

Coronary Artery Bypass Surgery

Elective CABS improves the likelihood of long-term survival in patients who have the following:

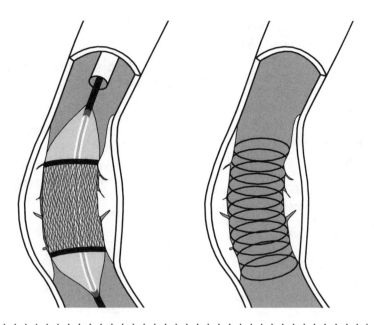

FIGURE 13.1   Percutaneous transluminal coronary angioplasty balloon catheter and two types of stents: latticed steel (left) and the coiled stent (right).

Reproduced with permission. T.E. Feldman. HOSPITAL PRACTICE *1998:33*(1)43. © 1998. The McGraw-Hill Companies. Illustration by Laura Duprey

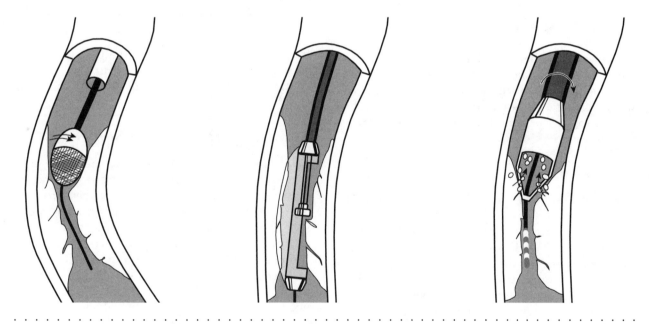

FIGURE 13.2   Intracoronary atherectomy procedures: rotational device (left), directional device (center), and extraction catheter (right).

Reproduced with permission. T.E. Feldman. HOSPITAL PRACTICE *1998:33*(1)43. © 1998. The McGraw-Hill Companies. Illustration by Laura Duprey

— Significant left main coronary artery disease

— Three-vessel disease

— Two-vessel disease with a proximal left anterior descending stenosis

— Two-vessel disease not amenable to PTCA and impaired left ventricular function (1)

For those experiencing failed angioplasty with persistent pain or hemodynamic instability, acute MI with persistent or recurrent ischemia **refractory** to medical therapy, **cardiogenic shock,** or failed PTCA with an area of myocardium still at risk, revascularization by CABS offers effective relief of angina pectoris and improves the patient's quality of life (1). The 1-year occlusion rate of the grafts is approximately 15% (45) with only a small additional annual occlusion rate for the ensuing years (31, 40). The CABS patient's postoperative education should include wound care, appropriate management of recurring symptoms, and risk factor modification.

Percutaneous Transluminal Coronary Angioplasty

In select cases of unstable angina, PTCA has an acute success rate of 84% to 90%. Following successful PTCA, restenosis occurs in approximately 25% of patients, almost always within the first 6 months (9, 25). There are several predictors of restenosis, listed here in order of importance:

1. Degree of acute residual stenosis after the PTCA is performed
2. The number of diseased vessels
3. The degree of reduction of the stenosis
4. The presence of acute dissection after PTCA
5. The presence of documented **variant angina** (16, 25, 29, 51)

Patients who have had PTCA in the setting of unstable angina should have close surveillance following hospital discharge and should be advised to seek prompt medical attention in the event of a recurrence of the symptoms that were occurring prior to their PTCA (16).

Stent Therapy

Although delivery of the stent has been 90% to 95% successful, acute closure and restenosis remain as serious limitations to both short- and long-term success (29, 61, 63). Because the major limitation of coronary stenting is the risk of stent **thrombosis,** chronic **anticoagulation therapy** is required that lowers the incidence of thrombosis and closure (47).

Acute closure rate has been reported to be 3% to 5% (10, 12). Restenosis rate ranges from 25% to 40% (29, 47, 63) with a 6-month **reocclusion** rate of 11% (39, 61). Predictors of stent restenosis are presented in table 13.1.

Graded Exercise Testing

The graded exercise test (GXT) is commonly used to establish functional status in patients who have undergone revascularization procedures. Although it is an integral component to the exercise program

TABLE 13.1
Predictors of Restenosis After PTCA/Stent
PTCA
— Degree of residual stenosis after PTCA — Number of diseased vessels — Degree of reduction of the stenosis — Presence or type of coronary dissection — Presence of documented variant angina
Stent
— Lesion eccentricity — Type of vessel stented (artery vs. vein) — Location of stent in vessel — Presence of multiple stents — Recurrence of unstable angina
All predictors are positively associated with risk for the revascularized vessel to reocclude. PTCA = percutaneous transluminal coronary angioplasty.

for all patients, the timing of the GXT is controversial. Standard administration procedures and contraindications to testing, discussed in chapter 6, should be followed.

Coronary Artery Bypass Surgery

Requiring all patients to have an exercise test after successful bypass surgery for the purpose of beginning a supervised and monitored exercise program is of questionable clinical benefit and an unnecessary financial burden (42). The patient's coronary anatomy is known, and unless there are surgical complications or postsurgical symptoms, the chance of detecting ischemia is extremely low. Also, because of the acute convalescent period, the patient may not be able to give a good physiological maximal effort, sacrificing test sensitivity.

A more opportune time for testing the patient is after incisional pain has resolved, blood volumes and hemoglobin concentrations have normalized, and skeletal muscular strength and endurance have improved from participating in low-level exercise and activity. At least 3 to 4 weeks postsurgery, the patient will be able to give a near-maximal physiological effort resulting in test results with greater diagnostic accuracy for assessing the patient's functional capacity, determining return to work status, or recommending the resumption of physically vigorous recreational activities. For patients whose surgical revascularization was not successful or who are experiencing symptoms suggestive of ischemia, a clinical exercise test before starting an exercise program is recommended. All testing procedures should follow professional guidelines (2-4) as noted in chapter 6.

Percutaneous Transluminal Coronary Angioplasty

Debate exists regarding the proper timing of stress testing in PTCA patients. Several reports of acute thrombotic occlusion associated with exercise testing shortly after successful PTCA have been described. Although no chest pain is reported during the test, ischemia within 1 hr after testing has been reported (11, 56). The mechanisms for the apparently abnormal test responses are unclear but are possibly related to the following:

1. Higher levels of **platelet aggregation** during exercise testing
2. An increase in **thromboxane** A2
3. Platelet activation and hyperreactivity increase during exercise
4. Increased arterial wall stress associated with increased coronary blood flow
5. The higher blood pressure that occurs during exercise, which may traumatize an already disrupted **intima** (11, 35, 43)

On the other hand, exercise testing of patients with PTCA has been accepted standard practice, particularly for those with incomplete revascularization (7). Gruentzig and coworkers (24) performed exercise testing within 2 days of successful angioplasty without complication. There is a large body of evidence supporting the use of early postprocedure exercise testing (1-2 days) to evaluate the functional status of the PTCA patient (8, 60).

Stent Therapy

Early testing after stent placement, especially if significant underlying **coronary dissection** is present, should be pursued with caution. If stress testing is clinically indicated during the early poststent period, vasodilator pharmacological stress (adenosine or dipyridamole) may be preferable to avoid the catecholamine elevations (and associated thrombogenicity) and the increased blood pressure associated with exercise (60).

As with CABS and PTCA, the need to test all patients after stent therapy before starting cardiac rehabilitation is debatable. For the successfully revascularized patient, an exercise test may be redundant and may not provide any further useful clinical information before the patient starts a supervised exercise program. It may be more prudent to forgo testing until the completion of cardiac rehabilitation, before return to work, or for the evaluation of symptoms or other suspected indications of restenosis.

For those patients who continue to experience symptoms after angioplasty or whose symptoms recur, then a diagnostic exercise test is recommended before starting or resuming exercise.

Opinions vary regarding the need for exercise testing of all revascularized patients before beginning a cardiac rehabilitation program. In the successfully revascularized, asymptomatic patient, a negative exercise test immediately following the procedure is likely. The patient may be too fatigued to give a good exercise effort but may be able to begin rehabilitation at a low level. The primary concerns with PTCA and stent are reocclusion and restenosis. Subsequent restenosis may not be detected immediately following the procedure. The patient can exercise in a supervised exercise program and be tested at a later date if symptoms recur or for assessment of functional capacity before return to work.

Exercise Prescription

Many body composition changes (loss of lean body mass) occur within the first few days of bed rest (64), supporting the need for early exercise intervention during hospital admission. Patients who perform typical ward activities and moderate, supervised ambulation do not suffer the magnitude of loss in lean body tissue as those who remain inactive. Early standing and low-level activities including range of motion and slow ambulation may be all that are required to deter postsurgical lean body mass loss while in the hospital (64).

After hospital discharge, many positive physiological adaptations occur in revascularized patients who participate in a supervised exercise program:

— Improved cardiac performance at rest and during exercise

— Improved exercise capacity of 20% to 25%

— Greater total work performed

— Improved angina-free exercise tolerance, much of which is attributable to peripheral muscular adaptations (5, 19, 23)

Patients in such a program gain the following:

— They more often achieve full working status.

— They have fewer hospital re-admissions.

— They are less likely to smoke at 6 months following completion of exercise therapy (5, 67).

When we compare the physiological and psychosocial outcomes between CABS, PTCA/stent, and MI patients at the beginning and completion of 12 weeks of cardiac rehabilitation, some group trends are apparent.

CABS patients may begin with lower functional capacities and lower ratings of quality of life and **self-efficacy** attributable to the surgical recuperative process, but they obtain greater improvements during the program and obtain similar or greater values than other cardiac patients at program completion. This may reflect lower rates of ischemia than MI patients, a greater confidence in their ability, and the potential psychological feeling that "something was done" about their heart disease and they are "cured." The PTCA/stent groups may have greater functional capacities and progress more rapidly than post-CABS and post-MI patients, including those PTCA patients who had an associated MI.

Exercise prescription for revascularization patients should follow the guidelines established by the American College of Sports Medicine (4), American Association of Cardiovascular and Pulmonary Rehabilitation (3), and American Heart Association (2). Special considerations include incisional discomfort, stability of sternum, and assurance of no incisional (sternum, arm, leg) infection. During the early phases of rehabilitation, it is important to assess the status of the incision and use caution when selecting activities for the patient.

Special Exercise Considerations

Although revascularized patients are equally knowledgeable about risk factors as post-MI patients, they are less compelled to make changes (20). Considerably more "large" lifestyle changes overall are initiated by post-MI patients than by revascularized patients (48). Patients undergoing revascularization may be less motivated to adhere to risk factor behavior change because of a perception of "being less sick or cured," which has a negative impact on compliance with risk factor modification (21, 32, 33, 65).

Revascularized patients may encounter, or anticipate, restrictions differently than other cardiac patients. The vast majority of PTCA patients are capable of resuming normal activities of daily living following hospital discharge. However, patients frequently perceive considerable restrictions after the procedure including all activities of daily living—leisure activities, sexual activity, and early return to work (14, 53, 58).

Depression remains prevalent in patients with coronary heart disease after major cardiac events (CABS and PTCA included). Cardiac rehabilitation does reduce the prevalence and severity of depression. Therefore, cardiac patients should be routinely screened and offered the benefits of comprehensive cardiac rehabilitation including psychosocial support and pastoral care (44).

Spouses may be more likely to seek information about the patient's psychological reactions and recovery, whereas patients are more likely to seek information about their physical condition and recovery (46). Patients tend to be more positive than spouses, who tend to be more fearful of the future (22). Also, patients and spouses differ in their views on the causes of coronary artery disease and about the responsibility for lifestyle changes and the management of health and stress (54). Therefore, it is important to assess both patient's and spouse's educational needs. Figure 13.3 lists important considerations in the rehabilitation of the post-PTCA patient.

Coronary Artery Bypass Surgery

One of the primary concerns for the CABS patient on entering outpatient cardiac rehabilitation is the state of incisional healing and sternal stability, **hypovolemia,** and low hemoglobin concentrations. During the initial patient interview, the rehabilitation professional needs to ensure that the surgical wound has no signs of infection, significant draining, or instability. Questions should focus on the following:

— Excessive or unusual soreness and stiffness

— Cracking or motion in the sternal region

— Whether the patient is sleeping at night

— How the patient's chest and leg incisions are responding to his or her current activities of daily living since discharge

Also, knowing how the patient performed during the inpatient program may help determine how soon he or she can begin the outpatient program and at what level the patient can begin exercising. For example, was the patient out of bed, upright, and walking soon after surgery without problems? If not, was it attributable to extreme physical discomfort, clinical or orthopedic difficulties, or a lack of motivation?

- Awareness of other cardiac risk factors
- Control of hypertension, obesity, and smoking
- Progressive exercise and weight reduction
- Counseling services for weight reduction, stress management, and smoking cessation
- Maintaining close contact between health professionals
- Identification of stressful factors
- Organizing and maintaining long-term follow-up records
- Reinforcing the noncurative nature of PTCA as a cardiac treatment modality.

Revascularization patients, and PTCA patients in particular, need to take a proactive approach to improve health outcomes.

FIGURE 13.3   Health care considerations for percutaneous transluminal coronary angioplasty (PTCA) patients.
Adapted from B. Gaw, 1992, "Motivation to change lifestyle following percutaneous transluminal coronary angioplasty," *Dimensions of Critical Care Nursing* 11: 68-74.

Historically, surgical patients did not begin cardiac rehabilitation for 4 to 6 weeks postsurgery or longer and avoided upper extremity exercise for even longer periods. Today, it is standard practice for patients to begin the outpatient program as soon as possible after discharge, many times within a week of surgery. For the uncomplicated revascularized patient, light upper extremity exercises are now prescribed, including range of motion exercises and light hand weights progressing to light resistive machinery and upper extremity ergometry.

Percutaneous Transluminal Coronary Angioplasty

The primary concern for the PTCA/stent patient is restenosis. At the patient's initial orientation session, questioning should address the presence of signs or symptoms indicative of their anginal equivalent. Education should include knowledge of symptoms, including anginal equivalents, management of angina (e.g., how to use nitroglycerin, going to the emergency department), precipitating factors (exertion or anxiety related), and care of the catheter insertion site. Patients with PTCAs and stents may begin the outpatient program as soon as they are discharged from the hospital or immediately following the procedure if it is performed on an outpatient basis.

Exercise training may alleviate the progression of coronary artery stenosis after PTCA by inhibiting smooth muscle cell proliferation, lowering serum lipids, and causing hemostatic changes (37). Aerobic training for 30 to 40 min, four to six times per week for 12 weeks, improves treadmill time and myocardial perfusion and reduces the restenosis rate at 3 months following PTCA (37). Furthermore,

there is improved sense of well-being, relief of depression, stress reduction, and sleep promotion (62).

As a result of angioplasty with improved techniques of revascularization, more patients with low-risk profiles are being referred to cardiac rehabilitation (i.e., patients with a greater exercise capacity, no evidence of ischemia, normal left ventricular function, and no arrhythmias). Specific examples include patients who are younger, have single vessel disease, and did not experience an MI before their PTCA. Regarding exercise prescription, these individuals may be treated similarly to apparently healthy individuals with the addition of education concerning the recognition of anginal equivalents, self-monitoring, self-care, and risk factor modification. Why do these lower risk patients need cardiac rehabilitation? Angioplasty patients commonly experience restenosis. Supervised exercise training and education will improve recognition of signs and symptoms associated with closure. Most importantly, angioplasty patients need instruction concerning appropriate exercise training, dietary modifications, medications, and general risk factor reduction to slow or reverse the coronary disease process (52, 55).

Because the PTCA patient remains on complete bed rest while the sheath is in situ for approximately 18 to 24 hr, the immobilization often causes back pain. Appropriate flexibility exercises that enhance ROM often help to resolve the low back pain.

Stent Therapy

Because stents are placed by using the same catheter procedure as in the PTCA, the same considerations exist. However, there is a greater risk for thrombosis following stent therapy. Consequently, patients are often placed on anticoagulant therapy for preventive purposes. Although there are no specific contraindications that would preclude exercise following recent stent placement, it is prudent to proceed with similar caution.

Exercise Recommendations

Over the past decade, the average length of hospital stay for cardiovascular patients has decreased and is still decreasing by more than one half day per year (52). Presently, the hospital stay for uncomplicated cases of CABS is usually 2 to 5 days and for PTCA/ stents 1 to 2 days. Even though cardiac rehabilitation begins as soon as possible during hospital admission, the shorter length of hospital stay has changed the inpatient program to basic range of motion exercises and ambulation, with the educational focus being on discharge planning—teaching about medications, home activities, and follow-up appointments. Educational topics previously covered in the inpatient setting are now the responsibility of the outpatient program. Moreover, it is imperative that cardiac rehabilitation professionals make every effort to enroll patients in an outpatient program. Practical application 13.1 details exercise prescription for the revascularized patient.

Cardiovascular Training

The initial exercise prescription is based on information gained from the patient's orientation interview to the outpatient cardiac rehabilitation program; patients are questioned concerning the presence of signs or symptoms, their activity while in the hospital, and their activity level since their return home from the hospital. Initially, the patients are closely observed and monitored to establish appropriate exercise intensities and durations that are within their tolerance. A starting program may include treadmill walking (5-10 min), cycle ergometry (5-10 min), combined arm and leg ergometry (5-10 min), and upper body ergometry (5 min). Initial intensities approximate 2 to 3 metabolic equivalents (METs; multiple of resting oxygen uptake of 3.5 ml $\cdot$ kg^{-1} $\cdot$ min^{-1}). The patient's heart rate, blood pressure, rating of perceived exertion (6), and signs and symptoms are monitored and recorded. Programs are gradually titrated during the initial sessions to a rating of perceived exertion of 11 to 14 in the absence of any abnormal signs or symptoms.

In general, exercise intensity is progressed by 0.5- to 1.0-MET increments (i.e., 0.5 mph or 2.0% grade on the treadmill or 12.5-25 W on the cycle) with the rate of progression based on the patient's symptoms, signs of overexertion, rating of perceived exertion, indications of any exercise-induced abnormalities, and prudent clinical judgment on the part of the cardiac rehabilitation staff. Those with greater exercise capacities (PTCA/stent patients with no MI) are started according to their exercise capacities and progressed more rapidly. The selection of exercise modality depends on the individual's program objectives. For example, those who are employed in a labor type occupation or perform a lot of upper extremity activities at home will spend a greater portion of their exercise time doing upper extremity exercises. If specific limitations preclude certain exercise modalities, program modifications are made that allow more time on tolerable equipment to obtain the greatest cardiovascular and muscular advantage.

For those patients who have an exercise test, standard recommended procedures for exercise prescriptions are followed (2-4).

Resistance Training

Muscular strength and endurance exercise training should be incorporated equally with cardiovascular endurance and flexibility exercise training during the early outpatient recovery period. Low-risk patients following revascularization can perform muscular strength and endurance exercise training safely and effectively (34). Depending on the patient's clinical and physical status, successful approaches for upper and lower extremity strength enhancement include 10 to 12 repetitions with a variety of types of equipment that may include elastic bands, hand weights, and various multistation machines. Usual guidelines include maintenance of regular breathing patterns (avoiding **Valsalva maneuvers**), selection of weights so that the last repetition of a set is moderately or somewhat hard, and progression when the perception of difficulty decreases. CABS patients may start range of motion exercises with light weights of 1 to 3 lb within 4 weeks of surgery as long as sternal stability is ensured and there is no excessive incisional discomfort. PTCA/stent patients may start resistance training immediately (26). Exercises should be selected that will strengthen muscle groups used during normal activities of daily living for lifting and carrying (1).

Weights are selected that allow the completion of 10 repetitions with the last three repetitions feeling moderately hard. Wrist weights with Velcro straps are used for those who cannot securely hold hand weights. Exercises are selected that use upper extremity muscle groups involved in routine lifting

Exercise Prescription for the Revascularized Patient

Training	Mode	Frequency	Intensity	Duration	Special considerations
Aerobic	Treadmill walking, cycling, combined arm and leg exercise, rowing, and stepper. Combination of above or others to ensure adequate use of major muscle groups and distribution between upper and lower extremities.	Daily	Asymptomatic (70-85% of HRmax; RPE 11-15) Symptomatic (below ischemic threshold; RPE 11-15)	At least 30 min. May be intermittent (e.g., 3 × 10 min) or continuous depending on patient tolerance.	Initially, need to be concerned with incisional discomfort in chest, arm, and leg of surgical patient. May need to restrict upper extremity exercises until soreness resolves. Also, for those with trans-catheter procedures, there may be some groin soreness at the catheter insertion site that may restrict certain physical movements.
Resistance	Elastic bands, hand weights, free weights, and multiple-station machines. Equipment selection is based on patient progress. A rational progression is the equipment in the order listed.	2 or 3 times per week	Select a weight where the last repetitions feel somewhat or moderately hard without inducing significant straining (bearing down and breath holding).	8-10 repetitions	For surgical patients, the initial upper extremity exercises may be range of motion without resistance—progressing initially with elastic bands or 1- to 3-lb increments. Further progression depends on sternal healing and stability. Exercises should be selected that use muscle groups involved in lifting and carrying.
Warm-up and cool-down exercises	Range of motion and flexibility exercises	Daily	Static stretching	5-10 min	Exercises should emphasize major muscle groups, especially lower back and posterior leg muscles.

HR = heart rate; RPE = rating of perceived exertion.

and carrying (figure 13.4). The lower extremity exercises are used for low functional capacity patients and are performed with weights that wrap around the ankle with Velcro straps. For low functional capacity patients who have difficulty with the exercises described, elastic bands of different thicknesses are used. Patients are progressed from hand weights to hydraulic resistance machines, again using resistances that result in a perception of difficulty of moderately hard for 2 or 3 sets of 10 to 12 repetitions.

The potential benefits of resistance training in the revascularized population include improving muscular strength and endurance and possibly attenuating the heart rate and blood pressure response to any given workload (less strain on the heart). A summary of resistance training is provided in figure 13.5.

Range of Motion

Each exercise sessions begins with a series of range of motion and flexibility exercises designed to main-

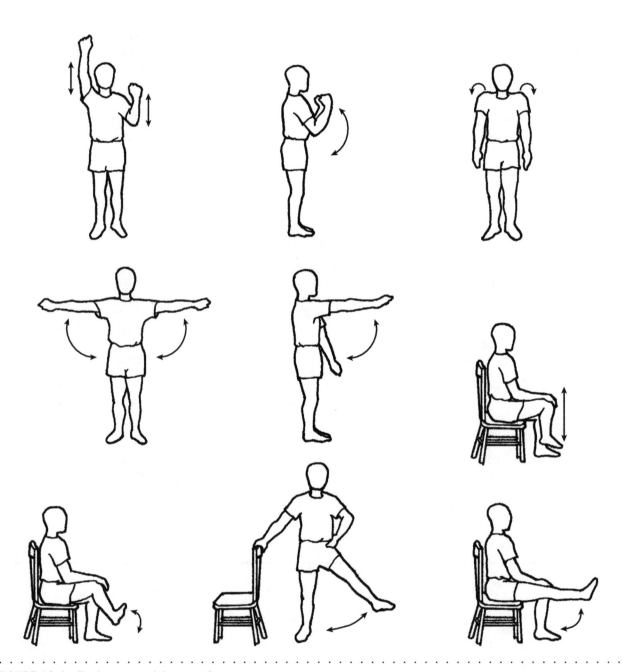

FIGURE 13.4　Resistance training exercises.

tain or improve the range of motion around joints and maintain or improve flexibility of major muscle groups (see figure 13.6). The exercises begin in the standing position (or seated in a chair, if difficulty standing) with the neck progressing downward to the shoulders and trunk and eventually to the lower extremities. The final seated stretches for the improvement of posterior leg muscles and lower back flexibility may be performed on the floor or in a chair for those with difficulty getting to the floor.

— Choose a weight that can be comfortably lifted for 12 to 15 repetitions.

— Avoid tight gripping during pushing, pulling, and lifting exercises.

— Do not hold breath during the activity. Exhale during the exertion phase and avoid straining.

— Perform two to three sets of each exercise and train three times per week.

— Rest 30 to 45 s between sets.

— Increase weight modestly (1-2 lb) after 15 to 20 repetitions of a given weight can be performed easily.

General resistance training guidelines for cardiac rehabilitation are presented. It is important to risk-stratify patients in order to determine eligibility.

FIGURE 13.5   Patient's guide for resistance training.

continued

FIGURE 13.6   An exercise for all seasons: Stretching.

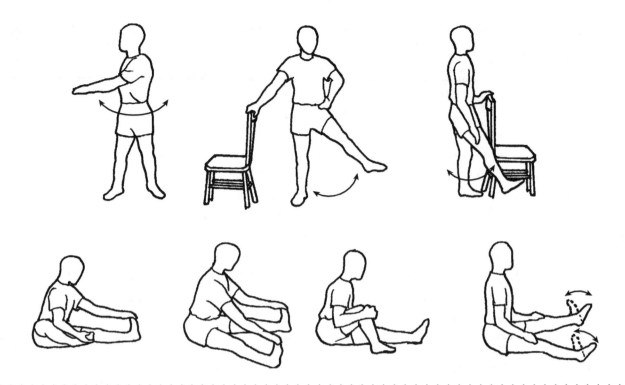

FIGURE 13.6   *(continued)*

Conclusion

Advances in coronary revascularization procedures and an aging population have led to a greater number of patients presenting for rehabilitation following CABS and PTCA with and without stent therapy. In addition to exercise programming, risk factor modification is essential to prevent reocclusion, particularly among PTCA patients. Furthermore, barring no new symptoms, the GXT may better serve its purpose of assessing functional status if it is postponed until later in the rehabilitation program. Provided that no untoward events occur over the course of rehabilitation, CABS and PTCA patients usually outperform their MI counterparts, achieving greater fitness improvements at a faster rate.

Case Study 13

Medical History

Patient: Mr. TW is a 65-year-old white male.

Procedures: Progress cardiac catheterization and off-pump two-vessel CABS.

History: Patient had two previous admissions within the last month for unstable angina.

Following PTCA and stent placement at the first admission, he had early recurrence of angina and an anterior-septal infarction. Having declined restudy, he had a trial on medical management. His clinical course was complicated by progressively limiting postinfarct angina.

Hospital Course

Cardiac catheterization demonstrated that the left anterior descending coronary artery had developed critical restenosis proximal to the stent. There was a long 60% stenosis attributable to

continued

dissection distal to the stent. Bypass surgery was advised. Two-vessel off-pump bypass was performed. His postoperative course was uneventful. He was discharged to home convalescence on postoperative day 3.

Diagnosis

Principle diagnosis: postinfarct angina

Secondary diagnosis: coronary artery disease, surgical procedures anteroseptal infarct, PTCA to the left anterior descending artery; diabetes mellitus; chronic obstructive pulmonary disease

Medications: Lopressor, Ecotrin, Albuterol inhaler, and Micronase with entrance into cardiac rehabilitation in 10 days

Exercise Prescription

Outpatient cardiac rehabilitation: Following CABS, the patient returned to cardiac rehabilitation 4 weeks postsurgery. The patient had an uncomplicated postsurgical course and had experienced no further symptoms since discharge. The patient was restarted in the program as follows.

Resting heart rate: 72
Resting blood pressure: 142/82

Initial Exercise Program:

Treadmill walking = 2.0 mph/0% grade for 10 min

Combined arm/leg ergometry = 100 W for 10 min

Upper body ergometry = 30 to 50 W for 5 min

Stepper = 6 METs for 5 min

Hydraulic resistance machine = 2 sets of 3 exercises for 10 repetitions

Patient completed 6 weeks in the program following surgery at the following workloads. Initially, 6 weeks were completed post-PTCA/stent therapy before surgery.

Treadmill walking = 2.5 mph/3% grade for 10 min

Combined arm/leg ergometry = 100 to 125 W for 10 min

Upper body ergometry = 100 W for 5 min

Rower = 50 to 75 W for 10 min

Hydraulic resistance machine = 2 sets of 3 exercises for 10 repetitions

Exercise heart rate: 90 to 96 beats $\cdot$ min^{-1}

Exercise rating of perceived exertion: 11

Remainder of program was uneventful. Patient completed a total of 12 weeks from his initial start and returned to his home walking program and activities of daily living.

Case Study 13 Discussion Questions

1. What differences may exist in the patient's exercise recommendations and educational guidelines for the first 6 weeks (pre-CABS) versus the latter 6 weeks (post-CABS)?

2. What kinds of exercise, if any, may need to be avoided after surgery and for how long?

3. At program completion, what type of home exercise would be recommended?

Glossary

angioplasty—Revascularization of a coronary artery or arteries by a balloon-tipped catheter that is inflated after being placed at the site of a coronary obstruction.

anticoagulation therapy—Pharmacological delaying or preventing of blood coagulation (clotting).

atherectomy—Procedure used for revascularization of an obstructed coronary artery consisting of a catheter tipped with either a metal burr that grinds a calcified atheroma (rotational atherectomy) or a rotating cup-shaped blade housed in a windowed cylinder that cuts or shaves the atheroma (directional atherectomy).

cardiogenic shock—Failure to maintain blood supply to the circulatory system and tissues because of inadequate cardiac output.

coronary dissection—Separation of tissue within the lining of a coronary artery.

culprit lesion—The primary obstruction responsible for decreased blood flow through a coronary artery.

distal—Away from the origin or centerline, as opposed to proximal.

eccentric lesion—A blockage that is equal distance away from the center of the artery—around the lining of the artery.

embolism—Obstruction of a blood vessel by foreign substances or a blood clot.

gastroepiploic—An artery with its origin in the stomach region used for coronary revascularization surgery.

heart and lung bypass—Device for maintaining the functions of the heart and lungs while either or both are unable to continue to function adequately.

hypovolemia—Diminished blood volume.

internal mammary—An artery with its origin in the chest region used for coronary revascularization surgery.

intima—Inner coat of a blood vessel.

ischemia—Local and temporary blood supply attributable to obstruction of the circulation.

MET—Metabolic equivalent of a particular physical activity expressed as multiples of resting energy expenditure.

platelet aggregation—Clumping together of platelets.

quality of life—Perception of life satisfaction.

refractory—Resilient or resistant to treatment.

reocclusion—To close again; reclosure.

restenosis—To become narrow or restricted again.

revascularization—Restoration of blood flow to a body part.

self-efficacy—Self-confidence.

stent—A stainless steel bridge, expanded by a balloon-tipped catheter, designed to hold open an area of stenosis within an artery.

sternotomy—The operation of cutting through the sternum.

thrombosis—The formation, development, or existence of a clot or thrombus within the vascular system.

thromboxane—Vasoconstrictor and platelet activation substance.

Valsalva maneuver—Expiration against a closed glottis resulting in an increased intrathoracic pressure and afterload on the heart.

variant angina—Angina pectoris occurring during rest; not necessarily preceded by exercise or an increase in heart rate.

References

1. ACC/AHA Guidelines for the Management of Patients with Acute Myocardial Infarction. A Report of the American College of Cardiology/American Heart Association Task Force on Practice Guidelines. *J Am Coll Cardiol* 1996; 28: 1328-1428.

2. AHA Medical/Scientific Statement, Position Statement. Cardiac rehabilitation programs: A statement for healthcare professionals from the American Heart Association. *Circulation* 1994; 90: 1602-1610.

3. American Association of Cardiovascular and Pulmonary Rehabilitation. *Guidelines for Cardiac Rehabilitation Programs.* 2nd ed. 1995; Champaign, IL: Human Kinetics.

4. American College of Sports Medicine. *ACSM's Guidelines for Exercise Testing and Prescription,* 6th ed. 2000; Baltimore: Lippincott, Williams & Wilkins.

5. Ben-Ari, E., J.J. Kellerman, E.Z. Fisman, A. Pines, B. Peled, and Y. Drory. Benefits of long-term physical training in patients after coronary artery bypass grafting—A 58-month follow-up and comparison with a non-trained group. *J Cardiopulm Rehabil* 1986; 6: 165-170.

6. Borg G. *Borg's Perceived Exertion and Pain Scales.* 1998; Champaign, IL: Human Kinetics.

7. Breisblatt, W.M., J.V. Barnes, F. Weiland, and L.J. Spaccavento. Incomplete revascularization in multivessel percutaneous transluminal coronary angioplasty: The role for stress thallium-201 imaging. *J Am Coll Cardiol* 1988; 11: 1183-1190.

8. Carlier, M., B. Meier, L. Finci, H. Karpuz, E. Nukta, and A. Righett. Early stress tests after successful coronary angioplasty. *Cardiology* 1993; 83: 339-344.

9. Cequier, A., R. Bonan, J. Crepeau, G. Cotte, P. De Guise, P. Joly, J. Lesperance, and D.D. Waters. Restenosis and progression of coronary atherosclerosis after coronary angioplasty. *J Am Coll Cardiol* 1988; 12: 49-55.

10. Cowley, M.J., A. Dorros, S.F. Kelsey, M. Van Raden, and K.M. Detre. Acute coronary complications associated with percutaneous transluminal angioplasty. *Am J Cardiol* 1984; 53: 12C-16C.

11. Dash, H. Delayed coronary occlusion after successful percutaneous transluminal coronary angioplasty: Association with exercise testing. *Am J Cardiol* 1982; 52: 1143-1144.

12. Ellis, S.G., G.S. Roubin, S.B. King, III, J.S. Douglas, W.S. Weintraum, R.G. Thomas, and W.R. Cox. Angiographic and clinical predictors of acute closure after native vessel coronary angioplasty. *Circulation* 1988; 77: 372-379.

13. Ewart, G.K., C.B. Taylor, L.B. Reese, and R.F. DeBusk. Effects of early postmyocardial infarction exercise test-

ing on self-perception and subsequent physical activity. *Am J Cardiol* 1983; 51: 1076-1080.

14. Faris, J., and N. Stotts. The effect of percutaneous transluminal coronary angioplasty on quality of life. *Prog Cardiovasc Nurs* 1990; 5: 132-140.

15. Feldman T.E., and R.F. Gunnar. Revascularization after CABS: Atherectomy and stenting. *Hosp Pract* 1998; 33: 43-64.

16. Foley, J.B., R.J. Chisholm, A.A. Common, A. Langer, and P.W. Armstrong. Aggressive clinical pattern of angina at restenosis following coronary angioplasty in unstable angina. *Am Heart J* 1992; 124: 1174-1180.

17. Foster, C., N.B. Oldridge, W. Dion, G. Forsyth, P. Grevenow, M. Hansen, J. Laughlin, C. Plichta, S. Rabas, R.E. Sharkey, and D.H. Schmidt. Time course of recovery during cardiac rehabilitation. *J Cardiopulm Rehabil* 1995; 15: 209-215.

18. Frick, M.H., P-T. Harjola, and M. Valle. Effect of aortocoronary grafts and native vessel patency in occurrence of angina pectoris after coronary bypass surgery. *Br Heart J* 1975; 37: 414-419.

19. Froelicher, V., D. Jensen, and M. Sullivan. A randomized trial of the effects of exercise training after coronary artery bypass surgery. *Arch Intern Med* 1985; 145: 689-692.

20. Gaw, B. Motivation to change lifestyle following percutaneous transluminal coronary angioplasty. *Dimens Crit Care Nurs* 1992; 11: 68-74.

21. Gaw-Ens, B., and G.P. Laing. Risk factor reduction behaviors in coronary angioplasty and myocardial infarction patients. *Can J Cardiovasc Nurs* 1994; 5: 4-12.

22. Gillis C. The family dimension of cardiovascular care. *Can J Cardiovasc Nurs* 1991; 2: 3-7.

23. Goodman, J.M., D.V. Pallandi, J.R. Reading, M.L. Plyley, P.P. Liu, and T. Kavanagh. Central and peripheral adaptations after 12 weeks of exercise training in post-coronary artery bypass surgery patients. *J Cardiopulm Rehabil* 1999; 19: 144-150.

24. Gruentzig, A.R., A. Senning, and W.E. Siegenthaler. Nonoperative dilatation of coronary artery stenosis: Percutaneous transluminal coronary angioplasty. *N Engl J Med* 1979; 301: 61-68.

25. Guiteras Val, P., M.G. Bourassa, P.R. David, R. Bonan, J. Crepeau, I. Dyrda, and J. Lesperance. Restenosis after successful percutaneous transluminal coronary angioplasty: The Montreal Heart Institute Experience. Presentation 1987; 60: 50B-55B.

26. Haennel, R.G., H.A. Quinney, and C.T. Kappagoda. Effects of hydraulic circuit training following coronary artery bypass surgery. *Med Sci Sports Exerc* 1991; 23: 158-165.

27. Hannan, E.L., M.J. Racz, D.T. Arani, B.D. McCallister, G. Walford, and T.J. Ryan. A comparison of short- and long-term outcomes for balloon angioplasty and coronary stent placement. *J Am Coll Cardiol* 2000; 36: 395-403.

28. Hillers, T.K., G.H. Guyatt, N.B. Oldridge, J. Crowe, A. Willan. L. Griffith, and D. Feeney. Quality of life after myocardial infarction. *J Clin Epidemiol* 1994; 47(11): 1287-1296.

29. Holmes, D.R., Jr., R.E. Vlietstra, H.C. Smith, G.W. Vetrovec, M.J. Cowley, D.P. Faxon, A.R. Gruentzig, S.F. Kelsey, K.M. Detre. M.J. Van Raden, and M.B. Mock. Restenosis after percutaneous transluminal coronary angioplasty (PTCA): A report from the PTCA registry on the National Heart, Lung, and Blood Institute. *Am J Cardiol* 1984; 53(Suppl.): 77C-81C.

30. Information and Research Exchange. Exercise for recent heart disease/surgery. *Nurse Practitioner* 1997; 22 (15): 232-235.

31. Irscoitz, S.B., D.R. Redwood, and L.E. Grauer. Long-term durability of patent saphenous vein aorto-coronary bypass grafts. *Am J Cardiol* 1974; 33: 146.

32. Jenkins, N., and L. Kotra-Ottoboni. Patient care aspects of percutaneous transluminal coronary angioplasty. In *Coronary Angioplasty*. 2nd ed. Ed. D. Clark. 1991; New York: Wiley-Liss.

33. Jensen K., L. Banwart, R. Venhaus, S. Popkess-Vawter, and S. Perkins. Advanced rehabilitation nursing care of coronary angioplasty patients using self-efficacy theory. *J Adv Nurs* 1993; 18: 926-931.

34. Kelemen, M.H. Resistive training safety and assessment guidelines for cardiac and coronary prone patients. *Med Sci Sports Exerc* 1989; 21: 675-677.

35. Kestin, A.S., P.A. Ellis, M.R. Barnard, A. Errichetti, B.A. Rosner, and A.D. Michelson. Effects of strenuous exercise on platelet activation state and reactivity. *Circulation* 1993; 88: 1502-1511.

36. King, S.B., III. Interventions in cardiology: What does and does not work. *Am J Cardiol* 2000; 86: 3H-5H.

37. Kubo, H., H. Hirai, and K. Machii. Exercise training and the prevention of restenosis after percutaneous transluminal coronary angioplasty (PTCA). *Ann Acad Med* 1992; 21: 42-46.

38. Lapin, E.S., J.A. Murray, R.A. Bruce, and L. Winterschield. Changes in maximal exercise performance in evaluation of saphenous vein bypass surgery. *Circulation* 1973; 47: 1164-1173.

39. Leon, M.B., Y. Almagor, R. Erbel, P.S. Teirstein, J. Perez, and R.A. Schatz. Subacute thrombotic events after coronary stent placement: Clinical spectrum and predictive factors. *Circulation* 1989; 80: II-174.

40. Lesperance, J., M.G. Bourassa, J. Saltiel, L. Campeau, and C.M. Grondin. Angiographic changes in aortocoronary vein grafts. Lack of progression beyond the first year. *Circulation* 1973; 48: 633.

41. Levine, M.J., B.M. Leonard, J.A. Burke, I.D. Nash, R.D. Safian, D.J. Diver, and D.S. Baim. Clinical and angiographic results of balloon-expandable intracoronary stents in right coronary artery stenoses. *J Am Coll Cardiol* 1990; 16: 332-339.

42. McConnell, T.R., T.A. Klinger, J.K. Gardner, C.A. Laubach, Jr., C.E. Herman, and C.A. Hauck. Cardiac

rehabilitation without exercise tests for post-myocardial infarction and post-bypass surgery patients. *J Cardiopulm Rehabil* 1998; 18: 458-463.

43. Mehta, J., P. Mehta, and C. Horalek. The significance of platelet-vessel wall prostaglandin equilibrium during exercise-induced stress. *Am Heart J* 1983; 105: 895-904.

44. Milani, R.V., C.J. Lavie, and M.M. Cassidy. Effects of cardiac rehabilitation and exercise training programs on depression in patients after major coronary events. *Am Heart J* 1996; 132: 726-732.

45. Morris, G.C., G.J. Reul, and J.F. Howell. Follow-up results of distal coronary artery bypass for ischemic heart disease. *Am J Cardiol* 1972; 29: 180.

46. Moser D.K., K.A. Dracup, and C. Marsden. Needs of recovering cardiac patients and their spouses compared views. *Int J Nurs Stud* 1993; 30: 105-114.

47. Nath, F.C., D.W.M. Muller, S.G. Ellis, U. Rosenschein, A. Chapekis, L. Quain, C. Zimmerman, and E.J. Topol. Thrombosis of a flexible coil coronary stent: Frequency, predictors and clinical outcome. *J Am Coll Cardiol* 1993; 21: 622-627.

48. Newton, K., E. Sivarajan, and J. Clarke. Patient perceptions of risk factor changes and cardiac rehabilitation outcomes after myocardial infarction. *J Cardiac Rehabil* 1985; 5: 159-168.

49. Oldridge, N., G. Guyatt, N. Jones, J. Crowe, J. Singer, D. Feeney, R. McKelvie, J. Runions, D. Streiner, and G. Torrance. Effects on quality of life with comprehensive rehabilitation after acute myocardial infarction. *Am J Cardiol* 1991; 67: 1084-1089.

50. Oldridge, N.B., and B.L. Rogowski. Self-efficacy and inpatient cardiac rehabilitation. *Am J Cardiol* 1990; 66: 362-365.

51. Oxford, J.L., A.P. Selwyn, P. Ganz, J.J. Popma, and C. Rogers. The comparative pathobiology of atherosclerosis and restenosis. *Am J Cardiol* 2000; 86(Suppl): 6H-11H.

52. Pashkow, F. Cardiac rehabilitation: Not just exercise anymore. *Cleve Clin J Med* 1996; 63: 116-123.

53. Pashkow, F.J. Issues in contemporary cardiac rehabilitation. *J Am Coll Cardiol* 1993; 21: 822-834.

54. Patterson, J.M. 1989. Illness beliefs as a factor in patient-spouse adaptation to treatment for coronary artery disease. *Fam Sys Med* 1989; 7: 428-443.

55. Popma, J.J., M. Sawyer, A.P. Selwyn, and S. Kinlay. Lipid-lowering therapy after coronary revascularization. *Am J Cardiol* 2000; 86(Suppl): 18H-28H.

56. Przybojewski, J.Z., and H.F.H. Welch. Acute coronary thrombus formation after stress testing following percutaneous transluminal coronary angioplasty: A case report. *S Afr Med J* 1985; 67: 378-382.

57. Rahimtoola, S.H. Postoperative exercise response in the evaluation of the physiologic status after coronary bypass surgery. *Circulation* 1982; 65: II-106–II-114.

58. Robertson, D., and C. Keller. Relationships among health beliefs, self-efficacy, and exercise adherence in patients with coronary artery disease. *Heart Lung* 1992; 21: 56-63.

59. Roskamm, H., A. Weisswange, C.H. Hahn, K.W. Jauch, M. Schmuziger, J. Peterson, P. Rentrop, and K. Schnellbacher. Hemodynamics at rest and during exercise in 222 patients with coronary heart disease before and after aorto-coronary bypass surgery. *Cardiology* 1977; 62: 247-260.

60. Samuels, B., J. Schumann, H. Kiat, J. Friedman, and D.S. Berman. Acute stent thrombosis associated with exercise testing after successful percutaneous transluminal angioplasty. *Am Heart J* 1995; 130: 1210-1222.

61. Schatz, R.A., D.S. Bain, M. Leon, S.G. Ellis, S. Goldberg, J.W. Hirshfeld, M.W. Cleman, H.S. Cabin, C. Walker, J. Stagg, M. Buchbinder, P.S. Teirstein, E.J. Topol, M. Savage, J.A. Perez, R.C. Curry, H. Whitworth, J.E. Sousa, F. Tio, Y. Almagor, R. Ponder, I.M. Penn, B. Leonard, S.L. Levine, R.D. Fish, and J.C. Palmaz. Clinical experience with the Palmaz-Schatz coronary stent. *Circulation* 1991; 83: 148-161.

62. Scriver, V.S., J. Crowe, A. Wilkinson, and C. Meadowcroft. A randomized controlled trial of the effectiveness of exercise and/or alternating air mattress in the control of back pain after percutaneous transluminal coronary angioplasty. *Heart Lung* 1994; 23: 308-316.

63. Serruys, P.W., H.E. Luijten, K.J. Beatt, R. Geuskens, P.J. de Feyter, M. von den Brand, J.H. Reiber, H.J. ten Katen, G.A. van Es, and P.G. Hugenholtz. Incidence of restenosis after successful coronary angioplasty: A time-related phenomenon. A quantitative angiographic study in 342 consecutive patients at 1, 2, 3, and 4 months. *Circulation* 1988; 77: 361-371.

64. Shaw, D.K., D.T. Deutsch, P.M. Schall, and R.J. Bowling. Physical activity and lean body mass loss following coronary artery bypass graft surgery. *J Sports Med Phys Fitness* 1991; 31: 67-74.

65. Shaw, R.E., F. Cohen, J. Fishman-Rosen, M.C. Murphy, S.H. Startzer, D.A. Clark, and R.K. Myler. Psychologic predictors of psychosocial and medical outcomes in patients undergoing coronary angioplasty. *Psychosom Med* 1986; 48: 582-597.

66. Tooth, L., and K. McKenna. Cardiac teaching: Application to patients undergoing coronary angioplasty and their partners. *Patient Educ Couns* 1995; 25: 1-8.

67. Waites, T.F., E.W. Watt, and G.F. Fletcher. Comparative functional and physiologic status of active and drop-out coronary bypass patients of a rehabilitation program. *Am J Cardiol* 1982; 51: 1087-1090.

14

Kimberly A. Bonzheim, MSA, FACSM
William Beaumont Hospital
Cardiac Rehabilitation and Exercise Laboratories
Royal Oak, MI

Kimberly A. Skelding, MD
William Beaumont Hospital
Cardiac Rehabilitation and Exercise Laboratories
Royal Oak, MI

Valvular Heart Disease

Valvular heart disease in adults is multifactorial and can be attributed to previous rheumatic fever, degenerative changes, congenital abnormalities, infection, and myocardial ischemia, among other causes. Valvular stenosis obstructs blood flow, and the chamber behind the narrow valve must produce extra work to sustain cardiac output. Stenosis develops or progresses gradually because the normal valve orifice is initially larger than necessary; therefore, the stenosis is usually severe before exercise symptoms occur. Valve leakage, also known as regurgitation, incompetence, or insufficiency, becomes problematic as the leakage backward becomes larger leading to less blood moving forward through the heart to the body. Leakage can be the result of ventricular dilation or abnormalities of the valve structure itself. It can be a chronic process or can occur acutely secondary to trauma to the valve or ischemia to one of its components. However, most valvular disease is mixed, with a degree of both regurgitation and stenosis. The circulatory abnormalities in valvular heart disease at rest are usually exacerbated with exercise.

During exercise, large amounts of blood are returned to the heart from the periphery. Even in mild valvular disease, symptoms that are nonexistent at rest or with daily activities can become quite severe with near-maximal exercise. In patients with severe aortic stenosis or regurgitation, acute left ventricular failure followed by acute pulmonary edema can result from exercise. In those with mitral disease, backflow of blood into the lungs can cause serious or even lethal pulmonary edema within minutes. Patients with mild valvular disease who do not have the normal increase in cardiac output during exercise can experience early overall muscle fatigue attributable to inadequate oxygen availability in the blood.

Scope

The near disappearance of acute rheumatic fever in the developed world, associated with an increase in life expectancy, has resulted in degenerative etiologies becoming the most common causes of valvular heart disease, with a dominance of aortic stenosis and mitral incompetence (25). The older average age of patients undergoing surgical repair explains the increasing concerns for comorbidities and the higher number of mixed (valvular and coronary artery bypass) surgeries. The improvement in surgical and interventional methods has widened the operative indications that are now, in certain circumstances, even considered for asymptomatic patients (11, 25).

The etiology of valvular heart disease in Western countries is strikingly different from that in developing countries, which have a greater incidence of rheumatic fever. This disease affects patients earlier, can decrease their life expectancy, and can prevent them from leading a normal life. Rheumatic heart disease continues to be a significant challenge for the World Health Organization.

Physiology and Pathophysiology

Critical to understanding valvular heart disease on a physiological, pathophysiological, and treatment basis is an understanding of the anatomy of the heart valves and their relation to heart function. Each heart valve has a unique architecture and consequent function. In general, heart valves are classified into one of two groups: **atrioventricular (A-V) valves** or **semilunar valves.**

The A-V valves, as described by their name, lie between the atria and the ventricles. The **tricuspid valve** is positioned on the right side of the heart, and the **mitral valve** is on the left. The blood flow-

ing between these two chambers normally is not under high velocity or flow. These valves are thin and filmy and can be easily damaged. They open and close passively, dependent on the blood flow and pressure gradients between the atrium and the ventricles during diastole. Papillary muscles, contiguous with the ventricular wall, are attached to the two cusps of the mitral valve and the three cusps of the tricuspid valve by string-like fibers called chordae tendineae. The function of the chordae is to pull the cusps inward toward the ventricles when closing and to prevent them from bulging too far into the atrium during ventricular contraction, as shown in figure 14.1.

The three half-moon-shaped thicker leaflets of the semilunar valves are exposed to a much higher velocity of blood flow than the A-V valves. Forceful ventricular contractions drive blood through the relatively small valvular opening of the **pulmonary valve** from the right ventricle into pulmonary circulation, and through the **aortic valve**, subject to the forceful contraction of the left ventricle, into systemic circulation. This high pressure is needed, as the pressure of the systemic circulation is tremendous compared with the pulmonary circulation in healthy adults. The high blood velocity through both these A-V valves decreases the chance of vegetations developing on the cusps in the setting of

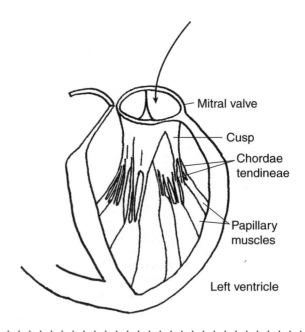

FIGURE 14.1   The anatomy of an atrioventricular valve highlighting the chordae tendineae that prevent the valve from bulging too far backward into the atria with a ventricular contraction.

Courtesy of Wm. Beaumont Hospital

bacteremia but increases the degree of wear and tear, especially to the leaflet edges.

The cross-sectional area of each valve is important in defining normal valve function. A decrease in cross-sectional area will increase the pressure gradient across the valve and increase the workload of the preceding chamber. The cross-sectional area and pressure gradients are delineated through **two-dimensional (2-D) echocardiography** and **Doppler echocardiography.**

The pathway of blood flow through the heart is germane to the understanding of valvular physiology. As shown in figure 14.2, deoxygenated venous blood returns from the periphery via the inferior and superior vena cavae. Blood first enters the right atrium. As the right atrium fills, the pressure exceeds the pressure in the right ventricle, and the tricuspid valve opens, allowing blood to flow into the right ventricle. The right ventricle, through active and passive filling which is calcium dependent, develops increased intracavitary pressure and the tricuspid valve is pulled shut, stopping the flow of blood. With the tricuspid valve shut, the right ventricle (systole) contracts, increasing the pressure in the right ventricle and forcing the pulmonary valve to open. The deoxygenated blood then flows into the pulmonary arteries and to the lungs for oxygenation. The now higher pressure in the pulmonary artery causes the pulmonary valve to snap shut to prevent blood from flowing back into the right ventricle.

Blood returning from the lungs via the pulmonic veins collects in the left atrium during systole when the mitral valve is closed, and the pressure increases with increased filling. This pressure increase in the left atrium easily opens the mitral valve, and the left ventricle begins to fill during diastole. As the blood flows from the left atrium to the left ventricle, the mitral valve begins to close. Just before the mitral valve closes, the left atrium contracts and forces another bolus of blood into the left ventricle, also known as the atrial kick. This atrial kick is lost in atrial fibrillation when the atrium does not contract in an organized manner and leads to decreased exercise tolerance and fatigue, especially in older adults. With the mitral valve closed, the pressure in the left ventricle increases with the onset of ventricular systole and the aortic valve is forced open to release the oxygenated blood into the arterial bed. To prevent backflow into the left ventricle, the aortic valve clamps shut during diastole.

Clinical Considerations

This chapter focuses on the mitral and aortic valves, because they are the most commonly affected by valvular abnormalities. Tricuspid and pulmonic valve disease is rare and seen only in select patient populations (e.g., intravenous drug users, children with congenital heart disease, pulmonary disease patients), and thus it is not discussed in this chapter.

Mitral Stenosis

Mitral stenosis is identified when the cross-sectional area of the mitral valve is less than 4 to 6 cm². In mild mitral stenosis, the valve area is greater than 2 cm², resulting in an increased stress on the left atrium to propel blood across this orifice. As the orifice decreases to less than 1 cm², critical mitral stenosis is diagnosed, and a pressure gradient of 20 cm² is required to move blood through the valve. The pressure required in the left atrium to propel this blood forward is great and is often transmitted to the pulmonary veins and capillaries, leading to symptoms.

Mitral stenosis is most often the result of previous **rheumatic fever,** an autoimmune disease that

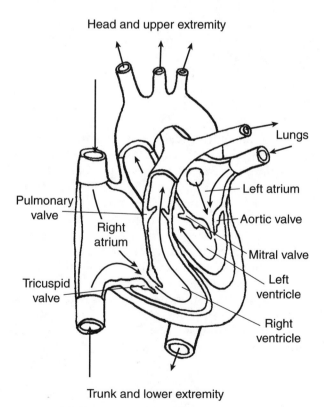

Head and upper extremity

Lungs

Pulmonary valve

Right atrium

Left atrium

Aortic valve

Mitral valve

Left ventricle

Tricuspid valve

Right ventricle

Trunk and lower extremity

FIGURE 14.2   Blood flow through the heart valves and chambers.

can cause large hemorrhagic, fibrinous, bulbous lesions to develop along the susceptible edges of the valve leaflets (30). These acute lesions will frequently become stuck together and form scar tissue over time that permanently fuses the valve leaflets together, impeding blood flow or preventing the valve from closing completely. In addition, mitral stenosis can be linked to congenital, degenerative, and inflammatory causes as well as rare occurrences secondary to genetic influence (38). Interestingly, any impedance to flow from the left atrium to the left ventricle (i.e., tumor, thrombus) can simulate mitral stenosis in symptoms and hemodynamics (45).

Signs and Symptoms

The most common presenting symptom is exertional dyspnea. This occurs secondary to increased pulmonary pressures, which are a result of elevated left atrium pressure required to propel the blood across the stenotic valve. In addition to exertion, episodes of atrial fibrillation, emotional stress, or a high output state during infection can trigger dyspnea.

Hemoptysis has been heralded as a classic finding in mitral stenosis. Abrupt bleeding occurs secondary to the rupture of small pulmonary veins that are engorged secondary to a transmitted elevated left atrial pressure (37). Frothy sputum and blood-tinged sputum are also seen. Patients can also present with chest pain, embolic events, or hoarseness caused by an enlarged left atrium compressing the left recurrent laryngeal nerve (40).

Diagnostic and Laboratory Evaluations

Two-dimensional (2-D) echocardiography is the most accurate way to evaluate the valve orifice size, physical appearance of the valve, and abnormal valvular motion. Subjective evaluation of left atrium size, valvular movement, and doming of the valve can also be achieved through this modality (34). The appearance of a stenotic mitral valve is often described as "fish mouth" appearing on echocardiography or intraoperatively, and there is marked thickening and fibrosis of the cusps and chordae. Associated Doppler imaging evaluates the transvalvular gradient from which the degree of stenosis can be calculated. Improved images can be obtained by using transesophageal echocardiography (see practical application 14.1) with the added ability to evaluate for a thrombus in the left atrium.

Electrocardiogram (ECG) characteristics evident in moderate to severe mitral stenosis include evidence of left atrial enlargement and a vertical axis. In severe cases, right ventricular hypertrophy, caused

PRACTICAL APPLICATION 14.1

Transesophageal Echocardiogram

A **transesophageal echocardiogram (TEE)** is a more sensitive test for identifying valvular abnormalities. It is typically recommended after a surface echo has been completed and the data are considered insufficient. The procedure requires the patient to swallow an endoscopic probe while under conscious sedation. The imaging probe is then positioned in the esophagus next to the heart. This higher frequency imaging probe provides better resolution of valvular structure, valve detail, left ventricular size, and function. A TEE provides better windows for imaging if the body habitus of a patient does not allow for an adequate view from the surface. A TEE is also useful to assess a mitral valve stenosis patient before valvuloplasty to better view potential clots in the left atrium, which can lead to significant mortality and morbidity rates. A TEE is a somewhat invasive procedure with potential risk of aspiration, esophageal perforation, and hemodynamic compromise.

from the transmittal of elevated pressures from the left atrium backward, can also be seen. On X-ray evaluation, the frontal view is often normal with abnormalities seen only on the lateral view. These abnormalities can include left atrial, right atrial, right ventricular, and pulmonary artery enlargement.

With the advent of surgical and percutaneous intervention, the natural history of mitral stenosis is rarely seen. From historical data, there is thought to be an asymptomatic time interval of 15 to 20 years to the onset of symptoms. After the onset of symptoms, the progression to severe symptoms occurs within approximately 3 years (32).

Treatment

Once identified, mitral stenosis should be followed closely because mild symptoms can continue for a protracted period of time. However, as symptoms begin to progress, invasive treatment strategies should be considered because typically a quick decline in function occurs in the setting of moderate disease (mitral valve orifice <1.5 cm^2). Surgical treatment strategies include open valvulotomy, in which the chest is opened to expose the heart and incise the valve orifice to enlarge the valve area under direct visualization. In the United States, valve replacement is more often performed because the patients are older, the valve annulus is usually calcified, and there is often a second valvular abnormality of the mitral valve or a different valve. Balloon valvuloplasty is a nonsurgical option with lower mortality rates, morbidity rates, and cost than the surgical options. In this procedure, a balloon catheter is advanced across the intra-atrial septum into the valve orifice and expanded, mechanically opening the valve orifice. Balloon valvuloplasty as well as open valvulotomy are palliative measures only. Coronary angiography is only performed if clinically indicated prior to surgical intervention to delineate the degree of coronary artery disease.

Medical therapy for mitral stenosis is multifactorial. Symptomatic relief as well as prophylaxis must be instituted. Symptomatically, patients improve with the use of diuretics and restriction of dietary sodium. In those patients with atrial fibrillation, aggressive rate control should be attempted with β-blockade plus digoxin, as often the left atrium size is enlarged and maintenance of normal sinus rhythm is unlikely. These patients are very sensitive to an increase in heart rate (HR) because they are dependent on a sufficient time for ventricular filling or diastole. Patients in atrial fibrillation or those with a history of embolic events should also receive oral anticoagulation to prevent future events. Antibiotic prophylaxis should be instituted in all patients with a history of rheumatic heart disease or evidence of mitral stenosis.

Mitral Regurgitation

Mitral regurgitation (MR) occurs when the valvular apparatus, the leaflets, papillary muscles, or chordal structures do not allow the valve cusps to close properly. Therefore, during left ventricular systole, there is blood flow both forward and backward through the partially opened mitral valve. This backward flow decreases cardiac output because a portion of the stroke volume is propelled into the left atrium, increasing left atrial pressure, which is transmitted backward through the pulmonary bed.

MR, like mitral stenosis, can be a result of dysfunction in any part of the valvular structure including the leaflets that can be damaged by rheumatic heart disease. The leaflets thicken, stiffen, become fibrotic, and are unable to coapt, leading to valvular leak. Acute MR can be the result of destruction of the leaflets secondary to infectious endocarditis or traumatic damage to the valve via a penetrating wound or blunt trauma, which can distort the valvular structure and cause acute MR.

Signs and Symptoms

Disease progression is slower in MR than in mitral stenosis. Patients usually only become symptomatic at the end stage of this disease process. Early symptoms may be very nonspecific and include weakness and fatigue attributable to low cardiac output. As the left ventricle fails, the patient can become more susceptible to atrial fibrillation and episodes of pulmonary edema as the left atrium size increase.

In contrast, patients with acute MR are often quickly symptomatic because they have not had the opportunity to compensate for the abruptly elevated left atrial pressure and often have acute pulmonary edema as well as symptoms of right heart failure (i.e., abdominal discomfort secondary to a congested liver, peripheral edema).

Diagnostic and Laboratory Evaluations

Echocardiography is often performed after a systolic heart murmur is heard at the base of the heart, and yearly thereafter. Two-dimensional echocardiography will help identify the cause of MR in many cases. Annulus calcification, vegetations, chordal structures, and often even papillary muscles can be evaluated with this modality. It is important to quantify the left atrium and ventricle sizes, which, if greatly enlarged, can suggest that the regurgitation is severe.

Left ventricular function can help with prognosis and determining surgical timing. It is important to realize that ejection fraction (EF) in the presence of MR may falsely appear normal. EF represents the percentage of blood ejected from the left ventricle; in the case of MR, blood is pumped forward through the aortic valve and backward through the diseased mitral valve. Therefore, an MR patient having a normal EF determined by echocardiography actually has moderate impairment of the left ventricle as part of the EF goes backward. Accordingly, a patient with an EF of even 40% to 45% has a severely impaired left ventricle when associated with MR. Doppler evaluation of the valve can determine severity of disease. If the mitral jet is greater than 8 cm^2, the MR is severe. A transesophageal echocardiogram (TEE) is preferred for evaluating the mitral valve and is considered the gold standard for presurgical evaluation.

Patients with mild MR can remain asymptomatic and stable for years (43). Regurgitation progresses, sometimes rapidly, in patients who develop other maladies such as infectious endocarditis or additional valvular disease or those with a genetic predisposition for collagen vascular disease. Severe MR leads to left ventricular dysfunction and failure. Its progression depends on comorbid conditions, degree of regurgitation, and premorbid left ventricular function.

Treatment

Acute MR does not allow the left atrium and ventricle to compensate gradually over time; therefore, patients are extremely ill and require urgent treatment. Patients with acute MR often present with respiratory distress secondary to pulmonary edema and need respiratory support, diuresis, and acute vasodilatory therapy with intravenous nitroprusside.

Chronic MR also benefits from afterload reduction, as in vasodilator therapy with angiotensin-converting enzyme inhibitors or hydralazine (41). Essentially, any treatment that can decrease the resistance to forward flow will make the patient feel better. Increasing forward flow will, over time, decrease left atrial pressure, decrease left ventricular volume, and improve cardiac output to some extent.

Further pathology resulting from chronic MR such as atrial fibrillation, embolic events, infective endocarditis, or congestive heart failure should be treated accordingly. Once the evaluation of MR has indicated that surgical intervention is necessary, the decision to repair or replace the valve is made. Ideally, it is more beneficial to repair the valve so that the subvalvular apparatus (i.e., papillary muscles, chordae) remains intact and can function normally. Older patients typically have a calcified, degenerative valve structure and more often require replacement. Mechanical valve replacement requires long-term anticoagulation therapy, and risks are involved with its use. Thromboembolism, infectious endocarditis, and valve dysfunction can occur with the placement of a mechanical valve. In addition, bioprosthetic valves have the risk of early degeneration.

Most patients have a great improvement in signs and symptoms after their operation. However, those patients with decreased left ventricular function can continue to be symptomatic and will need to continue with medical management of their heart failure.

Acute MR caused by papillary/chordal rupture or infectious endocarditis should be treated early and aggressively. Surgical mortality rate is higher in this patient population, but the options are limited.

Mitral Valve Prolapse

Mitral valve prolapse (MVP) occurs when the leaflets prolapse into the left atrium, which prevents the valve leaflets from joining together, causing a degree of MR. MVP is the most common valvular abnormality, occurring in 2% to 6% of the population (6), and it happens more frequently in women. It can occur secondary to primary dysfunction of any part of the valvular structure. Primary pathological changes of the mitral valve include the annulus, leaflets, chordae, or papillary muscles, and as these changes become more pronounced, the degree of MVP and consequently MR increases. MVP can also occur secondary to connective tissue diseases that would allow the structures pathological stretch or as a congenital defect.

Signs and Symptoms

Most patients with MVP are asymptomatic and are diagnosed after a murmur is found on examination. Dysfunction of the autonomic nervous system has also been linked to MVP as part of the MVP syndrome (7). Anxiety is often a concomitant symptom. As the MVP worsens and the degree of regurgitation increases, symptoms attributable to MR begin to occur.

Diagnostic and Laboratory Evaluations

An ECG may be normal, or in more severe MVP, atypical inferior changes may be seen secondary to the strain on the papillary muscles as the leaflets prolapse greatly into the left atrium. These patients may have any number of arrhythmias, but paroxysmal supraventricular tachycardia is the most common (21).

As in most valvular abnormalities, the best diagnostic testing is performed with 2-D echocardiography. The ultrasound images will show the mitral valve leaflets prolapsing into the left atrium like a sail caught in the wind when the left ventricle contracts in systole. The leaflets should be analyzed, because greatly thickened leaflets are more prone to infective endocarditis. Through Doppler evaluation, the degree of MR can be identified, which is prognostically important. Views not seen on 2-D echocardiography can usually be well delineated on TEE.

MVP has an excellent prognosis. Many people with this disorder will never progress to severe MR or become increasingly symptomatic (46). Ten percent to 15% of patients will progress to severe MR, most often attributable to rupture of the papillary muscles after years of strain (46). Although MVP is more common in women, when it occurs in men, the progression is often quicker. Some patients with MVP will develop infective endocarditis and will have secondary complications, including embolic events, that can be catastrophic.

Treatment

Asymptomatic patients with MVP alone require reassurance and a follow-up echocardiogram performed every 3 to 5 years. Symptomatic patients who experience palpitations, chest pain, anxiety, or fatigue may reduce symptoms by eliminating stimulants such as caffeine, alcohol, and cigarettes. Others respond well to β-blocker therapy.

Patients with MVP accompanied by MR should have this diagnostic evaluation yearly. Those patients with auscultative evidence of MVP and MR diagnosed on echocardiography require antibiotic prophylaxis for any dental, gastrointestinal, or urogenital procedures. Additionally, clinical judgment will dictate which patients' symptoms require further evaluation through stress testing, Holter monitoring, or angiography with treatment as needed. Moreover, as the often accompanying MR progresses, it is treated surgically as in isolated MR without evidence of prolapse.

Aortic Stenosis

Aortic stenosis (AS) is a narrowing of the aortic valve orifice leading to obstruction of the left ventricle's outflow tract. The aortic valve must be reduced by one fourth its normal size to significantly alter circulation. Aortic stenosis is usually a result of congenital or degenerative causes. Congenital AS is attributable to a malformation of the aortic valve in utero. This malformation can present itself as a unicuspid, bicuspid, or tricuspid valve (39). Unicuspid (one cusp) results in stenosis early in life and is often fatal. Bicuspid (two cusps) can also be identified early in life but more commonly is found on echocardiography after symptoms occur later in childhood or early adulthood. There is usually no stenosis seen at birth in bicuspid valves, and their stenosis is secondary to turbulence and increased flow that lead to degenerative changes on an accelerated course. The tricuspid valve often has abnormalities in the shape or size of the cusps, which like in bicuspid disease accelerate degenerative disease.

Acquired AS can be secondary to degenerative changes from wear over time or attributed to rheumatic heart disease. Age-related or senile calcific AS results from years of mechanical stress on a once-normal aortic valve. A calcified annular ring is usu-

ally seen in addition to calcified coronary arteries. Diabetes mellitus and hypercholesterolemia can escalate this process (12), and congenitally acquired diseases such as Paget's or familial hypercholesterolemia can lead to AS in adulthood.

Signs and Symptoms

Three signs and symptoms occur commonly in AS and lead to prognostic implications: chest pain, syncope, and heart failure. At the onset of symptoms, treatment must begin or the patient's status will decline rapidly with an average survival less than 2 to 3 years (6).

Syncope often occurs secondary to a compensatory mechanism. As AS progresses, there is a fixed cardiac output that can make its way through the aortic orifice. This leads to systemic vasodilation, which can reduce blood flow to the brain and cause the patient to pass out.

Heart failure is an ominous sign that occurs late in AS when the left ventricle no longer has the ability to hypertrophy. It begins to dilate, and the cycle of heart failure begins. The decrease in cardiac output causes a backup of blood into the left atrium and edema into the lungs. These patients are usually fatigued and have a decreased appetite and insomnia, as well as other stigmata of low cardiac output. Patients with AS and heart failure have a very poor prognosis if not surgically treated.

Diagnostic and Laboratory Evaluations

It is common for AS patients to have left ventricular hypertrophy on their ECG. Left atrial enlargement can also be seen. Arrhythmias and conduction blocks can be seen late in the disease attributable to calcific spicules protruding from the valve annulus into the conduction system. Disruption of the left atrium and ventricle architecture leads to atrial fibrillation and ventricular tachycardias.

A 2-D echocardiogram allows for measurement of the aortic valve, which should normally be between 1.6 and 2.6 cm. Visualization of the leaflets can help diagnose congenital defects and the degree of calcification. However, these views are best seen by transesophageal echocardiography. Both modes of testing are able to evaluate the left ventricle for prognostic information. Doppler interrogation of the valve will determine the left ventricle-aortic pressure gradient, providing further prognostic information.

Calcification of the aortic valve, increased left ventricular size, and poststenotic aortic enlargement can all be seen on X-ray or fluoroscopic evaluation. Angiography can provide information regarding AS, but there is a risk involved with injecting contrast into the left ventricle and crossing the aortic valve. The addition of this osmotic fluid (contrast) can lead to pulmonary edema or hemodynamic compromise. Moreover, crossing the often calcified aortic valve can disrupt small pieces of calcium or plaque, which can embolize to the brain or elsewhere.

Chest pain or angina usually occurs first, and without treatment these patients live a median of 5 years (22). Angina occurs either because there is truly concomitant coronary artery disease or because the enlarged musculature of the left ventricle, from pushing against the stenotic aortic valve, has inadequate vasculature to supply the additional muscle mass. Furthermore, because of the high intraventricular pressures with AS, very little blood flows through the coronary arteries during systole and must be overcompensated for in diastole. Unfortunately, the intraventricular pressures may still be too high during diastole, which compresses the endocardium, further inhibiting blood flow. Therefore, a patient may experience significant ischemia and angina.

Treatment

The treatment for asymptomatic AS comes in the form of patient counseling. Emphasis should be placed on reporting symptoms immediately. Follow-up is also very important, because serial echocardiography can provide vital information on the progression and prognosis for this disease process and should be performed annually or biannually depending on the severity and progression of disease. Patients with critical AS should not engage in strenuous physical exercise. In addition, AS patients should undergo antibiotic prophylaxis before dental, gastrointestinal, and urogenital procedures.

As the ventricle begins to fail, the treatment should follow standard treatment for heart failure with the caveat that diuretics should be used with caution. The left ventricle becomes very dependent on preload, or blood flow returning to the right heart, for its cardiac output as the AS increases in severity.

In adults with severe AS that becomes symptomatic, or in asymptomatic patients with progressive left ventricular dysfunction, valve replacement surgery is the procedure of choice. Valvulotomy and balloon valvuloplasty, although useful in mitral stenosis, are ineffective or have only a short-term benefit in adult patients with AS. Patients operated on soon after their onset of symptoms without evidence of left ventricular dysfunction do very well after surgery and tend to feel remarkably better (10).

Prior to their operation, most patients except the very young should undergo cardiac catheterization to assess the need for concomitant bypass grafting.

Restenosis is a major problem that occurs in half of the patients at 6 months (31) following balloon valvuloplasty. In younger patients the procedure is more durable, lasting for years; however, the damaged valve will inevitably become restenotic and require replacement.

Aortic Regurgitation

Aortic regurgitation (AR) is the leaking of the valve structure between the left ventricle and the aorta. Aortic regurgitation can be the result of an intrinsic valve disorder or a problem inherent to the aortic root. Thickening, fibrosis, and damage to the aortic valve can be the result of infective endocarditis, rheumatic heart disease, or trauma. Inherent congenital deformities and disease can also lead to AR. Dilation and enlargement of the aorta can cause dilation of the valve annulus and eventual regurgitation. This process can be secondary to congenital malformation and disease as well as degeneration with age (35).

Acute AR can be secondary to an acute aortic dissection involving the aortic root contiguous with the aortic valve, attributable to hypertension or congenital collagen malformation as in Marfan's syndrome. It can also be a manifestation of infective endocarditis leading to perforation and destruction of the valve.

Signs and Symptoms

Patients with AR can remain asymptomatic for years. The first symptoms may be secondary to heart failure and could include shortness of breath and fatigue. As the left ventricle becomes big and boggy, arrhythmias and extrasystolic beats can become symptomatically troublesome to the patient.

Acute AR can make a patient hemodynamically decompensate quickly. The left ventricle is not accustomed to the increase in volume, and as this pressure is relayed backward, pulmonary congestion and edema occur. Evidence of low cardiac output (e.g., hypotension, fatigue) will also present dramatically. As the forward stroke volume decreases, the HR increases in an often futile attempt to maintain cardiac output.

Diagnostic and Laboratory Evaluations

Findings on the ECG, as in other valvular abnormalities, are nonspecific. Depending on the state of the left ventricle, left ventricular hypertrophy, loss of

R waves across the precordial leads, and evidence of left ventricle dilation and failure, may be seen. Chest X-rays are also nonspecific and may show an enlarged left ventricle when the AR is chronic and severe.

A radionuclide ventriculogram can help with diagnosis. This nuclear test works by labeling the red blood cells with a nuclear tracer to track the nuclear tagged cells as they move through the heart. Left ventricular function is measured and regurgitant fraction quantified by taking count ratios of the red blood cells that remain in the ventricles after systole. The difference between left and right stroke volume determines the regurgitant fraction. A ratio of more than 2.0 signifies severe AR (23).

Echocardiography provides a subjective view of the aortic valve structure, which can help to determine the etiology of AR. Objective measurements of end-diastolic and systolic volumes and EF aid in the follow-up and prognosis. The flow or jet of AR can traverse the left ventricle and affect the mitral valve function as well as causing additional symptoms and decline of the ventricle. Doppler interrogation of the valve leads to the most sensitive measurements made in AR. The velocity across the valve orifice can lead to severity estimations and regurgitant orifice measurements.

Patients with chronic AR can be asymptomatic for years. Unfortunately, once symptoms present or evidence of left ventricular failure is seen, progression is quick. Death usually follows the development of angina or heart failure by 4 and 2 years, respectively. Acute AR progresses to death quickly if untreated. There is no time for the left ventricle to compensate for the extra regurgitant volume, and hemodynamic compromise occurs.

Treatment

Mild to moderate AR can be managed expectantly. Close follow-up of symptoms and echocardiographic evaluation of the left ventricle can help with the timing of surgery and prognosis. Antibiotic prophylaxis should be performed on all AR patients when undergoing dental, gastrointestinal, or urogenital procedures. Chronic AR leads to heart failure and the dilation of the left ventricle, which can predispose a patient to arrhythmias. In addition to treatment for those conditions, most patients will benefit from afterload reduction that decreases systemic vascular resistance to help restore cardiac output.

Surgical treatment is curative and should be performed prior to a progressive decrease in left ventricular dysfunction or at the early onset of symptoms. Those patients treated early will have the best outcomes, with a chance to recover left ventricle size and function; the chance for recovery will be permanently lost as the disease progresses over time.

Select AR patients with normal valve structure but a dilated annulus secondary to left ventricular or aortic root dilation can undergo surgical repair rather than replacement. In certain instances, the aortic root may be so dilated that it is replaced with a graft, and the native valve is without evidence of AR. Damage to the valve caused by trauma can also be treated with repair.

Graded Exercise Testing

The purpose of conducting exercise testing in patients with valvular heart disease is to obtain data on exercise tolerance, hemodynamics, arrhythmias, symptoms, concurrent coronary artery disease, and, ideally, ventricular function and oxygen uptake. In asymptomatic patients, an exercise test may uncover valuable clues to valve disease severity that have not been identified through the patient's daily activities. In contrast, a patient with a normal medical history may have a low functional capacity or exercise-induced symptoms that could be warning signs of underlying valvular disease. Exercise testing may also be used to help determine when surgery is indicated as well as to evaluate the patient's response to medical and surgical interventions (see practical application 14.2).

PRACTICAL APPLICATION 14.2

Client–Clinician Interaction

A clinical exercise physiologist working with a valvular heart disease patient should keep the patient's symptoms at the forefront of patient assessment. Frequent contact with the patient in a cardiac rehabilitation setting allows a clinician to watch for the new onset of or subtle changes in dyspnea, chest discomfort, palpitations, and fatigue that may signal underlying changes in their valvular disease progression or heart failure. The patient also needs to understand the importance of self-monitoring symptoms and the need to communicate changes with his or her physician.

A complete medical history and cardiac examination should be conducted before exercise testing to rule out severe aortic stenosis or other contraindications to exercise testing. Consider using a slow and more gradual test protocol (e.g., modified Bruce, Balke, ramp) in this often deconditioned population, which allows for a more gradual onset of symptoms or abnormal responses. A conservative tendency to terminate the exercise test is warranted. Frequent (usually every 1-2 min) and careful monitoring of the blood pressure (BP) during an exercise test is imperative. A hypotensive response (>10 mmHg decrease from baseline) (2), the absence of an appropriate increase in systolic BP, the slowing of the HR with increasing exercise, symptoms (e.g., dyspnea, chest discomfort, pallor, fatigue), and ectopy are all evidence of a limited cardiac output potentially from the diseased valve's dysfunction and are reasons to stop the test (3, 4). Pulmonary congestion induced by exercise may cause coughing rather than dyspnea, and the test should be terminated if the coughing becomes significant.

The exercise electrocardiogram has a higher false-positive rate in valvular patients because of hypertrophy and baseline abnormalities (17). Patients with aortic valve disease may develop left bundle branch block with exercise (9). Therefore, additional imaging (echocardiography or myocardial perfusion imaging) is recommended when underlying coronary artery disease is suspected. Stress echocardiography provides a unique noninvasive means to evaluate cardiovascular function and hemodynamic responses during increased cardiac work and a means to assess the exercise response of the left and right ventricular function, transvalvular and prosthetic valve gradients, and right ventricular systolic pressure.

In addition, assessment of left ventricular function can be obtained with alternative stress tests including radionuclide angiography and exercise or dobutamine stress echocardiography.

Maximal ventilatory oxygen uptake ($\dot{V}O_2$max) is the maximal amount of oxygen a person can extract from inspired air during dynamic exercise and indirectly represents a measure of the cardiovascular system's functional limits. Measured or estimated $\dot{V}O_2$max is an important criterion used by physicians to objectively determine degree of disability for treatment, and this measurement differentiates exaggerated symptoms and psychological impairments. Maximal oxygen uptake expressed in metabolic equivalents (3.5 ml of oxygen per kilogram per minute) is a valuable prognostic indicator. Achievement of 5 metabolic equivalents or less is associated with a poor prognosis in patients younger than 65.

Contraindications

Known moderate valvular disease is considered a relative contraindication for stress testing according to *ACSM's Guidelines for Exercise Testing and Prescription* (2). Associated symptoms (e.g., shortness of breath, chest discomfort, fatigue) and available information regarding changes in left ventricular size and function should be considered before the patient is tested. A discussion with the patient's physician may also be warranted to determine the patient's suitability and safety. In patients with severe symptomatic aortic stenosis, the risks of exercise testing (syncope and cardiac arrest) outweigh the benefits, and therefore, exercise testing remains contraindicated for these patients (2, 18).

Recommendations and Anticipated Responses

Following is specific information regarding the expected exercise responses in the presence of selected valvular heart disease conditions.

— Mitral stenosis: A normal or excessive HR response to exercise may be seen in patients with mitral stenosis. Because they cannot increase their stroke volume, their normal increase in cardiac output is blunted, and cardiac output may eventually decrease during exercise. This is usually accompanied by exercise-induced hypotension. An increase in pulmonary resistance causes an increase in myocardial oxygen demands. Consequently, chest discomfort and ST-segment depression may occur from the reduction in coronary perfusion or may be attributable to pulmonary hypertension. The ST-segment depression is the result of a decrease in coronary perfusion and the increase in myocardial oxygen demand attributable to right ventricular overload. The tachycardia shortens diastole and increases pulmonary blood flow during exercise, which aggravates the preexisting mitral stenosis and may cause pulmonary congestion.

— MR: Patients with mild to moderate MR and a normal left ventricle are able to maintain a normal cardiac output during exercise. Their HR, BP, and ECG responses are usually normal unless MR occurs suddenly during exercise (attributable to ischemic papillary muscle dysfunction), in which case a flat systolic BP response can occur. Severe MR patients have a decreased cardiac output and limited exercise capacity, and a hypotensive response can be expected. Because there is no significant increase in myocardial oxygen consumption, ST-segment depression is not typical; however, arrhythmias frequently occur.

— MVP: Primary MVP syndrome refers to a small subset of patients who are often sent to the stress testing laboratory for evaluation of symptoms (e.g., typical or atypical chest discomfort, palpitations). Asymptomatic MVP patients usually have no significant abnormalities or functional impairment with stress testing (5), unlike symptomatic patients, who have a relatively high prevalence of false-positive tests and ventricular ectopy (24). Therefore, myocardial perfusion imaging or echocardiography in conjunction with the electrocardiogram results is strongly recommended.

— AS: Exercise-induced syncope in patients with AS is an important and well-appreciated symptom. Four proposed mechanisms for effort syncope include carotid hyperreactivity, left ventricular failure, arrhythmia, and left ventricular baroreceptor stimulation (thought to be the most plausible cause) (17). An abrupt elevation of left ventricular systolic pressure with exercise without a corresponding increase in aortic pressure could allow the left ventricular baroreceptors to produce "a violent depressor reflex." The subsequent response could be bradycardia, peripheral vasodilation, and hypotension, which would all reduce coronary arterial flow and result in left ventricular dysfunction and arrhythmias.

A review of the literature (4) demonstrates that these potentially lethal complications are rare when exercise testing is performed with appropriate caution and monitoring in patients with mild to moderate AS. Exercise testing is used more aggressively in children with congenital AS to determine the need for surgical intervention.

Exercise testing in both pediatric and adult patients with AS should focus on frequent BP and continuous ECG monitoring with careful attention to the onset of symptoms. In the presence of an abnormal BP response, the patient should undergo a minimum 2-min cool-down walk to avoid the acute left ventricular volume overload that may occur when the patient is supine (17).

— AR: Patients with AR tend to maintain a normal exercise capacity longer than those with AS because of the lower oxygen cost required to maintain heart function with a volume overload rather than working against high pressures (AS). As the myocardium eventually fails, the HR slows and effective EF and forward stroke volume decrease. The ventricular diameter increases to hold the residual blood volumes, increasing the myocardial metabolic requirements, and as coronary flow is compromised, myocardial ischemia occurs.

Serial exercise testing is useful in AR progression by using the onset of ST-segment depression, a reduction of the HR response to each workload, and a decrease in $\dot{V}O_2$max as indicators of worsening left ventricular function. Significant AR will affect the regional wall motion of the left ventricle during exercise and adversely affects the accuracy of exercise echocardiography for the diagnosis of coronary artery disease in this patient population (44).

Exercise Prescription

Cardiac rehabilitation is recommended postprocedure to educate patients on the safety of exercise, symptoms, and risk factor modification (if appropriate). It will also increase their sense of confidence and well-being. Miller (26) reported that it takes 3 to 6 months to recover maximally, even with exercise training of gradually increasing intensity. Therefore, it is recommended that the patient ideally spend approximately 6 months in a rehabilitation setting to evaluate medical management and allow for a sufficient return of functional capacity and a more rapid recovery (see practical application 14.3).

Special Exercise Considerations

— MS: Mitral stenosis tends to limit the patient's ability to exercise, much more so than the other mitral valve afflictions. Exercise intensity and duration may need to be adjusted based on the patient's shortness of breath. Because of the increase in left atrial size and pressure, the patient can develop atrial fibrillation. Therefore, the best and safest method to determine an effective exercise prescription is an exercise test (26). If the results show mild mitral stenosis and the patient is asymptomatic, no further testing is needed on the initial work-up because these patients remain stable for many years (6). If the patient is symptomatic, the exercise guidelines should be set below the onset of significant symptoms, arrhythmias, or changes in hemodynamics. Patients who have greater than mild mitral stenosis should be counseled to avoid unusual physical stress.

— MR: If MR is mild without a change in left ventricular size, moderate exercise is well tolerated. Moderate MR patients with mild left ventricular and left atrium enlargement are limited to 50% of maximal functional capacity for 20 to 40 min. As the disease worsens, exercise training should be limited to low to moderate intensity (26). Research has identified that the regurgitant jet intensifies and left atrial pressures increase significantly, markedly increasing

Literature Review

The limited exercise training and cardiac rehabilitation patient data available for valvular heart disease patients support that exercise, when prescribed appropriately, can safely improve hemodynamics and functional capacity. Exercise is a useful therapy in this often deconditioned patient population. The landmark studies (28, 42) found that in aortic valve (AV) replacement patients, exercise training significantly increased functional capacity. Habel et al. (19) found no significant valve dysfunction or hemolysis after mitral valve (MV) replacement.

AV replacement patients may substantially increase their functional capacity after surgery, unlike MV surgery patients, who typically show little or no improvement (8). The contrast between these two groups is often attributable to a recovery of left ventricular function in AV patients that does not occur in the MV patients. Exercise capacity, when assessed by anaerobic threshold, shows that a third of AV replacement patients experience a significant increase especially if they were initially deconditioned (29).

In a recent study (20), researchers found that octogenarians undergoing aortic valve replacement (n = 47) and MV replacement or repair (n = 14), with concomitant coronary artery bypass grafting (n = 27), had safe outcomes. However, they typically had a longer hospital stay and more complications and required more home health care. The study estimated functional performance pre- and postoperatively and found a significant improvement following successful surgery. In addition, the patients who underwent surgical repair were more likely to resume independent activities and were less symptomatic. Earlier referral for surgery before significant heart failure develops and cardiac rehabilitation postsurgery may contribute to even better outcomes for these elderly adults.

arterial pressures, when intense static exercise is performed (e.g., powerlifting); therefore, this should be avoided (26).

— MVP: Patients with mild forms of MVP should be encouraged to lead a normal lifestyle and to exercise regularly (26). An exercise stress test is not necessary before prescribing exercise for an asymptomatic patient; however, a patient with exercise-induced symptoms, especially palpitations, could benefit from some additional testing (27).

MVP patients with mild mitral regurgitation may perform normal activities and exercise without restrictions unless they are prone to malignant arrhythmias, a family history, sudden death in the setting of MVP, embolic events, or syncope attributable to arrhythmias (26). Patients need to understand that their mitral valve disease could worsen over time and should be aware of warning signs such as a marked decrease in exercise capacity or the presence of symptoms or palpitations.

— AS: Patients with mild AS can safely participate in exercise training at a moderate intensity (26). For moderate to severe stenosis, an exercise test is needed to determine the threshold for cardiac output limitations (e.g., hypotension, symptoms, electrocardiogram changes, arrhythmias). The tar-

get HR range would then be set well below the threshold. Patients should be reevaluated every 6 to 12 months for signs of disease progression and its impact on exercise safety (26). Patients with severe AS should be advised to limit activity to relatively low levels (6).

— AR: AR patients who are asymptomatic, lead active lives, and have good echocardiogram results do not need an exercise test (26). They are not limited in physical activity or, in some cases, competitive athletics (6). Mild to moderate AR is very tolerable with exercise, and patients will continue to benefit from training. The patient should be reassessed every 3 to 6 months to document that his or her condition is stable and left ventricular size is unchanged (26). A patient who has more severe AR, is sedentary, and has atypical symptoms will benefit from exercise testing to evaluate functional capacity, symptoms, and hemodynamic effects of exercise to determine a safe training HR range. Isometric exercise in this patient population is contraindicated (6).

Exercise Recommendations

As a guide for determining exercise intensity in patients with valvular heart disease, an exercise stress test is strongly recommended for enhanced safety and training benefits. The test data allow for exercise intensity to be set at levels that are associated with normal hemodynamics, mild symptoms, and nonthreatening arrhythmias. Typically the best candidates for exercise therapy are patients with mild valve disease and normal ventricular function.

Exercise prescription guidelines are the same as those for other cardiac patients. If the patient has underlying coronary artery disease and potentially significant arrhythmias, close observation and ECG monitoring are recommended in cardiac rehabilitation. Low- and moderate-risk valve replacement patients without coronary artery disease or a history of significant ectopy probably do not need to be

telemetry monitored (1). Commonly, valve patients are deconditioned from before surgery or repair, so it is wise to start slowly and encourage the patient to become more active by adding exercise gradually to increase endurance. Practical application 14.4 discusses exercise prescription for patients with valvular heart disease.

If an exercise test is not available, a target HR range should be determined by symptoms, ECG monitoring at approximately 20 to 30 beats above resting HR, and a perceived exertion level of "fairly light to somewhat hard." To prescribe greater intensity levels, a stress test in this higher risk population is recommended.

PRACTICAL APPLICATION 14.4

Exercise Prescription

Special Considerations

The safety of resistance training should be evaluated based on the criteria used for other cardiac patients. Surgical patients should use range of motion exercises for the first 3 months to prevent unnecessary pulling on the sternum (33). Resistance training is contraindicated for severe stenotic or regurgitant valvular disease patients (33). Commonly used assessments of baseline strength should be avoided in patients who have undergone valve replacement and repair (1). Patients with a thoracic aortic dilation repair should avoid strength training or should be limited to low resistance levels (1). Pure isometric exercise is not recommended for valvular heart disease patients or any other cardiac patients, because the safety and efficacy have yet to be proven (33).

Aerobic Training

- Mode: Any appropriate aerobic exercise based on functional capacity. Guidelines for ergometers using the upper body should follow those established for coronary artery bypass patients. Stretching, flexibility activities, and range of motion exercises (10-15 repetitions to a perceived exertion of "fairly light" to "somewhat hard") can be started as an inpatient (33).

- Frequency: Three to five times per week; active on most days.

- Intensity: *With a stress test:* Initially 40% to 60% of HR reserve progressing up to 80% of HR reserve if clinically appropriate; a rating of perceived exertion "fairly light" to "somewhat hard." *Without a stress test:* 20 to 30 beats above resting HR. *Special considerations:* Intensity levels should be set 10 beats below the onset of symptoms with close attention to any change in symptom patterns that could signify underlying changes in valvular disease progression, heart failure, or angina.

- Duration: May start with short bouts of activity and progress to 30 to 40 min continuous.

Resistance Training

- Mode: Weight machines or free weights.

- Frequency: Two to three days per week.

- Intensity: A weight that can be lifted 10-15 repetitions to moderate fatigue or a rating of perceived exertion of "somewhat hard" (1, 14, 33). Moderate-risk patients can proceed to a perceived exertion of "hard" (33), whereas a low-risk patient may go to volitional fatigue after 10 to 15 repetitions (2, 14).

Patients with more progressed valvular disease need to clearly understand their HR guidelines, be able to rate their perceived exertion, and be aware of the potential consequences of exceeding their limits. Patients should be advised to report any new symptoms of chest discomfort, shortness of breath, lightheadedness, palpitations, or syncope to their physicians immediately.

A long preoperative course associated with muscle atrophy heightens the importance of adding a resistance training component to the exercise prescription. The safety of resistance training should be evaluated based on the criteria used for other cardiac patients with an emphasis on light weights and frequent repetitions. Contraindications include severe stenotic or regurgitant valvular disease, unstable angina, uncontrolled hypertension (>160/ 100 mmHg), uncontrolled arrhythmias, a recent episode of heart failure that has not yet been effectively treated, and hypertrophic cardiomyopathy (2, 14, 15). Left ventricular function and indicators of ischemia should also be taken into consideration when one is determining the safety of a resistance training program. Patients with poor left ventricular function and myocardial ischemia may have wall-

motion abnormalities or serious ventricular arrhythmias that occur during strength training (13, 36). Commonly used assessments of baseline strength should be avoided in patients who have undergone valve replacement and repair (1). Patients with a thoracic aortic dilation repair should avoid strength training or be limited to low resistance levels (1).

Conclusion

Physical activity recommendations for individuals with valvular heart disease depend on the valves involved, whether they are stenotic or regurgitant, the severity of the disease, and the presence of myocardial dysfunction or coronary artery disease. Before beginning an exercise program, patients would benefit from a complete medical evaluation including a cardiac exam. Graded exercise testing can provide important information about the individual's functional capacity and hemodynamic responses. Before the patient starts the exercise program, the clinician should be thoroughly acquainted with the individual's history and appropriate training level to ensure appropriate responses.

Case Study 14

Medical History and Diagnosis

Mr. GL is a 60-year-old African American male with a history of moderate aortic stenosis and an EF of 40%. Both were evaluated by 2-D echocardiography last week. He also has a history of hypercholesterolemia and a family history of coronary artery disease. The patient is referred to your laboratory by his cardiologist for an exercise stress test with myocardial perfusion imaging to evaluate his symptom of new-onset substernal chest discomfort on exertion. The patient states that he is surprised he is feeling chest pain because he really has not done much activity around the house over the last year.

Exercise Evaluation

The stress test was terminated because of an onset of chest discomfort with 1 mm of anterolateral ST depression. His systolic BP was flat, and he experienced two multiform ventricular couplets. His nuclear medicine images were positive for ischemia. His subsequent cardiac catheterization identified an 80% proximal left anterior descending (LAD) lesion and an 85% distal left circumflex lesion. The patient was sent for coronary artery bypass graft surgery and aortic valve replacement.

The patient has been sent to your outpatient cardiac rehabilitation program 1 month after bypass surgery. Now that he is feeling better, he is anxious to return to work after his long medical

continued

leave. He is certain cardiac rehabilitation is his ticket to better endurance and disease prevention. His wife is not so sure and could use some reassurance.

Before the patient enters your program, a symptom-limited stress test is required. The results are as follows:

Pretest Data

Resting ECG: normal sinus rhythm with left ventricular hypertrophy by voltage criteria

Resting HR: 59 beats $\cdot$ min^{-1}

Resting BP: 106/72 mmHg

Medications: Toprol, Lipitor, Coumadin, Lisinopril, aspirin

Protocol: modified Bruce terminated because of fatigue after 8 min

Peak HR: 114 beats $\cdot$ min^{-1}

Peak BP: 124/70 mmHg

Rating of perceived exertion: hard

No additional ST-segment changes were seen over the abnormal baseline with occasional multifocal premature ventricular contractions. The patient was asymptomatic.

Case Study 14   Discussion Questions

1. Is it safe to stress test this patient, and if so, what precautions would you take?

2. How would you reassure his wife regarding the benefits of his participation in a cardiac rehabilitation program?

3. Would you recommend placing him in a telemetry-monitored cardiac rehabilitation program?

4. How would you determine his starting target HR range?

5. Your patient would like to begin using his free weights at home again. He used to be able to bench 175 lb when he was in college. How would you handle this?

Glossary

aortic valve—A three-cusp semilunar valve that opens with the systolic contraction of the left ventricle allowing blood to be pumped throughout the body.

atrioventricular (A-V) valves—The tricuspid and mitral valves that control blood flow from the atrium to the ventricles.

Doppler echocardiography—A diagnostic test using ultrasound waves in a continuous or pulsing manner to assess velocity for determination of the gradient across the valve, from which the degree of stenosis can be calculated.

mitral valve—A two-cusp (bicuspid) atrioventricular valve that opens during diastole to allow blood to flow from the left atrium into the left ventricle.

mitral valve prolapse (MVP)—The most common form of valvular heart disease occurring in 2% to 6% of the population that usually has a benign prognosis; however, it can be the precursor to significant mitral regurgitation.

pulmonary valve—A three-cusp semilunar valve that opens with the systolic contraction of the right ventricle, allowing blood to be pumped to the lungs.

rheumatic fever—An autoimmune disease that begins with a streptococcal (specifically, group A hemolytic streptococci) infection that could present itself as a sore throat, ear infection, or scarlet fever and is responsible for the majority of valvular lesions worldwide, especially in less developed countries.

semilunar valves—The pulmonary and aortic valves that are exposed to the high velocities from rapid blood ejection from the ventricles into the lungs or the body, respectively.

transesophageal echocardiogram (TEE)—An internal echocardiogram obtained by a patient swallowing an endoscope that is positioned in the esophagus next to the heart; the endoscope provides images with better resolution of valvular structure and delineation of valve detail and left ventricular size and function.

tricuspid valve—A three-cusp atrioventricular valve that opens during diastole to allow blood to flow from the right atrium into the right ventricle.

two-dimensional (2-D) echocardiography—The use of ultrasound waves that create an image of the heart showing the valve orifice size, physical appearance of the valve, and abnormal valvular motion.

References

1. American Association of Cardiovascular and Pulmonary Rehabilitation. 1999. *Guidelines for Cardiac Rehabilitation and Secondary Prevention Programs.* 3rd ed., pp. 145-147. Champaign IL: Human Kinetics.

2. American College of Sports Medicine. 2000. *ACSM's Guidelines for Exercise Testing and Prescription.* 6th ed., p. 50. Baltimore: Lippincott, Williams & Wilkins.

3. Areskog NH. 1984. Exercise testing in the evaluation of patients with valvular aortic stenosis. *Clinical Physiology* 4: 201-208.

4. Atwood JE, Lawanishi S, Myers J, and Froelicher VF. 1988. Exercise and the heart: Exercise testing in patients with aortic stenosis. *Chest* 93: 1083-1087.

5. Barzilay J, Froone P, Gross M, et al. 1986. Exercise testing and physical fitness in mitral valve prolapse. *Journal of Cardiopulmonary Rehabilitation* 6: 465-468.

6. Bonow RO, Carbello B, deLeon AC Jr, et al. 1998. ACC/AHA Guidelines for the management of patients with valvular heart disease: Executive summary. A report from the American College of Cardiology/American Heart Association Task Force on Guidelines (Committee on Management of Patients with Valvular Heart Disease). *Circulation* 98: 1949-1984.

7. Boudoulas H, Lolibash AJ, Baker P, et al. 1989. Mitral valve prolapse and the mitral valve prolapse syndrome: A diagnostic classification and pathogenesis of symptoms. *American Heart Journal* 118(4): 796-818.

8. Carstens V, Behrenbeck D, Helger H. 1983. Exercise capacity before and after cardiac valve surgery. *Cardiology* 70: 41-49.

9. Cumming GR. 1987. Valvular and congenital heart disease in adults. In *Exercise Testing and Exercise Prescription for Special Cases,* ed. Skinner JS, p. 270. Philadelphia: Lea & Febiger.

10. Culliford AT, Gallowy AC, Colvin SB, et al. 1991. Aortic valve replacement for aortic stenosis in persons aged 80 years and over. *American Journal of Cardiology* 67(15): 1256-1260.

11. Delahaye JP. 1997. Acquired valvular heart diseases. In *Women & Heart Disease,* eds. Julian DG and Wenger NK, pp. 349-361. London: Martin Dunitz.

12. Deutscher S, Rockette HE, Krishnaswami V. 1984. Diabetes and hypercholesterolemia among patients with calcified aortic stenosis. *Journal of Chronic Disease* 37(5): 407-415.

13. Effron MB. 1989. Effects of resistive training on left ventricular function. *Medicine and Science in Sports and Exercise* 21: 694-697.

14. Fletcher GF, Balady G, Froelicher VF, et al. 1995. Exercise standards: A statement for healthcare professionals from the American Heart Association. *Circulation* 91: 580-615.

15. Franklin BA, Bonzheim K, Gordon S, et al. 1991. Resistance training in cardiac rehabilitation. *Journal of Cardiopulmonary Rehabilitation* 11: 99-107.

16. Froelicher V. 1999. The role of exercise testing in the evaluation of the patient with stable heart disease. In *Lifestyle Medicine,* ed. Rippe J, pp. 735-736. Malden, MA: Blackwell Science.

17. Froelicher V, Marcondes G. 1989. *Manual of Exercise Testing,* pp. 224-249. Chicago: Year Book Medical.

18. Gibbons RJ, Balady GJ, Beasley JW, et al. 1997. ACC/AHA Guidelines for exercise testing. A report of the American College of Cardiology/American Heart Association Task Force on practice guidelines (Committee on Exercise Testing). *Journal of the American College of Cardiology* 30(1): 260-311.

19. Habel-Verge C, Landry F, Desaulnier D, et al. 1987. L'entrainement physique apres un remplacement valulaire mitral. *Canadian Medical Association Journal* 136: 142-147.

20. Khan JH, McElhinney DB, Hall TS, Merrick SH. 1998. Cardiac valve surgery in octogenarians: Improving quality of life and functional status. *Archives of Surgery* 133: 887-893.

21. Kligfield P, Devereux RB. 1995. Arrhythmia in mitral valve prolapse. In *Cardiac Arrhythmia: Mechanisms, Diagnosis and Management,* eds. Podrid PR and Kowey PR, p. 1253. Baltimore: Williams & Wilkins.

22. Lester SJ, Heilbron B, Gin K, et al. 1998. The natural history and rate of progression of aortic stenosis. *Chest* 113(4): 1109-1114.

23. Manyari DE, Nolewajka AJ, Kostuk WJ. 1982. Quantitative assessment of aortic valvular insufficiency by radionuclide angiography. *Chest* 81(2): 170-176.

24. Massie B, Botvinick E, Shomes D, et al. 1978. Myocardial perfusion scintigraphy in patients with mitral valve prolapse. *Circulation* 57: 19-26.

25. Michel PL, Garbarz E, Messika-Zeitouri D, Roussin I. 2000. The best of valvular heart disease in 1999. *Archives des maladies du coeur et des vaisseaux* 93(1 Spec NO): 97-102.

26. Miller HS. 1999. Exercise training in special populations: Valvular heart disease. In *Cardiac Rehabilitation: A Guide to Practice in the 21st Century,* pp. 155-162. New York: Marcel Dekker.

27. Moir William T. 1989. Nonischemic cardiovascular disease. In *Exercise in Modern Medicine,* eds. Franklin BA,

Gordon S, and Timmis GC, p. 88. Baltimore: Williams & Wilkins.

28. Newell JP, Kappagoda CT, Stoker JB, Deverall PB, Watson DA, Linden RJ. 1980. Physical training after heart valve replacement. *British Heart Journal* 44: 638-649.

29. Niemela K, Ikakeimo M. Takkunen J. 1985. Determination of the anaerobic threshold in the evaluation of functional status before and following valve replacement. *Cardiology* 72: 165-173.

30. Olson LJ, Subramanian R, Ackermann DM, et al. 1987. Surgical pathology of the mitral valve: A study of 712 cases spanning 21 years. *Mayo Clinic Proceedings* 62(1): 22-34.

31. Otto CM, Michel MC, Kennedy JW, et al. 1994. Three year outcome after balloon aortic valvuloplasty: Insights into prognosis of valvular aortic stenosis. *Circulation* 89(2): 642-650.

32. Peterson KL. 1989. Timing of cardiac surgery in chronic mitral valve disease: Implications of natural history studies and left ventricular mechanics. *Seminars in Thoracic and Cardiovascular Surgery* 1(2): 106-117.

33. Pollock ML, Franklin BA, Balady GJ, et al. 2000. Resistance exercise in individuals with and without cardiovascular disease. AHA Science Advisory (Committee on Exercise, Rehabilitation, and Prevention, Council on Clinical Cardiology, American Heart Association); Position paper endorsed by the American College of Sports Medicine. *Circulation* 101: 828-833.

34. Rahimtoola SH. 1989. Perspective on valvular heart disease: An update. *Journal of the American College of Cardiology* 14(1): 1-23.

35. Rahimtoola SH. 1996. Aortic regurgitation. In *Valvular Heart Disease and Endocarditis, Atlas of Heart Disease*, Vol. 11, ed. Rahimtoola SH, pp. 200-215. St. Louis: Mosby.

36. Sagiv M, Hanson P, Besozzi M, et al. 1989. Left ventricular responses to upright isometric handgrip and deadlift in men with coronary artery disease. *American Journal of Cardiology* 55: 1298-1302.

37. Scarlat A, Bodner G, Livon M. 1986. Massive haemoptysis as the presenting symptom in mitral stenosis. *Thorax* 41(5): 413-414.

38. Schoen FJ, St. John Sutton, M. 1991. Contemporary pathologic considerations in valvular disease. In *Cardiovascular Pathology*, eds. Virmani R, Atikinson JF, Feuoglio JJ, pp 334-342. Philadelphia: Saunders.

39. Selzer A. 1987. Changing aspects of the natural history of valvular aortic stenosis. *New England Journal of Medicine* 317(2): 91-98.

40. Sharma NG, Kapoor CP, Mahambre L, et al. 1973. Ortner's Syndrome. *Journal of the Indian Medical Association* 60(11): 427-429.

41. Shimoyama H, Sabbah HN, Roman H, et al. 1995. Effects of long-term therapy with enalapril on severity of functional mitral regurgitation in dogs with moderate heart failure. *Journal of the American College of Cardiology* 25(3): 768-772.

42. Sire S. 1987. Physical training and occupational rehabilitation after aortic valve replacement. *European Heart Journal* 8: 1215-1220.

43. Supino P, Roman MJ, Devereux RB, et al. 1994. Natural history of the asymptomatic/mildly symptomatic patient with severe mitral regurgitation secondary to mitral valve prolapse and normal right and left ventricular performance. *American Journal of Cardiology* 74(4): 374-380.

44. Wahi S, Marwick TH. 1999. Aortic regurgitation reduces the accuracy of exercise echocardiography for diagnosis of coronary artery disease. *Journal of the American Society of Echocardiography* 12(11): 967-973.

45. Wrisley D, Giambartolomei A, Lee I, et al. 1991. Left atrial ball thrombus: Review of clinical and echocardiographic manifestations with suggestions for management. *American Heart Journal* 121(6 Pt 1): 1784-1790.

46. Zuppiroli A, Rinaldi M, Kramer-Fox R. 1995. Natural history of mitral valve prolapse. *American Journal of Cardiology* 75(15): 1028-1032.

Steven J. Keteyian, PhD
Henry Ford Heart and Vascular Institute
Detroit, MI

Thomas J. Spring, MS
Henry Ford Heart and Vascular Institute
Detroit, MI

Chronic Heart Failure

Heart failure (HF) is a varied clinical syndrome with a complex pathophysiology that is still being defined (29). It is "the pathophysiological state in which the heart is unable to pump blood at a rate commensurate with the requirements of metabolizing tissues" (11, p. 503).

The inability of the left ventricle (LV) to pump blood adequately is most often a failure of systolic function. **Systolic dysfunction,** or the inability of cardiac myofibrils to shorten against a load, leads to a reduced **ejection fraction.** However, some individuals with normal systolic contractile function can still present with symptoms of HF. In these patients, the disorder is not associated with an inability of the heart to contract; instead it is attributable to an abnormal increase in resistance to filling of the LV—referred to as diastolic dysfunction. Think of diastolic dysfunction as a stiff or noncompliant chamber that is partially unable to relax and expand as blood flows in during diastole.

Scope

The public health burden associated with HF is immense, in many ways exceeding other chronic health disorders (29). Approximately five million people are afflicted with the syndrome, and approximately 500,000 new cases occur each year. Because of the aging U.S. population and increased survival of patients with cardiovascular disorders, the prevalence of HF will increase two- to threefold over the next 10 years.

HF is the leading reason for hospitalizations in people 65 years of age and older and directly or indirectly contributes to 250,000 deaths annually. The 5-year mortality rate for a person newly diagnosed with HF is 50%. The economic burden imposed by HF in the United States is enormous, standing at about $20 billion annually.

Physiology and Pathophysiology

HF remains a final common denominator for many cardiovascular disorders. Although LV diastolic dysfunction can be a cause of HF, this chapter primar-

ily focuses on LV systolic dysfunction. In patients with HF attributable to LV systolic dysfunction, some degree of diastolic dysfunction is usually present as well.

Figure 15.1 depicts the complex abnormalities and changes that occur following loss of systolic function. When cardiac cells (myocytes) die because of infarction, chronic alcohol use, long-standing hypertension, disorders of the cardiac valves, viral infections, or yet unknown causes, a diminished LV systolic function results. Among all cases of LV systolic dysfunction, approximately 60% are caused by **ischemic heart disease** (i.e., coronary atherosclerosis). For this reason, it is common to hear these patients referred to as having an ischemic versus a nonischemic cardiomyopathy, where a **nonischemic cardiomyopathy** refers to some other disease process involving heart muscle (e.g., viral cardiomyopathy or alcoholic cardiomyopathy).

As figure 15.1 also shows, a variety of physiological adaptations and compensatory changes occur in response to LV systolic dysfunction. Most of the

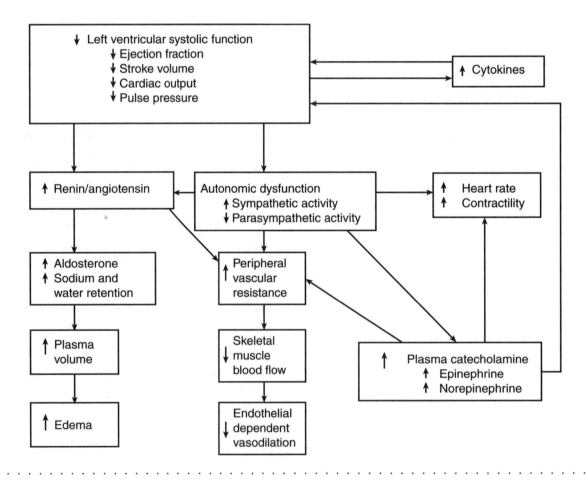

FIGURE 15.1   Schematic representation describing some of the main physiological and pathophysiological adaptations that occur at rest in patients with heart failure attributable to left ventricular systolic dysfunction.

current medical therapies used to treat HF are aimed at modifying one or more of these abnormalities.

Key characteristics unique to the pathophysiology of HF attributable to systolic dysfunction include the following:

1. A reduced ejection fraction
2. An increase in LV mass, pressure, diastolic, and systolic volumes
3. Sodium/water retention leading to edema secondary to activation of the renin-angiotensin–aldosterone system
4. An imbalance of the autonomic nervous system such that parasympathetic activity is inhibited and sympathetic activity is increased

Additionally, abnormalities of other hormones and chemicals such as a diminished production of nitrous oxide (endothelium-derived relaxing factor), increased endothelin-1, and increased cytokines (e.g., tumor necrosis factor-α) all contribute to adverse cardiac and vascular remodeling.

Substantial clinical evidence now indicates that many of these factors contribute to the remodeling of the LV, reshaping it from an elliptical to a spherical form. This change in shape itself contributes to a further loss in LV systolic function. Currently, several of the treatment strategies used in patients with HF interrupt this process, referred to as *reverse remodeling.*

In patients with isolated diastolic dysfunction, LV end-diastolic and systolic volumes are reduced, and systolic function (i.e., ejection fraction) may be normal or increased. Table 15.1 describes normal LV characteristics as well as those associated with LV systolic and diastolic dysfunction.

Clinical Considerations

Before a patient with HF can be cleared for exercise rehabilitation, his or her past and current medical history must be reviewed and functional status evaluated. Exercise testing provides important information about the patient's status; however, signs, symptoms, and medications must also be considered.

Signs and Symptoms

Clinically, patients with HF present with several key characteristics. Two of the classic symptoms are these:

1. Exercise intolerance as manifested by fatigue or shortness of breath on exertion
2. Peripheral edema, recent weight gain, or both

The former are usually associated with complaints of difficulty sleeping flat or awakening suddenly during the night to "catch my breath." Sudden awakening caused by labored breathing is referred to as **paroxysmal nocturnal dyspnea.**

Labored or difficult breathing while exerting oneself is called **dyspnea on exertion** (DOE), and difficulty breathing while lying supine or flat is referred to as **orthopnea.** Clinically, the severity of orthopnea is rated based on the number of pillows that are needed under a patient's head to prop him or her up sufficiently to relieve dyspnea. For example, so-called "two-pillow orthopnea" means that two pillows are needed under the head and shoulders of a patient in order for him or her to breathe comfortably while recumbent.

TABLE 15.1
Comparison of Typical Left Ventricle Characteristics

	End-diastolic volume (ml)	End-systolic volume (ml)	Stroke volume (ml)[a]	Ejection fraction (%)[b]
Normals	120	55	65	55
Systolic dysfunction	160	110	50	30
Diastolic dysfunction	85	35	50	60

[a] Stroke volume = (end diastolic volume) - (end systolic volume). [b]Ejection fraction = (stroke volume)/(end diastolic volume).

These signs and symptoms represent findings that the clinical exercise physiologist may need to evaluate to ensure that on any given day it is safe to exercise. For example, a patient's complaint of increased DOE or recent weight gain may or may not be clinically meaningful in HF patients, but such signs become especially concerning if associated with increased ankle edema or fluid accumulation in the lungs. The severity of ankle edema is typically evaluated on a scale of 1 to 3, as discussed in chapter 4. Lung sounds called "rales" are associated with pulmonary congestion and are best heard using the diaphragm portion of a stethoscope. Rales appear as a crackling noise during inspiration.

Normal first (S_1) and second (S_2) heart sounds are associated with the abrupt closure of the mitral/tricuspid valves and aortic/pulmonic valves, respectively (see chapter 4). In patients with HF attributable to LV systolic dysfunction, an abnormal third heart sound (S_3) can often be heard when the bell portion of the stethoscope is lightly placed on the chest wall over the apex of the heart. This S_3 sound occurs soon (120-200 ms) after S_2 and is most likely attributable to vibrations caused by the inability of the LV wall/chamber to appropriately accept incoming blood during the early, rapid stage of diastolic filling. Students are encouraged to listen to the audiotapes available in most medical libraries as a means to learn both normal and abnormal breath and heart sounds.

Diagnostic and Laboratory Evaluations

The diagnosis of HF attributable to LV systolic dysfunction, although based on the presence of signs and symptoms, also requires a reduced ejection fraction. Most often, ejection fraction is measured by using an **echocardiogram** machine; however, a radionuclide test or cardiac catheterization can assess this parameter as well. Normally, LV ejection fraction is greater than 50%. This means that at least half of the blood in the LV at the end of diastole is ejected into the systemic circulation during systole. In patients with HF attributable to systolic dysfunction, LV ejection fraction is typically reduced below 45%. Severe LV dysfunction may be associated with an ejection fraction of 30% or lower. The decrease in ejection fraction is qualitatively proportional to the amount of myocardium that is no longer functional. Chronic HF attributable to systolic dysfunction is also usually associated with an enlarged LV.

The diagnosis of HF attributable to **LV diastolic dysfunction** is much less exact. Although an echocardiogram can be used to evaluate character-

istics during diastole, quite often the diagnosis is made when the clinical syndrome of congestive HF (i.e., fatigue, dyspnea, and edema) requires hospitalization in the presence of a somewhat normal ejection fraction.

Routine evaluation or work-up of patients with HF now includes a graded exercise test with measured peak oxygen consumption ($\dot{V}O_2$peak) (figure 15.2). The 3-year risk of death for patients who achieve a $\dot{V}O_2$peak greater than 17 ml $\cdot$ kg^{-1} $\cdot$ min^{-1} is 19%, clearly lower than patients who achieve a peak value less than 14 ml $\cdot$ kg^{-1} $\cdot$ min^{-1} (figure 15.3) (50). Measured $\dot{V}O_2$peak helps determine which patients require aggressive medical therapy or possible cardiac transplant.

Weber and associates (63) observed that $\dot{V}O_2$peak is related to exercise cardiac output reserve and, as a result, classified patients using an A through D scale called the Weber class (table 15.2). Although most clinicians traditionally use New York Heart Association (NYHA) Functional Class (see chapter 4, table 4.2) to describe a patient's clinical status, use of that system can be problematic because it represents a

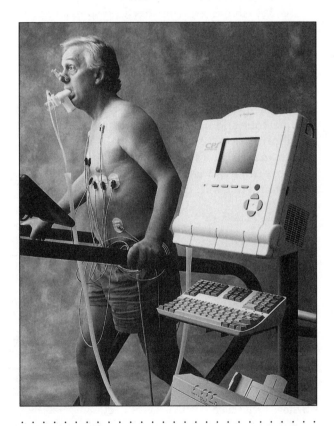

FIGURE 15.2 Example of automatic gas collection system to assess expired air during cardiopulmonary exercise testing.

Courtesy of Medical Graphics Corporation.

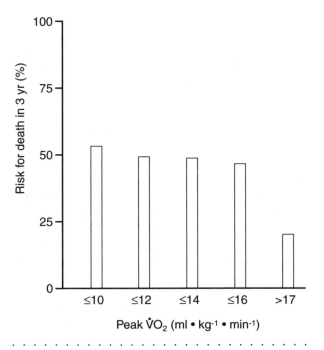

FIGURE 15.3   Three-year risk for death based on achieved peak V̇O₂.

subjective assessment open to interpretation. Clinician variability can be minimized when the Weber class is used, because V̇O₂ represents a quantifiable and reproducible parameter.

Although much of the diagnosis of HF relies on the use of echocardiography and patient symptoms, other laboratory tests such as measuring brain natriuretic peptide may be used. In patients with recent onset HF caused by a large anteroseptal myocardial infarction, large increases in serum troponin and creatine phosphokinase are typically observed in the blood (see chapter 12). Additionally, in these patients' electrocardiogram (ECG), changes such as Q waves in leads V1 through V4 are usually evident.

In some patients, a cardiac catheterization is performed to determine if, and to what extent, ischemic heart disease contributed to the problem. Ischemic heart disease is the underlying cause for HF in about 60% of cases.

Treatment

Over the past 15 years, great strides have been made in the medicines used to treat patients with HF. These advances have led to fewer HF-related deaths and hospitalizations, less symptoms, and increased exercise tolerance. In patients refractory to optimal medical therapy and who demonstrate a deteriorating clinical state consistent with a 1-year survival rate less than 50%, cardiac transplant is a consideration.

This section summarizes the guidelines for the medical treatment of patients with HF attributable to LV systolic dysfunction by the Heart Failure Society of America (29), the American College of Cardiology (1), and the American Heart Association. The current medical therapy for these patients includes angiotensin-converting enzyme (ACE) inhibitors, β-adrenergic receptor blockers, diuretics, and, possibly, digoxin. ACE inhibitors reduce 5-year mortality rate by approximately 30% among patients with diagnosed HF. They also improve exercise tolerance, alter the rate of HF progression, influence structural remodeling, and decrease future hospitalizations.

The single most important contribution to the medical management of patients with HF in the past 5 years has been the use of β-blocker therapy. Initially thought to be contraindicated, β-blockers are now known to be safe, to reduce morbidity and mortality rates (~30%), and to improve resting ejection fraction by approximately 7 percentage points.

Congestion and fluid overload remain important complications in many patients with HF. To address this, diuretic therapy is commonly used. And the role

TABLE 15.2	
Weber A Through D Scale	
Weber class	**Peak V̇O₂ (ml · kg^{-1} · min^{-1})**
A	>20
B	16-20
C	10-16
D	<10

that sodium restriction plays in minimizing fluid congestion cannot be overemphasized. Sodium restriction can help reduce the need for diuretics, and a dietary plan that aggressively restricts sodium intake may actually allow for the discontinuation of diuretic therapy.

The use of digoxin for patients with HF, although still unresolved, has been in practice for decades. Although the controversy applies mostly to HF patients with normal sinus rhythm, this agent has been shown to reduce hospitalizations by 27%. The use of digoxin in patients with HF and atrial fibrillation is not challenged. Others agents that may be used to treat patients with HF attributable to systolic dysfunction include antiplatelet and anticoagulation therapies, angiotensin receptor blockers, and an aldosterone antagonist.

In some instances, and despite aggressive attempts to optimize medical therapy, a patient's clinical condition continues to deteriorate such that special medications or mechanical left ventricular assist devices are required to augment circulation and maintain adequate blood flow. Without these efforts, many such patients die in months if not weeks. Alternately, cardiac transplantation is considered, which is a surgical procedure that is now accepted as standard therapy for patients with end-stage HF that is refractory to maximal medical therapy. An overview of this procedure is given in practical application 15.1.

An important part of the care for any chronic disease remains secondary prevention. For the exercise professional working with HF patients, this includes counseling to help manage behavioral habits known to exacerbate the condition. For example, in patients with an ischemic cardiomyopathy, every attempt should be made to help stabilize existing coronary atherosclerosis through aggressive risk factor management and prevent further loss of cardiac myocytes attributable to reinfarction.

For the patient with HF attributable to other causes such as illegal substances or alcohol abuse, the exercise physiologist should support healthy behaviors and be alert for signs of relapse that may require referral to a specialist. For all patients with HF, observing a low-sodium diet is an important step toward preventing congestion and fluid overload, thus reducing chances of subsequent hospitalizations. Consistent with this, compliance with all prescribed medications, especially diuretics aimed at removing excess fluid, is an important variable that an exercise professional can assess during clinical appointments or prior to exercise class. See practical application 15.2.

Graded Exercise Testing

The use of exercise testing in patients with HF provides an enormous amount of useful information. Not only is information gathered on severity of illness and 3-year survival, but response to medications, response to an exercise training program, and information to guide exercise training intensity are obtained as well.

Medical Evaluation

The method for exercise testing patients with HF differs little from the testing used in patients with other types of heart disease. Although the majority of the exercise tests conducted in these patients use a steady-state (3 min per stage) protocol like the modified Bruce or Naughton (55), a ramp protocol can be performed with a stationary cycle. With this method, external work rate is increased 10 to 15 W every minute (50, 51). A ramp protocol generally results in less variable data during submaximal exercise because of the more gradual increments in work rate that it provides. $\dot{V}O_2$peak is approximately 10% to 15% lower when measured with cycle ergometry versus a treadmill (34, 51, 53).

Because an accurate measure of $\dot{V}O_2$ is needed in these patients, the use of prediction equations to estimate $\dot{V}O_2$ is discouraged because they tend to overpredict functional capacity (22). Measured exercise capacity using a cardiopulmonary cart (figure 15.2) is preferred, and such equipment is available in either the cardiac noninvasive or the pulmonary function laboratory of most hospitals. In addition to measuring $\dot{V}O_2$peak, determining **ventilatory derived lactate threshold** (V-LT) is helpful. An adequate discussion of this parameter is provided elsewhere (22); however, an often used approach when determining V-LT is the V-slope method (6).

Contraindications

Fifteen years ago, standard teaching in most medical schools was that moderate or harder physical activity should be avoided or withheld in patients with HF. It was thought that the increased hemodynamic stress that exercise placed on an already weakened heart would further worsen heart function. As a result, most guidelines listed HF as an absolute contraindication to exercise testing. Today, however, patients with stable HF routinely undergo cardiopulmonary exercise testing to evaluate cardiorespiratory function.

All other contraindications to exercise testing still apply to patients with HF, including acute myocardial infarction and malignant arrhythmia (chap-

Cardiac Transplantation

Cardiac transplantation is an effective therapeutic alternative for persons with end-stage HF. Each year, approximately 3000 to 4000 such transplants are performed worldwide, with 1-and 3-year survival rates approximating 83% and 74%, respectively (31). In the vast majority of patients undergoing cardiac transplant, the atria of the recipient's heart are attached at the level of the atria of the donor's heart.

After surgery, many patients with cardiac transplant continue to experience exercise intolerance. In fact, $\dot{V}O_2$peak is typically between 14 and 22 mL $\cdot$ kg^{-1} $\cdot$ min^{-1} among untrained patients. For this and other reasons, it is common to enroll these patients in a home-based or supervised cardiac rehabilitation program as soon as 4 weeks after surgery. An expected increase in $\dot{V}O_2$peak ranges between 15% and 25% (39, 42).

There is an increasing probability of developing accelerated atherosclerosis in the coronary arteries of the donor heart. For this and other reasons, the traditional risk factors for ischemic heart disease such as hypertension, obesity, diabetes, and hyperlipidemia are aggressively treated.

Patients with cardiac transplant represent a unique physiology, in that the donated heart they received is decentralized. This means that except for the parasympathetic, postganglionic nerve fibers that are left intact, all other cardiac autonomic fibers are severed. Regeneration of these fibers may occur after 1 year in some patients, but it is likely best to assume that decentralization, for the most part, is permanent.

Because of the decentralized myocardium, the cardiovascular response of these patients to a bout of acute exercise differs from normally innervated people. At rest, HR is elevated at 90 to 100 beats $\cdot$ min^{-1}; this is because there is no longer parasympathetic (i.e., vagal nerve) influence on the sinoatrial node. When exercise begins, HR changes little because (a) there is no parasympathetic input that can be withdrawn and (b) there are no sympathetic fibers to directly stimulate the heart. During latter exercise, HR slowly increases because of an increase in norepinephrine in the blood. At peak exercise, HR is lower and both cardiac output and stroke volume are approximately 25% lower than age-matched controls.

Because of the absence of parasympathetic input, the decline of HR in recovery takes longer than normal. This is most observable during the first 2 min of recovery, during which HR may stay at or near the value achieved at peak exercise. Systolic blood pressure recovers in a generally normal fashion after exercise (21), which is why this measure can be used to assess adequacy of recovery.

The number of people needing a cardiac transplant each year far exceeds the available number of donors. As a result, many patients die while awaiting a heart, whereas others receive mechanical left ventricular assist devices or enroll in experimental programs testing new medications or other devices. The magnitude of this donor shortage will only increase over the next 15 years, as the number of patients with end-stage HF continues to increase.

---

ter 6). Interestingly, arrhythmias are common in patients with HF; therefore, information detailing a patient's history should be communicated to the person supervising an exercise test. Please note that the use of exercise testing to assess myocardial ischemia can be problematic, because many patients with HF present with ECG findings that invalidate or reduce the sensitivity of the test (e.g., left bundle branch block, left ventricular hypertrophy, and nonspecific ST-wave changes attributable to digoxin therapy).

Recommendations and Anticipated Responses

Compared with healthy normal people, patients with HF exhibit differences in their cardiorespiratory and peripheral responses at rest and during exercise (table 15.3) (3, 36, 43). Resting stroke volume and cardiac output are both lower in patients with HF versus controls (stroke volume: ~50 vs. ~75 ml $\cdot$ beat^{-1}; cardiac output: <2.5 vs. >2.5 L $\cdot$ min^{-1} $\cdot$ m^{-2}, respectively). Resting heart rate (HR) is increased (HF = 75-105 beats $\cdot$ min^{-1} vs. controls = 60-80 beats $\cdot$ min^{-1}) and systolic blood pressure may be reduced, attributable to both the underlying LV systolic dysfunction and the use of afterload-reducing agents such as ACE inhibitors.

During submaximal exercise, patients with HF exhibit a higher HR, a lower stroke volume response and attenuated increases in cardiac output and $\dot{V}O_2$

Client–Clinician Interaction

This chapter discusses several important issues relative to supervising exercise training in patients with HF. One of these issues includes the exercise professional's responsibility to ensure that on any given day, the patient is free of any signs or symptoms that might indicate that it is necessary to withhold exercise.

To accomplish this involves not only that the patient verbalize to you any problems she might be having, but also that you take the initiative to ask the patient key questions. Following are examples of these types of questions:

— How did you sleep over the weekend? Any more waking up during the night short of breath?

— How has your body weight been over the past week?

— It's been hot the past couple days; any increased difficulty breathing?

— Do you ever get that swelling in your ankles any more?

Although seemingly harmless, each enables you to assess change in HF-related symptoms. With time and experience, you will develop many of your own "pearls" about when and how to assess these patients.

Patients with heart failure may not tolerate the first few days of their exercise rehabilitation well. Therefore, verbal encouragement and guidance on your part are important. Explain that it is common to expect an improved ability to perform routine activities of daily living more comfortably. For some patients, this alone may be quite fulfilling. Other patients may wish to improve their exercise capacity to the point that they can resume activities that they previously had to avoid. Although such a goal may be realistic, be sure to emphasize that it is prudent to advance their training volume in a progressive manner. Trying to do too much too soon may actually lead to disappointment, if their functional improvement does not keep pace with their self-assigned interests.

Finally, when working with patients with HF, emphasize the importance of regular attendance to their exercise regimen. If they need to miss exercise because of personal or medical reasons, let them know that you look forward to seeing them back in class when they feel better.

---

compared with persons without HF (13, 57, 65). To compensate, the extraction of oxygen in exercising muscles is higher in patients with HF than in persons without HF (62). Also during exercise, plasma norepinephrine, an endogenous **catecholamine,** is released directly to the heart by sympathetic post-ganglionic fibers at increased levels. And there is a disproportionate increase in plasma norepinephrine levels as well, with the magnitude of the increase generally related to the severity of the illness (41).

Increasing blood flow to metabolically active skeletal muscles is part of a complex interplay between blood pressure and vasoconstriction/vasodilation response to exercise. Patients with HF have a reduced exercise-induced vasodilation attributable to both the increased plasma norepinephrine (41) and lower levels of a local chemical called endothelium-derived relaxing factor (16, 44).

Compared with healthy normal persons, at peak exercise patients with HF exhibit a lower power output (45% decrease), lower cardiac output (40% decrease), lower $\dot{V}O_2$ (40-50% decrease), lower stroke volume (50% decrease), and lower HR (20% decrease) (13, 14, 30, 52, 57). Depending on severity of illness, $\dot{V}O_2$peak typically ranges from 8 to 21 ml $\cdot$ kg^{-1} $\cdot$ min^{-1}. For patients not taking a β-adrenergic blocking agent, peak HR generally ranges from 132 to 150 beats $\cdot$ min^{-1}. This blunted or attenuated response of HR to exercise is referred to as **chronotropic incompetence** and represents a hallmark feature of HF. In 1999, Robbins and coworkers (55) showed that, like $\dot{V}O_2$peak, chronotropic incompetence during exercise is a powerful and independent predictor of mortality rate in patients with HF.

Despite the diminished peak exercise capacity of these patients, the degree of exercise intolerance is not related to the magnitude of LV systolic dysfunction. Among 234 patients tested in our laboratory we observed no relationship (r = 0.14) between ejection fraction (range = 8-40%) and $\dot{V}O_2$peak (range = 7.5-34.3 ml $\cdot$ kg^{-1} $\cdot$ min^{-1}).

TABLE 15.3

Characteristics of Patients With Heart Failure Attributable to Systolic Dysfunction

	Resting	Submaximal exercising	Peak exercising
CARDIORESPIRATORY			
Ejection fraction	↓	↓	↓
Cardiac output	↓	↓	↓
Stroke volume	↓	↓	↓
Heart rate	↑	↑	↓
Oxygen consumption	↑↔	↓	↓
Arterial-mixed venous oxygen difference	↑	↑	↑↔
$\dot{V}O_2$ at ventilatory-derived lactate threshold	N/A	↓	N/A
PERIPHERAL			
Arterial blood lactate	↔	↑	↓
Total systemic vascular resistance	↑	↑	↑
Blood flow in active muscle	↔	↓	↓
Skeletal muscle mitochondrial density	↓	N/A	N/A
Skeletal muscle oxidative enzymes	↓	N/A	N/A
Reliance on anaerobic metabolism	N/A	↑	↔

Note. ↑ = increased response compared with healthy subjects; ↓ = decreased response compared with healthy subjects; ↔ = similar responses compared with healthy subjects; NA = not applicable.

This suggests that other factors besides LV ejection fraction alone contribute to the exercise intolerance experienced by these patients. Although this may seem a bit perplexing, in that abnormalities in tissues other than the damaged heart are involved, this in fact occurs. Two such noncardiac mechanisms that limit exercise capacity are as follows:

1. An inability to sufficiently dilate peripheral vasculature as a means to increase blood flow to the metabolically active muscles (64)
2. Histological and biochemical abnormalities within the skeletal muscle itself (19, 41, 58)

In HF patients, there is a decrease in the percentage of type 1 muscle fibers, diminished oxidative enzymes, and decreased capillary density. As a result, and when compared to normals, patients with HF rely on anaerobic pathways to produce energy earlier during exercise.

Exercise Prescription

For patients with HF attributable to LV systolic dysfunction, continuously ECG monitored exercise is probably advised for 9 to 12 sessions (2), with the intention that such a practice will help identify abnormal findings and advoid clinical events. After demonstrating for 2 to 3 weeks that they are able to tolerate a three time per week exercise program, patients can begin a one to two time per week home exercise program. As patients continue to improve and demonstrate no complications to exercise after 12 to 24 supervised sessions, they can transfer to an unmonitored or all home-based exercise program.

Special Exercise Considerations

Despite the increased attention given to using moderate exercise training in the treatment plan of patients with HF, few guidelines exist for prescribing exercise in these patients. Prudent eligibility criteria for exercise training in patients with HF might be as follows:

— Ejection fraction less than 40%
— NYHA class II or III
— Receiving standard drug therapy of ACE inhibitors, β-adrenergic blockade, or diuretics for at least 4-6 weeks
— Absence of any other cardiac or noncardiac problems that would limit participation in exercise

Table 15.4 (3) suggests relative and absolute contraindications to exercise training in patients with HF.

The ejection fraction cutoff of less than 40% was chosen to remain consistent with the majority of the exercise studies conducted in these patients to date. By no means do we intend to infer that patients with an ejection fraction greater than 40% do not benefit from an exercise training regimen. On the contrary,

these patients improve as well. Likewise, in addition to enrolling stable NYHA class II and III patients in an exercise program, it may be possible to include carefully selected and motivated class IV patients who are free of pulmonary congestion. We previously reported a case detailing physiological outcomes in an ambulatory class IV patient who underwent 24 weeks of exercise training while receiving continuous dobutamine therapy (32). Such patients typically require more supervision during exercise.

Another important exercise consideration for patients with HF is compliance. In July 2000 we retrospectively reviewed the records of all patients enrolled in the Henry Ford Hospital cardiac rehabilitation program between August 1, 1977, and June 30, 2000. We observed that in patients who suffered a myocardial infarction, nearly 75% of both men and women completed the program. These percentages were lower in patients with HF and no myocardial infarction, in that 71% of men and only 53% of women completed the program.

One reason related to their lower compliance is the fact that they often have comorbidities or other illnesses that interrupt or prevent regular program attendance. For example, arrhythmias, pneumonia, adjustments in medications, and hospitalizations

Literature Review

Although exercise is an important component in the treatment of HF today, it was not until the late 1980s and early 1990s that sufficient research showed that patients could safely derive benefit. Lee et al. (45), Conn et al. (15), and Squires et al. (56) were the first to describe the structured use of exercise training in patients with HF. Improvements in functional capacity and NYHA functional class were observed. In 1988, two uncontrolled trials (4, 59) both demonstrated improvements in functional capacity, as measured by $\dot{V}O_2$peak. Since that time, more than a dozen randomized controlled trials showed that exercise training safely results in a 15% to 25% increase in $\dot{V}O_2$peak (table 15.5). Worth noting is the fact that most of the patients in these trials were taking ACE inhibitors, diuretics, and digoxin, which means that the gain in exercise capacity is in addition to what was derived from standard medical therapy at the time.

Likewise, improvements in submaximal exercise tolerance were observed as well, as evidenced by 20% to 25% improvements in V-LT. The mechanisms responsible for the increases in peak exercise capacity and V-LT are likely several, given that abnormalities in central transport (i.e., HR and stroke volume), nutritive blood flow to the active skeletal muscles, and skeletal muscle histology and biochemistry all contribute to exercise intolerance.

However, no large-scale trial exists to prove that regular exercise training improves survival or lowers recurrent hospitalizations in patients with HF. A 1999 study using a small sample of 99 patients randomized to either 1 year of exercise training or no exercise did show an approximate 67% decrease in risk for subsequent deaths and hospitalizations among exercise trained patients (7). Health-related quality of life was also improved in these patients versus the no-exercise group. Although this information is certainly promising, larger trials are needed to confirm these findings before the use of exercise therapy is standardized for all eligible patients.

TABLE 15.4

Relative and Absolute Contraindications

Relative contraindications	Absolute contraindications
1. 1.4–2.3 kg increase in body mass over previous 5-7 days	1. Signs or symptoms of worsening or unstable heart failure over previous 3-5 days
2. Concurrent continuous or intermittent dobutamine therapy	2. As defined by existing evidenced-based guidelines, abnormal blood pressure, early ischemic changes, or unexpected life-threatening arrythmia
3. Abnormal increase or decrease of blood pressure with exercise	3. Uncontrolled metabolic disorder (e.g., hypothyroidism, diabetes)
4. New York Heart Association Functional Class IV	4. Acute systemic illness or fever
5. Complex ventricular arrhythmia at rest or appearing with exertion	5. Recent embolism or thrombophlebitis
6. Supine resting heart rate $\geq$100 beats $\cdot$ min^{-1}	6. Active pericarditis or myocarditis
7. Preexisting comorbidity or behavioral disorder	7. Third-degree heart block without pacemaker
8. Implantable cardiac defibrillator with heart rate limit set below the target heart rate for training	8. Significant uncorrected valvular disease except mitral regurgitation due to HF related LV dilation
9. Extensive myocardial infarction resulting in left ventricular dysfunction within previous 4 weeks	

can all affect regular participation. Additionally, some patients with HF often do not tolerate their first few exercise sessions well, experiencing fatigue later in the day. To help overcome the poorer compliance in this patient population, be cognizant of these issues and ask patients to avoid trying to schedule too much on the days they exercise. Also, inform patients to expect interruptions in their exercise therapy caused by both HF- and non-HF-related issues. When they feel better, they should make every attempt to return to class.

Exercise Recommendations

Exercise training for patients with HF should be designed to improve cardiovascular and muscular endurance, muscular strength, and flexibility. Training for cardiorespiratory endurance is an obvious strategy for patients with HF; however, muscular strength and flexibility training improves functional capacity and fosters independence.

Cardiovascular Training

Unique issues pertinent to prescribing exercise in patients with HF are described next including mo-

dality, intensity, duration, and frequency of exercise. (See chapter 7 for general cardiorespiratory exercise prescription information.)

Mode To improve cardiorespiratory fitness, select activities that engage large muscle groups such as stationary cycling or walking. Benefits in exercise capacity or tolerance gained from either of these modalities transfer fairly well to routine activities of daily living. Consistent with this, consider including some upper body ergometry activities to improve function in the upper limbs as well. The exercise prescription for resistance training is described later in this section.

In terms of comparing one modality to another, a review of the exercise training trials involving patients with HF indicates that similar increases in $\dot{V}O_2$peak occur with both walking and cycle exercise (7, 8, 12, 17, 18, 26-28, 33, 37, 39). Many of these studies are shown in table 15.5.

Frequency Similar to people without HF, a regular exercise regimen of four to five times per week is required. As mentioned, patients should be counseled to remain regular with their exercise habits and,

Summary of Exercise Recommendations for Patients With Heart Failure

Screening

A limited physical examination should be performed to identify acute signs or symptoms that would prevent participation in exercise. These might include an increase in body mass greater than 1.4 kg during the previous 24 to 48 hr, lung sounds (i.e., rales) consistent with pulmonary congestion, complaints of increased difficulty sleeping while lying flat, sudden awakening during the night because of labored breathing, or increased swelling in the ankles or legs. Additionally, a graded exercise test is usually warranted to both identify any important exercise-induced arrhythmias and develop a target HR range that allows for determining a safe exercise intensity.

Exercise Program

As tolerated, modulate type, frequency, duration, and intensity of activity to progressively attain an exercise energy expenditure of 700 to 1000 kcal · week^{-1}.

Cardiovascular Training

— Mode: Walking and cycling to start, progressing to arm ergometry if interested and tolerated.

— Frequency: Four to five times per week.

— Intensity: Using the HR reserve method, progressively increase training HR from 60% to 70% of HR reserve. Titrate intensity such that rating of perceived exertion is maintained between 11 and 14. Also, if gas exchange was performed as part of an exercise test completed before the patient started rehabilitation, be sure that exercise training HR is four or more beats below HR at V-LT.

— Duration: Progressively increase, as tolerated, from 10 min or more per session up to 40 min per session.

Resistance Training

— Mode: Using free weights, fixed machines, or elastic bands, incorporate six to eight regional body exercises that focus on specific muscle groups. Improving lower limb strength and endurance represents a primary goal for these patients.

— Frequency: One to two times per week.

— Intensity: Approximately 60% of 1-repetition maximum.

— Duration: One set of 12 to 15 repetitions for each of the involved exercises.

Range of Motion

— Mode: Static stretching.

— Frequency: Before and after each aerobic or resistance training workout.

— Duration: Five minutes before and 5 to 10 min after each workout, with 10 to 30 s devoted to the major muscle groups and joints.

Special Considerations

— Many patients with HF are inactive and possess a low tolerance for activity. Be sure to progressively increase exercise dose in a manner that is individualized for every patient.

— For the first week or so of exercise training, patients may be a bit tired later in the day. To compensate for this, ask them to temporarily limit the amount of other, home-based activities.

— Exercise training programs for patients with HF are often interrupted because of both cardiac- and non-cardiac-related reasons. Inform patients that this is not uncommon and encourage them to regularly exercise when they feel well.

Summary of Exercise Recommendations for Patients With Cardiac Transplant

Screening

Because patients with cardiac transplant present with a decentralized heart, use of a graded exercise test before they begin an exercise program for the purpose of developing an HR range is not indicated. However, such a test will help screen for exercise-induced arrhythmias, quantify exercise tolerance, and serve as a marker to assess exercise outcomes at a future date. Ehrman et al. (20) showed that an exercise ECG does not detect ischemic heart disease well in this patient group. If ischemic heart disease is suspected in a cardiac transplant patient, an exercise test with cardiac imaging is preferred.

Exercise Program

As tolerated, modulate type, frequency, duration, and intensity of activity to progressively attain an exercise energy expenditure of 700 to 1000 kcal $\cdot$ week^{-1}.

Cardiovascular Training

— Mode: Walking and cycling to start, progressing to arm ergometry if interested and tolerated.

— Frequency: Four to five times per week.

— Intensity: Titrate intensity such that rating of perceived exertion is maintained between 11 and 14 (38); this level of exertion results in a significant increase in $\dot{V}O_2$peak.

— Duration: Progressively increase, as tolerated, from 10 min or more per session up to 40 min per session.

Resistance Training

— Mode: Using free weights, fixed machines, or elastic bands, incorporate six to eight regional body exercises that focus on specific muscle groups. Improving lower limb strength and endurance represents a primary goal for these patients.

— Frequency: One to two times per week

— Intensity: Approximately 60% of 1-repetition maximum.

— Duration: One set of 12 to 15 repetitions for each of the involved exercises.

Range of Motion

— Mode: Static stretching.

— Frequency: Before and after each aerobic or resistance training workout.

— Duration: Five minutes before and 5 to 10 min after each workout, with 10 to 30 s devoted to the major muscle groups and joints.

Special Considerations

— Loss of muscle mass and bone mineral content occurs in patients on long-term corticosteroids such as prednisone. As a result, resistance training programs play an important role in partially restoring bone health and muscle strength (10). However, high-intensity resistance training should be avoided because of increased risk of fracture in bone that has compromised bone mineral content.

— Although isolated cases of chest pain or angina associated with accelerated graft atherosclerosis have been reported, decentralization of the myocardium essentially eliminates angina symptoms in most patients.

continued

— A regular exercise program performed within the first few months after surgery may result in an exercise HR during training that is equal to or exceeds the peak HR achieved during an exercise test taken before a patient started training. This response is not uncommon and supports using rating of perceived exertion as the method to guide intensity of effort.

— Marked increases in body fat leading to obesity sometimes occur in cardiac transplant patients. This is likely caused by both long-term corticosteroid use and restoration of appetite following illness. Exercise modalities chosen for training should consider any possible limitation caused by excessive body mass.

— Because of the sternotomy, postoperative range of motion in the thorax and upper limbs may be limited for several weeks.

— Although definitive evidence is lacking, calf discomfort is experienced during walking in approximately 15% of cardiac transplant patients. This symptom may be related to the use of cyclosporine.

if their program is interrupted because of personal or medical reasons, to plan on restarting the program as soon as possible. After being in an exercise program for several weeks, it is not uncommon for patients to comment that they notice less DOE and less fatigue during routine daily activities. Part of this includes educating patients about the detraining effect that occurs, which means informing them that many of the benefits they noticed will fade when exercise is stopped.

Intensity   Exercise intensity can be established using variables derived from an exercise test. In patients with HF, different ranges of exercise intensity have been used to improve cardiorespiratory fitness (table 15.5). As you can see, relatively similar gains in cardiorespiratory fitness (i.e., $\dot{V}O_2$peak) are evident regardless of whether a light (40-50%), moderate (60-70%), or high (70-80%) intensity regimen was used to guide exercise intensity.

Using the HR reserve method, for the first few exercise sessions guide exercise intensity at about the 60% level. Titrate this based on a patient's subjective feelings using the rating of perceived exertion scale set at 11 to 14. One reason to start these patients at this lower intensity level is to allow them an opportunity to adjust to the exercise. As mentioned, fatigue later in the day is not uncommon, so restricting exercise intensity at first may help minimize this.

As a patient demonstrates an improved ability to tolerate therapy, increase the intensity of effort to 70% of HR reserve. Again, be sure that a patient's rating of perceived exertion is maintained between 11 and 14. LV function may decrease further during exercise when patients with LV systolic dysfunction

are exercised above V-LT (52). Therefore, it is prudent to keep HR in the 60% to 70% range, which is likely at or below V-LT for most patients.

Duration   Most exercise trials involving patients with HF increased training duration up to 30 to 60 min of continuous exercise. On occasion, it may be necessary to start an individual patient with two to three bouts of exercise that are each 4 to 8 min in duration. Interval training such as this has been investigated in patients with HF by Myers and co-workers (47, 48), using a work/rest ratio during cycle ergometry of 30:60. This model allows for greater work rates to be placed on the muscles without a sustained higher intensity stress on the nervous, endocrine, and cardiorespiratory systems.

Resistance Training

The role and benefit of resistance training in patients with HF have not been widely investigated. To date, less than 10 studies have addressed the topic, most showing that mild to moderate resistance training is tolerated and improves muscle strength 20-45% in stable patients with HF. Given that loss of muscle strength and endurance is common in these patients (49, 61, 62), it is appropriate to assume that improvements in these measures through resistance training are helpful. However, no specific guidelines addressing resistance training in patients with HF currently exist.

Instead, resistance training recommendations in these patients are drawn from the scientific advisory statement published in 2000 (54). Generally, exercise or lift intensity should be set at 50% to 60% of 1-repetition maximum, with the patient doing one set of 10 to 12 then progressing to one set of 15 repe-

TABLE 15.5

Review of the Major Exercise Training Trials

Authors	Training intensity	Training mode	Peak $\dot{V}O_2$ (mL · kg^{-1} · min^{-1}) (before/after)
Coats et al. (12)	60-80% peak HR	Bike ergometer	T: 13.2±0.9/15.6±1.0 C: not given
Kiilavuori et al. (40)	50-60% of peak $\dot{V}O_2$	Bike ergometer	T: 20.7±2.1/23.9±3.0 C: 18.9±1.6/19.1±1.6
Belardinelli et al. (8)	40% peak $\dot{V}O_2$	Bike ergometer	T: +12% C: –5%
Hambrecht et al. (27)	70% peak $\dot{V}O_2$	Bike ergometer	T: 17.5±5.1/23.3±4.2 C: 17.9±5.6/17.9±5.6
Keteyian et al. (33)	60% initial, then 70-80% of peak $\dot{V}O_2$	Bike ergometer, walk, row	T: 16.0±3.7/18.5±4.4 C: 14.7±4.2/15.2±4.8
Kiilavuori et al. (39)	50-60% of peak $\dot{V}O_2$	Bike ergometer, walk, swim, row	T: 19.3±1.6/21.7±2.5 C: 18.3±1.3/18.2±1.5
Dubach et al. (18)	70-80% peak $\dot{V}O_2$	Bike ergometer, walk	T: 19.4±3.0/25.1±4.8 C: 18.8±3.9/19.8±4.3
Keteyian et al. (37)	50-60% initial, then 70-80% of peak $\dot{V}O_2$	Bike ergometer, walk, row	T: 16.1±0.7/18.4±0.9 C: 14.6±0.8/15.3±0.9
Belardinelli et al. (7)	60% peak $\dot{V}O_2$	Bike ergometer	T: 15.7±2.0/19.9±1.0 C: 15.2±2.0/16.0±2.0

Note. HR = heart rate; T = treatment, before/after training; C = control, before/after training.

titions. The specific exercises should address the individual needs of the patient but will likely include all of the major muscle groups. As patients improve, load should be increased 5% to 10%. However, before each training session, patients should be evaluated for any signs or symptoms of HF or excessive muscle/joint discomfort.

A summary of the exercise prescriptions used for patients with HF and for patients with cardiac transplant is given in practical applications 15.4 and 15.5 respectively.

Physiological Adaptations and Maladaptations

Improvements with exercise training in patients with HF can be seen not only in functional and cardiorespiratory measures (practical application 15.5) but also in the skeletal muscle and other organ systems as well. Benefits of exercise in patients with HF are widespread and warrant further investigation.

Myocardial Function

Exercise training results in a modest increase in peak cardiac output (8-15%), attributable to increases in both peak HR and peak stroke volume. Interestingly, the 4% to 8% increase in peak HR (i.e., partial reversal of chronotropic incompetence) accounts for up to 40% to 50% of the increase in $\dot{V}O_2$peak (2, 28, 33, 37, 43). This training adaptation of an increase in peak HR differs from that observed in healthy individuals, where no increase in peak HR is expected with training (22).

Despite initial concerns to the contrary, these changes in central transport occur without concomitant harmful effects to the myocardium. In fact, numerous trials, some up to 1 year in duration, show that regular training does not lead to disproportionate worsening of LV ejection fraction or an increase in LV size (17, 23, 24). This absence of a training-induced maladaptation within the LV was observed with both moderate intensity and higher intensity training protocols.

Skeletal Muscle

In response to exercise training, there are improvements in the skeletal muscle. These include an improved ability to dilate the small blood vessels that nourish the metabolically active muscles, which leads to an approximate 25% to 30% increase in local blood flow (27, 28). The mechanisms responsible for this include an exercise training–induced decrease in plasma norepinephrine (37) and other vasoconstrictor agents (9), as well as increased levels of endothelium-derived relaxing factor (25).

Although only a few studies investigated the effects of exercise training on intrinsic characteristics of the skeletal muscle, it appears there may be a benefit at this level as well. Specifically, the volume density of mitochondria and the enzymes involved with aerobic metabolism (e.g., cytochrome C oxidase) is improved 19% to 40% (27). There is also a mild shift in fiber type, involving a small increase in type 1 (26). Finally, Barnard et al. (5) showed a 31% increase in single-repetition leg extension strength following 8 weeks of combined aerobic and resistance training, compared with no change in a group of HF patients who performed aerobic training only.

Other Organ Systems

As mentioned in the Physiology and Pathophysiology section of this chapter, patients with HF also have abnormalities of the autonomic nervous system, such as an increase in sympathetic activity and a decrease in parasympathetic activity. This im-balance likely makes the heart more prone to arrhythmia and less responsive to stressful situations.

Several studies show that exercise training favorably alters both autonomic and hormonal function in patients with HF (12, 40, 60). There is down-regulation of sympathetic nervous system and an increase toward more parasympathetic activity (+50%). Additionally, the levels of circulating norepinephrine in the blood both at rest and during exercise are decreased with training (12, 37).

Finally, various characteristics of ventilation are improved with exercise training, all contributing to the fact that less dyspnea is reported. Specifically, the relationship between minute ventilation and CO_2 produced is changed, such that during submaximal exercise less CO_2 is exhaled for any given amount of minute ventilation.

Conclusion

HF, today's fastest growing cardiac-related diagnosis, is at near-epidemic proportions in the United States. And over the next 10 years, much research will investigate and define to what extent exercise alters both the pathophysiology of the disease and the associated clinical outcomes. Concerning clinical outcomes, in October 2002 over 40 medical centers in the U.S. and Canada, in conjunction with the National Heart, Lung and Blood Institute, began a five-year clinical trial evaluating the effectiveness of up to 2.5 years of exercise training on mortality and hospitalizations in patients with stable HF. Based on current research we know that, among eligible patients with stable HF, regular exercise training improves exercise tolerance and quality of life. Therefore, healthcare practitioners involved in the care of such patients should consider regular exercise training when developing their treatment strategies.

Case Study 15

Medical History

Mr. WT is a 55-year-old African American male who was initially hospitalized in 1996. The patient stated he had no prior history of diabetes, angina, or myocardial infarction. He did have a positive history for cocaine use, as recently as 1 week prior to hospitalization. He had a strong family history of ischemic heart disease and denied alcohol and tobacco use. He worked as a computer programmer and was sedentary during leisure.

continued

Diagnosis

At the time of admission, he complained of being unable to lie flat without being short of breath, which had increased over the prior several days. Cardiac catheterization revealed three-vessel coronary artery disease, which included a 20% lesion in the left main, a 95% stenosis in the left anterior descending artery, total occlusion of the circumflex, and a 70% narrowing in the proximal right coronary artery. An echocardiogram performed during the same admission demonstrated a left ventricular ejection fraction of 30%. Diagnosis was ischemic cardiomyopathy and the patient was discharged from the hospital and scheduled to see a cardiologist in the outpatient clinical later in the week. Medications at the time of hospital discharge were lisinopril, simvastatin, lopressor, digoxin, and furosemide.

Exercise Test Results

Before the patient enrolled in cardiac rehabilitation, a graded exercise test was completed. Resting blood pressure was 138/96 mm Hg. Resting ECG showed sinus rhythm with a rate of 72 beats $\cdot$ min^{-1}. Occasional ventricular premature beats were noted at rest, along with left ventricular hypertrophy. An old anterior-lateral infarction pattern was present, age indeterminate.

The patient exercised for 6 minutes on a treadmill to a $\dot{V}O_2$ of 17.2 mL $\cdot$ kg$^{-1}\cdot$ min^{-1}. Chest pain was denied and exercise was stopped because of fatigue and dyspnea. Peak HR was 123 beats $\cdot$ min^{-1}, peak blood pressure was 158/98 mm Hg, and 1.0 mm of additional ST-segment depression was observed in lead V_6. Isolated ventricular premature beats were again observed.

Other Procedures

The patient was seen in the outpatient clinic 6 days after discharge from the hospital. Blood lipids were obtained and revealed total cholesterol of 227 mg $\cdot$ dL^{-1}, high-density lipoprotein cholesterol of 27 mg $\cdot$ dL^{-1}, and low-density lipoprotein cholesterol of 172 mg $\cdot$ dL^{-1}. A dobutamine echocardiogram was performed, which showed improved contractility. Based on this test and the cardiac catheterization results, coronary artery bypass surgery was recommended. The patient refused surgery, opting for medical management, lifestyle changes, and cardiac rehabilitation.

Development of Exercise Prescription

Mr. WT's goal was to make important and aggressive changes in his lifestyle. This included increasing his activity levels from being inactive to exercising 3 to 5 days per week. During rehabilitation he was able to tolerate 30 min of exercise without complication. The upper HR limit for exercise intensity was set at 104 beats $\cdot$ min^{-1}, which represented 4 beats below the HR observed at the V-LT obtained during his exercise test.

Case Study 15 Discussion Questions

1. Given Mr. WT's prior history of being inactive during leisure, what is the likelihood that he will still be following a regular exercise program 3 years in the future? What steps can you take now to ensure long-term compliance?

2. Which symptoms did Mr. WT complain of at the time of hospitalization that were consistent with the diagnosis of cardiomyopathy or heart failure?

continued

Case Study 15   *(continued)*

> 3. What target levels would you recommend that Mr. WT achieve for his blood lipid levels?
>
> 4. Based on data from Mr. WT's exercise test, did he show chronotropic incompetence? Is this a common or uncommon finding in patients with HF?
>
> 5. Given the $\dot{V}O_2$peak measured during his exercise test, what magnitude of improvement, if any, would you expect after 12 weeks of exercise training? How would this compare to a sedentary, apparently healthy person who also undergoes 12 weeks of training?

Glossary

catecholamine—Chemicals released in the body that are major elements in the response to stress and exercise. Two catecholamines of interest are epinephrine and norepinephrine. Both exert, among other properties, a positive inotropic and chronotropic effect on cardiac function.

chronotropic incompetence—Blunted or attenuated response of HR to exercise such that peak HR is less than 85% of age-predicted; a hallmark feature of HF.

dyspnea on exertion (DOE)—Labored or difficulty breathing during exertion.

echocardiogram—An investigation of the heart and great vessels with ultrasound technology as a means to diagnose cardiovascular abnormalities.

ejection fraction—The percentage of end-diastolic blood volume that is pumped out of the ventricle during systole.

heart failure—The pathophysiological state in which an abnormality of cardiac function is responsible for failure of the heart to pump blood at a rate commensurate with the requirements of metabolizing tissues.

ischemic heart disease—A pathological condition in which blood flow to the myocardium is reduced below the demand and a lack of oxygen delivery to cardiac tissue results (i.e., coronary atherosclerosis).

LV diastolic dysfunction—Clinically, diagnosis is less exact than systolic dysfunction. Quite often diagnosis is made when the clinical syndrome of congestive HF (fatigue, dyspnea, and orthopnea) requires hospitalization in the presence of a relatively normal ejection fraction.

LV systolic dysfunction—Ejection fraction reduced below 45% (severe considered <30%) as measured by echocardiogram or another quantitative measure.

nonischemic cardiomyopathy—Disease process involving cardiac muscle that is not related to ischemic heart disease; may be attributable to viral infection or ethanol abuse.

orthopnea—Labored or difficult breathing while lying flat or supine.

paroxysmal nocturnal dyspnea—Sudden awakening caused by labored or difficult breathing.

ventilatory derived lactate threshold (V-LT)—The point where a nonlinear increase in blood lactate occurs during exercise; when determined with ventilatory parameters is sometimes referred to as *ventilatory threshold*.

References

1. American College of Cardiology/American Heart Association. Guidelines for the Evaluation and Management of Chronic Heart Failure in the Adult: Executive Summary. *J Am Coll Cardiol* 38: 2101-2113, 2001.

2. American Association of Cardiovascular and Pulmonary Rehabilitation. *Guidelines for Cardiac Rehabilitation and Secondary Prevention Programs*. Champaign, IL: Human Kinetics, 1999.

3. Afzal, A., C.A. Brawner, S.J. Keteyian. Exercise training in heart failure. *Prog Cardiovasc Dis* 41: 175-190, 1998.

4. Arvan, S. Exercise performance of the high risk acute myocardial infarction patient after cardiac rehabilitation. *Am J Cardiol* 62: 197-201, 1988.

5. Barnard K.L., K.J. Adams, A.M. Swank, M. Kaelin, M.R. Kushnik, D.M. Denny. Combined high intensity and aerobic training in congestive heart failure patients. *J Strength Conditioning Res* 14: 383-388, 2000.

6. Beaver, W.L., K. Wasserman, B.J. Whipp. A new method for detecting anaerobic threshold by gas exchange. *J Appl Physiol* 60: 2020-2027, 1986.

7. Belardinelli, R., D. Georgiou, G. Cianci, A. Purcaro. Randomized, controlled trial of long-term moderate exercise training in chronic heart failure. *Circulation* 99: 1173-1182, 1999.

8. Belardinelli R., D. Georgiou, V. Scoccoo, et al. Low intensity exercise training in patients with heart failure. *J Am Coll Cardiol* 26: 975-982, 1995.

9. Braith, R.W., M.A. Welsch, M.S. Feigenbaum, et al. Neuroendocrine activation in heart failure is modified by endurance exercise training. *J Am Coll Cardiol* 34: 1170-1175, 1999.

10. Braith, R.W., R.M. Mills, M.A. Welsch, J.W. Keller, M.L. Pollock. Resistance training restores bone mineral density in heart transplant recipients. *J Am Coll Cardiol* 28: 1471-1477, 1996.

11. Colucci, W.S. and E. Braunwald. Pathophysiology of heart failure. In *A Textbook of Cardiovascular Medicine*. Braunwald, E., D.P. Zipes, and P. Libby (Eds.). Saunders. Philadelphia, 2001, p. 503.

12. Coats, A.J.S., S. Adamopoulos, A. Radaelli. Controlled trial of physical training in chronic heart failure. *Circulation* 85: 2119-2131, 1992.

13. Cohen-Solal, A., J.M. Chabernaud, R. Gourgon. Comparison of oxygen uptake during bicycle exercise in patients with chronic heart failure and in normal subjects. *J Am Coll Cardiol* 16: 80-85, 1990.

14. Colucci, W.S., J.P. Ribeiro, M.B. Rocco. Impaired chronotropic response to exercise in patients with congestive heart failure. *Circulation* 80: 314-323, 1989.

15. Conn, R.H., R.S. Williams, A.G. Wallace. Exercise responses before and after physical conditioning in patients with severely depressed left ventricular function. *Am J Cardiol* 49: 296-300, 1982.

16. Drexler, H., D. Hayoz, T. Munzel, et al. Endothelial function in chronic congestive heart failure. *Am J Cardiol* 69: 1596-1601, 1992.

17. Dubach, P., J. Myers, G. Dziekan, et al. Effect of exercise training on myocardial remodeling in patients with reduced left ventricular function after myocardial infarction. *Circulation* 95: 2060-2067, 1997.

18. Dubach, P., J. Myers, G. Dziekan, et al. Effect of high intensity exercise training on central hemodynamic responses to exercise in men with reduced left ventricular function. *J Am Coll Cardiol* 29:1591-1598, 1997.

19. Duscha, B.D., W.E. Kraus, S.J Keteyian, et al. Capillary density of skeletal muscle. *J Am Coll Cardiol* 33: 1956-1963, 1999.

20. Ehrman, J.K., S.J. Keteyian, A.B. Levine, et al. Exercise stress tests after cardiac transplantation. *Am J Cardiol* 71: 1372-1373, 1993.

21. Ehrman, J.K., S.J. Keteyian, R. Shepard, et al. Cardiovascular responses of heart transplant recipients to graded exercise testing. *J Appl Physiol* 73: 260-264, 1992.

22. Foss, M.L., S.J. Keteyian. *Fox's Physiological Basis for Exercise and Sport* (6th ed). Boston: McGraw-Hill, 1998.

23. Giannuzzi, P., L. Tavazzi, P.L. Temporelli, et al. Long-term physical training and left ventricular remodelling after anterior myocardial infarction: Results of the exercise in anterior myocardial infarction (EAMI) trial. *J Am Coll Cardiol* 22: 1821-1829, 1993.

24. Giannuzzi, P., P.L. Temporelli, U. Corra, et al. Attenuation of unfavorable remodelling by exercise training in post infarction patients with left ventricular dysfunction: Results of the exercise in left ventricular dysfunction (ELVD) trial. *Circulation* 96: 1790-1797, 1997.

25. Hambrecht, R., E. Fiehn, C. Weigl, et al. Regular physical exercise corrects endothelial dysfunction and improves exercise capacity in patients with chronic heart failure. *Circulation* 98: 2709-2715, 1998.

26. Hambrecht, R., E. Fiehn, J. Yu, et al. Effects of endurance training on mitochondrial ultrastructure and fiber type distribution in skeletal muscle of patients with stable chronic heart failure. *J Am Coll Cardiol* 29: 1067-1073, 1997.

27. Hambrecht, R., J. Niebauer, E. Fiehn, et al. Effects on cardiorespiratory fitness and ultrastructural abnormalities of leg muscles. *J Am Coll Cardiol* 25: 1239-1249, 1995.

28. Hambrecht, R., S. Gielen, A. Linke, et al. Effects of exercise training on left ventricular function and peripheral resistance in patients with chronic heart failure. *JAMA* 283: 3095-3101, 2000.

29. Heart Failure Society of America. HFSA guidelines for management of patients with heart failure caused by left ventricular systolic dysfunction. *J Cardiac Failure* 5: 358-382, 1999.

30. Higginbotham, M.B., K.G. Morris, E.H. Conn, et al. Determinations of variable exercise performance among patients with severe left ventricular dysfunction. *Am J Cardiol* 51: 52-60, 1983.

31. Hosenpud, J.D., L.E. Bennett, B.M. Keck, M.M. Boucek, R.J. Novick. The registry of the international society for heart and lung transplantation: Seventeenth official report-2000. *J Heart Lung Transplant* 19: 909-931, 2000.

32. Kataoka, T., S.J. Keteyian, C.R.C. Marks, et al. Exercise training in a patient with congestive heart failure on continuous dobutamine. *Med Sci Sports Exerc* 26: 678-681, 1994.

33. Keteyian, S.J., A.B. Levine, C.A. Brawner, et al. Exercise training in patients with heart failure. A randomized, controlled trial. *Ann Intern Med* 124: 1051-1057, 1996.

34. Keteyian, S.J., C.A. Brawner. Disparities in gas exchange response during cycle and treadmill exercise in heart failure patients. *Med Sci Sports Exerc* 27(Suppl): S159, 1995.

35. Keteyian, S.J., C. Brawner. Cardiac transplant. In *ACSM's Exercise Management for Persons With Chronic Diseases and Disabilities*. Durstine, J.L. and Moore, G.E. (Eds.). Human Kinetics. Champaign, IL, 2003, pp. 70-75.

36. Keteyian, S.J., C.A. Brawner, J.R. Schairer. Exercise testing and training of patients with heart failure due to left ventricular systolic dysfunction. *J Cardiopulm Rehabil* 17: 19-28, 1997.

37. Keteyian, S.J., C.A. Brawner, J.R. Schairer, et al. Effects of exercise training on chronotropic incompetence in patients with heart failure. *Am Heart J* 138: 233-240, 1999.

38. Keteyian, S.J., J. Ehrman, F. Fedel, K. Rhoads. Heart rate-perceived exertion relationship during exercise in orthotopic heart transplant patients. *J Cardiopulm Rehabil* 10: 287-293, 1990.

39. Kiilavuori, K., A. Sovijarvi, H. Naveri, et al. Effect of physical training on exercise capacity and gas exchange in patients with chronic heart failure. *Chest* 110: 985-991, 1996.

40. Kiilavuori, K., L. Toivonen, H. Naveri, H. Leinonen. Reversal of autonomic derangements by physical training in chronic heart failure assessed by heart rate variability. *Eur Heart J* 16: 490-496, 1995.

41. Kinugawa, T., K. Ogino, H. Kitamura, et al.. Response of sympathetic nervous system activity to exercise in patients with congestive heart failure. *Eur J Clin Invest* 221: 542-546, 1991.

42. Kobashigawa, J.A., D.A. Leaf, N. Lee, et al. A controlled trial of exercise rehabilitation after heart transplantation. *N Engl J Med* 340: 272-277, 1999.

43. Kokkinos, P.F., W. Chouair, P. Graves, V. Papademetrius, S. Ellahham. Chronic heart failure and exercise. *Am Heart J* 140: 21-28, 2000.

44. Kubo, S.H., T.S. Rector, A.J. Bank, et al. Endothelial-dependent vasodilatation is attenuated in patients with heart failure. *Circulation* 84: 1586-1596, 1991.

45. Lee, A.P., R.P. Ice, R. Blessey, et al. Long term effects of physical training in coronary patients with impaired ventricular function. *Circulation* 60: 1519-1526, 1979.

46. Mancini, D.M., G. Walter, N. Reichek, et al. Contribution of skeletal muscle atrophy to exercise intolerance and altered muscle metabolism in heart failure. *Circulation* 85: 1364-1373, 1992.

47. Meyer, K., L. Samek, M. Schwaibold, et al. Interval training in patients with severe chronic heart failure: Analysis and recommendations for exercise of procedures. *Med Sci Sports Exerc* 29: 306-312, 1997.

48. Meyer, K., M. Schwaibold, S. Westbrook, et al. Effects of short-term exercise training and activity restriction on functional capacity in patients with severe chronic congestive heart failure. *Am J Cardiol* 78: 1017-1022, 1996.

49. Minotti, J.R., I. Christoph, R. Oka, M.W. Weiner, L. Wells, B.M. Massie. Impaired skeletal muscle function in patients with congestive heart failure. *J Clin Invest* 88: 2077-2082, 1991.

50. Myers, J., L. Gullestad, R. Vagelos, et al. Cardiopulmonary exercise testing and prognosis in severe heart failure: 14 ml/kg/min revisited. *Am Heart J* 139: 78-84, 2000.

51. Myers, J., N. Buchanan, D. Walsh, et al. Comparison of the ramp versus standard exercise protocols. *J Am Coll Cardiol* 17: 1334-1342, 1991.

52. Normandin, E.A., D.N. Camaione, B.A. Clark, III, et al. A comparison of conventional versus anaerobic threshold exercise prescription methods in subjects with left ventricular dysfunction. *J Cardiopulm Rehabil* 13: 110-116, 1993.

53. Page E., A. Cohen-Solal, G. Jondeau, et al. Comparison of treadmill and bicycle exercise in patients with chronic heart failure. *Chest* 106: 1002-1006, 1994.

54. Pollock, M.L., B.A. Franklin, G.J. Balady, et al. Resistance exercise in individuals with and without cardiovascular disease. *Circulation* 101: 828-833, 2000.

55. Robbins, M., G. Francis, F.J. Pashkow, et al. Ventilatory and heart rate responses to exercise. *Circulation* 100: 2411-2417, 1999.

56. Squires, R.W., C.J. Lavie, T.R. Brandt, et al. Cardiac rehabilitation in patients with severe ischemic left ventricular dysfunction. *Mayo Clin Proc* 62: 997-1002, 1987.

57. Sullivan, M.J., D.J. Knight, M.B. Higginbotham, F.R. Cobb. Relation between central and peripheral hemodynamics during exercise in patients with chronic heart failure. *Circulation* 80: 769-781, 1989.

58. Sullivan, M.J., H.J. Green, F.R. Cobb. Skeletal muscle biochemistry and histology in ambulatory patients with long-term heart failure. *Circulation* 81: 518-527, 1991.

59. Sullivan, M.J., M.B. Higginbotham, F.R. Cobb. Exercise training in patients with severe left ventricular dysfunction. *Circulation* 78: 506-515, 1988.

60. Toepher, M., K. Meyer, P. Maier, et al. Influence of exercise training and restriction of activity on autonomic balance in patients with severe congestive heart failure. *Clin Sci* 91(Suppl): 116, 1996.

61. Toth, M.J., S.S. Gottlieb, M.L. Fisher, et al. Skeletal muscle atrophy and peak oxygen consumption in heart failure. *Am J Cardiol* 79: 1267-1269, 1997.

62. Volterrani, M., A.L. Clark, P.F. Ludman, et al. Predictors of exercise capacity in chronic heart failure. *Eur Heart J* 15: 801-809, 1994.

63. Weber, K.T., G.T. Kinasewitz, J.S. Janicki. Oxygen utilization and ventilation during exercise in patients with chronic cardiac failure. *Circulation* 65: 1213-1223, 1982.

64. Wilson, J.R., J.L. Martin, D. Schwartz, et al. Exercise intolerance in patients with chronic heart failure: Role of impaired nutritive flow to skeletal muscle. *Circulation* 69: 1079-1087, 1984.

65. Zhang, Y.Y., K. Wasserman, K.E. Sietsmea, et al. O_2 uptake kinetics in response to exercise. *Chest* 103: 735-741, 1993.

A.S. Contractor, MD
Director, Cardiac Rehabilitation
Cumballa Hill Hospital and Heart Institute
Bombay, India

Neil F. Gordon, MD, PhD, MPH
President and CEO, Intervent USA, Inc.
Savannah, GA

Hypertension

Hypertension is arbitrarily defined as a level of blood pressure (BP) at which a person has an increased risk of developing a morbid cardiovascular event or will clearly benefit from medical therapy. Perhaps the best operational definition is the level at which the benefits (minus the risks and costs) of action exceed the risks and costs (minus the benefits) of inaction (23). The Sixth Report of the Joint National Committee on Prevention, Detection, Evaluation, and Treatment of High Blood Pressure (JNC VI) defines hypertension as having a **systolic** BP of 140 mmHg or greater, having a **diastolic** BP of 90 mmHg or greater, or taking antihypertensive medication (21). The classification of BP for adults (aged 18 years or older) is shown in table 16.1.

In more than 95% of cases, the **etiology** of hypertension is unknown and it is called essential, **idiopathic,** or primary hypertension. Secondary hypertension is **systemic** hypertension with a known cause. Table 16.2 lists the causes of secondary hypertension and their relative frequency. (Although essential hypertension and secondary hypertension are the major classifications of hypertension, several other descriptive terms are used to define various types of hypertension.) Isolated systolic hypertension is defined as systolic BP of 140 mmHg or more and diastolic BP less than 90 mmHg. Malignant hypertension is the syndrome of markedly elevated

TABLE 16.1

Classification of Blood Pressure (BP) for Adults Aged 18 Years and Older[a]

Category	Systolic BP (mmHg)	Diastolic BP (mmHg)
Optimal	<120	<80
Normal	<130	<85
High-normal	130-139	85-89
Stage 1 hypertension[b]	140-159	90-99
Stage 2 hypertension[b]	160-179	100-109
Stage 3 hypertension[b]	≥180	≥110

[a]Not taking antihypertensive drugs and not acutely ill. When systolic and diastolic BPs fall into different categories, the higher category should be selected to classify the individual's BP status. [b]Based on the average of two or more readings taken at each of two or more visits after an initial screening.

TABLE 16.2

Prevalence of Various Forms of Hypertension in the General Population and in Specialized Referral Clinics

Diagnosis	General population (%)	Specialty clinic (%)
Essential hypertension	92-94	65-85
SECONDARY/RENAL HYPERTENSION		
Parenchymal	2-3	4-5
Renovascular	1-2	4-16
SECONDARY/ENDOCRINE HYPERTENSION		
Primary aldosteronism	0.3	0.5-12
Cushing's syndrome	<0.1	0.2
Pheochromocytoma	<0.1	0.2
Oral contraceptive induced	0.5-1	1-2
Miscellaneous	0.2	1

Reprinted, from A. Fauci, et al., 1998, *Harrison's principles of internal medicine* (New York: McGraw-Hill). Reproduced with permission of the McGraw-Hill Companies.

BP associated with **papilledema.** In these cases, the diastolic BP is usually greater than 140 mmHg. **White coat hypertension** is the situation in which a person's BP is elevated when measured by a physician or other healthcare personnel but is normal when measured outside of a healthcare setting.

Scope

Hypertension affects more than 50 million Americans, and its prevalence increases with age. In 1996, hypertension was listed as a primary or contributing cause of death in about 202,000 of the more than

2,000,000 U.S. deaths that year (4,10). It afflicts approximately 65% of the population in the 65- to 74-year-old group. According to the National Health and Nutrition Examination Survey III (NHANES III 1988-94), the estimated prevalence of high BP for U.S. adults ages 20 to 74 was 24.4% for non-Hispanic white males and 19.3 for females; 35% for non-Hispanic black males and 34.2% for females; and 25.2% for Mexican-American males and 22% for females (4, 10, 21). Compared with whites, blacks develop hypertension at an earlier age, and it is more severe at any decade of life.

Awareness of hypertension increased from the NHANES II (1976-1980) survey to the phase 1 NHANES III (1988-1991) survey, but it has plateaued since then and did not show an increase in the phase 2 NHANES III (1991-1994) survey. Similar trends are seen in the treatment data and the percentage of patients whose hypertension is under control (table 16.3). If hypertension awareness, treatment, and control rates had continued the trend established between 1976-1980 and 1988-1991, there would have been an increase in 1991-1994 in awareness to 76.2%, in treatment to 59.6%, and in control to 31.2% instead of the levels shown in table 16.3 (21).

The total estimated direct and indirect cost of hypertensive disease for 2002 was $47.2 billion. This figure includes health expenditures (direct costs, which include the cost of physicians and other professionals, hospital and nursing home services, medications, home health, and other medical durables) and lost productivity resulting from **morbidity** and mortality (indirect costs) (4).

Physiology and Pathophysiology

A variety of systems are involved in the regulation of blood pressure: renal, hormonal, vascular, peripheral and central adrenergic systems. BP is the product of cardiac output (CO) and total peripheral resistance (TPR): BP = CO × TPR. The pathogenic mechanisms leading to hypertension must lead to increased total TPR by inducing vasoconstriction, to increased CO, or to both. Hypertension is frequently associated with a normal cardiac output and elevated total peripheral (vascular) resistance.

Essential hypertension tends to cluster in families and represents a collection of genetically based diseases and syndromes with a number of underlying inherited biochemical abnormalities. Factors considered to be important in the genesis of essential hypertension include genetic factors, salt sensitivity, inappropriate renin secretion by the kidneys, and the environment. Environmental factors that have been implicated in the development of hypertension include obesity, physical inactivity, and alcohol and salt intake.

Although secondary hypertension forms a very small percentage of cases of hypertension, it is important to recognize these cases because they can often be improved or cured by surgery or specific medical therapy. Nearly all the secondary forms of hypertension are renal or **endocrine** hypertension. Renal hypertension is usually attributable to a derangement in the renal handling of sodium and fluids leading to volume expansion or an alteration

TABLE 16.3

Trends in the Awareness, Treatment, and Control of High Blood Pressure in Adults, United States, 1976-1994[a]

	NHANES II (1976-1980)	NHANES III (phase 1; 1988-1991)	NHANES III (phase 2; 1991-1994)
Awareness	51	73	68
Treatment	31	55	53
Control[b]	10	29	27

Note. NHANES = National Health and Nutrition Examination Survey.
[a]Data are percentage of adults aged 18 to 74 years with systolic blood pressure 140 mmHg or greater, diastolic blood pressure 90 mmHg or greater, or taking antihypertensive medication. [b]Systolic blood pressure less than 140 mmHg and diastolic blood pressure less than 90 mmHg.

in renal secretion of vasoactive materials resulting in a systemic or local change in arteriolar tone. Endocrine hypertension is usually attributable to an abnormality of the adrenal glands.

Untreated hypertension leads to premature death, the most common cause being heart disease, followed by stroke and renal failure. Hypertension damages the **endothelium,** which predisposes the individual to **atherosclerosis** and other **vascular pathologies.** In the presence of **hyperlipidemia** and a damaged endothelium, atherosclerotic plaque develops, whereas in its absence, the **intima** thickens. Increased **afterload** on the heart caused by hypertension may lead to left **ventricular hypertrophy** and is an important cause of **congestive heart failure.** Hypertension-induced vascular damage can lead to stroke and transient ischemic attacks as well as end-stage renal disease. A meta-analysis of nine studies, involving 420,000 individuals, revealed that prolonged increases in usual diastolic BP of 5 and 10 mmHg were associated with at least 34% and 56% increases in stroke risk and with at least 21% and 37% increases in coronary heart disease (CHD) risk, respectively (23).

Clinical Considerations

The clinical evaluation of a person with hypertension should be aimed at assessing secondary forms of hypertension, assessing factors that may influence therapy, determining if target organ damage is present, and identifying other risk factors for cardiovascular disease. It is also vital to determine an accurate pretreatment baseline blood pressure.

Signs and Symptoms

Hypertension is often referred to as the "silent killer," because most patients do not have specific symptoms related to their high BP. Headache is popularly considered a symptom of hypertension, although it occurs only in severe hypertension; most commonly, such headaches are localized to the **occipital** region and are present on awakening in the morning. Dizziness, palpitations, and easy fatigability are other complaints related to elevated BP. Some symptoms such as **epistaxis, hematuria,** and blurring of vision are attributable to underlying vascular disease, which may have led to the hypertension.

Diagnostic and Laboratory Evaluations

An accurate BP reading is the most important part of the diagnostic evaluation. A person is classified

as hypertensive based on the average of two or more BP readings, at each of two or more visits after an initial screening visit. Figure 16.1 lists the recommended techniques for BP measurement.

1. Patients should be seated in a chair with their backs supported and their arms bared and supported at heart level. Patients should refrain from smoking or ingesting caffeine during the 30 min preceding the measurement.

2. Measurement should begin after at least 5 min of rest.

3. The appropriate cuff size must be used to ensure accurate measurement. The bladder within the cuff should encircle at least 80% of the arm. Many adults will require a large adult cuff.

4. Two or more readings separated by 2 min should be averaged. If the first two readings differ by more than 5 mmHg, additional readings should be obtained and averaged.

FIGURE 16.1   Recommended techniques for blood pressure measurement.
From *Archives of Internal Medicine,* 1997157:2413-46.

Measurement of BP at home by the patient or family or automated ambulatory monitoring helps verify the diagnosis of hypertension and response to treatment. Advantages of self-measurement of BP include distinguishing sustained hypertension from "white-coat hypertension" (21). Table 16.4 provides follow-up recommendations based on the initial set of BP measurements.

Laboratory tests for hypertension should include urinalysis, **hematocrit,** blood chemistry (sodium, potassium, creatinine, and lipid profile), and an electrocardiogram (21). These routine tests will help determine the presence of target organ damage and other risk factors. From an exercise prescription point of view, the electrocardiogram is an important test, because it may reveal the presence of arrhythmias, baseline ST-segment changes, and the individual's resting heart rate. Other tests that can be of value, depending on indications, include **creatinine clearance,** microscopic urinalysis, chest X-ray, echocardiogram, and serum calcium, phosphate, and uric acid.

Treatment

The goal of prevention and management of hypertension is to reduce morbidity and mortality rates

TABLE 16.4

Recommendations for Follow-Up Based on Initial Blood Pressure (mmHg)[a] Measurements for Adults

Systolic	Diastolic	Follow-up recommended[b]
<130	<85	Recheck in 2 years
130-139	85-89	Recheck in 1 year[c]
140-159	90-99	Confirm within 2 months[c]
160-179	100-109	Evaluate or refer to source of care within 1 month
≥180	≥110	Evaluate or refer to source of care immediately or within 1 week depending on clinical situation

[a]If systolic and diastolic categories are different, follow recommendations for shorter follow-up. [b]Modify the scheduling of follow-up according to reliable information about past blood pressure measurements, other cardiovascular risk factors, or target organ disease. [c]Provide advice about lifestyle modifications.

by the least intrusive means possible. This may be achieved through lifestyle modification alone or in combination with pharmacological treatment. Lifestyle modifications (figure 16.2) help in controlling BP as well as other risk factors for cardiovascular disease.

A review of randomized controlled trials of over 6 months duration analyzing the effect of weight reduction in reducing BP found a decrease of 5.2/5.2 mmHg and 2.8/2.3 mmHg in hypertensive and normotensive participants, respectively (12). Weight reduction enhances the effects of antihypertensive medications and positively affects other cardiovascular risk factors, such as diabetes and **dyslipidemia.** The Dietary Approaches to Stop Hypertension trial found that a diet rich in fruits, vegetables, and fat-free or low-fat dairy products reduced BP by 11.4/5.5 mmHg and 3.5/2.1 mmHg in hypertensive and normotensive persons, respectively (5).

Patients should be questioned in detail about their current alcohol consumption. Excessive alcohol consumption is a risk factor for high BP. The JNC VI report recommends that those who drink alcohol should be counseled to limit their daily intake to no more than 1 oz (30 ml) of ethanol, that is, 24 oz (720 ml) of beer, 10 oz (300 ml) of wine, or 2 oz (60 ml) of 100-proof whiskey (21).

Epidemiological data demonstrate a positive association between sodium intake and level of BP. Patients with essential hypertension may be classified as salt sensitive and salt resistant, based on the absolute changes in BP that originate from dietary

- Lose weight if overweight.
- Limit alcohol intake to no more than 1 oz (30 ml) of ethanol (e.g., 24 oz of beer, 10 oz of wine, or 2 oz of 100-proof whiskey) per day or 0.5 oz of ethanol per day for women and lighter weight people.
- Increase aerobic physical activity (30-45 min most days of the week).
- Reduce sodium intake to no more than 100 mmol · day^{-1} (2.4 g of sodium or 6 g of sodium chloride).
- Maintain adequate intake of dietary potassium (approximately 90 mmol · day^{-1}).
- Maintain adequate intake of dietary calcium and magnesium for general health.
- Stop smoking and reduce intake of dietary saturated fat and cholesterol for overall cardiovascular health.

FIGURE 16.2 Lifestyle modifications for hypertension prevention and management.
From *Archives of Internal Medicine*, 1997157:2413-46.

salt intake (32). African Americans, older people, and patients with hypertension or diabetes are more sensitive to changes in dietary sodium chloride than are others in the general population. A recent review of randomized controlled trials of 6 months or longer duration in adults over the age of 45 years found a small but statistically significant effect of lowering

BP through salt reduction. Salt reduction resulted in pooled net systolic/diastolic BP changes of –2.9/ –2.1 mmHg in hypertensive individuals and –1.3/ –0.8 mmHg in normotensive individuals (12). The level of BP reduction was related to the level of salt reduction. One study revealed that a 100 mmol · day⁻¹ (approximately 6 g of sodium chloride or 2.4 g of sodium per day) reduction in sodium intake yielded a 5.8 mmHg systolic/2.8 mmHg diastolic BP reduction in hypertensive subjects and 2.3/ 1.4 mmHg BP reduction in normotensive subjects. The average American sodium consumption is more than 150 mmol · day⁻¹, and a moderate sodium restriction to no more than 100 mmol · day⁻¹ is recommended in the JNC VI report (21).

A high dietary potassium intake may protect against developing hypertension and improve BP control in patients with hypertension. An adequate intake of potassium (approximately 90 mmol · day⁻¹), preferably from food sources such as fresh fruits and vegetables, should be maintained (21).

The decision to initiate pharmacological therapy should be guided by the degree of BP elevation, the presence of target organ damage, and the presence of clinical cardiovascular disease (CVD) or other cardiovascular risk factors. The presence of CVD risk factors is assessed during the initial evaluation of the patient with hypertension. Their presence independently modifies the risk for future cardiovascular disease. Once the clinician has determined the individual's blood pressure, the presence of risk factors, and the presence of target organ damage and clinical cardiovascular disease (table 16.5), the individual's risk group can be determined as shown in table 16.6. This classification into risk groups helps guide therapeutic decisions.

According to the JNC VI guidelines, if there are no specific indications for another type of drug, a diuretic or β-blocker should be chosen as initial therapy, unless contraindicated. This recommendation is based on the findings of several randomized controlled trials that showed a reduction in morbidity and mortality rates with these agents (21). However, the guidelines recommend specific drug therapy when hypertension is complicated by the presence of associated risk factors, target organ damage, and cardiovascular disease.

Graded Exercise Testing

Hypertension is one of the major risk factors for CHD, and a graded exercise test is a useful screening tool for hypertensive patients. However, the American College of Sports Medicine (ACSM) does not recommend mass exercise testing to determine those individuals at high risk for developing hypertension as a result of an exaggerated BP response (15). Hypertensive individuals with an additional coronary risk factor, males more than 45 years of age, and females more than 55 years of age should perform an exercise test with electrocardiogram monitoring before starting a vigorous exercise program (15).

Medical Evaluation

A thorough medical history should assess the duration and severity of hypertension and symptoms and

---

TABLE 16.5
Components of Cardiovascular Risk Stratification in Patients With Hypertension

Major risk factors	Target organ damage / clinical cardiovascular disease
Smoking	Heart diseases
Dyslipidemia	— Left ventricular hypertrophy
	— Angina or prior myocardial infarction
Diabetes mellitus	— Prior coronary revascularization
	— Heart failure
Age >60 years	
	Stroke or transient ischemic attack
Sex (men and postmenopausal women)	
	Nephropathy
Family history of cardiovascular disease: women <65 years or men <55 years	Peripheral arterial disease
	Retinopathy

TABLE 16.6

Risk Stratification and Treatment

Blood pressure stages (mmHg)	Risk group A (no risk factors; no TOD/CCD)	Risk group B (at least one risk factor, not including diabetes; no TOD/CCD)	Risk group C (TOD/CCD and/or diabetes, with or without other risk factors)
High-normal (130-139/85-89)	Lifestyle modification	Lifestyle modification	Drug therapy
Stage 1 (140-159/90-99)	Lifestyle modification (up to 12 months)	Lifestyle modification (up to 6 months)	Drug therapy
Stages 2 and 3 (≥160/≥100)	Drug therapy	Drug therapy	Drug therapy

Note. TOD = target organ damage; CCD = clinical cardiovascular disease.

Client–Clinician Interaction

Patients with hypertension are at a heightened risk for atherosclerotic cardiovascular disease. Atherosclerotic cardiovascular disease is by far the leading cause of death in hypertensive patients. Therefore, in addition to assisting the patient with an appropriate exercise prescription, the clinician should attempt to educate the patient about atherosclerotic cardiovascular disease and its risk factors.

In view of the hypertensive patient's heightened risk for atherosclerotic cardiovascular disease, education also should be provided about factors that may help minimize the risk for exercise-related cardiac complications. These include the importance of an adequate warm-up and cool-down and the warning symptoms and signs of an impending cardiac event.

Drug therapy is often needed to optimize hypertension management and to facilitate cardiovascular disease risk reduction. Consequently, hypertensive patients are often receiving treatment with one or more medications. The clinician should educate the patient about the effect, if any, of specific medications on exercise performance and training. The clinician also should emphasize to the patient the importance of taking medications as prescribed and not discontinuing drug therapy without notifying his or her personal physician. If the clinician believes that the patient may be experiencing medication-related adverse effects, the clinician should refer the patient to his or her personal physician.

The clinician is likely to interact with the patient on many occasions throughout the course of a year. Therefore, it is likely that the clinician will have an opportunity to measure the patient's BP on many occasions. When discussing the patient's BP recordings, the clinician must strike a balance between not alarming the patient about minor day-to-day fluctuations in BP and expressing appropriate concern about excessive elevations in systolic or diastolic BP. However, if the clinician believes that the patient's BP is not under adequate control, the clinician should consult with the patient's physician.

signs, if any. It should include questions concerning the individual's risk factors for CHD and stroke, and symptoms and signs of CHD, heart failure, renal disease, and endocrine disorders. Information should be obtained about the past and present use of medications and about lifestyle habits. A comprehensive medical history will help in the treatment of primary hypertension and in the diagnosis and treatment of causes of secondary hypertension.

Contraindications

The American College of Cardiology/American Heart Association and ACSM guidelines on exercise testing state that severe arterial hypertension, defined as systolic BP greater than 200 mmHg or diastolic BP greater than 110 mmHg at rest, is a relative contraindication to exercise testing (15, 17).

Recommendations and Anticipated Responses

Standard exercise testing methods and protocols may be used for persons with hypertension (see chapter 6) (18). Before the hypertensive person undergoes graded exercise testing, it is important to obtain a detailed health history and baseline BP in both the supine and standing positions. The individual should be taking his or her usual antihypertensive medications when exercise testing is performed for the purpose of exercise prescription. Certain medications, especially β-blockers, affect BP at rest and during exercise and may also affect the heart rate response to exercise.

Abnormal BP Response   Blood pressure is a product of CO and TPR. Normally during exercise, the TPR decreases but the increase in CO more than compensates for the decrease in TPR and systolic BP increases. Diastolic BP usually remains the same or may decrease slightly because of the decrease in TPR. However, hypertensive patients often experience an increase in diastolic BP both during and after exercise. They are unable to reduce TPR to the same extent as normotensives (persons with normal BP). It is thought that impaired endothelial function in the early stage and, later, a reduced lumen-to-wall thickness could be responsible for the increased resistance during exercise (16).

Indications for Terminating Graded Exercise Test   A significant decrease in systolic BP (≥20 mmHg) from baseline or a failure of the systolic BP to increase with an increase in exercise intensity is an indication for terminating an exercise test. An excessive increase in BP, defined as a systolic BP of greater than 260 mmHg or diastolic BP greater than 115 mmHg, is an indication for terminating an exercise test (15).

Predictive Value of BP Response   An exaggerated BP response to graded exercise testing in normotensives has been associated with an accentuated future risk of developing hypertension. An exaggerated response can be arbitrarily defined as a level of BP higher than that expected for the individual being tested. Data from the Framingham Heart Study (34) showed that an exaggerated diastolic BP response to exercise is predictive of risk for new-onset hypertension in normotensive men and women and that an elevated recovery systolic BP is predictive of the future development of hypertension in men. This is one of the few studies on the predictive value of exercise BP in which there were large numbers of men and women. After multivariate adjustment, an exaggerated diastolic BP response during stage 2 of the Bruce treadmill protocol was observed to have the strongest association with new-onset hypertension in both men (odds ratio = 4.16) and women (odds ratio = 2.17). The study defined an exaggerated exercise BP response as either a systolic or diastolic BP above the 95th percentile of sex-specific, age-predicted values, during stage 2 of the Bruce protocol. These values are shown in table 16.7.

Studies at the Cooper Clinic (26) and Mayo Clinic (1) both revealed that an exaggerated exercise BP response had an odds ratio of 2.4 for predicting future hypertension. The subjects were mostly men, and the average follow-up time was about 8 years. The Mayo Clinic study also found that exercise hypertension was a significant predictor for total cardiovascular events but not for death or any individual cardiovascular event. In a review article, Benbassat and Froom (7) found that the prevalence of hypertension on follow-up among normotensive subjects with a hypertensive response to exercise testing was 2.06 to 3.39 times higher than that among subjects with a normotensive response. However, this predictive value was limited because 38.1% to 89.3% of those with a hypertensive response to exercise did not have hypertension on follow-up, and a normotensive response only marginally reduced the risk of future hypertension. After 17 years of follow-up in 4907 men, Filipovsky et al. (14) found that the exercise-induced increase of systolic BP was a risk factor for death from cardiovascular as well as noncardiovascular causes independent of resting BP. Similar findings have been shown in other studies (24, 27).

However, some studies showed no additional prognostic information regarding total mortality rate and cardiovascular events from exercise BP readings (13). According to the ACSM, mass exercise testing is not advocated to determine those individuals

TABLE 16.7

Sex-Specific, Predicted 95th Percentile Values for Systolic and Diastolic Blood Pressure (BP) at Stage 2 of Exercise (on the Bruce Protocol) for Different Age Groups

Age	Exercise Systolic BP	Exercise Diastolic BP
MEN		
20-24	190	93
25-29	193	97
30-34	196	101
35-39	198	103
40-44	201	105
45-49	204	106
50-54	208	107
55-59	211	107
60-64	214	107
65-69	218	106
WOMEN		
20-24	165	92
25-29	169	95
30-34	173	98
35-39	177	100
40-44	181	102
45-49	186	103
50-54	190	104
55-59	195	104
60-64	199	103
65-69	204	102

Reprinted, by permission, from J.P. Singh, M.G. Larson, T.A. Manolio, C.J. O'Donnell, M. Lauer, J.C. Evans, D. Levy, 1999, "Blood pressure response during treadmill testing as a risk factor for new-onset hypertension," *Circulation* 99: 1831-1836.

Literature Review

Reviews of studies on the BP-lowering effects of exercise training in hypertensive people have shown vastly differing results, perhaps because of the different study inclusion criteria and study designs. Recently, Petrella (28) reviewed the English language literature from January 1966 to January 1998. The search included articles in the MEDLINE database, with the MESH headings of blood pressure, hypertension, exercise, and exertion. A total of 39 studies were identified, and these showed reductions in BP of 13/8 mmHg in hypertensive patients. Most of the studies reviewed suffered from at least one major flaw, such as a lack of a valid control group, absence of blinded BP measurements, and the presence of cointervention. The ACSM, in its position stand on physical activity, physical fitness, and hypertension, reviewed 40 studies assessing the BP-lowering effects of endurance exercise training in individuals with hypertension. They found that exercise training reduced systolic BP by approximately 11 mmHg from an initial mean systolic BP of 153 mmHg and reduced diastolic BP by approximately 9 mmHg from an initial mean value of 99 mmHg (2). Hagberg et al. (19) analyzed 25 studies that examined the BP-lowering effects of endurance exercise on individuals with essential hypertension and found an average reduction of 10.8/8.2 mmHg in BP. However, when more stringent criteria are used to review studies, and when ambulatory BP monitoring is included, the results are not quite as impressive.

Ebrahim and Smith (12) conducted a review from 1966 to 1995 but included only randomized trials of exercise-only intervention among adults age 45 years or older with and without hypertension and with at least 6 months follow-up. They found eight trials that fit the criteria. These showed nonsignificant pooled net systolic/diastolic BP changes of –0.8/–3.7 mmHg and –0.2/+0.1 mmHg in hypertensive and normotensive participants, respectively.

It has been shown that 24-hr ambulatory BP monitoring may be more predictive of target organ damage than casual resting measures. In a recent study, Seals et al. (32) found that ambulatory-determined 24-hr levels of BP were unchanged with training in postmenopausal women. Previous trials have led to conflicting results, and even in trials showing a significant reduction in ambulatory BP, the magnitude of reduction was generally less than that observed in casual BP (2).

Recently, it has been hypothesized that exercise training–induced BP and plasma lipid improvements in hypertensives may be genotype dependent. Hagberg et al. (19) found evidence to support the possibility that ACE, apoE, LPL PvuII, and Hind III **genotypes** may identify hypertensives likely to reduce BP the most with exercise training.

The incidence of modifiable CHD risk factors is higher in hypertensive individuals. Exercise training has been shown to have a beneficial effect on obesity, lipid profiles, insulin resistance, and glucose intolerance (2). Dengel et al. (11) showed that a 6-month intervention of aerobic exercise and weight loss substantially reduced BP, improved insulin sensitivity by 39%, and resulted in a 50% reduction in the number of metabolic abnormalities associated with the insulin resistance syndrome in obese, hypertensive, middle-aged men. Brown et al. (9) found that an aerobic exercise program of 7 days improved insulin sensitivity in African American hypertensive women independent of changes in fitness levels, body composition, or body weight. Subjects performed aerobic exercise on the treadmill or stationary bicycle for 50 min on 7 consecutive days. The exercise intensity corresponded to 65% of their maximum heart rate reserve. Blair et al. (8) found that hypertensive men who were more fit had lower death rates compared with less fit men. Between 1970 and 1981, these authors tested 1832 men who reported a history of hypertension but were otherwise healthy. Mortality surveillance was conducted on the group through 1985. The inverse relation between fitness and all-cause mortality held even after investigators adjusted for the influence of age, serum cholesterol, resting systolic BP, body mass index, current smoking, and length of follow-up. Most studies show a beneficial effect of exercise on CHD risk factors, even if the exercise is not enough to increase fitness level or decrease body weight.

at high risk for developing hypertension in the future as a result of an exaggerated exercise BP response (2).

Exercise Prescription

Exercise training has been recommended as one of the important lifestyle modifications for the prevention and management of hypertension (21). When compared with active and fit individuals, those who are sedentary have a 20% to 50% increased risk of developing hypertension. Endurance exercise training by individuals who are at high risk for developing hypertension will reduce the increase in BP that occurs with age (15).

Although regular aerobic exercise has been shown to reduce BP, the mechanisms responsible for this remain largely unknown. There is evidence that exercise training is associated with a decrease in plasma norepinephrine levels, which may be responsible for the decrease in BP (2).

The kidneys play an important role in BP regulation. It is believed that exercise training may decrease BP by improving renal function in patients with essential hypertension. Another postulated mechanism is that regular physical activity causes favorable changes in arterial structure, which would presumably reduce peripheral vascular resistance (35).

Hyperinsulinemia has been postulated to raise BP via renal sodium retention, sympathetic nervous activation, and induction of vascular smooth-muscle hypertrophy (22, 30). Hypertension and hyperinsulinemia, along with insulin resistance, increased triglycerides, decreased high-density lipoprotein, and glucose intolerance, often are clustered together to form what has been called Syndrome X, or the metabolic syndrome. Even a single bout of exercise has a well-known insulinlike effect and dramatically increases skeletal muscle glucose transport. Exercise training increases insulin sensitivity, which can decrease serum insulin and BP (6).

PRACTICAL APPLICATION 16.3

Exercise Prescription

ACSM recommendations (15) for aerobic exercise programming for patients with hypertension include the following:

- Frequency: Three to seven days per week.
- Intensity: 40% to 70% $\dot{V}O_2$max (typically, this corresponds to 40-70% heart rate reserve, 50-80% maximal heart rate, and a rating of perceived exertion of 11-14).
- Duration: 30 to 60 min of continuous or intermittent (minimum of 10-min bouts) aerobic activities.

The frequency, duration, and intensity of aerobic exercise should be modulated to achieve a weekly energy expenditure of 700 (initial goal) to 2000 (long-term goal) kcal · week^{-1}. The patient should not exercise if resting systolic BP exceeds 200 mmHg or diastolic BP exceeds 115 mmHg. The clinician should be aware that β-blockers attenuate the heart rate response to exercise; α-blockers, calcium channel blockers, and vasodilators may cause postexertion hypotension (an adequate cool-down may be especially important for patients taking these medications); and certain diuretics may cause a decrease in serum potassium levels, thereby predisposing the patient to arrhythmias.

When performing resistance training, hypertensive patients generally should adhere to the American Heart Association's guidelines (29) for patients with cardiovascular disease (which has been endorsed by the ACSM). These include the following:

- Frequency: Two to three days per week
- Sets: One
- Reps: 10 to 15
- Stations/devices: 8 to 10 exercises that condition the major muscle groups

Special Exercise Considerations

Those with more marked elevations in BP (>180/110) should add endurance training to their treatment only after initiating drug therapy. Individuals should not be allowed to exercise on a given day if their resting systolic BP is more than 200 mmHg or diastolic BP is more than 115 mmHg.

Strength or resistance training is not recommended as the only form of exercise training for persons with hypertension. With the exception of circuit weight training, resistance training has not consistently been shown to lower BP. Resistance training is recommended as a component of a well-rounded fitness program but not when done independently.

Exercise Recommendations

The exercise recommendations for individuals with hypertension should take into account their medical history, current BP levels, presence of cardiovascular disease, and its risk factors. The program developed should include cardiovascular endurance training, resistance training, and flexibility exercises.

Cardiovascular Training

The ACSM, in its position statement on physical activity, physical fitness, and hypertension, states that cardiovascular exercise training at somewhat lower intensities (40-70% of $\dot{V}O_2$max) appears to lower BP as much, or more, than exercise at higher intensities (2). The mode (large muscle activities), frequency (3-7 days per week), and duration (30-60 min of continuous or intermittent aerobic activity) are similar to those recommended for healthy adults (see chapter 7) (15).

The ACSM now views exercise/physical activity for health and fitness in the context of an exercise dose continuum. That is, there is a dose response to exercise by which benefits are derived through varying quantities of physical activity ranging from approximately 700 to 2000 or more kilocalories of effort per week (3).

Resistance Training

In the past, hypertensive patients have been discouraged from participating in resistance training because of fear of overloading an already compromised **myocardium** (25). These fears were increased by a study performed by McDougall et al. (25), who recorded pressures in excess of 400/200 mmHg in

weight lifters during high-intensity resistance exercise. Subsequently, it has been hypothesized that circuit weight training would be more appropriate for a hypertensive population. Harris and Holly (20) evaluated a circuit weight training program in male subjects with BP between 140/90 and 160/99 mmHg. Subjects exercised at approximately 79% of their maximum heart rate, 3 days a week, for 9 weeks. They improved their muscular strength and cardiovascular endurance and lowered their diastolic BP from 96 to 91 mmHg. Resting heart rate and systolic BP did not change. Figure 16.3 lists guidelines for a circuit weight training program.

Resistance training for hypertensives should involve lower resistance with higher repetitions. The recommendations are to do one set of 8 to 10 different exercises that condition the major muscle groups, 2 to 3 days a week. Each set should consist of 10 to 15 repetitions (29). Circuit weight training has also been found to be beneficial for hypertensives and is defined as lifting a weight equal to 40% to 60% of 1-repetition maximum for 10 to 20 reps in a 30- to 60-s period. After a rest of 15 to 45 s, the person moves to the next exercise (25). BP should be monitored frequently before, during, and after resistance training during the initial few weeks of participation. The improvement in strength from a resistance training program will help hypertensive persons better perform both occupational and leisure tasks and will enhance their quality of life (25).

Range of Motion

Flexibility exercises should be included in the exercise routine. They should include a variety of upper and lower body range-of-motion activities, which should be performed on at least 2 to 3 days of the week. At least four repetitions per muscle group should be done, and each static stretch should be held for 10 to 30 s (3). The goal of these exercises is to reduce the risk of musculoskeletal injury and improve the individual's flexibility.

Conclusion

Hypertension affects more than 50 million Americans and is one of the leading causes of death. In more than 95% of cases, the etiology of hypertension is unknown. Hypertension is one of the major risk factors for coronary artery disease and is often found clustered with other CVD risk factors. Lifestyle modification helps control BP and reduces other

1. Select 8 to 10 exercises to get a well-balanced program.

2. Establish a conservative 1-repetition maximum (1RM) in each exercise or a 10RM on three to four key exercises.

3. Use 40% to 60% of 1RM.

4. Do 10 to 15 reps in 30 to 60 s.

5. Do two or three sets of each exercise in a circuit pattern, alternating between upper and lower body and moving from large to small muscle group exercises.

6. Begin with a 45-s rest between sets and gradually reduce to 15 to 30 s.

7. Train 2 or 3 days a week.

8. Machine weights are preferable for hypertensive patients.

9. Emphasize full range of motion and proper posture during all exercises.

10. Warm up with stretching and 10 to 15 min of moderate aerobic exercise (11-13 on rating of perceived exertion scale).

11. Perceived exertion during the circuit should be 11 to 14.

12. Avoid straining or heavy lifting; emphasize aerobic circuit training.

13. Use proper breathing technique to avoid the Valsalva maneuver.

14. Maintain a firm hand grip, but not too tight, to avoid a pressor response that may cause excessive increases in BP.

15. When you can comfortably complete two to three circuits with a given load, increase the load by 2.5 to 10.0 lb.

16. Be sure to adhere to the medical regimen as prescribed by your physician.

FIGURE 16.3 Guidelines for a circuit weight training program.

From *Archives of Internal Medicine*, 1997, 157:2413-46.

CVD risk factors. Exercise training helps prevent the development of hypertension as well as reduces the BP of those with hypertension. Chronic endurance training has shown to reduce both systolic and diastolic BP by about 10 mmHg. Studies have shown that cardiovascular exercise training at somewhat lower intensities (40-70% of $\dot{V}O_2$max) appears to lower BP as much as, or more than, exercise at higher intensities. The mode (large muscle activities), frequency (3-7 days per week), and duration (30-60 min of continuous or intermittent aerobic activity) are similar to those recommended for healthy adults. Resistance training for hypertensives should involve lower resistance with higher repetitions. Circuit training has found to be extremely beneficial for hypertensive individuals.

Case Study 16

Medical History and Diagnosis

Mr. NJ, a 64-year-old Latino male, comes to your fitness center to start an exercise program. He is apparently healthy but has a 15-year history of hypertension. His goal is to improve his health and fitness. At present he is sedentary and has a body mass index of 36. His medications include aspirin 325 mg daily and atenolol 50 mg daily. He has a family history of premature CHD, his triglycerides are elevated (375 mg · dl^{-1}), his resting BP is 148/88 mmHg, and his resting heart rate is 60 beats · min^{-1}. Mr. NJ drinks three to four glasses of wine with dinner each evening.

continued

Case Study 16 (*continued*)

Exercise Test Results

During a recent maximal exercise test, he exercised for 6 min using the Bruce treadmill protocol and achieved a maximum heart rate of 130 beats · min^{-1} and BP of 240/90 mmHg. He did not develop any significant arrhythmias or ST-segment changes.

Case Study 16 Discussion Questions

1. Is it necessary for Mr. NJ to exercise at a high intensity to help manage his hypertension?
2. How will atenolol (a β-blocker) affect the prescription of a target heart rate range for Mr. NJ?
3. What is an appropriate aerobic exercise prescription for Mr. NJ?
4. Can Mr. NJ participate in a resistance training program?
5. What additional lifestyle modifications should be discussed with Mr. NJ to optimize his BP control?

Glossary

afterload—The resistance against which the heart's pumping chamber or ventricle works.

atherosclerosis—Disease in which deposits of cholesterol and other lipids are formed within the intima of large and medium-sized arteries.

congestive heart failure—Failure of the heart to maintain the cardiac output resulting in congestion of blood in other bodily organs.

creatinine clearance—A kidney function test that indicates the amount of creatinine cleared.

diastolic—The pressure remaining in the arteries after cardiac contraction.

dyslipidemia—Any abnormality of serum lipids.

endocrine—Referring to glands that secrete hormones into the bloodstream.

endothelium—A thin layer of cells that line the inner surface of blood vessels.

epistaxis—Bleeding from the nose.

etiology—Cause.

genotype—The resultant expression of specific genes.

hematocrit—The percentage by volume of packed red blood cells in a sample of blood.

hematuria—Red blood cells in the urine.

hyperlipidemia—Elevated lipid levels in the blood.

idiopathic—Having no known cause.

intima—The inner layer of blood vessels containing endothelial cells.

morbidity—Manifestations of disease other than death.

myocardium—The heart muscle.

occipital—The posterior part of the skull.

papilledema—Swelling of the optic disk in the eye caused by severe hypertension.

systemic—The arterial system supplying the body.

systolic—The pressure generated in the arteries by contraction of the heart muscle.

vascular pathologies—Manifestations of disease in blood vessels.

ventricular hypertrophy—Muscle thickening in a pumping chamber of the heart.

References

1. Allison TG, Cordeiro MAS, Miller TD, Daida HD, Squires RW, Gau GT. Prognostic significance of exercise-induced systemic hypertension in healthy subjects. Am J Cardiol 1999; 83: 371-375.

2. American College of Sports Medicine position stand. Physical activity, physical fitness, and hypertension. Med Sci Sports Exerc 1993; 25(10): i-x.

3. American College of Sports Medicine position stand. The recommended quantity and quality of exercise for developing and maintaining cardiorespiratory and muscular fitness, and flexibility in healthy adults. Med Sci Sports Exerc 1998; 30: 975-991.

4. American Heart Association. 1999 Heart and Stroke Statistical Update. 1998; American Heart Association, Dallas.

5. Appel LJ, Moore TJ, Obarzanek E, et al. for the DASH collaborative research group. A clinical trial of the effects of dietary patterns on blood pressure. N Engl J Med 1997; 336: 1117-1124.

6. Barnard RJ, Wen SJ. Exercise and diet in the prevention and control of the metabolic syndrome. Sports Med 1994; 18: 218-228.

7. Benbassat J, Froom P. Blood pressure response to exercise as a predictor of hypertension. Arch Intern Med 1986; 146: 2053-2055.

8. Blair SN, Kohl HW III, Barlow CE, Gibbons LW. Physical fitness and all-cause mortality in hypertensive men. Ann Med 1991; 23: 307-312.

9. Brown MD, Moore GM, Korytkowski MT, McCole SD, Hagberg JM. Improvement of insulin sensitivity by short-term exercise training in hypertensive African-American women. Hypertension 1997; 30: 1549-1553.

10. Burt VL, Whelton P, Roccella EJ, Brown C, Cutler JA, Higgins M, Horan MJ, Labarthe D. Prevalence of hypertension in the US adult population. Results From the Third National Health and Nutrition Examination Survey, 1988-1991. Hypertension 1995; 26(1): 60-69.

11. Dengel DR, Hagberg JM, Pratley RE, Rogus EM, Goldberg AP. Improvements in blood pressure, glucose metabolism, and lipoprotein lipids after aerobic exercise plus weight loss in obese, hypertensive middle-aged men. Metabolism 1998; 47: 1075-1082.

12. Ebrahim S, Smith GD. Lowering blood pressure: A systematic review of sustained effects of non-pharmacological interventions. J Public Health Med 1998; 20: 441-448.

13. Fagard R, Staessen J, Amery A. Exercise blood pressure and target organ damage in essential hypertension. J Human Hypertens 1991; 5: 69-75.

14. Filipovsky J, Ducimetiere P, Safar ME. Prognostic significance of exercise blood pressure and heart rate in middle-aged men. Hypertension 1992; 20: 333-339.

15. Franklin B, Whaley M, Holey E. American College of Sports Medicine's Guidelines for Exercise Testing and Prescription (6th ed.). 2000; Lippincott, Williams & Wilkins, Baltimore.

16. Franz IW. Blood pressure measurement during ergometric stress testing. Z Kardiol 1996; 85(Suppl 3): 71-75.

17. Gibbons RJ, Balady GJ, Beasley JW, Bricker JT, Duvernoy WFC, Froelicher VF, Mark DB, Marwick TH, McCallister BD, Thompson PD, Winters WL, Jr, Yanowitz FG. ACC/AHA guidelines for exercise testing: A report of the American College of Cardiology/American Heart Association Task Force on Practice Guidelines (Committee on Exercise Testing). J Am Coll Cardiol 1997; 30: 260-315.

18. Gordon NF. Hypertension. In ACSM's Exercise Management for Persons with Chronic Diseases and Disabilities. 1997; American College of Sports Medicine, Human Kinetics, Champaign, IL, pp. 59-63.

19. Hagberg JM, Ferrell RE, Dengel DR, Wilund KR. Exercise training-induced blood pressure and plasma lipid improvements in hypertensives may be genotype dependent. Hypertension 1999; 34: 18-23.

20. Harris KA, Holly RG. Physiological response to circuit weight training in borderline hypertensive subjects. Med Sci Sports Exerc 1987; 19: 246-252.

21. Joint National Committee on Prevention, Detection, Evaluation, and Treatment of High Blood Pressure. The Sixth Report of the Joint National Committee on Detection, Evaluation, and Treatment of High Blood Pressure. Arch Intern Med 1997; 157: 2413-2446.

22. Kaplan NM. The deadly quartet. Arch Intern Med 1989; 149: 1514-1520.

23. Kaplan NM. Systemic hypertension: Mechanisms and diagnosis. In Braunwald E, ed. Heart Disease: A Textbook of Cardiovascular Medicine. 1997; Saunders, Philadelphia.

24. Kjeldsen SE, Mundal R, Sandvik L, Erikssen G, Thaulow E, Erikssen J. Exercise blood pressure and fatal myocardial infarction. J Hypertens 1994; 12(Suppl 3): 77.

25. LaFontaine T. Resistance training for patients with hypertension. Strength Condition 1997; 19: 5-9.

26. Matthews CE, Pate RR, Jackson KL, Ward DS, Macera CA, Kohl HW, Blair SN. Exaggerated blood pressure response to dynamic exercise and risk of future hypertension. J Clin Epidemiol 1998; 51(1): 29-35.

27. Mundal R, Kjeldsen SE, Sandvik L, Erikssen G, Thaulow E, Erikssen J. Exercise blood pressure predicts cardiovascular mortality in middle-aged men. Hypertension 1994; 24: 56-62.

28. Petrella RJ. How effective is exercise training for the treatment of hypertension? Clin J Sport Med 1998; 8: 224-231.

29. Pollock M, Franklin B, Balady G, Chaitman B, Fleg J, Fletcher B, Limacher M, Pina I, Stein R, Williams M, Bazzare T. AHA science advisory. Resistance exercise in individuals with and without cardiovascular disease. Circulation 2000; 101: 828-833.

30. Reaven GM. Role of insulin resistance in human disease. Diabetes 1988; 37: 1595-1607.

31. Reisin E. Nonpharmacologic approaches to hypertension. Weight, sodium, alcohol, exercise, and tobacco considerations. Med Clin North Am 1997; 81: 1289-1303.

32. Seals DR, Silverman HG, Reiling MJ, Davy KP. Effect of regular aerobic exercise on elevated blood pressure in postmenopausal women. Am J Cardiol 1997; 80: 49-55.

33. Singh JP, Larson MG, Manolio TA, O'Donnell CJ, Lauer M, Evans JC, Levy D. Blood pressure response during treadmill testing as a risk factor for new-onset hypertension. The Framingham Heart Study. Circulation 1999; 99: 1831-1836.

34. Tanaka H, Reiling MJ, Seals DR. Regular walking increases peak limb vasodilatory capacity of older hypertensive humans: Implications for arterial structure. J Hypertens 1998; 16: 423-428.

35. Williams GH. Approach to the patient with hypertension. In E Braunwald, A Fauci, S Hauser, J Jameson, D Kasper, D Longo eds. Harrison's Principles of Internal Medicine. 1998; McGraw-Hill, New York.

Richard M. Lampman, PhD
Director of Research, Department of Surgery
St. Joseph Mercy Hospital
Adjunct Associate Professor
Department of Physical Medicine and Rehabilitation
University of Michigan Medical School
Ann Arbor, MI

Seth W. Wolk, MD
Vascular Surgeon
Michigan Vascular and Heart Institute
Department of Surgery
St. Joseph Mercy Hospital
Clinical Assistant Professor, Department of Surgery
University of Michigan Medical School
Ann Arbor, MI

Peripheral Arterial Disease

Peripheral arterial disease (PAD) is a form of vascular disease characterized by narrowing of the major arteries supplying blood to the lower extremities. PAD is a marker of systemic atherosclerosis, and as such it is associated with coronary and cerebrovascular disease (32). It has been estimated that approximately 9.6% of cardiovascular events are reported in people with PAD, and about 17,400 deaths occur each year (62).

Scope

Age and sex influence the prevalence of PAD. A recent population study from the Netherlands found the prevalence of advanced PAD for individuals over the age of 55 years to be 16.9% and 20.5% in women and men, respectively (77). The prevalence rate of PAD in women lags behind men by 10 years, with male/female ratios varying from 1 to 8 with increasing age (68).

A common symptom of PAD is **intermittent claudication (IC).** Intermittent claudication produces symptoms of limping, lameness, and pain that occur in individuals while performing mild activity (i.e., walking). Symptoms of IC are caused by a lack of oxygen to the skeletal muscle (ischemia). When PAD progresses to symptoms of intermittent claudication, the prevalence of PAD in those over 50 years has been reported to be 2% to 7% for men and 1% to 2% for women, and more than 5% to 20% in men over the age of 70 years (55). In an elderly population of patients with comorbidities, the prevalence of PAD was high in those with cardiovascular risk factors, being 61.4% in diabetics, 21.4% in those with ischemic heart disease, and 13.4% in those with renal failure compared with 4.1% in controls (1).

The incidence of PAD in a population is difficult to assess because many individuals ignore symptoms and do not seek medical treatment. The incidence of asymptomatic PAD, diagnosed with noninvasive testing, is three to four times more frequent than symptomatic PAD, and when PAD has advanced to critical ischemia, the incidence is estimated to be between 0.05% and 0.1% of the population (118). There appears to be a general increase in the incidence of IC up to the age of 60 years (81).

When IC is present, patients have a two to four times increase in mortality rates, primarily from cardiovascular disease (64), and it has been estimated that 9.6% of cardiovascular events occur in patients with IC (62). The life expectancy is poor, with a mortality rate of 3% to 5% per year once IC is present and 20% per year when IC is advanced to critical ischemia, with coronary heart disease accounting for 50%, stroke for 15%, and vascular disease in the stomach for 10% of deaths (22).

Physiology and Pathophysiology

The development of atherosclerosis affecting the intima of the peripheral arterial circulation sets the stage for clinically significant PAD. Atherosclerosis of the arterial wall has been hypothesized to develop in response to specific phospholipids contained in low-density lipoprotein that have become trapped and then oxidized in the endothelium (86). **Endothelial cell** aberrations appear associated with abnormal oxidative stresses involving increased endothelial superoxide anion, which leads to endothelium-produced nitric oxide or the accelerated degradation of **endothelium-derived relaxing factors** such as nitric oxide (71, 87, 113). Vasomotor tone is abnormal in the atherosclerotic vessel because of endothelial dysfunction secondary to increased oxidant degradation of endothelium-derived

nitric oxide or prostaglandin endoperoxide, PGH2 (45, 114). Reducing a major risk factor, such as hypercholesterolemia, can normalize vascular superoxide anion production and improve endothelium-dependent vascular relaxation (88).

The percentage stenosis of the artery as well as the flow velocity across the lesion accounts for the degree of the hemodynamic abnormality in the distal extremity (14, 54). Flow velocity varies inversely with peripheral resistance. Blood flow at rest may not be influenced by a critical arterial stenosis (i.e., 50-90% stenosis), but the lesion can impair adequate blood flow to muscles to sufficiently meet metabolic requirements with exercise (124).

Regional blood flow of leg muscles, oxygen consumption rate, and oxygen extraction fraction have been shown to be normal at rest in those with PAD, but of these three variables, muscle regional blood flow and oxygen consumption rates were higher in the presence of PAD compared with the absence of PAD (20, 111). There appears to be a delayed **hyperemia** in the ischemic muscle following exercise and therefore a lag of postexercise metabolism that is related to the severity of the disease.

Metabolic and neural changes suggested to be associated with PAD include the following:

1. Impaired energy utilization of claudicating skeletal muscle
2. Impaired resynthesis of high-energy phosphate substrate following exercise
3. Low concentrations of adenosine triphosphate and phosphocreatine (91), leading to severe IC pain
4. Loss of muscle fibers leading to abnormalities of oxidative adenosine triphosphate synthesis
5. Reduced metabolic efficiency of the muscle rather than a direct consequence of inadequate blood flow alone
6. Oxidative stress involving mitochondrial injury
7. Peripheral nerve damage that occurs in the presence of ischemia (27, 49), leading to both sensory impairment and motor weakness (24)

More studies are needed to discern whether these metabolic and fiber changes are related to PAD or merely to muscle disuse in those with IC. Evidence does suggest that muscle dysfunction and poor exercise tolerance are characteristic of those with IC. Generalized atrophy may result from a loss of muscle fibers, and weakness may be attributable to denervation of the skeletal muscle. Oxidative metabolism may be inefficient. Unfortunately, these abnormalities impair a patient's ability to ambulate efficiently.

Major risk factors for developing PAD are similar to those for coronary heart and cerebrovascular atherosclerosis. Epidemiological studies show that the incidence of this disease increase with the following factors:

— Age
— Smoking
— Non-insulin-dependent diabetes mellitus
— Impaired glucose intolerance
— Hypertension
— Low levels of high-density lipoprotein cholesterol
— Hypercholesterolemia
— Hypertriglyceridemia
— High blood levels of **apolipoprotein** B
— High low-density lipoprotein and low high-density lipoprotein cholesterol
— High **homocysteine** and fibrinogen
— Increased blood viscosity
— Elevated **C-reactive protein** levels (12, 16, 51, 81, 99, 118)

Men, primarily those who smoke cigarettes, appear to have the highest incidence of PAD. Now that women have begun smoking at higher rates, the incidence of PAD in women who smoke is on the increase. It has been estimated that approximately 25% of women between the ages of 55 and 74 years have symptoms of PAD disease. Especially at risk are those who are menopausal, smoke cigarettes, and have diabetes mellitus (43). Heavy smokers are at a fourfold risk of developing IC (64).

Diabetes mellitus is a major risk factor associated with the development of PAD (81). Hyperinsulinemia to an oral glucose challenge has been shown to be an independent risk factor for PAD in the nondiabetic population (94), and smoking appears to exacerbate the postglucose hyperinsulinemic response (93). Although diabetics have a higher incidence of PAD than those with normal glucose tolerance, elevated systolic blood pressure and triglyceride levels in those with diabetes may account for the higher incidence of PAD (74).

PAD has also been closely linked to hypertension and hyperlipidemia (64), to abnormal plasma viscosity (92) and plasma levels of fibrinogen (63), and to hyperhomocysteinemia (61).

Clinical Considerations

Impaired exercise capacity resulting from IC can severely limit an individual's ability to meet the occupational, social, and personal demands of daily life. As IC progresses, symptoms occur at progressively shorter periods of activity. This progression can threaten or impair functional independence and diminish subjective quality of life (65). It is estimated that 25% of those with IC will need vascular intervention and that less than 5% will require a major amputation. However, the atherosclerotic process can progress to obliterate previously nondiseased arteries in both lower extremities (118). When present, IC is associated with a 20% to 40% increase in mortality rates, primarily from cardiovascular disease (64).

Signs and Symptoms

PAD is characterized by an initial asymptomatic phase followed by a period of symptomatic IC and finally a phase of critical limb ischemia (ulceration, tissue loss, and **gangrene**). Acute arterial occlusion may occur at any point along the continuum. The prognosis for limb-threatening ischemia is reasonably good. The need for vascular intervention arises in only 25% of men with IC. Only about 5% of patients with IC will require a major amputation of a leg or a foot. The atherosclerotic process can, however, progress to previously nondiseased arteries, and initially unilateral disease may become bilateral (118). Furthermore, because atherosclerosis is a systemic disease, those with PAD have increased cardiovascular morbidity rates (e.g., myocardial infarction, cerebrovascular events, and rupture of aortic aneurysms) and mortality rates when compared with a normal population.

To characterize PAD clinically, a variety of symptom severity scales have been devised. Two popular scales are the Fontaine stages and Rutherford categories (104). The Fontaine stages are as follows:

— I, asymptomatic
— IIa, mild **claudication**
— IIb, moderate-severe claudication
— III, ischemic rest pain
— IV, ulceration or gangrene

The Rutherford categories are these:

— 0, asymptomatic
— 1, mild claudication
— 2, moderate claudication

— 3, severe claudication
— 4, ischemic rest pain
— 5, minor tissue loss
— 6, major tissue loss

Advanced PAD is manifested by exercise-induced **ischemia** of the lower limb resulting in symptoms of moderate to severe muscular pain referred to as intermittent claudication (IC). Rest usually relieves symptoms of pain associated with IC. The pathogenesis of IC is related primarily to restrictive or occlusive atherosclerotic ischemia but may be caused by endothelial dysfunction or thrombosis proximal to the exercising muscle.

Diagnostic and Laboratory Evaluations

The diagnosis of advanced PAD can be made clinically by palpation of the patient's peripheral pulses, but a lack of a pulse can be attributable to causes other than PAD.

The ankle/brachial index (ABI) is an excellent noninvasive testing method for detecting asymptomatic PAD. A stenosis of at least 50% is important hemodynamically and results in a decreased resting ABI of 0.15 (7). Postexercise ABI may be a useful method to identify severe ischemia. An ABI can also help detect the severity of PAD and has been shown to be associated with symptoms and cardiovascular mortality, as shown in table 17.1. A resting ABI of less than 0.5 is a significant predictor (relative risk of 2.3) for progression of PAD (23).

Noninvasive testing in a population produces about a five times higher prevalence of PAD than would be expected. Hiatt, Hoag, and Hamman (51) demonstrated that to best diagnose PAD, an abnormal dorsalis pedis and posterior tibial ABI in the same leg at rest should be present (see table 17.2). Skin microcirculatory screening and classification

have been reported to be clinically useful to detect critical ischemia that will result in amputation (117).

An ABI at rest for a normal untrained individual has been reported to be 1.08 ± 0.08 and for those who are trained, 1.15 ± 0.05; these values fell 1 min following acute strenuous exercise to 0.70 ± 0.06 and 0.80 ± 0.08, respectively, in untrained and trained subjects (21). Those individuals with IC usually present with a resting ABI ranging between 0.50 and 0.90 (55). An ABI less than 0.50 is usually associated with severe limits on activities of daily living. An ABI of less than 0.30 is associated with more severe ischemic symptoms and a poor prognosis for limb salvage.

Murabito et al. (81) have developed an IC risk profile based on the Framingham Heart Study. This profile is an office survey tool that allows for the identification of high-risk patients and can also be used as a basis for educating patients on modifying risk factors for IC. The severity of ipsilateral PAD can be obtained through the San Diego Claudication Questionnaire (15).

Tests performed in the noninvasive laboratory to evaluate patients with PAD regarding hemodynamic and functional parameters include the following:

— **Doppler ultrasonography**
— Pulse volume recording
— Segmental blood pressure measurement
— Transcutaneous oximetry
— Color-assisted ultrasound imaging
— Exercise testing
— Postocclusive reactive hyperemia testing
— Near-infrared spectroscopy

Recent evidence suggests that elevated blood levels of C-reactive protein can predict future risk of developing symptomatic PAD (89).

TABLE 17.1

The Association of Symptoms and Cardiovascular Morbidity Rate (Over 3 Years) to the Ankle/Brachial Index (ABI) in Patients with Intermittent Claudication (IC)

ABI	Symptoms with walking	Cardiovascular morbidity rate (%)
>0.7-<0.9	None to slight IC	12
0.5-0.7	Debilitating IC	33
<0.3-0.5	Severe IC	60

TABLE 17.2
· · · · · ·
Diagnostic Criteria for Determining the Prevalence of Peripheral Arterial Disease (PAD) in Population Studies Using Specific Abnormal Cutoff Points for the Ankle/Brachial Index (ABI)

Criteria used to detect PAD	Detection of PAD and association with cardiovascular risk factors
Abnormal resting ABI in both dorsalis pedis and posterior tibial arteries in the same lower extremity	Higher frequency of PAD and an absent pulse compared with normal subjects or those with one abnormal vessel or an abnormal ABI after exercise
	Increased risk of PAD with increasing age, non-insulin-dependent diabetes, smoking, hypertension, and hypercholesterolemia
Abnormal resting ABI in one lower extremity	Increased risk of PAD with age and smoking alone
Abnormal ABI in either lower extremity only after exercise	Not associated with any cardiovascular risk factor except male sex

---

Treatment

Comorbidities, especially silent myocardial ischemia, are very common in those with PAD (106). Because these maladies greatly increase the risk for cardiovascular complications, they should be thoroughly evaluated if suspected and should receive aggressive medical or surgical treatment. It has been suggested that patients with a metabolic equivalents (METs) level less than 5 METs are at a higher risk for cardiac complications if they undergo high-risk vascular surgery (10). The risk for those with a low metabolic equivalent level, who engage in routine low-intensity exercise therapy, has yet to be determined and may not be a factor in the presence of adequate treatment for comorbidities. Medications to treat hypertension, dyslipidemia, carbohydrate intolerance, left ventricular hypertrophy, and coronary artery disease are often all used in this patient population because many patients with IC have many concurrent diseases. Exercise therapy can be of great importance because it can profoundly improve many of these medical conditions, and once a patient with PAD becomes involved in regular exercise, medications for some or all of these diseases can usually be reduced or even discontinued.

The goals of therapy for patients with IC are to relieve symptoms; prevent progression of atherosclerosis, which may result in severe ischemia, gangrene, and limb loss; prevent other cardiovascular or cerebrovascular events; and improve functional capacity and quality of life (50). Treatment options for patients with IC include exercise therapy, medical intervention, **angioplasty,** and surgery (table 17.3). Aggressive risk factor reduction for atherosclerosis should be a high clinical priority. Exercise therapy, caloric restriction for those overweight, and increased dietary fiber in those who have abnormal lipid blood levels are important lifestyle and dietary changes that can favorably affect comorbidities (44, 101). Patients with IC are often depressed (82). Therefore, the patient's psychological status should be considered when he or she is being treated for incapacitating IC, especially when the patient is undergoing smoking cessation. Surgical treatments and angioplasty are costly, can increase morbidity rates, do not address comorbidities, and do not treat associated mental or physical problems of poor functional capacity and weakness. Criteria for possible surgical intervention when a lower limb is in jeopardy include symptoms of severe ischemia at rest over a 2-week period, a systolic blood pressure at the ankle less than 50 mmHg, or a systolic blood pressure less than 30 mmHg at the big toe.

Graded Exercise Testing

Exercise stress testing provides baseline hemodynamic information not present at rest and can provide helpful information in the treatment of

TABLE 17.3
· · · · · · ·
Treatment Options

MEDICAL TREATMENT

Medication	Outcomes	Comments
Pentoxifylline (an oral hemorheologic agent)	May be beneficial to patients with ischemic rest pain	Unclear whether this drug was efficacious for those with only IC (67)
Cilostazol, a drug with vasodilating and antiplatelet properties	Significant statistical improvement was found in walking distance	Clinical relevance has been questioned because patients experienced side effects of the drug such as headache, abnormal stools, diarrhea, and dizziness (78)
Carnitine or propionyl-L-carnitine	Treatment for patients with IC resulted in increased claudication-limited exercise (50)	
Vasodilators	Treatment for IC has been shown to be ineffective (13)	
Verapamil	An optimal dose has been shown to increase mean pain-free walking distance by 29% and maximal walking distance by 49% in patients with stable IC (6)	

SURGICAL INTERVENTIONS

Technique	Outcomes	Comments
Central reconstructive vascular surgery	May be associated with both short-term and long-term morbidity and mortality rates of the surgical procedure	Usually not performed in the majority of patients with PAD because their major symptom of IC can be successfully improved with daily exercise therapy and, if necessary, medical treatment and weight reduction
Surgical revascularization or PTA	Recommend that patients follow a physical training program after arterial reconstruction surgery for IC to preserve the high metabolic capacity in the peripheral skeletal muscles (57)	Justifiable only if they are relatively safe and effective and will be durable for certain arterial vessels
PTA, laser-assisted angioplasty, and **atherectomy** for appropriate arteries	PTA may be most effective for an aortoiliac stenosis (119)	May be performed in the presence of severe ischemia (46)
Endoluminal stent placement	Patients with IC had improved hemodynamics	Major complication rate of 11% (83)
Surgical bypass	Normalizing the patient's ABI	Risks of both short- and long-term morbidity

RISK FACTOR REDUCTION PROGRAMS		
	Outcomes	**Comments**
Smoking cessation		Smoking is a major risk factor for PAD (a two- to threefold increase)
Diet and weight reduction	Carrying an extra weight of 5 kg or more significantly reduces a PAD patient's claudication distance (123)	Obesity is directly related to claudication distance and is a risk factor for cardiovascular events
Serum lipid and glucose normalizing	Lipid-normalizing therapy has been shown to reduce the risk in the development or worsening of IC (90)	Diabetes is highly associated with PAD and with the progression of vascular disease (35)
	Medication, diet, and exercise will help normalize elevated LDL cholesterol (<100 mg $\cdot$ dl^{-1}), hypertriglyceridemia, and low levels of HDL cholesterol (35)	Because abnormal rheological factors are thought responsible for reduced blood flow in the legs of patients with IC, worsening of their ischemic condition, and promoting progression in clinical symptoms, treatments to normalize hemostatic and rheological factors may prevent further deterioration by PAD (110)
Routine exercise	Exercise therapy has been shown to improve glucose homeostasis probably by reducing insulin resistance (101)	

PHYSICAL EXERCISE THERAPY		
Technique	**Outcomes**	**Comments**
	Effectively improves a patient's ability to walk significantly farther before the onset of ischemic pain (113)	The following were shown important for the greatest improvement in distance to walk pain:
		Frequency: At least three sessions per week
		Intensity: At an exercise level close to maximum pain during training
		Time: 30 min or more of walking per exercise session
		Mode of exercise: walking
		Length of program: greater than 6 months (39).

Note. IC = intermittent claudication; PAD = peripheral arterial disease; PTA = percutaneous transluminal angioplasty; ABI = ankle/brachial index; HDL = high-density lipoprotein; LDL = low-density lipoprotein.

those with advanced peripheral atherosclerosis. Exercise-induced IC symptoms and cardiopulmonary measures are important parameters for devising an individualized exercise prescription and can be adequately assessed through treadmill testing (3). Exercise stress testing also provides objective evidence that exercise therapy was beneficial to a pa-

tient, and routine testing is helpful in optimally and safely increasing exercise intensity levels when patients undergo exercise therapy.

Prior to engaging in vigorous exercise training, a patient should undergo an exercise stress test to evaluate the status of suspected or established myocardial disease (3). This test can also be used to evaluate functional capacity in clinical and research applications.

In those who are without known heart disease or are stable with respect to cardiovascular disease and plan to begin a low-level exercise therapy program (walking at their normal walking speed or exercising at equivalent intensity on some mechanical device), an initial progressive exercise stress test, with multilead electrocardiogram monitoring for detecting cardiac ischemia, may not be necessary. A vascular assessment should include resting and postexercise testing on a treadmill for ABI, peak heart rate and blood pressure, and time and distances to the onset of claudication and to the maximal claudication pain. Peak oxygen consumption measures are usually impractical to obtain because of claudication symptoms that cause a patient to terminate a test prematurely before arriving at a peak cardiovascular response level. Total body and local metabolic parameters are important to measure in research protocols if appropriate testing equipment is available.

Medical Evaluation

After seeing his or her primary care physician, a patient suspected of having advanced PAD should see a vascular surgeon for a more thorough vascular evaluation before embarking on an exercise program. The vascular surgeon's evaluation to confirm advanced PAD will include a medical history, a physical examination, measures of ankle and toe pressures, an ABI for each leg (figure 17.1), a determination of which arteries are involved, an assessment of cardiovascular risk factors, and an assessment of the degree of disability and severity of symptoms. Smoking history, systolic and diastolic blood pressures, lipid and lipoprotein profiles, diabetes status, hemoglobin, serum creatine, height, weight, and body mass index (calculated) should also be included in the initial evaluation. For special research projects, a hypercoagulability screen can be performed, and homocysteine, lipoprotein a [LP(a)], and C-reactive protein levels can be measured.

Contraindications

In general, indications and contraindications to exercise stress testing for those with PAD are similar to those for cardiac patients (3, 34). If a patient does not have clinical signs and symptoms of cardiac disease, he or she does not need to have an exercise stress test to evaluate for myocardial ischemia before engaging in low-intensity exercise therapy. In this case, the exercise training therapy intensity should be kept similar to the person's exercise effort while maintaining a normal walking speed. Guidelines for contraindications to exercise testing and training can be found in the *ACSM's Guidelines for Exercise Testing and Prescription* (3).

Recommendations and Anticipated Responses

Because the functional capacity of patients with PAD is often low, and their physical performance is usually restricted by their symptoms of IC, they present special needs when they undergo evaluation of their cardiac status and functional capacity in response

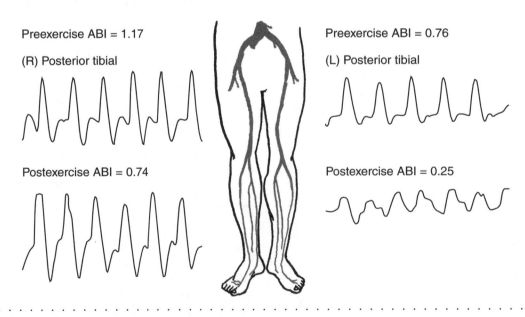

Preexercise ABI = 1.17

(R) Posterior tibial

Postexercise ABI = 0.74

Preexercise ABI = 0.76

(L) Posterior tibial

Postexercise ABI = 0.25

FIGURE 17.1   Right and left lower extremity Doppler velocity tracings and ankle/brachial index (ABI) at rest before and immediately following acute exercise.

to acute exercise. If the major objective is to evaluate the patient's cardiac status (i.e., the presence of myocardial ischemia), the test should be strenuous enough to elicit a pressure rate product (PRP = systolic blood pressure $\times$ heart rate) close to 3000. This double product (PRP) noninvasively estimates myocardial oxygen uptake. A patient with PAD is usually unable to perform an appropriate amount of skeletal muscular work with leg exercise alone to elicit an adequate PRP because of IC symptoms, rather than angina symptoms. Such a patient may need to perform arm work (e.g., arm ergometry) together with leg work to achieve an adequate myocardial stress for diagnostic testing of myocardial ischemia (8). The sensitivity for detecting myocardial ischemia in those with IC might be enhanced by including thallium scintigraphy with leg and arm ergometry. An alternative method to arm and leg ergometry is to induce a cardiac challenge by using a pharmacological agent such as ergonovine. Although this pharmacological method can be used to help evaluate cardiac status, it would not provide important information regarding hemodynamic measures and claudication indexes to exercise.

Exercise stress testing can be used to evaluate changes in the ABI following acute exercise (see figure 17.1). Functional impairment can be assessed by maximal treadmill walking time, pain-free walking time (**initial claudication distance**), and walking time to severe claudication (absolute claudication time). It has been recommended that for research, a run-in phase that consists of two or three treadmill tests be performed to familiarize patients to treadmill testing (68). The end point of testing is determined by the patient's perception of the severity of pain and may vary from test to test until a patient becomes familiar with the pain sensation. During testing, the patient should not be allowed to use handrail support because it influences the reliability of test results (41).

Either a level walking protocol or a progressive workload treadmill protocol can be used clinically to test for claudication pain and for hemodynamic measurements of the lower limbs in response to exercise in patients with PAD (42). Internationally accepted settings for a constant workload are a speed of 2 mph (3.2 km $\cdot$ hr^{-1}) and a grade of 12% (best for those with claudication distance between 50 and 150 m) (68).

Although a single-load test results in the workload and walking time following a linear function, a graded test usually follows a progressive incremental workload with respect to time. A graded exercise test starts at a horizontal level, and the slope is incrementally increased at specific time intervals throughout the test. The speed may vary but is usually kept constant because a patient with IC may find it difficult to adjust to speed changes. A typical protocol for a graded treadmill testing protocol is a constant speed of 3.2 km $\cdot$ hr^{-1}, 0% grade for 3 min, with a 3.5% increase in grade every 3 min (68).

Other testing protocols may have some utility. Gardner et al. (43), reported that a progressive stair-climbing testing protocol was also reliable for clinically evaluating claudication and hemodynamic responses to exercise in patients with stable IC. Furthermore, similar peak oxygen consumption (O$_2$peak) values were found in this study between testing protocols when patients with IC underwent graded walking, level walking, or progressive stair climbing. Stair climbing, while resulting in similar metabolic, claudication, and peripheral hemodynamic parameters, may not be the test of choice if cardiac parameters are of interest. This is true especially if patients are given exercise prescriptions involving walking, because fewer metabolic demands are placed on the cardiovascular system with stair climbing (41). For diagnosing lower limb arterial insufficiency in an office setting, 30 s of heel raising has been shown to cause changes in ankle pressure 1 min postexercise expressed as percentage change similar to those induced by treadmill exercise (4).

Advantages and disadvantages of different testing protocols for patients with PAD are shown in table 17.4. Basically, the mode of testing depends on whether the objective is diagnosis for advanced cardiac disease, IC symptoms, or to prescribe individualized exercise therapy.

$\dot{V}O_2$peak in patients with IC can be predicted by using a multiple regression equation with measures of time to maximal claudication pain and maximal heart rate taken during an incremental graded exercise test (120). Practical application 17.1 discusses methods for evaluating daily physical activity and quality of life in people with PAD.

Exercise Prescription

Exercise therapy is recommended as an important treatment for individuals with PAD, and the benefits include improved functional capacity, modification of risk factors, greater ambulation speed, longer ambulation distance to the onset of severe claudication, and improved psychological status. Muscle function and structural changes occur, mitochondrial bodies increase, oxidative enzymes increase, and free fatty acid oxidation metabolism

TABLE 17.4

Exercise Testing Protocols for Patients With Peripheral Arterial Disease

Mode of testing	Advantages	Disadvantages
Treadmill test	Common form of physiological stress	Cardiac evaluation is limited because of IC symptoms
— Single-stage test	Good test for those who are very deconditioned	May not be strenuous enough for some patients
— Multistage continuous	Results in high test–retest correlation coefficients	Sometimes difficult for patients to adjust to work increments
Ramp protocols — Treadmill — Cycle ergometry	Involves a steady increase in cardio-respiratory and hemodynamic responses	Lacks norms for physiological and hemodynamic responses in those with IC
Stair stepping	Progressive test is reliable for evaluating symptoms of IC	Lacks norms for physiological responses
Cycle ergometry	Good test for those with limitations that restrict weight bearing	Unfamiliar method of exercise. IC symptoms may differ from those obtained while walking.
Cycle and arm ergometry	Good test for detecting and evaluating myocardial ischemia in those with IC	Difficult to measure blood pressure responses
Heel raising	Good test for office setting without adequate testing equipment	Unknown workloads for test reproducibility.

Note. IC = intermittent claudication.

increases as muscles adapt to exercise training. Most importantly, the morbidity and mortality rates associated with exercise training are very low in those with IC (29, 30, 47, 53, 97). Womack et al. (121) showed that exercise rehabilitation for patients with PAD improved their walking economy as assessed by a decreased slow component of the oxygen cost of ambulation.

Patients may be in a supervised clinical setting or an unsupervised home exercise program depending on their needs, especially as it applies to comorbidities (59, 69, 89). Home exercise programs are usually very conducive to a patient's schedule and are less costly than hospital-based programs. This latter point is important, because many patients are retired and on fixed incomes. Patients may benefit equally by either approach as long as they closely follow an individualized and structured exercise protocol. If a patient is clinically without known cardiac disease or has cardiac disease but is otherwise stable cardiovascularly, he or she can safely exercise at home. Even if close cardiac monitoring is ini-tially required, most patients can eventually exercise at home on a routine basis. See practical application 17.2 for a review of the literature about peripheral arterial disease.

Special Exercise Considerations

Patients' desires to improve their medical status and their willingness to comply closely to their program are important issues to address when counseling them on exercise training therapy. Expected physiological, hemodynamic, and psychological outcomes need to be discussed so patients will know how they will individually benefit from engaging in routine exercise training. An important consideration is that a patient's primary care physician and vascular surgeon highly support and encourage long-term exercise training therapy. Cognitive factors such as stress, motivation to participate in exercise training, and the belief that the therapy is beneficial have been shown to relate to improvements in walking distance (102). Rather than a physician suggesting exercise

Methods for Evaluating Daily Physical Activity and Quality of Life

Either an accelerometer or a pedometer has been reported to reliably estimate physical activity on two consecutive days (109). It is important that patients provide information regarding their quality of life and functional status, because their symptoms of IC usually prevent them from engaging in normal daily activities of living (51). Daily functional status can be assessed by validated questionnaires such as the Walking Impairment Questionnaire (which tests ability to walk), the Physical Activity Recall (habitual physical activity level), and the Medical Outcomes Study Short Form 20 or 36 (physical, social, and role functioning, well-being, and overall health) (96). Most scores on these questionnaires have been shown to improve with exercise therapy, more so with treadmill walking than strength training, and these scores continue to improve with exercise therapy over time (51).

The quality of life assessment tool, the Short Form-36, is considered a valid survey instrument for evaluating those with IC and many other patient populations (112). The Sickness Impact Profile has been shown to be a sensitive method for evaluating the overall dysfunction in those with IC (5). The McMaster Health Index Questionnaire, administered to patients with IC, revealed a significant impairment in general health and physical, social, and emotional function in patients compared with controls (9). Furthermore, because this study found that these measures were not associated with treadmill performance, the many ramifications of the disease itself rather than the functional level of a patient need to be considered when a treatment is provided.

A 6-min walk test, performed outside the laboratory by those with IC, has some clinical utility, because this test has been shown to produce highly reliable measurements that are associated with the functional and hemodynamic severity of IC (79). For more physically functional and otherwise healthy patients, using a 6-min walk test can provide objective information for evaluating the progress of exercise therapy without expensive laboratory equipment. Furthermore, patients can perform this test on their own and without sophisticated testing equipment.

Literature Review

Williams et al. (120) reported that a structured exercise program along with risk factor reduction for 45 patients with IC resulted in an increase of 122% to 450% in patients' walking distances. Of those patients who smoked, 88% quit smoking after embarking on the exercise program. Important in this report was that exercise therapy was not associated with any morbidity or mortality. Follow-up of these patients after 1 year showed a significant reduction in hyperlipidemia (both total serum cholesterol and triglycerides levels), and after 2 years, 84% had maintained or improved their walking distances compared with their distances at exit from the structured exercise program. Eighty-three percent of those who quit smoking after starting the exercise training program remained nonsmokers at follow-up. Others have reported benefits of exercise training and, in general, found routine walking therapy to prolong pain-free walking distance. Feinberg et al. (33) studied the effects of 12 weeks of exercise therapy (without an attempt made to modify risk factors for atherosclerosis) in 19 patients with IC. Major endpoints were absolute systolic ankle pressure, ABI, maximum walking time, claudication pain time, and the ischemic window (the area under the curve of the exercise-induced decrease of the ankle pressure and its recovery recorded over time). The authors found 659% and 846% increases for maximum walking time and claudication pain time, respectively. No change occurred in the absolute ankle pressure or the ABI and, importantly, the average ischemic window measures decreased by 58.7%.

Ten published reports with good scientific methodological criteria, but most with small sample size, were used by Robeer et al. (100) in an attempt to establish the effect of exercise rehabilitation therapy in older (age 60-76 years) patients with IC. Scientific criteria examined included the study population, the intervention used, outcome variables examined, and data analysis and presentation. This information was used to develop a "weighted scale" for the methodological quality of

continued

randomized clinical trials employing exercise therapy to evaluate their efficacy. Included in this analysis was the identification of outcome predictors for exercise therapy, but this measure was found in only one of the studies. Of the 10 published studies considered to be of good research design, there was a 28% to 210% improvement in pain-free/maximum walking distance/time. These investigators were unable to determine outcome predictors (i.e., optimal frequency, type, mode, intensity, and duration of exercise or the value of supervision) for exercise training and, therefore, recommended additional clinical studies be performed to develop optimal rehabilitation programs for patients with IC. Another report from these investigators using these 10 published studies concluded that even after they reviewed studies of good scientific methodology, a definition of an optimal exercise program, the effect of adherence to an exercise program long term, and how long beneficial effects of exercise last still remain unclear (11).

An earlier review of the literature by Ernst and Fialka (30) resulted in a recommendation that exercise be performed regularly for at least 2 months and at a high intensity of effort to be beneficial. Continued improvements in functional status, however, have been noted to continue over a 24-week period of adherence to a structured walking exercise program (95). A model that predicts the outcome of exercise training for those with IC by fitting multistate transition models using autoregressive logistic regression was proposed by de Vries et al. (19). Important covariate parameters that predict outcome of supervised exercise were time, age, ABI, and duration of IC.

Patients can markedly improve their functional capacity from participating in either a highly structured home exercise program or a formal supervised exercise program in a clinical setting (59), which especially can improve their perception of their health and well-being (89). An unsupervised home walking program can also improve the ability of patients with IC to walk longer distances without pain (98). This is an important point, because it is often more conducive for daily exercise and less expensive for patients to exercise at home rather than to travel to a hospital setting for supervised exercise therapy, even for a short time.

Different modes of training for patients with IC have been demonstrated to show some crossover improvements but are mostly specific to the exercise therapy (e.g., treadmill walking, stair climbing, biking) performed over time in a training program. Thus, if a patient is physically training by exercising on a stair climber machine or a motor-driven treadmill, both modes can improve exercise capacity. But for testing purposes, the training effect will be most apparent on the apparatus used for training (60). This has implications for evaluating progress in physical fitness levels, because testing should be performed on an apparatus (e.g., bicycle ergometer, treadmill, and stair climber) similar to that used for exercise therapy.

training to the patient without providing explicit directions, the patient should be referred to a specialist (e.g., exercise physiologist or physical therapist) to receive an individualized, progressive, and structured exercise training program. This individualized approach provides a program of exercise therapy that affords the lowest possible risks of complications and takes into account comorbidities and medications for these conditions that may necessitate close supervision by the patient's primary care physician or vascular surgeon. Merely giving a simple recommendation to exercise will, in most cases, not help a patient to engage in appropriate exercise therapy, and recidivism will be high. As the patient becomes more physically fit, he or she may proceed to a higher level of exercise training effort, a decision made by the patient and his or her primary care

physician, vascular surgeon, and exercise physiologist or physical therapist.

Patients with IC are often frightened to push themselves physically for fear of damaging their painful claudicating lower limb skeletal muscles. They need to be assured that walking through their claudication pain while exercise training will not cause lasting harm to their muscles, and will ultimately result in better muscle function. In those with IC, no pain with exercise will result in little physiological gain in improvements in their pain-free walking distance over time with exercise training. Patients should be clearly told that if they have chest pain (possibly angina pectoralis) they should stop exercising and, if it persists, go to a local emergency room immediately. However, patients should be encouraged to tolerate pain in the legs attributable to claudication as best as possible, and they should be continuously reassured that exercise will not harm the skeletal muscles. Patients need to understand that routine exercise is one of the most beneficial therapies that they can do for their vascular condition as well as for reducing risk factors of their underlying general atherosclerosis.

Proper foot care is very important, especially in those with diabetes mellitus, to prevent blisters and possible infections. Properly fitting footwear and socks that do not bunch up are crucial to prevent soft tissue injury or foot ulcers. Daily inspection of the toes and plantar surfaces of the feet is essential because it is important to detect any abnormality early. Patients should be advised to return to their physicians immediately should any changes occur in their feet.

Resistance activities can supplement walking or other cardiovascular activities. These exercises can be properly selected for strengthening muscles of major muscles groups. Selection of muscle groups should be based on those needed to perform activities of daily living. The resistance training effort should be easy to moderate, using free weights and dumbbells or a wide variety of weight resistance machines, and should include 8 to 10 repetitions performed one to three times with slightly slower eccentric contractions.

Although patients with IC will usually exercise at a low intensity, it is recommended that they begin their aerobic exercise session with a very low walking intensity for 2 to 5 min (warm-up) and after the exercise session (cool-down). These warm-up and cool-down sessions may not be as important until a patient can perform at least 10 to 15 min of continuous exercise.

In those with severe symptoms of IC brought on by exercise, transient ischemia followed by reperfusion occurs following a bout of acute strenuous exercise. This physiological response has led to some controversy as to whether exercise is contraindicated for patients with advanced PAD (see table 17.5). Total **antioxidant** capacity and renal tubular function have been shown to be abnormal following acute exercise in patients with IC (72).

Increased thrombin formation was reported to occur with acute exercise in patients with PAD compared with healthy subjects (84). This suggests that acute exercise, also resulting in catecholamine release along with local muscle ischemia, could possibly increase the preexisting prothrombic potential of a diseased arterial wall. Others have reported that major vascular endothelial injury does not occur following acute exercise. This was concluded because

concentrations of plasma markers for endothelial damage did not change even though the median ABI changed from 0.96 before exercise to 0.59 following treadmill exercise testing in patients with symptomatic IC (122). If there is a major concern about the possibility of an ischemic-reperfusion injury, drug therapy may provide some protection against systemic vascular endothelial injury following acute exercise; pentoxifylline has been shown to reduce the post-1-hr increase seen in urinary albumin, expressed as a creatinine ratio (115).

Exercise Recommendations

An aerobic exercise program can be very beneficial for PAD patients. Although one's overall blood flow in the diseased extremity may not change a great deal, patients will experience physiological adaptations that promote a better utilization of oxygen and increase their walking time and most importantly, their quality of life. Walking is the mode of exercise training most effective for reducing IC pain symptoms in patients with advanced PAD. In respect to duration, the goal is to reach 20 to 30 min of activity. However, with PAD, continuous exercise for this period is difficult, and starting with intermittent exercise is strongly suggested. The ability to achieve continuous exercise for the suggested period of time depends on the severity of disease and the patient's pain threshold. The intensity of the walking pace is also directly related to the severity of disease and should start at or below one's normal walking pace and progress to tolerable pain. Last, in regard to frequency in PAD patients, it appears that progressing to daily activity should be the goal for optimal results. Enhanced functional performance can be seen soon after the person embarks on a routine exercise program, but it usually takes 6 months of exercise training to optimize IC pain-relief walking distances. See practical application 17.3 for information about prescribing exercise for patients with peripheral arterial disease.

Mode

Walking is an excellent training method for individuals with IC. It is a rhythmic, dynamic aerobic activity, the most common weight-bearing activity that is both familiar and, in most cases, enhances the muscles most affected by PAD (80). Walking regularly, with the duration and intensity progressed systematically, develops and sustains an individual's physical fitness functional capacity and usually specifically trains muscle having the most severe claudicating symptoms. Intermittent walking programs, starting at a low intensity (approximately the

TABLE 17.5

Markers Showing Systemic Effects of Ischemia-Reperfusion With a Single Bout of Exercise in Patients With Intermittent Claudication (IC)

Markers	Change with acute exercise in IC
ENDOTHELIAL DAMAGE	
Von Willebrand's factor[a]	NC
FREE RADICALS AND NEUTROPHIL ACTIVATION	
Systematic neutrophil count[a]	Increased
Neutrophil elastase	Increased
Neutrophil hydrogen peroxide	NC
Soluble P-selectin	NC
Monocytes	NC Platelets
THROMBIN FORMATION AND FIBRIN DEGRADATION	
Thrombin–antithrombin III complex	Increased
D-dimer[a]	Increased[b]
Tissue plasminogen activator (t-PA)[a]	Increased[b]
Plasminogen activator inhibitor-I antigens[a]	NC
t-PA activity	Increased[b]
Plasmin-α 2-antiplasmin complex	Increased[b]
Plasma catecholamines	Increased[b]
OTHER	
Urinary microalbumin excretion	Increased
Neutrophil deformability	Increased
Plasma thromboxane	Increased[b]
Interleukin-8	Increased

Note. NC = no change.

[a]Levels at rest are markedly higher in patients with IC compared with controls. [b]Acute exercise in controls results in a similar increase.

Edwards et al. (25); Kirkpatrick et al. (66); Lewis et al. (72); Mustonen et al. (84); Turton et al. (116); Woodburn et al. (122)

patient's normal walking speed) and progressing to a more brisk pace slowly over weeks, are an excellent method to improve a patient's fitness level without markedly increasing the risk of adverse events (i.e., musculoskeletal or joint injuries) (69).

Strength training alone or in combination with treadmill training for patients with IC has been com-pared to determine whether one modality of train-ing was superior (48). These authors reported that patients with PAD and IC symptoms showed im-provements in peak walking times with treadmill exercise or with weight training for skeletal muscles of the legs. The treadmill group, however, showed an improvement in $\dot{V}O_2$peak, but no changes were

PRACTICAL APPLICATION 17.3

Exercise Prescription

Aerobic Exercise Training

— Mode of training: intermittent walking program (69).

— Frequency of training: three to five days per week but preferably daily with a goal of expending 2000 kcal · week^{-1}.

— Intensity: initially at normal walking pace with progressive efforts toward maximum pain while walking.

— Time: 20-30 min of walking per session.

— Method: intermittent walking bouts of 2 to 5 min followed by a 1- to 2-min recovery period performed repetitively to accomplish 20 to 30 min of walking. Progression accomplished by increasing the walking bouts by 1 min per week.

— Clinical follow-up: 2 to 6 months following exercise training.

Resistance Training

— Muscle group: specifically those of lower extremities and generally major muscle groups throughout the body.

— Repetitions: 8-10 repetitions over the full range of muscle function with slower eccentric muscle contractions.

— Sets: one to three sets per session.

— Frequency: three times per week.

— Progression of effort: no more than 10% according to readiness.

observed in $\dot{V}O_2$peak or claudication onset time for those participating in weight training alone. These results demonstrate the importance of using walking as a major exercise modality for those with IC, showing that resistance training can serve as an adjunct treatment for strengthening major muscle groups in the legs.

Frequency

At the onset of exercise therapy, frequency should be a minimum of three times per week. As the individual adapts to the exercise program and there is minimal muscle soreness 24 hr postexercise, the frequency should gradually increase (i.e., an additional day every 1-2 weeks). Ideally, the goal for PAD patients is to walk daily and eventually reach a caloric expenditure of 2000 kcal · week^{-1}.

Intensity

The exercise intensity prescribed for patients undergoing therapy has implications when considering time to onset of claudication pain and time to maximal claudication pain but appears not to alter the dissipation of pain during recovery (38). This is one reason why intermittent exercise training protocols

can work well for patients with IC. Furthermore, the intensity of effort is not associated with the decrease in ABI with exercise but is related to the recovery time for measures of the ABI (105). These findings are important for prescribing exercise training based on a patient's perception of pain, but patients need to be encouraged to tolerate as much IC discomfort as possible. Because intermittent exercise protocols allow recovery periods interspersed throughout exercise sessions, they are ideal for patients with IC (69). Intermittent exercise therapy can also be very effective in increasing total work accomplished and can be followed more closely by patients with advanced PAD and associated IC symptoms.

The intensity of effort can initially be set at a normal walking pace because this effort is well within a patient's capability and is one that the patient feels safe in performing. Setting intensity according to a target heart rate often proves impractical for patients with IC because many patients are older and find this approach difficult to master because of the difficulty in palpating their pulse as well as monitoring their pulse using a wristwatch. If a patient is interested in monitoring his or her exercise heart rate electronically, many accurate heart rate counters are commercially available. PAD progresses with age, and those who are older (mean age of 75.5 years), compared with those younger (mean age of 60.4 years), may show greater impairment in peripheral hemodynamic measurements but do so without exaggerated heart rate or blood pressure responses during a progressive treadmill walking test to maximal leg pain (37). This would have implications for training if exercise intensity were based on age-predicted exercise heart rate responses rather than IC symptoms.

An optimal threshold of exercise intensity is not known for those with PAD and most likely varies from one patient to the next depending, in most part, on the severity of disease and related symptoms. The intensity need not be high, especially when a patient is embarking on an exercise program. The focus should be on the total time (preferably 20-30 min) of exercise per session. The initial intensity of training chosen for those with PAD should be similar to their normal ambulating pace. This approach to intensity of effort has many advantages, as patients with IC are usually very deconditioned because of their IC pain symptoms, which often prevent them from even minimal activity. Having patients with IC exercise at their normal walking pace usually allows them to begin and progress in an exercise training program well within their capabilities. As a patient's physical fitness conditioning improves, an exercise stress test is recommended for evaluating cardiac status and claudicating indexes before their intensity of exercise effort is markedly increased.

Duration

The duration of activity is very dependent on the severity of PAD. However, the goal would be to achieve 20 to 30 min of continuous walking. Because of the uniqueness of exertional IC that occurs with PAD, exercise needs to be intermittent. Generally, a good starting point is to determine an intensity that the individual can maintain for 2 to 5 min followed by a recovery phase of 1 to 2 min. Begin at one's normal walking pace, or slightly less than normal and progress to a pace that is equivalent to one's maximum tolerable pain. Irrespective of the initial walking time, the duration of activity per bout of exercise should ideally be increased by 1 min per week.

Physiological Adaptations and Maladaptations

Some studies have shown that exercise rehabilitation for those with IC does not increase blood flow

(26, 58, 70, 75, 76, 111, 125), but others reported that exercise effectively improved leg blood perfusion (2, 28, 52, 73). A review of studies using exercise training therapy for those with IC reported improvements in walking tolerance, but blood flow improvements alone could not account for the total improvement in pain-free walking distances (39). Patients following a 6-month exercise rehabilitation program showed a 115% increased distance to the onset of claudication pain, and this improvement was independently related to the 27% increase found in blood flow (40).

Another mechanism by which exercise therapy may improve walking distance in those with IC is by enhancing skeletal muscle oxidative metabolism without altering anaerobic metabolism (73). The benefits of exercise conditioning for patients with IC appear more likely attributable to an improvement in calf muscle oxidative metabolism rather than to changes in skeletal muscle blood flow.

Improvement in pain-free walking distance following exercise therapy has been attributed to the following factors:

1. Improved biomechanics of ambulation resulting in decreased metabolic demands (107, 109, 121)
2. Increased collateral circulation resulting in increased peripheral blood flow (2, 28, 108, 110)
3. A reduction in blood viscosity (31, 36)
4. An increase in blood cell filterability and a decrease in red cell aggregation (36)
5. Regression of atherosclerosis (52)
6. Increased extraction of oxygen and metabolic substrates resulting from improvements in skeletal muscle oxidative metabolism (17, 56, 111)
7. Increased pain tolerance (52, 125)

Of these reported factors associated with improved walking distance, it appears unlikely that an increased blood flow secondary to increased collateral circulation is the major contributing factor (18, 26, 85, 103, 111).

Although some evidence suggests that a low-grade inflammatory response may occur with acute strenuous exercise, regular exercise training for those with IC appears not to cause long-term endothelial inflammation, improves blood **rheology** properties, and may attenuate progression of atherosclerosis. This suggests that exercise therapy is a viable treatment option for those with IC.

Conclusion

This chapter provides a review of information regarding PAD with accompanying IC and the benefits of exercise training for patients with IC symptoms without jeopardy of losing a lower extremity. Limb salvage is usually not the primary goal of treatment because IC is the primary symptom that responds well to conservative measures. In that IC is an expression of systemic atherosclerosis, control of underlying metabolic and cardiovascular comorbidities should clearly be a priority to reduce cardiac and cerebrovascular morbidity and mortality rates. Intensive risk factor reduction therapy for inhibiting the progression of atherosclerosis, especially abstinence from tobacco products, is a major goal. Importantly, exercise therapy can improve the functional capacity of these patients and may also prevent other cardiovascular diseases. Our current knowledge regarding the benefits of exercise comes primarily from studies involving patients who are highly motivated, fairly young, and with stable comorbidities. Studies still need to be conducted investigating the frequency, intensity, and time of training necessary to gain optimal benefits of exercise therapy for those with IC. Because our population is aging, efficacy of exercise therapy in those over 65 years of age, especially in a home-based setting, needs to be further investigated.

Although the conservative treatments of risk factor reductions and exercise have been shown to be efficacious for patients with IC, third party payers presently reimburse none of these therapeutic services involving exercise therapy. Insurance companies need to recognize the value of vascular rehabilitation services for cost containment, and strongly support the availability of these services in health care facilities. As an important aspect of the overall medical process, healthcare personnel treating patients with IC need to continue analysis of therapy outcomes to further delineate and define appropriate and optimal interventions.

Case Study 17

Medical History

Mrs. RU is a 61-year-old African-American female who came to the Vascular Clinic complaining of right leg pain symptoms for more than 8 months. She complains of intermittent tingling sensations in the right foot as well as significant right buttock and thigh pain with ambulation. She is presently able to walk approximately 50 ft before the onset of painful leg symptoms. The symptoms make her feel moderately disabled. She denies exertional angina or any major orthopedic problems that would prevent her from engaging in a mild exercise program.

Her past history is significant for coronary artery bypass surgery 10 years earlier. She underwent a subsequent coronary artery bypass surgery revision and mitral valve placement approximately 7 years later. A recent cardiac echo showed severely depressed ventricular function with an ejection faction estimated to be about 35%.

Her risk factors for atherosclerosis include a history of hypertension as well as insulin-dependent diabetes mellitus. She smoked cigarettes until 1996 but at that time quit and has maintained smoking cession. She is also hyperlipidemic, both hypertriglyceridemic and hypercholesterolemic. Her family history is positive for coronary artery disease.

Her physical examination shows that she is a healthy-appearing middle-aged female. Her blood pressure is 118/70 and her pulse is 80 and regular. Her femoral, popliteal, and pedal pulses are absent on the right extremity. The right foot is warm with mild delayed capillary refilling. There is no evidence of ischemic tissue loss or pedal edema. On the left extremity, the femoral popliteal and posterior tibial pulses are 2$^+$.

Exercise Testing

She undergoes a constant load treadmill test in the vascular laboratory. Results of testing show the resting ABI on the right to be 0.75, which dropped to 0.51 with acute exercise. The right first digit

continued

pressure is 15 mmHg. The left ABI and left first digit pressures are normal. Doppler velocity waveform analysis is consistent with suprainguinal arterial occlusive disease of the right leg. The arterial Doppler study shows abnormal right lower extremity wave forms and pressures.

Diagnosis

The medical history and exercise test results are consistent with a diagnosis of PAD of her right leg. Her current weight is 160 lb. She is approximately 25 to 30 lb overweight for a woman of 5 ft 5 in. She is encouraged to lose approximately 0.25 to 0.50 lb per week to reduce her body weight to a goal of 130 lb. The recommended balanced diet includes a high-fiber, low-cholesterol, and low saturated fat diet with approximately 1700 to 1800 kcal per day.

Exercise Prescription

Intermittent exercise is recommended, involving walking in the halls of her apartment complex at her normal walking pace. Although she is allowed to exercise on a stationary cycle ergometer, she is encouraged to choose walking as often as possible to improve metabolism in the gastrocnemius and soleus muscles. The exercise bouts were initially set for 2 to 5 min followed by a 1- to 2-min recovery period (slower walking or sitting). She is instructed not to extend herself beyond the point of mild intermittent claudication pain. She is instructed to repeat the exercise recovery cycle, if possible, 6 to 15 times to ensure 25 to 30 min of exercise per training session. She is to progress by increasing the bouts by 1 min per week. Because IC symptoms rather than cardiac dysfunction limited her exercise performance, the patient is instructed to closely monitor any chest discomfort also while exercising.

   The patient markedly increased her ability to ambulate. She initially was only able to walk half the length of the hallway and eventually was able to walk at a very slow pace continuously for 20 to 30 min. The patient reported that she felt like she had more vigor and strength and did not experience such severe claudication symptoms at this level of effort. She denied experiencing any cardiac symptoms throughout her exercise training program. Because this protocol seemed to be working

Parameters[a]	Pretraining	Posttraining
ASP	88	87
ABI	0.75	0.73
ICD (m)	16	134
ACD (m)	120	165
Total cholesterol (mg · dl^{-1})	248	205
HDL cholesterol (mg · dl^{-1})	34	40
Triglycerides (mg · dl^{-1})	190	138

Note. ASP = ankle systolic blood pressure (mmHg); ABI = ankle/brachial index; ICD = initial claudication distance; ACD = absolute claudication distance; HDL = high-density lipoprotein
[a]Hemodynamics of the most severely affected leg.

continued

well and because she had relatively impaired cardiac function (low ejection fraction), she was maintained on this exercise regime. After 6 months, she had decreased her body weight to 133 lb. Shown in the preceding table are baseline and follow-up (6 months) parameters.

Case Study 17 Discussion Questions

1. Because Ms. RU has poor cardiac function and a low functional capacity, how could a daily intermittent exercise protocol be adjusted to optimize her duration of effort?

2. With such a low cardiac function, would progression of effort be restricted by claudication symptoms or by cardiac factors?

3. As Ms. RU loses weight, should the intensity of effort be significantly increased?

4. Will physiological adaptations to exercise training be most pronounced in the cardiopulmonary or the peripheral vascular system?

5. How often should Ms. RU's exercise prescription be revised, what components (frequency, intensity, duration) may be altered, and what other safe modes of exercise might be recommended?

Glossary

absolute claudication distance—The distance walked when the patient stops ambulating because of severe pain.

angioplasty—Reconstitution or recanalization of a blood vessel; may involve balloon dilation, mechanical stripping of intima, forceful injection of fibrinolytics, or placement of a stent.

antioxidant—An agent that inhibits oxidation and thus prevents the deterioration of other materials through oxidative processes.

apolipoprotein—The protein component of lipoprotein complexes that is a normal constituent of plasma chylomicrons, high-density lipoprotein, low-density lipoprotein, and very low density lipoprotein in humans.

atherectomy—Any removal by surgery or specialized catheterization of an atheroma in an artery.

claudication—Cramping symptoms and the feeling of weakness of the legs induced by walking.

C-reactive protein—A β-globulin found in the serum of various persons with certain inflammatory, degenerative, and neoplastic diseases.

Doppler ultrasonography—Application of the Doppler effect in ultrasound to detect movement of scatterers (usually red blood cells) by the analysis of the change in frequency of the returning echoes.

endothelial cell—One of the squamous cells forming the lining of serous cavities, blood, and lymph vessels and the inner layer of the endocardium.

endothelial derived relaxing factors—Diffusible substances produced by endothelial cells that cause vascular smooth muscle relaxation; nitric oxide is one such substance.

gangrene—Necrosis of body tissues caused by obstruction, loss, or diminution of blood supply.

homocysteine—A homolog of cysteine.

hyperemia—Increased amount of blood in a part or organ.

initial claudication distance—The distance walked when the patient first perceived claudication pain.

intermittent claudication—Recurrent cramping symptoms at regular intervals when walking.

ischemia: Local anemia caused by arterial narrowing with restriction of the blood supply.

percutaneous transluminal angioplasty—Intra-arterial catheterization through the skin to enlarge a narrowed lumen with a balloon-tipped catheter; may include positioning of an intravascular stent.

rheology—The study of the deformation and flow of liquids and semisolids.

References

1. Al Zahrani, H.A., H.M. Al Bar, A. Bahnassi, and A.A. Abdulaal. 1997. The distribution of peripheral arterial disease in a defined population of elderly high-risk Saudi patients. *International Angiology,* 16(2): 123-128.

2. Alpert, J., O. Larson, and N. Lassen. 1969. Exercise and intermittent claudication. Blood flow in the calf muscle during walking studied by the zenon-133 clearance methods. *Circulation,* 39: 353-359.

3. American College of Sports Medicine. ACSM's Guidelines for Exercise Testing and Prescription. 1995, Williams & Wilkins, Media, PA.

4. Amirhamzeh, M.M., H.J. Chant, J.L. Rees, L.J. Hands, R.J. Powell, and W.B. Campbell. 1997. A comparative study of treadmill tests and heel raising exercise for peripheral arterial disease. *European Journal of Vascular & Endovascular Surgery,* 13(3): 301-305.

5. Arfvidsson, B., J. Karlsson, A.G. Dahllof, K. Lundholm, and M. Sullivan. 1993. The impact of intermittent claudication on quality of life evaluated by the Sickness Impact Profile technique. *European Journal of Clinical Investigation,* 23(11): 741-745.

6. Bagger, J.P., P. Helligsoe, F. Randsbaek, H.H. Kimose, and B.S. Jensen. 1997. Effect of Verapamil in intermittent claudication: A randomized, double-blind, placebo-controlled, cross-over study after individual dose-response assessment. *Circulation,* 95: 411-414.

7. Baker, J.D., and D. Dix. 1981. Variability of Doppler ankle pressure with arterial occlusive disease: An evaluation of ankle index and brachial-ankle pressure gradient. *Surgery,* 89: 134-137.

8. Barbadimos, A.N., and L.R. Zohman. 1999. Intravenous dipyridamole thallium imaging V combined arm-leg cycle stress testing of patients unable to exercise on the treadmill. *American Journal of Physical Medicine & Rehabilitation,* 78(2): 111-116.

9. Barletta, G., S. Perna, C. Sabba, A. Catalano, C. O'Boyle, and G. Brevetti. 1996. Quality of life in patients with intermittent claudication: Relationship with laboratory exercise performance. *Vascular Medicine,* 1(1): 3-7.

10. Bartels, C., J.F. Bechtel, V. Hossmann, and S. Horsch. 1997. Cardiac risk stratification for high-risk vascular surgery. *Circulation,* 95(11): 2473-2475.

11. Brandsma, J.W., B.G. Robeer, S. van den Heuvel, B. Smit, C.H. Wittens, and R.A. Oostendorp. 1998. The effect of exercises on walking distance of patients with intermittent claudication: A study of randomized clinical trials. [published erratum appears in Physical Therapy 1998 May;78(5):547]. *Physical Therapy,* 78(3):278-286; discussion 286-288. [published erratum appears in *Physical Therapy* 1998 78(5): 547]

12. Buchwald, H., H.R. Bourdages, C.T. Campos, P. Nguyen, S.E. Williams, and J.R. Boen. 1996. Impact of cholesterol reduction on peripheral arterial disease in the Program on the Surgical Control of the Hyperlipidemias (POSCH). *Surgery,* 120(4): 672-679.

13. Coffman, J.D., and J.A. Mannick. 1972. Failure of vasodilator drugs in arteriosclerosis obliterans. *Annals of Internal Medicine,* 76(1): 35-39.

14. Coffman J.D. 1988. Pathophysiology of obstructive arterial disease. *Herz,* 13(6): 343-350.

15. Criqui, M.H., J.O. Denenberg, C.E. Bird, A. Fronek, M.R. Klauber, and R.D. Langer. 1996. The correlation between symptoms and non-invasive test results in patients referred for peripheral arterial disease testing. *Vascular Medicine,* 1(1): 65-71.

16. Criqui, M.H., J.O. Denenberg, R.D. Langer, and A. Fronek. 1997. The epidemiology of peripheral arterial disease: Importance of identifying the population at risk. *Vascular Medicine,* 2(3): 221-226.

17. Dahllof, A.G., P. Bjorntorp, J. Holm, and T. Schersten. 1974. Metabolic activity of skeletal muscle in patients with peripheral arterial insufficiency. *European Journal of Clinical Investigation,* 4(1): 9-15.

18. Dahllof, A.G., J. Holm, T. Schersten, and R. Sivertsson. 1976. Peripheral arterial insufficiency, effect of physical training on walking tolerance, calf blood flow, and blood flow resistance. *Scandinavian Journal of Rehabilitation Medicine,* 8(1): 19-26.

19. de Vries, S.O., V. Fidler, W.D. Kuipers, and M.G. Hunink. 1998. Fitting multistate transition models with autoregressive logistic regression: Supervised exercise in intermittent claudication. *Medical Decision Making,* 18(1): 52-60.

20. Depairon M., and M. Zicot. 1996. The quantitation of blood flow/metabolism coupling at rest and after exercise in peripheral arterial insufficiency, using PET and 15-0 labeled tracers. *Angiology,* 47(10): 991-999.

21. Desvaux, B., P. Abraham, D. Colin, G. Leftheriotis, and J.L. Saumet. 1996. Ankle to arm index following maximal exercise in normal subjects and athletes. *Medicine & Science in Sports & Exercise,* 28(7): 836-839.

22. Dormandy, J., M. Mahir, G. Ascady, F. Balsano, P. De Leeuw, P. Blombery, M.G. Bousser, D. Clement, J. Coffman, A. Deutshinoff, et al. 1989. Fate of the patient with chronic leg ischaemia. A review article. *Journal of Cardiovascular Surgery,* 30(1): 50-57.

23. Dormandy, J.A., and G.D. Murray. 1991. The fate of the claudicant: A prospective study of 1969 claudicants. *European Journal of Vascular Surgery,* 5(2): 131-133.

24. Eames, R.A., and L.S. Lange. 1967. Clinical and pathological study of ischaemic neuropathy. *Journal of Neurological Neurosurgery and Psychiatry,* 30(2): 215-226.

25. Edwards, A.T., A.D. Blann, V.J. Suarez-Mendez, A.M. Lardi, and C.N. McCollum. 1994. Systemic response in patients with intermittent claudication after treadmill exercise. *British Journal of Surgery,* 81(12): 1738-1741.

26. Ekroth, R., A.G. Dahllof, B. Gundevall, J. Holm, and T. Schersten. 1978. Physical training with intermittent claudication: Indications, methods, and results. *Surgery,* 84(5): 640-643.

27. England J.D., J.G. Regensteiner, S.P. Ringel, M.R. Carry, and W.R. Hiatt. 1992. Muscle denervation in peripheral arterial disease. *Neurology,* 42(5): 994-999.

28. Ericsson, B., K. Haeger, and S.E. Lindell. 1970. Effect of physical training on intermittent claudication. *Angiology,* 21(3): 188-192.

29. Ernst, E. 1992. Exercise: The best therapy for intermittent claudication? *British Journal of Hospital Medicine,* 48(6): 303-304, 307.

30. Ernst, E., and V. Fialka. 1993. A review of the clinical effectiveness of exercise therapy for intermittent claudication. *Archives of Internal Medicine,* 153(20): 2357-2360.

31. Ernst, E.E., and A. Matrai. 1987. Intermittent claudication, exercise, and blood rheology. *Circulation,* 76(5): 1110-1114.

32. Federman, D.G., J.T. Trent, C.W. Froelich, J. Demirovic, and R.S. Kirsner. 1998. Epidemiology of peripheral vascular disease: A predictor of systemic vascular disease. *Ostomy Wound Management,* 44(5): 58-62, 64, 66 passim.

33. Feinberg, R.L., R.T. Gregory, J.R. Wheeler, S.O. Snyder, Jr., R.G. Gayle, F.N. Parent, III, and R.B. Patterson.1992. The ischemic window: A method for the objective quantitation of the training effect in exercise therapy for intermittent claudication. *Journal of Vascular Surgery,* 16(2): 244-250.

34. Fletcher, G.F., G. Balady, V.F. Froelicher, L.H. Hartley, W.L. Haskell, and M.L. Pollock. 1995. Exercise standards: A statement for healthcare professionals from the American Heart Association Writing Group. *Circulation,* 91(2): 580-615.

35. Fowkes, F.G., E. Housley, R.A. Riemersma, C.C. Macintyre, E.H. Cawood, R.J. Prescott, and C.V. Ruckley. 1992. Smoking, lipids, glucose intolerance, and blood pressure as risk factors for peripheral atherosclerosis compared with ischemic heart disease in the Edinburgh Artery Study. *American Journal of Epidemiology,* 135(4): 331-340.

36. Gallasch, D., C. Diehm, C. Dofer, T. Schmitt, A. Stage, and H. Morl. 1985. Effect of physical training on the blood flow properties in patients with intermittent claudication. *Klinische Wochenschrift,* 63(12): 554-559.

37. Gardner, A.W. 1993. Claudication pain and hemodynamic responses to exercise in younger and older peripheral arterial disease patients. *Journal of Gerontology,* 48(5): M231-M236.

38. Gardner, A.W. 1993. Dissipation of claudication pain after walking: Implications for endurance training. *Medicine & Science in Sports & Exercise,* 25(8): 904-910.

39. Gardner, A.W., and E.T. Poehlman. 1995. Exercise rehabilitation programs for the treatment of claudication pain. A meta-analysis. *Journal of American Medication Association,* 274(12): 975-980.

40. Gardner, A.W., L.I. Katzel, J.D. Sorkin, L.A. Killewich, A. Ryan, W.R. Flinn, and A.P. Goldberg. 2000. Improved functional outcomes following exercise rehabilitation in patients with intermittent claudication. *Journal of Gerontology (Series A) Biological Sciences & Medical Sciences,* 55(10): M570-M577.

41. Gardner, A.W., J.S. Skinner, C.X. Bryant, and L.K. Smith. 1995. Stair climbing elicits a lower cardiovascular demand than walking in claudication patients. *Journal of Cardiopulmonary Rehabilitation,* 15(2): 134-142.

42. Gardner, A.W., J. S. Skinner, N.R. Vaughan, C.X. Bryant, and L.K. Smith. 1992. Comparison of three progressive exercise protocols in peripheral vascular occlusive disease. *Angiology,* 43(8): 661-671.

43. Gerhard, M., P. Baum, and K.E. Raby. 1995. Peripheral arterial-vascular disease in women: Prevalence, prognosis, and treatment. *Cardiology,* 86(4): 349-355.

44. Hall, J.A., and J. Barnard. 1982. The effects of an intensive 26-day program of diet and exercise on patients with peripheral vascular disease. *Journal of Cardiac Rehabilitation,* 2: 569-574.

45. Harrison, D.G., and Y. Ohara. 1995. Physiologic consequences of increased vascular oxidant stresses in hypercholesterolemia and atherosclerosis: Implications for impaired vasomotion. *American Journal of Cardiology,* 75(6): 75B-81B.

46. Hertzer, N.R. 1991. The natural history of peripheral vascular disease. Implications for its management. *Circulation,* 83(2 Suppl.): I12-I19.

47. Hiatt, W.R., J.G. Regensteiner, and E.E. Wolfel. 1993. Special populations in cardiovascular rehabilitation. Peripheral arterial disease, non-insulin-dependent diabetes mellitus, and heart failure. *Cardiology Clinics,* 11(2): 309-321.

48. Hiatt, W.R., E.E. Wolfel, R.H. Meier, and J.G. Regensteiner. 1994. Superiority of treadmill walking exercise versus strength training for patients with peripheral arterial disease. Implications for the mechanism of the training response. *Circulation,* 90(4): 1866-1874.

49. Hiatt, W.R., J.G. Regensteiner, E.E. Wolfel, M.R. Carry, and E.P Brass. 1996. Effect of exercise training on skeletal muscle histology and metabolism in peripheral arterial disease. *Journal of Applied Physiology,* 81(2): 780-788.

50. Hiatt, W.R. 1997. Current and future drug therapies for claudication. *Vascular Medicine,* 2(3): 257-262.

51. Hiatt, W.R., S. Hoag, and R.F. Hamman. 1995. Effect of diagnostic criteria on the prevalence of peripheral arterial disease. The San Luis Valley Diabetes Study. *Circulation,* 91(5): 1472-1479.

52. Hiatt, W.R., J.G. Regensteiner, M.E. Hargarten, E.E. Wolfel, and E.P. Brass. 1990. Benefit of exercise conditioning for patients with peripheral arterial disease. *Circulation,* 81(2): 602-609.

53. Hiatt, W.R., A.T. Hirsch, J.G. Regensteiner, and E.P. Brass. 1995. Clinical trials for claudication. Assessment of exercise performance, functional status, and clinical end points. Vascular Clinical Trialists. *Circulation,* 92(3): 614-621.

54. Higgins D., W.P. Santamore, P. Walinsky, and P. Nemir, Jr. 1985. Hemodynamics of human arterial stenoses. *International Journal of Cardiology,* 8: 177-192.

55. Hilleman, D.E. 1998. Management of peripheral arterial disease. *American Journal of Health-System Pharmacy,* 55(19 Suppl. 1): S21-S27.

56. Holm, J., A.G. Dahllof, P. Bjorntorp, and T. Schersten.1973. Enzyme studies in muscles of patients with intermittent claudication. Effect of training. *Scandinavian Journal of Clinical & Laboratory Investigation,* 31(Suppl. 128): 201-205.

57. Holm, J., A.G. Dahllof, and T. Schersten. 1975. Metabolic activity of skeletal muscle in patients with peripheral arterial insufficiency. Effect of arterial reconstructive surgery. *Scandinavian Journal of Clinical & Laboratory Investigation,* 35(1): 81-86.

58. Johnson, E.C., W.F. Voyles, H.A. Atterbom, D. Pathak, M.F. Sutton, and E.R. Greene. 1989. Effects of exercise training on common femoral artery blood flow in patients with intermittent claudication. *Circulation,* 80 (5 Pt 2): III59-III72.

59. Jonason, T., I. Ringqvist, and A. Oman-Rydberg. 1981. Home-training of patients with intermittent claudication. *Scandinavian Journal of Rehabilitation Medicine,* 13(4): 137-141.

60. Jones, P.P., J.S. Skinner, L.K. Smith, F.M. John, and C.X. Bryant. 1996. Functional improvements following StairMaster vs. treadmill exercise training for patients with intermittent claudication. *Journal of Cardiopulmonary Rehabilitation,* 16(1): 47-55.

61. Kang, S.S., P.W. Wong, and M.R. Malinow. 1992. Hyperhomocysteinaemia as a risk factor for occlusive vascular disease. *Annual Review of Nutrition,* 12: 279-298.

62. Kannel, W.B. 1996. The demographics of claudication and the aging of the American population. *Vascular Medicine,* 1(1): 60-64.

63. Kannel W.B., R.B. D'Agostino, and A.J. Belanger. 1992. Update on fibrinogen as a cardiovascular risk factor. *Annals of Epidemiology,* 2: 457-466.

64. Kannel, W.B., and D.L. McGee. 1985. Update on some epidemiologic features of intermittent claudication: The Framingham Study. *Journal of the American Geriatrics Society,* 33(1): 13-18.

65. Khaira, H.S., R. Hanger, and C.P. Shearman. 1996. Quality of life in patients with intermittent claudication. *European Journal of Vascular & Endovascular Surgery,* 11(1): 65-69.

66. Kirkpatrick, U.J., M. Mossa, A.D. Blann, and C.N. McCollum. 1997. Repeated exercise induces release of soluble P-selectin in patients with intermittent claudication. *Thrombosis & Haemostasis,* 78(5): 1338-1342.

67. Kokesh, J., A. Kazmers, and R.E. Zierler. 1991. Pentoxifylline in the nonoperative management of intermittent claudication. *Annals of Vascular Surgery,* 5(1): 66-70.

68. Labs, K.H., J.A. Dormandy, K.A. Jaeger, C.S. Stuerzebecher, and W.R. Hiatt. Basel PAD Clinical Methodology Group. 1999. Transatlantic Conference on clinical trial guidelines in peripheral arterial disease: Clinical trial methodology. *Circulation,* 100: e75-e81.

69. Lampman, R.M. 1997. Exercise prescription for chronically ill patients. *American Family Physician,* 55(6): 2185-2192.

70. Larsen, O.A., and N.A. Lassen. 1966. Effect of daily muscular exercise in patients with intermittent claudication. *Lancet,* 2(7473): 1093-2006.

71. Laursen, J.B., S. Rajagopalan, Z. Galis, M. Tarpey, B.A. Freeman, and D.G. Harrison. 1997. Role of superoxide in angiotensin II-induced but not catecholamine-induced hypertension. *Circulation,* 95(3): 588-593.

72. Lewis, D.R., A. Day, J.Y. Jeremy, P.V. Newcombe, S.T. Brookes, R. Baird, F.C. Smith, and P.M. Lamont. 1999. Vascular surgical society of Great Britain and Ireland: Systemic effects of exercise in claudicants are associated with neutrophil activation. *British Journal of Surgery,* 86(5): 699-700.

73. Lundgren, F., A.G. Dahllof, K. Lundholm, T. Schersten, and R. Volkmann. 1989. Intermittent claudication-surgical reconstruction or physical training? A prospective randomized trial of treatment efficiency. *Annals of Surgery,* 209(3): 346-355.

74. MacGregor, A.S., J.F. Price, C.M. Hau, A.J. Lee, M.N. Carson, and F.G. Fowkes. 1999. Role of systolic blood pressure and plasma triglycerides in diabetic peripheral arterial disease. The Edinburgh Artery Study. *Diabetes Care,* 22(3): 453-458.

75. Mannarino E., L. Pasqualini, S. Innocente, V. Scricciolo, A. Rignanese, and G. Ciuffetti. 1991. Physical training and antiplatelet treatment in stage II peripheral arterial occlusive disease: Alone or combined? *Angiology,* 42(7): 513-521.

76. Mannarino, E., L. Pasqualini, M. Menna, G. Maragoni, and U. Orlandi. 1989. Effects of physical training on peripheral vascular disease: A controlled study. *Angiology,* 40(1): 5-10.

77. Meijer, W.T., A.W. Hoes, D. Rutgers, M.L. Bots, A. Hofman, and D.E. Grobbee. 1998. Peripheral arterial disease in the elderly: The Rotterdam Study. *Arteriosclerosis, Thrombosis & Vascular Biology,* 18(2): 185-192.

78. Money, S.R., J.A. Herd, J.L. Isaacsohn, M. Davidson, B. Cutler, J. Heckman, and W.P. Forbes. 1998. Effect of cilostazol on walking distances in patients with intermittent claudication caused by peripheral vascular disease. *Journal of Vascular Surgery,* 27(2): 267-275.

79. Montgomery, P.S., and A.W. Gardner. 1998. The clinical utility of a six-minute walk test in peripheral arterial occlusive disease patients. *Journal of the American Geriatrics Society,* 46(6): 706-711.

80. Morris, J.N., and A.E. Hardman. 1997. Walking to health. *Sports Medicine,* 23(5): 306-332. [published erratum appears in *Sports Medicine* 1997 24(2): 96.]

81. Murabito, J.M., R.B. D'Agostino, H. Silbershatz, and W.F. Wilson. 1997. Intermittent claudication. A risk profile from The Framingham Heart Study. *Circulation,* 96(1): 44-49.

82. Murphy, T.P. 1998. Medical outcomes studies in peripheral vascular disease. *Journal of Vascular & Interventional Radiology,* 9(6): 879-889.

83. Murphy, T.P., A.A. Khwaja, and M.S. Webb. 1998. Aorto-iliac stent placement in patients treated for intermittent claudication. *Journal of Vascular & Interventional Radiology,* 9(3): 421-428.

84. Mustonen, P., M. Lepantalo, and R. Lassila. 1998. Physical exertion induces thrombin formation and fibrin degradation in patients with peripheral atherosclerosis. *Arteriosclerosis, Thrombosis & Vascular Biology,* 18(2): 244-249.

85. Myhre, K., and D.G. Sorlie. 1982. Physical activity and peripheral atherosclerosis. *Scandinavian Journal of Social Medicine,* (Suppl.) 29: 195-201.

86. Navab, M., J.A. Berliner, A.D. Watson, S.Y. Hama, M.C. Territo, A.J. Lusis, D.M. Shih, B.J. Van Lenten, J.S. Frank, L.L. Demer, P.A. Edwards, and A.M. Fogelman. 1996. The Yin and Yang of oxidation in the development of the fatty steak. A review based on the 1994 George Lyman Duff Memorial Lecture. *Arteriosclerosis, Thrombosis, and Vascular Biology*, 16(7): 831-842.

87. Ohara, Y., T.E. Peterson, and D.G. Harrison. 1993. Hypercholesterolemia increases endothelial superoxide anion production. *Journal of Clinical Investigation*, 91(6): 2546-2551.

88. Ohara, Y., T.E. Peterson, H.S. Sayegh, R.R. Subramanian, J.N. Wilcox, and D.G. Harrison. 1995. Dietary correction of hypercholesterolemia in the rabbit normalizes endothelial superoxide anion production. *Circulation*, 92(4): 898-903.

89. Patterson, R.B., B. Pinto, B. Marcus, A. Colucci, T. Braun, and M. Roberts. 1997. Value of a supervised exercise program for the therapy of arterial claudication. *Journal of Vascular Surgery*, 25(2): 312-319.

90. Pedersen, T.R., J. Kjekshus, K. Pyorala, A.G. Olsson, T.J. Cook, T.A. Musliner, J.A. Tobert, and T. Haghfelt. 1998. Effect of simvastatin on ischemic signs and symptoms in the Scandinavian simvastatin survival study (4S). *American Journal of Cardiology*, 81(3): 333-335.

91. Pernow, B., B. Saltin, R. Wahren, R. Cronestrand, and S. Ekestroom. 1975. Leg blood flow and muscle metabolism in occlusive arterial disease of the leg before and after reconstructive surgery. *Clinical Science and Molecular Medicine*, 49(3): 265-275.

92. Poredos, P., and B. Zizek. 1996. Plasma viscosity increase with progression of peripheral arterial atherosclerotic disease. *Angiology*, 47(3): 253-259.

93. Price, J.F., A.J. Lee, and F.G. Fowkes. 1996. Hyperinsulinaemia: A risk factor for peripheral arterial disease in the non-diabetic general population. *Journal of Cardiovascular Risk*, 3(6): 501-505.

94. Reaven, G.M. 1988. Banting lecture: Role of insulin resistance in human disease. *Diabetes*, 37: 1595-1607.

95. Regensteiner, J.G., J.F. Steiner, and W.R. Hiatt. 1996. Exercise training improves functional status in patients with peripheral arterial disease. *Journal of Vascular Surgery*, 23(1): 104-115.

96. Regensteiner, J.G., A. Gardner, and W.R. Hiatt. 1997. Exercise testing and exercise rehabilitation for patients with peripheral arterial disease: Status in 1997. *Vascular Medicine*, 2(2): 147-155.

97. Regensteiner, J.G., and W.R. Hiatt. 1995. Exercise rehabilitation for patients with peripheral arterial disease. *Exercise & Sport Sciences Reviews*, 23: 1-24.

98. Regensteiner, J.G., T.J. Meyer, W.C. Krupski, L.S. Cranford, and W.R. Hiatt. 1997. Hospital vs home-based exercise rehabilitation for patients with peripheral arterial occlusive disease. *Angiology*, 48(4): 291-300.

99. Ridker, P.M., M. Cushman, M.J. Stampfer, R.P. Tracy, and C.H. Hennekens. 1998. Plasma concentration of C-reactive protein and risk of developing peripheral vascular disease. *Circulation*, 97(5): 425-428.

100. Robeer, G.G., J.W. Brandsma, S.P. van den Heuvel, B. Smit, R.A. Oostendorp, and C.H. Wittens. 1998. Exercise therapy for intermittent claudication: A review of the quality of randomized clinical trials and evaluation of predictive factors. *European Journal of Vascular & Endovascular Surgery*, 15(1): 36-43.

101. Rosfors, S., S. Bygdeman, B.B. Arnetz, G. Lahnborg, L. Skoldo, P. Eneroth, and A. Kallner. 1989. Longterm neuroendocrine and metabolic effects of physical training in intermittent claudication. *Scandinavian Journal of Rehabilitation Medicine*, 21(1): 7-11.

102. Rosfors, S., B.B. Arnetz, S. Bygdeman, L. Skoldo, G. Lahnborg, and P. Eneroth. 1990. Important predictors of the outcome of physical training in patients with intermittent claudication. *Scandinavian Journal of Rehabilitation Medicine*, 21(3): 135-137.

103. Ruell, P.A., E.S. Imperial, F.J. Bonar, P.F. Thursby, and G.C. Gass. 1984. Intermittent claudication. The effect of physical training on walking tolerance and venous lactate concentration. *European Journal of Applied Physiology & Occupational Physiology*, 52(4): 420-425.

104. Rutherford, R.B., J.D. Baker, C. Ernst, K.W. Johnston, J.M. Porter, S. Ahn, and D.N. Jones. 1997. Recommended standards for reports dealing with lower extremity ischemia: Revised version. *Journal of Vascular Surgery*, 26(3): 517-538.

105. Sakurai, T., M. Masushita, N. Nishikimi, and Y. Nimura. 1997. Effect of walking distance on the change in anklebrachial pressure index in patients with intermittent claudication. *European Journal of Vascular & Endovascular Surgery*, 13(5): 486-490.

106. Salmasi, A.M., A. Nicolaides, A. Al-Katoubi, T.N. Sonecha, P.R. Taylor, S. Serenkuma, and H.H. Eastcott. 1991. Intermittent claudication as a manifestation of silent myocardial ischemia: A pilot study. *Journal of Vascular Surgery*, 14(1): 76-86.

107. Saltin, B. 1981. Physical training in patients with intermittent claudication. In *Physical conditioning and cardiovascular rehabilitation*, ed. L.S. Cohen, M.B. Mock, and I. Ringqvist, 181-196. New York: Wiley.

108. Sanne, H., and R. Sivertsson. 1968. The effect of exercise on the development of collateral circulation after experimental occlusion of the femoral artery in the cat. *Acta Physiologica Scandinavica*, 73(3): 257-263.

109. Schoop, W. 1973. Mechanism of beneficial action of daily walking training of patients with intermittent claudication. *Scandinavian Journal of Clinical Laboratory Investigation*, (Suppl. 128) 31: 197-199.

110. Skinner, J.S., and D.E. Strandness. 1967. Exercise and intermittent claudication II: Effect of physical training. *Circulation*, 36: 23-29.

111. Sorlie, D., and K. Myhre. 1978. Effects of physical training in intermittent claudication. *Scandinavian Journal of Clinical & Laboratory Investigation*, 38(3): 217-222.

112. Stewart, A.L., S. Greenfield, R.D. Hays, W.H. Rogers, S.D. Berry, E.A. McGlynn, and J.E. Ware, Jr. 1989. Functional status and well-being of patients with chronic conditions. Results from the Medical Outcomes Study. *Journal of American Medical Association*, 262(7): 907-913. [published erratum appears in *Journal of American Medical Association*, 262(18): 2542]

113. Tan, K.H., de Cossart, L., Edwards, P.R. 2000. Exercise training and peripheral vascular disease. *British Journal of Surgery*, 87(5): 553-562.

114. Tesfamariam, B. 1994. Free radicals in diabetic endothelial cell dysfunction. *Free Radical Biology & Medicine*, 16(3): 383-391.

115. Tsang, G.M., K. Sanghera, P. Gosling, F.C. Smith, I.S. Paterson, M.H. Simms, and C.P. Shearman. 1994. Pharmacological reduction of the systemically damaging effects of local ischaemia. *European Journal of Vascular Surgery*, 8(2): 205-208.

116. Turton, E.P., J.I. Spark, K.G. Mercer, D.C. Berridge, P.J. Kent, R.C. Kester, and D.J. Scott. 1998. Exercise-induced neutrophil activation in claudicants: A physiological or pathological response to exhaustive exercise. *European Journal of Vascular & Endovascular Surgery*, 16(3): 192-196.

117. Ubbink, D.T., G.H. Spincemaille, R.S. Reneman, and M.J. Jacobs. 1999. Prediction of imminent amputation in patients with non-reconstructible leg ischemia by means of microcirculatory investigations. *Journal of Vascular Surgery*, 30(1): 114-121.

118. Verhaeghe, R. 1998. Epidemiology and prognosis of peripheral obliterative arteriopathy. *Drugs*, 56(Suppl. 3): 1-10.

119. Whyman, M.R., F.G. Fowkes, E.M. Kerracher, I.N. Gillespie, A.J. Lee, E. Housley, and C.V. Ruckley. 1997. Is intermittent claudication improved by percutaneous transluminal angioplasty? A randomized controlled trial. *Journal of Vascular Surgery*, 26: 551-557.

120. Williams, L.R., M.A. Ekers, P.S. Collins, and J.F. Lee. 1991. Vascular rehabilitation: Benefits of a structured exercise/risk modification program. *Journal of Vascular Surgery*, 14(3): 320-326.

121. Womack, C.J., D.J. Sieminski, L.I. Katzel, A. Yataco, and A.W. Gardner. 1997. Improved walking economy in patients with peripheral arterial occlusive disease. *Medicine & Science in Sports & Exercise*, 29(10): 1286-1290.

122. Woodburn, K.R., A. Rumley, A. Murtagh, and G.D. Lowe. 1997. Acute exercise and markers of endothelial injury in peripheral arterial disease. *European Journal of Vascular & Endovascular Surgery*, 14(2): 140-142.

123. Wyatt, M.G., P.M. Scott, D.J. Scott, K. Poskitt, R.N. Baird, and M. Horrocks. 1991. Effect of weight on claudication distance. *British Journal of Surgery*, 78(11): 1386-1388.

124. Young, D.F., N.R. Cholvin, R.L. Kirkeeide, and A.C. Roth. 1977. Hemodynamics of arterial stenoses at elevated flow rates. *Circulation Research*, 41(1): 99-107.

125. Zetterquist, S. 1970. The effect of active training on the nutritive blood flow in exercising ischemic legs. *Scandinavian Journal of Clinical Laboratory Investigation*, 25: 101-111.

Kerry J. Stewart, EdD
Division of Cardiology
Johns Hopkins School of Medicine
Bayview Medical Center
Baltimore, MD

Pacemakers and Internal Cardiac Defibrillators

The heart of the average person beats about 100,000 times per day. Each contraction results from an electric impulse that is initiated in the sinoatrial (SA) node, passes through the atrioventricular (AV) node, and is spread through the ventricles. An artificial pacemaker maintains a normal heart rate when the intrinsic electrical circuitry of the heart fails. The most common indications for pacemaker implantation are a heart rate that is too slow (symptomatic bradycardia) or fails to increase appropriately with exercise (chronotropic incompetence) or an electrical pathway that is blocked (conduction system disease). Rhythm disorders that involve the SA node are classified under the broad term **sick sinus syndrome.** This includes the inability to generate a heartbeat or increase the heart rate in response to the body's changing circulation demands. Heart block may cause a loss of **AV synchrony,** a term that refers to the sequence and timing of the atria and ventricles. Normally, the ventricles contract a fraction of a second after they have been filled with blood following an atrial contraction. Asynchrony may not allow the ventricles to fill with enough blood before contracting. Depending on the patient's specific condition, the artificial pacemaker may replace SA node signals that are delayed or blocked along the pathway between the upper and lower heart, maintain a normal timing sequence between the upper and lower heart chambers, and ensure that the ventricles contract at a sufficient rate. The second revision of the American College of Cardiology/American Heart Association "Guidelines for Implantation of Cardiac Pacemakers and Anti-arrhythmia Devices" provides an extensive review of the scientific literature and recommendations for which bradyarrhythmias and tachyarrhythmias may optimally be treated with a pacemaker (10). Table 18.1 shows pacemaker types that are used as therapy for different medical conditions (17).

The need for pacing can occur at any age. Although some infants require a pacemaker from birth, about 85% of the population needing a pacemaker are over the age of 65 years, with an equal distribution among men and women. About 115,000 pacemakers are implanted in the United States each year. This number is likely to increase because of the growing number of elderly persons in the population. In the 1950s, external pacemakers were used to treat symptomatic bradycardia. The 1960s produced AV pacemakers that provided a more physiological method of pacing. Today, with miniaturized electronic circuitry, pacemakers improve the quality of life by optimizing the hemodynamic state at rest and can produce an appro-priate heart response to meet the physiological demands of exercise. Because of newer technology, many patients can maintain or even begin an exercise program after pacemaker implantation. Therefore, it is important for the clinical exercise physiologist to understand how pacemakers work to emulate normal cardiac rate, conduction, and rhythm in response to physiological and metabolic needs. It is also important to know about pacemaker programmed settings, how these settings may effect exercise capacity and the exercise prescription, and how to determine if the patient's response to exercise is appropriate. The clinical exercise physiologist should communicate observations of the patient's responses to exercise to the pacemaker physician, who can reprogram the pacemaker to optimal settings.

Pacing System

A pacing system consists of separate but closely integrated components that stimulate the heart to contract with precisely timed electrical impulses. The pacemaker (also known as the pulse generator) is a small metal case that contains the circuitry which controls the electrical impulses, along with a battery. Modern pacemaker batteries are the lithium-

---

TABLE 18.1

Pacemaker Therapies for Different Medical Conditions

Pacemaker varieties	Treatment conditions
Rate responsive for chronotropic incompetence	Patients who need to sustain a heart rate that matches their metabolic needs to their daily lifestyle or condition.
Mode switch for managing atrial arrhythmias in patients with bradyarrhythmia	Many patients with sinus node dysfunction and atrioventricular (AV) block will experience atrial fibrillation. Mode switch therapy reduces symptoms of atrial fibrillation during dual chamber pacing.
Rate drop response for neurocardiogenic syncope	Patients with carotid sinus syndrome and vasovagal syncope are being treated for their symptoms by preventing their heart rate from falling below a prescribed level.
Ablate and pace for atrial fibrillation	Patients with drug-refractory atrial fibrillation have been shown to benefit from ablation of the atrioventricular junction and implantation of a pacing system to maintain an appropriate heart rate.
Search AV for patients with intermittent or intact AV conduction	Intrinsic AV activation is generally preferred to a ventricular-paced contraction because it provides improved hemodynamics and extended pacemaker longevity.

Adapted from Medtronics Web site, http://www.medtronic.com/brady/clinician/therapies/clinther.html (17).

type and can last for many years, depending on the extent to which the pacemaker is actually used. In some cases, the patient is dependent on the pacemaker to always provide the cardiac rhythm and conduction. In other cases, the pacemaker acts as a backup and will fire intermittently when the sinus node fails to produce an appropriate rate or when the conduction system fails to transmit the impulses. The pacing leads are insulated wires connected to a pacemaker. The leads carry the electrical impulse to the heart and the heart's own responses back to the pacemaker. These leads are very flexible to accommodate both the moving heart and the body. Depending on the type of pacemaker that is implanted, there may be one or two leads. Each lead has at least one electrode that can deliver energy from the pacemaker to the heart and sense information about the heart's electrical activity.

Pacemaker Implantation

This surgery, performed by surgeons or cardiologists, typically takes an hour or less and is often done as an outpatient procedure. Most patients receive a local anesthetic and remain awake during the surgery. In some cases, general anesthesia and a brief hospital stay are required. The pulse generator is usually implanted below the collarbone just beneath the skin. The leads are threaded into the heart through a vein located near the collarbone. The tip of each lead is then positioned inside the heart. In some cases, the pulse generator is positioned in the abdomen and the pacemaker leads are attached to the outside of the heart. After implantation, the pacemaker can be adjusted with an external programming device. The device works by sending radio frequency signals to the pacemaker via a transmitter that is placed on the chest.

Temporary External Pacemakers

An external pacemaker pulse generator is a temporary device that is commonly used in emergency and critical care settings, after open heart surgery, or until a permanent pacemaker can be implanted. This device is used outside the body as a temporary substitute for the heart's intrinsic pacing. Many of these devices have adjustments for impulse strength, duration, R-wave sensitivity, and other pacing variables.

Permanent Pacemaker Types

There are two basic types of permanent pacemakers—single-chamber and dual-chamber. Both

monitor the heart and send out pacing signals as needed to meet physiological demands.

1. A **single-chamber pacemaker** usually has one lead to carry signals to and from either the right atrium or, more commonly, the right ventricle. This type of pacemaker can be used for a patient whose SA node sends out signals too slowly but whose conduction pathway to the lower heart is intact. A single-chamber pacemaker is also used if there is a slow ventricular rate in the setting of permanent atrial fibrillation. In this case, the tip of the lead is usually placed in the right ventricle.

2. A **dual-chamber pacemaker** has two leads, with the tip of one lead positioned in the right atrium and the tip of the other lead located in the right ventricle. This type of pacemaker can monitor and deliver impulses to either or both of these heart chambers. A dual-chamber pacemaker is used when the SA node is unreliable and the conduction pathway to the lower chamber is partly or completely blocked. When pacing does occur, the contraction of the atria is followed closely by a contraction in the ventricles, resulting in timing that mimics the heart's natural way of working.

Pacemakers are categorized by a standardized coding system developed by The North American Society of Pacing and Electrophysiology and the British Pacing and Electrophysiology Group. Figure 18.1 shows the coding system for pacemaker functions. The letters refer to the chamber paced, the chamber sensed, what the pacemaker does when it senses an event, and other programmable features. This example is a ventricular demand pacemaker that is also rate responsive. The first V indicates that the ventricle is paced. The second V indicates that the pacemaker is programmed to sense for an impulse in the ventricle. The I indicates that when the pacemaker senses the patient's own ventricular impulse, the pacemaker is inhibited. The R indicates that the pacemaker is "rate responsive" or "rate adaptive." A sensor in the pacemaker senses physical activity and adjusts the patient's pacing rate according to the level of activity.

Figure 18.2 shows the code for a dual-chamber pacemaker that is also rate responsive. The D stands for dual. Because of two leads, a dual-chamber pacemaker can pace both the atria and the ventricles. Likewise, the pacemaker can sense in the atria and ventricles. The third D indicates that the pacemaker can either be inhibited or triggered. The pacemaker will watch for atrial activity, and if it detects none, it will pace the atrium. After an appropriate AV time

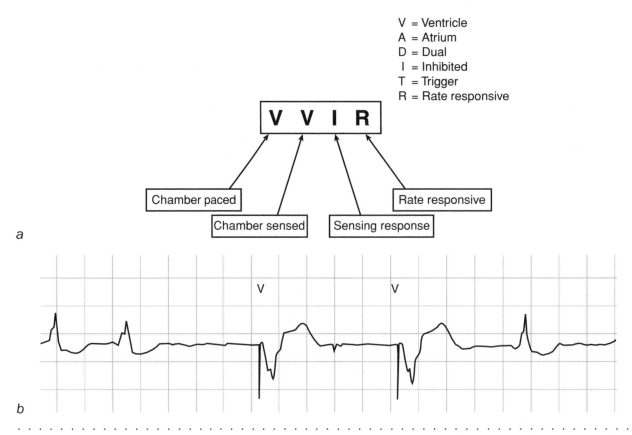

FIGURE 18.1   *(a)* Coding for a ventricular demand pacemaker that is also rate responsive. The first V indicates that the ventricle is paced. The second V indicates that the pacemaker senses in the ventricle. The I indicates that the pacemaker will be inhibited. The R indicates that the pacemaker is rate responsive. *(b)* VVI operation during atrial fibrillation. Ventricular pacing (V) occurs at the programmed lower rate limit of 60 beats · min^{-1} when the intrinsic ventricular activity falls below that level. Intrinsic ventricular activity at a faster rate inhibits ventricular pacing.

interval, the pacemaker will watch for a ventricular depolarization. If this is sensed, the pacemaker will be inhibited. If there is no ventricular activity, the pacemaker will pace the ventricle.

Exercise Physiology

The physiology of exercise for patients with pacemakers is the same as for other patients. The difference is how their physiology interacts with the device. For patients who cannot provide an appropriate cardiac output response to exercise, modern pacemakers attempt to increase the cardiac output to meet changing physiological demands.

The increase in oxygen uptake from rest to maximal exercise follows the formula: oxygen uptake is equal to cardiac output multiplied by arteriovenous oxygen difference. From rest to maximal exercise, oxygen uptake can increase 700% to 1200%, arteriovenous oxygen difference by 200% to 400%, and cardiac output by 200% to 400%. Cardiac output is

equal to heart rate multiplied by stroke volume. With exercise, stroke volume can increase by 15% to 20%, whereas heart rate can increase by 200% to 300%. Thus, heart rate is the most important component for increasing cardiac output and is most closely related to metabolic demands. Although AV synchrony contributes to cardiac output, this factor is more important at rest and less important with exercise.

Physiological Pacing

The term physiological pacing refers to the maintenance of the normal sequence and timing of the contractions of the upper and lower chambers of the heart. AV synchrony provides a higher cardiac output without increasing myocardial oxygen uptake. Dual-chamber pacemakers attempt to provide this physiological beneficial function. The pacemaker senses the patient's sinus node and, in complete heart block, sends an impulse to the ventricle following an appropriate AV timing interval. Although

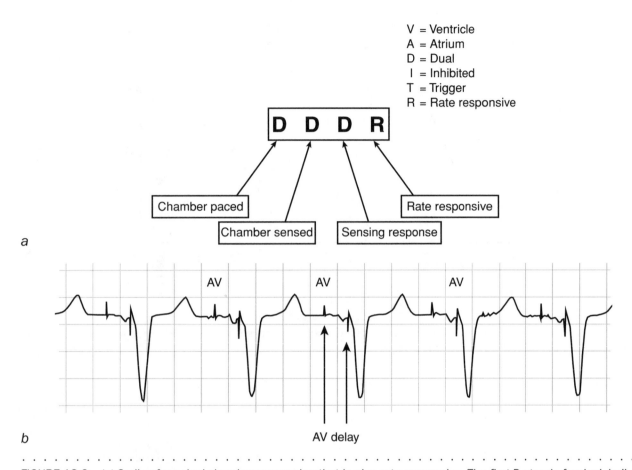

a

b

FIGURE 18.2   *(a)* Coding for a dual chamber pacemaker that is also rate responsive. The first D stands for dual, indicating pacing in both the atria and the ventricles. The second D indicates sensing capability in both the atria and ventricles. The third D indicates inhibited or triggered, and the R stands for rate responsive. *(b)* DDD operation. Atrial pacing (A) occurs at the programmed lower rate limit of 75 beats · min^{-1}. Because of complete heart block, the pacemaker tracks the atrial rate to pace the ventricle (V) at the same rate after a programmed AV delay.

the specific change in cardiac output depends on many factors, the optimal AV delay in normal persons to produce the maximum cardiac output is about 150 ms from the beginning of atrial depolarization. The efficiency of cardiac work decreases with a shorter or longer AV interval. In normal subjects, the AV interval shortens with increased heart rate. The pacemaker can also set the AV interval based on heart rate. A dual-chamber pacer can also initiate an atrial impulse in sick sinus syndrome. AV synchrony augments ventricular filling and cardiac output, improves venous return, and assists in valve closure. The loss of atrial function increases atrial pressure and pulmonary congestion. The benefit of AV synchrony is independent of any measure of left ventricular function. The maintenance of normal AV synchrony allows for improved hemodynamic responses with a more normal increase in cardiac output (13). Because of a higher cardiac output at any given level of work with synchronous pacing, the

arteriovenous oxygen difference is narrower and the serum lactate is lower. Thus, synchronous pacing results (figure 18.3) in less anaerobic metabolism at the same level of work (15). AV synchrony and stroke volume provide their most important contributions to cardiac output at rest, whereas an increase in heart rate is the predominant factor contributing to cardiac output during exercise (figure 18.4).

Mode Switching in Dual-Chamber Pacemakers

Many patients with sinus node dysfunction and AV block will develop atrial arrhythmias, with the most common arrhythmia being atrial fibrillation. To prevent the dual-chamber pacemaker from tracking or matching every atrial impulse with a ventricular pacing pulse, **mode switching** controls the ventricular rate. The highest rate at which the pacemaker will respond in terms of matching the sinus rate with

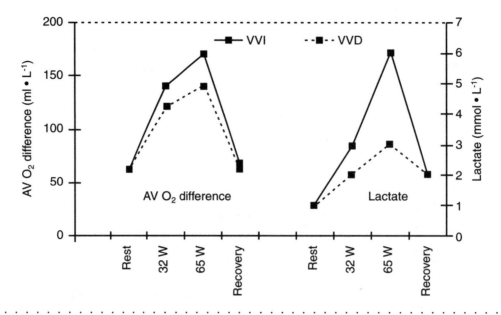

FIGURE 18.3   Synchronous pacing (VDD) results in less anaerobic metabolism at the same level of work compared with nonsynchronous pacing (VVI). AV, atrioventricular.

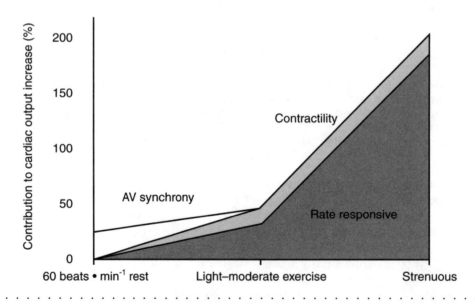

FIGURE 18.4   The relative contributions of atrioventricular (AV) synchrony, stroke volume (contractility), and heart rate to cardiac output at rest and exercise. Heart rate is the most important factor contributing to cardiac output during exercise.

a ventricular response is known as the **maximal tracking rate.** Mode switching will temporarily revert to a nontracking mode so that irregular or excessive atrial activity does not drive the ventricles to a very high rate. The mode-switching feature is programmable and, depending on the specific pacemaker model, can be adjusted for optimal performance in any given patient.

Rate Responsive Pacing

The development of **rate responsive pacing** (also called rate adaptive rate or rate modulated) has dramatically changed the application of pacing with regard to physical activity. The rate responsive function is used when the native sinus node cannot increase heart rate to meet metabolic demands. Increasing heart rate in response to exercise is probably the single most important factor for increasing cardiac output and oxygen uptake. A sudden increase in exercise requires the heart rate to quickly adjust to the workload. Rate responsive pacemakers can sense the body's physical need for increased cardiac output and produce an appropriate cardiac rate in patients with **chronotropic incompetence.** The highest rate at which the pacemaker will pace the ventricle in response to a sensor driven rate is known as the maximal sensor rate. When rate-modulated pacing is compared with non-rate-modulated pacing (figure 18.5), we see that exercise capacity is very limited without an appropriate increase in heart rate (6).

Several physiological and metabolic changes occur during exercise as the demand for energy increases:

— Movement that produces vibration and acceleration

— Respiration

— Heat that raises body temperature

— Electricity activity that produces electrocardiographic and electromyographic changes

— Carbon dioxide

— Lactic acid, which reduces blood pH

— Changes in intracardiac pressure

Various sensors have been developed to detect these changes and, based on computer algorithms, generate the electrical impulses that are used to pace the heart. The development of optimal sensors and algorithms for rate-modulated pacing systems must meet several requirements:

— The sensor should rapidly detect acceleration and deceleration of physiological changes.

— The response should be proportionate to the exercise workload and metabolic demands.

— The response should be sensitive to both exercise and nonexercise requirements such as posture, anxiety and stress, vagal maneuvers, circadian variations, and fever.

— The response should be specific and not be falsely triggered.

The most common rate responsive pacemakers detect motion in response to physical activity. Vibration sensors use a piezoelectric crystal located in the pulse generator to detect forces generated during movement. These forces are transmitted to the sensor through connective tissue, fat, and muscle. Acceleration sensors detect body movement in anterior and posterior directions (8). The circuitry is also located in the pulse generator. Because the sensor is not in direct contact with the pacemaker case, there is no reaction to vibration or pressure.

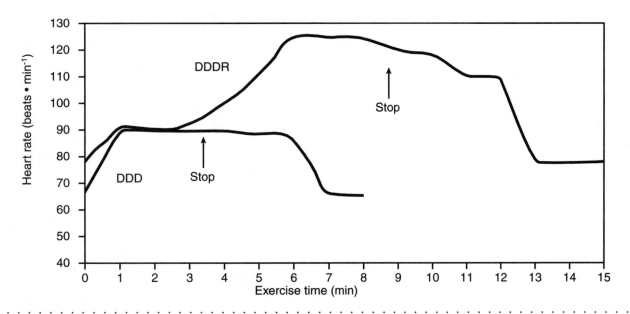

FIGURE 18.5   A comparison of rate-modulated (DDDR) versus non-rate-modulated pacing (DDD) during treadmill exercise. Without an appropriate increase in heart rate, exercise capacity is very limited in a patient with chronotropic incompetence.

Adapted from Bodenhamer and Grantham (6).

Producing an appropriate heart rate in response to certain work tasks poses a technological challenge. A simple task of walking up and down stairs can produce different heart rate responses, based on the type of sensor used to sense motion. Compared with a normal heart rate response, an accelerometer produces similar results going up stairs but overestimates the metabolic demand going down stairs. The vibration sensors produce a heart rate that is too low for stepping up and a rate that is too high for stepping down. This occurs because the vibration of walking down stairs is greater than walking up stairs, even though the metabolic demand is greater when walking up stairs (4). In contrast to single sensors, dual-sensor rate response provided by activity and minute ventilation may help to overcome these types of problems, as observed in appropriate heart rate responses while patients are ascending and descending stairs (2). Recent advances in pacemaker technology may allow the use of combined sensors and advance algorithms to improve rate performance over a single sensor system (16).

Graded Exercise Testing

It is desirable that patients with pacemakers capable of rate modulation undergo exercise testing to ensure appropriate rate responses (12). Exercise capacity and quality of life are improved by appropriately programmed rate-responsive pacemakers compared with fixed rate units (19). These devices can be programmed to more closely match the needs of the patient. The primary pacemaker settings can be adjusted to optimize responses to physical activity:

— Sensitivity of the sensor
— Responsiveness to a physiological change
— Rate at which the cardiac rate changes
— Minimum rate at rest and maximal rate at peak activity

Exercise testing is used to guide the adjustment of these settings to improve exercise capacity and reduce symptoms. Exercise testing helps to establish **upper rate limits** and adjust the sensitivity and responsiveness of the sensor. Exercise testing is also used to determine the **anginal threshold,** if any. It would not be prudent to pace the heart rate beyond the point at which ischemia would occur.

Several different approaches to exercise testing can be used. These include using informal or formal protocols with or without real-time electrocardiogram (ECG) monitoring and with or without determination of optimal rate responsive parameters. The patient's health status and lifestyle, the type of pacemaker, and the facilities and experience of personnel will also determine the specific approach to exercise testing. For many patients, informal exercise testing is a reasonable and less expensive alternative to formal treadmill testing (12).

Empiric adjustment of the rate response parameters is common. With informal testing, the patient walks at a self-determined casual and brisk pace, usually for about 3 min each. The sensor-driven cardiac rate can be determined by examining the ECG. Because pacemakers are capable of storing an electronic record of pacemaker activity, the physician, using a special computer, can also interrogate the pacemaker to examine a histogram display of the heart rate response during the walk. The optimal pacemaker rate is determined empirically. For casual exercise, the target is often 10 to 20 beats $\cdot$ min^{-1} above the **lower rate limit.** For brisk exercise, the target can be 20 to 50 beats $\cdot$ min^{-1} above the lower rate limit. This approach to exercise testing is best suited for less active patients who are unlikely to reach their upper rate limit. By examining a display of the sensed atrial rate as measured by an event counter in the pacemaker, or by measuring the heart rate by ECG and asking the patient about symptoms, the physician makes a clinical judgment whether the patient is chronotropically competent. If not, the pacemaker will need to be programmed to elicit an appropriate response.

Formal exercise testing allows for a chronotropic evaluation that seeks to optimally match the pacemaker-augmented response of the chronotropically incompetent patient to the metabolic requirements of the body (21). Formal exercise testing is typically best for active patients likely to reach the programmed **maximum sensor rate.** Programming the upper rate of rate-adaptive pacing improves exercise performance and exertional symptoms during both low and high exercise workloads compared with a standard nominal value of 120 beats $\cdot$ min^{-1} (7).

With formal exercise testing, the protocol selected requires careful consideration. Many different protocols for exercise testing exist. Nevertheless, many of the traditional protocols such as the Bruce and Naughton protocols are designed to test for coronary artery disease. Their usefulness in defining optimal programming for rate-responsive pacemakers may be limited. A widely used protocol for assessing patients with pacemakers is the **chronotropic assessment exercise protocol** (20, 21) shown in table 18.2.

The advantage of this protocol is that the workload gradually increases to mimic the range of ac-

TABLE 18.2
Chronotropic Assessment Exercise Protocol

Stage	Speed	Grade	Cumulative time	Metabolic equivalents
1	1.4	2	2	2.0
2	1.5	3	4	2.8
3	2.0	4	6	3.6
4	2.5	5	8	4.6
5	3.0	6	10	5.8
6	3.5	8	12	7.5
7	4.0	10	14	9.6
8	5.0	10	16	12.1
9	6.0	10	18	14.3
10	7.0	10	20	16.5
11	7.0	15	22	19.0

tivities of daily living. It allows for a more complete assessment of how the pacemaker responds at the lower metabolic equivalent (MET) ranges where patients typically spend most of their time. The chronotropic assessment exercise protocol has five stages at a lesser exercise intensity than the Bruce produces in the second stage (22). Because the Bruce protocol increases by 2 to 3 METs during each 3-min stage, it would be difficult to assess the patient's work capacity and to assess the ability of the **pacemaker sensor** to provide an adequate hemodynamic response.

Exercise Prescription

Dual-chamber pacemakers are in greater use today. Clinical exercise physiologists need to be familiar with the normal behavior of these devices during exercise. Figure 18.6 shows DDDR operation. The rate at which the sensor-driven heart rate increases follows algorithms that are programmed into the pacemaker. Among the key parameters are the slope of the heart rate increase and decline and the sensitivity of the sensor. With increased physical activity, the pacemaker will follow the sinus rate up to a maximal tracking rate. The activity sensor can be programmed to allow a further increase in the paced rate to the maximal sensor rate in response to physical activity. If the patient continues to exercise, the pacemaker may reach its maximal tracking rate or maximum sensor rate. When this occurs, the pacemaker will not further increase the heart rate. If the patient's native sinus rate continues to increase beyond this point, the pacemaker will switch to an AV block mode because the sinus rate now exceeds the rate at which the pacemaker will permit tracked ventricular pacing. The pacemaker will first switch to a Wenckebach-type block to gradually slow the ventricles with 2:1 AV block ensuing if the sinus rate continues to rise. This feature protects against nonexercise sinus tachycardia that might otherwise force the pacemaker to produce ventricular tachycardia. At higher levels of exercise, the metabolic demands will be high, but 2:1 block may occur and slow the ventricular rate. In this situation, the development of 2:1 block is a normal feature of the pacemaker but can cause symptoms because of the sudden drop in heart rate. If this occurs, the patient may be exercising too hard or the maximal rate is set too low. If the pacemaker is also rate adaptive, the sensor setting may be too low. The exercise physiologist should record and communicate episodes of abrupt decreases in heart rate to the patient's

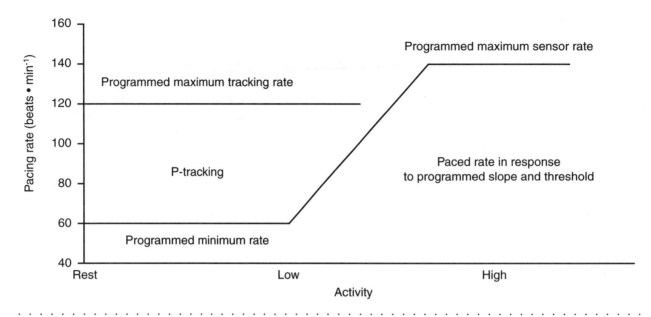

FIGURE 18.6   DDDR pacing and physical activity. The pacemaker follows the sinus rate to a maximal-tracking rate. In response to physical activity, the sensor-driven response can drive the rate to the maximal sensor rate. The paced rate increases in response to programmed slope and threshold settings.

pacemaker physician so that programmed settings can be evaluated for possible change.

Several of the exercise modalities that are commonly prescribed in cardiac rehabilitation and adult fitness programs may pose a particular challenge in some patients with activity sensors (3). In the example shown in figure 18.7, because most of the increase in work is accounted for by raising the slope rather than speed on a treadmill, there is little change in the generated forces that would be detected by the vibration sensor. Thus, the heart rate determined by the vibration sensor is too slow for the metabolic demand of the increased work. In this case, the accelerometer sensor is better able to provide a heart rate that more closely matches an appropriate rate response.

The clinical exercise physiologist should also be aware of how a vibration sensor responds to outdoor and stationary cycling (figure 18.8). The response of this type of sensor is particularly relevant to cardiac rehabilitation because stationary cycling is the dominant form of exercise in many programs. This may explain why some patients complain of unusual fatigue and shortness of breath during stationary cycling but not other types of exercise such as treadmill walking.

Patients with an artificial pacemaker will require long-term surveillance by their physicians to ensure optimal adjustment of the programming for their individual needs, to maximize the life expectancy of the pacemaker through adjustment of pacemaker

output settings, and to identify and treat complications. Pacemaker follow-up relies on clinical, electrocardiographic, and device assessment. Other tests may include exercise testing, Holter monitoring, and echocardiography. The device assessment requires a specialized programmer to verify pacemaker functions. In many cases, interrogation of the pacemaker over the telephone is done periodically to provide useful information about selected functions of the pacemaker when a more complete test is not deemed necessary. The clinical exercise physiologist can play an important role in the overall evaluation of the patient by providing feedback to the physician about heart rate, blood pressure, and symptomatic responses to exercise. The case study at the end of the chapter illustrates the role of the clinical exercise physiologist in the management of the patient with a pacemaker.

Special Considerations

Exercise prescription requires special attention to the type of pacemaker that is implanted. With fixed-rate pacemakers, the cardiac output and arterial pressure are increased by stroke volume (1). Target heart rate cannot be used to guide exercise intensity. Instead the patient follows ratings of perceived exertion (RPE). It is also important to monitor the blood pressure to ensure that an appropriate tensity. When the sinus node is normal, it is desirable to have the pacemaker "track" native sinus activity by pacing

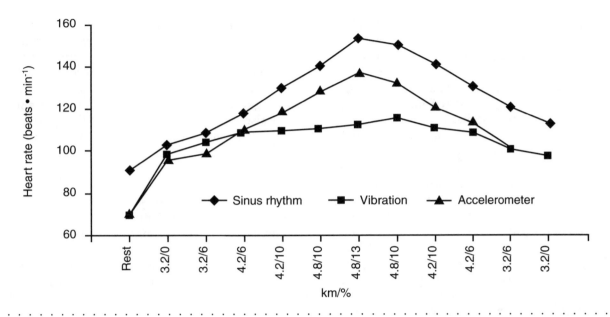

FIGURE 18.7 Generated forces. The diamonds show a normal sinus response at rest and with increased treadmill work. Work is increased primarily by raising the slope rather than speed. At rest, sinus rate is about 20 beats above pacemaker rate. This difference is maintained throughout the test. The accelerometer sensor (triangles) is able to produce a heart rate that better matches with the workload compared with the vibration sensor (squares).

Adapted from Alt and Matula (3).

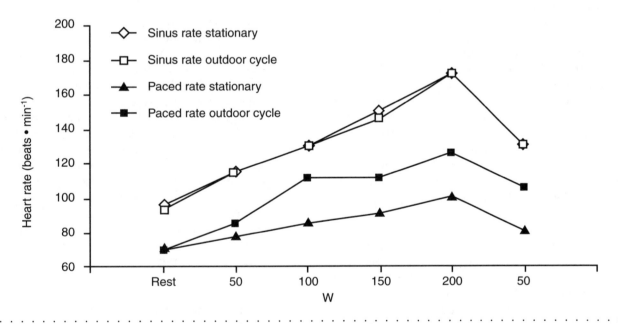

FIGURE 18.8 Vibration sensor. The upper lines represent a normal sinus rate with outdoor cycling (open squares) and stationary cycling (open diamonds) at increasing workloads and recovery. The lower lines represent the sensor. The difference of 20 beats at rest is maintained throughout the test during outdoor cycling (filled squares). During stationary cycling (filled triangles), the paced cardiac rate is considerably slower than both the sinus rate and outdoor cycling. This is because there is less body motion and vibration with stationary cycling.

Adapted from Alt and Matula (3).

the ventricle after an appropriate AV delay (11). In all cases, the target heart rate must be lower than the anginal threshold in a patient with ischemia (5). Tailoring the exercise prescription and modifying the pacemaker's response rate based on cardiopulmonary stress testing that determines the anaerobic threshold have been shown to provide functional advantages for patients in cardiac rehabilitation (9).

Because rate responsive pacemakers will mediate the heart rate response to exercise, the type of sensor must be taken into account when exercise is prescribed (18). Sensors that detect movement may respond slowly to stationary cycling and increased treadmill slope. Again, the RPE and MET equivalents are extremely useful in establishing the exercise prescription. With modern pacemakers, pacing occurs only when needed. In many patients, such as those with normal sinus function with intermittent heart block, the exercise prescription can be written the same as for most other patients. Regarding heart rate monitoring, one study found that the use of dry-electrode heart rate monitors that transmit a signal to a monitor such as those worn on the wrist had no adverse effect on pacemaker function (14).

Exercise Recommendations

Patients with pacemakers can derive benefits from an exercise program similar to those gained by other individuals. The area of greatest concern when prescribing exercise is the issue of exercise intensity. Because of the variety of pacemakers, types of sensors that are used to detect an increase in activity, and mode of exercise prescribed, there can be a great deal of variability in an appropriate heart rate response. Therefore, when patients with pacemakers start exercising, they should be monitored to make sure that the pacemaker is responding appropriately.

Mode

Generally all forms of exercise are acceptable in patients with pacemakers, except activities that can cause direct contact with the pacemaker. Therefore, contact sports such as football, soccer, and hockey are generally not recommended. All forms of aerobic exercise are generally acceptable and most likely carry the greatest benefit in regard to improving overall health and decreasing risk factors for cardiovascular disease. When a patient with a pacemaker performs any form of aerobic exercise, rate responsive pacemakers should be evaluated to see that they are increasing the heart rate appropriately relative to the intensity of exercise. Pacemakers that rely on vibration or accelerometer sensors to detect body motion during exercise may not produce an adequate response for activities such as stationary cycling and increased treadmill slope. Unusual shortness of breath and fatigue may indicate a lack of rate responsive pacing and need to be monitored during different forms of activity. Weight training may also be acceptable, although it is important to make sure that weights or bars do not come in contact with the pacemaker.

Frequency

The frequency of activity is based on the goals of the program. If someone is interested in improving his or her health and exercising at an intensity less than 60% of maximal aerobic capacity, daily activity is recommended. If the individual is able and willing to exercise at a higher intensity (60-85% of maximal aerobic capacity), 3 to 5 days a week are recommended.

Intensity

The intensity should be in the recommended range of 40% to 85% of maximal aerobic capacity but is primarily dependent on comorbid conditions (e.g., angina, chronic heart failure). Also, upper limits of the pacemaker (tracking and sensing) can influence upper exercise intensity levels. Because heart rate will not increase in a patient with a fixed-rate pacemaker, RPE and MET equivalents need to be used to evaluate exercise intensity. Additionally, blood pressure should be monitored to show appropriate increases with increasing workload. A patient with a fixed-rate pacemaker should not exceed an exercise intensity above the point that blood pressure begins to plateau with increasing workload. With dual-chamber and rate responsive pacemakers, heart rate can be used to determine exercise intensity and should be used along with RPE and METs. Knowledge of maximal tracking or sensing rates will determine upper intensity level. Patients should be monitored closely and activities should be chosen based on the ability of the pacemaker to adjust heart rate with increasing metabolic demands.

Duration

The duration of activity is similar to the general recommended guidelines to promote health and fitness (20-60 min) and is dependent on the goals. Ideally, the duration should be adjusted so that the individual achieves at least 1000 kcal energy expenditure per week.

Special Considerations

If a patient goes into second-degree type 1 block (Wenckebach) while exercising, it is most likely caused by the patient's native sinus rate exceeding the pacemaker's maximal tracking or sensor rate, and the intensity of exercise should be reduced. If the exercise professional notices a decrease in heart rate well below the patient's tolerable limits, this information should be forwarded to the pacemaker physician so that programmed settings can be evaluated. Practical application 18.1 provides more information about living with a pacemaker.

Automatic Internal Cardioverter Defibrillators

Individuals who survive ventricular fibrillation or other life-threatening ventricular tachycardias are at high risk for recurrence. Although medications that stabilize heart rhythm are available, medications are not entirely effective and often produce serious side

---

Patient Education About Living With a Pacemaker

The clinical exercise physiologist is often a primary source of patient education and will be asked about precautions that pacemaker patients should be aware of in their day-to-day lives. This section addresses some common questions and issues.

Sports and Recreational Activities

Many active patients, after appropriate medical clearance, can travel, drive, bathe, shower, swim, resume sexual activities, return to work, walk, hike, garden, golf, fish, or participate in other similar activities. However, contact sports that includes jarring, banging, or falling such as playing football, baseball, and soccer should be avoided. Also, patients should avoid hunting if a rifle butt is braced against the implant site.

Work Activities

Most office equipment is unlikely to generate the type of electromagnet interference that can affect a pacemaker. However, equipment with large magnets should not be carried if they are held near the pacemaker. Those patients working with heavy industrial or electrical equipment need to consult with their physician about resuming work because this equipment often produces high levels of electromagnetic interference that can affect pacemaker function.

Home Activities

People with pacemakers can participate in most activities of daily living and can be reassured that most home electrical devices will not interfere with pacemaker operation. However, some precautions are recommended. Cellular phones should be kept at distance of at least 6 in. between the phone and the pacemaker site, 12 in. for phones transmitting above 3 W. Also, the patient should hold the cell phone to the ear opposite the pacemaker site and should not carry a phone in a pocket or on a belt within 6 in. of the pacemaker. Ordinary cordless, standard desk, and wall telephones are considered safe. The patient should not lift or move large speakers because their large magnets may interfere with the pacemaker. Most general household electrical appliances like televisions and blenders and outdoor tools such as electric hedge clippers, leaf blowers, and lawn mowers do not usually interfere with pacemakers. However, the patient should avoid tools where the body comes into close contact with electric spark-generating components such as a chain saw. Also, caution is advised when working near the coil, distributor, or spark plug cables of a running engine. The safe approach is to turn off the engine before making any adjustments.

Travel

Most people with pacemakers can travel but should tell airport security personnel that they have a pacemaker or other implanted medical devices before going through security systems. Although airport security systems will not affect the pacemaker, the pacemaker's metal case could trigger the metal detection alarm. It is unlikely that the pacemaker will set off or be affected by home, retail, or library security systems.

effects. Increasingly, automatic internal cardioverter defibrillators (AICDs) are being used to control life-threatening ventricular arrhythmias. An AICD is a battery-driven implanted device, similar to a pacemaker, that is programmed to detect and then stop a life-threatening ventricular arrhythmia by delivering an electrical shock directly to the heart. Some models provide tiered therapy in that they have the capability of providing antitachycardia pacing, cardioversion, and defibrillation, as needed.

Most modern AICDs are implanted beneath the skin and muscle of the chest or abdomen, and electrodes that sense the heart rhythm and deliver the shock are inserted into the heart through veins. Nevertheless, the site and placement of the electrode wires will vary, depending on the patient and model of AICD used. In some cases, electrode patches are sewn to the surface of the heart, whereas some patients may receive electrodes that are placed under the skin of the chest near the heart. Rapid technological advances have produced devices that serve as both a pacemaker and an AICD. Microchips inside the device record rhythms and shocks to be used to determine optimal therapy.

Special Considerations

In many cases, the failure of the heart's intrinsic pacing and conduction system is associated with comorbid conditions such as myocardial infarction and chronic heart failure. Many patients with artificial pacemakers or AICDs are elderly and often have limited exercise capacity. The exercise prescription must consider not only the indications for the pacemaker and the type implanted but also the limitations to exercise associated with comorbidities. Besides monitoring the patient for appropriate heart rate responses, the exercise physiologist must pay close attention to signs and symptoms that might occur with increased heart rate such as exercise-induced angina, failure of blood pressure to increase or decrease, and marked shortness of breath.

Exercise Recommendations

The American College of Cardiology/American Heart Association "Guidelines for Implantation of Cardiac Pacemakers and Antiarrhythmia Devices" (10) provide recommendations for AICD therapy. These guidelines emphasize the need for the physician to establish limitations on the patient's specific physical activities. The guidelines also refer to policies on driving that advise the patient with an AICD to avoid operating a motor vehicle for a minimum

of 3 months and preferably 6 months after the last symptomatic arrhythmic event to determine the pattern of recurrent ventricular fibrillation/tachycardia. After appropriate evaluation and observation, many patients with AICD can participate in exercise programs. In most cases, the guidelines for exercise prescription are similar to those for any other patient with cardiovascular disease and should consider the patient's underlying diagnoses, medications, and symptoms. An exercise stress test is essential for establishing an appropriate exercise prescription. The prescribed target heart rate should be at least 20 beats below the heart rate cutoff point at which the device will shock. The exercise prescription must also take into account the existence of an **angina threshold** or exercise-induced hypotension, because many of these patients have severe coronary artery disease and poor left ventricular function. Furthermore, many patients with AICDs take β-blockers to limit heart rate to control symptoms and to prevent firing of the device. The benefits of pacing are

— alleviation or prevention of symptoms,

— restoration or preservation of cardiovascular function,

— restoration of function capacity,

— improved quality of life,

— enhanced survival, and

— participation in exercise training programs with many forms of physical activity.

Conclusion

Because of the increased prevalence of pacemakers and AICDs, clinical exercise physiologists and cardiac rehabilitation professionals need to know how pacemakers function and their limitations. There are basically two types of pacemakers: single- and dual-chamber units. Knowledge of the universal coding system is required to understand appropriate pacemaker function (figures 18.1 and 18.2). Initial pacemakers operated at fixed rates and were primarily used for patients who were symptomatic because of bradycardia or high-degree AV blocks. Because of the inability to increase heart rate with exertion in fixed rate pacemakers, rate responsive pacemakers have been developed, so that cardiac output can appropriately increase under physical activity. Motion sensors (vibration and accelerometers) are used in rate responsive pacemakers and each one has advantages and disadvantages. In addition, dual-sensor (activity and ventilation) pacemakers have been developed to further enhance a normal heart rate response with exertion. In patients with rate responsive pacemakers, graded exercise testing should be used to ensure an appropriate increase in heart rate and to allow for adjustment if the unit is not properly functioning. Graded exercise testing allows for optimal programming of the pacemaker to provide maximal hemodynamic benefit and quality of life.

When prescribing exercise training with rate responsive pacemakers, the clinical exercise physiologist must make sure that heart rate increases appropriately with exertion. In some activities where there is not a great deal of change in body movement (cycling, uphill walking), vibration sensors or accelerometers are not able to detect a difference. Therefore, it is important to determine the type of activity an individual is planning on doing before implanting a pacemaker, if possible. In addition, when prescribing different modes of activity, the clinical exercise physiologist must consider the type of pacemaker. Ideally, patients with rate responsive pacemakers should be monitored to determine if the physiological response is acceptable with exertion (i.e., heart rate, ECG, blood pressure, RPE, and METs). In addition, patients with pacemakers should avoid contact sports that carry a risk of direct contact with the pacemaker. Overall, there is not a great deal of limitation in prescribing exercise in pacemaker patients, other than making sure that there is an appropriate physiological response (increase in heart rate and/or blood pressure) with increasing levels of physical exertion.

The use of AICDs is increasing to control life-threatening ventricular arrhythmias. Before patients with AICDs start an exercise program, an exercise test is recommended to determine the safety of exercise and rule out any other underlying diagnoses. When one is prescribing exercise, the major concern with patients with ACIDs is to avoid reaching the critical heart rate that will cause the device to shock. It is recommended that training heart rate stay 20 beats $\cdot$ min^{-1} below the preset heart rate that produces a shock. Otherwise, there are no other specific limitations in prescribing exercise in this select population.

Case Study 18

Medical History

Mrs. JD is a 64-year-old female African American referred to cardiac rehabilitation 6 weeks following implantation of a DDDR pacemaker with a vibration sensor. The indication for the pacemaker was marked sinus bradycardia and chronotropic incompetence. Her primary complaint was episodes of shortness of breath, undue fatigue, lightheadedness, and weakness at rest and during exertion. She had no other significant cardiac history except for mild hypertension. Mrs. JD also has mild arthritis in her hands and knees, and her body mass index is 29. Presently, her symptoms at rest are resolved but she complains of early exertional fatigue and shortness of breath while doing housework and taking short walks in her neighborhood. Her resting heart rate is paced at 60 beats $\cdot$ min^{-1}.

Exercise Prescription

The pacemaker settings relevant to her exercise prescription were a lower rate limit of 60 beats $\cdot$ min^{-1}, programmed maximum tracking rate to 110 beats $\cdot$ min^{-1}, and a programmed sensor rate to 110 beats $\cdot$ min^{-1}. Mrs. JD was given a standard exercise prescription consisting primarily of stationary cycling for 15 min and walking on a treadmill for 15 min. Exercises with hand-held weights and flexibility exercises were prescribed for warm-up. Because she did not have an exercise stress test before starting the cardiac exercise program, her exercise intensity for aerobic exercise was set at 12 to 14 on the Borg RPE scale.

continued

Case Study 18 *(continued)*

Response to Initial Exercise Training Session

On the stationary cycle, Mrs. JD complained of shortness of breath and early fatigue, stopping at 4 min. Her heart rate peaked at 75 beats · min^{-1}. On the treadmill, she was able to walk for 10 min at 1.5 mph, reaching a peak heart rate of 98 beats · min^{-1}. Her main complaint was shortness of breath.

Referral for Pacemaker Assessment

Because of limited exercise tolerance, Mrs. JD was referred to her cardiologist for a pacemaker evaluation. Her physician administered an office-based walking protocol during which Mrs. JD complained of shortness of breath after 3 min of "brisk" walking. Her heart rate reached a peak of 102 beats · min^{-1}. The pacemaker was programmed to a maximum tracking rate of 110 beats · min^{-1} and a maximum sensor rate of 130 beats · min^{-1}, and the threshold of the sensor was lowered to be more sensitive to body movements.

Adjustment of the Exercise Prescription

In cardiac rehabilitation, the exercise prescription was changed to two bouts of 15 min each of treadmill walking with 2 min rest between bouts. The warm-up was unchanged. Mrs. JD tolerated 2.0 mph at 0% grade on the treadmill as prescribed, reporting only mild leg fatigue and shortness of breath. Her peak heart rate reached 122 beats · min^{-1}.

Case Study 18 Discussion Questions

1. Why was the maximum tracking rate left unchanged when the pacemaker settings were adjusted?

2. Why was stationary cycling dropped as an exercise modality for this patient?

3. What would be the initial choice for progressing the exercise intensity on the treadmill as the patient improves her fitness—an increase in speed, an increase in grade, or both?

Glossary

A-V synchrony—The sequence and timing of the atria and ventricles during systole.

angina threshold—Point at which the supply of oxygen is less than the demand, leading to ischemia and producing symptoms of angina pectoris. Generally observed during physical or mental exertion in patients with significant coronary artery disease.

chronotropic assessment exercise protocol—Treadmill protocol used to determine if heart rate response is appropriate throughout the length of the exercise test.

chronotropic incompetence—Lack of an appropriate increase in heart rate with physical exertion. Considered an abnormal response if peak heart rate doesn't reach two standard deviations of one's age-predicted maximum heart rate, assuming the patient was highly motivated and not on medications to blunt heart rate response (i.e., β-blockers, calcium channel blockers).

dual-chamber pacemaker—Pulse generator that can pace or sense in the atrium or ventricle.

lower rate limit—The rate at which the pulse generator begins pacing in the absence of intrinsic activity.

maximum sensor rate—The maximum rate for a rate responsive pacemaker that can be achieved under sensor control.

maximum tracking rate—The maximum rate at which the pulse generator will respond to atrial events.

mode switching—A programmed feature of dual-chamber pacemakers that prevents tracking or matching every atrial impulse with a ventricular pacing pulse to prevent tracking of rapid atrial rates to the ventricle.

pacemaker sensor—Sensor incorporated into the pulse generator that detects a physiological stimulus in order to control the heart rate to match physiological demands.

rate responsive pacing—Refers to a pacemaker that changes the rate by sensing a physiological stimulus. Other terms to describe this type of pacemaker are rate modulated, adaptive rate, or sensor driven.

sick sinus syndrome—Syndrome in which the sinus node is not functioning at an appropriate rate, leading to sinus bradycardia, pauses, arrest, or exit block. Syncopal episodes can be caused by this abnormality.

single chamber pacemaker—Pulse generator that can pace or sense in the atrium or ventricle.

upper rate limit—The highest rate at which ventricular pacing will track 1:1 each sensed atrial event.

References

1. Alexander T, Friedman DB, Levine BD, Pawelczyk JA, Mitchell JH: Cardiovascular responses during static exercise. Studies in patients with complete heart block and dual chamber pacemakers. Circulation 89: 1643-7, 1994

2. Alt E, Combs W, Willhaus R, Condie C, Bambl E, Fotuhi P, Pache J, Schömig A: A comparative study of activity and dual sensor: Activity and minute ventilation pacing responses to ascending and descending stairs. Pacing Clin Electrophysiol 21: 1862-8, 1998

3. Alt E, Matula M: Comparison of two activity-controlled rate-adaptive pacing principles: Acceleration versus vibration. Cardiol Clin 10: 635-58, 1992

4. Bacharach DW, Hilden TS, Millerhagen JO, Westrum BL, Kelly JM: Activity-based pacing: Comparison of a device using an accelerometer versus a piezoelectric crystal. Pacing Clin Electrophysiol 15: 188-96, 1992

5. Barold SS, Barold HS: Optimal cardiac pacing in patients with coronary artery disease. Pacing Clin Electrophysiol 21: 456-61, 1998

6. Bodenhamer RM, Grantham RN: Mode selection: The therapeutic challenge: Adaptive-Rate Pacing. St. Paul, Cardiac Pacemakers, 1993, pp 19-52

7. Carmouche DG, Bubien RS, Kay GN: The effect of maximum heart rate on oxygen kinetics and exercise performance at low and high workloads. Pacing Clin Electrophysiol 21: 679-86, 1998

8. Erdelitsch-Reiser E, Langenfeld H, Millerhagen J, Kochsiek K: New concept in activity-controlled pacemakers: Clinical results with an accelerometer-based rate adaptive pacing system. Pacing Clin Electrophysiol 15: 2245-9, 1992

9. Greco EM, Guardini S, Citelli L: Cardiac rehabilitation in patients with rate responsive pacemakers. Pacing Clin Electrophysiol 21: 568-75, 1998

10. Gregoratos G, Cheitlin M, Conill A, Epstein A, Fellows C, Ferguson TJ, Freedman R, Hlatky M, Naccarelli G, Saksena S, Schlant R, Silka M: ACC/AHA guidelines for implantation of cardiac pacemakers and antiarrhythmia devices: A report of the American College of Cardiology/American Heart Association Task Force on Practice Guidelines (Committee on Pacemaker Implantation). J Am Coll Cardiol 31: 1175-1209, 1998

11. Harper GR, Pina IL, Kutalek SP: Intrinsic conduction maximizes cardiopulmonary performance in patients with dual chamber pacemakers. Pacing Clin Electrophysiol 14: 1787-91, 1991

12. Hayes DL, Von Feldt L, Higano ST: Standardized informal exercise testing for programming rate adaptive pacemakers. Pacing Clin Electrophysiol 14: 1772-6, 1991

13. Holmes DR: Hemodynamics of cardiac pacing, in Furman S, Hayes DL, Holmes DR (eds): A practice of cardiac pacing (ed 2). Mount Kisco, NY, Futura, 1989, pp 167-191

14. Joglar JA, Hamdan MH, Welch PJ, Page RL: Interaction of a commercial heart rate monitor with implanted pacemakers. Am J Cardiol 83: 790-2, A10, 1999

15. Kruse I, Arnman K, Conradson TB, Rydén L: A comparison of the acute and long-term hemodynamic effects of ventricular inhibited and atrial synchronous ventricular inhibited pacing. Circulation 65: 846-55, 1982

16. Leung SK, Lau CP, Tang MO, Leung Z, Yakimow K: An integrated dual sensor system automatically optimized by target rate histogram. Pacing Clin Electrophysiol 21: 1559-66, 1998

17. Medtronic: Therapies for Medical Conditions. Minneapolis, Medtronic, 1999

18. Sharp CT, Busse EF, Burgess JJ, Haennel RG: Exercise prescription for patients with pacemakers. J Cardiopulm Rehabil 18: 421-31, 1998

19. Sulke N, Dritsas A, Chambers J, Sowton E: Is accurate rate response programming necessary? Pacing Clin Electrophysiol 13: 1031-44, 1990

20. Wilkoff B, Corey J, Blackburn G: A mathematical model of the chronotropic response to exercise. J Electrophysiol 3: 176-180, 1989

21. Wilkoff BL, Miller RE: Exercise testing for chronotropic assessment. Cardiol Clin 10: 705-17, 1992

22. Wood MA, Stambler BS, Ellenbogen KA: Patient management: Optimal programming of adaptive-rate pacemakers: Adaptive-rate pacing. St. Paul, Cardiac Pacemakers, 1993, pp 86-110

Chronic Obstructive Pulmonary Disease

Michael J Berry, PhD
Department of Health and Exercise Science
Wake Forest University
Winston-Salem, NC

C. Mark Woodard, MS, MHA, MBA
Department of Health and Exercise Science
Wake Forest University
Winston-Salem, NC

Chronic obstructive pulmonary disease (COPD) is defined by the American Thoracic Society as a disease characterized by the presence of airflow obstruction that is attributable to either chronic bronchitis or emphysema (13, 14). **Chronic bronchitis** is a clinical diagnosis for patients who have chronic cough and sputum production. It is formally defined by the American Thoracic Society as the presence of a productive cough most days during three consecutive months in each of two successive years (14, 146). The cough is a result of hypersecretion of mucus, which in turn is the result of an enlargement of the mucus-secreting glands. In contrast to the clinical diagnosis for chronic bronchitis, **emphysema** is a pathological or anatomical diagnosis marked by abnormal permanent enlargement of the respiratory **bronchioles** and the alveoli, that is, the airspaces distal to the terminal bronchioles. It is accompanied by destruction of the lung **parenchyma** without obvious fibrosis (146, 148, 158). Most patients with COPD have both chronic bronchitis and emphysema with the relative extent of each varying among patients (see figures 19.1 and 19.2). The World Health Organization's International Classification of Disease (ICD) codes used by nosologists to classify COPD are 490, 491, 492, 494, 495, and 496.

Patients with COPD experience acute **exacerbations,** or periods of worsening symptoms. The

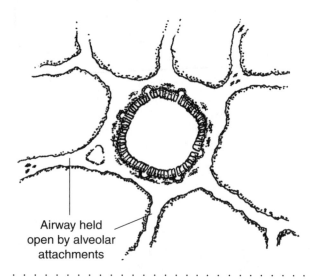

FIGURE 19.1   A normal airway that has little inflammation or mucus plugging and is being held open by parenchymal lung tissue.

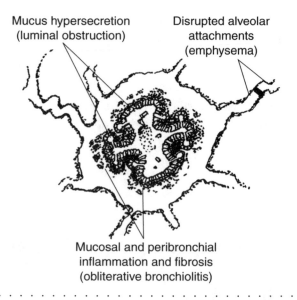

FIGURE 19.2   An obstructed airway that has significant inflammation and mucus plugging. Also shown is a loss of alveolar attachments, thus making airway collapse more likely.

pathogenesis of an exacerbation is not well understood, and it may be difficult to define clinically. These exacerbations can lead to **respiratory failure** and are a major cause of hospitalizations in the United States. The risk factors for the development of COPD are listed in table 19.1.

At times, asthma (ICD code 493) has been subsumed under the rubric of COPD. It is characterized by inflammation and hyperresponsiveness of the tracheobronchial tree to a variety of stimuli (13). Although asthma patients experience exacerbations or attacks, they are interspersed with symptom-free periods where there is complete or near-complete reversal of airway narrowing. In contrast, most patients with chronic bronchitis do not exhibit significant reversibility of their airway narrowing and present with residual symptoms between exacerbations. Whereas patients with COPD and asthma share similar clinical characteristics, the pathology of the two syndromes differs considerably, suggesting that they are different diseases (101). There is accumulating evidence to suggest that COPD is a multiple–organ system disease with ventilatory impairments and muscle dysfunction contributing to the exercise intolerance seen in these patients (15). Because of the differences between COPD and asthma, they should be considered separately.

Scope

COPD was the fourth leading cause of death in 1999, resulting in 124,181 deaths. After the numbers are adjusted for age, the death rate for COPD has increased 4.1% compared with 1998. Additionally, COPD was the only leading cause of death, among the top five causes, that showed an increase (117).

In addition to being the fourth leading cause of death in the United States, COPD is a major cause of morbidity and disability and a major healthcare cost. There were 553,000 hospitalizations in 1995 for which COPD was given as the diagnosis at the time of discharge. The average length of stay for COPD hospitalizations was 6.3 days, with the mean cost of each of these hospitalizations being $10,684.00 (5, 117). In 1995, 16,087,000 physician office visits were attributable to COPD. Respiratory tract infections in COPD patients have been shown to have a major impact on utilization of healthcare resources (70). Between 1990 and 1992, 800,000 Americans reported that emphysema caused them to limit their activity (117). Serres et al. (136) reported that COPD patients have lower levels of physical activity than age-matched controls.

Physiology and Pathophysiology

The anatomical, physiological, and pathological abnormalities associated with COPD often result in debilitation for the COPD patient. Because of the direct insult that cigarette smoke, the primary risk

TABLE 19.1

Chronic Obstructive Pulmonary Disease (COPD) Risk Factors

Risk factor	Scope	Comments
MAJOR RISK FACTORS		
Cigarette smoking	15% of smokers developing clinically significant COPD (77, 140).	The age at which smoking is started, total pack years, and current smoking status are all predictive of COPD mortality rates.
α_1-antitrypsin deficiency	Less than 1% of COPD cases in the United States.	Genetic deficiency affects integrity of the alveolar walls (29, 43).
OTHER RISK FACTORS Exposure to passive smoke		
Air pollution		
Hyperresponsive airways		
Occupational factors		
Sex		Higher prevalence of COPD in men than in women (140).
Race		Death rates are highest for white males (106).
Socioeconomic status		

factor for the development of COPD, has on the lungs, it has long been thought that the lungs were the primary organs affected by COPD. However, recent research suggests that the disease process itself or certain aspects of the disease process not only adversely affect the lungs but may also adversely affect skeletal muscle.

Lungs

Cigarette smoking affects the large airways (**bronchi**), the small airways (bronchioles), and the pulmonary parenchyma. The pathological conditions that develop are a result of the effect that cigarette smoke has on each of these structures. Additionally, the degree of airway reactivity of each individual patient will also have an effect. Within the large airways, cigarette smoke causes the bronchial mucus glands to become enlarged and the gland ducts to become dilated. Excessive cough and sputum production are a result of these factors and are the characteristic symptoms of chronic bronchitis (157). These alterations in the large airways have very little effect on airflow or **spirometry.** The airflow ob-

struction that is characteristic of COPD does not occur until additional damage is incurred by the small airways and the lung parenchyma. The changes in the smaller airways include mucus plugging, inflammation, and an increase in the smooth muscle (see figure 19.1). These changes decrease the cross-sectional area of the airways and can have a profound effect on airflow and the COPD patient's spirometry (157). The spirometry of a COPD patient is characterized by reductions in expiratory flow rates including the **forced expiratory volume in 1 s (FEV$_1$),** the FEV$_1$/**forced vital capacity (FVC)** ratio and the mid-expiratory flow rate.

With emphysema there are destructive changes to the alveolar walls, and the net effect of these changes is twofold. First, destruction of the alveolar walls results in a loss of the tethering or supportive effect that the alveoli have on the smaller airways. This tethering effect helps keep the airways open during expiration. Without this alveolar support, the smaller airways are likely to collapse during expiration, thus adding further to the airway obstruction (see figure 19.2). The second effect of destruction of the alveolar walls is to diminish the elastic recoil of the lungs, which in turn decreases the force that moves air out of the lungs. The combination of these two effects reduces airflow and increases the amount of work the respiratory muscles must perform to meet the ventilatory demands of the body.

The combined effects of airway obstruction and the reduced expiratory driving force increase the time needed for expiration. If inspiration occurs before the increased expiratory time requirement can be met, then the normal end-expiratory lung volume will not reached. This results in an increased functional residual capacity and **hyperinflation** of the lungs. Furthermore, the diaphragm will assume a shorter more flattened position. Because the diaphragm is a skeletal muscle, it operates according to the length–tension relationship (109), whereby the tension developed by skeletal muscle is a function of its resting length. As a muscle is shortened or lengthened beyond its optimal length, the potential for tension development decreases (68). Because the diaphragm is shortened with hyperinflation, it has less force-generating potential (44, 124, 145). There is some evidence to suggest that the diaphragm adapts to these chronic changes by shortening the optimal length of its fibers (142). As such, each fiber would have the potential to generate its maximal force at its new length (58). Despite these adaptive changes, there is still evidence to suggest that COPD patients have a decreased diaphragmatic pressure-generating capacity.

In normal healthy individuals, the end-expiratory lung volume is decreased by approximately 200 to 400 ml with moderate exercise (76, 138). In contrast, patients with COPD demonstrate an increase in the end-expiratory lung volume with exercise, leading to dynamic hyperinflation of the lungs (150). This resulting dynamic hyperinflation leads to further diaphragm weakness and may contribute to **dyspnea** and a reduced exercise tolerance. Because the diaphragm is a skeletal muscle, it has been hypothesized that positive adaptations may result from training of the diaphragm muscle (96, 132).

Skeletal Muscle

In addition to the damaging effect that COPD has on the lungs, there is evidence to suggest the presence of skeletal muscle dysfunction in COPD patients (36). This skeletal muscle dysfunction may contribute to the exercise intolerance seen in COPD patients. It has been reported that COPD patients have diminished peripheral muscle strength (20, 40, 48, 54, 69, 73). Patients with COPD have been found to have a 20% to 30% reduction in quadriceps strength compared with age-matched controls (48,

69, 73). These decreases in strength are accompanied by a reduction in muscle cross-sectional area (20, 54). Other studies have also reported that the muscle mass in patients with COPD is reduced (53, 134, 171). Presently, it is not known which muscle fiber type is most affected by the disease. An analysis of muscle biopsies from patients with moderate COPD showed no changes in the proportions of fiber types. However, significant atrophy of type II fibers was correlated with weight loss (82). In contrast to these findings, studies with advanced COPD patients have reported a reduction in the proportion of type II fibers compared with control subjects (84, 166). This finding is consistent with the report of reduced oxidative enzyme activities in these patients (104, 105). Additionally, this reduction in type II fibers has been shown to be accompanied by a corresponding increase in type IIb fibers (166). Both chronic **hypoxemia** (78) and a lack of physical activity (166) may contribute to the changes in fiber types. Chronic steroid use has been suggested as a contributor to the muscle weakness (47, 48). Other possible contributors to the skeletal muscle abnormalities seen in COPD patients include chronic **hypercapnia,** inflammation, nutritional depletion, and comorbid conditions that may also affect skeletal muscle function (15).

Disease Progression

As previously mentioned, one hallmark of patients with COPD is a reduction in airflow, which is most prominent during maximal efforts. This reduction is typically quantified by using the results of pulmonary function tests, with the FEV_1 being one of the standards used to assess disease severity and monitor disease history. Additionally, the FEV_1 has been shown to be a strong predictor of mortality rate from COPD (156). In healthy nonsmokers, the FEV_1 declines by 20 to 30 ml per year (89, 147, 154). In both men and women smokers, there are increased rates of decline in FEV_1 (34, 93, 154). The rate of decline is both age- and sex-dependent, with the greatest rates occurring in men between the ages of 50 and 70 (34). Individuals who have characteristics compatible with an emphysematous form of COPD have a rate of decline in the FEV_1 of 70 ml per year (31).

In those individuals who quit smoking, the decline of FEV_1 is less pronounced than the decline observed in those who continue to smoke (34, 93). In fact, ex-smokers show rates of decline of FEV_1 similar to those of nonsmokers, and in younger ex-smokers the FEV_1 has actually been shown to increase following smoking cessation (34). Although complete cessation of smoking has beneficial effects on the decline of the FEV_1, the effect of attempting to quit smoking and relapsing on FEV_1 decline is equivocal. Sherrill et al. (139) reported that the rate of decline of the FEV_1 is steeper in ex-smokers who resume smoking compared with those who continue to smoke. More recently, Murray et al. (115) reported that attempts to quit smoking in patients with mild COPD slow the rate of FEV_1 compared with those who continue to smoke.

Signs and Symptoms

The diagnosis of COPD is made based on patient history, a physical examination, and the results of laboratory and radiographic studies. The diagnosis is suspected in patients who have a history of smoking and present with an acute respiratory illness or respiratory symptoms such as a productive cough. It has been reported that a smoking history of 70 or more **pack years** is suggestive of COPD (17). The acute respiratory illness is characterized by increased cough, purulent sputum production, wheezing, dyspnea, and occasional fever. With progression of the disease, the interval between these illnesses decreases (14).

In the early disease stages, slowed expiration and wheezing are noted during the physical examination. Additionally, breath sounds are decreased, heart sounds may become distant, and course crackles may be heard at the base of the lungs (14).

Diagnostic and Laboratory Evaluations

Results from pulmonary function tests are necessary for establishing a diagnosis of COPD and for determining the severity of the disease; however, they cannot be used to distinguish between chronic bronchitis and emphysema. The FEV_1, FVC, FEV_1/FVC, and single-breath diffusing capacity are the primary pulmonary function tests recommended to aid in the diagnosis of COPD (12). In patients with COPD, the results from all these tests are less than what would be predicted for a person of similar age, sex, and stature. Other recommended tests include lung volume measurements and determination of arterial blood gas levels. Lung volume measurements often reveal an increase in total lung capacity, functional residual capacity, and residual volume. Arterial blood gases may reveal hypoxemia in the absence of hypercapnia in the early stages of the disease, with a worsening of hypoxemia and hypercapnia presenting in the later stages of the disease (14).

Because emphysema is defined in anatomic terms, the chest **roentgenogram** can sometimes be used to differentiate between emphysema and chronic bronchitis. Whereas patients with chronic bronchitis often have a normal chest roentgenogram, the roentgenogram of patients with advanced emphysema may reveal large lung volumes, hyperinflation, a flattened diaphragm, and vascular attenuation (14). Computed tomography has greater sensitivity and specificity than the chest roentgenogram and can be used for both qualitative and quantitative assessment of emphysema. Because the additional information gained from computed tomography will rarely alter therapy, it is infrequently used in the routine care of COPD patients. Although computed tomography is not recommended for routine use with COPD patients, it is the best way of recognizing emphysema and probably has a significant role in recognizing localized emphysema that is amenable to surgical treatment (158).

The American Thoracic Society proposes staging patients with COPD into distinct categories based on the degree of airflow obstruction. This organization suggests using the FEV_1 as the staging criterion. As such, patients with an FEV_1 greater than or equal to 50% of predicted are categorized as stage 1 or as having mild disease. Those with an FEV_1 between 35% and 49% of predicted are categorized as stage 2 or as having moderate disease. Finally, those with an FEV_1 less than 35% of predicted are categorized as stage 3 or as having severe disease (12, 14). The majority of COPD patients are categorized as having mild disease.

Clinical Considerations

Although it was originally believed that COPD had minimal effects on health-related quality of life and exercise capacity in this group of patients (14), recent evidence suggests that these patients do have a substantially compromised health-related quality of life (61) and a reduced exercise capacity (35). As the severity of the disease progresses, quality of life and exercise capacity of these patients continue to decrease. Additionally, increased healthcare costs are associated with caring for the more severely diseased patients (14). Practical Application 19.1 discusses malnutrition in chronic obstructive pulmonary disease.

Treatment

Once a diagnosis of COPD has been made, a multifaceted approach to the treatment and management of the patient should be adopted. Comprehensive treatment of the patient should include smoking

Malnutrition in Chronic Obstructive Pulmonary Disease

Malnutrition is a problem for as many as 25% of all COPD patients (30, 83). Additionally, as many as 50% of hospitalized COPD patients exhibit protein and calorie malnutrition (83). In a review of 90 COPD patients, researchers found that those patients who required hospitalization and mechanical ventilation demonstrated the most severe nutritional decrements (62). Whereas the causes of malnutrition in COPD patients have not been clearly defined (168), the results of malnutrition have. Weight loss in COPD patients has been shown to be a predictor of mortality rate (170). Additionally, prolonged malnutrition results in deleterious changes to the diaphragm muscle such that its ability to generate force is decreased (87, 98). This fact coupled with the fact that COPD patients exhibit dysfunction in other skeletal muscles suggests the need for nutritional support in these patients. It has been demonstrated that when COPD patients are given sufficient calories in excess of their needs, they will gain weight and this weight gain is accompanied by significant improvements in ventilatory and peripheral muscle strength (169). Given the need for nutritional intervention and the positive outcomes that can result, the clinical exercise physiologist should consult with a nutritionist when working with underweight COPD patients.

cessation, oxygen therapy, pharmacological therapy, and pulmonary rehabilitation (which includes exercise training).

Because smoking is a major cause of COPD, smoking cessation is a major therapy in the treatment of COPD patients. It is only one of two interventions that has been shown to improve patient survival (14).

The second therapy that has been shown to improve survival in patients with COPD is long-term oxygen therapy. Both the British Medical Research Council study (112) and the National Heart, Lung and Blood Institute's Nocturnal Oxygen Therapy Trial (119) showed that patients receiving long-term oxygen therapy experienced a significant reduction in mortality rates. Other benefits of long-term oxygen therapy include a reduction in **polycythemia** (97), decreased pulmonary artery pressure (1-3), and improved neuropsychiatric function (75).

Acute administration of supplemental oxygen has been shown to preserve exercise tolerance in hypoxemic patients. Whether patients with COPD undergo an acute bout of exercise with administration of supplemental oxygen (28, 41, 45, 114, 123) or are trained with supplemental oxygen (49, 123, 133, 172), the benefits are significant. The goal of oxygen therapy is to reverse hypoxemia and prevent tissue **hypoxia** (14). In order for COPD patients to realize benefits from supplemental oxygen therapy during training, they must demonstrate hypoxemia during training. If oxygen is to be prescribed for COPD patients, the goal is to maintain the partial pressure of oxygen in arterial blood above 60 mmHg or the percentage saturation above 90. Therefore, the delivery method and the dosage of oxygen, or the

flow rate, need to be considered. COPD patients needing supplemental oxygen during exercise will often use a liquid oxygen supply. These systems, although more expensive, are lightweight and easily refilled from larger stationary sources. Oxygen concentrators cannot be used during exercise because of their weight and need for an electrical supply.

Pharmacological therapy in patients with COPD is aimed at inducing bronchodilation, decreasing the inflammatory reaction, and managing and preventing respiratory infections (14). **Bronchodilator** therapy includes the use of β_2-agonists, anticholinergic agents, and theophylline. The use of β_2-agonists may result in tremors, anxiety, palpitations, and arrhythmias. Because of these problems, careful dosing and monitoring of patients with known cardiovascular disease are necessary (14).

Once a patient develops persistent symptoms, anticholinergic agents such as ipratropium bromide may be prescribed because their effect is more intense and of a longer duration. Additionally, they may have less potentially deleterious side effects than β_2-agonists.

Theophylline, one of the methylxanthines, is a third agent that may be used to induce bronchodilation. In addition to its bronchodilator effects, theophylline will increase cardiac output, will decrease pulmonary vascular resistance, and may have anti-inflammatory effects (153, 173). Despite the beneficial effects of theophylline, its popularity has declined because of its toxicity (79, 137) and potential to adversely interact with other drugs (131).

Whereas the administration of corticosteroids to treat asthma is common, the role of these drugs in the management of COPD patients has not been well established. It appears that patients with COPD who experience an exacerbation may benefit from corticosteroid therapy (6), whereas those patients with stable COPD are less likely to experience benefits from corticosteroid therapy (113).

Pulmonary rehabilitation is defined as a multidimensional continuum of services directed at persons with pulmonary disease and their families. It is usually delivered by an interdisciplinary team of specialists, with the goal of achieving and maintaining the individual's maximum level of independence and functioning in the community (63). These services typically include patient assessment, patient education, exercise training, psychosocial interven-tion, and patient follow-up (8). The various components of each of these services are listed in table 19.2. The goals of pulmonary rehabilitation are to decrease airflow limitations, improve exercise capacity or physical function, prevent and treat secondary medical complications, decrease respiratory symptoms, and improve the patient's quality of life. As a result of participating in a comprehensive pulmonary rehabilitation program, patients have demonstrated improvements in quality of life (67, 90), sense of well-being (16), self-efficacy (130), and functional capacity (130). Additionally, functional status has been shown to be a strong predictor of survival in patients with advanced lung disease

---

TABLE 19.2

Various Components of the Different Services Offered in Pulmonary Rehabilitation

Patient assessment	Patient training and education	Exercise training	Psychosocial interventions	Patient follow-up
Medical history	Anatomy and physiology	Mode, duration, frequency, and intensity	Identification of support systems	Outcome measurements of physical function and health-related quality of life
Pulmonary function tests	Pathophysiology of lung disease	Upper and lower extremity endurance training	Treatment of depression	Support groups
Symptom assessment	Description and interpretation of assessment tests	Upper and lower extremity strength training	Treatment of anxiety	Maintenance programs
Physical function assessment	Breathing retraining	Inspiratory muscle training	Anger management	
Nutritional assessment	Bronchial hygiene	Flexibility and posture	Sexuality issues	
Activities of daily living assessment	Medication information	Orthopedic limitations	Adaptive coping styles	
Educational assessment	Symptom management	Home exercise plans	Adherence to lifestyle modifications	
Psychosocial assessment	Activities of daily living and energy conservation		Relapse prevention	
	Nutrition			
	Psychosocial issues			
	Smoking cessation			

Guidelines for Pulmonary Rehabilitation Programs. 2nd ed. American Association of Cardiovascular and Pulmonary Rehabilitation. (8)

following pulmonary rehabilitation (27). It is difficult to determine which of the specific components of pulmonary rehabilitation is responsible for these improvements, because they are all integrally related. The American Thoracic Society recommends that patients with COPD be referred to a pulmonary rehabilitation program after they have been placed on optimal medical therapy and still demonstrate the following (14):

1. Severe symptoms
2. Several emergency room or hospital admissions within the previous year
3. Diminished functional status that limits their activities of daily living
4. Impairments in quality of life

Although patients who are referred to pulmonary rehabilitation typically have severe disease, recent research indicates that patients with mild and moderate disease will benefit from participation in the exercise component of a pulmonary rehabilitation program similarly to those with severe disease (23, 163).

Graded Exercise Testing

Exercise testing is an integral component in the evaluation of patients with COPD. In patients with mild or moderate disease, symptoms generally do not present until an increased demand is placed on the respiratory system, such as with exercise. In patients with severe disease, the functional capacity is reduced to such a level that even simple activities of daily living may impose a challenge to the respiratory system. Most patients with moderate to severe COPD have a reduced exercise capacity as a result of a reduced ventilatory capacity in the face of an increased ventilatory demand (18, 51, 86). Because exercise places an increased demand on the respiratory system, exercise testing provides an objective evaluation of the functional capacity of the COPD patient. Additionally, exercise testing can be used to detect COPD and cardiovascular disease, follow the course of the disease, detect exercise hypoxemia, determine the need for supplemental oxygen during exercise training, evaluate the response to treatment, and prescribe exercise (151). Practical application 19.2 describes client–clinician interaction for patients with COPD.

Medical Evaluation

During the exercise tests, the minimum monitoring should include measurement of blood pressure,

a 12-lead electrocardiogram (ECG), analysis of arterial oxygen saturation, and measurement of dyspnea. These should be measured before the start of the test, continuously throughout the test, and at the termination of the test. Blood pressure should be measured with the patient's arm relaxed and the manometer mounted at eye level. Automatic monitors for blood pressure measurement during exercise are available, and their use has been found acceptable (71). Placement for the 12-lead ECG is typically the Mason-Likar. The gold standard for the measurement of arterial oxygen saturation is co-oximetry using arterial blood. If this is unavailable, the use of a pulse oximeter is acceptable as long as the pulse oximeter has been validated during exercise. At oxygen saturations greater than 90%, these devices have a high degree of reliability (118). However, as oxygen saturation drops below 90%, their reliability worsens (118). Because of the problems with precisely defining the degree of hypoxemia with these instruments, it is probably best to use these devices to qualify whether desaturation is occurring and then to correct it with supplemental oxygen. The final variable that should be monitored during the exercise test in patients with COPD is dyspnea. A number of scales are available for use that have been validated and proven reliable (4, 102, 103). One particular dyspnea scale of interest to the exercise specialist is the Borg scale. This scale has been adapted (figure 19.3) for use during exercise testing and training with COPD patients (26).

If the equipment is available, gas exchange and ventilatory measurements should be obtained. These measures provide valuable information that can be used to more accurately prescribe the exer-

0	Nothing at all
0.5	Very, very weak shortness of breath
1	Very weak shortness of breath
2	Weak shortness of breath
3	Moderate shortness of breath
4	Somewhat strong shortness of breath
5	Strong shortness of breath
6	
7	Very strong shortness of breath
8	
9	
10	Very, very strong shortness of breath
	Maximal shortness of breath

FIGURE 19.3 Rating of perceived dyspnea chart.

Client–Clinician Interaction

Oftentimes, the first interaction between the clinical exercise physiologist and the COPD patient is when the patient has been referred for pulmonary rehabilitation. Unfortunately, patients with COPD are often referred to pulmonary rehabilitation only after they have experienced an exacerbation of their disease or when their dyspnea has become so oppressive that they are severely disabled. As a result, these patients are often anxious, scared, frustrated, and depressed. It is important for the clinical exercise physiologist to be aware of these problems when working with the COPD patient and to present a positive, yet realistic, picture of the benefits of exercise training for the COPD patient.

Dyspnea, the primary symptom of COPD, often results in a vicious cycle of fear and anxiety followed by inactivity that results in deconditioning, which results in further dyspnea. Unless this cycle can be broken, COPD patients are destined to lose their independence and become dependent on others to meet their most basic needs. COPD patients need to be made aware that they can learn to live with their dyspnea and that exercise can help to reduce the intensity of dyspnea and the distress associated with dyspnea. These patients should be taught strategies that will help them to manage their dyspnea on a daily basis. These strategies include such things as monitoring the effects of various medications on dyspnea and avoiding factors that can result in dyspnea—such as stress.

The clinical exercise physiologist must also be aware that COPD patients often experience exacerbations or periods of worsening symptoms. As a result, these patients may not be able to exercise at their prescribed intensity or may miss exercise sessions completely. If possible, the patient should be encouraged to continue exercising even if at a much lower intensity. If this is not possible, the patient should be encouraged to resume exercising once he or she has recovered from the exacerbation.

The successful clinical exercise physiologist can effectively interact with each person on a one-to-one basis. Such a professional is sensitive to the individual needs of the patient and is able to tailor each patient's program to meet that patient's individual needs. As a result, the patient is able to take control of his or her disease with less fear and anxiety.

cise intensity, to evaluate the effectiveness of an exercise intervention, and to provide information regarding the extent of the lung disease (107, 135). Of special concern when one is measuring gas exchange and ventilatory parameters in COPD patients is the patient requiring supplemental oxygen. These patients should be tested on an elevated fraction of inspired oxygen such that it equates with the flow rate established for the use of supplemental oxygen. Most commercially available gas exchange measurement systems have established procedures that allow for the use of elevated fractions of inspired oxygen during exercise testing and conversions of oxygen flow rates to inspired oxygen fractions.

Contraindications

Exercise testing has been shown to be very safe, even in high-risk populations, with one or fewer deaths per 10,000 tests (65). To minimize risk to patients, recommended guidelines from the American College of Sports Medicine (ACSM) (11), the American Association of Cardiovascular and Pulmonary Rehabilitation (8), and the American Heart Association (64) should be closely followed. In general, the procedures for testing patients with COPD follow those for testing other at-risk populations.

Before conducting the exercise test, the clinician should review information from a medical exam and history to identify contraindications to testing, as observed in the previously cited guidelines. An additional concern with COPD patients is the patient with accompanying pulmonary hypertension. Some experts advise caution with these individuals because of the risk of serious cardiac arrhythmias or even sudden death while testing (8).

Recommendations and Anticipated Responses

The exercise mode, the test protocol, and the monitoring equipment are all fundamental considerations in exercise testing. The exercise mode should be one that will increase total body oxygen demands by requiring the use of a large muscle group. The most common exercise testing modalities for the COPD patient are the treadmill and the bicycle ergometer. One exercise mode that should not be used to test COPD patients is arm ergometry. This is because patients with severe COPD often use the accessory muscles of inspiration for breathing at rest. As such, any additional burden placed on these muscles could result in significant symptoms and distress for the patient (38).

The testing protocol should start at a work rate that can easily be accomplished by the patient, should have increments in the work rate that are progressively difficult, and should last a total duration of 8 to 12 min. The initial stages should be of such an intensity to allow the patient an adequate amount of time to warm up and become accustomed to the exercise bout. The work rate increments should be small and based on characteristics of the patient (e.g., sex, size, severity of disease, and previous level of physical activity). It has been observed that the incremental rate of work rate will affect the exercise responses of COPD patients (46). For example, the peak work rate achieved for a given level of oxygen consumption will be greater when the work rate is increased quickly. Unfortunately, because of severe deconditioning and extreme shortness of breath that some patients with COPD experience when performing even mild physical activity, it may not be possible to have a test that lasts the minimum recommended time duration.

The responses of the COPD patient to an exercise test will vary depending on the severity of the disease. In patients with mild disease, the results of their exercise test are consistent with those of normal individuals or may demonstrate abnormalities indicative of cardiovascular disease or deconditioning (55). Exercise responses in moderate and severe COPD patients compared with age-matched healthy controls are shown in table 19.3. Peak oxygen uptake and peak work rates are usually reduced in COPD patients with moderate or severe disease (37, 104, 121, 141, 164). Concomitant with the lower peak oxygen consumption is a lower peak heart rate (HRpeak) and a greater heart rate reserve (predicted HRpeak minus measured HRpeak). Oxygen pulse is also low in COPD patients because they terminate exercise at a low work rate. In normal and deconditioned individuals and in cardiac patients, the ventilatory reserve (maximal voluntary ventilation minus the peak minute ventilation) is high at peak exercise. In contrast, the patient with COPD has a low ventilatory reserve, and, in some cases, peak minute ventilation is equal to or even greater than the maximal voluntary ventilation (162). Addition-

TABLE 19.3
Exercise Test Responses in Patients With Chronic Obstructive Pulmonary Disease Compared With Normal Healthy Subjects

Parameter	Finding
Peak work rate	Decreased
Peak oxygen consumption	Decreased
Peak heart rate	Decreased
Peak ventilation	Decreased
Heart rate reserve	Increased
Ventilatory reserve	Decreased
Arterial partial pressure of oxygen	Decreased
Arterial oxygen saturation	Decreased
Lactate threshold	Occurs at a lower work rate
Ventilatory threshold	Absent

ally, the peak minute ventilation is lower than predicted (37). The partial pressure of oxygen in the arterial blood and the percentage saturation of hemoglobin in the arterial blood are often low at maximal exercise in the COPD patient with moderate or severe disease (14). Although it was traditionally thought that patients with COPD did not develop anaerobiosis and did not demonstrate a lactate threshold during incremental exercise, recent research suggests that these patients can develop a significant anaerobiosis and do demonstrate a lactate threshold, although these occur at relatively low work rates (37, 152). Because of their ventilatory impairment, patients with moderate and severe COPD will not show a disproportionate increase in minute ventilation with the development of anaerobiosis (152). Thus, the detection of a ventilatory threshold may not be possible in these patients.

Exercise testing is an important tool in assessing the patient with COPD. However, the test and equipment used must be designed to meet the needs of the patient and the clinician administering the test. Additionally, because of the abnormal responses of the COPD patient, care must be exercised when interpreting the results of these tests.

Exercise Prescription

Participation in a pulmonary rehabilitation program that includes at least 4 weeks of exercise training can result in improvements that are clinically significant for COPD patients (91). The improvements realized by patients who participate in exercise rehabilitation are in the realm of quality of life,

specifically in the relief of dyspnea and an improvement in patients' perceptions of how well they can cope with their disease. Practical application 19.3 reviews the literature about exercise and COPD.

Special Exercise Considerations

Dyspnea or shortness of breath and a reduced exercise capacity are two of the most common complaints of COPD patients. In addition to the diaphragm muscle, the accessory muscles of inspiration are activated during exercise. These accessory muscles of inspiration include the scalene, sternocleidomastoid (52), and serratus anterior (128). Even at low work rates, unsupported arm exercise results in greater levels of dyspnea compared with lower extremity exercise in patients with COPD (38). It has been hypothesized that arm exercise requires the use of the accessory muscles of inspiration, thereby decreasing their participation in ventilation and increasing the work of the diaphragm. This may explain, in part, why patients with COPD complain of dyspnea when performing activities of daily living with their upper extremities (155). Thus, it appears that strategies aimed at improving the function of these accessory muscles of inspiration, such as resistance training, could benefit COPD patients.

Skeletal muscle dysfunction may also contribute to the reduced exercise tolerance seen in COPD patients (36). Muscle strength has been found to be a significant contributor to symptom intensity during exercise in COPD patients (73). Additionally, quadriceps muscle strength has been shown to be positively correlated with both the 6-min walk distance and maximal oxygen consumption in COPD

PRACTICAL APPLICATION 19.3

Literature Review

Recently, the American College of Chest Physicians and the American Association of Cardiovascular and Pulmonary Rehabilitation released evidence-based guidelines for pulmonary rehabilitation (9). This document contains recommendations for pulmonary rehabilitation, and the scientific evidence supporting these recommendations are reviewed. Lower extremity exercise training received a grade of A. This grade reflects the fact that there is strong scientific evidence to support the use of lower extremity exercise training in COPD patients. This evidence is from the results of well-designed, well-conducted, controlled (both randomized and nonrandomized) trials with statistically significant results that support the use of lower extremity exercise training. The results of controlled randomized clinical trials that have included lower extremity exercise training as part of an intervention are shown in the *Results of Controlled Randomized Clinical Trials Examining the Efficacy of Lower Extremity Exercise Training* table.

continued

Results of Controlled Randomized Clinical Trials
Examining the Efficacy of Lower Extremity Exercise Training

Author	Increased exercise capacity	Increased peak oxygen consumption	Improved quality of life
Ambrosino et al. (7)	Yes	Not measured	Not measured
Berry et al. (22)	Yes	No	Not measured
Booker (25)	No	Not measured	Yes
Busch and McClements(32)	Yes	Not measured	No
Cambach (33)	Yes	Not measured	Yes
Goldstein et al. (67)	Yes	Not measured	Yes
Jones et al. (85)	Yes	Not measured	Yes
Lake et al. (92)	Yes	Not measured	Yes
Larson et al. (94)	Yes	Yes	Yes
McGavin et al. (110)	Yes	No	Yes
Reardon et al. (127)	Yes	No	Not measured
Ries et al. (130)	Yes	Yes	No
Strijbos et al. (149)	Yes	Not measured	Yes
Toshima et al. (159)	Yes	Not measured	No
Weiner et al. (165)	Yes	Not measured	Not measured
Wijkstra et al. (167)	No	Not measured	Yes

Shown in this table are the effects of lower extremity exercise training on submaximal and maximal exercise capacity, peak oxygen consumption, and quality of life. In nearly all of the studies, exercise capacity, as evaluated from time on the treadmill or a timed distance walk, was found to improve following lower extremity exercise training. Whether improvements in exercise capacity will translate into improvements in domains such as physical function and activities of daily living has yet to be determined. As recently pointed out at a workshop convened by the National Institutes of Health investigating the efficacy of pulmonary rehabilitation, limiting the evaluation of interventions to outcomes such as timed walks or physiological measures was myopic and provided incomplete measures of medical outcomes (63). The conclusion from this august group was that the success of therapeutic interventions should be based on a variety of medical outcomes such as health-related quality of life, respiratory symptoms, frequency of exacerbations, activities of daily living, cost–benefit relationships, use of healthcare resources, and mental, social, and emotional function (63).

patients (69, 73). Based on these observations, it appears that strength training may also prove beneficial for the rehabilitation of patients with COPD.

Exercise Recommendations

In general, four approaches are recommended for improving respiratory and skeletal muscle dysfunctions. These include lower extremity aerobic exercise training, **ventilatory muscle training,** upper extremity strength training, and whole-body strength training. Practical application 19.4 discusses exercise prescription for patients with COPD.

Cardiovascular Training

Despite the strong evidence supporting the use of lower extremity exercise as a therapeutic intervention for patients with COPD, guidelines for the prescription of lower extremity exercise in these patients have not been well defined. More specifically, the intensity at which these patients should perform lower extremity exercise is unclear and debatable. The ACSM recommends that the mode of exercise should be any aerobic exercise that involves large muscle groups, such as walking or cycling. The minimal recommendation for frequency is three to five

Exercise Prescription

Aerobic	Recommendation	Special considerations
Mode	Typical activities include walking and stationary cycling. However, other common forms of activity may be appropriate (e.g., stair stepping, rowing) if they do not promote excessive dyspnea (shortness of breath).	Exercise intensity recommendations are not clearly defined and should be determined individually based on what is tolerable by the patient when using a dyspnea scale. Dyspnea scale of 3 (moderate) is equal to 50% of peak oxygen consumption, and 6 (severe to very severe) is equal to 85% of peak oxygen consumption. Significant increases in exercise capacity have been observed when subjects exercise at their gas exchange threshold.
Frequency	Three to five times per week.	Upper body aerobic exercise may result in greater dyspnea and should be monitored.
Duration	Work up to a minimal duration of 20 to 30 min.	Exercise compliance and increased risk of injury should be considered when determining exercise intensity.
Intensity	≥50% of peak oxygen consumption.	All COPD patients should be closely monitored, especially at near-maximal intensities.

Resistance	Recommendations	Special considerations
Mode	Free weights or weight machines for upper and lower body exercises.	Respiratory muscle weakness is common in pulmonary patients and may contribute to dyspnea. Therefore, resistance training of the accessory muscles of respiration should be strongly suggested. Inspiratory muscle training may be considered in patients who remain symptomatic despite optimal therapy and is considered an adjunctive exercise therapy.
Frequency	Two to three times per week.	Optimal frequency appears to be dependent on the muscle group being exercised. For the spine, 1 to 2 days per week will yield optimal results. For the upper and lower extremities, two to three times per week will yield optimal results.
Intensity	One set of 8 to 15 repetitions to fatigue per exercise.	The intensity should be such that fatigue results after 8 to 15 repetitions. For more frail persons, the higher end of this range should be used. Resistance should be increased as strength increases. Each repetition should consist of 2 to 3 s of a concentric contraction (lifting) and 4 to 6 s of an eccentric contraction (lowering). One set should be performed. There is no strong evidence to suggest that a greater number of sets will yield greater increases in strength (60).

times per week with a minimal duration goal of 20 to 30 min of continuous activity (11). With respect to exercise intensity, the ACSM provides four very disparate recommendations (11). The first strategy is to have patients exercise at 50% of their peak oxygen consumption. The rationale for this recommendation is that this is the minimal intensity recommended for apparently healthy individuals. Because patients with COPD are deconditioned, training at this intensity should improve their exercise capacity. However there is some research to suggest that training at lower exercise intensities may not lead to physiological improvements (160). Another potential problem with this approach is the method used to monitor the exercise intensity. Because there is a linear relationship between %HRpeak and %$\dot{V}O_2$peak, it has been suggested that a given percentage of HRpeak can be used to monitor the exercise intensity (11). Results from a recent investigation demonstrate that the relationship between %HRpeak and %$\dot{V}O_2$peak described by the ACSM cannot be used with COPD patients (143). It appears that when one prescribes exercise based on the relationship between %HR and %$\dot{V}O_2$peak in COPD patients, heart rate intensity is underpredicted. Therefore, if this approach to monitoring exercise intensity is used, errors in the exercise prescription can result.

A second approach to prescribing exercise intensity is to have patients exercise at an intensity above the anaerobic threshold (11). The rationale behind this suggestion is that training at an intensity that induces metabolic acidosis will reduce minute ventilation after training, thus allowing for an increase in ability of these patients to perform heavy work (37).

The third approach recommended by the ACSM for prescribing exercise intensity in COPD patients is to have them exercise at near-maximal or maximal levels. Because many COPD patients are limited by their ventilatory system and not the cardiovascular system, it has been thought that these patients should be able to exercise at levels approaching those achieved at maximum levels during graded exercise tests (126). Noncompliance and an increased risk of injuries are two potential problems that could result when exercise is prescribed at high intensities. Therefore, if either the second or third approach to exercise prescription approach is used, close monitoring of the patient is recommended so that injuries can be prevented and problems with noncompliance can be avoided.

A final strategy recommended by the ACSM for prescribing exercise intensity in COPD patients deals more with monitoring the exercise intensity rather than prescribing a specific intensity. This approach uses ratings of perceived dyspnea to define exercise intensity (11). It has been shown that COPD patients are able to regulate and monitor their exercise intensity using dyspnea ratings obtained during an incremental exercise test (80). If rating of perceived dyspnea is to be used to monitor exercise intensity, a target dyspnea of 3 (moderate on the Borg scale) (26) corresponds to an intensity of approximately 50% of peak oxygen consumption and 6 (severe to very severe) corresponds to an intensity of approximately 85% of peak oxygen consumption (59).

Ventilatory Muscle Training   Ventilatory muscle training is recommended for COPD patients to increase ventilatory muscle strength and endurance with the ultimate goal of improving exercise capacity, relieving the symptoms of dyspnea, and improving health-related quality of life. Three strategies have been used to train the ventilatory muscles:

1. Voluntary isocapnic **hyperpnea**
2. Inspiratory resistive loading
3. Inspiratory threshold loading

With voluntary isocapnic hyperpnea, the patient is instructed to breathe at as high a level of minute ventilation as possible for 10 to 15 min. With this technique, the patient is hyperventilating and, therefore, a rebreathing circuit must be used to maintain **isocapnia.** This rebreathing circuit is complex and not portable, and the patient requires constant monitoring to ensure isocapnia when using this device. Because of these problems, this type of training has not been used or studied extensively.

During **inspiratory resistive loading,** the patient breathes through inspiratory orifices of smaller and smaller diameter while attempting to maintain a normal breathing pattern. A potential problem with the use of this device is the patient may slow his or her breathing frequency in an attempt to decrease the sensation of effort. As a result of the change in the breathing pattern, the load on the inspiratory muscles is reduced such that a training response may not occur.

With **inspiratory threshold loading,** the patient breathes through a device that only permits air to flow through it once a critical inspiratory pressure has been reached. These devices are small, do not require supervision, and avoid the problems associated with changing breathing patterns.

Results from studies examining the efficacy of ventilatory muscle training in COPD patients are equivocal. Table 19.4 shows the results of randomized controlled clinical trials examining the effects

TABLE 19.4

Results of Randomized Clinical Trials Examining the Efficacy of Ventilatory Muscle Training

Reference	Type of training	Outcomes (compared with a control group)
Belman and Shadmehr (19)	Resistive loading	Improved inspiratory muscle strength and endurance
Berry et al. (22)	Threshold loading coupled with general exercise conditioning	No change in inspiratory muscle strength No change in 12-min walk distance No change in dyspnea ratings
Bjerre-Jepsen et al. (24)	Resistive loading	No change in inspiratory muscle endurance No change in exercise tolerance
Chen et al. (39)	Resistive loading coupled with standard pulmonary rehabilitation	Improved inspiratory muscle endurance Improved inspiratory muscle strength No change in maximal or constant load cycle exercise
Dekhuijzen et al. (50)	Resistive loading coupled with standard pulmonary rehabilitation	Improved inspiratory muscle strength No improvement in maximal work capacity Increased 12-min walk distance
Falk et al. (57)	Resistive loading	Decreased dyspnea Improved submaximal exercise time
Goldstein et al. (66)	Threshold loading coupled with standard pulmonary rehabilitation	Improved inspiratory muscle endurance No change in inspiratory muscle strength No change in exercise tolerance
Guyatt et al. (72)	Resistive loading	No improvement in inspiratory muscle strength or endurance No improvement in 6-min walk distance No improvement in health-related quality of life
Harver et al. (74)	Resistive loading	Improved inspiratory muscle strength Decreased dyspnea
Larson et al. (95)	Threshold loading	Improved inspiratory muscle strength Improved inspiratory muscle endurance Improved 12-min walk distance No improvement in health-related quality of life
Lisboa et al. (99)	Threshold loading	Improved inspiratory muscle strength and endurance Decreased dyspnea Increased 6-min walk distance
Lisboa et al. (100)	Threshold loading	Decreased dyspnea Improved 6-min walk distance
McKeon et al. (111)	Resistive loading	No change in inspiratory muscle strength Increased inspiratory muscle endurance No increase in maximal cycle exercise, 12-min walk distance, or treadmill walking
Noseda et al. (120)	Resistive loading	Increased inspiratory muscle endurance No change in maximal or constant load cycle exercise
Pardy et al. (122)	Resistive loading	Improved 12-min walk distance Improved submaximal exercise endurance
Preusser et al. (125)	Threshold loading	Improved inspiratory muscle strength and endurance Improved 12-min walk distance
Wanke et al. (161)	Threshold loading coupled with general exercise conditioning	Improved inspiratory muscle strength and endurance Improved maximal exercise capacity
Weiner et al. (165)	Threshold loading coupled with general exercise conditioning	Improved inspiratory muscle strength and endurance Improved 12-min walk distance Improved submaximal exercise time

of inspiratory resistive loading and inspiratory threshold loading. Of the 18 studies presented, inspiratory muscle strength was found to increase in 9 of them and inspiratory muscle endurance was found to increase in 10. These results suggest that inspiratory muscle training does not add significantly to a program of general exercise conditioning in patients with COPD.

In the evidence-based guidelines for pulmonary rehabilitation (9), inspiratory muscle training received a grade of B. This grade reflected the fact that the scientific evidence from both observational and controlled clinical trials provided inconsistent results. Because of this grade, it was recommended that inspiratory muscle training not be considered an essential component of pulmonary rehabilitation. However, in patients who have decreased respiratory muscle strength and breathlessness and who remain symptomatic despite optimal therapy, it may be considered as an adjunctive exercise therapy.

Specific recommendations regarding the intensity, frequency, or duration of training for inspiratory muscle training have not been developed. The majority of studies that have reported improvements in inspiratory muscle function have had patients perform inspiratory muscle training at a minimum of 30% of their maximal inspiratory pressure. The duration of this training has been for at least 15 min and the frequency is at least three times per week. These appear to be the minimal requisites of an exercise prescription if inspiratory muscle strength and endurance are to be improved.

Resistance Training

Based on the results of preliminary studies that have evaluated upper, lower, and whole body strength training in COPD patients, it appears as if strength training may offer distinct advantages over other forms of exercise training. As such, it should be included in a comprehensive exercise rehabilitation program. Presently there are no clear recommendations as to the optimal strength training prescriptions that should be used with COPD patients. Until such recommendations are put forth, COPD patients should adhere to existing recommendations for strength training in older adults (10, 56).

Upper Body Upper extremity strength training has been proposed as a training modality to help reduce dyspnea in COPD patients. It has been suggested that ventilatory muscle fatigue and dyspnea may occur when COPD patients use their upper extremities to perform activities of daily living. Fatigue results because of the additional work that the accessory muscles of inspiration must perform in helping support the arms during such activities (38). To date, three studies have specifically examined the efficacy of upper extremity resistance training with COPD patients. A summary of these studies is shown in table 19.5.

From these studies, it appears that patients with COPD can tolerate and benefit from a training program consisting of upper extremity strength exercise. It has not been conclusively demonstrated that this type of training will improve activities of daily living or physical function. However, preliminary results support the recommendation that upper extremity resistance training be included as part of a comprehensive rehabilitation program (9).

These preliminary studies do not provide clear recommendations on the specific exercises that would benefit this population or on the resistance or number of repetitions that will provide the optimal benefits for these patients. It is hypothesized that the exercises should involve the accessory muscles of inspiration, that is, the muscles involved in shoulder elevation. With respect to the amount of resistance used and number of sets and repetitions to be completed, the ACSM guidelines (10) and the recommendations of Evans (56) should be followed.

Whole Body Strength training has been advocated as a means of ameliorating the problems associated with the skeletal muscle dysfunction seen in COPD patients. Because COPD patients demonstrate atrophy of type I and type II fibers and decreases in strength, it is not unreasonable to suspect that resistance training may benefit these patients. Despite a strong rationale supporting resistance training in COPD patients, there is a lack of knowledge regarding strength training and its effects in this group of patients. To date, only two studies have systematically investigated the effects of strength training in patients with COPD.

In a randomized controlled study, the effects of strength training in a group of COPD patients was evaluated (144). Those randomized to the strength training group completed three sets of 10 repetitions of single arm curls, single leg extensions, and single leg presses. The resistance was progressively increased during the 8 weeks of training. Muscular strength and cycling endurance time were found to increase in the experimental group. Dyspnea and mastery of activities of daily living, measures of health-related quality of life, were found to improve in the experimental group following training. These results provided the first evidence supporting the use of whole-body strength training in patients with COPD.

TABLE 19.5

Results of Trials Examining the Efficacy of Upper Extremity Resistance Training

Reference	Type of training	Outcomes
Couser et al. (42)	Arm ergometry at 60% of maximal workload. Unsupported arm exercise that consisted of bilateral shoulder abduction and extension for 2 min. Weight was added as tolerated. Both groups performed leg cycle ergometry at 60% of maximal workload.	Following training, minute ventilation and oxygen consumption were lower during arm elevation. Respiratory muscle strength did not change following training. No differences were reported between the two groups.
Martinez et al. (108)	Arm ergometry at a workload that engendered an RPE of 12 to 14 and an RPD of 3. Unsupported arm exercise that consisted of five shoulder and upper arm exercises for up to 3.5 min. Weight was added as tolerated. Both groups performed leg cycle ergometry at a workload that engendered an RPE of 12 to 14 and an RPD of 3 and inspiratory muscle training.	No difference in improvements in 12-min walk, cycle ergometer test, or respiratory muscle function. Task-specific unsupported arm exercise tests improved in the group that performed unsupported arm exercise. Decreased oxygen consumption in unsupported arm exercise tests in the group that performed unsupported arm exercise.
Ries et al. (129)	Gravity resistance exercises that included five low-resistance, high-repetition exercises to improve arm and shoulder endurance. Proprioceptive neuromuscular facilitation that included lower frequency progressive resistance training with weights to improve arm and shoulder strength and endurance. Both groups participated in standard pulmonary rehabilitation that included walking.	Compared with a control group, both training groups improved on training test specific to the exercise modality. Patients reported subjective improvement in ability to perform activities of daily living using the upper extremities. No change in performance of cycle ergometry tests, simulated activities of daily living tests, or ventilatory muscle endurance tests.

Note. RPE = rating of perceived exertion; RPD = rating of perceived dyspnea.

More recently, the addition of strength training to a program of aerobic training was evaluated in patients with COPD (21). The strength training program consisted of one upper extremity and three lower extremity exercises. Both resistance and the number of sets patients completed during each session were progressively increased during the 12 weeks of training. Significant increases in muscular strength and muscle mass were found in the patients in the resistance training program. However, changes in peak work rate, distance walked in 6 min, and quality of life were not significantly different between the two groups. Based on these results, it was concluded that the addition of strength training to an aerobic training program did not result in additional improvements in exercise capacity or quality of life. Although this study did not provide direct support for the use of strength training in COPD patients, it did confirm that the peripheral muscles in COPD patients show structural adaptations to strength training. It should be noted that the strength and muscular deficiencies seen in the COPD patients in this investigation were not completely corrected with the training regimen used. Thus, a more intense training program or a longer period of training may be needed in this group of patients to realize the potential benefits of strength training.

Conclusion

Chronic obstructive pulmonary disease is a common condition that affects a large number of older individuals. The disease process spans several decades and eventually results in significant morbidity and mortality rates. Research suggests that exercise can be used as an effective therapeutic intervention in patients with COPD. The review presented here supports the notion that participation in an exercise program will decrease

dyspnea and increase exercise capacity—two of the most common complaints of COPD patients. Despite the positive findings from previous research, a number of questions regarding exercise and the COPD patient remain unanswered. We hope that future research will provide these answers.

Case Study 19

Medical History

Mr. DM is a 69-year-old white male who complains of shortness of breath on exertion and occasionally at rest. The patient does not report any symptoms suggestive of myocardial ischemia. Additional findings from the medical history include treatment for hypertension and prostate cancer diagnosed within the past 5 years. The patient quit smoking cigarettes approximately 3 years ago, and reports a 102 pack year smoking history (average of 2 packs per day for 51 years). The patient was admitted to a local hospital for an exacerbation of respiratory symptoms approximately 4 months before enrolling in the exercise program. The remainder of the medical history is unremarkable. The patient does not use supplemental oxygen at the time of enrollment into the exercise program; oxygen saturation at rest via pulse oximetry is 95%.

The patient's score on the Chronic Respiratory Disease Questionnaire, a measure of health-related quality of life, dyspnea subscale is 5 (on a 1-7 scale), corresponding to "some shortness of breath" when performing activities of daily living. The patient also reports a sedentary lifestyle, rarely walking outside of the home and not participating in any sport or recreational activities.

Results of the preexercise medical exam reveal the following:

— Height and weight of 70 in. and 222 lb (body mass index = 31.8)
— Resting heart rate of 85 and blood pressure of 144/98
— Enlarged anteroposterior chest diameter and decreased breath sounds, prolonged expiration, and wheezes
— Regular pulse with no murmurs, gallops, or bruits noted
— Normal hearing and vision, absence of edema in lower extremities, and good mobility

Upon enrollment into the exercise program, the patient reports the following medications:

Atrovent inhaler, eight puffs twice a day (anticholinergic bronchodilator)
Doxapram HCL, 50 mg three times a day (respiratory stimulant)
Furosemide, 40 mg four times a day (diuretic)
Hytrin, 2 mg four times a day (α_1-selective adrenoceptor-blocking agent)
Prednisone, 5 mg four times a day (corticosteroid)
Proventil, 0.5% twice a day (β_2-adrenergic bronchodilator)
Serevent inhaler, two puffs twice a day (β_2-adrenergic bronchodilator)
Theo-Dur, 300 mg twice a day (methylxanthine derivative)
Ventolin inhaler, two puffs twice a day (β_2-adrenergic bronchodilator)

continued

Pulmonary function testing reveals a forced vital capacity of 5.31 L (127% of predicted); an FEV_1 of 1.60 L (49% of predicted); an FEV_1/FVC ratio of 30%; and a maximal voluntary ventilation of 77 L (60% of predicted). After administration of 200 μg of albuterol via metered dose inhaler, the FVC improved by 80 ml and the FEV_1 improved by 20 ml. Blood gas analysis was not performed.

Diagnosis

— Stage 2 (moderate) obstructive lung disease with shortness of breath on exertion
— Obesity
— Hypertension
— Physical deconditioning

Exercise Test Results

The patient performs a graded exercise test on the treadmill with continuous 12-lead ECG monitoring and blood pressure assessments. Ratings of perceived dyspnea are assessed with the Borg scale (1-10), oxygen saturation is assessed via pulse oximetry, and respired gas analysis is performed with a metabolic cart. Resting data include heart rate 88, blood pressure 144/100, and oxygen saturation 94%. The resting ECG is essentially normal, with mild nonspecific T-wave flattening noted in the lateral chest leads.

The patient is only able to complete the first stage of the graded exercise test using a modified Naughton protocol, walking for 2 min at 1.5 miles and 1.0% grade. Heart rate is 125 beats · min^{-1} (83% of age-predicted maximum) and blood pressure is 194/100 at maximal exercise. The patient reports a dyspnea rating of 5, corresponding to "strong shortness of breath" on the Borg scale. No ECG changes consistent with ischemia are noted, and the patient does not report chest tightness, pain, or pressure. Rare premature ventricular contractions (PVCs) are observed during exercise. Oxygen saturation at maximal exercise decreases to 85%, and the peak oxygen consumption is 14.7 ml · kg^{-1} · min^{-1}. The test is terminated because of shortness of breath and oxygen desaturation.

Other Outcome Measures

The patient also performs a 6-min walk for distance and a hands-over-head task for time before beginning the exercise program. The distance covered during the 6-min walking trial is 948 feet; oxygen saturation decreases to 85% and the patient reports a shortness of breath rating on the Borg dyspnea scale of 7 (severe shortness of breath or very hard breathing).

The hands-over-head task is designed to assess upper body strength and susceptibility to dyspnea when the patient uses the upper extremities. The task involves removing and replacing 10-lb weights along a row of six pegs positioned at shoulder height. The patient completes this task in 57.4 s with a dyspnea rating of 3 (moderate shortness of breath) and an oxygen saturation of 86%. The average time for subjects in our rehabilitation program to complete this task is 50.3 s.

Development of Exercise Prescription

The primary consideration in prescribing exercise for this patient with chronic obstructive lung disease is his ability to maintain adequate oxygen saturation. The oxygen saturation values from the graded exercise test, the 6-min walk, and hands-overhead tasks indicate that the patient should

continued

be prescribed supplemental oxygen for use when exercising. In this case, the clinical exercise specialist can serve as patient advocate by providing the primary care physician or pulmonologist with documentation that supports the need for supplemental oxygen. During exercise, oxygen flow rate should be adjusted to maintain a minimum oxygen saturation of 90% or greater.

The lack of ECG changes suggestive of myocardial ischemia during the treadmill test does not preclude the presence of coronary artery disease. Coronary artery disease is common in patients with COPD, and the diagnostic sensitivity of treadmill testing is improved if the patient can attain a maximal or near-maximal level of exertion. In this case, the patient achieves a heart rate corresponding to approximately 83% of predicted. Signs and symptoms of myocardial ischemia should be carefully monitored during exercise training in this population.

The exercise prescription for this patient includes the following components:

— Aerobic training through walking to improve functional capacity, perception of dyspnea, and ability to perform activities of daily living

— Upper body strength training exercises with dumbbells (biceps curl, triceps extension, shoulder flexion, shoulder abduction, and shoulder shrugs) to increase muscular strength and lean body mass

— Stretching exercises three times weekly, performed after walking to improve joint range of motion and mobility

— Frequency of three times weekly in supervised setting to maximize training effects, minimize fatigue and risk of injury, and maximize compliance

— Intensity of aerobic training exercise at dyspnea rating of 3 to 5 on Borg dyspnea scale and intensity of strength training at two sets of each exercise with a weight that allows 12 to 15 repetitions of the movement to maximize training effects, minimize risk of untoward cardiovascular/pulmonary events, and maximize compliance

— Duration of 30 min per session (interval training may be required, especially early in the training program) to maximize training effects, minimize fatigue and risk of injury, and maximize compliance

Case Study 19   Discussion Questions

1. How would the results of this patient's graded exercise test be expected to differ from those of a healthy age-matched nonsmoker?

2. What improvements can be expected in the graded exercise test and the other outcome measures as a result of this patient participating in a program of exercise rehabilitation using the previously described exercise prescription?

3. How are the results of this patient's pulmonary function tests different from that of a healthy age-matched nonsmoker?

4. What physiological factors would account for these pulmonary function test differences?

5. Would involvement in an exercise program result in improvements in these pulmonary function tests? Why or why not?

Glossary

α₁-antitrypsin (AAT)—Protein produced in the liver and found in the lungs that inhibits neutrophil elastase.

α₁-antitrypsin deficiency—A genetic disorder characterized by abnormally low levels of α₁-antitrypsin, thereby predisposing an individual to emphysema.

bronchi—Large airways of the lungs.

bronchioles—Small airways of the lungs.

bronchodilator—A drug that relaxes the smooth muscles surrounding the bronchi and bronchioles.

chronic bronchitis—Disease characterized by the presence of a productive cough most days during three consecutive months in each of two successive years.

chronic obstructive pulmonary disease (COPD)—Presence of airflow obstruction that is attributable to either chronic bronchitis or emphysema.

dyspnea—Shortness of breath.

elastin—Structural protein found in the walls of the alveoli.

emphysema—Disease characterized by abnormal permanent enlargement of the respiratory bronchioles and the alveoli.

exacerbation—A period of worsening symptoms.

forced expiratory volume in 1 s (FEV₁)—The maximum amount of air that can be exhaled in 1 s; may be expressed as an absolute value, a percentage of the forced vital capacity, or a percentage of a predicted value.

forced vital capacity (FVC)—The maximum amount of air that can be exhaled after a maximal inspiration.

hypercapnia—An increased arterial carbon dioxide content.

hyperinflation—An overinflated lung resulting in a greater functional residual capacity and total lung capacity.

hyperpnea—More rapid and deeper breathing than what is normal.

hypoxemia—A decreased arterial oxygen content.

hypoxia—A state of oxygen deficiency.

inspiratory resistive loading—The act of inspiring air against a resistance greater than normal.

inspiratory threshold loading—The act of inspiring after attaining and proceeding at a predetermined inspiratory pressure (threshold point).

isocapnia—Normal arterial carbon dioxide levels.

pack years—Number of packs of cigarettes smoked per day multiplied by the number of years smoked; for example, if a person smoked two packs a day for 20 years, she or he would have a 40 pack year history.

parenchyma—The essential or primary tissue of the lungs.

polycythemia—An abnormally elevated level of red cells in the blood.

respiratory failure—Failure of the respiratory system to keep gas exchange at an acceptable level.

roentgenogram—A photograph made with X-rays.

spirometry—The measurement of respiratory gases using a spirometer.

ventilatory muscle training—Specific exercises that are used to increase respiratory muscular strength.

References

1. Abraham, A.S., R.B. Cole, and J.M. Bishop. Reversal of pulmonary hypertension by prolonged oxygen administration to patients with chronic bronchitis. *Circ. Res.* 23: 147-157, 1968.

2. Abraham, A.S., R.B. Cole, I.D. Green, R.B. Hedworth-Whitty, S.W. Clarke, and J.M. Bishop. Factors contributing to the reversible pulmonary hypertension of patients with acute respiratory failure studies by serial observations during recovery. *Circ. Res.* 24: 51-60, 1969.

3. Abraham, A.S., R.B. Hedworth-Whitty, and J.M. Bishop. Effects of acute hypoxia and hypervolaemia singly and together, upon the pulmonary circulation in patients with chronic bronchitis. *Clin. Sci.* 33: 371-380, 1967.

4. Adams, L., N. Chronos, R. Lane, and A. Guz. The measurement of breathlessness induced in normal subjects: Validity of two scaling techniques. *Clin. Sci.* 69: 7-16, 1985.

5. Agency for Health Care Policy and Research. Clinical Classifications for Health Policy Research: Hospital Inpatient Statistics, 1995. HCUP-3 Research Note. 1999, 16-17.

6. Albert, R.K., T.R. Martin, and S.W. Lewis. Controlled clinical trial of methylprednisolone in patients with chronic bronchitis and acute respiratory insufficiency. *Ann. Intern. Med.* 92: 753-758, 1980.

7. Ambrosino, N., P.L. Paggiaro, M. Macchi, M. Filieri, G. Toma, F.A. Lombardi, F.D. Cesta, A. Parlanti, A.M. Loi, and L. Baschieri. A study of short term effect of rehabilitative therapy in chronic obstructive pulmonary disease. *Respiration* 41: 40-44, 1981.

8. American Association of Cardiovascular and Pulmonary Rehabilitation. Guidelines for Pulmonary Rehabilitation Programs, 2nd ed. Champaign, IL, Human Kinetics. 1998.

9. American College of Chest Physicians/American Association of Cardiovascular and Pulmonary Rehabilitation Pulmonary Rehabilitation Guidelines Panel. Pulmonary rehabilitation: joint ACCP/AACVPR evidence-based guidelines. ACCP/AACVPR Pulmonary Rehabilitation Guidelines Panel. *Chest* 112: 1363-1396, 1997.

10. American College of Sports Medicine. The recommended quantity and quality of exercise for developing and maintaining cardiorespiratory and muscular fitness in healthy adults. *Med. Sci. Sports Exerc.* 22: 265-274, 1990.

11. American College of Sports Medicine. ACSM's Guidelines for Exercise Testing and Prescription, 5th ed. Baltimore, Williams & Wilkins. 1995.

12. American Thoracic Society. Evaluation of impairment/disability secondary to respiratory disorders. *Am. Rev. Respir. Dis.* 133: 1205-1209, 1986.

13. American Thoracic Society. Standards for the diagnosis and care of patients with chronic obstructive pulmonary disease (COPD) and asthma. *Am. Rev. Respir. Dis.* 136: 225-244, 1987.

14. American Thoracic Society. Standards for diagnosis and care of patients with chronic obstructive pulmonary disease. *Am. J. Respir. Crit. Care. Med.* 152: S77-S152, 1995.

15. American Thoracic Society and European Respiratory Society. Skeletal muscle dysfunction in chronic obstructive pulmonary disease. *Am. J. Respir. Crit. Care Med.* 159: S1-40, 1999.

16. Atkins, C.J., R.M. Kaplan, R.M. Timms, S. Reinsch, and K. Lofback. Behavioral exercise programs in the management of chronic obstructive pulmonary disease. *J. Consult. Clin. Psychol.* 52: 591-603, 1984.

17. Badgett, R.G., D.J. Tanaka, D.K. Hunt, M.J. Jelley, L.E. Feinberg, J.F. Steiner, and T.L. Petty. Can moderate chronic obstructive pulmonary disease be diagnosed by historical and physical findings alone? *Am. J. Med.* 94: 188-196, 1993.

18. Belman, M.J. Exercise in chronic obstructive pulmonary disease. *Clin. Chest Med.* 7: 585-597, 1986.

19. Belman, M.J., and R. Shadmehr. Targeted resistive ventilatory muscle training in chronic obstructive pulmonary disease. *J. Appl. Physiol.* 65: 2726-2735, 1988.

20. Bernard, S., P. LeBlanc, F. Whittom, G. Carrier, J. Jobin, R. Belleau, and F. Maltais. Peripheral muscle weakness in patients with chronic obstructive pulmonary disease. *Am. J. Respir. Crit. Care Med.* 158: 629-634, 1998.

21. Bernard, S., F. Whittom, P. LeBlanc, J. Jobin, R. Belleau, C. Berube, G. Carrier, and F. Maltais. Aerobic and strength training in patients with chronic obstructive pulmonary disease. *Am. J. Respir. Crit. Care Med.* 159: 896-901, 1999.

22. Berry, M.J., N.E. Adair, K.S. Sevensky, A. Quinby, and H.M. Lever. Inspiratory muscle training and whole-body reconditioning in chronic obstructive pulmonary disease. *Am. J. Respir. Crit. Care Med.* 153: 1812-1816, 1996.

23. Berry, M.J., W.J. Rejeski, N.E. Adair, and D. Zaccaro. Exercise rehabilitation and chronic obstructive pulmonary disease stage. *Am. J. Respir. Crit. Care Med.* 160: 1248-1253, 1999.

24. Bjerre-Jepsen, K., N.H. Secher, and A. Kok-Jensen. Inspiratory resistance training in severe chronic obstructive pulmonary disease. *Eur. J. Respir. Dis.* 62: 405-411, 1981.

25. Booker, H.A. Exercise training and breathing control in patients with chronic airflow limitation. *Physiotherapy* 70: 258-260, 1984.

26. Borg, G.A. Psychophysical bases of perceived exertion. *Med. Sci. Sports Exerc.* 14: 377-381, 1982.

27. Bowen, J.B., J.J. Votto, R.S. Thrall, M.C. Haggerty, R. Stockdale-Woolley, T. Bandyopadhyay, and R.L. ZuWallack. Functional status and survival following pulmonary rehabilitation. *Chest* 118: 697-703, 2000.

28. Bradley, B.L., A.E. Garner, D. Billiu, J.M. Mestas, and J. Forman. Oxygen-assisted exercise in chronic obstructive lung disease. The effect on exercise capacity and arterial blood gas tensions. *Am. Rev. Respir. Dis.* 118: 239-243, 1978.

29. Brantly, M., T. Nukiwa, and R.G. Crystal. Molecular basis of alpha-1-antitrypsin deficiency. *Am. J. Med.* 84: 13-31, 1988.

30. Braun, S.R., N.L. Keim, R.M. Dixon, P. Clagnaz, A. Anderegg, and E.S. Shrago. The prevalence and determinants of nutritional changes in chronic obstructive pulmonary disease. *Chest* 86: 558-563, 1984.

31. Burrows, B., J.W. Bloom, G.A. Traver, and M.G. Cline. The course and prognosis of different forms of chronic airways obstruction in a sample from the general population. *N. Engl. J. Med.* 317: 1309-1314, 1987.

32. Busch, A.J., and J.D. McClements. Effects of a supervised home exercise program on patients with severe chronic obstructive pulmonary disease. *Phys. Ther.* 68: 469-474, 1988.

33. Cambach, W., R.V. Chadwick-Straver, R.C. Wagenaar, K.A. van, and H.C. Kemper. The effects of a community-based pulmonary rehabilitation programme on exercise tolerance and quality of life: A randomized controlled trial. *Eur. Respir. J.* 10: 104-113, 1997.

34. Camilli, A.E., B. Burrows, R.J. Knudson, S.K. Lyle, and M.D. Lebowitz. Longitudinal changes in forced expiratory volume in one second in adults. Effects of smoking and smoking cessation. *Am. Rev. Respir. Dis.* 135: 794-799, 1987.

35. Carter, R., B. Nicotra, W. Blevins, and D. Holiday. Altered exercise gas exchange and cardiac function in patients with mild chronic obstructive pulmonary disease. *Chest* 103: 745-750, 1993.

36. Casaburi, R. Skeletal muscle function in COPD. *Chest* 117: 267S-271S, 2000.

37. Casaburi, R., A. Patessio, F. Ioli, S. Zanaboni, C.F. Donner, and K. Wasserman. Reductions in exercise lactic acidosis and ventilation as a result of exercise training in patients with obstructive lung disease. *Am. Rev. Respir. Dis.* 143: 9-18, 1991.

38. Celli, B.R., J. Rassulo, and B.J. Make. Dyssynchronous breathing during arm but not leg exercise in patients with chronic airflow obstruction. *N. Engl. J. Med.* 314: 1485-1490, 1986.

39. Chen, H., R. Dukes, and B.J. Martin. Inspiratory muscle training in patients with chronic obstructive pulmonary disease. *Am. Rev. Respir. Dis.* 131: 251-255, 1985.

40. Clark, C.J., L.M. Cochrane, E. Mackay, and B. Paton. Skeletal muscle strength and endurance in patients with mild COPD and the effects of weight training. *Eur. Respir. J.* 15: 92-97, 2000.

41. Cotes, J.E., and J.C. Gilson. Effect of oxygen on exercise ability in chronic respiratory insufficiency. *Lancet* 1: 872-876, 1956.

42. Couser, J.I., F.J. Martinez, and B. Celli. Pulmonary rehabilitation that includes arm exercise reduces metabolic and ventilatory requirements for simple arm elevation. *Chest* 103: 37-41, 1993.

43. Crystal, R.G. The alpha 1-antitrypsin gene and its deficiency states. *Trends. Genet.* 5: 411-417, 1989.

44. Danon, J., W.S. Druz, N.B. Goldberg, and J.T. Sharp. Function of the isolated paced diaphragm and the cervical accessory muscles in C1 quadriplegics. *Am. Rev. Respir. Dis.* 119: 909-919, 1979.

45. Dean, N.C., J.K. Brown, R.B. Himelman, J.J. Doherty, W.M. Gold, and M.S. Stulbarg. Oxygen may improve dyspnea and endurance in patients with chronic obstructive pulmonary disease and only mild hypoxemia. *Am. Rev. Respir. Dis.* 146: 941-945, 1992.

46. Debigare, R., F. Maltais, M. Mallet, R. Casaburi, and P. LeBlanc. Influence of work rate incremental rate on the exercise responses in patients with COPD. *Med. Sci. Sports Exerc.* 32: 1365-1368, 2000.

47. DeCramer, M., B.V. de, and R. Dom. Functional and histologic picture of steroid-induced myopathy in chronic obstructive pulmonary disease. *Am. J. Respir. Crit. Care Med.* 153: 1958-1964, 1996.

48. DeCramer, M., L.M. Lacquet, R. Fagard, and P. Rogiers. Corticosteroids contribute to muscle weakness in chronic airflow obstruction. *Am. J. Respir. Crit. Care Med.* 150: 11-16, 1994.

49. Degre, S., R. Sergysels, R. Messin, P. Vandermoten, P. Salhadin, H. Denolin, and A. De Coster. Hemodynamic responses to physical training in patients with chronic lung disease. *Am. Rev. Respir. Dis.* 110: 395-402, 1974.

50. Dekhuijzen, P.N., H.T. Folgering, and H.C. Van. Target-flow inspiratory muscle training during pulmonary rehabilitation in patients with COPD. *Chest* 99: 128-133, 1991.

51. Dillard, T.A. Ventilatory limitation of exercise. Prediction in COPD. *Chest* 92: 195-196, 1987.

52. Druz, W.S., and J.T. Sharp. Activity of respiratory muscles in upright and recumbent humans. *J. Appl. Physiol.* 51: 1552-1561, 1981.

53. Engelen, M.P., A.M. Schols, W.C. Baken, G.J. Wesseling, and E.F. Wouters. Nutritional depletion in relation to respiratory and peripheral skeletal muscle function in out-patients with COPD. *Eur. Respir. J.* 7: 1793-1797, 1994.

54. Engelen, M.P., A.M. Schols, J.D. Does, and E.F. Wouters. Skeletal muscle weakness is associated with wasting of extremity fat-free mass but not with airflow obstruction in patients with chronic obstructive pulmonary disease. *Am. J. Clin. Nutr.* 71: 733-738, 2000.

55. Epstein, S.K., and B.R. Celli. Cardiopulmonary exercise testing in patients with chronic obstructive pulmonary disease. *Cleve. Clin. J. Med.* 60: 119-128, 1993.

56. Evans, W.J. Exercise training guidelines for the elderly. *Med. Sci. Sports Exerc.* 31: 12-17, 1999.

57. Falk, P., A.M. Eriksen, K. Kolliker, and J.B. Andersen. Relieving dyspnea with an inexpensive and simple method in patients with severe chronic airflow limitation. *Eur. J. Respir. Dis.* 66: 181-186, 1985.

58. Farkas, G.A., and C. Roussos. Adaptability of the hamster diaphragm to exercise and/or emphysema. *J. Appl. Physiol.* 53: 1263-1272, 1982.

59. Faryniarz, K., and D.A. Mahler. Writing an exercise prescription for patients with COPD. *J. Respir. Dis.* 11: 638-644, 1990.

60. Feigenbaum, M.S., and M.L. Pollock. Prescription of resistance training for health and disease. *Med. Sci. Sports Exerc.* 31: 38-45, 1999.

61. Ferrer, M., J. Alonso, J. Morera, R.M. Marrades, A. Khalaf, M.C. Aguar, V. Plaza, L. Prieto, and J.M. Anto. Chronic obstructive pulmonary disease stage and health-related quality of life. *Ann. Intern. Med.* 127: 1072-1079, 1997.

62. Fiaccadori, E., C.S. Del, E. Coffrini, P. Vitali, C. Antonucci, G. Cacciani, I. Mazzola, and A. Guariglia. Hypercapnic-hypoxemic chronic obstructive pulmonary disease (COPD): Influence of severity of COPD on nutritional status. *Am. J. Clin. Nutr.* 48: 680-685, 1988.

63. Fishman, A.P. Pulmonary rehabilitation research. *Am. J. Respir. Crit. Care Med.* 149: 825-833, 1994.

64. Fletcher, G.F., G. Balady, V.F. Froelicher, L.H. Hartley, W.L. Haskell, and M.L. Pollock. Exercise standards. A statement for healthcare professionals from the American Heart Association. *Circulation* 91: 580-615, 1995.

65. Gibbons, L.W., T.L. Mitchell, and V. Gonzalez. The safety of exercise testing. *Prim. Care* 21: 611-629, 1994.

66. Goldstein, R., J. De Rosie, S. Long, T. Dolmage, and M.A. Avendano. Applicability of a threshold loading device for inspiratory muscle testing and training in patients with COPD. *Chest* 96: 564-571, 1989.

67. Goldstein, R.S., E.H. Gort, D. Stubbing, and G.H. Guyatt. Randomised controlled trial of respiratory rehabilitation. *Lancet* 344: 1394-1397, 1994.

68. Gordon, A.M., A.F. Huxley, and F.J. Julian. The variation in isometric tension with sarcomere length in vertebrate muscle fibres. *J. Physiol. (Lond.)* 184: 170-192, 1966.

69. Gosselink, R., T. Troosters, and M. DeCramer. Peripheral muscle weakness contributes to exercise limitation in COPD. *Am. J. Respir. Crit. Care Med.* 153: 976-980, 1996.

70. Greenberg, S.B., M. Allen, J. Wilson, and R.L. Atmar. Respiratory viral infections in adults with and without chronic obstructive pulmonary disease. *Am. J. Respir. Crit. Care Med.* 162: 167-173, 2000.

71. Griffin, S.E., R.A. Robergs, and V.H. Heyward. Blood pressure measurement during exercise: A review. *Med. Sci. Sports Exerc.* 29: 149-159, 1997.

72. Guyatt, G., J. Keller, J. Singer, S. Halcrow, and M. Newhouse. Controlled trial of respiratory muscle training in chronic airflow limitation. *Thorax* 47: 598-602, 1992.

73. Hamilton, A.L., K.J. Killian, E. Summers, and N.L. Jones. Muscle strength, symptom intensity, and exercise

capacity in patients with cardiorespiratory disorders. *Am. J. Respir. Crit. Care Med.* 152: 2021-2031, 1995.

74. Harver, A., D.A. Mahler, and J.A. Daubenspeck. Targeted inspiratory muscle training improves respiratory muscle function and reduces dyspnea in patients with chronic obstructive pulmonary disease. *Ann. Intern. Med.* 111: 117-124, 1989.

75. Heaton, R.K., I. Grant, A.J. McSweeny, K.M. Adams, and T.L. Petty. Psychologic effects of continuous and nocturnal oxygen therapy in hypoxemic chronic obstructive pulmonary disease. *Arch. Intern. Med.* 143: 1941-1947, 1983.

76. Henke, K.G., M. Sharratt, D. Pegelow, and J.A. Dempsey. Regulation of end-expiratory lung volume during exercise. *J. Appl. Physiol.* 64: 135-146, 1988.

77. Higgins, M.W., and T. Thom. Incidence, prevalence, and mortality: Intra- and inter-country differences. In M.J. Hensley and M.J. Saunders, eds., Clinical Epidemiology of Chronic Obstructive Pulmonary Disease. New York, Marcel Dekker. 1990, 23-43.

78. Hildebrand, I.L., C. Sylven, M. Esbjornsson, K. Hellstrom, and E. Jansson. Does chronic hypoxaemia induce transformations of fibre types? *Acta Physiol. Scand.* 141: 435-439, 1991.

79. Holford, N., P. Black, R. Couch, J. Kennedy, and R. Briant. Theophylline target concentration in severe airways obstruction—10 or 20 mg/L? A randomised concentration-controlled trial. *Clin. Pharmacokinet.* 25: 495-505, 1993.

80. Horowitz, M.B., B. Littenberg, and D.A. Mahler. Dyspnea ratings for prescribing exercise intensity in patients with COPD. *Chest* 109: 1169-1175, 1996.

81. Hughes, J.R., M.G. Goldstein, R.D. Hurt, and S. Shiffman. Recent advances in the pharmacotherapy of smoking. *J. Am. Med. Assoc.* 281: 72-76, 1999.

82. Hughes, R.L., H. Katz, V. Sahgal, J.A. Campbell, R. Hartz, and T.W. Shields. Fiber size and energy metabolites in five separate muscles from patients with chronic obstructive lung diseases. *Respiration* 44: 321-328, 1983.

83. Hunter, A.M., M.A. Carey, and H.W. Larsh. The nutritional status of patients with chronic obstructive pulmonary disease. *Am. Rev. Respir. Dis.* 124: 376-381, 1981.

84. Jakobsson, P., L. Jorfeldt, and A. Brundin. Skeletal muscle metabolites and fibre types in patients with advanced chronic obstructive pulmonary disease (COPD), with and without chronic respiratory failure. *Eur. Respir. J.* 3: 192-196, 1990.

85. Jones, D.T., R.J. Thomson, and M.R. Sears. Physical exercise and resistive breathing training in severe chronic airways obstruction: Are they effective? *Eur. J. Respir. Dis.* 67: 159-165, 1985.

86. Jones, N.L., G. Jones, and R.H. Edwards. Exercise tolerance in chronic airway obstruction. *Am. Rev. Respir. Dis.* 103: 477-491, 1971.

87. Kelsen, S.G., M. Ference, and S. Kapoor. Effects of prolonged undernutrition on structure and function of the diaphragm. *J. Appl. Physiol.* 58: 1354-1359, 1985.

88. Killian, K.J., P. LeBlanc, D.H. Martin, E. Summers, N.L. Jones, and E.J. Campbell. Exercise capacity and ventilatory, circulatory, and symptom limitation in patients with chronic airflow limitation. *Am. Rev. Respir. Dis.* 146: 935-940, 1992.

89. Knudson, R.J., R.C. Slatin, M.D. Lebowitz, and B. Burrows. The maximal expiratory flow-volume curve. Normal standards, variability, and effects of age. *Am. Rev. Respir. Dis.* 113: 587-600, 1976.

90. Lacasse, Y., G.H. Guyatt, and R.S. Goldstein. The components of a respiratory rehabilitation program: A systematic overview. *Chest* 111: 1077-1088, 1997.

91. Lacasse, Y., E. Wong, G.H. Guyatt, D. King, D.J. Cook, and R.S. Goldstein. Meta-analysis of respiratory rehabilitation in chronic obstructive pulmonary disease. *Lancet* 348: 1115-1119, 1996.

92. Lake, F.R., K. Henderson, T. Briffa, J. Oppenshaw, and A.W. Musk. Upper-limb and lower-limb exercise training in patients with chronic airflow obstruction. *Chest* 97: 1077-1082, 1990.

93. Lange, P., S. Groth, G.J. Nyboe, J. Mortensen, M. Appleyard, G. Jensen, and P. Schnohr. Effects of smoking and changes in smoking habits on the decline of FEV_1. *Eur. Respir. J.* 2: 811-816, 1989.

94. Larson, J.L., M.K. Covey, S.E. Wirtz, J.K. Berry, C.G. Alex, W.E. Langbein, and L. Edwards. Cycle ergometer and inspiratory muscle training in chronic obstructive pulmonary disease. *Am. J. Respir. Crit. Care Med.* 160: 500-507, 1999.

95. Larson, J.L., M.J. Kim, J.T. Sharp, and D.A. Larson. Inspiratory muscle training with a pressure threshold breathing device in patients with chronic obstructive pulmonary disease. *Am. Rev. Respir. Dis.* 138: 689-696, 1988.

96. Leith, D.E., and M. Bradley. Ventilatory muscle strength and endurance training. *J. Appl. Physiol.* 41: 508-516, 1976.

97. Levine, B.E., D.B. Bigelow, R.D. Hamstra, H.J. Beckwitt, R.S. Mitchell, L.M. Nett, T.A. Stephen, and T.L. Petty. The role of long-term continuous oxygen administration in patients with chronic airway obstruction with hypoxemia. *Ann. Intern. Med.* 66: 639-650, 1967.

98. Lewis, M.I., G.C. Sieck, M. Fournier, and M.J. Belman. Effect of nutritional deprivation on diaphragm contractility and muscle fiber size. *J. Appl. Physiol.* 60: 596-603, 1986.

99. Lisboa, C., V. Munoz, T. Beroiza, A. Leiva, and E. Cruz. Inspiratory muscle training in chronic airflow limitation: Comparison of two different training loads with a threshold device. *Eur. Respir. J.* 7: 1266-1274, 1994.

100. Lisboa, C., C. Villafranca, A. Leiva, E. Cruz, J. Pertuze, and G. Borzone. Inspiratory muscle training in chronic

airflow limitation: Effect on exercise performance. *Eur. Respir. J.* 10: 537-542, 1997.

101. Magnussen, H., K. Richter, and C. Taube. Are chronic obstructive pulmonary disease (COPD) and asthma different diseases? *Clin. Exp. Allergy* 28 Suppl 5: 187-194, 1998.

102. Mahler, D.A., R.A. Rosiello, A. Harver, T. Lentine, J.F. McGovern, and J.A. Daubenspeck. Comparison of clinical dyspnea ratings and psychophysical measurements of respiratory sensation in obstructive airway disease. *Am. Rev. Respir. Dis.* 135: 1229-1233, 1987.

103. Mahler, D.A., D.H. Weinberg, C.K. Wells, and A.R. Feinstein. The measurement of dyspnea. Contents, interobserver agreement, and physiologic correlates of two new clinical indexes. *Chest* 85: 751-758, 1984.

104. Maltais, F., P. LeBlanc, F. Whittom, C. Simard, K. Marquis, M. Belanger, M.J. Breton, and J. Jobin. Oxidative enzyme activities of the vastus lateralis muscle and the functional status in patients with COPD. *Thorax* 55: 848-853, 2000.

105. Maltais, F., A.A. Simard, C. Simard, J. Jobin, P. Desgagnes, and P. LeBlanc. Oxidative capacity of the skeletal muscle and lactic acid kinetics during exercise in normal subjects and in patients with COPD. *Am. J. Respir. Crit. Care Med.* 153: 288-293, 1996.

106. Mannino, D.M., C. Brown, and G.A. Giovino. Obstructive lung disease deaths in the United States from 1979 through 1993. An analysis using multiple-cause mortality data. *Am. J. Respir. Crit. Care Med.* 156: 814-818, 1997.

107. Marciniuk, D.D., and C.G. Gallagher. Clinical exercise testing in chronic airflow limitation. *Med. Clin. North Am.* 80: 565-587, 1996.

108. Martinez, F.J., P.D. Vogel, D.N. Dupont, I. Stanopoulos, A. Gray, and J.F. Beamis. Supported arm exercise vs unsupported arm exercise in the rehabilitation of patients with severe chronic airflow obstruction. *Chest* 103: 1397-1402, 1993.

109. McCully, K.K., and J.A. Faulkner. Length-tension relationship of mammalian diaphragm muscles. *J. Appl. Physiol.* 54: 1681-1686, 1983.

110. McGavin, C.R., S.P. Gupta, E.L. Lloyd, and G.J.R. McHardy. Physical rehabilitation for the chronic bronchitis: Results of a controlled trial of exercises in the home. *Thorax* 32: 307-311, 1977.

111. McKeon, J.L., J. Turner, C. Kelly, A. Dent, and P.V. Zimmerman. The effect of inspiratory resistive training on exercise capacity in optimally treated patients with severe chronic airflow limitation. *Aust. N. Z. J. Med.* 16: 648-652, 1986.

112. Medical Research Council Working Party. Long term domiciliary oxygen therapy in chronic hypoxic cor pulmonale complicating chronic bronchitis and emphysema. *Lancet* 1: 681-686, 1981.

113. Mendella, L.A., J. Manfreda, C.P. Warren, and N.R. Anthonisen. Steroid response in stable chronic obstructive pulmonary disease. *Ann. Intern. Med.* 96: 17-21, 1982.

114. Miller, W.F., and H.F. Taylor. Exercise training in the rehabilitation of patients with severe respiratory insufficiency due to pulmonary emphysema. *South Med. J.* 55: 1216-1221, 1962.

115. Murray, R.P., N.R. Anthonisen, J.E. Connett, R.A. Wise, P.G. Lindgren, P.G. Greene, and M.A. Nides. Effects of multiple attempts to quit smoking and relapses to smoking on pulmonary function. *J. Clin. Epidemiol.* 51: 1317-1326, 1998.

116. National Emphysema Treatment Trial Group. Rationale and design of the national emphysema treatment trial: A prospective randomized trial of lung volume reduction surgery. *Chest* 116: 1750-1761, 1999.

117. National Heart, Lung and Blood Institute. National Institutes of Health. *Morbidity and Mortality: 1998 Chartbook on Cardiovascular, Lung and Blood Diseases*. Bethesda, MD, National Institutes of Health. 1998, 1-128.

118. Nickerson, B.G., C. Sarkisian, and K. Tremper. Bias and precision of pulse oximeters and arterial oximeters. *Chest* 93: 515-517, 1988.

119. Nocturnal Oxygen Therapy Trial Group. Continuous or nocturnal oxygen therapy in hypoxemic chronic obstructive lung disease: A clinical trial. *Ann. Intern. Med.* 93: 391-398, 1980.

120. Noseda, A., J.P. Carpiaux, W. Vandeput, T. Prigogine, and J. Schmerber. Resistive inspiratory muscle training and exercise performance in COPD patients. A comparative study with conventional breathing retraining. *Bull. Eur. Physiopathol. Respir.* 23: 457-463, 1987.

121. Oelberg, D.A., B.D. Medoff, D.H. Markowitz, P.P. Pappagianopoulos, L.C. Ginns, and D.M. Systrom. Systemic oxygen extraction during incremental exercise in patients with severe chronic obstructive pulmonary disease. *Eur. J. Appl. Physiol.* 78: 201-207, 1998.

122. Pardy, R.L., R.N. Rivington, P.J. Despas, and P.T. Macklem. Inspiratory muscle training compared with physiotherapy in patients with chronic airflow limitation. *Am. Rev. Respir. Dis.* 123: 421-425, 1981.

123. Pierce, A.K., P.N. Paez, and W.F. Miller. Exercise training with the aid of a portable oxygen supply in patients with emphysema. *Am. Rev. Respir. Dis.* 91: 653-659, 1965.

124. Polkey, M.I., D. Kyroussis, C.H. Hamnegard, G.H. Mills, M. Green, and J. Moxham. Diaphragm strength in chronic obstructive pulmonary disease. *Am. J. Respir. Crit. Care Med.* 154: 1310-1317, 1996.

125. Preusser, B.A., M.L. Winningham, and T.L. Clanton. High- vs low-intensity inspiratory muscle interval training in patients with COPD. *Chest* 106: 110-117, 1994.

126. Punzal, P.A., A.L. Ries, R.M. Kaplan, and L.M. Prewitt. Maximum intensity exercise training in patients with chronic obstructive pulmonary disease. *Chest* 100: 618-623, 1991.

127. Reardon, J., E. Awad, E. Normandin, F. Vale, B. Clark, and R.L. ZuWallack. The effect of comprehensive outpatient pulmonary rehabilitation on dyspnea. *Chest* 105: 1046-1052, 1994.

128. Reid, D.C., J. Bowden, and P. Lynne-Davies. Role of selected muscles of respiration as influenced by posture and tidal volume. *Chest* 70: 636-640, 1976.

129. Ries, A.L., B. Ellis, and R.W. Hawkins. Upper extremity exercise training in chronic obstructive pulmonary disease. *Chest* 93: 688-692, 1988.

130. Ries, A.L., R.M. Kaplan, T.M. Limberg, and L.M. Prewitt. Effects of pulmonary rehabilitation on physiologic and psychosocial outcomes in patients with chronic obstructive pulmonary disease. *Ann. Intern. Med.* 122: 823-832, 1995.

131. Rodrigo, C., and G. Rodrigo. Treatment of acute asthma. Lack of therapeutic benefit and increase of the toxicity from aminophylline given in addition to high doses of salbutamol delivered by metered-dose inhaler with a spacer. *Chest* 106: 1071-1076, 1994.

132. Rollier, H., A. Bisschop, G. Gayan-Ramirez, R. Gosselink, and M. DeCramer. Low load inspiratory muscle training increases diaphragmatic fiber dimensions in rats. *Am. J. Respir. Crit Care Med.* 157: 833-839, 1998.

133. Rooyackers, J.M., P.N. Dekhuijzen, H.C. Van, and H.T. Folgering. Training with supplemental oxygen in patients with COPD and hypoxaemia at peak exercise. *Eur. Respir. J.* 10: 1278-1284, 1997.

134. Schols, A.M., P.B. Soeters, A.M. Dingemans, R. Mostert, P.J. Frantzen, and E.F. Wouters. Prevalence and characteristics of nutritional depletion in patients with stable COPD eligible for pulmonary rehabilitation. *Am. Rev. Respir. Dis.* 147: 1151-1156, 1993.

135. Schwaiblmair, M., T. Beinert, M. Seemann, J. Behr, M. Reiser, and C. Vogelmeier. Relations between cardiopulmonary exercise testing and quantitative high-resolution computed tomography associated in patients with alpha-1-antitrypsin deficiency. *Eur. J. Med. Res.* 3: 527-532, 1998.

136. Serres, I., V. Gautier, A. Varray, and C. Prefaut. Impaired skeletal muscle endurance related to physical inactivity and altered lung function in COPD patients. *Chest* 113: 900-905, 1998.

137. Shannon, M. Predictors of major toxicity after theophylline overdose. *Ann. Intern. Med.* 119: 1161-1167, 1993.

138. Sharratt, M.T., K.G. Henke, E.A. Aaron, D.F. Pegelow, and J.A. Dempsey. Exercise-induced changes in functional residual capacity. *Respir. Physiol.* 70: 313-326, 1987.

139. Sherrill, D.L., P. Enright, M. Cline, B. Burrows, and M.D. Lebowitz. Rates of decline in lung function among subjects who restart cigarette smoking. *Chest* 109: 1001-1005, 1996.

140. Sherrill, D.L., M.D. Lebowitz, and B. Burrows. Epidemiology of chronic obstructive pulmonary disease. *Clin. Chest Med.* 11: 375-387, 1990.

141. Shuey, C.B.J., A.K. Pierce, and R.L.J. Johnson. An evaluation of exercise tests in chronic obstructive lung disease. *J. Appl. Physiol.* 27: 256-261, 1969.

142. Similowski, T., S. Yan, A.P. Gauthier, P.T. Macklem, and F. Bellemare. Contractile properties of the human diaphragm during chronic hyperinflation. *N. Engl. J. Med.* 325: 917-923, 1991.

143. Simmons, D.N., M.J. Berry, S.I. Hayes, and S.A. Walschlager. The relationship between %HRpeak and %VO$_2$peak in patients with chronic obstructive pulmonary disease. *Med. Sci. Sports Exerc.* 32: 881-886, 2000.

144. Simpson, K., K. Killian, N. McCartney, D.G. Stubbing, and N.L. Jones. Randomised controlled trial of weight-lifting exercise in patients with chronic airflow limitation. *Thorax* 74: 70-75, 1992.

145. Smith, J., and F. Bellemare. Effect of lung volume on in vivo contraction characteristics of human diaphragm. *J. Appl. Physiol.* 62: 1893-1900, 1987.

146. Snider, G.L. Chronic obstructive pulmonary disease: A definition and implications of structural determinants of airflow obstruction for epidemiology. *Am. Rev. Respir. Dis.* 140: S3-S8, 1989.

147. Snider, G.L., L.J. Faling, and S.I. Rennard. Chronic bronchitis and emphysema. In J.F. Murray and J. A. Nadel, eds., Textbook of Respiratory Medicine. Philadelphia, Saunders. 1994, 1342.

148. Snider, G.L., J. Kleinerman, W.M. Thurlbeck, and Z.K. Bengali. The definition of emphysema. Report of a National Heart, Lung, and Blood Institute, Division of Lung Diseases workshop. *Am. Rev. Respir. Dis.* 132: 182-185, 1985.

149. Strijbos, J.H., D.S. Postma, R.V. Altena, F. Gimeno, and G.H. Koeter. A comparison between an outpatient hospital-based pulmonary rehabilitation program and a home-care pulmonary rehabilitation program in patients with COPD. *Chest* 109: 366-372, 1996.

150. Stubbing, D.G., L.D. Pengelly, J.L. Morse, and N.L. Jones. Pulmonary mechanics during exercise in subjects with chronic airflow obstruction. *J. Appl. Physiol.* 49: 511-515, 1980.

151. Sue, D.Y. Exercise testing in the evaluation of impairment and disability. *Clin. Chest Med.* 15: 369-387, 1994.

152. Sue, D.Y., K. Wasserman, R.B. Moricca, and R. Casaburi. Metabolic acidosis during exercise in patients with chronic obstructive pulmonary disease. Use of the V-slope method for anaerobic threshold determination. *Chest* 94: 931-938, 1988.

153. Sullivan, P., S. Bekir, Z. Jaffar, C. Page, P. Jeffery, and J. Costello. Anti-inflammatory effects of low-dose oral theophylline in atopic asthma. *Lancet* 343: 1006-1008, 1994.

154. Tager, I.B., M.R. Segal, F.E. Speizer, and S.T. Weiss. The natural history of forced expiratory volumes. Effect of cigarette smoking and respiratory symptoms. *Am. Rev. Respir. Dis.* 138: 837-849, 1988.

155. Tangri, S., and C. R. Woolf. The breathing pattern in chronic obstructive lung disease during the performance of some common daily activities. *Chest* 63: 126-127, 1973.

156. Thomason, M.J., and D.P. Strachan. Which spirometric indices best predict subsequent death from chronic obstructive pulmonary disease? *Thorax* 55: 785-788, 2000.

157. Thurlbeck, W.M. Pathophysiology of chronic obstructive pulmonary disease. *Clin. Chest Med.* 11: 389-403, 1990.

158. Thurlbeck, W.M., and N.L. Muller. Emphysema: Definition, imaging, and quantification. *Am. J. Roentgenol.* 163: 1017-1025, 1994.

159. Toshima, M.T., R.M. Kaplan, and A.L. Ries. Experimental evaluation of rehabilitation in chronic obstructive pulmonary disease: Short term effects on exercise endurance and health status. *Health Psychol.* 9: 237-252, 1990.

160. Vallet, G., S. Ahmaidi, I. Serres, C. Fabre, D. Bourgouin, J. Desplan, A. Varray, and C. Prefaut. Comparison of two training programmes in chronic airway limitation patients: Standardized versus individualized protocols. *Eur. Respir. J.* 10: 114-122, 1997.

161. Wanke, T., D. Formanek, H. Lahrmann, H. Brath, M. Wild, C. Wagner, and H. Zwick. Effects of combined inspiratory muscle and cycle ergometer training on exercise performance in patients with COPD. *Eur. Respir. J.* 7: 2205-2211, 1994.

162. Wasserman, K., J.E. Hansen, D.Y. Sue, and B.J. Whipp. Principles of Exercise Testing and Interpretation. Philadelphia, Lea & Febiger. 1986, 1-274.

163. Wedzicha, J.A., J.C. Bestall, R. Garrod, R. Garnham, E.A. Paul, and P.W. Jones. Randomized controlled trial of pulmonary rehabilitation in severe chronic obstructive pulmonary disease patients, stratified with the MRC dyspnoea scale. *Eur. Respir. J.* 12: 363-369, 1998.

164. Wehr, K.L., and R.L.J. Johnson. Maximal oxygen consumption in patients with lung disease. *J. Clin. Invest.* 58: 880-890, 1976.

165. Weiner, P., Y. Azgad, and R. Ganam. Inspiratory muscle training combined with general exercise reconditioning in patients with COPD. *Chest* 102: 1351-1356, 1992.

166. Whittom, F., J. Jobin, P.M. Simard, P. LeBlanc, C. Simard, S. Bernard, R. Belleau, and F. Maltais. Histochemical and morphological characteristics of the vastus lateralis muscle in patients with chronic obstructive pulmonary disease. *Med. Sci. Sports Exerc.* 30: 1467-1474, 1998.

167. Wijkstra, P.J., E.M. TenVergert, R.V. Altena, V. Otten, D.S. Postma, and G.H. Koeter. Long term benefits of rehabilitation at home on quality of life and exercise tolerance in patients with chronic obstructive pulmonary disease. *Thorax* 50: 824-828, 1995.

168. Wilson, D.O., R.M. Rogers, and D. Openbrier. Nutritional aspects of chronic obstructive pulmonary disease. *Clin. Chest Med.* 7: 643-656, 1986.

169. Wilson, D.O., R.M. Rogers, M.H. Sanders, B.E. Pennock, and J.J. Reilly. Nutritional intervention in malnourished patients with emphysema. *Am. Rev. Respir. Dis.* 134: 672-677, 1986.

170. Wilson, D.O., R.M. Rogers, E.C. Wright, and N.R. Anthonisen. Body weight in chronic obstructive pulmonary disease. The National Institutes of Health Intermittent Positive-Pressure Breathing Trial. *Am. Rev. Respir. Dis.* 139: 1435-1438, 1989.

171. Wuyam, B., J.F. Payen, P. Levy, H. Bensaidane, H. Reutenauer, B.J. Le, and A.L. Benabid. Metabolism and aerobic capacity of skeletal muscle in chronic respiratory failure related to chronic obstructive pulmonary disease. *Eur. Respir. J.* 5: 157-162, 1992.

172. Zack, M.B., and A.V. Palange. Oxygen supplemented exercise of ventilatory and nonventilatory muscles in pulmonary rehabilitation. *Chest* 88: 669-674, 1985.

173. Ziment, I. Pharmacologic therapy of obstructive airway disease. *Clin. Chest Med.* 11: 461-486, 1990.

Brian W Carlin, MD
Assistant Professor of Medicine
Drexel University College of Medicine
Philadelphia, PA
Allegheny General Hospital
Pittsburgh, PA

David Seigneur, MS
Director, Cardiopulmonary Rehabilitation
Allegheny General Hospital
Pittsburgh, PA

Asthma

Asthma represents a continuum of a disease process characterized by inflammation of the airway wall. This inflammation characterizes all forms of asthma, whether mild, moderate, or severe in nature. No two patients with asthma are alike, and thus therapeutic interventions for patients with asthma must be individualized.

Asthma is characterized by variable **airflow obstruction** and affects many people worldwide. An operational definition of asthma is a chronic inflammatory disorder of the airways that leads to **airway hyperreactivity,** airflow limitation, and respiratory symptoms including wheezing, coughing, and **dyspnea** occurring particularly at night or in the early morning. These symptoms are usually associated with widespread but variable airflow obstruction that is often reversible either spontaneously or with treatment. The inflammation may also be associated with an increase in the existing bronchial hyperresponsiveness of the airway wall to a variety of stimuli (39).

Scope

Asthma affects at least 17 million Americans (up to 5% of the U.S. population) (11). Incidence rates of up to 3.9% per year are noted, with the highest incidence among younger ages. Most studies suggest that the prevalence of asthma in the world has been increasing over the last several decades by 5% per year (21, 36, 46). Most childhood asthma begins in infancy (before the age of 3), and viral infections are proposed to be a critical component in its development (51). The incidence is apparently higher in some patient populations, with up to 23% of inner-city African Americans noted to have asthma compared with 5% of Caucasians (31). In addition, asthma morbidity in Hispanics in the United States is highest in the northeast (26).

Despite the availability of good medical therapy, the morbidity and mortality rates associated with asthma, both in the United States and worldwide, have increased, particularly within the African American population. The reasons for this increase in morbidity and mortality rates are unclear but likely include variability of the pathophysiology of the disease process itself from one patient to the next, influence of environmental factors on the development and progression of the disease, inability of a patient to effectively access health care, and inability of the patient to comply with the recommended treatment regimen.

Physiology and Pathophysiology

Knowledge about the pathophysiology of asthma has increased significantly over the last decade. Various cellular mediators are responsible for the development of the inflammation associated with asthma (8). The **CD4 lymphocyte** (Th2 subgroup) is currently believed to promote inflammation by the **eosinophils** and **mast cells** (24) with subsequent infiltration of these cells into the airway wall resulting in edema formation (30). The airway lumen is also structurally changed, becoming occluded by a plug composed of mucus (produced secondary to glandular hypersecretion and epithelial cell injury) (8). Abnormal collagen deposition with hypertrophy and hyperplasia of goblet cells, submucosal gland cells, smooth muscle cells, and blood vessel cells also occurs. Dilation of the blood vessels occurs. Subsequent airway remodeling occurs with thickening of the basement membrane attributable to **subepithelial fibrosis** (16, 44). Figure 20.1 shows airway remodeling.

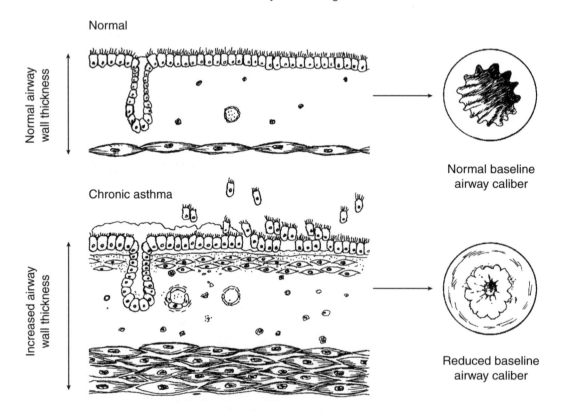

FIGURE 20.1   Illustrations of normal tissue, swelling, and remodeling.

Recent developments have further elucidated the mechanisms responsible for this airway remodeling. During lung development, repair, and inflammation, local production of various inflammatory mediators (e.g., **cytokines** and growth factors) mediates growth control and creates an **epithelial-mesenchymal trophic unit.** In asthma, the bronchial epithelium is highly abnormal, with structural changes involving separation of columnar cells from their basal attachments and functional changes including increased expression and release of proinflammatory cytokines, growth factors, and mediator-generating enzymes. Beneath this damaged subepithelium, an increased number of subepithelial myofibroblasts deposit collagens, causing basement membrane thickening. The extent of the epithelial damage in asthma may be the result of impaired epidermal growth factor receptor–mediated repair. Impaired epithelial repair cooperates with the Th2 environment, which leads to inflammation to alter the communication with the epithelial-mesenchymal trophic unit. This leads to myofibroblast activation, excessive matrix deposition, and production of mediators that propagate and amplify the remodeling responses throughout the airway wall (25). The end effect of this airway remodeling causes chronic scarring of the airway with subsequent irreversible changes within the airway wall. This would eventually cause the patient to have persistent and permanent symptoms.

Significant airway remodeling in the peripheral airways can occur even in those patients with clinically mild asthma. Such patients can be classified as "mild persistent," in terms of symptoms and lung function, yet have substantial remodeling and thus be at risk for complications associated with the ongoing remodeling responses.

Asthma is currently classified into various forms based on clinical and pulmonary function criteria (table 20.1) (39). Most patients with asthma are bothered by symptoms infrequently, although in some, progression to a more severe long-standing illness may occur. In addition, some patients may have airflow obstruction that is fixed and cannot be reversed with **bronchodilator** medication. This fixed obstruction may be the result of airway wall structural alterations (e.g., deposition of collagen beneath the epithelial basement membrane) and may be secondary to the inadequately treated airway inflammation during the earlier stages of the disease process (29). Asthma severity varies not only from one patient to the next but also within the same patient from time to time. Escalation from a mild degree of asthma to a severe **exacerbation** can occur at any time. Death from asthma can occur in patients regardless of the stage of their disease. In fact, one study showed that up to one third of the deaths attributable to asthma were in patients with mild asthma (43).

Clinical Considerations

Asthma is characterized by reversible airway obstruction and airway hyperresponsiveness. It is the most frequent chronic disease of childhood and is often missed as an initial diagnosis of a child who has recurrent episodes of "bronchitis" or "colds." A child who has asthma may have a remission of symptoms in the mid-teens only to have recurrence of symptoms later in life. In some instances, adults may develop asthma (e.g., often following a viral respiratory tract infection). Potential triggers to asthma include those that enhance the inflammatory response (e.g., allergens, viral infections, occupational

TABLE 20.1

Classification of Asthma Severity: Clinical Features Before Treatment

	Days with symptoms	Nights with symptoms	PEF or FEV$_1$
Severe persistent	Continual	Frequent	≤60%
Moderate persistent	Daily	≥5/month	>60% – ≤80%
Mild persistent	3-6/week	3-4/month	≥80%
Mild intermittent	≤2/week	≤2/month	≥80%

Note. PEF = Peak expiratory flow rate; FEV$_1$ = forced expiratory volume in 1 s.

exposures) and those that activate **bronchospasm** (e.g., exercise, irritants, emotions, aspirin) (10). Exposure to such triggers as well as house dust mites, cats, dogs, or cockroaches may also precipitate or potentiate symptoms.

Signs and Symptoms

The symptoms associated with asthma (dyspnea, wheezing, mucus production, and chest tightness) are a result of the pathophysiological (airway inflammation and narrowing) changes noted previously. Variable clinical presentations occur from one patient to the next with approximately 15% of asthmatics failing to appreciate any discomfort even after an acute 20% reduction in their **forced expiratory volume in 1 s (FEV_1)** (45). In some instances, the clinical symptoms may only develop after exercise (exercise-induced bronchospasm) and may be manifest simply as a decline in the exercise tolerance of that person. In addition, the nature and language of symptom perception of asthma and the associated neural pathways may vary from one patient to the next (6). A high index of suspicion in a patient who presents with dyspnea will help lead to the correct diagnosis of asthma. Nocturnal symptoms (awakening with cough, dyspnea, wheezing) or cough should also prompt consideration of the diagnosis.

Various factors are associated with the development of an acute exacerbation of asthma. Respiratory infections can lead to sensitization to aeroallergens, increase in cytokine production, stimulation of sensory nerves, and loss of epithelium-derived relaxing factors. Such reactions exacerbate airway inflammation and enhance airway responsiveness. Exercise and cold air inhalation can also induce the inflammatory response.

Diagnostic and Laboratory Evaluations

Pulmonary function tests assist in the diagnosis of asthma. Episodes of asthma symptoms are usually associated with widespread but variable airflow obstruction that is often reversible either spontaneously or with treatment. Obstruction is defined by **spirometry** as a decrease in the FEV_1 to less than 80% predicted and a decrease in the FEV_1/**forced vital capacity (FVC)** ratio to less than 65%. Reversibility is defined by spirometry as an increase in FEV_1 equal to or greater than 12% and at least 200 ml after using a short-acting inhaled β_2-agonists (e.g., albuterol). Normal or increased lung volumes are noted. One component of the spirometry testing is a plot comparing the rate of airflow noted during exhalation with the volume of the lung at which that par-

ticular flow was noted (referred to as the flow–volume loop). Such a tracing (figure 20.2) can be helpful to differentiate between airway obstruction secondary to asthma (which will often show an improvement in the flow rates following the administration of a bronchodilator) or to emphysema (which will show no improvement in flow rates).

In some instances, normal spirometry may be present and the airway obstruction may only be noted following provocation testing. Nonspecific triggers (e.g., methacholine, histamine) can be used (in an aerosol form) to try to determine if there is any decline in the FEV_1 or FVC following aerosol administration. Such testing, called **bronchoprovocation** (or methacholine) testing, can easily be performed in most pulmonary function laboratories. A greater than 20% decline in FEV_1 following testing is abnormal and likely indicates the presence of asthma.

Other studies can be useful to help substantiate the diagnosis of asthma. The chest roentgenogram may show hyperinflation (increase in the retrosternal airspace, flattening of the diaphragms) of the **lung parenchyma** without **parenchymal infiltrates** and is helpful to rule out other causes for dyspnea (e.g., pneumonia, pneumothorax, congestive heart failure). Other laboratory studies (including the electrocardiogram, blood chemistries, or hematocrit) are not helpful to either confirm or rule out a diagnosis of asthma.

Treatment

The general goals of asthma therapy are noted in practical application 20.1. These goals provide criteria that the clinician and patient should use to evaluate the patient's particular response to therapy. It is necessary to first determine the patient's personal goals of therapy, then share the general goals of asthma therapy with the patient, and finally agree on the goals that will be the foundation for the treatment plan.

Assessment of the severity of the disease is an important step to guide further therapy. Severity can be classified based on the symptoms, nocturnal component, physical activity limitations, and spirometric measures (39). Patients with intermittent asthma have brief (less than twice weekly) symptomatic periods and are asymptomatic between episodes. They have no limitation in physical activity. Patients with mild, persistent asthma have symptoms present more than twice weekly, awaken with asthma symptoms more than two nights per month, and may have a decline in exercise tolerance. Spirometry is

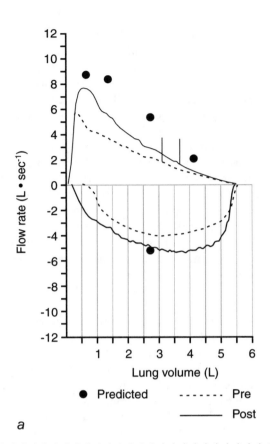

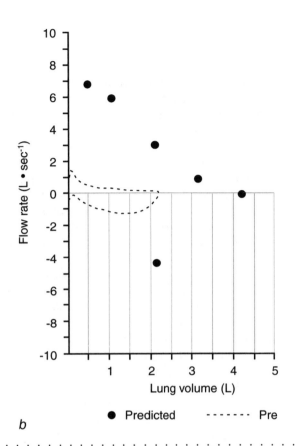

a   *b*

FIGURE 20.2 Flow volume tracings of *(a)* an asthma patient and *(b)* an emphysema patient. The *x*-axis represents the volume of the lung (measured in liters), and the *y*-axis represents the flow rate (measured in liters per second). The dots represent the predicted normal flow-volume trading for a patient who is the same age and height as the subject; the dotted line represents the flow-volume tracing prior to the administration of a bronchodilator; and the solid line represents the flow-volume tracings following the administration of a bronchodilator. Note the difference in the shape of each flow–volume loop before and after bronchodilator administration. *(a)* Significant improvement in the flow rate following administration of a bronchodilator. *(b)* Significant decrement in the flow rate (compared with the predicted normal) and no improvement is noted after administration of a bronchodilator.

PRACTICAL APPLICATION 20.1

General Goals of Asthma Therapy

1. Prevent chronic asthma symptoms and asthma exacerbations during the day and night.

2. Maintain normal activity levels.

3. Have normal or near-normal lung function.

4. Be satisfied with the asthma care received.

5. Have no or minimal side effects while receiving optimal medications.

normal in each of these patient groups. Those patients who have moderate persistent asthma have symptoms daily, awaken once per week with symptoms, and have a decline in exercise tolerance noted several times per week. The FEV_1 in this group ranges from 60% to 80% of predicted. Those patients with severe, persistent asthma have continuous symptoms, awaken frequently at night with symptoms, and have frequent interference with exercise. The FEV_1 in this group is below 60% of predicted.

Depending on the severity of the patient's asthma, various medical regimens have been recommended (39). Several types of medications are used:

— Short- and long-acting β-agonists

— Oral and inhaled corticosteroids

— Theophyllines

— Leukotriene modifiers
— Cromones

The β-agonists and theophyllines act primarily to reduce smooth muscle constriction, whereas the corticosteroids, leukotriene modifiers, and cromones reduce inflammation.

Based on the severity of asthma, specific treatment guidelines have been developed. For a patient with intermittent asthma, intermittent use of an inhaled β-agonist may be all that is necessary. For a patient with mild, persistent asthma, use of an inhaled corticosteroid (low dose) or oral leukotriene modifier should be considered. For a patient with moderate, persistent asthma, use of an inhaled corticosteroid (low to moderate dose) daily combined with an inhaled long-acting β-agonist should be considered. For a patient with severe, persistent asthma, use of a high-dose inhaled corticosteroid (or possibly an oral corticosteroid) and an inhaled long-acting β-agonist should be considered. Should the patient have symptoms persisting throughout the treatment regimen, an inhaled short-acting β-agonist should be used. Each patient's therapy must be individualized given the overall severity of his or her asthma and in many instances requires various combinations of the previously mentioned medications.

Graded Exercise Testing

It is important to confirm the presence of exercise-induced bronchospasm by exercise testing, because in many instances the patient's history may not be sufficient for an accurate diagnosis. An adolescent basketball player may develop unusual fatigue or shortness of breath, which may be perceived as a lack of conditioning or lack of willingness to exercise when, in reality, he or she may actually have exercise-induced bronchospasm. Because exercise-induced bronchospasm may affect the maximal exercise capacity of athletes by increasing ventilatory cost and decreasing maximal ventilatory capacity (and may explain, in part, the day-to-day variability in exercise performance capacity of some athletes), even highly trained athletes should be screened for asthma (22). Practical application 20.2 describes client–clinician interaction regarding exercise testing for a patient with asthma.

Medical Evaluation

A thorough history and physical examination of each patient undergoing exercise testing should be performed. This should include those aspects of the history as mentioned previously including symptoms such as presence of cough, shortness of breath at rest or with exercise, and wheezing. In addition, exercise capability and performance of activities of daily living should also be assessed. The presence of an elevation in respiratory rate at rest and active bronchospasm should be assessed. After reviewing the medical history and physical examination of the patient, the clinician will have a general idea of that individual's functional capacity and can then develop an appropriate exercise testing protocol.

PRACTICAL APPLICATION 20.2

Client–Clinician Interaction

To appropriately determine the level of exercise capability that a patient with asthma might be able to attain (as well as to follow during training sessions), the examiner should review the following features of the clinical assessment:

1. Presence or absence of symptoms (e.g., cough, wheezing, shortness of breath) while at rest or with exercise. A history of exercise limitation (attributable to the previously mentioned symptoms) in an otherwise asymptomatic patient should alert the examiner to the possibility of exercise-induced bronchospasm.

2. Use of medication prior to exercise (e.g., inhaled β-agonist, inhaled corticosteroid, oral leukotriene modifier).

3. Correct use of the medication (particularly when using a metered dose inhaler).

4. Correct use of warm-up and cool-down periods during exercise.

A correct diagnosis of asthma (or exercise-induced bronchospasm) must be made and followed with appropriate use of medications prior to exercise to allow the patient to optimize his or her exercise capabilities.

Measurement of oxygen consumption, carbon dioxide production, and anaerobic threshold by metabolic cart testing during progressive incremental exercise provides a detailed assessment of an individual's response to maximal symptom-limited exercise and is helpful for further decisions regarding the exercise prescription.

Contraindications

For most patients with asthma, there are few limitations or contraindications to exercise. Should a patient have acute bronchospasm, chest pain, or an increased level of shortness of breath above that usually experienced, then exercise testing should be withheld at that time. Certainly, if the patient has severe exercise debilitation or other confounding medical problems (e.g., unstable angina, debilitating orthopedic condition), then exercise testing and training may not be feasible.

Recommendations and Anticipated Responses

The exercise test is performed to a symptom-limited maximum (on either a bicycle ergometer or a treadmill). The work intervals should be short (2-min stages), and the increase in workload should be approximately 1 metabolic equivalent from stage to stage in those with significant asthma. Ideally, a protocol should be chosen that will elicit a patient's maximal effort between 8 and 12 min. Spirometry assessment at baseline (preexercise) and at 5-min intervals (up to 30 min) following exercise cessation should be included. A decline of the FEV_1 by 15% compared with the preexercise FEV_1 confirms the diagnosis of exercise-induced bronchospasm.

Exercise capacity should be assessed at the onset of the training period and at the end of the training period (e.g., 6-8 weeks later). Various types of measurements can be made to assist in this assessment.

Measurement of cardiopulmonary parameters (including **maximal oxygen consumption,** expired oxygen, and carbon dioxide levels) affords the most comprehensive type of testing available. During such testing, oxyhemoglobin saturation can be measured by pulse oximetry. Such testing is expensive and time consuming and may not be readily available. A submaximal exercise test performed on either a cycle ergometer or a treadmill is easier to perform and less costly. Finally, either a 6- or 12-min walk test is the most easily administered of any of the tests, is inexpensive, and is reproducible. Regardless of the method of testing used, exercise capability before and after a comprehensive exercise program should be formally assessed.

Exercise Prescription

Given the great variability of the pathophysiological processes among patients with asthma, the response to exercise varies widely as well (20, 41, 47). A patient with mild persistent asthma may successfully compete at the Olympic level, whereas a patient with severe persistent asthma may not be able to walk across the room without significant dyspnea. On the other hand, a patient may have mild persistent asthma and be unable to perform any type of meaningful exercise because of extreme dyspnea. In addition, the individual patient's exercise ability will vary from time to time depending on the current status of the asthma, especially during an exacerbation when exercise tolerance may be severely limited (34, 41). In many instances, patients with asthma have adjusted their activity level to avoid dyspnea. Practical application 20.3 reviews the literature concerning asthma and exercise training.

PRACTICAL APPLICATION 20.3

Literature Review

A wide variety of physiological outcomes might be expected as a result of exercise training for patients with asthma. No effect, either adverse or beneficial, has been reported to occur in regard to static lung function measurements (spirometry) including bronchial hyperreactivity related to exercise training. Several physiological changes have been noted to occur following a training program, including increases in maximal oxygen uptake, oxygen pulse, and anaerobic threshold. Significant reductions in blood lactate level, carbon dioxide production, and minute ventilation at maximal exercise have also been shown to occur. Subjective responses also have been noted, particularly a reduction in perceived breathlessness at equivalent workloads, following exercise training. This latter response could be attributable to a central nervous system "desensitizing" effect, a decrease in minute ventilation at submaximal workloads, or an increase in the endorphin levels without a concomitant reduction in ventilatory chemosensitivity.

continued

A variety of training schedules have been used for patients with asthma. Various types of exercise including gym, games, distance running, swimming, cycling, altitude training, and treadmill running (to name just a few) have all been shown to improve exercise capability. The frequency of exercise training varied from study to study, ranging from once weekly to daily for periods of time ranging from 20 min up to 2 hr per training session. Training periods of 6 to 8 weeks were generally used. The intensity of exercise also varied from a gradual increase in exercise endurance to short, heavy increases (12).

In one study, 26 adults with mild to moderate asthma (FEV_1 63%) underwent a 10-week supervised rehabilitation program with emphasis on individualized physical training. Daily exercise (swimming) for 2 weeks was followed by twice weekly exercise. Exercise training intensity was measured by a target heart rate during the first 2 weeks and then by perceived sense of exertion as measured by a Borg scale (1-10 scale) during the latter 8 weeks. Each subject was encouraged to exercise to a Borg level of 7 to 8. All subjects were able to perform high-intensity exercise (8-90% of their maximum predicted heart rate), and improvements in cardiovascular conditioning and walk distance were observed after the program. An abatement in asthma symptoms and decreased anxiety were noted following the training period (19).

Ongoing exercise following the initial training program has been shown to be effective for patients with asthma. Of 58 patients who had previously undergone a 10-week rehabilitation program, 39 reported continuation of regular exercise. Cardiovascular conditioning (as measured by a 12-min walk distance) and lung function values remained unchanged in all patients. There was, however, a significant decrease in the number of emergency department visits over the 3-year period compared with the year prior to entry into the rehabilitation program in these 39 patients. A decrease in asthma symptoms was noted only in a subgroup of patients (n = 26) who exercised one or two times per week. Continued exercise following a supervised rehabilitation program is helpful for patients with mild to moderate asthma (18).

Patients with exercise-induced bronchospasm should be encouraged to undergo exercise training as well. Instruction on preventive strategies allowing adequate control of airway inflammation and bronchoconstriction is important for these patients. For each patient, the clinician should consider triggers for the development of asthma under such situations as being outside and exercising on a day with a high ozone concentration or high allergen counts in the atmosphere, or the development of symptoms following ingestion of certain foods within an hour or two prior to exercise (e.g., milk products, vegetables). As discussed previously, an adequate warm-up period and use of inhaled or oral medications prior to exercise should be stressed.

Comprehensive rehabilitation programs for patients with asthma include much more than just exercise training. Components of the initial patient assessment should include patient interview, medical history, diagnostic testing, symptoms and physical assessment, nutritional evaluation, activities of daily living assessment, educational and psychosocial history, and goal development. Actual program content should include education regarding the disease process, triggers of asthma, self-management of the disease (medication use, warning signs and symptoms associated with exacerbations, peak flow monitoring, metered dose inhaler technique, importance of exercise warm-up and cool-down), activities of daily living, psychosocial intervention, and dietary intake and nutrition counseling. Follow-up and evaluation of outcomes are of vital importance as part of the rehabilitation process. Questionnaires used to assess the asthma patient's quality of life (measuring variables such as symptoms, emotions, exposure to environmental stimuli, and activity limitation) have been well validated and should be used as part of this assessment process (27). In addition, cost of medications and equipment, time lost from work or school, and utilization of healthcare resources (e.g., emergency department visits, calls to the patient's physician) can be assessed as part of the follow-up.

Not all patients with asthma are candidates for such comprehensive rehabilitation programs. Most patients with asthma can be managed quite effectively with a combination of medications and general exercise recommendations. In those patients who have moderate or severe persistent asthma, those who have failed medical therapy and have had a significant decline in their performance of the activities of daily living, or those in whom the disease process has drastically adversely affected their lifestyle, comprehensive rehabilitation offers an effective means to improve overall quality of life.

The widespread acceptance, over many years, that patients with asthma cannot and should not exercise has led to many recommendations that these patients avoid exercise. Many parents have unnecessarily restricted their children who have asthma from exercise because the fear that exercise may make the asthma "worse." Attempts have been made to more completely educate patients and their families about the importance of exercise and how such exercise can be performed safely. The use of β-agonist or leukotriene modifier therapy prior to exercise as well as the avoidance of conditions known to precipitate that person's asthma are important mainstays of the treatment and should be used aggressively when asthma plays a role in exercise intolerance.

A variety of cardiopulmonary and metabolic responses to exercise in patients with asthma have been noted. From a cardiopulmonary perspective, treadmill exercise for patients with exercise-induced bronchospasm without prior treatment increases ventilation/perfusion inequality, physiological dead space, and arterial blood lactate levels (5). From a metabolic response perspective, a "blunted" sympathoadrenal response to exercise (7, 49), an alteration in potassium homeostasis, and an excessive secretion of growth hormone have all been shown to occur. The role that each of these metabolic responses may play in the exercise limitation noted in patients with asthma is unknown. Again, given the wide variety of pathophysiological processes present in each patient who has asthma, a wide variety of cardiopulmonary responses to exercise can be expected.

One of the most confounding variables noted in patients with asthma who are attempting to exercise is the effect of dyspnea on exercise capability. The decision to exercise is often weighted against discontinuation of exercise because of the increasing levels of dyspnea experienced by the patient. In one study, "harmful anticipation" significantly increased the perception of visceral changes associated with exercise (38). A wide variability between the degree of airway obstruction, exercise tolerance, and the severity of breathlessness has been noted in several studies (35, 42), but this only accounts for up to 63% of the variance in breathlessness that asthmatics rated during progressive incremental exercise. This complexity concerning the development of symptoms and exercise tolerance might be one reason that the diagnosis of exercise-induced bronchospasm is obscured. In one study that screened 503 children with asthma, an average of one child in each classroom had previously undiagnosed asthma (the symptoms of which were often absent or attributed

to other causes) (48). Given these variable responses to exercise from a cardiopulmonary, metabolic, and symptom perspective, no unified conclusions can be made regarding such effects in patients with asthma as a whole. Individual assessment of each patient is thus important when one is trying to determine the degree (and thus subsequent effects) of exercise intolerance present.

Special Exercise Considerations

Exercise-induced bronchospasm occurs in 50% to 100% of patients with asthma (15). The mechanisms behind exercise-induced bronchospasm are thought to relate to the consequences of heating and humidifying large volumes of air during exercise. Cooling and drying of the airways lead to inflammatory mediator release. The airways are narrowed by the bronchial smooth muscle and cellular abnormalities similar to those discussed previously in response to these mediators (3, 4).

The symptoms of exercise-induced bronchospasm are similar to those of asthma, in general (e.g., dyspnea, chest pain, cough, wheezing), but are associated with short periods of intense physical activity. The typical response of a patient who has exercise-induced bronchospasm involves a 10-min period of bronchodilation at the beginning of exercise followed by progressive bronchospasm peaking up to 10 min following the completion of exercise. Spontaneous resolution of the exercise-induced bronchospasm symptoms occurs over the ensuing 60 min (28, 37). In some, a "refractory" period occurs during exercise, enabling the patient to "run through" the asthma episode.

Patients can prevent exercise-induced bronchospasm through a variety of means. An appropriate warm-up prior to the actual exercise is an important nonpharmacological method to be used. A warm-up period of 15 min (continuous exercise) at up to 60% of the maximal oxygen consumption can significantly decrease postexercise bronchoconstriction in moderately trained athletes. Interval warm-up may be used, but it has been noted in some instances (based on eight 30-s runs at 100% maximal oxygen consumption with a 1.5-min rest in between trials) not to significantly reduce postexercise bronchoconstriction. Thus, at least 15 min of moderate intensity exercise should precede significant exercise for active persons with asthma (15).

Pharmacological methods used to prevent exercise-induced bronchospasm include the use of inhaled β-agonist medications and leukotriene modifiers. Although short-acting β-agonists (e.g., albuterol sulfate) have been useful to ameliorate exercise-induced bronchoconstriction, longer acting agents (e.g., salmeterol xinafoate) have recently been shown to protect against exercise-induced bronchoconstriction as well (40). Leukotriene modifiers (e.g., montelukast) given as a single daily oral dose have also been shown to prevent exercise-induced bronchoconstriction. Tolerance to the medication and rebound worsening of lung function after discontinuation of treatment were not seen with the use of montelukast (17, 33). Inhaled corticosteroids and cromolyn sodium have also been shown to be useful in preventing exercise-induced bronchospasm (23, 32). Given the widely variable pathophysiological components of exercise-induced bronchospasm, therapy must be individually tailored in most instances.

Exercise Recommendation

Comprehensive supervised rehabilitation is helpful for patients with asthma. Control of airway inflammation and bronchoconstriction by use of an appropriate medical regimen is of primary importance for each person. Without adequate control of these pathological processes, exercise capability will not improve significantly. Improvements in aerobic capacity, muscle strength, and endurance are important goals achieved through an appropriate exercise prescription. The exercise prescription should be based on objective measurements of the exercise capabilities and individualized for each patient. This comprehensive approach can often improve exercise tolerance and quality of life for each patient.

Depending on coexisting medical and orthopedic conditions, patients can undertake exercise training through a variety of modalities, such as stationary cycling, treadmill, walking, and swimming. In addition, if weakness of a specific muscle group is noted, then exercises addressing that weakness should be undertaken (e.g., for weakness of the trapezius or biceps, specific weight training targeted at this muscle group). Practical application 20.4 provides an exercise prescription for patients with asthma.

Cardiovascular Training

For cardiorespiratory fitness, the exercise training should be 20 to 60 min in duration completed 3 to 5 days per week (1). The mode of exercise should take into consideration the patient's interests, past exercise experience, and availability of equipment while acknowledging the effects of the surrounding environment. Exposure to cold air, low humidity, or air pollutants is more likely to induce bronchospasm. Intermittent exercise, lower intensity sports, and the presence of warm, humid air (swimming) are generally better tolerated. In addition, if weakness of a specific muscle group is noted, then resistance exercises to address that weakness can be instituted (e.g., using dumbbell weights, elastic bands, or machine weights).

There is presently no consensus as to the optimal level of intensity at which a patient with asthma should train (13, 14). The intensity prescription for such a patient should be based on clinical and exercise test data in addition to the patient's specific goals. Training can then be initiated at an exercise intensity just below the anaerobic threshold for those patients able to tolerate such a level of exercise. Such involved testing may not be available in

---

Exercise Prescription

1. Assess patient's underlying respiratory status and goals for exercise.

2. Assess maximum level of exercise.

3. If maximum level of exercise has been determined by measurement of oxygen consumption and carbon dioxide production (cardiopulmonary exercise testing), begin exercise prescription at an initial intensity level just below the anaerobic threshold.

4. If such measurements are unavailable, begin exercise at a level of exercise at which the patient is comfortable performing for 5 min.

5. Instruct the patient to continue exercise for 20 to 60 min per session.

6. Have the patient perform sessions three to five times per week.

7. Increase exercise intensity by 5% with each session.

8. When maximal level of intensity is attained, increase exercise duration by 5%.

many areas. Submaximal workload can be determined without a metabolic cart by using either a treadmill or a cycle ergometer. Training in these patients can then be initiated at an intensity of 50% to 85% of heart rate reserve (maximal heart rate minus resting heart rate from the evaluation). Again, this intensity is below anaerobic threshold for most individuals (2). For patients with more limiting asthma, a target intensity based on perceived dyspnea (such as a Borg scale) may be more appropriate (9).

Resistance Training

Breathing exercises to strengthen the respiratory muscles have been used by some, but their overall effectiveness is controversial. Deep diaphragmatic breathing was used in 67 patients with asthma and significantly decreased the use of medical services and the intensity of asthma symptoms. However, there was no significant change in overall physical activity as measured by an inventory scale (21). Inspiratory muscle training that used a threshold inspiratory muscle training device in a double-blind sham trial showed a significant increase in inspiratory muscle strength (as expressed by the maximum inspiratory pressure measured at residual volume)

and respiratory muscle endurance. The training group also had a significant reduction in the amount of asthma symptoms, number of hospitalizations, and absence from work or school compared with the sham group (50).

Conclusion

Asthma represents a complex process involving airway inflammation and bronchoconstriction. Environmental risk factors (such as indoor allergens, viral infections) or other triggers (such as exercise, cold air) can initiate an allergic response, resulting in airway inflammation and airway hyperresponsiveness. Airflow limitation results, and the patient develops symptoms such as chest tightness, wheezing, and shortness of breath. Exercise limitations and decreased levels of fitness frequently are noted in patients with asthma but in many instances are not considered to be important for some time following the development of symptoms. Exercise limitations and fitness levels can be improved in those patients treated with an appropriate medication and exercise regimen.

Case Study 20

Medical History

Ms. JR is a 22-year-old Caucasian college senior. Throughout high school she was active in competitive sports including soccer, swimming, and field hockey. On occasion throughout high school she would develop an increase in shortness of breath and a cough. Her primary care physician told her that she had bronchitis and that she should not worry about this. On entry into college she continued with competitive soccer and swimming. At the end of a long run during soccer games she would note an increase in cough and slight wheeze. She would not note any symptoms following her swimming practice. She continued to exercise but noticed an increase in coughing and wheezing over the ensuing year.

Diagnosis and Test Results

Her parents became concerned regarding her discomfort and tried to convince her not to exercise because "it makes you feel much worse and could be dangerous." With such ongoing symptoms, she withdrew from soccer. She sought advice of the college physician, who told her she might have asthma given the symptoms of wheezing. Spirometry was performed, which revealed an FEV_1 of 3.09 L (96% predicted), an FVC of 3.54 L (95% predicted), a **peak expiratory flow rate (PEFR)** of 6.97 L $\cdot$ s^{-1} (95% predicted), and an FEV_1/FVC ratio of 87%. Given these results, showing "normal" pulmonary function, she was told that she did not have asthma but rather bronchitis and was told to continue her exercise (after a course of antibiotics).

continued

She continued to swim but would note that at the end of a training session she was slightly more short of breath than usual and had a heaviness over her anterior chest region. She sought the advice of another physician, who ordered an exercise test with the measurement of expired gases during progressive incremental bike exercise. Spirometry was performed at 15, 30, and 60 min following the exercise test. Maximal oxygen consumption was 3.13 L · min^{-1} (52.2 ml · kg^{-1} · min^{-1}). Flow rates were as follows:

— FEV$_1$ (L): preexercise = 3.09; 15 min postexercise = 2.87; 30 min postexercise = 2.20; 60 min postexercise = 2.24
— FVC (L): preexercise = 3.54; 15 min postexercise = 3.32; 30 min postexercise = 2.97; 60 min postexercise = 3.03
— PEFR \ (L · sec^{-1}): preexercise = 6.97; 15 min postexercise = 6.00; 30 min postexercise = 5.25; 60 min postexercise = 5.26

Treatment and Exercise Prescription

As a result of these studies, a diagnosis of asthma (exercise induced) was made. The patient was started on a short-acting β-agonist (albuterol sulfate) administered 30 min before exercise. She was instructed to warm up for 15 min with low- to moderate-intensity exercise or swimming before starting a high-intensity swim practice. Exercise tolerance subsequently improved while exercise-associated symptoms became rare (for the most part abated).

Case Study 20   Discussion Questions

1. Why was the initial diagnosis of asthma not entertained?
2. How was the actual diagnosis of asthma (exercise induced) made? What tests should be useful in this determination?
3. How did the recommendations improve her exercise tolerance? Why was swimming initially better tolerated than soccer?
4. Discuss the intensity, frequency, and duration of exercise training for patients with asthma.
5. How would the development of asthma symptoms at the end of a 3-hr practice session influence the choice of medication (e.g., short-acting vs. long-acting β-agonist)?

Glossary

airflow obstruction—The blockage of the flow of air out of the lung that can occur secondary to narrowing of the airway lumen.

airway hyperreactivity—The ability of the airway wall to be sensitive to various inhalants.

asthma—A continuum of disease processes characterized by inflammation of the airway wall.

bronchodilator—A substance that will dilate (open) a bronchus.

bronchoprovocation—A type of pulmonary function testing in which a particular medication (e.g., methacholine) is aerosolized in an attempt to induce bronchospasm.

bronchospasm—Spasmodic contraction of the smooth muscle of the bronchi, as occurs in asthma.

CD4 lymphocyte—A type of white blood cell that is part of the body's immunological system.

cytokine—A generic term for nonantibody proteins released by one cell population on contact with a specific antigen, which act as intercellular mediators.

dyspnea—The perception of shortness of breath.

eosinophils—A granular leukocyte that contains vasoactive amines.

epithelial-mesenchymal trophic unit—The structural unit composed of epithelial cells and mesenchymal cells present in the bronchus responsible for the development of the bronchus and the repair of the bronchus following injury.

exacerbation—Worsening of an underlying condition.

forced expiratory volume in 1 s (FEV₁)—The amount of air able to be exhaled forcefully in 1 s (measured by pulmonary function testing).

forced vital capacity (FVC)—The total amount of air able to be exhaled forcefully (measured by pulmonary function testing).

lung parenchyma—The essential (or functional) elements of the lung.

lymphocytes—Any of the mononuclear, nonphagocytic leukocytes found in the blood, lymph, or lymphoid tissues that are the body's immunologically competent cells.

mast cells—Connective tissue cells that are important in the body's defense mechanisms needed during injury or infection.

maximal oxygen consumption—The maximum amount of oxygen consumed (or used) by the body, usually measured under conditions of maximal exercise.

parenchymal infiltrates—The deposition or diffusion in lung tissue of substances not normal to it.

peak expiratory flow rate (PEFR)—The highest flow rate (exhalation of gas from the lung) a patient can generate during a forceful expiration.

spirometry—The measurement of the breathing capacity of the lungs.

subepithelial fibrosis—The structural changes noted beneath the epithelial layer of the bronchus resulting in scar tissue formation in this area.

References

1. American College of Sports Medicine. ACSM's Guidelines for Exercise Testing and Prescription (sixth edition). 2000; Lippincott, Williams, & Wilkins, Baltimore.

2. American College of Sports Medicine Position Stand. The recommended quantity and quality of exercise for developing and maintaining cardiorespiratory and muscular fitness, and flexibility in healthy adults. Med Sci Sports Exerc 1998; 30: 975-91.

3. Anderson, S.D., and E. Daviskas. The mechanism of exercise-induced asthma is . . . . J Allergy Clin Immunol 2000; 106: 453-9.

4. Anderson, S.D., and K. Holzer. Exercise induced asthma: Is it the right diagnosis in elite athletes? J Allergy Clin Immunol 2000; 106: 419-28.

5. Anderson, S.D., M. Silverman, and S.R. Walker. Metabolic and ventilatory changes in asthmatic patients during and after exercise. Thorax 1972; 27: 718-25.

6. Banzett, R.B., J.A. Dempsey, D.E. O'Donnell, and M.Z. Wambolt. NHLBI Workshop Summary. Symptom perception and respiratory sensation in asthma. Am J Respir Crit Care Med 2000; 162: 1178-82.

7. Barnes, P.J., M.J. Brown, M. Silverman, et al. Circulating catecholamines in exercise and hyperventilation induced asthma. Thorax 1981; 36: 435-40.

8. Barnes, P.J., K.F. Chung, and C.P. Page. Inflammatory mediators of asthma: An update. Pharmacol Rev 1998; 50: 515-96.

9. Borg, G. Perceived exertion as an indicator of somatic stress. Scand J Rehabil Med 1970; 2: 92-8.

10. Busse, W.W. Inflammation in asthma: The cornerstone of the disease and target therapy. J Allergy Clin Immunol 1998; 102: S17-22.

11. Centers for Disease Control and Prevention. Forecasted state-specific estimates of self-reported asthma prevalence—United States. MMWR 1998; 47: 1022-5.

12. Clark, C.J., and L.M. Cochrane. Physical activity and asthma. Curr Opin Pulmon Med 1999; 5: 68-75.

13. Clark, C.J., and L.M. Cochrane. Assessment of work performance in asthma for determination of cardiorespiratory fitness and training capacity. Thorax 1988; 43: 745-98.

14. Cochrane, L.M., and L.J. Clark. Benefits and problems of a physical training programme for asthmatic patients. Thorax 1990; 45: 345-51.

15. Cypcar, D., and D.F. Lemanske. Asthma and exercise. Clin Chest Med 1994; 15: 351-68.

16. Ebina, M., H. Yaegashi, R. Chiba, et al. Hyperreactive site in the airway tree of asthmatic patients recorded by thickening of bronchial muscles: A morphometric study. Am Rev Respir Dis 1990; 141: 1327-32.

17. Edelman, J.M., J.A. Turpin, E.A. Bronsky, et al. Oral montelukast compared with inhaled salmeterol to prevent exercise-induced bronchoconstriction. Ann Intern Med 2000; 132: 97-104.

18. Emtner, M., M. Finne, and G. Stalenheim. A three year followup of asthmatic patients participating in a 10-week rehabilitation program with emphasis on physical training. Arch Phys Med Rehabil 1998; 78: 539-44.

19. Emtner, M., M. Herela, and G. Stalenheim. High intensity physical training in adults with asthma. Chest 1996; 109: 323-30.

20. Garfinkel, S.K., S. Kesten, K.R. Chapman, et al. Physiologic and nonphysiologic determinants of aerobic fitness in mild to moderate asthma. Am Rev Respir Dis 1992; 145: 741-5.

21. Girodo, M., K.A. Ekstrand, and G.J. Metivier. Deep diaphragmatic breathing: Rehabilitation exercise for the asthmatic patient. Arch Phys Med Rehabil 1992; 73: 717-20.

22. Helenius, I., and T. Haahtela. Allergy and asthma in elite summer sport athletes. J Allergy Clin Immunol 2000; 106: 444-52.

23. Henriksen, J.M., and R. Dahl. Effect of inhaled budesonide alone and in combination with low-dose terbutaline in children with exercise induced asthma. Am Rev Respir Dis 1983; 128: 993-7.

24. Holgate, S.T. The cellular and mediator basis of asthma in relation to natural history. Lancet 1997; 350(Suppl II): 5-9.

25. Holgate, S.T., D.E. Davies, P.M. Lackie, S.J. Wilson, S.M. Puddicombe, and J.L. Lordan. Epithelial-mesenchymal interactions in the pathogenesis of asthma. J Allergy Clin Immunol 2000; 105: 193-204.

26. Homa, D.M., D.M. Mannino, and M. Lara. Asthma mortality in U.S. Hispanics of Mexican, Puerto Rican, and Cuban Heritage, 1990-1995. Am J Respir Crit Care Med 2000; 161: 504-9.

27. Juniper, E.F., G.H. Guyatt, P.J. Ferrie, et al. Measuring quality of life in asthma. Am Rev Respir Dis 1993; 147: 832-8.

28. Kowabori, I., W.E. Pierson, L.C. Loveday, et al. Incidence of exercise induced asthma in children. J Allergy Clin Immunol 1996; 58: 447-55.

29. Laitinen, A., and L.A. Laitinen. Airway morphology: Endothelium/basement membrane. Am J Respir Crit Care Med 1994; 150: 514-7.

30. Laitinen, L.A., A. Laitinen, T. Haahtela. Airway mucosal inflammation even in patients with newly diagnosed asthma. Am Rev Respir Dis 1993; 147: 697-704.

31. Lang, D.M., and M. Polansky. Patterns of asthma mortality in Philadelphia 1969 to 1991. N Engl J Med 1994; 331: 1542-6.

32. Lee, T.M., M.J. Brown, L. Nagy, et al. Exercise induced release of histamine and neutrophil chemotactic factor in atopic asthmatics. J Allergy Clin Immunol 1982; 70: 73-81.

33. Leff, J.A., W.W. Busse, D. Pearlman, et al. Montelukast, a leukotriene-receptor antagonist, for the treatment of mild asthma and exercise induced bronchospasm. N Engl J Med 1998; 339: 147-52.

34. Ludwick, S.K., J.W. Jones, T.K. Jones, et al. Normalization of cardiopulmonary endurance in severely asthmatic children after bicycle ergometry therapy. J Pediatr 1986; 109: 446-51.

35. Mahler, D.A., K. Faryniarz, T. Lentine, et al. Measurement of breathlessness during exercise in asthmatics: Predictor variables, reliability, and responsiveness. Am Rev Respir Dis 1991; 144: 39-44.

36. Mannino, D.M., D.M. Homa, C.A. Pertowski, et al. Surveillance for asthma—United States, 1960-1995. MMWR 1998; 47: 1-27.

37. McFadden, E.R. Jr, and F.A. Gilbert. Exercise induced asthma. N Engl J Med 1994; 330: 1362-7.

38. Meyer, R., B. Froner-Herwig, and H. Sporkel. The effect of exercise and induced expectations on visceral perception in asthmatic patients. J Psychosom Res 1990; 34: 454-60.

39. Murphy S. National Asthma Education and Prevention Program. Expert Panel Report II: Guidelines for the Diagnosis and Management of Asthma (NIH Publication 1997:97-4051). 1997; U.S. Department of Health and Human Services, Bethesda, MD.

40. Nelson, J.A., L. Strauss, M. Skowronski, et al. Effect on long term salmeterol treatment on exercise induced asthma. N Engl J Med 1998; 339: 141-6.

41. Orenstein, D.M., M.E. Reed, F.T. Grogan, et al. Exercise conditioning in children with asthma. J Pediatr 1985; 106: 556-60.

42. Peel, E.T., C.A. Soutar, and A. Seaton. Assessment of variability of exercise tolerance limited by breathlessness. Thorax 1988; 43: 960-4.

43. Robertson, C.F., A.R. Rubinfeld, and G. Bowes. Pediatric asthma deaths in Victoria: The mild at risk. Pediatr Pulmonol 1992; 13: 95-100.

44. Roche, W.R., R. Beasley, J.H. Williams, and S.T. Holgate. Subepithelial fibrosis in the bronchi of asthmatics. Lancet 1989; i: 520-4.

45. Rubenfeld, A., and M. Pain. Perception of asthma. Lancet 1976; 1(7965): 882-4.

46. Senthilselvan, A. Prevalence of physician diagnosed asthma in Saskatchewan, 1981-1990. Chest 1999; 114: 388-92.

47. Strunk, R.C., D. Rubin, L. Kelly, et al. Determination of fitness in children with asthma: Use of standardized tests for functional endurance, body fat composition, flexibility, and abdominal strength. Am J Dis Child 1988; 142: 940-4.

48. Tsanakas, J.N., R.D. Milner, O.M. Bannister, et al. Free running asthma screening test. Arch Dis Child 1988; 63: 261-5.

49. Warren, J.B., R.J. Keynes, M.J. Brown, et al. Blunted sympathoadrenal response to exercise in asthmatic subjects. Br J Dis Chest 1982; 76: 147-50.

50. Weiner, P., R. Axgad, R. Ganam, and R. Weiner. Inspiratory muscle training in patients with bronchial asthma. Chest 1992; 102: 1357-61.

51. Zeiger, F.S., C. Dawson, and S. Weiss. Relationships between duration of asthma and asthma severity among children in the childhood asthma management program (CAMP). J Allergy Clin Immunol 1999; 103: 376-87.

Steven R. Boas, MD
Children's Memorial Hospital
Chicago, IL

Michael J. Danduran, MS
Children's Memorial Hospital
Chicago, IL

Cystic Fibrosis

Cystic fibrosis (CF) is a genetic disorder that affects the respiratory, digestive, and reproductive systems. Excessively viscid mucus causes obstruction of passageways including pancreatic and bile ducts, intestines, and bronchi. Also, the sodium and chloride contents of sweat are increased.

Scope

CF is the most common life-shortening genetic disease in the Caucasian population. Currently, more than 23,000 patients have CF in the United States, with approximately 1000 new patients diagnosed each year (26). Two thirds of all individuals with CF are less than 18 years of age. The median survival age is 31 years of age, although the oldest reported individual with CF was 73 years old. CF is inherited as an autosomal recessive disorder affecting approximately 1 in 3000 live births in the Caucasian population and 1 in 17,000 in the African American population. Individuals with other ethnic backgrounds are also affected but in decreased frequencies. The estimated total cost to treat CF in the United States was more than $900 million in 1995, which represents a cost per CF patient of almost $40,000 per year (25).

Physiology and Pathophysiology

The gene for CF is located on chromosome 7 and results in the altered production of a protein called the **cystic fibrosis transmembrane conductance regulator (CFTR),** a protein that functions as a chloride channel regulated by cyclic adenosine monophosphate. More than 1000 unique mutations of the CF gene have been identified, although one type, δ F508, accounts for more than 70% of CF genes in the United States.

The primary role of CFTR appears to be as a chloride channel, although other functions for CFTR have been proposed as well. The abnormal CFTR leads to abnormal sodium chloride and water movement through the cell membrane. When this occurs in the lungs, abnormal thick and dry mucus ensues, resulting in bronchial airway obstruction, bacterial infection, and inflammation. As this "vicious cycle" continues, lung tissue is progressively destroyed with eventual respiratory failure (figure 21.1). Lung disease accounts for more than 95% of the morbidity and mortality associated with CF. However, with aggressive intervention and early diagnosis, survival has been extended, with adults living well into their 30s and 40s.

The average age at time of diagnosis is approximately 6 months (26). Typically, one or more symptoms are affected in CF and can lead to its diagnosis including the respiratory, gastrointestinal, sinus, and sweat gland systems. The underlying common theme to all these systems is the cellular abnormality of ion transport necessary for proper function of epithelial structures.

Respiratory System

At birth, the lungs are normal on a histological basis. As the vicious cycle of infection, inflammation, and impaired mucus clearance ensues, the lungs become colonized with bacteria, with 60% of all patients with CF growing *Pseudomonas aeruginosa* later in life (26). Other bacteria such as *Staphylococcus aureus* occur in more than 40% of patients, whereas others occur less frequently. Although a young child may present with acute signs of respiratory infection, most children with CF present at the time of diagnosis with signs of chronic infection including cough, sputum pro-

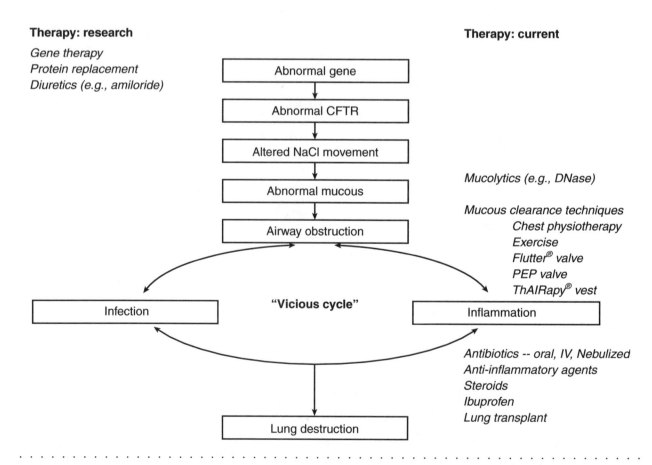

FIGURE 21.1   The "vicious cycle" of cystic fibrosis. CFTR = cystic fibrosis transmembrane conductance regulator.

duction, wheeze, fever, and failure to thrive. Many infants or young children with CF have been previously misdiagnosed as having asthma, bronchitis, allergies, pneumonia, or bronchiolitis. Chest radiographs may indicate the presence of acute or chronic changes such as infiltrates, **bronchiectasis** (chronically overdilated airways), or hyperlucency. When pulmonary function is assessed in older children (>5 years) at the time of diagnosis, evidence of airways obstruction—reduced forced expiratory volume in 1 s (FEV_1) or forced expiratory flow between 25% and 75% of forced vital capacity (FEF_{25-75})—or hyperinflation (elevated residual volume and right ventricle/total lung capacity ratio) may exist. Exercise tolerance may become significantly compromised as well compared with normative values. Ultimately, the progressive loss of lung tissue and airways obstruction lead to respiratory failure. The time course for this progression is variable, with some adults with CF experiencing very little lung damage and some children experiencing extensive lung disease.

Gastrointestinal and Nutritional Systems

In more than 90% of individuals with CF, exocrine **pancreatic insufficiency** is present, resulting in malabsorption of important nutrients including fat and protein. Malabsorption can lead to frequent fatty stools (steatorrhea), malodorous stools, and abdominal pain. The combination of the need for increased caloric intake (attributable to increased resting energy expenditure, cough, and infection) and poor utilization of nutrients via malabsorption often leads to malnutrition or a constant struggle to maintain body weight. Additionally, other gastrointestinal organs can be affected, resulting in liver disease, endocrine pancreatic insufficiency (CF-related diabetes mellitus), and gall bladder disease.

Sinuses

The development of **pansinusitis** and **nasal polyposis** is common for individuals with CF. For many, this finding may be inconsequential, although some individuals may find it difficult to breath through their nose. Additionally, pansinusitis with associated bacterial colonization may contribute to the extent of lung disease, with some individuals requiring aggressive medical intervention (e.g., antibiotics, nasal irrigation, and endoscopic surgery).

Sweat Glands

Although all epithelial cells will demonstrate the chloride transport defect, the sweat glands are the organ on which the diagnostic test was based. The basis of the **sweat test** (i.e., pilocarpine iontophoresis analysis) rests on the presence of very high salt content in the sweat of individuals with cystic fibrosis. A sweat chloride concentration of greater than $60 \text{ mEq} \cdot \text{dl}^{-1}$ is considered diagnostic for CF.

Clinical Considerations

The clinical manifestations of CF are variable with differing involvement of the pulmonary and gastrointestinal organ systems. Comprehensive evaluation that includes assessment of the signs and symptoms, diagnostic studies, and pulmonary function testing helps determine the severity of disease.

Signs and Symptoms

CF is usually diagnosed by the presence of classic signs and symptoms (table 21.1). Because of the expansive nature of the disease many systems are affected, which often requires that patient care be coordinated by a CF care team that may include pulmonologists, gastroenterologists, nurses, respiratory therapists, physical therapists or exercise clinicians, a social worker, a nutritionist, a psychologist, a genetic counselor, and pulmonary function technologists. Nevertheless, respiratory and gastrointestinal support are the mainstay of therapy for children with CF.

Diagnosis and Laboratory Results

Many states have instituted a newborn screening for CF allowing for earlier detection. A positive sweat test and/or genetic mutation analysis can confirm the diagnosis of CF. Additional laboratory testing should be performed when the diagnosis of CF is considered:

- Sputum culture (positive for *P. aeruginosa* or other CF bacteria)
- Chest radiograph
- Static and dynamic lung assessment if age appropriate
- Blood sampling for complete cell count
- Liver function
- Nutritional parameters (e.g., total protein, albumin)
- Renal function (e.g., blood urea nitrogen, creatinine)
- Fat-soluble vitamins (e.g., vitamins A and E)
- Glucose

TABLE 21.1

Clinical Signs and Symptoms of Cystic Fibrosis

System	Signs and Symptoms
Respiratory	Chronic productive cough, pneumonia, wheezing, hyperinflation, exercise intolerance, *Pseudomonas aeruginosa* bronchitis
Gastrointestinal/nutritional	Steatorrhea, failure to thrive, biliary cirrhosis, intestinal obstruction, abdominal pain, malodorous stools
Sinuses	Chronic sinusitis, nasal polyps
Sweat glands	Salty taste, recurrent dehydration, chronic metabolic acidosis
Other	Infertility, pubertal delay, digital clubbing, family history

Assessment of **static pulmonary function,** as defined by the properties of the lung at rest or baseline, is essential in the acute and chronic management of individuals with CF. Simple spirometry, as well as assessment of lung volumes, diffusion capacity, and bronchodilator responsiveness, assists in the detection of an acute **pulmonary exacerbation.** Although some individuals with CF have little or mild lung disease, most people affected with CF will demonstrate varying degrees of airway obstruction with signs of hyperinflation. Additionally, almost 30% of people with CF will demonstrate signs of airway hyperreactivity when exposed to a bronchodilator. Declines in FEV_1 or indexes of smaller airway function (e.g., FEF_{25-75}, FEF_{50}, FEF_{75}) over time may serve as warning signs of acute or chronic lung deterioration.

In conjunction with static pulmonary function assessment, **dynamic pulmonary function,** as defined by lung function in response to changing physiological state (e.g., work, exercise, physiologic stress), also plays an important role as a diagnostic tool and in monitoring the patient's clinical condition. In fact, aerobic exercise tolerance has been strongly correlated with long-term survival in patients with CF (2, 61, 64). When healthcare workers assess the lungs under a measurable stress (e.g., exercise), ventilatory limitations and impairment in other parameters such as oxygen saturations that depend on dynamic lung function may become apparent that were not noted when the patient was at rest. Regular assessment of exercise tolerance in patients with CF is an integral component of their medical care. The exercise clinician who treats and assesses children with CF should comprehensively

understand both dynamic and static lung function measurements and the role that exercise testing serves in the management of this population.

Treatment

Current treatment for CF is complex. Specialized CF care centers have emerged that can offer the multiple-specialty care necessary for these individuals. Both preventive and acute management are required to optimize health for individuals with CF. Respiratory and gastrointestinal support are the mainstay of therapy for children with CF.

Treatment of the pulmonary component can be best viewed in terms of addressing the vicious cycle of infection and inflammation (figure 21.1). Current strategies are designed to intervene in this process at multiple levels, thus minimizing the progressive loss of lung tissue. A combination of mucolytic agents and daily mucus clearance techniques can help maintain good pulmonary hygiene. Because bacterial colonization leads to a brisk inflammatory response (e.g., cough, sputum production, increased work of breathing), use of antibiotics becomes necessary. The choice of oral, nebulized, or intravenous antibiotics is determined by the severity of the acute exacerbation. Exercise as part of the therapeutic medical regimen is standard care in most CF specialized centers. As progressive lung destruction and deterioration occur, the final choice of therapy is lung transplantation with a 3-year survival rate of approximately 60%.

In addition to the pulmonary aspects of CF, important attention must be given to the nutritional aspects of CF. For the 90% of individuals with CF who are pancreatic insufficient, the use of pancreatic enzymes can help in the utilization of nutrients. Fat-soluble vitamins are also given as supplements. The use of a high-fat, high-calorie diet to ensure the consumption of sufficient calories is standard care for most patients with CF.

Although children with CF have a tendency to lose salt while exercising, especially in extremely hot, humid weather, most children will consume sufficient salt. Ready access to a salt shaker or salty snacks (e.g. pretzels, potato chips) along with liberal fluid intake usually suffices. In rare instances, salt tablets may be required and should be prescribed under a CF physician's direction. See practical application 21.1 on thermoregulation.

Exercise is routinely recommended for all individuals with CF, regardless of pulmonary status. In conjunction with regular chest physiotherapy, exercise can enhance clearance of mucus from the bronchial tree (7). However, exercise alone does not appear to be as effective as standard chest physiotherapy. The positive psychological well-being of the individual with CF also appears to be an important benefit of regular exercise (48).

Graded Exercise Testing

The importance of a complete exercise evaluation prior to the development of an exercise prescription for individuals with CF cannot be underestimated. This evaluation can help identify potential risk factors for this specific population. Although many individuals with CF have limitations in exercise performance, physical activity remains a vital part of the therapeutic management plan. An effective exercise program should help optimize all aspects of fitness including overall well-being, both physical and psychological. The exercise clinician plays an important role in conjunction with the patient, parent, and CF medical team in establishing realistic goals and in developing an achievable exercise program to enhance the individual's quality of life (see practical application 21.2).

Medical Evaluation

A thorough medical history is necessary to identify potential risk factors that may limit exercise performance in individuals with CF. The history should focus on factors that may be present in the

---

PRACTICAL APPLICATION 21.1

Thermoregulation

Fluid management during exercise poses unique challenges for the individual with CF. Persons with CF tend to lose more salt in their sweat per surface area than do their non-CF counterparts. Thus, the sodium and chloride levels in the bloodstream often decrease after exercise, whereas these levels are maintained in those without CF (66). In addition, individuals with CF tend to underestimate their fluid needs during exercise. In one study, patients with CF lost twice as much body weight as healthy subjects when drinking fluid only when thirsty (10). When subjects were forced to drink fluids every 20 min, no weight loss was seen. Although children with CF will drink more fluid if it is flavored, they still can become dehydrated if they rely solely on the thirst mechanism. A recent study demonstrated that when the salt content of the beverages exceeded $50 \text{ mmol} \cdot \text{L}^{-1}$, individuals with CF consumed sufficient fluid to avoid dehydration (51). The authors noted that at this level of salt concentration, the drinks are perceived as less palatable. Thus, a physiological mechanism of the higher salt ingestion somehow triggered the thirst mechanism independent of taste alone.

Clinical signs of early dehydration should be discussed with all individuals who plan to exercise. These signs include light-headedness, heat intolerance, flushed skin, decreased urine output, concentrated (dark yellow) urine production, nausea, headaches, and muscle cramps.

Adequate hydration needs to be stressed for those who will exercise in warm, humid climates. During sports and activity, consumption of 4 oz of fluid every 20 min is a good general rule. For children who cannot readily quantify fluid amounts, eight gulps of fluid equals approximately 4 oz. Additionally, it is important for children to drink fluids after exercise as well to help offset any weight loss. The type of fluid chosen is not critical, because even the sports beverages contain less than $50 \text{ mmol} \cdot \text{L}^{-1}$ of salt. Thus, fluid intake should be paired with the consumption of salty snacks (e.g., pretzels, potato chips) to ensure the thirst mechanism becomes activated. Salt tablets are rarely indicated and should only be used under medical supervision.

Client–Clinician Interaction

Motivation may be the most powerful factor in determining whether an exercise program will succeed. Motivation is unique for each individual. Because two thirds of the patients with CF are children or teenagers, motivational issues are especially important given adolescent issues of self-image, fitting in with peer groups, and establishing physical abilities. Although a complete understanding of the pathophysiological process associated with CF is important, the appropriate client–clinician interaction may be even more important in ensuring successful exercise testing and program satisfaction. The following table is designed to assist the exercise clinician in motivating individuals with CF to give maximal efforts during testing and to adhere to exercise prescriptions.

Situation	Motivational tip
Clinical testing	— Make the testing experience fun for the younger children (e.g., a sticker for each stage of the test completed).
	— Explain all procedures in detail.
	— Forewarn the child what he or she will feel during testing (e.g., breathlessness, muscle fatigue, and cough).
	— Listen to the child's questions and concerns.
	— Pick an apparatus (treadmill, bike) the individual feels would allow him or her the greatest success.
	— Use positive motivational phases such as "you can do it," "you're almost there," "we are so proud of you," or "only one more minute."
	— Minimize use of phrases that influence decisions such as "do you need to stop?" or "do you have to quit?"
Exercise adherence	— Enhance a child's accountability by allowing him or her to play an active role in planning.
	— Have the individual set goals in conjunction with the exercise clinician.
	— Establish a partnership and assist in the development of the exercise program with the child.
	— Use incentive systems or find a training partner for the child to assist in his or her program.
	— Address the child's or parents' concerns (e.g., increased weight loss, not being able to keep up with friends, poor body image) associated with exercise programs and facilities.
	— Acknowledge the concerns and barriers to adherence to the program and work to find alternative strategies.
	— Understand that what is "successful" will be unique for individuals with CF (e.g., completion of a marathon, learning to in-line skate, walking for 20 min with supplemental oxygen).
	— Communicate frequently and address concerns before they lead to noncompliance.

person with CF which can alter the pulmonary–cardiovascular–peripheral systems necessary for effective oxygen delivery and utilization during exercise.

The most important consideration prior to testing a child with CF is to determine the subject's level of pulmonary disease. Prior pulmonary function data can help predict which child is likely to experience oxyhemoglobin desaturation with exercise testing. An FEV_1 of less than 50% of predicted, a diffusing capacity of the lungs of less than 65% of predicted, or a low resting **oxyhemoglobin saturation** places the child with CF at much greater risk of oxygen desaturation during exercise (36, 43, 54). A history of wheezing, chest tightness, or chest pain during exercise may indicate the presence of exercise-induced bronchoconstriction, which is seen in 22% to 55% of CF patients (49, 79, 83). Additional considerations, such as a history of pneumothorax or **hemoptysis** (coughing up blood) at rest or during exercise, should be reviewed. A history of nocturnal headaches or cyanosis may suggest advanced lung disease with associated **hypoxemia.** Because exercising in altitude may exaggerate hypoxemia, it should be determined whether the exercise testing or training program will occur at altitude (8, 76). Few cardiovascular limitations exist for individuals with CF. A history of pulmonary hypertension or cor pulmonale would require consultation with a cardiologist prior to testing. Signs and symptoms of right-side heart failure should be sought (e.g., edema, venous congestion, hypoxemia). Peripheral factors such as **scoliosis, kyphosis,** and tight hamstrings are commonly present in individuals with CF and may reduce the mechanical efficiency during exercise.

Other organ systems can be affected by CF and become an issue during acute exercise. Liver disease with associated ascites (abdominal distention) may interfere with respiratory muscle effectiveness, whereas liver-related bleeding disorders may be exacerbated with increased blood pressure during exercise. A tendency toward dehydration should be determined, because excessive salt loss with physical exertion is commonly seen in CF (66, 67). The climate, especially one that is warm and humid, in which the individual will be exercising for both testing and training is an important factor. Because salt is lost more readily in individuals with CF and the thirst mechanism is less reliable in preventing dehydration, care must be taken to avoid dehydration. The extent of malnutrition and body composition (e.g., lean muscle mass) should be noted, because this may alter the mechanical load applied during exer-

cise testing. Finally, the use of validated physical activity questionnaires or diaries can help determine how physically active the individual is prior to developing a precise prescription. Multiple options for assessing activity are available and vary in format (e.g., recall questionnaires, activity diaries). The appropriateness of each tool varies for younger individuals; however, some are designed specifically for use in the pediatric population (e.g., previous day physical activity recall) (69).

Contraindications

Although no absolute contraindications to exercise testing exist, special considerations for the individual with a history of pulmonary hypertension, acute hemoptysis, pneumothorax, oxygen dependence, exercising at altitude, a bleeding disorder secondary to liver disease, and severe malnutrition need to be made. Monitoring during testing should include continuous pulse oximetry, electrocardiogram, and vital sign assessment. Supplemental oxygen and a short-acting bronchodilator (e.g., albuterol by inhaler) should be available to all patients with CF during or following testing as needed.

Recommendations

Following a complete history, the assessment of exercise capacity can be made. Typical cardiopulmonary responses in CF patients are listed in table 21.2. The precise protocol and testing location depend on multiple factors including the desired goals for training, the age of the child, the medical considerations of the disease, and the resources available. Finally, a clearly defined goal for testing should be determined so that individualized exercise programs can be designed. Such goals may include determining the heart rate at which oxyhemoglobin desaturation occurs, monitoring for improvement in response to medical therapy, or comparison to prior performance.

Aerobic Testing Numerous reproducible exercise protocols exist for testing maximal exercise performance in children: the Godfrey, McMaster, and James protocols for bicycle testing and the Bruce, modified Bruce, and Balke protocols for treadmill testing (60, 73). Monitoring of pulse oximetry, electrocardiogram, and blood pressure response with exercise should be considered in all patients with CF, especially those with more advanced disease. The information provided by a maximal aerobic test is extremely comprehensive and allows the exercise clinician to develop an appropriate exercise prescription. However, this form of testing requires sophisticated exercise equipment and highly trained

TABLE 21.2	
Cardiopulmonary Assessment for Individuals With Cystic Fibrosis	
STATIC PULMONARY FUNCTION (REST OR BASELINE)[a]	
Parameter	**Change**
Spirometry: FEV_1, FEF_{25-75}	Decreased
Lung volumes: RV, FRC, RV/TLC	Increased
Diffusion capacity: DL_{CO}	Decreased
Oxyhemoglobin saturation: SpO_2	Decreased
DYNAMIC CARDIAC AND PULMONARY FUNCTION (IN RESPONSE TO EXERCISE OR STRESS)[b]	
Aerobic capacity: PWC, $\dot{V}O_2$max	Decreased
Breathing response: $\dot{V}_E$, $\dot{V}_E$/MVV	Increased
Gas exchange: $\dot{V}CO_2$, EtCO$_2$	Increased
Blood pressure response	Normal
Heart rate at rest	Increased
Heart rate during peak exercise	Decreased

Note. FEV_1 = Forced expiratory volume in 1 s; FEF_{25-75} = forced expiratory flow between 25% and 75% of the forced vital capacity; RV = residual volume; FRC = functional residual capacity; DL_{CO} = diffusion capacity of the lung by the carbon monoxide technique; SpO_2 = pulse oximetry; PWC = peak work capacity; $\dot{V}O_2$max = oxygen consumption; $\dot{V}_E$ = minute ventilation; $\dot{V}$/MVV = ratio of minute ventilation to maximal voluntary ventilation; $\dot{V}CO_2$ = carbon dioxide production; EtCO$_2$ = end tidal carbon dioxide.

[a]Static pulmonary function is decreased and declines with advancing lung disease. [b]Abnormal parameters tend to follow extent of lung disease; aerobic performance weakly correlated with static lung function parameters.

technical staff and can result in a significant financial cost to the patient.

A submaximal aerobic test may be useful in determining whether exercise desaturation or breathlessness takes place and can help verify the effectiveness of exercise prescriptions established from maximal tests. The level of cooperation required for a submaximal test is less than that of a maximal test, making this test easier to perform for the young child or child with significant ventilatory limitations. Gas exchange parameters are not routinely required for this test. A treadmill or bicycle similar to that used during a maximal test is the ideal equipment for the submaximal test. Traditional submaximal protocols require the subject to exercise for 5 to 8 min at increasing workloads until about 75% of the age-predicted maximal heart rate is reached. However, a child with CF may not be able to reach the theoretical age-predicted maximal heart rate because of ventilatory limitations. Exercise at 75% of age-predicted maximal heart rate for children with

significant respiratory compromise may actually cause the child to approach or exceed his or her maximal capacity (63). Prior heart-rate data from a maximal test, if available, as well as a thorough history to determine the level of disease severity, may be helpful in determining the appropriate workload needed for a submaximal test. Heart rate, blood pressure, and pulse oximetry should be monitored during testing. The lowered technical demands, financial costs, and ease of multiple testing (e.g., tracking performance during treatment or over time) make the submaximal test an attractive alternative to the maximal test.

Formal exercise tests are not always accessible for children with CF. Over the past few years, several walking and running tests have been developed in an attempt to better mimic "real-life" work and to offer simple-to-administer testing protocols. Two-, 6- and 12-min walk tests have been used for individuals with CF (38, 40, 75). These protocols allow the subject to walk over a set time period at his or her own pace while heart rate and oxygen saturations are monitored. Total distance traveled, the development of exercise breathlessness, and oxygen desaturation are recorded and can be compared over time with prior tests. Outcome variables have been relatively well correlated to standardized maximal tests for individuals with mild to severe CF lung disease. Shuttle tests have also been used for individuals with CF (16, 17). In one version, the subject walks (or runs) at increasing speeds (set by an audio signal) over a set course. The individual continues with the test until he or she is unable to continue or keep the pace. Heart rate, pulse oximetry, and distance traveled are monitored during this test. Although both the walk tests and shuttle tests are relatively simple to perform, a limitation is that they depend on patient effort, which makes motivation by the test administrator essential. A 3-min step test has been developed as a modification of the Master two-step exercise test used in adult cardiac testing (9). The subject steps on a single step set a standard height (6 in.) at a rate of 30 steps per minute for 3 min. Cadence can be controlled by a metronome. The test is complete after 3 min or when the subject is unable to continue. The total number of steps can be tabulated along with the extent of heart-rate increase, oxygen desaturation, and sensation of breathlessness (71). All of these noninvasive tests are easy to perform and do not require sophisticated equipment. The utility of these tests for assessment of aerobic fitness in milder patients with CF remains unknown.

Muscular Endurance Although many tests have been proposed to assess muscular endurance in healthy children (84), relatively few protocols have been used for children with CF. The Wingate Anaerobic Test (WAnT) has been the most widely used test to assess both short-term mechanical power or strength (measured by peak power) as well as leg muscle endurance over a brief time period (measured by mean power). The WAnT was designed to measure nonoxidative muscle function with peak power indicative of muscle strength and mean power providing information about anaerobic muscle endurance. The test consists of a 30-s all-out sprint on a cycle ergometer against a fixed resistance. Determination of resistance depends on lean muscle mass, but a standard starting point is 75 g of resistance per kilogram of body weight (45). The test is demanding, but children with CF have been able to complete it (15, 18). Although sophisticated equipment is available to perform the WAnT, a mechanical cycle ergometer (e.g., Monarck or Fleisch) can be adapted for the WAnT. Alternative protocols for testing muscle endurance in children with CF include use of an isokinetic cycle ergometer and cycling at supramaximal levels. Both of these protocols have been used in the research setting for testing children with CF (52, 78) but are not readily accessible outside the academic exercise laboratory. Measurements of respiratory and peripheral muscle fatigue have been used as research tools to assess both respiratory and peripheral muscle function in individuals with CF (50, 53).

Muscular Strength Strength of both respiratory muscle and peripheral muscle groups has been assessed in individuals with CF. Peak inspiratory pressure determination specifically measures the muscles used for inspiration and consists of the subject's inspiring a breath of air at residual volume against an occluded airway. The greatest inspiratory subatmospheric pressure that can be developed is recorded. Similarly, peak expiratory pressure measures the strength of the abdominal and accessory muscles of breathing and consists of the subject's exhaling forcefully against an occluded airway, usually at total lung capacity. These maneuvers are relatively easy to perform (74). The equipment required is usually part of a standard body plethysmography system. Alternatively, handheld direct reading manometers or electronic pressure transducers and recorders can be used.

Peripheral muscle strength assessment has not traditionally been used for children with CF. However, standard techniques including use of dynamometers, cable tensiometers, isokinetic muscle testing, and free weights can be applied to test specific muscle groups. The age of the subject will often determine the choice of test. For very young children, a child's own body weight can be used as a resistance tool per the testing criteria of the President's Council on Physical Fitness and Sport (e.g., push-ups, pull-ups) (31). The WAnT also provides some measurement of peripheral muscle strength, as discussed previously. Expected muscular endurance and strength responses in children with CF are listed in table 21.3.

Body Composition Because individuals with CF tend to be lower in both body weight and height than individuals without CF, assessment of lean body mass becomes important. From a clinical perspective, monitoring body composition is an important part of nutritional assessment. Typical body composition responses in CF relative to age and gender matcher peers are listed in table 21.4. For the exercise clinician, documenting body composition is essential in determining workloads needed for testing that

---

TABLE 21.3
Muscle Endurance and Strength for Individuals with Cystic Fibrosis

Parameter	Change
MUSCULAR ENDURANCE[a]	
WAnT mean power; isokinetic cycle ergometry power	Decrease
Muscle efficiency	Decrease
MUSCULAR STRENGTH[b]	
Respiratory muscle strength (PI_{max}, PE_{max})	Decrease
Peripheral muscle strength (dynamometer; grip strength, WAnT peak power)	Decrease

Note. WAnT = Wingate Anaerobic Test; PI_{max} = peak inspiratory pressure; PE_{max} = peak expiratory pressure.
[a]Muscle endurance decreases as lung disease progresses and may reflect impaired nutritional status and intrinsic cellular deficiencies. [b]Decreases as disease progresses and may reflect nutritional status and loss of mechanical efficiency.

TABLE 21.4
Body Composition and Flexibility Assessment for Individuals With Cystic Fibrosis

Parameter	Change
BODY COMPOSITION[a]	
Body weight	Decrease
Lean muscle mass	Decrease
Body mass index	Decrease
Percent body fat (BIA, skinfold assessment)	Decrease
FLEXIBILITY[b]	
Peripheral muscle flexibility (hamstrings, quadriceps)	Decrease
Posture—extent of kyphosis	Increase

Note. BIA = bioelectric impedance analysis.
[a]Decreases reflect increased caloric expenditure with advanced lung disease, poor oral intake, and release of cachectic mediators. [b]Reflects deconditioning associated with decreased activity and/or advanced disease.

---

depend on the amount of muscle mass present as well as for scaling absolute exercise data per muscle mass. Additionally, monitoring other interventions (nutritional or exercise) depends on the distribution of muscle and fat mass. Many techniques exist for determining fat distribution in healthy adults. Some of the underlying assumptions of these techniques are in question for children, for individuals with chronic lung disease, and in conditions associated with electrolyte disturbances. Currently, single-site (triceps) or multiple-site skinfold assessment has been the most commonly used technique for monitoring body fat in children with CF. Skinfold calipers (e.g., Harpenden, Lange) provide an inexpensive means for determining body composition. Use of pediatric reference equations is mandatory (56). An alternative technique for body composition assessment uses bioelectrical impedance analysis through commercially available systems. Both skinfold and bioelectrical impedance assessments are easy to perform, inexpensive, and reproducible in the hands of a trained technician. Although the validity of these techniques for assessment of body composition in children with CF is still in question, these techniques remain the current procedures of choice. A common practice by some CF specialty centers is to use both techniques as a means of establishing internal reliability of measurements.

Flexibility For most individuals with CF, flexibility is not a major limiting factor in exercise performance. As lung disease advances, thoracic kyphosis ensues with associated mechanical inefficiencies seen with exercise (46). Some of these postural changes can be associated with tight hamstrings leading to potential exercise limitations and injury (46). Early identification of these abnormalities through routine assessment of large muscle group range of motion, as part of an exercise assessment, can lead to the formation of stretching programs and stabilization of abnormal posture.

Anticipated Responses

Individuals with CF have impaired exercise tolerance as demonstrated by a reduced maximal oxygen consumption and peak work capacity compared with healthy children (34, 35, 44, 57, 81). The ratio of minute ventilation to the **maximal voluntary ventilation,** a marker of ventilatory limitation, may exceed 100% (normal 70-80%) and worsen as CF lung disease progresses (19, 24). Individuals with CF demonstrate expiratory airflow limitation as evidenced by tidal loop analysis during exercise (5). End tidal carbon dioxide, another marker of ventilatory limitation, increases with exercise and is related to the severity of lung disease (21, 24, 58). Alveolar ventilation appears normal in patients with mild lung disease with a compensatory increase in the tidal volume (21). But as disease severity increases, alveolar hypoventilation becomes evident as the tidal volume approaches and is limited by the vital capacity (55). As this occurs, breathing frequency increases as a compensatory factor but does not adequately provide the minute ventilation necessary for the increased exercise intensity. Gas exchange can also be compromised as evidenced by the lack of increase in the diffusion capacity of the lungs following exercise (87). The cardiovascular system is generally able to keep up with the oxygen demands of the exercising muscle and only becomes compromised with advanced disease. Heart rate and blood pressure responses to exercise appear normal, although a lower peak heart rate is seen as the disease progresses (34).

Muscular efficiency in children with CF can be reduced by up to 25%. The reduced efficiency may reflect altered aerobic pathways at the mitochondrial level (29). The muscular strength of CF patients appears reduced when compared with healthy controls as measured by hamstring and quadriceps force generation or grip strength (28, 37). When muscular strength is normalized by lean body mass, differences do not exist for hamstring or quadriceps strength but persist for grip strength, making conclusions about muscle strength in individuals with CF difficult. Studies using the WAnT have demonstrated decreased anaerobic performance in individuals with CF that is related to muscle mass quantity (15). Overall oxygen cost of work appears elevated during exercise for individuals with CF (30). Additionally, it appears that energy metabolism during exercise is abnormal in children with CF (13, 85).

Exercise Prescription

As lung disease progresses, lung function may become significantly compromised for individuals with CF. Once the high-risk individual is identified, specific precautions may be needed before the initiation of an exercise prescription. Use of supplemental oxygen while exercising, for those individuals who are prone to oxyhemoglobin desaturation, has been beneficial in allowing for successful exercise training and recovery (23, 59, 78). With encouragement from parents and medical personnel, the use of supplemental oxygen should not discourage safe participation in exercise training. Although the clinical guidelines for exercise testing per the American College of Sports Medicine discourage testing in individuals with FEV_1 less than 60% of predicted, with appropriate medical direction, children with significant lung disease associated with CF can safely participate in a formal exercise program (88). Additional concerns exist for the child who will be exercising at high altitude. For these individuals, it would be helpful to determine oxygen saturation during a maximal exercise challenge at sea level to determine the risk of desaturation at high altitude (8). Practical application 21.3 reviews the literature regarding exercise training for patients with CF.

Because children with CF are susceptible to dehydration with prolonged exercise, especially in warm climates, fluid intake should be carefully monitored and encouraged. Fluid intake every 20 to 30 min should suffice. Use of sport rehydration beverages may result in greater consumption of fluid compared with water, although both are acceptable. (See practical application 21.1 on thermoregulation.)

Literature Review

Despite the pathophysiological manifestations of CF, a number of individuals with CF have performed at very high athletic levels, accomplishing many of the same athletic endeavors as their non-CF counterparts. The short-term benefits of exercise for individuals with CF include the therapeutic aspects of enhanced mucus clearance, improved cardiopulmonary fitness, and positive psychological well-being. Furthermore, the psychosocial aspects of improved self-esteem and greater sense of accomplishment are also associated with regular physical activity. The long-term benefits are more difficult to define, because the natural course of CF is complex and multifactorial. All aspects of fitness appear to show positive benefits in conjunction with an exercise training program. Because many of the training programs used in individuals with CF have varied by duration, intensity, and modality, it is difficult to establish causal relationships between training and disease progression. Exercise tolerance and long-term survival have been correlated with one another, but a causal relationship has not been established (64).

Cardiorespiratory Benefits

Some benefits in static pulmonary function have been seen after exercise training. In a group of individuals with advanced lung disease, increases in FEV_1 were seen in response to an in-patient bicycling program consisting of 20 min per session at 75% of prestudy maximal intensity (41). Additionally, improvements were observed in forced vital capacity (FVC), FEV_1, FEF_{25-75}, and peak expiratory flow following intensive exercise (1-hr swim, 2.5-km jog, several hours of hiking per day) during a 17-day elective stay in a pediatric rehabilitation hospital in the mountains of Austria (86). Despite these findings, most exercise training programs have not shown increases in spirometric indexes but rather have resulted in a slower deterioration of lung function when compared with the nonexercising group.

Improvements in dynamic lung function and other parameters dependent on dynamic lung function following exercise training are well documented as evidenced by a lower resting heart rate, improved maximal oxygen consumption, increased maximal heart rate, increased physical work capacity, enhanced ventilatory threshold, and improved maximal minute ventilation. Exercise programs during a hospitalization for an infectious exacerbation can serve as an adjunct to traditional modalities including chest physiotherapy and bronchial drainage and have been associated with improvements in peak oxygen consumption and peak work capacity (1, 20, 72). Formal supervised training programs including running, cycling, and swimming programs, as well as structured camps, have helped to maximize compliance with exercise. However, length of participation and intensity of training vary in these studies, with greater training effects seen in the more intense programs (32, 33, 39, 65).

Muscular Strength and Endurance

The benefits of exercise training on anaerobic function such as muscular strength and endurance have received less attention. Increased ventilatory muscle endurance by up to 56% was seen after a training program that involved swimming and canoeing (50). Exercise programs focusing on respiratory muscle training have also shown training adaptations as demonstrated by increased peak inspiratory pressures (4, 77). Limited data are available on the peripheral skeletal muscle adaptations to exercise. Both weight training and home cycling for 6 months increased muscle strength in individuals with CF, although long-term adherence to these programs is of concern (39, 82).

Body Composition and Nutrition

Children with CF undergoing regular exercise training (e.g., swimming, biking, running, weight lifting) are capable of increasing body mass despite increased caloric needs (6, 20, 42). Nutritional supplementation has been associated with improved aerobic exercise tolerance and respiratory muscle strength in some small case reports (22, 80).

continued

Psychological Well-Being

The long-term psychosocial benefits of exercise have been fairly well established. The Quality of Well-Being scale, designed to measure daily functioning, has been shown to correlate with exercise capacity in individuals with CF (47, 68). Furthermore, improvements in self-concept and well-being have been shown to be associated with involvement in CF summer camps (48, 86). Whether summer camp participation is analogous to exercise training is speculative but may provide some insight into the psychological benefits of regular physical activity.

Exercise Influence on CF Disease

The science of exercise immunology has received tremendous interest over the past decade. The ability of exercise to alter the immune system is now well recognized (62), with several epidemiological studies suggesting that heavy acute or chronic exercise is associated with an increased risk of upper respiratory tract infections (62, 70). This notion is consistent with the belief among the general public and athletes that heavy exertional activities predispose an individual to illnesses, whereas moderate levels offer a protective effect.

The role of exercise in preventing the deterioration of lung function or occurrence of complications associated with CF is unclear. Although studies have linked exercise tolerance with long-term prognosis, the preventive benefits of exercise were not established (61, 64). Although some studies have shown a beneficial effect of chronic exercise on the immune system, this has not been established in individuals with CF (12, 11). Whether the improved prognosis associated with exercise tolerance in CF is mediated by an exercise-enhanced immune modulation is an attractive but speculative theory.

Special Exercise Considerations

Conditions associated with CF such as CF-related diabetes, exercise-induced asthma, and liver disease require special consideration. Specific guidelines from the CF physician should be made for these conditions before an exercise program is initiated. Hypoglycemia and acute bronchospasm can be relatively easy to prevent. Although severe CF-related liver disease is uncommon, its presence can result in a bleeding tendency. Individuals who have either enlarged visceral organs (e.g., liver, spleen) or a bleeding tendency should avoid contact sports. For this subpopulation, appropriate choices for exercise and physical activity are track and field events, swimming, and dance. Practical application 21.4 describes exercise in individuals with advanced lung disease.

Exercise Recommendations

The goals of an exercise program for individuals with CF should include enhancing physical fitness, reducing the severity or recurrence of disease, and ensuring safe and enjoyable participation. To maximize compliance with any exercise program, activity should be selected carefully to enhance cardiopulmonary fitness and other exercise goals as described subsequently.

Because two thirds of all individuals with CF are in the pediatric and adolescent age group, exercise prescriptions should accommodate the special needs of younger participants. Children, especially under the age of 10, often do not respond to a formal structured exercise program. Children respond well when an exercise program matches a child's muscular development, strength, and coordination with age-appropriate activities. Additionally, a gradual progression in the level of physical activity should allow for attainment of exercise goals while minimizing the risk of injury and noncompliance. This concept of gradual progression depends on the fitness parameter that is being addressed and is described next. The use of "play" consisting of games and diversionary tactics may be most beneficial in meeting the preceding criteria while maintaining good compliance and teaching an active lifestyle. Older children (>10 years) may be able to undergo a more structured program built on the mode, intensity, duration, and frequency of exercise.

Although not specifically designed for children, the recommendations for exercise prescription in practical application 21.5 follow the guidelines established by the American College of Sports Medicine (88). These guidelines were developed to address each component of fitness. The guidelines have been adapted for adults with CF as well as for the pediatric

Exercise in Individuals With Advanced Lung Disease

Traditionally, individuals with CF and severe obstructive pulmonary disease have not received the attention of an exercise clinician because of an extremely conservative approach to their exercise participation. Fear of exercise-induced hypoxia leading to pulmonary hypertensive episodes, cardiac ischemia, and dyspnea as well as the perception of limited beneficial effects of training have all been cited as deterrents to regular physical activity. Although concerns of hypoxia exist for individuals with advanced CF, appropriate exercise prescriptions can be safely administered.

The first consideration for an individual with advanced lung disease who wishes to participate in regular physical activity is to determine whether exercise will induce oxyhemoglobin desaturation. A baseline maximal aerobic exercise challenge should be administered with monitoring of pulse oximetry and electrocardiogram. If the subject completes the challenge without desaturation (defined as a value <90% on room air), then supplemental oxygen is not required. Although an individual may desaturate to less than 90% during the challenge, the point at which this occurs becomes critical. Because most aerobic exercise prescriptions use submaximal intensity levels (60-75% of maximal), one should determine if the subject desaturated at this submaximal level. If yes, then supplemental oxygen needs to be administered for subsequent exercise training, whereas it is not required for those who do not desaturate. If supplemental oxygen is required, a repeat exercise challenge should be administered after an appropriate recovery period to document that the subject remains normoxic during exercise. The level of oxygen supplementation required should be recorded.

Special considerations for the person requiring oxygen for exercise should be made. The choice of activity may need to be modified to allow for the presence of oxygen tanks. Activities using stationary modalities (e.g., treadmills, bicycles) may be more appropriate for this group. Many health clubs are capable of accommodating individuals with these special needs.

Individuals with advanced lung disease are capable of demonstrating the beneficial effects of exercise training. Improved gas exchange, ventilation, aerobic tolerance, peripheral muscle adaptations, and sense of well-being have all been documented following exercise training (41, 59).

Appropriate exercise prescriptions can be developed for even the most debilitated individual with CF. This population has not traditionally received the benefits of interacting with an exercise clinician. One could argue that individuals with severe CF have the most gains to make from exercise that would help reestablish functional ability. By understanding the particular needs of this population, the exercise clinician can have a major impact in helping to achieve this goal.

---

population (27). Including the parent in the development of an exercise program is vital to its success, because many parents falsely perceive potentially negative consequences of exercise for their child with CF (e.g., weight loss) (14).

Cardiorespiratory Training

The main objective of cardiorespiratory training is to improve aerobic capacity. Higher levels of aerobic fitness have been associated with better quality of life and improved survival rates. The components of the exercise prescription are reviewed here to optimize the client's exercise training program.

Mode   No specific activity has been identified as optimal for patients with CF and those with more severe lung dysfunction. Choice of modality depends on the subject's personal preference and need not be costly. Cardiopulmonary benefits have been seen

with a multitude of activities, some of which require little or no equipment (e.g., walking, jogging). Treadmills, bicycles, or rowing machines can all be incorporated into a successful exercise program. For those patients who experience desaturation, exercising in an environment where supplemental oxygen is available may limit some choices but should not prohibit participation in an exercise program.

Frequency   To ensure improvements in the cardiorespiratory conditioning of individuals with CF, 3 to 5 days of exercise per week appear optimal. Exercising beyond five times per week may lead to an increased risk of injury, especially if the intensity of exercise is too high. If more than five times per week is desired, then the use of cross-training (e.g., strength training, stretching) is advised to allow for adequate muscle recovery. Signs of increased fatigue and staleness may be a result of over-

PRACTICAL APPLICATION 21.5

Exercise Prescription

	DISEASE SEVERITY: MILD[a] TO MODERATE			DISEASE SEVERITY: SEVERE[b]		
	Aerobic	**Anaerobic**	**Flexibility**	**Aerobic**	**Anaerobic**	**Flexibility**
Type	Any enjoyable aerobic activity: biking, swimming, walking, jogging, team sports.	Sprinting, shuttle runs, plyometrics, weightlifting with supervision.	Stretching and yoga can be used for relaxation and improved flexibility.	Supplement oxygen if needed, which may limit choices; stationary ergometers work well.	Increasing disease state = increased risk. Light weights can be used to maintain muscle tone.	Stretching to enhance chest wall mobility, body relaxation, and general flexibility.
Frequency	3-5 days per week	3-5 days per week	2-7 days per week	3-5 days per week	1-2 days per week	2-7 days per week
Intensity	70-80% of measured maximum.	10-12 repetitions, low resistance.	Pain-free, 10-30 s per stretch.	Measuring a true maximum is vital because pulmonary limitations limit peak heart rate to 70-80%.	Very light resistance, limit activities that would induce a **Valsalva maneuver.**	Pain free, 10-30 s per stretch.
Duration	20-60 min	20-30 min	10 min	20-60 min	10-20 min	10 min
Progression	No more than 10% in any given 2-week period.	Work upper and lower body, progress when 10-12 reps are no longer challenging.	Natural progression as flexibility improves.	No more than 10% in any given 2-week period.	Minimal progression, maintaining aerobic activities will reduce muscle wasting.	Natural progression as flexibility improves.
Retest	Yearly: cycle or treadmill ergometer.	Yearly: Wingate grip strength, shuttle run ±1-repetition maximum.	Yearly: sit-and-reach or alternative tests.	Yearly: cycle or treadmill ergometer. 6- to 12-min walk.	Wingate or alternative.	Yearly: sit-and-reach.

[a]Mild to moderate disease severity: Minimal risk of desaturation, no sport or activity restrictions; forced expiratory volume in 1 s (FEV_1) >50%, and a diffusion capacity (DL_{CO}) >65%. [b]Severe disease severity: Increased risk of desaturation, which may require supplemental oxygen; FEV_1 <50% or DL_{CO} <65%.

training and should prompt a reduction in exercise frequency.

Intensity Exercise intensity should range from 70% to 85% of the measured maximum heart rate, with lower intensities used for beginners. If a maximum cardiopulmonary test is not performed, then the general rule of using a maximum heart rate of 200 beats · min^{-1} for intense running and 195 for cycling can be applied for children and adolescents. Once a child has completed puberty, the formula of 220 minus age can be applied as it would for adults. Thus, a heart rate of 140 to 170 beats · min^{-1} is a reasonable estimate of the heart rate that should be

attained for optimal cardiopulmonary benefit. The use of a steady-state protocol (e.g., treadmill or cycle ergometer) can ensure that the appropriate workload is established. This is easily done by choosing a submaximal work load and having the subject exercise for 10 min while measuring heart rate throughout. This appropriate intensity of exercise is necessary for obtaining positive gains in cardiorespiratory fitness.

Duration Exercise sessions should last 20 to 60 min with the option of two abbreviated sessions providing similar benefits. Attention span may play a role in the child's ability to perform an activity for longer than 10 min. For the younger patient, varying the exercise sessions by interspersing different activities may minimize boredom and enhance compliance (e.g., 5 min on the bike followed by 15 jumping jacks followed by 5 min on the treadmill at various speeds and grades).

Progression   Because too rapid a progression in the exercise dose may cause the patient to lose enthusiasm for the activity, special attention should be given to advances in the exercise prescription. As in adults, no more than a 10% increase in activity duration should occur between any 2-week period during the exercise program. Frequency can gradually be increased from 3 times per week for the beginner to 5 times per week over a 3-month period. Finally, the progression of an individual's program should be based on the goals and the individual's particular needs.

Muscular Strength and Endurance Training
Individuals with CF may benefit from resistance training through both a generalized increase in muscle strength and a decrease in residual air trapped in pulmonary dead space. Although additional research is needed to quantify the added benefits of resistance training, special considerations are discussed next.

Mode   Free weights, weight machines, or resistance against one's own body weight can all be used to enhance peripheral muscle strength and endurance, provided proper direction is given to individuals with CF. Plyometrics, sprinting, cycling, and other modalities that require high-intensity, short-burst duration can also develop muscle strength and endurance. Individuals with CF can perform plyometrics in an age-appropriate manner if proper technique and supervision are given. Other modalities that address multiple muscle groups are ideal for enhancing muscular strength and endurance. Respiratory muscle training should be done in consultation with a clinician trained in respiratory disorders.

Frequency   A frequency of three to five times a week is usually appropriate. Care should be taken to allow for adequate muscle recovery. By alternating major muscle groups when training, one can minimize muscle injury while maximizing training effects. Subtle signs of overuse injuries (e.g., muscle soreness, joint pain) should prompt a reduction in the frequency.

Intensity   Muscle strength and endurance can be optimized through high-repetition, low-intensity resistance training or through other modalities as

earlier described. The American Academy of Pediatrics Committee on Sports Medicine does not recommend high-intensity resistance training for children because of the potential of musculoskeletal injury, epiphyseal fractures, ruptured intervertebral disks, and growth plate injury until a child reaches full maturation (Tanner stage 5) (3). However, strength training programs that use lower intensity weights and modalities can be permitted if the planned program is appropriate for the child's stage of maturation (3). Whether adults with CF can safely participate in weight lifting is controversial. Lifting heavy weights can be associated with a Valsalva maneuver with resultant increased thoracic pressures. In the susceptible patient with CF (e.g., prior history of **pneumothorax**, advanced lung disease), this increased thoracic pressure may result in a spontaneous pneumothorax. Consultation with a CF clinician before initiation of a weight-lifting program is strongly encouraged. Once an individual is cleared for participation, a resistance of 50% to 60% of 1-repetition maximum is generally used. Three sets of 12 or more repetitions should produce favorable strength gains.

Duration   In children and adolescents with CF, the duration of each session depends on the number of muscle groups exercised. Generally, 10 to 30 min of properly performed activities can increase muscular strength and endurance.

Progression   In individuals with CF, the progression of a strength program should be slow. Repetitions or resistance should only be increased when the muscle has adapted to the current workload. The progression for increased resistance should occur when the individual is able to perform 8 to 12 repetitions without fatigue to the muscle. For non-weight-lifting activities, activity should progress by no more than 10% each 2-week period.

Flexibility Training
Generalized flexibility exercises should be an adjunct to any exercise program. Adequate range of motion is essential for minimizing risk of skeletal injury and ensuring healthy aging.

Mode   Because stretching can be performed with little to no equipment (e.g., floor mat), it is one of the easiest aspects of fitness to address. Stretching may provide a protective mechanism against injury, and it can also be a source of tension release and relaxation. Stretching exercises should focus on large muscle groups and should be included before and after activity as part of an effective warm-up and cool-down.

Frequency   Stretching exercises should be considered a routine component of any exercise program. Stretching can occur before, during, and after an exercise session depending on the activity chosen. A stretching program of 2 or more days per week can yield positive results with a decreased tightening of the hamstrings and quadriceps and more efficient use of respiratory muscles during exercise. Stretching for relaxation or tension relief can be performed daily. For individuals with specific flexibility issues (e.g., posture abnormalities, tight hamstrings), a more comprehensive stretching program may be needed.

Intensity   Proper stretching technique will help ensure that the intensity used for stretching is appropriate. A proper stretch often feels like a gentle pull in the muscle and should not be forced. Proper breathing including exhaling prior to the stretch and inhaling afterward will help minimize injury. Finally, slowly releasing a stretch back into a neutral position will allow the muscle to recover.

Duration   The American College of Sports Medicine guidelines suggest that a stretch should last between 10 and 30 s (88). The use of progressive stretching that includes a 10- to 30-s stretch followed by an additional 10 to 30 s has been proposed as well.

Progression   The flexibility of an individual will gradually improve as the muscle adapts to an increased stretch. Slow progression of a stretching program should occur over a 5-week period. The length at which a stretch is held can be gradually progressed from an initial 10 to 30 s to 40 to 50 s by the end of 5 weeks.

Conclusion

Individuals with CF appear to benefit from exercise training programs. Both static and dynamic pulmonary function either improve or have a slower rate of deterioration following training programs. Improved muscle strength and endurance following exercise programs are well documented. Body weight can be successfully maintained or increased during exercise intervention. Improvements in psychological well-being have also been observed. Finally, the potential impact of exercise on prognosis may make exercise a quite attractive therapeutic modality.

Case Study 21

Medical History and Diagnosis

Mr. EM is a 15-year-old Caucasian male who was diagnosed with cystic fibrosis at 8 months of age secondary to recurrent respiratory infections and failure to thrive. Mr. EM has done relatively well with intermittent respiratory infections requiring antibiotic and hospital therapy. Because of his inability to consume adequate calories, a gastrostomy tube was placed to allow for supplemental nocturnal nutrition. Mr. EM is very self-conscious about his gastrostomy tube and finds it difficult to partake in any physical activity including school physical education classes.

Exercise Evaluation

An exercise evaluation was performed as part of Mr. EM's medical care. An activity questionnaire revealed that Mr. EM enjoys individualized sports as opposed to team sports. Although he owns a bicycle and treadmill, he admits to not using them to any degree because of boredom and lack of motivation.

- Pulmonary function tests: FEV_1 = 68% of predicted; FEF_{25-75} = 42% of predicted
- Resting pulse oximetry = 95% on room air
- Residual volume = 229% of predicted
- Diffusing capacity of the lungs = 82% of predicted

continued

Maximal graded ergometry test (Godfrey protocol) results:

— Physical work capacity = 120 W (68% of predicted)
— $\dot{V}O_2$peak = 32.1 ml $\cdot$ kg^{-1} $\cdot$ min^{-1} (59% of predicted)
— Peak end tidal CO_2 = 42 mmHg
— Lowest exercise oxygen saturation = 92%
— Ratio of minute ventilation to the maximal voluntary ventilation = 104%
— Resting heart rate = 88 beats $\cdot$ min^{-1}
— Peak heart rate = 190 beats $\cdot$ min^{-1}
— Body composition: weight = 48.6 kg (10th percentile for age); height = 162 cm (5th percentile for age)
— Body mass index = 18.5 kg $\cdot$ m^{-2}
— Percent body fat (bioelectrical impedance analysis) = 15%
— Flexibility: good posture; mild hamstring tightness

Exercise Prescription

Overall, Mr. EM has moderate obstructive pulmonary disease with mild ventilatory limitations to exercise. Although oxygen saturations decrease with exercise, he can safely participate in aerobic activities without adverse effect (but probably not at altitude). Overall, nutrition is only fair with lower than expected lean body mass. From a psychological perspective, Mr. EM appears to be typical of teenagers with a lack of motivation for exercise and a concern over body image.

The following program was developed to improve cardiorespiratory fitness and to effectively use the calories being given by his nocturnal feedings.

— Mr. EM was encouraged to select individualized activities where he could potentially excel (e.g., tennis, cross country, track, dance), three times per week, 20 to 30 min, moderate intensity (goal heart rate of 140 beats $\cdot$ min^{-1}), progression to four to five times a week at 20 to 30 min within 6 months.
— As a component of each exercise session, 15 min of stretching during warm-up and cool-down that targets large muscle groups with emphasis on hamstring flexibility was taught.
— Personal motivation strategies were used to enhance compliance with program.
— Mr. EM will have a friend join the local YMCA where both can play tennis.
— He will initiate a jogging program, working toward trying out for a cross country team at school.
— The clinician will discuss with parents their role in encouraging Mr. EM and the use of an appropriate reward system (e.g., will pay for membership to YMCA and running shoes if Mr. EM attends).
— Exercise clinician will undertake frequent telephone follow-up with Mr. EM and formal reevaluation in 6 months by to monitor progress, reestablish new goals, and offer motivation.

Case Study 21 Discussion Questions

1. How does Mr. EM's gastrostomy tube pose a challenge in designing an exercise prescription?
2. What is the role of the exercise clinician in optimizing compliance with the exercise program?
3. What anticipatory counseling should be offered prior to initiation of an exercise program?
4. What steps should the exercise clinician take should Mr. EM not meet his goals or prove to be noncompliant?

Glossary

bronchiectasis—Chronic dilatation of a bronchus usually associated with secondary bacterial infection.

cystic fibrosis transmembrane conductance regulator (CFTR)—a protein responsible for moving salt across cell membranes.

dynamic pulmonary function—Lung function in response to changing physiological conditions.

FEF$_{25-75}$—Forced expiratory flow occurring between 25% and 75% of the forced vital capacity; assessed via a spirometer; reflects lower airway flow.

FEV$_1$—Forced expiratory volume in 1 s; marker of airway obstruction; an individual maximally inhales and then exhales into a spirometer with the first second recorded.

hemoptysis—Expectoration of blood arising from the respiratory system; for individuals with CF, reflects further infection or advancing disease.

hypoxemia—Insufficient oxygenation of the blood; assessed by arterial blood gas or pulse oximetry.

kyphosis—Excessive angulation of the spine resulting in an increased anteroposterior diameter of the chest cavity; humpback; may reflect chronic pulmonary disease.

maximal voluntary ventilation—Amount of air maximally breathed in expressed as liters per minute.

nasal polyposis—Growths of tissue in the nose that may block the air passage through the nostril; not life threatening.

oxyhemoglobin saturation—Percentage of hemoglobin bound to oxygen; assessed noninvasively by pulse oximeter or invasively via arterial blood gas sampling.

pancreatic insufficiency—Inadequate exocrine function of the pancreas resulting in little or no production of pancreatic enzymes needed for digestion (i.e., lipase, amylase, protease); results in nutrient malabsorption.

pansinusitis—Chronic inflammation and infection involving all sinuses; commonly seen in individuals with CF.

pneumothorax—An acute collection of air in the pleural space; results in collapse of the affected lung; common in advancing CF lung disease.

pulmonary exacerbation—An episode of worsening lung disease caused by increased infection and inflammation.

scoliosis—Lateral curvature of the spine.

static pulmonary function—Properties of the lung measured at rest or baseline.

sweat test—Diagnostic test used for CF; usually involves stimulation of the skin's sweat glands by chemical (i.e., pilocarpine) and electrical means (i.e., iontophoresis); an elevation of greater than 60 mEq · dl^{-1} is highly suggestive of CF.

Valsalva maneuver—Forced exhalation with the glottis, nose, and mouth closed resulting in increased intrathoracic pressure, slowing of the heart rate, and decreased return of blood to the heart.

References

1. Alison JA, Donnelly PM, Lennon M, Parker S, Torzillo P, Mellis C, Bye PTP. The effect of a comprehensive, intensive inpatient treatment program on lung function and exercise capacity in patients with cystic fibrosis. Phys Ther 1994; 74: 583-593.

2. Alison J, Duong B, Robinson M, Regnis J, Donnelly P, Bye PTP. Level of aerobic fitness and survival in adults with cystic fibrosis. Am J Respir Crit Care Med 1997; 155 (Part 2): A642.

3. American Academy of Pediatrics: Committee on Sports Medicine Policy Statement. Strength training, weight and power lifting, and body building by children and adolescents. Pediatrics 1990; 86: 801-803.

4. Asher MI, Pardy RL, Coates AL, Thomas E, Macklem PT. The effects of inspiratory muscle training in patients with CF. Am Rev Respir Dis 1982; 126: 855-859.

5. Babb TG. Mechanical ventilatory constraints in aging, lung disease, and obesity: Perspectives and brief review. Med Sci Sports Exerc 1999; 31 (Suppl 1): S12-S22.

6. Bakker W. Nutritional state and lung disease in cystic fibrosis. Neth J Med 1992; 41: 130-136.

7. Baldwin DR, Hill AL, Peckham DG, Knox AJ. Effect of addition of exercise to chest physiotherapy on sputum expectoration and lung function in adults with cystic fibrosis. Respir Med 1994; 88: 49-53.

8. Balfour-Lynn IM, Carr SB, Madge S, Prasad A, MacAlister L, Laverty A, Dinwiddie R. Effect of altitude on exercise testing in children with cystic fibrosis. Proceedings of the 20th European Working Group on Cystic Fibrosis, 1995. Brussels.

9. Balfour-Lynn IM, Prasad SA, Laverty A, Whitehead BF, Dinwiddie R. A step in the right direction: Assessing exercise tolerance in children with cystic fibrosis. Pediatr Pulmonol 1998; 25: 278-284.

10. Bar-Or O, Blimkie CJ, Hay JA, MacDougall JD, Ward DS, Wilson WM. Voluntary dehydration and heat intolerance in cystic fibrosis. Lancet 1992; 339: 696-699.

11. Boas SR, Danduran MJ, McBride AL, McColley SA, O'Gorman MRG. Post exercise immune correlates in children with and without cystic fibrosis. Med Sci Sports Exerc 2000; 32: 1997-2004.

12. Boas SR, Danduran MJ, McColley SA, Beeman K, O'Gorman MRG. Immune modulation following aerobic exercise in children with cystic fibrosis. Int J Sports Med 2000; 21: 294-301.

13. Boas SR, Danduran MJ, McColley SA. Energy metabolism during anaerobic exercise in children with cystic fibrosis and asthma. Med Sci Sports Exerc 1999; 31: 1242-1249.

14. Boas SR, Danduran MJ, McColley SA. Parental attitudes about exercise in cystic fibrosis. Int J Sports Med 1999; 20: 334-338.

15. Boas SR, Joswiak ML, Nixon PA, Fulton JA, Orenstein DM. Factors limiting anaerobic performance in adolescent males with cystic fibrosis. Med Sci Sports Exerc 1996; 28: 291-298.

16. Bradley J, Howard J, Wallace E, Elborn S. Validity of a modified shuttle test in adult cystic fibrosis. Thorax 1999; 54: 437-439.

17. Bradley J, Howard J, Wallace E, Elborn S. Reliability, repeatability, and sensitivity of the modified shuttle test in adult cystic fibrosis. Chest 2000; 117: 1666-1671.

18. Cabrera ME, Lough MD, Doershuk CF, DeRivera GA. Anaerobic performance—assessed by the Wingate test—in patients with cystic fibrosis. Pediatr Exerc Sci 1993; 5: 78-87.

19. Canny GJ. Ventilatory response to exercise in cystic fibrosis. Acta Paediatr Scand 1985; 74: 451-452.

20. Cerny FJ, Cropp GJA, Bye MR. Hospital therapy improves exercise tolerance and lung function in cystic fibrosis. Am J Dis Child 1984; 138: 261-265.

21. Cerny FJ, Pullano TP, Cropp GJA. Cardiorespiratory adaptations to exercise in cystic fibrosis. Am Rev Respir Dis 1982; 126: 217-220.

22. Charge TD, Drury D, Pianosi P, Kopelman H, Coates AL. Nutritional rehabilitation and changes in respiratory strength, function, and maximal exercise capacity in cystic fibrosis. Am Rev Respir Dis 1991; 143 (Suppl): A300.

23. Coates AL. Oxygen therapy, exercise, and cystic fibrosis. Chest 1992; 101: 2-4.

24. Cropp GJ, Pullano TP, Cerny FJ, Nathanson IT. Exercise tolerance and cardiorespiratory adjustments at peak work capacity in cystic fibrosis. Am Rev Respir Dis 1982; 126: 211-216.

25. Cystic Fibrosis Foundation, Homeline Newsletter; Bethesda, MD, January 1997.

26. Cystic Fibrosis Foundation, Patient Registry, 1997; Annual Data Report, Bethesda, MD, September 1998.

27. Danduran MJ, Boas SR. Fibrosi cistica ed esercizio fisico. Il Fisioterapista 2000; 6: 36-41.

28. Darbee J, Watkins M. Isokinetic evaluation of muscle performance in individuals with cystic fibrosis. Pediatr Pulmonol 1987; 3 (Suppl): 140-141.

29. deMeer K, Jeneson JAL, Gulmans VAM, van der Laag J, Berger R. Efficiency of oxidative work performance of skeletal muscle in patients with cystic fibrosis. Thorax 1995; 50: 980-983.

30. deMeer K, Gulmans VAM, van der Laag J. Peripheral muscle weakness and exercise capacity in children with cystic fibrosis. Am J Respir Crit Care Med 1999; 159: 748-754.

31. Docherty D. Field tests and test batteries. In: *Measurement in Pediatric Exercise Science*. D Docherty (Ed). Champaign, IL: Human Kinetics, 1996, pp. 285-327.

32. Dunlevy CL, Douce FH, Hill E, Baez S, Clutter J. Physiological and psychological effects of low-impact aerobic exercise on young adults with cystic fibrosis. J Cardiopulm Rehabil 1994; 14: 47-51.

33. Edlund LD, French RW, Herbst JJ, Ruttenberg HD, Ruhling RO, Adams TD. Effects of a swimming program on children with cystic fibrosis. Am J Dis Child 1986; 140: 80-83.

34. Freeman W, Stableforth DE, Cayton RM, Morgan MDL. Endurance exercise capacity in adults with cystic fibrosis. Respir Med 1993; 87: 541-549.

35. Godfey S, Mearns M. Pulmonary function and response to exercise in cystic fibrosis. Arch Dis Child 1971; 46: 144-151.

36. Goldring RM, Fishman AP, Turino GM, Cohen HI, Denning CR, Andersen DH. Pulmonary hypertension and cor pulmonale in cystic fibrosis of the pancreas. J Pediatr 1964; 65: 501-524.

37. Griffiths DM, Miller L, Flack E, Connett GJ. Reduced grip strength in children with cystic fibrosis. Neth J Med 1999; 54: S62.

38. Guillen MAJ, Posadas AS, Asensi JRV, Moreno RMG, Rodriguez MAN, Gonzalez AS. Reproducibility of the walking test in patients with cystic fibrosis. An Esp Pediatr 1999; 51: 475-478.

39. Gulmans VAM, deMeer K, Brackel HJL, Faber JAJ, Berger R, Helders PJM. Outpatient exercise training in children with cystic fibrosis: Physiological effects, perceived competence, and acceptability. Pediatr Pulmonol 1999; 28: 39-46.

40. Gulmans VAM, van Veldhoven NHMJ, de Meer K, Helders PJM. The six-minute walking test in children with cystic fibrosis: Reliability and validity. Pediatr Pulmonol 1996; 22: 85-89.

41. Heijerman HGM, Bakker W, Sterk PJ, Dijkman JH. Oxygen-assisted exercise training in adult cystic fibrosis patients with pulmonary limitation to exercise. Int J Rehabil Res 1991; 14: 101-115.

42. Heijerman HGM. Chronic obstructive lung disease and respiratory muscle function: The role of nutrition and exercise training in cystic fibrosis. Respir Med 1993 B: 49-51.

43. Henke KG, Orenstein DM. Oxygen saturation during exercise in cystic fibrosis. Am Rev Respir Dis 1984; 129: 708-711.

44. Hjeltnes N, Stanghelle JK, Skyberg D. Pulmonary function and oxygen uptake during exercise in 16 year old boys with cystic fibrosis. Acta Paediatr Scand 1984; 73: 548-553.

45. Inbar O, Bar-Or O, Skinner JS. 1996. *The Wingate Anaerobic Test*. Champaign, IL: Human Kinetics.

46. Jenesma M, Concannon D, Gallagher CG. An evaluation of spinal posture and hamstring muscle flexibility in cystic fibrosis adults. Pediatr Pulmonol 1998; 17: 350.

47. Kaplan RM, Anderson JP, Wu AW, Mathews WC, Kozin F, Orenstein DM. The quality of well-being scale. Application in AIDS, cystic fibrosis, and arthritis. Med Care 1989; 27 (Suppl): 27-43.

48. Kaplan TA, McKey RM, Toraya N, Moccia G. Impact of CF summer camp. Clin Pediatr 1992; 31: 161-167.

49. Kaplan TA, Moccia G, McKey RM. Unique pattern of pulmonary function after exercise in patients with cystic fibrosis. Pediatr Exerc Sci 1994; 6: 275-286.

50. Keens TG, Krastius IRB, Wannemaker EM, Levison H, Crozier DN, Bryan AC. Ventilatory muscle endurance training in normal subjects and patients with cystic fibrosis. Am Rev Respir Dis 1977; 116: 853-860.

51. Kriemler S, Wilk B, Schurer W, Wilson WM, Bar-Or O. Preventing dehydration in children with cystic fibrosis who exercise in the heat. Med Sci Sports Exerc 1999; 31: 774-779.

52. Lands LC, Heigenhauser GJ, Jones NL. Analysis of factors limiting maximal exercise performance in cystic fibrosis. Clin Sci 1992; 83: 391-397.

53. Lands LC, Heigenhauser GJF, Jones NL. Respiratory and peripheral muscle function in cystic fibrosis. Am Rev Respir Dis 1993; 147: 865-869.

54. Lebecque P, Lapierre JG, Lamarre A, Coates AL. Diffusion capacity and oxygen desaturation effects on exercise in patients with cystic fibrosis. Chest 1987; 91: 693-697.

55. Levison H, Cherniack RM. Ventilatory cost of exercise in chronic obstructive pulmonary disease. J Appl Physiol 1968; 25: 21-27.

56. Lohman, T. *Advances in Body Composition Assessment.* Champaign, IL: Human Kinetics, 1992.

57. Loutzenhiser JK, Clark R. Physical activity and exercise in children with cystic fibrosis. J Pediatr Nurs 1993; 8: 112-119.

58. Marcotte JE, Grisdale RK, Levison H, Coates AL, Canny GJ. Multiple factors limit exercise capacity in cystic fibrosis. Pediatr Pulmonol 1986; 2: 274-281.

59. Marcus CL, Bader D, Stabile M, Wang CI, Osher AB, Keens TG. Supplemental oxygen and exercise performance in patients with severe cystic fibrosis. Chest 1992; 101: 52-57.

60. McKone EF, Barry SC, FitzGerald MX, Gallagher CG. Reproducibility of maximal exercise ergometer testing in patients with cystic fibrosis. Chest 1999; 116: 363-368.

61. Moorcroft AJ, Dodd ME, Webb AK. Exercise testing and prognosis in adult cystic fibrosis. Thorax 1997; 52: 291-293.

62. Nieman DC. The immune response to prolonged cardio-respiratory exercise. Am J Sports Med 1996; 24: S98-S103.

63. Nixon PA, Orenstein DM. Exercise testing in children. Pediatr Pulmonol 1988; 5: 107-122.

64. Nixon PA, Orenstein DM, Kelsey SF, Doershuk CF. The prognostic value of exercise testing in patients with cystic fibrosis. N Engl J Med 1992; 327: 1785-1788.

65. Orenstein DM, Franklin BA, Doershuk CF, Hellerstein HK, Germann KJ, Horowitz JG, Stern RC. Exercise conditioning and cardiopulmonary fitness in cystic fibrosis: The effects of a three-month supervised running program. Chest 1981; 80: 392-398.

66. Orenstein DM, Henke KG, Costill DL, Doechuk CF, Lemon PJ, Stern RC. Exercise and heat stress in cystic fibrosis patients. Pediatr Res 1983; 17: 267-269.

67. Orenstein DM, Henke KG, Green CG. Heat acclimation in cystic fibrosis. J Appl Physiol 1984; 57: 408-412.

68. Orenstein DM, Nixon PA, Ross EA, Kaplan RM. The quality of well-being in cystic fibrosis. Chest 1989; 95: 344-347.

69. Pate RR, Barabowski T, Dowda M, Trost S. Tracking of physical activity in young children. Med Sci Sports Exerc 1996; 28: 92-96.

70. Peters EM, Bateman ED. Ultramarathon running and upper respiratory tract infections: An epidemiologic survey. South Afr Med J 1983; 64: 582-584.

71. Prasad SA, Randall SD, Balfour-Lynn IM. Fifteen-count breathlessness score: An objective measure for children. Pediatr Pulmonol 2000: 30: 56-62.

72. Rachinsky SV, Kapranow NI, Tatochenko VK, Simonowa OI, Turina JE. Submaximal physical loads in cystic fibrosis. Acta Univ Carol 1990; 36: 198-200.

73. Rowland TM. Aerobic exercise testing protocols. In: *Pediatric Laboratory Exercise Testing: Clinical Guidelines.* TW Rowland (Ed). Champaign, IL: Human Kinetics, 1991, pp. 19-41.

74. Ruppel GL. Spirometry and related test. In: *Manual of Pulmonary Function Testing.* GL Ruppel (Ed). St. Louis: Mosby, 1998, pp. 52-53.

75. Ruter K, Staab D, Magdorf K, Kleinau I, Paul K, Hetzer R, Wahn U. The 12-minute walk test as assessment for lung transplantation in CF patients. Neth J Med 1999; 54 (Suppl): S58.

76. Ryujin DT, Samuelson WM, Marshall BC. Oxygen saturation in adult cystic fibrosis patients during exercise at 1500 meters above sea level. Pediatr Pulmonol 1998; 17 (Suppl): 331.

77. Sawyer EH, Clanton TL. Improved pulmonary function and exercise tolerance with inspiratory muscle conditioning in children with cystic fibrosis. Chest 1993; 104: 1490-1497.

78. Shah AR, Gozal D, Keens TG. Determinants of aerobic and anaerobic exercise performance in cystic fibrosis. Am J Respir Crit Care Med 1998; 157: 1145-1150.

79. Silverman M, Hobbs FDR, Gordon IRS, Carswell F. Cystic fibrosis, atopy, and airways lability. Arch Dis Child 1978; 53: 873-877.

80. Skeie B, Askanazi J, Rothkopf MM, Rosenbaum SH, Kvetan V, Ross E. Improved exercise tolerance with long term parenteral nutrition in cystic fibrosis. Crit Care Med 1987; 15: 960-962.

81. Stanghelle JK, Hjeltnes N, Michalsen H, Bangstad HJ, Skyberg D. Pulmonary function and oxygen uptake during exercise in 11 year old patients with cystic fibrosis. Acta Paediatr Scand 1986; 75: 657-661.

82. Strauss GD, Osher A, Wang C, Goodrich E, Gold F, Colman W, Stabile M, Dobrenchuk A, Keens TG. Variable weight training in cystic fibrosis. Chest 1987; 92: 273-276.

83. van Haren EHJ, Lammers JWJ, Festen J, van Herwaarden CL. Bronchial vagal tone and responsiveness to histamine, exercise and bronchodilators in adult patients with cystic fibrosis. Eur Respir J 1992; 5: 1083-1088.

84. Van Praagh E, Franca NM. Measuring maximal short-term power output during growth. In: *Pediatric Anaerobic Performance*. E Van Praagh (Ed). Champaign, IL: Human Kinetics, 1998, pp. 151-190.

85. Ward SA, Tomezsko JL, Holsclaw DS, Paolone AM. Energy expenditure and substrate utilization in adults with cystic fibrosis and diabetes mellitus. Am J Clin Nutr 1999; 69: 913-919.

86. Zach M, Oberwaldner B, Hausler F. Cystic fibrosis: Physical exercise versus chest physiotherapy. Arch Dis Child 1982; 57: 587-589.

87. Zelkowitz PS, Giammona ST. Effects of gravity and exercise on the pulmonary diffusing capacity in children with cystic fibrosis. J Pediatr 1969; 74: 393-398.

88. Zwiren LD, Manos TM. Exercise testing and prescription considerations throughout childhood. In: *ACSM's Resource Manual for Guidelines for Exercise Testing and Prescription*. JL Rottman (Ed). Baltimore: Williams & Wilkins, 1998, pp. 507-515.

22

John R. Schairer, DO
Advanced Cardiovascular Health Specialists
Livonia, MI

Steven J. Keteyian, PhD
Henry Ford Heart and Vascular Institute
Detroit, MI

Cancer

Both the words *cancer* and *malignancy* are commonly used for the medical term **neoplasm.** Neoplasm describes a disease characterized by the uncontrolled growth of aberrant cells with local tissue invasion and/or systemic **metastasis.** Cancer is unique compared with other diseases because it represents a collection of diseases that can originate in any organ system, can spread to other organ systems, and has multiple etiologies. Cancer affects all nationalities, races, and ages, as well as both men and women. The treatment can vary with each cancer and can include chemotherapy, radiation therapy, biotherapy, or surgery, individually or in combination.

Any discussion about the role of exercise for patients with cancer is multidimensional. Physical inactivity is often cited as a risk factor for developing cancer, and exercise is now increasingly used as an adjunctive treatment for the exercise intolerance that often occurs attributable to the disease and its treatment. Also, exercise has been hypothesized to enhance immunity and alter function within the endocrine system, which could be beneficial in primary prevention and as an adjunct to treatment. For the purpose of our discussion, we focus primarily on the three most common cancers: lung, breast, and colon/rectum.

Scope

Cancer is the major cause of morbidity and mortality throughout the world. In the United States, cancer ranks as the second leading cause of death, behind cardiovascular disease. It is estimated that 1.3 million new cases of cancer are diagnosed in the United States annually. Forty percent of Americans will develop cancer during their lifetime, and 560,000 Americans die annually (8). Unlike heart disease and stroke, which have declined since the 1960s, the age-adjusted cancer mortality rate is increasing (6). In 1930, the age-adjusted death rate from cancer was 143 per 100,000 population. It rose to 163 in 1970, and was 171 in 1989 (3). Within the next 25 years, cancer will become the leading cause of death in the United States.

The annual economic burden to care for the 100 million Americans who will develop cancer in their lifetime is estimated to be more than $35 billion for hospital costs and physician services and another $12 billion in lost productivity (11, 35). However, the story is not completely bleak. With continued improvement in diagnosis and treatment, cancer patients are living longer. The 5-year survival rate for all cancers has increased to almost 60%. There are 8.2 million Americans alive today who have survived cancer.

Cancer is found throughout the world, with about a threefold difference between those countries having the highest frequency of cancer and those with the least. Geographic variation can be as much as 100-fold for specific cancers. For example, esophageal cancer mortality is very high in South Africa, China, and Iran. It is much less frequent in the United States, except for those areas that have a high incidence of alcoholism. The most common cancers in the Western countries are lung, large bowel, and breast, whereas in southeast China, nasopharyngeal cancer is the most common malignancy.

Haenszel et al. (29) and Muir et al. (44) demonstrated that migrating populations tend to acquire the cancer incidence profile of their new country of residence, suggesting that genetics is less important in the genesis of cancer than environmental influences. Results from epidemiological studies show that more than two-thirds of cancer deaths might be prevented through lifestyle modification (18). One-third of cancer deaths are due to cigarette smoking, and another one-third are attributed to alcohol use, specific sexual practices, pollution, and dietary factors.

There are differences between men and women in the type and frequency of cancer and the likelihood of dying of cancer. For nearly all cancers, the incidence rates are higher in men than in women. The exceptions are thyroid, gallbladder, and, of course, breast and uterine cancer. Among men, the most common cancers in order of prevalence are prostate, lung, and colon/rectum. With women, the most common cancers are breast, colon/rectum, and lung. If death from cancer is considered, the order of the cancers changes. More men die of lung cancer followed by colon/rectum cancer and finally prostate cancer. Among women, lung cancer is now responsible for more deaths than breast cancer and colon/rectum.

Age also effects the distribution and frequency of cancer. Among patients under the age of 15, the most common cancers are leukemia, brain, and endocrine. For patients over age 75, the most common cancers are lung, colon, and breast. Cancer incidence increases with age. Between birth and age 39 years, 1 in 58 men and 1 in 52 women will develop cancer. The incidence of cancer in men and women between ages 60 to 79 years is 1 in 3 and 1 in 4, respectively (15).

Physiology and Pathophysiology

"The final common pathway in virtually every instance is a cellular genetic mutation that converts a well-behaved cellular citizen of the body into a destructive renegade that is unresponsive to the ordinary checks and balances of the normal community of cells" (13).

The stem cell theory was developed to explain the sequential triggering mechanisms that cause cells to begin to specialize and develop new structures and discrete functions (figure 22.1). The stem cell is pluripotent, which means it is an uncommitted cell with various developmental options still open. The process by which the **pluripotent stem cell** is able to develop special functions and structures within an organ system is called differentiation. Thus, some stem cells are triggered to differentiate and become hair cells, and some cells become cardiac myocytes. The pluripotent stem cell also has the capacity for self-renewal. However, once it becomes a committed stem cell, it no longer has this feature and is destined to develop along its specialized pathway of differentiation. The best example is a pluripotent hematopoietic stem cell, with its capacity to form red blood cells, and granulocytes and other white blood cells.

The stem cell model for cancer proposes that tumors arise from carcinogenic-causing events occurring within the stem cells of a particular tis-

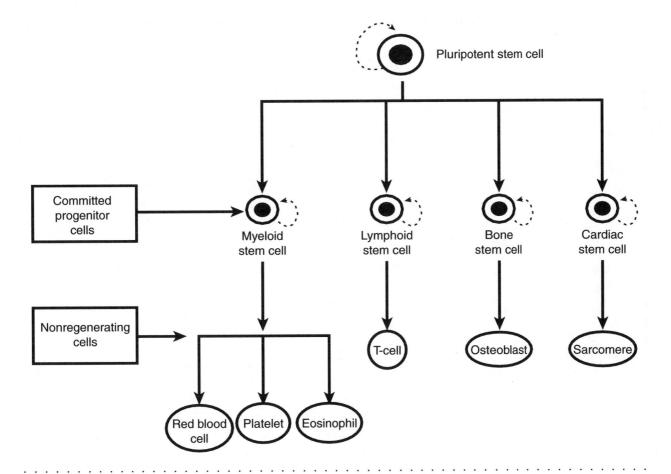

FIGURE 22.1   Stem cell sequence theory. This schematic presentation of the stem cell theory depicts how the uncommitted pluripotent stem cell develops or differentiates into several committed, nonregenerating cells with specialized functions.

sue. It is believed that a cancer-causing insult produces a defect in the control of normal stem cell function, resulting in abnormalities in self-renewal, differentiation, and proliferation. In other words, the normal quality and quantity control for cell function and growth is lost.

The carcinogenic event for many cancers is unknown, but four broad categories have been identified:

1. Environment
2. Heredity
3. Oncogenes
4. Hormones

Environmental factors are implicated in 60% to 90% of cancers, partly because of the known association between certain agents and the development of cancer (table 22.1). For example, lung cancer is more prevalent among miners and those who work with asbestos, chromate, or uranium. Exposure to certain

solvents is associated with leukemia, and excessive exposure to sunlight is responsible for a higher incidence of skin cancer in farmers. Other environment factors such as a diet high in fat and cigarette smoking are well-known carcinogens.

Support for hereditary or genetic factors in cancer comes primarily from animal studies. In humans, genetic factors have been implicated in colon (e.g., familial polyposis), breast, and stomach cancers. The presence of breast cancer in a female relative more than doubles a woman's likelihood of developing the disease.

Genes that act at the cellular level causing uncontrolled proliferation of previously normal cells are called oncogenes. Possibly activated by heredity or the environment (e.g., radiation, chemicals, viruses), oncogenes produce a defect in the control of normal stem cell function. Finally, some hormones are thought to possess carcinogenic potential. For example, the ovary, breast, and uterus are hormonally sensitive; therefore, estrogen may play a role in the cancers of these organs in women.

TABLE 22.1
Environmental Causes of Cancer

Agent	Cancer
Alcohol	Oropharynx, larynx, esophagus, liver
Arsenic and arsenic compounds	Lung, skin
Asbestos	Lung, gastrointestinal
Benzene	Leukemia, Hodgkin's
Beryllium and beryllium compounds	Lung, leukemia
Cadmium and cadmium compounds	Prostate
Chromium compounds	Lung
Ethylene oxide	Leukemia
Nickel compounds	Nasal, lung, leukemia
Radon	Lung
Strontium	Lung, leukemia
Tobacco	Mouth, pharynx, larynx, esophagus, lung, pancreas, bladder
Ultraviolet ray	Skin cancer (melanoma)
Uranium	Lung, leukemia
Vinyl chloride	Liver

The Immune System

Any discussion about cancer would be incomplete if it did not address the body's **immune system.** The immune system is responsible for mediating the interaction between an individual's "internal" and "external" environment. The overall function of the immune system is to rid the body of infectious agents and malignant cells. It can identify the malignant cell because of its genetic changes and respond. The immune system can also inhibit subsequent formation of a tumor by countering factors responsible for its growth. The important role that exercise plays in modulating the immune system has only recently been studied.

The immune system is divided into two major categories: innate and acquired responses (table 22.2). The innate immune system response is nonspecific and immediate, beginning within minutes of an insult. It occurs without "memory" for the eliciting stimulus. This process is called inflammation. The innate system is composed of the monocyte/macrophage system, neutrophils, and natural killer (NK) cells, all of which represent our first line of defense against cancer. These cells are capable of lysing tumor cells without having been exposed to them previously.

The adaptive or acquired immune system is characterized by an antigen-specific response to a foreign antigen or pathogen and generally takes several days or longer to materialize. A key feature of acquired immunity is memory for the antigen, such that subsequent exposure leads to a more rapid and often more vigorous response. The primary cell type for this task is the cytotoxic T lymphocyte. Practical application 22.1 discusses acute and chronic exercise and immunity.

TABLE 22.2

Components of the Immune System

Immune system component	Description

Innate immune system
  — Monocyte
  — Macrophage
  — Neutrophils
  — Natural killer cells
	Nonspecific response
	  — Mature to become macrophages
	  — Nonspecific killing response of tumor cells by phagocytosis and cytolysis, a process by which cell-specific molecules are injected through pores in the cell membrane and rupturing the cell membrane

Acquired or adaptive immune system
  — Cytotoxic T-lymphocytes
	Antigen-specific response
	  — Requires tumor antigens in association with class I major histocompatibility antigens

PRACTICAL APPLICATION 22.1

Acute and Chronic Exercise and Immunity

Several excellent review articles discuss the impact of exercise on immune system function (22, 36, 52, 67, 70, 87, 88). The most popular is the inverted J hypothesis (see following figure). This model suggests that there is a dose of exercise/physical activity that enhances function of the immune system and reduces cancer incidence. Note that prolonged, high-intensity, or exhaustive training is associated with immune function that is even below that of sedentary individuals. Additionally, overtraining or intense competition may lead to immunosuppression. Clinically, it seems that moderate to high levels of physical activity are associated with decreased incidence and/or mortality rates for various cancers (66, 74). However, presently it is not known if exercise plays its greatest role on the immune system in preventing cancer, eliminating early cancer, or potentiating the treatment of cancer after it has occurred.

Exercise bouts of low to moderate intensity (<60% $\dot{V}O_2$max) and moderate duration (<60 min per bout) exert less stress on the immune system than do prolonged highly intense sessions (>95% $\dot{V}O_2$max and >90 min) (12, 46, 52). Macrophages, NK cells, and neutrophils appear to be the most responsive to the effects of exercise, in terms of both number and function (27, 54, 65) (see following table).

Regardless of intensity, acute exercise induces a profound leukocytosis that includes an increase in neutrophils. Although the number of neutrophils is increased with moderate exercise during periods of high-intensity training, neutrophil function has been reported to be suppressed. Pyne et al. (65, 64) reported that elite swimmers undertaking intensive training have a significantly lower neutrophil oxidative activity at rest than do controls, and that cell function is further suppressed during strenuous exercise. Suppression of neutrophil function during periods of heavy training probably partly explains the increased risk for upper respiratory tract infections among some competitive athletes.

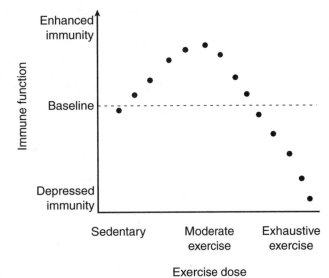

continued

407

INNATE IMMUNE SYSTEM		
Component	**Effect of acute exercise**	**Effect of chronic exercise**
NK cells	Immediate increase in cell count and cytolytic activity.	Increase in resting NK cell count and activity, both in blood and (depressed for 2-24 hr postexercise) spleen.
Macrophages	Immediate increase in monocyte and macrophage count.	Response is unclear. Resting monocyte count unchanged.
	Adherence unchanged. Increased phagocytosis with moderate activity.	May cause adaptations that alter exercise response.
Neutrophils	Large and sustained increase with moderate exercise.	Function is suppressed during periods of strenuous activity.
	Most PMN functions decrease significantly after strenuous exercise.	

ACQUIRED IMMUNE SYSTEM		
Component	**Effect of acute exercise**	**Effect of chronic exercise**
T lymphocytes	Moderate activity enhances cell proliferation, with depressed levels 30 min postexercise.	Regular, moderate exercise enhances cell proliferation.
	Vigorous activity causes a transient decrease in proliferation.	

Note. NK = natural killer; PMN = polymorphonuclear leukocytes.

In response to acute exercise, regardless of the intensity or duration, the number of monocytes in peripheral blood increases transiently (85). Both acute and moderate exhaustive exercise have been shown to enhance a variety of macrophage functions including chemotaxis (25, 43), adherence (49, 57), and phagocytic (15, 22) and cytotoxic (14, 86) activity. There are relatively few data regarding the effects of chronic, regular exercise on macrophage function. Although acute exercise may be a potent stimulus for macrophage activity, chronic training may actually diminish this response (21, 56, 57). High-intensity exercise over several days can decrease the number of macrophage cells by more than 50% (86).

NK cells are very responsive to exercise, such that in recovery, cell activity and circulating numbers are transiently increased (23, 27). Following high-intensity exercise, NK cell cytotoxic activity is increased 40% to 100%. This has been observed in marathon runners compared with sedentary controls (53) and elite cyclists during the summer (intense training period) versus the winter months (low training period) (77). Several studies using moderate endurance activity over 8 to 15 weeks reported no significant elevations in NK cell activity (7, 50), suggesting that endurance activity must take place over a long period of time (i.e., years) before a response is observed.

High-intensity exercise is associated with increased numbers of cytotoxic T lymphocytes on the order of 50% to 100% immediately after exercise. As with NK cells, this increase is transient and resolves in 30 min (47).

During recovery from high-intensity aerobic-type exercise, subjects experience a sustained **neutrophilia** and **lymphocytopenia** (28, 54) and possible compromised host protection. NK cell activity and T-cell function are reduced as well. These changes are believed to be attributable to excessive elevations in cortisol (45, 48, 51). Moderate exercise results in a lower cortisol response and is associated with a more favorable immune response.

Clinical Considerations

Because of the diversity of organ system involvement, there is no one specific part of the history and physical examination that focuses on cancer. As a result, it is incumbent on the clinician to perform a complete and accurate history and physical examination.

Signs and Symptoms

Early on, the symptoms of cancer are usually nonspecific such as weight loss, fatigue, nausea, and malaise. It takes an astute clinician to make the diagnosis of cancer at this time, and yet early detection is key to maximizing the patient's chance for survival (see figure 22.2).

Cancer's seven early warning signs, as compiled by the American Cancer Society. Notice that the first letter of each word spells the word *caution*.

Change in bowel or bladder habits
A sore that does not heal
Unusual bleeding or discharge
Thickening or lump in breast or elsewhere
Indigestion or difficulty in swallowing
Obvious change in a mole or wart
Nagging cough or hoarseness

FIGURE 22.2   Cancer's seven early warning signs.

Used with permission from the American Cancer Society.

Later on, the patient will develop symptoms specific for the involved organ, such as shortness of breath in lung cancer or jaundice attributable to biliary obstruction in pancreatic carcinoma. By this time though, prognosis is much poorer. Social history is also important, revealing occupational exposure to carcinogens or habits such as smoking or ethanol ingestion. The family history may reveal familial predisposition to a cancer, one that requires closer surveillance in the future. The review of systems may reveal symptoms indicating that the cancer has already metastasized.

Diagnostic and Laboratory Evaluations

Because the prevention of cancer is not always possible, the earliest detection of the disease is the next best strategy to reduce cancer mortality rates. To help accomplish this, the American Cancer Society recommends a series of screening procedures and evaluations, which are shown in table 22.3.

Once cancer is suspected, the first diagnostic principle is that adequate tissue must be obtained to establish the diagnosis. Because the therapy used for each type and subtype of cancer is often unique, every effort must be made to obtain appropriate tissue samples—even if treatment is delayed for a short time. The process of obtaining a sample of tissue is called a **biopsy.**

The second diagnostic principle is to determine the extent of spread of the cancer, also known as **staging.** In leukemia, this can be accomplished through routine history and physical examination, laboratory tests, chest X-ray, and bone marrow biopsy. With solid tumors, a biopsy and an extensive examination are often needed. The degree to which the cancer has spread is reflected in its stage, which guides the type of treatment most appropriate for the patient. An example of a simplified staging system is shown in table 22.4. Each cancer will have a staging system unique to itself, one that takes into consideration pathogenic features, the modes of spread, and the curability of the disease.

Treatment

There are four treatment options for cancer. The selection of which treatment option to use will depend on the type, location, and stage of the cancer. Treatment options include the following:

1. Surgery
2. Radiation therapy
3. Chemotherapy
4. Biotherapy

Surgery is the oldest and most definitive treatment for cancer. There are two types of surgery: curative and palliative. **Curative surgery** is the primary treatment for about one-third of cancers that are small and not yet metastasized. If the tumor is removed along with a small amount of surrounding normal tissue, the chance for a cure is good.

In **palliative surgery,** a large tumor mass is removed to make the patient more comfortable, to relieve obstruction of vital organs, and to reduce the tumor burden. For example, a colon cancer may be removed to prevent bowel obstruction or an ovarian cancer may be removed to prevent obstruction of a ureter. Decreasing tumor burden may also make the tumor more susceptible to radiation or chemotherapy. Palliative surgery does not usually change overall survival.

TABLE 22.3
American Cancer Society Recommendations for Early Detection

Test	Sex	Age	Frequency
Flexible sigmoidoscopy	M & F	50 and over	Every 3-5 years
Fecal occult blood	M & F	50 and over	Every year
Digital rectal exam	M & F	40 and over	Every year
Prostate exam	M	50 and over	Every year
Prostatic specific antigen	M	50 and over	Every year
Papanicolaou (Pap) test	F	18 and over	Every year (after 3 normal exams, may be performed less often)
Pelvic exam	F	18-40	Every 1-3 years
	F	Over 40	Every year
Endometrial tissue	F	Menopause	
Breast self-exam	F	20-40	Every month
Breast clinical exam	F	20 and over	Every 3 years
	F	Over 40	Every year
Mammography	F	40-49	Every 1-2 years
	F	Over 49	Every year
Chest X-ray			Not recommended
Sputum cytology			Not recommended

Note. The American Cancer Society recommends a series of cancer screening procedures. Although not all experts agree with all aspects of these recommendations, they serve as a useful guide to screen patients for cancer. M = males; F = females.

Used with permission from the American Cancer Society.

TABLE 22.4
Tissue–Nodal–Metastasis (TNM) Classification System for Breast Cancer

Tumor size (T)		Nodal involvement (N)		Metastasis (M)	
Ts	in situ	N0	No nodal metastasis	M0	No distant metastasis
T1	<2 cm	N1	Movable axillary nodes	M1	Distant metastasis
T2	2-5 cm	N2	Fixed axillary nodes		
T3	>5 cm	N3	Internal mammary nodes		

These TNM categories are combined to give the stage (e.g., stage I = T1 N0 M0).

Radiation therapy is thought to stop the growth of malignant cells by damaging the DNA within the cell. Radiation can be applied either from implanted internal sources (brachytherapy) or from external machines. Most radiation therapy is applied in small fractions, usually between 180 and 250 rad · day^{-1}. Radiation therapy can be used alone or in conjunction with surgery or chemotherapy. Radiation is good for the localized tumor that has not metastasized or those that are difficult to reach with surgery (e.g., within the brain).

Chemotherapy uses chemical agents that kill rapidly growing cells. Cells that grow the fastest are best treated by chemotherapy, which frequently can result in a cure. However, some cancer cells may become resistant to chemotherapeutic agents. Using several drugs at the same time (i.e., combination chemotherapy) is one way to minimize the resistance.

Cancer cells possess distinct surface protein antigens that are targets for antibody-directed or cell-mediated immunity. **Biotherapy** stimulates the body's immune response to these protein antigens. Biotherapy also involves the production of antibodies outside the body, which are then administered to the cancer patient in an attempt to destroy the tumor. Biotherapy includes bone marrow transplantation and the use of cytokines such as α-interferon and interleukin-5.

Graded Exercise Testing

Before beginning an exercise program, cancer patients should be screened for coronary artery disease and cancer-related issues. Patients with cancer have the same incidence of coronary artery disease as the general population. The American College of Sports Medicine (4) recommends that men over the age of 40, women over the age of 50, and individuals with two or more risk factors undergo exercise testing before beginning an exercise program. Practical application 22.2 describes client–clinician interaction for patients with cancer.

PRACTICAL APPLICATION 22.2

Client–Clinician Interaction

Part of this individualized approach to using exercise in patients with cancer requires that you, as the clinical exercise professional, accomplish two things. First, become familiar with the type of therapy or therapies that your patient is undergoing. This may require some extra reading when you encounter a therapy you are unfamiliar with. Also, talk with oncologists and surgeons about the agents or interventions used to treat cancer. Such discussions should include the clinical presentation of expected side effects and the natural history of the disease. Ultimately, your ability to interact with patients in a learned fashion will be improved.

Second, develop a skills set that advances your ability to ask questions about how well your patient is tolerating therapy and his or her disease. Such skills usually come from working with many patients, improving your abilities to evaluate the interaction between exercise, cancer, and the cancer treatment. Because part of any exercise program includes regular follow-up, either in an exercise program or by telephone to their home, you are in a unique position to establish patient confidence and assess clinical status. Obviously, any concerns can be brought to the attention of the patient's doctors, making your role an important part of long-term patient surveillance.

Clinical features that you should pay special attention to might be sudden loss of exercise tolerance over several days to a week or so, increased shortness of breath with exertion, an inordinate increase in anxiety or depression as manifested by difficulty falling to sleep or disinterest in social contact, and sudden changes in nutritional status. These are issues that you can "work in" to the routine follow-up phone calls or clinic visits you might be using to evaluate exercise compliance and progress. In fact, there may be times when the information you gather dictates that exercise be withheld for a period of time, while a specific treatment protocol runs its course.

Because more evidence is needed to describe how exercise improves function and whether exercise lowers risk for cancer recurrence or future hospitalizations, it is likely best to keep a patient's attention focused on how exercise can help him or her lessen fatigue, regain control, and improve quality of life. This means using exercise to help patients lead a more comfortable life, one that allows them to perform the activities they enjoy without symptoms or limitations.

Medical Evaluation

The history and physical examinations are important assessments of the patient prior to an exercise program. The history should include both non-cancer and cancer considerations. Noncancer considerations include other medical problems such as age, diabetes, hypertension, fitness level, and orthopedic problems. Cancer issues of importance include type and stage of cancer, type of treatment, side effects of therapy, psychological status, and timing of tests and therapy. The physical examination should attempt to identify acute signs or symptoms that would prevent participation in exercise. These might include bone tenderness indicative of metastatic lesions; gait instability secondary to neuropathy attributable to chemotherapy or central nervous system involvement; and complications such as wound healing, immune suppression, and propensity to bleeding that occur because of the treatment used. Other complications that may require a modification of the exercise plan include nausea, vomiting, fatigue, and weakness.

Contraindications

In addition to the usual contraindications for exercise and exercise testing in patients free of cancer, there are additional considerations unique to patients with cancer. These include nutritional status; complications such as metastatic lesions to bones of the pelvis, back, or legs; metabolic considerations such as electrolyte abnormalities and dehydration; and instability secondary to neuropathy resulting from chemotherapy or central nervous system involvement. Cancer treatment can result in wounds from surgery, immune suppression, a propensity to bleeding, and other side effects such as fatigue, nausea, vomiting, and weakness. All of these factors must be taken into consideration when a patient enters an exercise program and must continually be reevaluated throughout the program. Cancer-related contraindications to exercise are shown in figure 22.3.

Recommendations and Anticipated Responses

Patients with cancer are frequently deconditioned because of both the disease process and the treatments. The average maximal work capacity in patients with malignancy is reduced to 3 to 5 metabolic equivalents or a peak $\dot{V}O_2$ of 11 to 21 ml $\cdot$ kg^{-1} $\cdot$ min^{-1}. Exercise is often limited by general or leg fatigue. Peak heart rate may be reduced as well. The ideal exercise test should last 8 to 12 min. Protocols beginning at lower work rates and progressing more slowly should be considered (e.g., modified Bruce, Naughton-Balke, modified Balke).

- Hemoglobin <10.0 g $\cdot$ dl^{-1}
- White blood cells <3,000 $\cdot$ μl^{-1}
- Neutrophil count <0.5 $\times$ 10$^9 \cdot$ml^{-1}
- Platelet count <50 $\times$ 10$^9 \cdot$ml^{-1}
- Fever >38° C (100.4° F)
- Unsteady gait (ataxia)
- Cachexia or loss of >35% of premorbid weight
- Limiting dyspnea with exertion
- Bone pain
- Severe nausea

FIGURE 22.3   Contraindications to exercise for patients with cancer.

Exercise Prescription

The exercise program for cancer patients does not require electrocardiographic monitoring; however, some supervision and instruction about heart rate monitoring and proper exercise technique are desirable. Initially, the exercise prescription should be reviewed with the patient, and the patient should be taught both how to take his or her pulse and the symptoms of an adverse response to exercise. The patient should also be instructed in the types of exercise best suited to his or her cancer. The actual exercise program can be performed at home, in a healthcare facility, or outdoors. If resistance training is being incorporated, the patient may benefit from temporarily enrolling in a medical fitness center. See practical application 22.3 for a review of the literature about exercise training for cancer patient.

Special Exercise Considerations

The exercise professional working with the patient with cancer needs to assess the risk–benefit relationship case by case. Also, the fluctuating clinical status of the individual cancer patient needs to be considered. The patient may experience setbacks attributable to the progression of the cancer and its symptoms. If this occurs, exercise goals need to be modified accordingly. At some point it may even be necessary to suspend exercise permanently, if the risk–benefit relationship no longer justifies exercise as a component of the treatment.

Because cancer is a constellation of diseases, it is also important to know which cancer the patient has and to what extent it has spread. Some patients may have cancers that involve the bones, in which

Literature Review

The role of exercise in treatment of disease is divided into primary prevention, secondary prevention, and adjunctive therapy. Primary prevention looks at whether the disease can be prevented by exercise. The role exercise may play in primary prevention is promising. Secondary prevention refers to prevention of additional disease once the original disease has been treated. At this time, there is no evidence that exercise helps prevent recurrence of cancer. Adjunctive therapy refers to treatment in addition to, or potentiating, usual treatment. Here again, exercise clearly plays a role.

Clinically, people who exercise generally have a healthier lifestyle. They tend to avoid carcinogens such as a diet high in fat, alcohol, and tobacco, and they tend to maintain their weight. The latter is important, because obesity is linked to an increased risk for breast, colon, rectum, prostate, endometrial, and kidney cancers.

Data show that higher levels of physical activity lower overall cancer mortality (58, 60, 75). The American Cancer Society concluded in a 1996 report (2) that "there is the potential to reduce the overall incidence, morbidity, and mortality from certain kinds of cancer if successful physical activity and dietary interventions are implemented" (p. 325). In fact, for the nonsmoker, dietary and physical activity interventions are the most important modifiable determinants of cancer risk.

Determining a relationship between physical activity and cancer risk is hindered because it is not known in what time frame exercise needs to occur to decrease the cancer risk. Exercise may help block initiators of cancer, in which case exercise done consistently at a relatively young age may be most beneficial. This is the rationale behind studies that investigate if participation in high school and college athletics reduces the risk of cancer during adulthood. Alternatively, exercise may counter promoters of cancer cell replication, so that exercise during a later phase of the neoplastic process may be preferred to decrease the development of clinically significant disease (69, 73). Therefore, the point at which exercise occurs during one's lifespan may be an important factor relative to its effect on cancer development.

One example is breast cancer. Large doses of estrogens, when given continuously, can cause breast carcinoma in susceptible strains of mice. There are four phases in a woman's life when estrogen plays a role: menarche, first pregnancy, menopause, postmenopause. It is not certain at which point or points in a woman's life cycle that exercise exerts its greatest anticancer effect.

Several studies indicate that breast cancer risk is directly related to the cumulative number of ovulatory menstrual cycles (30, 36, 62, 76, 80). Intense exercise delays menarche, which can be thought of as favorable, because the risk of breast cancer is increased twofold in women who experience menarche under the age of 12 versus at age 13 or older. Epidemiological data indicate that for every year that menarche is delayed, breast cancer risk is reduced by 5% to 15% percent (33, 76). Moderate levels of activity have been shown to increase the risk of anovulatory cycles threefold. Delayed menarche and anovulatory cycles decrease the woman's exposure to estrogen and progesterone (9, 26, 81). The first full pregnancy induces differentiation of the breast and may change the sensitivity of the breast to both endogenous and exogenous risk factors (64). And women who experience natural or artificially induced menopause before the age of 45 have a markedly reduced risk of breast cancer compared with women whose menopause occurred after the age of 55 (74). Postmenopausal women generally have an increased body mass index, which is a significant risk factor for breast cancer (31, 59, 68). In postmenopausal women, fat tissue is the primary source of estrogen. Therefore, obesity increases the woman's exposure to estrogen.

The role of exercise in preventing a site-specific cancer is not fully understood. The relationship between colorectal cancer and physical activity has been studied the most (37, 78). Forty-eight studies with 40,000 patients demonstrated a 1.3- to 3.6-fold increased risk for colon cancer in inactive individuals. There appears to be no association between activity level and rectal cancer. Decreased bowel transit time because of physical activity may account for the decrease in colon cancer incidence but no change in rectal cancer. Twenty-six studies with 108,321 women demonstrated that both occupational and leisure time activity reduces breast cancer risk by about 30%. The review of the data in the studies evaluating the possible protective effect of exercise in preventing endometrial, ovarian, prostate, and lung cancer is quite favorable but inconsistent at this time.

continued

How much physical activity is needed to reduce the risk of cancer? The definitive answer is not known at this time. Data from Blair et al. (10) indicate that the reduction in cancer risk occurred primarily between the very low fitness group and the moderately fit group, with no further decrease in risk among the more fit subjects. Paffenbarger et al. (60) reported a decrease in all-cause mortality for alumni expending 2000 or more kcal · week^{-1}, but these authors did not break out cancer deaths in this group. Lee (39) also reported that exercising at a moderate intensity, greater than 4.5 metabolic equivalents for 6 to 8 hr per week, decreased the incidence of lung cancer.

The influence of exercise in patients with cancer can be divided into three areas: fatigue, psychosocial, and therapy-related side effects. It has been shown that improving one of these areas can have a positive domino effect on other aspects of quality of life (23).

Fatigue

Having the energy to carry out activities of daily living as well as occupational, leisure, and social activities is inherent to many quality of life measures (42, 61, 72). Up to 70% of cancer patients recovering from surgery, or receiving chemotherapy or radiotherapy, report loss of energy. This symptom persists in up to one third of cancer survivors for years after cessation of treatment (71).

There are many potential causes of fatigue in patients with cancer, including preexisting conditions, the direct effect of the cancer, symptoms related to cancer, effects from the treatment used for the cancer, and the demands of dealing with cancer. One frequently underestimated factor contributing to fatigue is loss of physical fitness as a result of bed rest. When patients experience fatigue they are often told to "take it easy" and "get plenty of rest," further limiting their level of daily activities and perpetuating deconditioning and exercise intolerance. The exercise capacity of the cancer patient is markedly reduced, often with a $\dot{V}O_2$peak of 11 to 21 ml · kg · min^{-1} (71). However, one-third or more of the decline in functional capacity is likely attributable to the consequences of physical inactivity (4). The rapid decline is seen in multiple systems but is most dramatic in the cardiopulmonary system. Additional consequences of physical inactivity include limited joint mobility, osteoporosis, impaired balance, and lowered pain threshold (5). Impaired physical fitness is an important contributing factor that lessens quality of life in patients with cancer.

Psychosocial

Exercise also has an immediate mood-elevating effect (71) and can help cancer victims psychologically, breaking the vicious cycle of physical inactivity and tissue loss. Because physical mobility is associated with health and well-being, rehabilitation efforts that incorporate exercise are associated with increased energy levels and decreased fatigue, which in turn improve comfort, concentration, appetite, and sleep habits (10, 34, 59).

Dependence on others for assistance with routine activities of daily living often intensifies the psychological responses to disease and treatment in patients with cancer (42). By fostering functional independence, exercise improves self-concept, self-esteem, confidence, self-image, sense of personal worth, control, and self acceptance while at the same time providing a means for attaining a sense of control over one's life (1, 20, 41, 84). Exercise also improves quality of life by decreasing feelings of depression, tension, anxiety, anger, hostility, helplessness, and pessimism (10, 16, 24, 79). Winningham (82) reported that these changes in psychological state give patients hope by improving their feelings of well-being and their ability to cope and adapt to stress. With 8.2 million cancer survivors alive today, emphasis on quality of life issues is paramount.

Cancer Therapy Side Effects

Thirty-six studies (14) have examined the relationship between exercise and quality of life following cancer diagnosis and have consistently demonstrated beneficial effects on a wide range of quality of life outcomes, regardless of exercise prescription, cancer rate, or cancer treatment. Dimeo et al. (18, 19) reported the impact of an exercise training program on fatigue in 80 patients undergoing high-dose chemotherapy. The patients were randomized to one group that performed supine bicycling for 30 min per day during each day of their hospitalization for chemotherapy. A second

continued

group received chemotherapy only. After 7 weeks, physical performance (+34%) and hemoglobin concentration were significantly greater for the training group. Heart rate and lactate concentration measured at a submaximal workload were also significantly reduced in the training group. Interestingly, the training group demonstrated a reduced duration of neutropenia and thrombocytopenia, severity of diarrhea, severity of pain, and duration of hospitalization. Conversely, at the time of hospital discharge, fatigue and somatic complaints had increased significantly in the nonexercising group but not in the training group. Finally, the training group had a significant change in measures of obsessive–compulsive traits, fear, interpersonal sensitivity, and phobic anxiety. These changes were not seen in the nonexercising group.

To counter the effects of inactivity, Winningham developed the Winningham Aerobic Interval Training Program, a 10-week aerobic interval program that alternated higher and lower intensity exercise. Cancer patients were entered into the study at a work rate that induced a heart rate of 60% to 85% of heart rate reserve, calculated from the highest rate achieved at pretest. Exercise sessions were conducted three times per week. Winningham et al. (82) and MacVicar et al. (41,42) reported a 40% increase in functional capacity, improved feelings of internal control, improved psychological status,(42) and lessened tension/anxiety, depression/dejection, and fatigue. Exercise decreased the incidence of nausea, weight gain, and muscle wasting (40, 83, 84).

case exercise of that extremity may cause pain or result in injury. Patients with metastatic disease to the pelvis or legs may benefit from exercise programs that allow them to sit down, such as a stationary bicycle or chair exercises. Water aerobics may also reduce stress on the skeletal system and allow exercise to continue.

Because of their malignancy or possibly because of their treatment, many patients will experience symptoms of extreme fatigue, shortness of breath, nausea, or vomiting. These symptoms must be taken into account when one is planning an exercise program. The type of cancer treatment and the treatment protocol are also important. For example, chemotherapy and radiation therapy both use protocols dictating how and when the treatment is administered. Any exercise program needs to be flexible enough to work around these schedules. Surgery has both an in-hospital and out-of-hospital recovery phase that must be considered. Finally, treatment-related side effects such as fatigue, depression, and nausea will likely influence, at the very least, intensity of exercise.

Cancer patients are faced with many factors that may require their exercise program be modified. Intravenous devices preclude some stretching exercises. Pool activities are acceptable in patients with indwelling central venous catheters, continent urinary devices, or colostomies but are contraindicated for patients with intravenous catheters, nephrostomy tubes, and urinary bladder catheters.

Stationary bicycle activities are frequently used for patients recovering from breast cancer or thoracic surgery. This mode also works well in patients with ataxia, central venous catheters, or lymphedema. Patients with primary or secondary bone cancer should avoid high-impact exercises. High-intensity exercise should be avoided during all cancer treatments.

Resistance training to maintain or enhance muscle strength is important and should be performed on machines rather than free weights to avoid any potential for bruising or bone fractures. Low-resistance, high-repetition workouts are recommended. Patients undergoing blood tests should avoid resistance training for 36 hr before blood is drawn.

As your patient's clinical exercise physiologist, you should take care to remember several things:

— Many patients with cancer are inactive and experience mood disturbances such as anxiety and depression that are themselves associated with little interest for exercise. Be sure to progressively increase exercise dose in a manner that is individualized for every patient.

— For the first week or so of exercise training, patients may be a bit tired later in the day. To compensate for this, ask them to limit the amount of other, home-based activities for the first few days.

— Exercise therapy for patients with cancer is often interrupted by a variety of treatment-related obligations and complications. Before even starting an exercise regimen, inform the patient that setbacks and interruptions are not uncommon. Instead of not exercising at all or stopping the exercise program, patients should plan around interruptions as best as they can and continue to adhere to their program whenever possible.

— Learn about the different types of treatments that your patients are receiving.

— Consider developing an "exercise buddy" system, which matches up cancer survivors who are now exercising with patients who are just starting out. Such group support from patients with similar medical problems may help improve short-term adherence.

Exercise Recommendations

It is generally recommended that exercise intensity be modified in patients with cancer, starting at 50% to 55% of heart rate reserve. Interval or intermittent exercise training on a cycle ergometer at an intensity of 60% to 75% of peak heart rate may also be well tolerated in ambulatory patients and can significantly improve functional capacity. Patients with cancer who present with severely impaired fitness may initially benefit from interrupted programs that incorporate several bouts of shorter duration exercise. As shown in table 22.5, patients with activity level categories of 0 and 1 can probably start with 15 to 20 min of continuous exercise. Patients in category 2 may need to start with periods of 5 to 10 min. When in doubt about where to start, use the 50% rule: Ask the patient how far he or she can walk before becoming too tired and start at half the distance or time (82). To build endurance, encourage long slow distances as opposed to faster short distances. Exercise frequency is ideally set at three to five times per week, but this can vary based on the individual patient. Patients undergoing radiation therapy or chemotherapy have complex schedules and require flexibility relative to planning which days to exercise. Increase a patient's exercise program between 30 s and 2 min each day, while maintaining a rating of perceived exertion of between 11 and 13. Practical application 22.4 provides an exercise prescription summary for patients with cancer.

Remember, pain and fatigue are common symptoms for cancer patients. You must be alert to distinguish the normal pain and fatigue of exercise from that caused by progression of disease, deconditioning, aging, or an exercise program that is too strenuous. Pain at the initiation of the exercise program is a sign to start at a lower level of intensity.

Finally, keep in mind that one purpose of an exercise program is to improve fitness, whether the patient has cardiovascular disease, has cancer, or is disease free. The components of fitness include strength, endurance, and flexibility. Therefore, each exercise program should incorporate aerobic training, resistance training, and range of motion exercises to achieve maximum fitness.

Physiological Adaptations and Maladaptations

Physical activity improves the ability of muscle cells to produce energy, whereas inactivity results in de-

---

TABLE 22.5
Categorizing Performance Levels

Activity level	Category	Exercise duration	Exercise frequency
Active, no limitations	0	15-20 min	Daily
Ambulatory, decreased leisure activity, can perform self-care	1	15-20 min	Daily
Ambulatory >50% of time, moderate fatigue, limited assistance with ADL	2	15-20 min	Daily
Ambulatory <50% of time, fatigue with mild exertion, requires assistance with ADL	3	5-10 min	Two sessions daily
Confined to bed	4	No exercise	—

Note. ADL = activities of daily living.

Exercise Prescription Summary

Aerobic Training

— Mode: Exercises involving large muscle groups are best. Walking is generally recommended except for those patients with skeletal involvement of the back, pelvis, or legs. For these patients, a non-weight-bearing activity may be preferred.

— Frequency: Three to five times per week. Exercise regimen must be flexible to accommodate any scheduled chemotherapy and radiation treatments or associated side effects.

— Intensity: The patient should progressively increase training heart rate, as tolerated, from 50% to 70% of heart rate reserve. The clinician should titrate exercise intensity so that rating of perceived exertion is between 11 and 13. High-intensity exercise should be avoided during cancer treatment protocols.

— Duration: Progressively increase duration from 15 min to 30 or more min per session. If needed, divide total exercise time into two to three intervals.

Resistance Training

— Mode: Standardized machines rather than free weights are preferred.

— Frequency: One to two times per week. Patients undergoing laboratory tests should avoid resistance training for 36 hr prior to testing.

— Intensity: Approximately 50% to 60% of 1-repetition maximum.

— Duration: One set of 12 to 15 repetitions for each of the desired muscle groups.

Range of Motion

— Mode: Static stretching.

— Frequency: Before and after aerobic and/or resistance training workouts.

— Intensity: The patient should stretch to the point of challenge, but not to the point of discomfort, and should relax into the stretch.

— Duration: Five minutes before and 5 to 10 min after each workout, with 10 to 30 s devoted to the major muscle groups and joints.

creased muscle function and exercise tolerance. In patients with cancer, investigators have found muscle fiber atrophy and necrosis. These changes are partly responsible for the fatigue that patients with cancer experience when they exert themselves (40). Exercise can counter these effects and stimulate the body to maintain or improve its abilities to extract oxygen for energy production.

A recent study by Segal et al. (66) randomized 123 women with breast cancer to either 6-months of regular exercise or a no-exercise control group. Those randomized to exercise trained five times per week. Although exercise capacity, as measured by $\dot{V}O_2$peak, was not different after training in the exercise group versus controls, physical function as assessed by patient questionnaire was improved. In 2000, Courneya et al. (14) reviewed all exercise interventional trials conducted between 1998 and 2000.

Eight trials were identified, with the majority of the findings showing improved $\dot{V}O_2$peak, muscle strength/function, or quality of life.

Conclusion

Cancer is a constellation of diseases. It can begin in any organ and spread to other organ systems. Its treatment includes surgery, radiation therapy, chemotherapy, and, more recently, biotherapy. Quality of life issues are extremely important among patients with cancer. Exercise benefits these patients primarily through quality of life issues, such as reducing fatigue, reducing dependence on others, and countering some of the side effects of cancer therapy. The role of exercise in preventing cancer or altering the natural history of the disease remains unknown.

Medical History

Mr. CB is a 68-year-old Caucasian male with metastatic renal cell carcinoma. He was a millwright for the city water department but is now retired. He has a medical history of dyslipidemia (total cholesterol = 178 mg · dl^{-1}, high-density lipoprotein cholesterol = 24 mg · dl^{-1}, triglyceride = 774 mg · dl^{-1}), hypertension, and myocardial infarctions in June and August 1997. He also has a history of atrial fibrillation. In January 1999 his body mass was 83 kg, blood pressure was 110/70 mmHg, and heart rate was 72 beats · min^{-1}. Current medications are metoprolol, isosorbide dinitrate, Solu-Cortef, fenofibrate, and aspirin.

Diagnosis

In 1994 during a follow-up visit for impotency in the Department of Urology, urine tests detected microscopic hematuria. The following year, still having difficulties of impotency, he complained of gross hematuria and right flank discomfort. An intravenous pyelogram suggested a renal mass. Results of a computed tomography scan showed a right renal mass that invaded the right kidney and the inferior vena cava. He underwent a right radical nephrectomy in November 1995 for stage IIIA renal cell carcinoma. Pathology showed invasion of the renal capsule, as well as the renal vein and inferior vena cava.

Exercise Test Results

His most recent electrocardiogram showed normal sinus bradycardia with evidence of a previous inferior wall myocardial infarction. Resting heart rate was 53 beats · min^{-1}. An exercise stress test completed in August 1997 was mildly suggestive of ischemia, with a $\dot{V}O_2$peak of 17.6 ml · kg^{-1} · min^{-1}. Peak heart rate was 92 beats · min^{-1} and exercise was discontinued because of mild to moderate grade angina. His cardiac status is now stable with Canadian Cardiovascular Society grade 2 angina.

Other Procedures

A follow-up computed tomography scan in April 1997 showed metastases in the lower lobe of the left lung. These nodules were considered too small for biopsy. In May 1997, he complained of radiating pain from the right buttock to his knee. A bone scan identified widespread bone metastases. Magnetic resonance imaging showed a mass in the left posterior lateral aspect of the lumbar vertebrae. He underwent 14 days of radiotherapy in June 1997, during which his back discomfort improved. As part of a clinical trial, in August 1997 he received two courses of interferon and chemotherapy (suramin and tamoxifen).

In May 1998, CB underwent surgical decompression and excision of an L3 vertebrae tumor. He began immunotherapy with α-interferon in July 1998. Repeat magnetic resonance imaging and bone scans through January 1999 showed the disease to be stable.

Development of Exercise Prescription

During chemotherapy in 1997, Mr. CB initiated an exercise program. He exercised 3 to 4 days a week at a perceived exertion of 11 to 12 (Borg scale, 6-20) at a prescribed heart-rate range of 72 to

continued

82 beats · min⁻¹. This heart-rate range was free of any electrocardiographic evidence of ischemia. Exercise modalities included treadmill, dual-action bike, and rower. Exercise therapy was tolerated well, without complications or symptoms.

Case Study 22   Discussion Questions

1. What effect does metoprolol have on heart rate response to exercise? Would you alter how you go about guiding exercise intensity for patients taking this drug?

2. Fatigue and mood disturbances are common complaints of patients with cancer, often attributable both to the disease itself and the treatments used to manage the disorder. Explain whether exercise should be used or withheld in and around those times when a patient is undergoing therapy. What clinical features and symptoms would lead you to consider (or not consider) exercise during this time? How might you quantify if, in fact, a patient is or is not responding to an exercise regimen?

3. This patient asks you to help him design and start a resistance training program. Do his test results indicate that this type of training be incorporated into his exercise regimen? Explain your answer and any concerns you have with respect to resistance training in patients with cancer.

Glossary

biopsy—A surgical procedure whereby a sample of tissue is obtained.

biotherapy—Stimulation of the body's immune response system to cancer-specific protein antigens.

chemotherapy—Use of chemical agents to kill rapidly growing cancer cells.

curative surgery—Surgery aimed at complete removal of tumor along with a small amount of surrounding normal tissue.

immune system—System that mediates the body's interaction between internal and external environments. Helps rid the body of infectious agents and malignant cells.

lymphocytopenia—A reduction in the number of lymphocytes in the circulating blood.

metastases—The spread of a disease process from one part of the body to another, as in the appearance of neoplasms in parts of the body remote from the site of the primary tumor.

neoplasm—Abnormal tissue that grows by cellular proliferation more rapidly than normal and continues to grow after the stimuli that initiated the growth ceases. Structural organization and function of neoplastic tissue are partially or completely different from the normal tissue.

neutrophilia—An increase in neutrophilic leukocytes in blood or tissue.

palliative surgery—Surgery aimed at removal of tumor to make patient more comfortable, relieve organ obstruction, or reduce tumor burden.

pluripotent stem cell—Uncommitted cell with various developmental options pending.

radiation therapy—Therapy meant to stop growth of malignant cells by damaging RNA within the cells.

staging—A system used to classify the extent and spread of cancer.

References

1. Aistars, J. Fatigue in the cancer patient: A conceptual approach to a clinical problem. *Oncol. Nurs. Forum* 14: 25-34, 1987.

2. American Cancer Society. 1996 Advisory Committee on Diet, Nutrition, and Cancer Prevention: Guidelines on diet, nutrition, and cancer prevention: Reducing risk of cancer with healthy food choices and physical activity. *Cancer J. Clin.* 46: 325-341, 1996.

3. American Cancer Society. *Cancer Facts & Figures.* Atlanta, GA: American Cancer Society, 1993.

4. American College of Sports Medicine: *Guidelines for Exercise Testing and Prescription* (3rd ed.). Philadelphia: Lea & Febiger, 1986.

5. Åstrand, P., and K. Rodahl: *Textbook of Work Physiology* (3rd ed.). New York: McGraw-Hill, 1986.

6. Bailar, J., and E. Smith. Progress against cancer? *N. Engl. J. Med.* 314: 1226, 1986.

7. Baslund, B., K. Lyngberg, V. Andersen, J. Halkjaer-Kristensen, M. Hansen, M. Klokker, and B.K. Pedersen.

Effect of 8 wk of bicycle training on the immune system of patients with rheumatoid arthritis. *J. Appl. Physiol.* 75: 1691-1695, 1993.

8. Bennett, J.C., F. Plum, and F. Cecil. *Textbook of Medicine* (20th ed.). Philadelphia: Saunders, 1996.

9. Bergner, M. Quality of life, health status, and clinical research. *Med. Care.* 27: S148-S156, 1989.

10. Blair, S.N., H.W. Kohl, R.S. Paffenbarger, et al. Physical fitness and all-cause mortality. A prospective study of healthy men and women. *JAMA* 262: 2395-2401, 1989.

11. Blesch, K.S., J.A. Paice, R. Wickham, N. Harte, D.K. Schnoor, S. Purl, K. Rehwalt, P.L. Kopp, S. Manson, S.B. Coveny, M. McHale, and M. Cahill. Correlates of fatigue in people with breast or lung cancer. *Oncol. Nurs. Forum* 18: 81-87, 1991.

12. Brown, J. The national economic burden of cancer: An update. *J. Natl. Cancer Inst.* 82: 1811-1814, 1990.

13. Butterworth, E., and S.L. Nehlsen-Cannarella. The effects of high-versus moderate-intensity exercise on natural killer cell cytotoxic activity. *Med. Sci. Sports Exerc.* 25: 1126-1134, 1993.

14. Courneya, K.S., J.R. Mackey, and L.W. Jones. Coping with cancer—Can exercise help? *Phys. Sportsmed.* 28: 49-73, 2000.

15. Davis J.M., M.I. Kohut, D.A. Jackson, L.M. Hertler-Colbert, E.P. Mayer, and A. Ghafar. Exercise effects on lung tumor metastases and *in vitro* alveolar macrophage anti-tumor cytotoxicity. *Am. J. Physiol.* 274: R1454-R1459, 1998.

16. De La Fuente, M., M.I. Martin, and E. Ortega. Changes in the phagocytic function of peritoneal macrophages from old mice after strenuous physical exercise. *Comp. Immunol. Microbiol. Infect. Dis.* 13: 189-198, 1990.

17. Decker, W.A., J. Turner-McGlade, and K.M. Fehir. Psychosocial aspects and the physiological effects of a cardiopulmonary exercise program in patients undergoing bone marrow transplantation (BMT) for acute leukemia (AL). *Transplant. Proc.* 2: 3608-3609, 1989.

18. Dimeo, F.C., S. Fetscher, W. Lange, R. Mertelsmann, and J. Keul. Effects of aerobic exercise on the physical performance and incidence of treatment-related complications after high-dose chemotherapy. *Blood* 90: 3390-3394, 1997.

19. Dimeo, F.C., R.D. Stieglitz, U. Novelli-Fischer, S. Fetscher, and J. Keul. Effects of physical activity on the fatigue and psychologic status of cancer patients during chemotherapy. *Cancer* 85: 2273-2277, 1999.

20. Doll, R., and R. Peto. The causes of cancer: Quantitative estimates of avoidable risks of cancer in the United States today. *J. Natl. Cancer Inst.* 66: 1191-1308, 1981.

21. Eide, R. The relationship between body image, self-image and physical activity. *Scand. J. Social Med.* 29 (Suppl.): 109-112, 1982.

22. Fehr, H., G. Lotzerich, and H. Michna. Influence of physical exercise on peritoneal macrophage functions: Histo-chemical and phagocytic studies. *Int. J. Sports Med.* 9: 77-81, 1988.

23. Fehr, H., G. Lotzerich, and H. Michna. Human macrophage function and physical exercise: Phagocytic and histochemical studies. *Eur. J. Appl. Physiol.* 58: 613-617, 1989.

24. Ferrell, B., M. Grant, B. Funk, N. Garcia, S. Otis-Green, and M. Schaffner. Quality of life in breast cancer. *Cancer Patient.* 4: 331-340, 1996.

25. Fobair, P.R., T. Hoppe, J. Bloom, R. Cox, A. Varghese, and D. Spiegel. Psychosocial problems among survivors of Hodgkin's disease. *J. Clin. Oncol.* 4: 805-814, 1986.

26. Forner, M.A., M.E. Collazos, C. Barriga, M. De La Foente, A.B. Rodriguez, and E. Ortega. Effect of age on adherence and chemotaxis capacities of peritoneal macrophages: Influence of physical activity stress. *Mech. Ageing Dev.* 75: 179-189, 1994.

27. Frisch, R., G. Wyshak, N. Albright, and T. Albright. Lower prevalence of breast cancer and cancers of the reproductive system among former college athletes compared to non-athletes. *Br. J. Cancer* 2: 885-891, 1985.

28. Gabriel, H., L. Schwartz., P. Born, and W. Kindermann. Differential mobilization of leukocyte and lymphocyte subpopulations into the circulation during endurance exercise. *Eur. J. Appl. Physiol.* 65: 529-534, 1992.

29. Garrey, W.E., and W.R. Bryan. Variations in white blood cell counts. *Physiol. Rev.* 15: 597-638, 1935.

30. Haenszel, W. Cancer mortality among the foreign born in the United States. *J. Natl. Cancer Inst.* 26: 37-132, 1961.

31. Henderson, B., R. Ross, and H. Judd. Do regular ovulatory cycles increase breast cancer risk? *Cancer* 56: 1206-1208, 1985.

32. Hershcopf, R., and H. Bradlow. Obesity, diet, endogenous estrogens, and the risk of hormone-sensitive cancer. *Am. J. Clin. Nutr.* 45: 283-289, 1987.

33. Hoffman-Goetz, L. Exercise, natural immunity, and tumor metastasis. *Med. Sci. Sports Exerc.* 26: 157-163, 1994.

34. Hsieh, C., D. Trichopoulos. and K. Katsouyanni. Age at menarche, age at menopause, height and obesity as risk factors for breast cancer: Association and interactions in an international case control study. *Int. J. Cancer* 46: 796-800, 1990.

35. Kampert, J.B., S.N. Blair, C.E. Barlow, et al. Physical activity, physical fitness, and all-cause and cancer mortality: A prospective study of men and women. *Am. Epidemiol.* 6: 452-457, 1996.

36. Kohl, H., R. LaPorte, and S. Blair. Physical activity and cancer. *Sports Med.* 6: 222-237, 1988.

37. La Vecchia, C., A. Decarli, S. Di Pietro, S. Franceschi, E. Negri, and F. Parazzini. Menstrual cycle patterns and the risk of breast disease. *Eur. J. Cancer Clin. Oncol.* 21: 417-422, 1985.

38. Lee, I-M. Physical activity, fitness, and cancer. In: *Physical Activity, Fitness and Health. International Proceedings and*

Consensus Statement. C. Bouchard, R.J. Shephard, and T. Stephens (Eds.). Champaign, IL: Human Kinetics, 1994, pp. 814-831.

39. Lee, I-M. Exercise and physical health: Cancer and immune function. Res. Q. Exerc. Sport 66: 286-291, 1995.

40. Lee, I-M., H.D. Sesso, and R.S. Paffenbarger. Physical activity and risk of lung cancer. Int. J. Epidemiol. 28: 620-625, 1999.

41. MacVicar, M.G., and M.L. Winningham. Promoting the functional capacity of cancer patients. Cancer Bull. 38: 235-239, 1986.

42. MacVicar, M.G., M.L. Winningham, and J.L. Nickel. Effects of aerobic interval training on cancer patients' functional capacity. Nurs. Res. 38: 348-351, 1989.

43. McSweeney, A.J., I. Grant, R.K. Heaton, K.M. Adams, and R.M. Timms. Life quality of patients with chronic obstructive pulmonary disease. Arch. Intern. Med. 142:473-478, 1982.

44. Michna, H. The human macrophage system: Activity and functional morphology. Bibl. Anat. 31: 1-38, 1988.

45. Muir, C., and J. Waterhouse (Eds.). Cancer Incidence in Five Continents (Vol 5). Lyon, France: IARC, 1987.

46. Munck, A., P.M. Guyre, and N.J. Holbrook. Physiological functions of glucocorticoids in stress and their relation to pharmacological actions. Endocr. Rev. 5: 25-44, 1984.

47. Nehlsen-Cannarella, S.L., D.C. Nieman, J. Jessen, L. Chang, G. Gusewitch, G.G. Blix, and E. Ashley. The effects of acute moderate exercise on lymphocyte function and serum immunoglobulin levels. Int. J. Sports Med. 12: 391-398, 1991.

48. Nieman, D.C. Exercise, upper respiratory infection, and the immune system. Med. Sci. Sports Exerc. 26: 128-139, 1994.

49. Nieman, D.C., and S.L. Nehlsen-Cannarella. The immune response to exercise. Sem. Hematol. 31: 166-179, 1994.

50. Nieman, D.C., J.C. Ahle, D.A. Henson, et al. Indomethacin does not alter natural killer cell response to 2.5 hours of running. J. Appl. Physiol. 79: 748-755, 1995.

51. Nieman, D.C., L.S. Berk, M. Simpson-Westerberg, K. Arabatzis, W. Youngberg, S.A. Tan, and W.C. Eby: Effects of long endurance running on immune system parameters and lymphocyte function in experienced marathoners. Int. J. Sports Med. 10: 317-323, 1989.

52. Nieman, D.C., K.S. Buckley, D.A. Henson, et al. Immune function in marathon runners versus sedentary controls. Med. Sci. Sports Exerc. 27: 986-992, 1995.

53. Nieman, D.C., D.A. Henson, G. Gusewitch, B.J. Warren, R.C. Dotson, D.E. Butterworth, and S.L. Nehlsen-Cannarella. Physical activity and immune function in elderly women. Med. Sci. Sports Exerc. 25: 823-831, 1993.

54. Nieman, D.C., S. Simandle, D.A. Henson, et al. Lymphocyte proliferation response to 2.5 hours of running. Natl. J. Sports Med. 16: 404-408, 1995.

55. Ortega, E., M.E. Collazos, C. Barriga, and M. De La Fuente. Stimulation of the phagocytic function in guinea pig peritoneal macrophages by physical activity stress. Eur. J. Appl. Physiol. 64: 323-327, 1992.

56. Ortega, E., M.A. Forner, and C. Barriga. Exercise-induced stimulation of murine macrophage chemotaxis: Role of corticosterone and prolactin as mediators. J. Physiol. (Lond.) 498: 729-734, 1997.

57. Ortega, E., M.A. Forner, C. Barriga, and M. De La Fuente. Effect of age and of swimming-induced stress on the phagocytic capacity of peritoneal macrophages from mice. Mech. Ageing Dev. 70: 53-63, 1993.

58. Paffenbarger, R., R. Hyde, A. Wing, C. Hsieh. Physical activity, all-cause mortality and longevity of college athletes. N. Engl. J. Med. 314: 605-613, 1986.

59. Paffenbarger, R., R. Hyde, A. Wing, D. Jung, and J. Kampert. Influence of changes in physical activity and other characteristics on all-cause mortality. Med. Sci. Sports Exerc. 23: S82, 1991.

60. Paffenbarger, R., J. Kampert, H. Chang. Characteristics that predict risk of breast cancer before and after menopause. Am. J. Epidemiol. 112: 258-267, 1980.

61. Pedersen, B.K. Effect of 8 wk of bicycle training on the immune system of patients with rheumatoid arthritis. J. Appl. Physiol. 75: 1691-1695, 1993.

62. Pickard-Holley, S. Fatigue in cancer patients. Cancer Nurs. 14: 13-19, 1991.

63. Pike, M., M. Krailo, and B. Henderson. Hormonal risk factors, breast tissue age and the age incidence of breast cancer. Nature 302: 767-770, 1983.

64. Pyne, D.B. Regulation of neutrophil function during exercise. Sports Med. 17: 245-258, 1994.

65. Pyne, D.B., M.S. Baker, P.A. Fricker, W.A. McDonald, and W.J. Nelson. Effects of an intensive 12-wk training program by elite swimmers on neutrophil oxidative activity. Med. Sci Sports Exerc. 27: 536-542, 1995.

66. Segal, R., D. Johnson, J. Smith, S. Colletta, J. Guyton, et al. Structured exercise improves physical functioning in women with stages I and II breast cancer: Results of a randomized controlled trial. J. Clin. Oncol. 19: 657-665, 2001.

67. Shephard, R.J. Exercise in the prevention and treatment of cancer: An update. Sports Med. 15: 258-280, 1993.

68. Shephard, R.J. Physical activity and the healthy mind. Can. Med. Assoc. J. 128: 525-530, 1983.

69. Shephard, R.J., and N.P. Shek. Cancer, immune function, and physical activity. Can. J. Appl. Physiol. 20: 1-25, 1995.

70. Shephard, R.J., and N.P. Shek. Associations between physical activity and susceptibility to cancer. Sports Med. 26: 193-315, 1998.

71. Shephard, R.J., and N.P. Shek. Cancer, immune function, and physical activity. Can. J. Appl. Physiol. 20: 1-25, 1995.

72. Simopoulos, A. Obesity and carcinogenesis: Historical perspective. *Am. J. Clin. Nutr.* 45: 271-276, 1987.

73. Spitzer, W.O. State of science of 1987: Quality of life and functional status as target variables for research. *J. Chronic Dis.* 40: 465-471, 1987.

74. Sternfeld, B. Cancer and the protective effect of physical activity: The epidemiological evidence. *Med. Sci. Sports Exerc.* 24: 1195-1209, 1992.

75. Thompson, H.J. Effects of exercise intensity and duration on the induction of mammary carcinogenesis. *Cancer Res. Suppl.* 54: 1608-1635, 1994.

76. Thompson, W. Exercise and health: Fact or hype? *South. Med. J.* 87: 567-574, 1994.

77. Trichopoulos, D., B. MacMahon., and P. Cole. The menopause and breast cancer risk. *J. Natl. Cancer Inst.* 48: 605-613, 1972.

78. Tvede, N., J. Steensberg, B. Baslund, J. Halkjaer-Kristensen, and B.K. Pedersen. Cellular immunity in highly-trained elite racing cyclists and controls during periods of training with high and low intensity. *Scand. J. Sports Med.* 1: 163-166, 1991.

79. U.S. Department of Health and Human Services. *Physical Activity and Health: A Report of the Surgeon General.* Atlanta: U.S. Department of Health and Human Services, Centers for Disease Control and Prevention, National Center for Chronic Disease Prevention and Health Promotion, 1996.

80. Vezina, M.L., and R.H. Ruegger. The psychology of running: Implications for nursing and health. *Nurs. Forum* 19: 109-121, 1980.

81. Vihko, R., and D. Apter. Endocrine characteristics of adolescent menstrual cycles: Impact of early menarche. *J. Steroid Biochem. Mol. Biol.* 20: 231-236, 1984.

82. Winningham, M.L. Effects of a bicycle ergometry program on functional capacity and feelings of control in women with breast cancer (dissertation). Columbus, OH: Ohio State University, 1993.

83. Winningham, M.L. Walking program for people with cancer: Getting started. *Cancer Nurs.* 14: 270-276, 1991.

84. Winningham, M.L., and M.G. MacVicar. The effect of aerobic exercise on patient reports of nausea. *Oncol. Nurs. Forum* 15: 447-450, 1988.

85. Winningham, M.L., M.G. MacVicar, M. Bondoc, J.I. Anderson, and J.P. Minton. Effect of aerobic exercise on body weight and composition in patients with breast cancer on adjuvant chemotherapy. *Oncol. Nurs. Forum* 16: 683-689, 1989.

86. Woods, J.A., and J.M. Davis. Exercise, monocyte/macrophage function, and cancer. *Med. Sci. Sports Exerc.* 26: 147-156, 1994.

87. Woods, J.A., J.M. Davis, E.P. Mayer, A. Ghaffar, and R.R. Pate. Exercise increases inflammatory macrophage antitumor cytotoxicity. *J. Appl. Physiol.* 75: 879-886, 1993.

88. Woods, J.A., J.M. Davis, J.A. Smith, and D.C. Nieman. Exercise and cellular innate immune function. *Med. Sci. Sports Exerc.* 31: 57-66, 1999.

Barbara Smith, RN, PhD
School of Nursing
University of Alabama at Birmingham
Birmingham, AL

James Raper, CFNP, DNS
School of Nursing and School of Medicine
University of Alabama at Birmingham
Birmingham, AL

Michael Saag, MD
School of Medicine
University of Alabama at Birmingham
Birmingham, AL

23

Human Immunodeficiency Virus

Human immunodeficiency virus (HIV) infection and **acquired immunodeficiency syndrome (AIDS)** are conditions that evolve from the initial infection to the time of death. Throughout this cycle, a person becomes increasingly debilitated, often to the degree that even low-level physical activity is difficult. Exercise training can help to attenuate physical functioning loss and may affect the progress of the disease cycle. This chapter reviews the disease and the recommendations for exercise in this population.

The case definition of AIDS has evolved over time, reflecting changes in technology and better understanding of the clinical manifestations of HIV infection. The first case definition was published in 1982 by the Centers for Disease Control and Prevention (CDC). The list of AIDS-defining-diseases at that time was drawn from the clinical experience of providers treating primarily homosexual men with AIDS.

First identified in 1983, the HIV-type 1 is a **retrovirus** that infects human cells bearing the CD4+ surface marker. The preferred CD4+ site is the **T-helper lymphocyte.** Once infection is established, a complex array of pathogenic mechanisms occurs,

423

typically over many years, and ultimately results in the depletion of CD4+ lymphocytes from the normal range of 500 to 1400 cells · mm⁻³. As the number of CD4+ declines, persons infected with HIV are predisposed to **opportunistic infections,** malignancies, **wasting,** and other complications of AIDS (3, 21, 42).

Once HIV was established as the causative virus and the HIV antibody test became available in 1985, the case definition was expanded from the 1983 definition to include clinical conditions less closely associated with immune suppression. These included the following:

— Disseminated histoplasmosis, a type of fungus

— Chronic isosporiasis, an intestinal protozoa

— Certain non-Hodgkin's lymphomas, cancer of lymph tissue (11)

In 1987 the definition of AIDS was further expanded to include the following:

— Wasting (loss of 10% of body weight in ≤3 months)

— HIV encephalopathy (inflammation of the brain)

— Extrapulmonary (outside the lung) tuberculosis (12, 13)

In addition, these expanded definitions allowed for a presumptive diagnosis to accommodate changing diagnostic and treatment practices.

In 1993, the CDC issued a revised classification system for defining AIDS using three CD4+ cell ranges and three clinical categories (9). This formed a matrix of nine mutually exclusive categories (table 23.1). The most important changes were the inclusion of all patients with fewer than 200 CD4+ cells · mm⁻³ or a CD4+ count of less than 14% of total lymphocytes. These were the first laboratory criteria for defining AIDS. In addition, the following were added to the list of AIDS-defining clinical conditions:

— Recurrent bacterial pneumonia

— Invasive cervical cancer

— Pulmonary tuberculosis

Finally, in January 2000, the CDC made the most recent revision to the HIV case definition by combining reporting criteria for HIV infection and AIDS to a single case definition.

Scope

Since 1981, when the first case of AIDS was reported, this infection has presented a significant burden for the U.S. healthcare system. A total of 774,467 cases of AIDS and 448,060 AIDS-related deaths were reported in the United States up to the year 2000 (14). It is estimated that more than 1 million people are infected with HIV in the United States, or 1 person in 250 (46), suggesting that many do not know that they are infected.

Determining those at greatest risk of contracting HIV is important to determine if screening needs to be performed (figure 23.1). Along with the traditional high-risk groups of men who have sex with men and intravenous drug users, other high-risk groups include recipients of blood transfusions prior to 1985 including those with **hemophilia,**

TABLE 23.1
· · · · · · ·
1993 Centers for Disease Control Classification System/Matrix for HIV-1 Infection

CD4+ cells · mm⁻³ categories	Clinical category A	Clinical category B	Clinical category C
>500 cells · mm⁻³ (>29%)	A1	B1	C1
200-499 cells · mm⁻³ (14-28%)	A2	B2	C2
<200 cells · mm⁻³ (<14%)	A3	B3	C3

Note. Shaded areas show the expanded AIDS surveillance case definition. These are reportable as AIDS cases in the United States and its territories, effective January 1, 1993. Category A includes asymptomatic, or persistent, generalized lymphadenopathy, or acute HIV infection. Category B includes symptomatic or not A or C. Category C includes AIDS-defining or indicator condition.

- Persons with other sexually transmitted diseases
- Persons in high-risk categories
  - Injection drug users
  - Men who have sex with men
  - Hemophiliacs
  - Sexual partners of persons with known HIV-1 infection or those at high risk
- Persons who consider themselves at risk or request the test
- Pregnant women
- Patients with tuberculosis
- Persons with an occupational exposure (e.g., deep penetrating wound, splashed with blood)
- Hospitalized persons age 15-24 where seroprevalence is >1% or case rate is >1/1000
- Persons who perform exposure-prone invasive procedures (surgeons)
- Blood, semen, and organ donors
- Persons with clinical or laboratory findings suggestive of HIV-1 infection

FIGURE 23.1   Indications for HIV-1 testing.

children born to HIV-infected women, and persons who engage in impulsive sexual behavior or with many partners. Increased HIV transmission by heterosexuals has resulted in a higher infection risk for sexually active adolescents and young adults, the mentally ill, and abusers of alcohol and other mood-altering drugs. Although other avenues of transmission (e.g., insect bites, saliva transfer, sweat) have been suggested, none is known to transmit the virus.

Since the early 1980s, the demographic characteristics of those newly infected with HIV have changed. Data show a disproportionate impact on minority communities as reflected in AIDS incidence rates that are higher among African Americans and Hispanics/Latinos than among Caucasians (14). Minority women account for 80% of all newly reported AIDS cases among women. For a more detailed description of the demographic characteristics of HIV/AIDS, access the CDC web site at www.CDC.gov.

Worldwide, the demographic characteristics of HIV/AIDS vary from country to country. The World Health Organization (WHO) estimated there are more than 33 million people worldwide living with HIV infection with approximately 2.5 million deaths

to date. Most of these have occurred in Africa. For a more detailed description of the worldwide demographic characteristics of HIV/AIDS, or the demographic characteristics of a specific country, access the WHO web site at www.who.org.

Physiology and Pathophysiology

The natural history of HIV infection from the time of transmission to death is 8 to 10 years in an individual not receiving antiretroviral therapy. HIV infection progresses through five stages (29):

1. Acute infection (seroconversion)
2. Early disease (CDC A1, B1, C1)
3. Middle disease (CDC A2, B2, or C2)
4. Late disease or AIDS (CDC A3, B3, or C3)
5. Advanced disease

Stage 1—Acute Infection

Primary HIV infection and seroconversion (the change from **HIV negative** to **HIV positive** antibody status) occur over a period of a few weeks to months. During this stage, there is a high concentration of HIV in plasma and a precipitous decline in the CD4+ cell count, indicating a decline in function of the immune system. Figure 23.2 depicts the process of HIV replication. CD4+ cell destruction may result from direct infection of the CD4+ cell by HIV or by the body's immune system response to the infection.

Eventually, the immune system gains control of viral replication, the HIV concentration in the blood (i.e., **viremia**) is reduced, and gradually viral levels stabilize (37, 40). Although some persons may not experience any symptoms, the majority will have mild to moderate flulike symptoms during this period.

Stage 2—Early Disease

Stabilization of the level of plasma viral load signifies the beginning of clinical latency or what is known as the early disease stage. At this time, plasma levels of CD4+ cells are >500 cells · mm^{-3}. During this period, the patient is clinically asymptomatic and generally has no findings on physical examination except for persistent generalized lymphadenopathy (enlargement of lymph nodes). At this time the lymph tissue serves as the major reservoir for

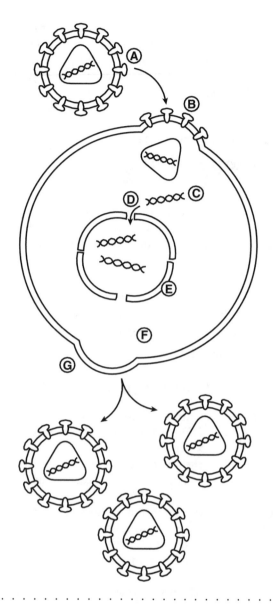

FIGURE 23.2   Replication of the HIV-1 virion. (A) HIV-1 virion with binding sites for CD4+ receptor. (B) Virion fuses (binds) with CD4+ receptors on cell surface and uncoating of virus core occurs. (C) Once genetic material is in the cell reverse transcription occurs. (D) Importation of viral genetic material into the cell nucleus occurs and it is then incorporated into the cell's genetic material. (E) Translation and transport occurs. (F) Assembly of new virions occurs. (G) New virus buds from the cell, matures, and the process is repeated.

HIV. This results in a relatively lower number of viral particles in peripheral blood than seen in stage 1. This stage is the longest, often lasting many years. Chronic immune activation and persistent viral replication characterize this period, but the rates at which HIV cell death and replacement occur are nearly in balance. However, over time the CD4+ cell count gradually declines at the rate of approximately 50 to 70 cells · mm^{-3} per year.

Stage 3—Middle Disease

This stage begins when CD4+ cells decrease to between 200 and 499 cells · mm^{-3} plasma. Although most patients remain asymptomatic during the middle stage of disease, many lesions or disorders of the skin and mucous membrane become evident. Patients with middle stage HIV disease, when left untreated, have a 20% to 30% chance of developing an AIDS-defining illness (i.e., moving to stage 4) or dying within 18 to 24 months (34).

Stage 4—Late Disease

Movement into this stage occurs when the plasma CD4+ cell count falls below 200 cells · mm^{-3}. At this point, the patient meets the CDC case definition of AIDS. During the late disease stage CD4+ cells are between 50 and 199 cells · mm^{-3} and the risks associated with developing an AIDS-defining illness increase (9). HIV usually increases during this stage of disease and correlates with a sharp decrease in CD4+ cell count (48). In addition, patients with late stage disease are at increased risk of developing opportunistic infections and malignancies that are included in the 1993 CDC list of AIDS-defining conditions.

Stage 5—Advanced Disease

When HIV infection is uncontrolled and CD4+ cells decrease to less than 50 cells · mm^{-3}, the final stage of the disease process is reached. It is characterized by a period of increasing opportunistic infection risk, certain cancers, wasting, and neurological complications. Each of these is associated with severe immunosuppression, and often several opportunistic diseases coexist. Without aggressive antiretroviral treatment and aggressive treatment of the opportunistic infection, death is imminent.

Clinical Considerations

The diagnosis and treatment of HIV infection and AIDS depend on clinical evaluation and careful consideration of treatment options. This section reviews these topics.

Signs and Symptoms

Primary HIV infection (stage 1) is often unrecognized clinically. However, many patients develop symptoms of viral illness (16). Following are the five most commonly described symptoms of primary HIV infection:

1. Fever
2. Sore throat
3. Fatigue
4. **Myalgia**
5. An average weight loss of 5 kg (59)

During the early disease stages (stages 1-2), up to 5% of patients will develop one or more of the following signs and symptoms:

1. Mild to moderate lymphadenopathy: increase in size of lymph nodes
2. Seborrheic dermatitis: dry, scaly skin
3. Psoriasis
4. Pruritic folliculitis: itchy red bumps
5. Fungal infection of the fingernails and toenails
6. Exaggerated responses to insect bites
7. Molluscum contagiosum: bumps under the skin containing a cheesy-looking substance
8. Taste distortion
9. Ulcers of the mouth

Muscle and joint pain, headache, and fatigue are commonly present during the middle disease stages (stages 2-4) and are related to higher levels of HIV virus in the blood. Other commonly reported middle stage signs and symptoms include these:

1. Recurrent fever blisters
2. Herpes simplex ulcerations
3. Shingles (i.e., varicella-zoster virus)
4. Removable white plaques on mucosal surfaces
5. Recurrent diarrhea
6. Skin problems
7. Intermittent fever
8. Wasting

During the late and advanced disease stages (stages 4-5), symptom progression often intensifies and includes multiple systems. AIDS defining opportunistic conditions include these:

1. Invasive cervical cancer
2. Disseminated **histoplasmosis**
3. **Kaposi's sarcoma**

4. *Mycobacterium avium* **complex**
5. *Pneumocystis carinii* **pneumonia (PCP)**
6. Tuberculosis
7. Wasting

Opportunistic infections are the primary cause of death in people with AIDS. *P. carinii* pneumonia is the primary diagnosis and affects up to 52% of patients with AIDS. Kaposi's sarcoma is the next most prevalent (26%). Cytomegalovirus retinitis is the leading cause of blindness in these patients and causes various other end-organ disease resulting in painful swallowing, bloody diarrhea with abdominal pain, and weakness of the lower extremities. Various endocrine disorders and neurological symptoms are common in late and advanced stage disease (14).

Diagnostic and Laboratory Evaluations

A blood test is required to conclusively diagnose HIV infection. Two techniques are used most commonly:

1. Enzyme immunoassay method
2. Western blot method

The enzyme immunoassay (EIA; also referred to as enzyme-linked immunosorbent assay, or ELISA) is the most commonly used of these tests. It measures the presence of antibodies to HIV viral proteins. A single positive EIA is classified as HIV-reactive but not positive for HIV infection until it is confirmed by one or more subsequent tests. It is very important to thoroughly explain to patients the need for the confirmatory subsequent test when they have a reactive EIA test, because this test is not 100% accurate. The Western blot method is the most commonly used retest method. Various organizations have developed and published criteria for interpreting Western blot reactivity (10, 17, 70). Reactivity patterns that do not fit these criteria are classified as indeterminate.

Ongoing evaluation of disease severity is important to determine appropriate treatment strategies. Several studies demonstrate the value of using CD4+ cell count as the best predictive marker of relative risk for the development of HIV-related opportunistic disease (66). However, CD4+ count only provides a surrogate marker for clinical progression (15, 18). Rapid decay of plasma HIV RNA concentration (or viral load) occurs in patients treated with potent **antiretroviral medications** (24, 45), and the degree of this effect is inversely related to disease progression (38, 39, 46). Decisions regarding the initiation of or changes in antiretroviral therapy are guided by plasma HIV RNA concentration, the CD4+ cell count, and the clinical condition of the patient. To establish the overall clinical condition and reveal existing comorbidities, the patient often undergoes the laboratory evaluations outlined in table 23.2.

Treatment

The management of HIV infection has evolved very rapidly. Therefore, the U.S. Department of Health and Human Services Panel of Clinical Practices for the Treatment of HIV recommends that the care of patients with HIV infection be supervised by an expert. Those working with HIV-infected individuals must have access to the latest available recommendations from the International AIDS Society–USA Panel or the HIV/AIDS Treatment Information Service (6). For best treatment results, all patients should be treated with the most up-to-date therapy as soon as possible after seroconversion. Often this does not occur until patients have reached stage 2 and demonstrate related symptoms. Recommendations to treat asymptomatic persons must be based on the individual's willingness to accept therapy, the probability of adherence to the prescribed regimen, and the prognosis in terms of time to an AIDS-

defining complication as predicted by plasma HIV RNA concentration and the CD4+ cell count (6). Once the decision is made to initiate therapy, the goal is maximal viral suppression for as long as possible.

Three categories of antiretroviral agents currently available are the mainstays of medical therapy.

1. Nucleoside analogs or nucleoside analog reverse transcriptase inhibitors (NRTIs)
2. Non-nucleoside reverse transcriptase inhibitors
3. **Protease inhibitors**

Chapter 5 provides specific information about medications used to treat HIV infection.

Treatment with these agents is complex. Patient education and involvement are considered important for optimal adherence to the treatment regimen. Therapeutic results are evaluated primarily by observing plasma HIV-RNA concentration. An effective or therapeutic response occurs when there is a 10-fold decrease in HIV RNA copies at 8 weeks and no detectable virus 4 to 6 months after initiation of treatment. Failure of therapy at 4 to 6 months may be the result of nonadherence to the prescribed drug

---

TABLE 23.2
Laboratory Tests for Baseline Evaluation Following HIV-1 Diagnosis

Evaluation	Test rationale
CD4+ cells/mm	Level of CD4+ cells and HIV disease stage
HIV RNA	Amount of virus and later to assess effectiveness of treatment
PPD skin test	Tuberculosis (comorbidity)
CBC	Presence of anemia, overall level of specific blood cells
Serology (VDRL/RPR)	Presence of syphilis, an STD
Serum chemistry	General health, status of kidney, and liver function
Hepatitis B and C	Assess for these viruses that are common comorbidities
G-6-PD	Test for presence or absence of this genetic disease, because it predisposes individuals to hemolytic anemia following exposure to certain drugs used in the treatment of HIV
Pap smear	Assess health of the cervix and test for other STDs
Chest X-ray	Evaluate for the presence of PCP and other pulmonary complications

Note. PPD = purified protein derivative; CBC = complete blood count; VDRL = Venereal Disease Research Laboratories; RPR = rapid plasma reagin; STD = sexually transmitted disease; PCP = *Pneumocystis carinii* pneumonia.

regimen, inadequate potency of medications, suboptimal levels of antiretroviral agents, medication resistance, and other factors that are poorly understood at this time. Patients whose therapy fails ideally should be changed to new medications that are devoid of anticipated cross-resistance and for which clinical trial data support a high probability of suppressive viral response.

The dramatic improvements in mortality and morbidity rates associated with HIV infection observed over the past decade are attributable, in part, to the widespread use of highly active antiretroviral therapy or HAART (24, 28, 47). In addition, the development of guidelines for the use of prophylactic antimicrobial agents and vaccines to prevent opportunistic infections such as **cytomegalovirus,** *P. carinii* pneumonia, *M. avium* complex, and others have also contributed to improved outcome. The exercise professional working with these patients should understand that these patients often have a multitude of other comorbid conditions requiring the use of other medications. These include chemotherapeutic agents for cancer, psychotropic drugs, medications for gastrointestinal distress, and other drugs to treat the side effects of the HIV medications. Some of these medications may have a direct or indirect affect on the ability to perform exercise.

With declining morbidity and mortality rates associated with HIV infection, there is much optimism about the future for HIV-infected patients. However, healthcare providers must be aware of the following physiological effects of the HAART regimen that place patients at an increased risk for developing coronary heart disease (23):

— Body fat redistribution or lipodystrophy
— Impaired glucose metabolism
— Insulin resistance
— Elevated blood lipids related to protease inhibitors or NRTIs

These effects have led some physicians to alter their approach to therapy in HIV-infected individuals. However, to date little emphasis has been placed on the use of exercise to counteract these side effects.

Medical Evaluation

When an HIV-infected individual begins an exercise program, it is important to conduct a medical evaluation to detect conditions that might contraindicate exercise. A history of extreme fatigue, weakness or malaise, fever, chills, night sweats, or wasting indicates active infection and should be considered because exercise may worsen these conditions. Care should be taken to identify individuals who are at high risk for developing wasting, such as those with uncontrolled viremia or those who are already actively wasting. Although wasting is not an absolute contraindication for performing an exercise test, the cause should be identified and treated before the individual starts an aerobic exercise program. The clinical exercise professional should assess body weight at regular intervals for rapid or excessive weight loss. Increased caloric expenditure associated with exercise training could exacerbate weight loss. Additionally, those individuals with severe peripheral lipoatrophy (i.e., decrease or absence of fat, increased central obesity, and the development of a cervical fat pad) may need to be cautioned about the exacerbation of peripheral fat loss that may occur with aerobic training. Additional tips for working with HIV-infected people are provided in practical applications 23.1 and 23.2.

The physical examination should focus on temperature, resting heart rate, blood pressure, body weight, and indicators of cardiovascular disease (e.g.,

---

PRACTICAL APPLICATION 23.1

Client–Clinician Interaction

"Healthy" patients who are HIV positive and choose to exercise often do so on their own or at a commercial health club. Therefore, the initial interaction between the clinical exercise professional and an HIV-infected patient is likely to occur at the patient's first visit to an exercise facility, often for treatment of established cardiac disease. These patients are often concerned about who is going to know about their diagnosis because of the stigma associated with their disease. The clinical exercise professional must be aware of this concern and assure the patient that only those staff members who need to know about their diagnosis will be told. The patient can also be assured that no staff member will discuss medical information with other patients. Additionally, the clinical exercise professional should not display any hesitancy or prejudice toward these patients.

Universal Precautions

An HIV-positive patient cannot be identified by simple observation. Additionally, one cannot rely on an individual to provide his or her HIV status because some are not willing to reveal this information and others do not know that they are HIV positive. Finally, there are other diseases that are equally or more detrimental to health than HIV. These include hepatitis B and C. Individuals with these and other conditions are more likely than a person with HIV to transmit their disease to someone else. Applying the universal precautions will greatly reduce risk of HIV and other disease transmission. The universal precautions listed here are suggested for working with any patient, including those with HIV.

Procedure	Wash hands	Gloves	Gown	Mask	Eyewear
Talking					
Adjusting intravenous fluid rate or other noninvasive equipment					
Examining patient without touching blood, body fluids, or mucous membranes (e.g., checking blood pressure)	X				
Examining patient including touching blood, body fluids, or mucous membranes (e.g., placing and removing mouth pieces, vigorous skin preparation for electrodes during exercise testing)	X	X			
Blood draw	X	X			
Inserting a venous catheter	X	X			
Handling soiled linen or other waste	X	X			
Intubating during cardiopulmonary resuscitation	X	X	X	X	X
Suctioning	X	X	X	X	X
Surgical procedures	X	X	X	X	X

Adapted from the University of Alabama at Birmingham (UAB) Interdisciplinary Standard: Universal Precautions CDC Category: Standard Precautions. For more detail on this topic search the CDC Web site at www.CDC.gov.

murmurs, arrhythmias, edema). Other findings indicative of disease progression include findings of psychomotor slowing, eye movement abnormalities, hyperreflexia, and peripheral neuropathy.

It is important to consider any recent history of changes in body weight (i.e., gain or loss, period of time, and distribution of fat) or other symptoms of metabolic disorders. Fat redistribution, characterized by peripheral lipoatrophy in HIV-infected men and women, is noted with specific HAART regimens (7, 32, 52, 61, 72). Metabolic disorders such as impaired glucose metabolism and reduced insulin sensitivity may occur in individuals on active anti-retroviral therapy, both with and without the use of protease inhibitors (7, 58, 62, 68, 69).

Exercise Testing

Because a number of antiretroviral medications cause severe nausea and vomiting following ingestion, the graded exercise test should be scheduled at a time when the patient is not as likely to experience these symptoms. It is best to discuss this with the patient to clarify this issue. See chapter 6 for general exercise testing recommendations.

Recommendations and Anticipated Responses

Depending on the individual's physical status, the exercise test protocol should be carefully selected. For patients with poor balance or coordination secondary to fatigue or neuromuscular effects, a cycle ergometer may be better than treadmill walking. Standard test end points should be used along with standard electrocardiogram and blood pressure monitoring. As a substitute for maximal exercise testing, a submaximal exercise evaluation, such as the 6-min walk, can provide information regarding fitness status and improvement.

Many individuals who are in early, middle, or late stage disease can expect to have a normal blood pressure, pulse rate, and respiratory responses to a graded exercise test. Some HIV medications known to cause peripheral neuropathy may also increase the risk of autonomic neuropathy, which in turn increases the risk of an abnormal blood pressure response, an elevated resting heart rate, and attenuated exercise heart rate response.

In general, exercise testing reveals that HIV-infected individuals have low $\dot{V}O_2$max values (27, 64, 67). This is noted even when there is no evidence of a pulmonary opportunistic infection (27). Possible reasons for a reduced $\dot{V}O_2$max include skeletal muscle effects of the infection and medications, and loss of cardiorespiratory endurance secondary to an increasingly sedentary lifestyle.

Exercise Prescription

Exercise training can counter some of the debilitating effects of HIV. These patients can expect to improve their cardiorespiratory fitness, muscular strength and endurance, flexibility, and body composition by participating in a regular exercise program. Additionally, exercise training may help to improve psychological status, including a reduction in depression and anxiety and improvement of mood and fatigue. Finally, it does not appear that exercise training impairs immune status in stable patients with HIV infection. In fact, exercise training may help to retard the disease process and possibly improve immune status.

When an HIV-positive individual requests guidance during exercise training, the clinical exercise professional should provide assistance for at least the first several exercise sessions. Goals include teaching the patient how to exercise at an appropriate intensity and with proper technique to optimize the benefits of the exercise and to reduce the likelihood of injury or overtraining syndrome. Early su-

pervision may also increase the likelihood that an individual will adhere to an exercise training program. These patients have no special exercise facility requirements. Practical application 23.3 reviews the literature concerning aerobic exercise, resistance training, and disease prevention.

Special Exercise Considerations

There is little risk of infection to the clinical exercise professional conducting a graded exercise test or supervising exercise sessions for an HIV-infected person. However, before working with any clinical population, an individual should become familiar with universal precautions to reduce the risk of contracting an infectious disease (see practical application 23.2).

The CD4+ cell count should be evaluated within 3 months after the initiation of an exercise training program. This is particularly important for individuals who are not tested regularly. As stated previously, HIV-positive patients are at risk of extreme fatigue as a result of an exercise training program. Each patient should be regularly evaluated for indications of fatigue, especially in the early stages of a training program. An easy method to evaluate fatigue is to use the rating of perceived exertion scale at a standardized work rate at least once a week.

Nausea is a common side effect for HIV-infected patients who begin a new medication or have a change in the amount of medication they take. If this occurs in association with exercise, which happens most often in newly diagnosed patients, patients may become discouraged with exercise because they feel sick. The clinical exercise professional should be prepared to deal with this situation. This includes making the patient aware of this possibility and emphasizing that the medicine, and not exercise, is likely causing the problem. One can work with the patient and his or her physician to develop a schedule that takes into consideration the patient's eating, medication, and exercise schedules to prevent or reduce medication side effects during exercise training.

Fatigue is often experienced by the HIV-infected patient and is inversely related to the CD4+ count. Fatigue may be the result of the infection or treatment. The patient experiencing fatigue may become depressed and resist continued involvement in an exercise regimen. It is important to encourage these patients to continue to exercise but at a reduced level. The goal is to prevent deconditioning and the subsequent cascade of increasing fatigue and reduced activity levels. An exception to encouraging continued exercise is if the patient is experiencing fever or

Literature Review

Aerobic Exercise

Research using aerobic exercise unequivocally shows improved physical and psychological end points in many chronic disease states including cardiovascular disease, hypertension, Parkinson's disease, pulmonary disease, renal dialysis, and cancer (8, 19, 20, 22, 26, 35, 50). Studies indicate that strength and fat-free mass can be increased by resistance training in older men and women and in women at risk for diabetes (56, 57, 60). Despite the wealth of investigation on these and other clinical populations, only a few studies have documented the benefits of aerobic and resistive exercise in HIV-infected individuals (31, 51, 53, 64, 65, 67). The following table provides an overview.

Author	HIV+ subject number	Type of exercise	Length of program	Primary exercise results
LaPerriere et al. (31)	10 HIV+ exercisers and 6 HIV+ controls	Aerobic	10 weeks	Increased estimated $\dot{V}O_2$max
Perna et al. (49)	28 HIV+	Aerobic	12 weeks	Increased $\dot{V}O_2$peak
MacArthur et al. (33)	6 HIV+ exercisers completed	Exercise training	24 weeks	Increased estimated $\dot{V}O_2$max
Rigsby et al. (51)	13 HIV+ exercisers and 11 HIV+ controls	Aerobic/resistance	12 weeks, 3 times per week	Increased total time at 150 W and increased strength.
Roubenoff et al. (53)	24 HIV+ exercisers	Resistance	8 weeks training + 8 weeks ad libitum	Increased strength, increased lean mass
Roubenoff et al. (55)	10 HIV+ exercisers and no controls	Aerobic and resistance	16 weeks, 3 times per week	No aerobic measure, increased strength, increased lean mass
Smith et al. (64)	18 HIV+ exercisers and 30 HIV+ controls	Aerobic	12 weeks, 3 times per week	Increased time to exhaustion on treadmill
Stringer et al. (67)	HIV+ exercisers and HIV+ controls	Aerobic	6 weeks	Increased measured $\dot{V}O_2$max

Two investigations demonstrate that the time subjects could exercise on a cycle ergometer was greater among exercise-trained than nontrained HIV subjects (31, 51). However, each of these studies used relatively small samples of men and did not directly measure $\dot{V}O_2$. Stringer et al. (67) reported a modest improvement ($0.2 \text{ L} \cdot \text{min}^{-1}$) in directly measured $\dot{V}O_2$max after 6 weeks of heavy training in HIV-positive patients. Smith et al. (64) reported a significant increase in time on a treadmill and a nonsignificant increase of $\dot{V}O_2$max ($0.16 \text{ L} \cdot \text{min}^{-1}$).

Lower $\dot{V}O_2$max values in some HIV-infected individuals may be related to the effects of medication or the HIV virus on the cardiovascular, pulmonary, or skeletal muscle systems or to comorbid conditions associated with HIV infection. Johnson et al. (27) reported that a group of HIV-infected active duty servicemen had a lower $\dot{V}O_2$max compared with an otherwise similar group of non-infected controls. The authors concluded that the lower $\dot{V}O_2$max was attributable to coronary artery disease. Sinnwell et al. (63) used magnetic resonance imaging and found a decreased muscle

continued

phosphocreatine level following a graded exercise test in individuals on AZT, compared with non-AZT-treated HIV-infected individuals. These authors suggested that the skeletal muscle cells in HIV-infected patients on AZT rely heavily on anaerobic pathways to generate adenosine triphosphate because of the effects of the drug. Mannix et al. (36), also using magnetic resonance imaging, determined that there was a decrease in oxidative, and an increase in anaerobic, sources of adenosine triphosphate (ATP) in skeletal muscle cells of stable AIDS patients compared with healthy controls. They concluded that the decrease in oxidative sources of ATP contributes to poor exercise tolerance and suggested it might be attributable to HIV-induced skeletal muscle mitochondrial damage.

In summary, the first few studies of aerobic exercise training in HIV-infected individuals are promising because they have demonstrated that aerobic exercise is safe and indicate that it may be an effective health promotion strategy for individuals with HIV infection. However, the mechanism for the reduced exercise $\dot{V}O_2$max and its response to training requires further study.

Resistance Training

Only a few studies have described the use of resistance exercise in HIV-infected individuals. Roubenoff et al. (53) reported increased lean mass with the use of resistance training. Additionally, this study reported no decline in CD4+ cells or increase in HIV RNA. Roubenoff and colleagues (55) combined aerobic and resistance exercise during a 16-week intervention using 10 HIV-infected men. The authors noted an increased 1-repetition maximum with three of the four exercises tested (leg press, leg extension, and chest press). Body weight, lean mass, and bone mineral density, measured with the dual energy X-ray absorptiometry method, did not change; however, total body fat and abdominal fat decreased. These data suggest that resistance training is not harmful, and is potentially beneficial, for patients with HIV infection. There remains a need for more investigation in this area.

Disease Prevention

Exercise may indirectly reduce HIV progression to AIDS by reducing psychological depression. Reductions in depression are associated with a decreased likelihood to engage in behavior that may be detrimental to one's health. Because the risk of HIV infection is related to behavior, those who exercise may indirectly reduce their risk of HIV infection. However, direct evidence of the theory does not exist.

A longitudinal observational study of a large HIV/AIDS clinic reported that HIV-infected patients who exercised three or more times per week had a lower risk of progressing to AIDS or death from AIDS at 1-year follow-up (41). Regular exercise training of moderate intensity does not appear to activate an acute phase immune response. Also, it is reported that a single bout of exercise does not increase viral replication in HIV-infected adults (54). Additionally, exercise training studies that use HIV-infected patients report either no change or an improved CD4+ cell count. Thus, the use of moderate-intensity exercise that promotes general health, along with appropriate antiretroviral therapy and opportunistic infection prophylaxis, may reduce the severity of HIV infection and progression to AIDS.

HIV-infected individuals who may wish to engage in highly intense exercise can point to elite athletes such as Magic Johnson, Greg Louganis, and others who have participated in rigorous sporting events with no health consequences. Although there are no data to indicate that highly intense exercise increases the morbidity and mortality rates associated with HIV infection, infected individuals may be wise to avoid prolonged, highly intense exercise. This recommendation is based on data from healthy individuals which indicates that highly intense, prolonged exercise is associated with an increase in upper respiratory infections and the suppression of specific immune factors (5, 43, 44). Each of these may be detrimental to the health of an HIV-infected person. These individuals should be encouraged to participate in moderate aerobic and resistance exercise training because of myriad potential physiological and psychological benefits. More longitudinal studies are needed to demonstrate the role of moderate exercise in slowing the progression of HIV infection and delaying the onset of opportunistic infections.

active wasting. If that is the case, the patient should be immediately referred to his or her physician to evaluate and treat the underlying cause of the fever or wasting.

Exercise Recommendations

Recommendations for exercise prescription for HIV-infected patients are based on a combination of the American College of Sports Medicine's *Guidelines for Exercise Testing and Prescription* (1), studies of the effects of aerobic exercise on HIV-infected individuals, and general practical experience. Practical application 23.4 provides specifics about exercise training recommendations for these patients.

Cardiovascular Training

As with any individual, to provoke continued adaptations of the cardiovascular system and skeletal muscles of HIV-infected individuals, an appropriate exercise prescription must include an appropriate mode, frequency, duration, and intensity (1, 2, 4, 30).

The exercise intensity must provide the highest level of total work without inducing excessive fatigue. Most individuals can exercise at a heart rate corresponding to between 40% and 85% of their $\dot{V}O_2$max (or heart rate reserve) achieved during an exercise evaluation. Those in the later stages of disease, or exhibiting indications of fatigue, should exercise at the lower end of this intensity range (e.g., 40-50%). If a heart rate range cannot be used, the rating of perceived exertion scale can be substituted. Training workload should be increased as tolerated by the individual. It is best to increase workload in small, regular increments of duration initially until a desired duration (i.e., 30-60 min) is attained on 3 to 5 days per week. This can then be followed by increases in pace or intensity.

Exercise Prescription Summary

Mode of training	Frequency, intensity, duration	Considerations
AEROBIC		
Walking	3-5 days per week	Watch for indications of fatigue, which may signify or exacerbate disease progression.
Jogging	30-60 min in target zone	
Stationary cycling	40-85% peak $\dot{V}O_2$	Progress patients as tolerated.
Other aerobic exercises	40-85% heart rate reserve	Assess for excessive weight loss.
		Consider a 3-month CD4+ determination for those who do not have this regularly determined.
RESISTANCE		
Free weights	2-4 days per week	Do not allow patients to perform on consecutive days.
Resistance machines	2-3 sets per exercise	Focus on major muscle groups.
	8-12 repetitions per set	Use stretch cords as an alternative.
	~50% of 1RM	Progress patient toward 80% of 1RM.
RANGE OF MOTION		
Static	Perform daily	Work through major joints.
	Hold stretch for up to 30 s	Avoid pain during stretch.
	Work through major joints	May supplement with towels or stretch cords to enhance stretch.
		Relaxation exercises, yoga or t'ai chi may assist with increasing range of motion and reduce anxiety associated with the disease.

Note. 1RM = 1-repetition maximum.

Because aerobic exercise training can increase the rate of weight loss, individuals should be weighed at least biweekly. If the patient loses more than 1 kg (2.2 lb) for 3 weeks in a row, his or her healthcare provider should be consulted. Additionally, those individuals with severe peripheral lipoatrophy may need to be cautioned about the exacerbation of peripheral fat loss that may occur in association with aerobic training. Reductions in exercise training volume may be necessary to reduce the rate of fat loss.

Resistance Training

Patients infected by HIV who perform resistance training demonstrate positive skeletal muscle adaptations that enhance strength and endurance. Patients report that activities of daily living become less difficult and less fatiguing. Additionally, increases in cross-sectional area are reported and may prevent wasting. Standard principles of training specificity and mode, frequency, duration, and overload (intensity) should be followed (2, 4, 30, 71). See chapter 7 for general resistance training recommendations.

The following recommendations for resistance training are based on the American College of Sports Medicine's 1990 position statement, studies by Roubenoff et al., and general practice experience with resistance exercise in HIV-infected and other chronically ill populations (2, 51, 53). Strength training exercises should include those that work both the upper and lower extremities, focusing on the large muscle groups. Common exercises include leg press, squats, elbow flexion and extension, pulley/cable pull-downs, bench and military presses, leg extension and curl, upright row, low back extension, and bent knee curl-ups. Resistance training can be performed on 2 to 4 days of the week with a training bout every other day. It is prudent for individuals to abstain from consecutive days of resistance training to reduce the chance of excessive fatigue, immunosuppression, and opportunistic infection. These recommendations can be adapted by experienced exercise professionals to meet the individual needs of their clients.

Range of Motion

As with any person performing a well-rounded exercise program, the HIV-positive patient should be instructed to perform range of motion routinely. General recommendations for range of motion exercise are presented in chapter 7.

Conclusion

Although it may seem counterintuitive to recommend exercise training for HIV-infected patients because of the risk of fatigue, a properly designed and implemented exercise training program can provide benefits for these patients. A clinical exercise professional can provide guidance for these patients who wish to begin an exercise training program. Future investigations should focus on the effects of exercise training on disease progression and mortality rates.

Case Study 23

Medical History and Diagnosis

Mr. BP, a 39-year-old Caucasian, bisexual male was diagnosed with hypertension about 6 years ago. He has had difficulty adjusting to antihypertensive medications, and thus his hypertension is not well controlled. He was first diagnosed with HIV infection about 2.5 years ago following repeated bouts of upper respiratory infection, persistent generalized lymphadenopathy, and an unexplained weight loss of 15 lb in 3 months. Since his HIV diagnosis he has been diagnosed with clinical depression.

At the time of diagnosis his weight was 187 lb and blood pressure was 166/90, and his laboratory work revealed the following:

— 206 CD4+ cells $\cdot$ mm^{-3}
— 98,650 HIV RNA copies $\cdot$ ml^{-1}
— Total cholesterol = 209 mg $\cdot$ dl^{-1}

continued

— High-density lipoprotein cholesterol = 37 mg · dl^{-1}

— Triglycerides = 145 mg · dl^{-1}

He was placed on a HAART regime that included the protease inhibitor Indinavir, plus two NRTIs, AZT and ddI. He was also encouraged to see his internist regarding his blood pressure. Six months after beginning HAART his laboratory work was as follows:

— 389 CD4+ cells · mm^{-3}

— 1009 HIV RNA copies · ml^{-1}

— Total cholesterol = 215 mg · dl^{-1}

— High-density lipoprotein cholesterol = 39 mg · dl^{-1}

— Triglycerides = 160 mg · dl^{-1}

At 2.5 years after diagnosis, he continues on the same therapy. His weight has returned to an almost normal 199 lb, and his most recent blood pressure is 144/86. He reports that he is feeling better because his virus is "under control" and he is seeing a therapist to deal with his depression. He also complains about getting a "real paunch" in his abdominal area. Laboratory work at his second to last clinic visit revealed the following:

— 347 CD4+ cells · mm^{-3}

— Less than 100 HIV RNA copies · ml^{-1}

— Total cholesterol = 238 mg · dl^{-1}

— High-density lipoprotein cholesterol = 37 mg · dl^{-1}

— Triglycerides = 678 mg · dl^{-1}

Laboratory work at his last visit revealed the following:

— 329 CD4+ cells · mm^{-3}

— Less than 100 HIV RNA copies · ml^{-1}

— Total cholesterol = 273 mg · dl^{-1}

— High-density lipoprotein cholesterol = 34 mg · dl^{-1}

— Triglycerides = 1265 mg · dl^{-1}

At the second visit, his provider talked with him about increasing his exercise to help with his "paunch," to reduce his cholesterol and triglyceride levels, and to reduce his blood pressure. His physician also scheduled an exercise evaluation to determine his level of fitness and to attain some feedback from a clinical exercise physiologist regarding beginning an exercise program.

Exercise Evaluation

At the time of the exercise evaluation the patient's resting vitals were as follows:

— Heart rate = 72

— Blood pressure = 146/98

— Respiratory rate = 22

continued

The exercise evaluation resulted in the following maximal values:

— Heart rate = 184

— Blood pressure = 270/96

— Respiratory rate = 44

— Peak oxygen consumption = 36 ml $\cdot$ min^{-1} $\cdot$ kg^{-1}

His resting electrocardiogram revealed normal sinus rhythm with nonspecific ST-segment changes. No ST-segment changes or arrhythmias were noted during exercise.

Before beginning his exercise program, the patient was advised to revisit his internist and have his blood pressure and his antihypertensive medication reevaluated. The physician added a diuretic to the patient's medical regime. When he returned to the exercise facility the next week, his resting blood pressure was 132/90.

Development of Aerobic Exercise Prescription

The following exercise prescription was developed by the clinical exercise professional who performed the exercise evaluation.

— Exercise mode: Stationary cycling or walking/jogging on a treadmill or track for at least 20 min followed by 10 min of aerobic exercise on either a stair stepper, cross-country ski machine, or the elliptical trainer.

— Exercise frequency: Three to five days each week.

— Exercise session duration: 30 min, not including warm-up and cool-down.

— Exercise intensity/workload: Exercise at a workload that will produce a heart rate of between 139 and 167 (60-85% of heart rate reserve).

— Exercise progression: As cardiovascular and muscular adaptations occur in response to training, adjust the workload to maintain the heart rate within the initially prescribed range. Consider adding a few minutes per week to reach a duration goal of 45 to 60 min.

The patient attended an average of four exercise sessions per week and over the next 12 weeks increased his exercise session duration from 30 to 45 total minutes. He lost approximately 7 lb, and although he complained of still "having a bit of a paunch," he was feeling considerably better about how he looked. His therapist had indicated his depression was improving. His resting blood pressure was 132/84 and resting heart rate was 68. Follow-up laboratory work revealed the following:

— 317 CD4+ cells $\cdot$ mm^{-3}

— Less than 400 HIV RNA copies $\cdot$ ml^{-1}

— Total cholesterol = 249 mg $\cdot$ dl^{-1}

— High-density lipoprotein cholesterol = 39 mg $\cdot$ dl^{-1}

— Triglycerides = 746 mg $\cdot$ dl^{-1}

His follow-up exercise evaluation revealed a maximal heart rate of 188, maximal blood pressure of 220/84, and a maximal respiratory rate of 44. Peak oxygen consumption was measured at 39.8 ml $\cdot$ min^{-1} $\cdot$ kg^{-1}. His electrocardiogram revealed normal sinus rhythm with nonspecific ST-segment changes, and no changes were noted during exercise.

continued

Case Study 23 *(continued)*

Case Study 23   Discussion Questions

1. How can skeletal muscle wasting be avoided in individuals with HIV by using exercise?

2. Develop a resistance training exercise prescription for this case study patient.

3. What precautions must be taken to guarantee the patient's safety when performing an exercise evaluation on someone who has HIV infection and a CD4+ cells · mm^{-3} count of 125?

4. What are the risks of a technician contracting HIV during an exercise evaluation? How can any risk be reduced?

5. Why was the patient sent back to his internist?

6. What is likely to limit exercise capacity in individuals with HIV infection?

7. What are the five stages of HIV infection, and what are the several implications for exercise testing and training?

8. What might explain the changes noted in his blood lipid profile?

Glossary

acquired immunodeficiency syndrome (AIDS)—A disease caused by HIV. See the text for the CDC case definition.

antiretroviral medications—Medications demonstrated to be effective against the HIV virus, which is a retrovirus.

CD4+ cells—A membrane receptor found on T-helper lymphocytes (or T4 cells). It is the preferred target of HIV.

cytomegalovirus—One of a group of highly host-specific herpes viruses.

hemophilia—A hereditary hemorrhagic diathesis caused by deficiency of coagulation factor VIII. Characterized by spontaneous or traumatic subcutaneous and intramuscular hemorrhages.

histoplasmosis—Infection resulting from inhalation or, infrequently, ingestion of fungal spores. May cause pneumonia.

HIV negative—Describes an individual without antibodies to HIV viral proteins. An individual who has been recently infected with HIV may not have had an opportunity to have developed antibodies to the virus.

HIV positive—Describes an individual with antibodies to HIV viral proteins. Many times this is used to describe individuals who are infected but who have not yet developed an AIDS-defining condition or whose CD4+cells are greater than 200 cells · mm^{-3}.

Kaposi's sarcoma—Firm, subcutaneous, brown-black or purple lesions usually observed on the face, chest, genitals, oral mucosa, or viscera.

myalgia—Pain in a muscle or muscles.

***Mycobacterium avium* complex**—Complex that consists of two predominant species, *M. avium* and *Mycobacterium intracellulare*. More than 95% of infections in patients with AIDS are caused by *M. avium*, whereas 40% of infections in immunocompetent patients are caused by *M. intracellulare*.

nucleoside reverse transcriptase inhibitor (NRTI)—A specific type of antiretroviral medication.

opportunistic infections—Infections that are most commonly seen in individuals who are immunocompromised, such as individuals with late or advanced HIV-1 disease, cancer, or other immunocompromising conditions.

***Pneumocystis carinii* pneumonia (PCP)**—An AIDS-defining condition caused by the parasite *P. carinii*.

protease inhibitor—A specific type of antiretroviral medication.

retrovirus—Viruses containing both RNA-dependent and DNA-synthesizing material.

T-helper lymphocyte—Lymphocytes whose secretions and other activities coordinate the cellular and humoral immune responses.

viremia—Viral particles in the blood.

virion—A single, encapsulated piece of viral genetic material.

wasting—Involuntary loss of more than 10% of body weight.

References

1. American College of Sports Medicine. ACSM's Guidelines for Exercise Testing and Prescription: Sixth Edition. Philadelphia: Lippincott Williams & Wilkins, p. 368, 2000.

2. American College of Sports Medicine. The recommended quantity and quality of exercise for developing and maintaining cardiorespiratory and muscular fitness in healthy adults: A position stand. Med Sci Sports Exerc 22(2): 265-274, 1990.

3. Barre-Sinoussi F., Chermann J.C., Rey F. Isolation of a T-lymphotrophic retrovirus from a patient at risk for acquired immune deficiency syndrome (AIDS). Science 220: 868-871, 1983.

4. Brooks G.A., Fahey T.D., White T.P. Exercise Physiology: Human Bioenergetics and Its Applications: Second Edition. Mountain View, CA: Mayfield, 1996.

5. Cannon J.G. Exercise and resistance to infection. J Appl Physiol 74: 973-981, 1983.

6. Carpenter C.C.J., Fischl M.A., Hammer S.M., et al. Antiretroviral therapy for HIV infection in 1998: Updated recommendations of the International AIDS Society–USA Panel. JAMA 280(1): 78-88, 1998.

7. Carr A.S., Samaras K., Burton S., et al. A syndrome of peripheral lipodystrophy, hyperlipidaemia, and insulin resistance in patients receiving HIV protease inhibitors (abstract). AIDS 12(7): F51-58, 1998.

8. Casaburi R.A., Patessio A., Ioli F., Zanaboni S., Donner C.F., Wasserman K. Reductions in exercise in lactic acidosis and ventilation as a result of exercise training in patients with obstructive lung disease. Am Rev Respir Dis 143: 9-18, 1991.

9. Centers for Disease Control and Prevention. 1993 Revised classification system for HIV infection and expanded surveillance case definition for AIDS among adolescents and adults. MMWR 41 (RR-17), 1992.

10. Centers for Disease Control and Prevention. Interpretation and use of Western blot assay for serodiagnosis of HIV infection. MMWR S7: 1-7, 1989.

11. Centers for Disease Control and Prevention. Revision of the case definition of acquired immunodeficiency syndrome for national reporting—United States. MMWR 34: 373-375, 1985.

12. Centers for Disease Control and Prevention. Revision of the CDC surveillance case definition for acquired immunodeficiency syndrome. MMWR CDC Surveillance Summary 36 (1S): 1S-5S, 1987.

13. Centers for Disease Control and Prevention. Update: Acquired immunodeficiency syndrome—United States. MMWR 36 (31): 522-526, 1987.

14. Centers for Disease Control and Prevention. HIV/AIDS Surveillance Report. Volume 11. Atlanta: U.S. Public Health Service, pp. 1-42, 1999.

15. Choi S., S. Lagakos, R. Schooley, P. Volberding. CD4+ lymphocytes are an incomplete surrogate marker for clinical progression in persons with asymptomatic HIV infection taking zidovudine. Ann Intern Med 118(9): 674-680, 1993.

16. Cooper D.A., Gold J.A., MacLean P. Acute retrovirus infection: Definition of a clinical illness associate with seroconversion. Lancet 1: 137-140, 1985.

17. CRS. Serologic diagnosis of human immunodeficiency virus infection by Western blot testing. JAMA 260: 674-679, 1988.

18. De Gruttola V., M. Wulfson, M. Fischl, A. Tsiatis. Modeling the relationship between survival and CD4 lymphocytes in patients with AIDS and AIDS-related complex. J Acquir Immune Defic Syndr Hum Retrovirol 6(4): 359-365, 1993.

19. Dimeo F., Rumberger B.G., Kaul J. Aerobic exercise as therapy for cancer fatigue. Med Sci Sports Exerc 30(4): 475-478, 1998.

20. Fish A.F., Smith B.A., Frid D.J., Christmas S.K., Post D., Montalto N.J. Step treadmill exercise training and blood pressure reduction in women with mild hypertension. Prog Cardiovasc Nurs 12(1): 28-35, 1997.

21. Gallo R.C., Salahuddin S.Z., Popovic M. Frequent detection and isolation of cytopathic retroviruses (HTLV-III) from patients with AIDS and at risk for AIDS. Science 224: 500-503, 1984.

22. Goldberg A.P., Hagberg J., Delmez J.A., et al. The metabolic and psychological effects of exercise training in hemodialysis patients. Am J Clin Nutr 33: 1620-1628, 1980.

23. Henry K., Melroe H., Huebsch J., et al. Severe premature coronary artery disease with protease inhibitors. Lancet 351: 1328, 1998.

24. Hogg R.S., Heath K.V., Yip B., et al. Improved survival among HIV-infected individuals following initiation of antiretroviral therapy. JAMA 279: 450-454, 1998.

25. Hughes M.D., Johnson V.A., Hirsch M.S. Monitoring plasma HIV RNA levels in addition to CD4+ lymphocyte count improves assessment of antiretroviral therapeutic response. Ann Intern Med 126: 929-938, 1997.

26. Hurwitz A. The benefit of a home exercise regimen for ambulatory Parkinson's disease patients. J Neurosci Nurs 21(3): 180-184, 1989.

27. Johnson J.E., Anders G.T., Blanton H.M., et al. Exercise dysfunction in patients seropositive for the human immunodeficiency virus. Am Rev Respir Dis 141: 618-622, 1990.

28. Katlama C., Valantin M.A., Calvez V., et al. ALTIS PLUS: Long term d 4T-3TC with and without ritonavir. In Program and Abstracts of the 5th Conference on Retroviruses and Opportunistic Infections; February 1-5, 1998, Chicago, Abstract 376, 1998.

29. Kilby J.M., Saag M.S., eds. Natural History of HIV Disease: Second Edition. Baltimore: Williams & Wilkins, pp. 49-58, 1999.

30. Lamb D.R. Physiology of Exercise. New York: Macmillan, 1984.

31. LaPerriere A., Fletcher M.A., Antoni M.H., Klimas N.G., Ironson G., Schneiderman N. Aerobic exercise training in an AIDS risk group. Int J Sports Med 12(Suppl. 1): S53-S57, 1991.

32. Lui A., Karter D., Turett G. Another case of breast hypertrophy in a patient treated with indinavir. Clin Infect Dis 26: 1482, 1998.

33. MacArthur R.D., Levine S.D., Birk T.J. Supervised exercise training improves cardiopulmonary fitness in HIV-infected persons. Med Sci Sports Exerc 25(6): 684-688, 1993.

34. MacDonell K.B., Chimiel J.S., Poggensee L. Predicting progression to AIDS: Combined usefulness of CD4 lym-

phocyte counts and p24 antigenemia. Am J Med 89: 706-712, 1990.

35. MacVicar M.G., Winningham M.L., Nickel J. Effects of aerobic interval training on cancer patients' functional capacity. Nurs Res 38(6): 348-351, 1989.

36. Mannix E.T., Boska M.D., Ryder K.D., Newcomer B., Manfredi F., Farber M.O. Impaired oxidative energy metabolism in AIDS: A 31P MRS study. Official Publication of the Federation of American Societies for Experimental Biology 10(3): A375, Abstract 2172, 1996.

37. Mellors J.W., et al. Quantitation of HIV RNA in plasma predicts outcome after seroconversion. Ann Intern Med 122(8): 573-579, 1995.

38. Mellors J.W., Kingsley L.A., Rinaldo C.R. Quantitation of HIV RNA in plasma predicts outcome after seroconversion. Ann Intern Med 122: 573-579, 1995.

39. Mellors J.W., Rinaldo C.R., Gupta P. Prognosis in HIV infection predicted by the quantity of virus in plasma. Science 272: 1167-1170, 1996.

40. Mellors J.W., Rinaldo C.R., Jr., Gupta P., White R.M., Todd J.A., Kingsley L.A. Prognosis in HIV infection predicted by the quantity of virus in plasma. Science 272: 1167-1170, 1996.

41. Mustafa T., Sy F.S., Macera C.A., et al. Association between exercise and HIV disease progression in a cohort of homosexual men. Ann Epidemiol 9(2): 127-131, 1999.

42. Neumann M., Harrison J., Saltareli M. Splicing variability in HIV type 1 revealed by quantitative RNA polymerase chain reaction. AIDS Res Hum Retroviruses 10: 1531-1532, 1994.

43. Nieman D.C. Immune response to heavy exertion. J Appl Physiol 82: 1385-1394, 1997.

44. Nieman D.C., Johanssen L.M., Lee J.W., Cermak J., Arabatzis K. Infectious episodes in runners before and after the Los Angeles marathon. J Sports Med Phys Fitness 30: 316-328, 1990.

45. O'Brien T.R., Blattner W.A., Waters D. Serum HIV RNA levels and time to development of AIDS in the multicenter hemophilia cohort study. JAMA 276: 105-110, 1996.

46. O'Brien W.A., Hartigan P.A., Daar E.S. Changes in plasma HIV RNA levels and CD4+ lymphocyte counts predict both response to antiretroviral therapy and therapeutic failure. Ann Intern Med 126: 939-945, 1996.

47. Palella F.J., Delaney K.M., Mooreman A.C., et al. Declining morbidity and mortality among patients with advanced human immunodeficiency virus infection. N Engl J Med 338: 853-860, 1998.

48. Pantaleo G., Fauci A.S. New concepts in the immunopathogenesis of HIV infection. Annu Rev Microbiol 50: 825-854, 1996.

49. Perna FM, LaPerriere A, Klimas N, Ironson G, Perry A, Pavone J, et al. Cardiopulmonary and CD4 changes in response to exercise training in early symptomatic HIV infection. Med Sci Sports Exerc 31(7): 973-979, 1999.

50. Preusser B.A., Winningham M.L., Clanon T.L. High- vs low-intensity inspiratory muscle interval training in patients with COPD. Chest 106(1): 110-117, 1994.

51. Rigsby L.W., Dishman R.K., Jackson A.W., MaClean G.S., Raven P.B. Effects of exercise training on men seropositive for the human immunodeficiency virus-1. Med Sci Sports Exerc 24(1): 6-12, 1992.

52. Roth V.R., Kravcik S., Angel J.B. Development of cervical fat pads following therapy with human immunodeficiency virus type 1 protease inhibitors. Clin Infect Dis 27: 65-67, 1998.

53. Roubenoff R., McDermott A., Wood M., Suri J. Feasibility of increasing lean body mass in HIV-infected adults using progressive resistance training. Presented at the 12th International AIDS Conference, Geneva, Switzerland, 1999.

54. Roubenoff R., Skolnik P.R., Shevitz A., et al. Effect of a single bout of acute exercise on plasma human immunodeficiency virus RNA levels. J Appl Physiol 86(4): 1197-1201, 1999.

55. Roubenoff R., Weiss L., McDermott A., et al. A pilot study of exercise training to reduce trunk fat in adults with HIV-associated fat redistribution. AIDS 13:1373-1375, 1999.

56. Ryan A.S., Pratley R.E., Elahi D., Goldberg A.P. Resistive training increases fat-free mass and maintains RMR despite weight loss in postmenopausal women. J Appl Physiol 79(3): 818-823, 1995.

57. Ryan A.S., Treuth M.S., Rubin M.A., et al. Effects of strength training on bone mineral density: Hormonal and bone turnover relationships. J Appl Physiol 77(4): 1678-1684, 1994.

58. Saint-Marc T., Touraine J.L. Effects of metformin on insulin resistance and central adiposity in patients receiving effective protease inhibitor therapy. AIDS 13(8): 1000-1002, 1999.

59. Schacker T., Collier A.C., Hughes J. Clinical and epidemiologic features of primary HIV infection. Ann Intern Med 125(4): 125-257, 1996.

60. Schneider B.A., Roehrig K.L., Smith B.A., Sherman W.M. Resistive exercise: Strength, body composition, glucose tolerance and insulin action in African American women. Unpublished manuscript.

61. Shaw A.J., McLean K.A., Evans B. Disorders of fat distribution in HIV infection. Int J STD AIDS 9: 595-599, 1998.

62. Shikuma C.M., Waslien C., McKeague J., et al. Fasting hyperinsulinemia and increased waist-to-hip ratios in non-wasting individuals with AIDS. AIDS 13: 1359-1365, 1999.

63. Sinnwell T.M., Kumaraaswamy S., Soueidan S., et al. Metabolic abnormalities in skeletal muscle of patients receiving zidovudine therapy observed by 31P in vivo magnetic resonance spectroscopy. J Clin Invest 96: 126-131, 1995.

64. Smith B., Neidig J.L., Nickel J., Mitchell G.L., Para M.D., Fass R.J. Aerobic exercise training: Effects on physiological fatigue, dyspnea, increased weight and central fat in HIV-infected adults. AIDS, in press.

65. Spence D.W., Galantino L.A., Mossberg K.A., Zimmerman S.O. Progressive resistance exercise: Effect on muscle function and anthropometry of a select AIDS population. Arch Phys Med Rehabil 71: 645-649, 1990.

66. Stein D.S., Korvick J.A., Vermund S.H. CD4+ lymphocyte cell enumeration for prediction of clinical course of human immunodeficiency virus disease. J Infect Dis 165: 352-363, 1992.

67. Stringer W.W., Berezovskaya M., O'Brien W.A., Beck C.K., Casaburi R. The effect of exercise training on aerobic fitness, immune indices, and quality of life in HIV+ patients. Med Sci Sports Exerc 30(1): 11-16, 1998.

68. Visnegarwala F., Krause K.L., Musher D.M. Severe diabetes associated with protease inhibitor therapy. Ann Intern Med 127(11): 947, 1997.

69. Walli R., Herfort O., Michl G.M., et al. Treatment with protease inhibitors associated with peripheral insulin resistance and impaired oral glucose tolerance in HIV-infected patients. AIDS 12(15): F167-F173, 1998.

70. World Health Organization. Proposed WHO criteria for interpreting results from Western blot assays for HIV, HIV-2, and HTLV-I/HTLV II. Wkly Epidemiol Rec 37: 281-283, 1990.

71. Wilmore J.H., Costill D.L. Physiology of Sport and Exercise. Champaign, IL: Human Kinetics, 1994.

72. Wurtz R. Abnormal fat distribution and use of protease inhibitors. Lancet 351: 1735-1736, 1998.

24

Virginia B. Kraus, MD, PhD
Division of Rheumatology, Allergy, and Clinical Immunology
Medical Director of the Duke Center for Living
Arthritis Rehabilitation Program
Department of Medicine
Duke University Medical Center
Durham, NC

Diane Wiggin, MS, PT
Clinical Coordinator of the Duke Center for Living
Arthritis Rehabilitation Program
Department of Physical and Occupational Therapy
Duke University Medical Center
Durham, NC

Arthritis

Arthritis is a condition affecting the **synovial joint** and is characterized by inflammation, varying degrees of degeneration of joint structures, and pain. Finding the correct balance of rest and physical activity in the comprehensive management of arthritis is a challenge. Despite traditional beliefs that individuals with arthritis should avoid vigorous physical activity, regular physical activity reduces impairment and improves joint function without aggravating symptoms (81).

Scope

More than 15% (40 million) of Americans have some form of arthritis. The **prevalence** of arthritis increases with age, affecting 50% of persons age 65 years or older (68), and it is the most prevalent condition in both women and men over 65 followed by high blood pressure and hearing impairment (12, 122). Prevalence is higher for women (17%) than men (12%). In 1992, the total cost of arthritis and musculoskeletal conditions was equivalent to 2.5% of the gross national product, or $149 billion (124). By 2020, the prevalence of arthritis in the United States is estimated to reach 18%, owing primarily to the aging of the population (50, 124).

There are more than 100 different forms of arthritis. Osteoarthritis (OA) is the most common (12.1% or 21 million Americans) and has the highest annual **incidence.** Rheumatoid arthritis (RA) occurs in 1% (2 million) of Americans and consists of gout (1.6 million), **spondyloarthropathy** such as ankylosing spondylitis (AS) (0.4 million), and juvenile rheumatoid arthritis (30-50 thousand children under age 16) (68). Arthritis is also associated with connective tissue diseases including systemic lupus erythematosus, dermatomyositis, and scleroderma.

Arthritis adversely affects psychosocial and physical function and is the leading cause of disability in later life (73). The impact of arthritis on social functioning is impressive: 25% of people with arthritis never leave their home or do so only with help, and 18% never participate in social activities (6). The energy cost of daily activity increases with increasing impairment (42, 96) and contributes to prolonged inactivity. Inactivity negatively affects health by increasing the risk of cardiovascular disease, hypertension, diabetes, and obesity (88).

Physiology and Pathophysiology

There are genetic components to OA, RA, and AS (77). Each is distinct in presentation and clinical manifestations, yet they have common biological and physiological consequences. For instance, individuals with any type of arthritis have impaired exercise tolerance (41, 86, 102). The following are affected by the disease and contribute to functional limitation in individuals with arthritis.

— Flexibility
— Biomechanical efficiency
— Muscle strength
— Endurance
— Speed
— Proprioception (7, 54)

These impairments are usually more pronounced for women.

Muscle weakness is the longest recognized and best established correlate of lower limb functional limitation in individuals with knee OA (7). In addition, joint motion is limited (51, 57, 97). The most common causes of disability are impairment of knee flexion, hip extension, and hip external rotation (115). Limited joint motion impairs joint nutrition (48, 56). And, although the joint depends on dynamic loading for maintenance of joint function and joint nutrition, both chronic insufficient and excessive loads are deleterious to joints (16).

Osteoarthritis is a dynamic disease process characterized by the uncoupling of the normal balance between degradation and repair of the components of the **articular** cartilage and subchondral bone (16, 100).

Rheumatoid arthritis begins as an **autoimmune** inflammatory process of the synovial lining of the presenting joint and affects the hand joints, with possible involvement of feet, ankles, knees, elbows, and spine. Inflammation of the synovial lining results in erosion of articular cartilage and marginal bone with subsequent joint destruction.

Ankylosing spondylitis is a process of inflammation and erosion at the **entheses.** It involves the sacroiliac joint and the spine. This is followed by healing during which new bone is formed, leading eventually to spinal fusion.

There are three levels of classifying arthritis disease stages:

1. Acute: reversible signs and symptoms in the joint related to synovitis
2. Chronic: stable, but irreversible structural damage brought on by the disease process
3. Chronic with acute exacerbation of joint symptoms: results in increased pain and decreased range of motion and function and is often related to overuse or superimposed injury

Each stage has disease-specific presentations, treatment considerations, and goals. Table 24.1 provides specifics about the stages of each type of these arthritic conditions.

Clinical Considerations

The various arthritides can be differentiated based on whether symptoms arise from the joint or from a periarticular location, the number of joints involved, their location, whether the distribution is **symmetric** or **asymmetric,** and the **chronicity** of disease (9, 69). Pharmacological treatment of OA, RA, and AS varies dramatically, but all need to include exercise in their comprehensive management.

Signs and Symptoms

In the evaluation of individuals with musculoskeletal complaints, the history and physical examination are the most informative elements. Restricted movement of a joint and tenderness to palpation

TABLE 24.1
Types of Arthritis, Stages, and Related Impairments

Type of arthritis	Disease stage	Related impairments
Osteoarthritis	Acute	Joint pain—often insidious
	Chronic	Radiographic joint disease
	Chronic with exacerbation	Increased joint pain ± joint swelling
Rheumatoid arthritis	Acute	Disease in multiple joints with limited range of motion, joint stiffness, muscle weakness and fatigue
	Chronic	Irreversible joint deformity present
	Chronic with exacerbation	Increased joint pain ± joint swelling which may occur in one or more joints
Ankylosing spondylitis	Acute	Spinal pain and stiffness without significant decrease in mobility
	Chronic	Spinal ankylosis predominant with decreased spinal and thoracic mobility
	Chronic with exacerbation	Increased pain and stiffness of the back or peripheral joint

along the axis of joint movement are indicative of arthritis. This contrasts with tenderness around the joint, which is more indicative of periarticular soft tissue problems. The signs and symptoms of arthritis are as follows:

— Pain
— Stiffness
— Joint **effusion**
— **Synovitis**
— Deformity
— **Crepitus**

Joint pain may arise from pathological changes in the joint capsule, periarticular ligaments, **intraosseous** hypertension, muscle weakness, subchondral microfractures, **enthesopathy, bursitis,** and psychosocial factors including depression (14). Pain does not arise, however, from articular cartilage directly, because it is **aneural.** Stiffness from OA is intermittent and typically of short duration. Stiffness in those with RA and AS varies with the severity of inflammation. The prognosis of recent onset arthritis is aided by determining whether the duration of symptoms has exceeded 4 to 6 weeks (53).

Diagnostic and Laboratory Evaluations

The American College of Rheumatology has developed diagnostic criteria for the classification of hip, knee, and hand OA, RA, and AS (62). Table 24.2 summarizes these criteria.

Joint imaging is often used to help confirm a particular arthritis diagnosis. The plain radiograph, which detects bony changes, is the traditional imaging modality. Typical features of OA include **osteophyte** formation, **cysts,** and **subchondral sclerosis.** RA joints develop erosions at joint margins from the invasion of synovial tissue at the intersection of cartilage and bone. Early AS is characterized radiographically as a squaring of the corners of the vertebrae. Later there is evidence of bone formation, ossification, and/or thin vertically oriented outgrowths that bridge the disc space and limit spinal motion (108).

There are currently no specific diagnostic tests or markers of arthritis. However, several useful serum and synovial fluid laboratory studies, in combination with joint imaging, are often used to support a particular arthritis diagnosis. These include rheumatoid factor, which is often negative early in RA but becomes positive in about 80% of individuals within the first 2 years after disease onset (123). Nonspecific measures of systemic inflammation, such as the sedimentation rate and C-reactive protein, are elevated in conjunction with active synovitis in RA and can be mildly elevated in active AS. C-reactive protein is also predictive of OA status (98, 112), underscoring the fact that inflammation can be a manifestation of all three diseases. Synovial fluid analysis is helpful to differentiate the type of

TABLE 24.2
. . . .
American College of Rheumatology Diagnostic Criteria for Arthritis

Arthritis type	Distinguishing characteristics	Presentation
OA	Joint pain Crepitus Gel phenomenon	Pain worsens throughout the day Affects any traumatized joint Hands, hips, knees, lumbar, and cervical spine
RA	Hand pain Swelling Fatigue Prolonged morning stiffness	Metacarpal phalangeal joints affected Intraphalangeal joints of wrists
AS	Low back pain Low back stiffness	Early discovertebral bone erosion Later vertebral fusion via bone formation, ossification, and spinal fusion

Note. OA = osteoarthritis; RA = rheumatoid arthritis; AS = ankylosing spondylitis.

arthritis when joint swelling is present and fluid readily obtained. The leukocyte count increases from a normal value of 500 cells · mm^{-3} in a joint to 2000 cells · mm^{-3} in OA and 5000-15,000 cells · mm^{-3} in RA and AS. Macromolecules originating from joint structures and measurable in blood, synovial fluid, or urine reflect processes taking place locally in the joint (106).

Treatment

A comprehensive treatment strategy for arthritis should strive to counteract inactivity and reduce disability and handicap. This may include the use of the following:

— Pharmacological agents
— Braces and bandages
— Activity modification
— Exercise

Previously, the traditional standard of treatment for an arthritic joint was rest (17). Current practice has shifted toward joint mobilization because of substantial evidence of the beneficial effects and relative safety of exercise for individuals with arthritis (21, 63, 90, 110, 125). As depicted in figure 24.1, exercise interventions can interact at each stage in the progression from arthritis pathology to handicap and can help to mitigate the effects of the disease process.

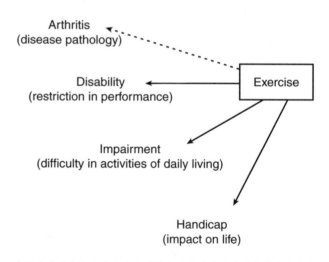

FIGURE 24.1   The World Health Organization classification of impairments, disabilities, and handicaps is presented to demonstrate the potential for a positive impact of exercise as an interactive and mitigating factor in this process. Strong evidence exists for benefits of exercise at the level of disability, impairment, and handicap with arthritis. Much remains to be learned about the effects of exercise on the level of disease pathology. The majority of studies show no worsening of arthritis with exercise, and a few suggest a beneficial effect on disease pathology itself.

Adapted from R. Shephard and P. Shek, 1997, "Autoimmune disorders, physical activity, and training, with particular reference to rheumatoid arthritis," *Exercise Immunol Rev* 3: 53-67.

The implementation of exercise as part of a comprehensive therapeutic management strategy is a challenge. Physicians rarely recommended exercise to their patients with arthritis, and if doctors do recommend exercise, little instruction is provided (28). Individuals with arthritis are typically poorly motivated to perform physical activity on their own. During leisure time, the period most amenable to efforts to increase physical activity, adults and children with arthritis are less active than healthy individuals (3, 61, 82). Thus, individuals with arthritis require education and encouragement to increase appropriate physical activities.

In addition to exercise, multiple nonpharmacological interventions are used in the rehabilitation of individuals with chronic arthritis (24):

— Education

— Physical and occupational therapy

— Braces and bandages

— Canes and other walking aids

— Shoe modification and orthotics

— Ice and heat modalities

— Weight reduction

— Avoidance of repetitive motion occupations

— Joint irrigation and joint surgery (in select circumstances)

The judicious use of rest is beneficial and generally recommended between days of dynamic exercise, particularly for those with systemic, multiple-joint disease. Joint replacement has markedly improved the quality of life for individuals with knee and hip arthritis. Surgery is usually reserved for unremitting pain or disability when conservative measures have failed. An estimated 350,000 joint replacements are performed every year in the United States (101), and more than 70% of total hip and knee replacements are for OA (35).

Pharmacological therapies can be administered orally, topically, or directly into joints. Topical agents include capsaicin cream and topical nonsteroidal anti-inflammatory drugs (NSAIDs). Oral agents frequently used to treat arthritis include analgesics in the form of acetaminophen, opioids, and oral NSAIDs. Dosing of an NSAID 1 hr before exercise may aid exercise compliance by minimizing pain and stiffness during activity. A variety of disease-modifying agents currently exist for RA (methotrexate, sulfasalazine, and the new tumor necrosis factor-α inhibitors, etanercept, infliximab, and adalimumab) but very few for OA or AS. Glucosamine and chondroitin sulfate are naturally occurring constituents of articular cartilage. Taken orally, these are reported to reduce OA pain and may slow progression of the disease.

In persons with OA, being overweight is associated with increased pain and disability (1, 26, 34) and may worsen disease activity (29). The inclusion of dietary intervention in combination with exercise may be necessary when weight loss is a goal for the arthritic person (79). Other dietary considerations for OA management may include adequate intake of vitamins D and C and selenium (74, 75), and for RA, n-3 fatty acids found in fish oil or cold water fish. These may decrease inflammation (20).

Exercise may have disease-modifying effects for arthritis because it decreases circulating CD4 (helper) lymphocyte counts, reduces the length of hospitalizations in RA patients (111), and reduces the number of inflamed joints in OA and RA (85). Exercise does not put an individual at increased risk for OA. In a few instances, however, exercise may promote arthritis. For instance, sports participation associated with an increased risk of injury may subsequently increase OA risk (65). However, in general, physical activity promotes joint health by promoting cartilage nutrition and weight loss, which decreases the risk of knee and hip OA.

Graded Exercise Testing

Individuals with RA have an increased risk of cardiovascular disease, possibly related to the inflammatory process (5, 27, 55). Individuals with arthritis tend to be more deconditioned than sedentary individuals without arthritis, increasing the risk of cardiovascular disease in general. A symptom-limited exercise test should be considered to screen for coronary artery disease, when appropriate, in those wishing to exercise and to guide their training (15). It is also commonly used for assessing cardiovascular status for surgical risk prior to joint replacement. However, joint symptoms and fatigue may adversely affect performance on an exercise test and may prohibit maximal testing (10, 25).

Medical Evaluation

The individual with arthritis should be screened for conditions that will guide the exercise prescription. Suggestions for questions that should be included as part of the exercise evaluation are listed in practical application 24.1 and reviewed in chapter 4.

Contraindications

Arthritis and musculoskeletal conditions are often listed as relative contraindications to graded exercise testing (15). However, one study found that of those

Client–Clinician Interaction

1. Individual's age, level of fitness, medications, personal goals, and lifestyle
2. Names of healthcare providers including primary care physician, rheumatologist, orthopedist, clinical exercise physiologist, and physical therapist
3. Rheumatologic diagnosis
4. Pattern of joint involvement: (a) symmetric or asymmetric, (b) upper or lower extremity involvement, (c) joints affected
5. Severity of disease activity (acute, chronic, or chronic with acute exacerbation)
6. Comorbidities (other medical conditions, including pulmonary disease, fibromyalgia, Raynaud's phenomenon, Sjogren's syndrome, osteoporosis)
7. Surgical history including joint replacements
8. Previous treatment (whether successful or not)
9. Presence of fatigue
10. Adequacy of footwear

with severe end-stage hip or knee arthritis attributable to OA or RA, 95% were capable of performing a symptom-limited exercise test using cycle ergometry methods (99). The majority also achieved a respiratory exchange ratio of greater than 1.0, indicating a metabolically maximal test. Approximately two thirds of subjects were capable of completing tests by pedaling with their legs, and the remainder performed the same task using their arms. Standard contraindications for exercise testing should also be followed (chapter 6).

Recommendations and Anticipated Responses

Cycle ergometry is generally the preferred mode because of reduced baseline exercise capacity and the high frequency of lower extremity impairment. Testing by arm ergometry may be necessary in those with severe lower extremity joint pain or with significant deformity that might cause the extremity to strike the flywheel (such as a **varus** or **valgus** deformity of the knee). The treadmill may be used in those with minimal or no functional disability.

Testing procedures for individuals with arthritis are similar to protocols recommended for elderly and deconditioned individuals (58). These should have small incremental changes in workload, for instance, increments of 10 to 15 W $\cdot$ min^{-1} on the cycle ergometer, or using the modified Naughton protocol using a treadmill (13). Submaximal testing does not provide optimal diagnostic information about cardiovascular disease but can be used to predict aerobic capacity (87).

Standard equations by the American College of Sports Medicine that estimate $\dot{V}O_2$ based on exercise testing are inaccurate in an elderly population with arthritis and tend to overestimate $\dot{V}O_2$peak. This in turn could lead to overprescribing exercise intensity for training (13). The following equation was developed to predict $\dot{V}O_2$peak in seniors with knee OA or cardiovascular disease: $\dot{V}O_2$ (ml $\cdot$ min^{-1} $\cdot$ kg^{-1}) = 0.0698 $\times$ speed (m $\cdot$ min^{-1}) + 0.8147 $\times$ grade (%) $\times$ speed (m $\cdot$ min^{-1}) + 7.533 ml $\cdot$ min^{-1} $\cdot$ kg^{-1}. This equation is valid in both men and women and requires that front handrails be used for support during the exercise test. This is an advantageous method of testing an arthritic individual with lower extremity disability in whom standard non-hand support methods of treadmill testing may be hazardous. In general, it is best to use cycle ergometry for testing because of the non-weight-bearing nature of the mode and its reduced reliance on balance. However, equations for estimating $\dot{V}O_2$ by cycle ergometry have not been validated for this population. The majority of those with arthritis are able to achieve maximal cardiovascular effort using cycle ergometry (99).

Exercise Prescription

Exercise benefits for the arthritic population include increased muscle strength, increased flexibility, improved sense of well-being, improved coping ability, weight control, enhanced quality of sleep, decreased blood pressure, and fewer heart attacks (90). The following are goals of exercise for the treatment of arthritis:

— Reduce inflammation and pain
— Prevent contractures and deformities
— Maintain or improve range of motion, muscle strength, and cardiovascular fitness

Immobilization and inactivity amplify the negative systemic and psychological manifestations that accompany arthritis (40). The effects of inactivity include rapid loss of strength (~3-8% per week), loss of endurance, negative nitrogen balance reflecting a loss of muscle protein, and loss of cartilage matrix components (11, 59, 66). Because cartilage is **avascular,** it depends on normal repetitive loading of the joint for its nutrition and normal physiological function (52). Moreover, joints with effusions may develop synovial ischemia attributable to elevated intra-articular pressure. Walking and cycling increase synovial blood flow in inflamed knees (56). The intra-articular oxygen partial pressure increases during joint movement in both healthy joints and those affected by OA; however, arthritic joints demonstrate less of an increase than normal joints (80).

Normal training heart rate ranges can be achieved by individuals with arthritis (78). Thus, standard heart-rate range development techniques can be used (see chapter 7). Practical application 24.2 reviews the literature about exercise and arthritis.

Supervised exercise training for individuals with arthritis most often occurs in a group setting (vs. personal training) and is well supported as a treatment modality (21, 32, 46, 70, 85, 91, 93, 119). An individual in an exercise program, whether supervised or unsupervised, requires education, skill acquisition and reinforcement, and regular monitoring by health professionals. A supervised group setting may be beneficial because of peer social support in a positive environment. Regular monitoring by health professionals helps to ensure accuracy and safety of exercise performance. Access to a physical therapist or clinical exercise physiologist who has previously evaluated the individual may help decrease anxiety and improve compliance (113).

However, unsupervised and independent exercise is important to the long-term success of an individual's exercise routine.

There are many exercise options for unsupervised programs. These include cardiovascular and strengthening exercises via an independent water exercise program, including water walking and independent joint range of motion exercises. Videotapes of exercise programs suited to individuals with arthritis are available through the Arthritis Foundation (see practical application 24.3). Other activities and exercises that can be performed throughout the day include chin tucks, corner pectoral stretches in a doorway, abdominal tightening, checking posture throughout the day in the mirror when in the bathroom, and extending walking time by taking the stairs or parking farther away from a destination. For individuals with RA and severe morning stiffness, active range of motion exercises performed within 15 min of bedtime can decrease morning stiffness (18). Walking in climate-controlled buildings such as fitness centers or shopping malls is an excellent option for cardiovascular exercise. The ability to perform a variety of cardiovascular exercises may help maintain interest and compliance. In addition, cross-training may prevent overuse injuries.

Range of motion exercises in a pool can be performed as a component of a supervised or an unsupervised exercise program. Following the Arthritis Foundation's low-intensity, recreational aquatic program (AFYAP), available nationally in YMCAs and in private facilities, increases hip range of motion and isometric strength when performed three times a week over 6 weeks (116). However, the AFYAP exercises may not be of sufficient intensity to increase strength and ROM in joints not affected by arthritis (116) and can be supplemented by strength training and joint ROM activity on land.

Special Exercise Considerations

The principles of exercise training vary according to the disease stage. In establishing an exercise prescription, it is also important to consider the success of various treatment modalities for particular joint impairments as well as the individual's affected areas, level of fitness, surgical history, comorbidities, medications, age, personal goals, and lifestyle.

Inflammation and joint degeneration associated with the disease process cause a cycle of decreasing function and increasing impairment. It is not yet clear whether therapeutic exercise alters the pathological process of arthritides. However, performed judiciously, exercise therapy can prevent, retard, or correct the mechanical and functional limitations that may occur throughout the course of the disease.

Exercise therapy can induce pain, the primary complaint and disabling factor related to arthritis. This can make it difficult to motivate an individual to maintain an exercise program. In an arthritic population that is generally older, is commonly sedentary, and may be using systemic steroids, avoidance of injury and pain is important to maintain

Literature Review

OA

The finding that the degree of severity of knee OA is associated with the level of cardiovascular deconditioning supports the concept that regular aerobic exercise should be performed by individuals with OA (104). Reviews of exercise for OA demonstrate positive effects on pain and disability as well as general safety (7, 118, 126). Sixteen distinct randomized clinical trials of exercise for OA were recently reviewed (7, 118). These demonstrate small to moderate beneficial effects of exercise therapy on pain, small beneficial effects on disability outcome measures, and moderate to large beneficial effects for patient global assessment. Several document improvement in knee OA pain with quadriceps-focused exercises (22). The largest controlled trial of exercise for knee OA included a home-based exercise intervention with limited supervision for 15 months following the initial 3 months of supervised exercise (32), which improved disability by arresting functional decline. Low-level exercise intensity (at 40% heart rate reserve) may be as effective as high-intensity cycling (at 70% heart rate reserve) in improving function, gait, and aerobic capacity and decreasing pain in OA patients (72). Therapeutic programs to strengthen knee extension and hip and ankle flexion in individuals with OA have resulted in strength gains as well as decreased pain and stiffness and improved mobility, strength, balance, gait, independence, and physical function (66, 107).

RA

With RA, joint mobility, muscle strengthening, and aerobic conditioning are equally important. These patients often have great improvements, likely because of their severe baseline disability (126). Isometric exercises moderately increase quadriceps femoris muscle strength (49, 71, 119). Studies demonstrate that dynamic exercise can increase muscle strength and aerobic capacity without detrimental effects on disease activity and in some cases can decrease disease activity (120). Greater benefits are observed with the use of dynamic aerobic exercise as opposed to static or isometric exercises for RA patients (30, 43, 114). Long-term effects of dynamic training over 4 to 8 years demonstrate better ability to perform activities of daily living than nonexercising control patients (92, 93). Nontraditional modalities such as dance and t'ai chi ch'uan may be beneficial to reduce depression, anxiety, fatigue, and tension (59, 94, 121). Combined sessions of intensive weight-bearing exercise and cycling may be more effective than lower intensity exercise (119). However, this may depend on the patient's level of disability. Water exercises, including seated immersion, have been shown to improve aerobic capacity and other physical and psychological measures in patients with RA (45, 85). Continuance of exercise after completing a supervised exercise program was associated with high physical activity levels and aerobic capacity at the start of the exercise program (84). However, home exercise programs have failed to establish significant improvements in physical impairments (23, 113). Training effects, such as improved strength, can be maintained as long as exercise continues. However, these benefits wane rapidly on termination of exercise (44).

AS

Investigations of exercise training effects in patients with AS are lacking. Daily exercise is considered vital to maintenance of spinal mobility, but long-term effects have not been studied (126). Significant short-term improvement in spinal and hip range of motion of AS patients who are enrolled in intensive physical therapy has been demonstrated, and the performance of regular moderate (2-4 hr · week^{-1}) exercise is associated with functional improvement for patients with AS (105).

Arthritis Information for the Patient and Clinician

Arthritis Foundation
  1330 West Peachtree Street
  Atlanta, GA 30309
  800-283-7800
  www.arthritis.org

Spondylitis Association of America
  P.O. Box 5872
  Sherman Oaks, CA 91413
  800-777-8189
  www.spondylitis.org

American College of Rheumatology/
Association of Rheumatology Health
Professionals
  60 Executive Park South, Suite 150
  Atlanta, GA 30329
  404-633-3777
  Fax: 404-633-1870
  www.rheumatology.org

American College of Sports Medicine
  P.O. Box 1440, Indianapolis, IN 46206-
    1440
  317-637-9200,
  Fax: 317-634-7817
  www.acsm.org

exercise compliance (8). The clinical exercise physiologist's role is to make recommendations that minimize these symptoms.

Attention to affected joints is important because they have decreased range of motion, instability, reduced muscle strength and flexibility, poor joint proprioception, and increased pain. The clinical exercise physiologist should query the individual and the medical record to determine which joint sites are affected. This may be refined with information from the individual's rheumatologist, primary care physician, or physical therapist. Impaired balance and increased fatigue are additional factors that must be considered when one is developing an exercise program. Site-specific exercise recommendations are listed in table 24.3.

Disease Staging

A primary consideration when one is developing an appropriate exercise prescription for an individual with arthritis is the disease stage. The focus of exer-

cise therapy for chronic stages of arthritis should be to maintain or improve function while minimizing or avoiding exacerbations. Most individuals referred for exercise therapy will be experiencing arthritis in a chronic stage. Table 24.4 lists the arthritis stages and their associated exercise-related particulars. Practical application 24.4 provides site-specific recommendations for the clinical exercise professional to follow in arthritic patients with disease-specific skeletal conditions.

Other Factors

Compliance with an exercise program can be a challenge in individuals with arthritis. Efforts on the part of the exercise professional to prevent musculoskeletal injury from exercise and to appreciate the specialized concerns of the individual with arthritis will facilitate overall enjoyment and compliance with the exercise program. Some special considerations for individuals with arthritis are described next.

Preventing Musculoskeletal Injury Secondary to Exercise   Because cardiovascular exercise involves high repetition of joint motion, there is a risk of overuse injuries. Fortunately, injuries attributable to supervised exercise are infrequent. It is estimated that there are 2.2 minor injuries per 1000 hr of exercise, and 0.48 major injuries per 1000 hr of exercise that lead to a reduction or discontinuation of exercise for at least 1 week (21).

Overuse of the soft tissue and bone may be minimized in this population by performing interval or cross-training during endurance exercise. Examples of this include the following:

1. Alternating cycle ergometry between 25% and 75% of the maximum work rate performed during a graded exercise test
2. Alternating between water walking and joint range of motion exercises in a pool setting
3. Walking and weight training
4. Walking and higher intensity cardiovascular exercise such as recumbent stair-stepping or cycling

Repeated loading of the lower limb during walking and other daily activity may contribute to the onset and progression of knee OA. In contrast, the development of strong knee extensors with quadriceps-strengthening exercise decreases impulse loading of the lower limb during walking by slowing the deceleration phase occurring prior to heel strike. Therefore, in a normal joint, adequate quadriceps strength in deceleration may help prevent knee injury and knee OA.

TABLE 24.3

Arthritis Site-Specific Recommendations

Site	Condition	Presentation	Recommendation
LOWER EXTREMITY			
Foot	Hallux valgus	Lateral deviation of the first digit of the foot leading to bursitis at the metatarsophalangeal joint ("bunion")	— Padding between first and second toe — Prescription for footwear that decreases need for extension of the joint during walking — If moderate to severe, avoid walking for exercise
Foot	Plantar fasciitis/heel spur	Sharp foot pain on weight bearing, especially in heel; worse with first steps upon wakening and with walking	— Avoid weight-bearing activities that increase pain — Wear shoes with supportive arch; may need soft or semirigid foot orthoses or heel cups — Stretch calf muscles throughout day — Wear night splints to maintain stretch in dorsiflexion
Leg	Musculotendinous inflammation of the anterior or posterior tibialis muscle (shin splints) or compartment syndrome	Aching pain in anterior lateral or posterior medial leg	— Avoid painful activities (usually walking) for as much as 2 weeks — May need improved footwear or orthotics — Evaluate training routine for sudden change that preceded onset — Acute onset with excruciating pain and area hard to the touch requires immediate medical attention—send to emergency room
Hip or knee	Hip OA, knee OA	— Gait deviation — Pain with weight bearing or stair climbing	— Patellar femoral OA, hip/groin pain without gait deviation or balance problems: rearward walking — Patellofemoral OA: leg press; no leg extension — Hip/groin pain: hip bridging, free-speed walking, stationary cycling — If gait deviation is caused by pain or decreased joint ROM, may need to use cane or rolling walker in the hand opposing the painful limb
Hip	Bursitis	— Lateral hip pain, may extend to lateral knee — Cannot tolerate lying or single-leg stance on affected side	— Control inflammation: ice lateral hip — Should see doctor to rule out other problems and for medication to control inflammation — Should see physical therapist

Site	Condition	Presentation	Recommendation
Knee	Valgus or varus deformity	Valgus deformity often called "knock knees"; varus deformity often called "bow-legged"	— Avoid weight-bearing exercise — Perform open chain strengthening — May need prescription for wedged insoles

UPPER EXTREMITY

Site	Condition	Presentation	Recommendation
Shoulder	Shoulder pathology, possibly adhesive capsulitis, tendinitis, or bursitis; may be secondary to OA	Pain with overhead or end range motion	— Perform shoulder ROM in pain-free range in pool with UE submerged — If pain disturbs function or sleep, should see doctor — Would benefit from physical therapy
Hand	OA of carpal metacarpal joint of the thumb	Pain in hand proximal to thumb	— Avoid gripping activity during exercise — Enlarge grips
Hand	Ulnar deviation in RA	Body of hand and fingers deviate to the small digit side of the hand	— Avoid gripping activity during exercise — Use large muscles and joints for functional activities

AXIAL SKELETON

Site	Condition	Presentation	Recommendation
Lumbar spine	Spinal stenosis	— Flexed low back when walking, standing, and sitting — Symptoms increase with extension (standing, looking overhead) — Often presents with claudication-type pain with walking	— Tolerates flexion exercises, seated cardiovascular exercise (recumbent bicycle or stair stepper) — Aquatic exercise — Rolling walker for household or community ambulation
Cervical spine	Atlantoaxial subluxation in patients with RA	— Facial sensory loss — Vertigo with cervical extension — Numbness/tingling of hands or feet — Difficulty walking — Loss of control of bowel or bladder — Transient loss of consciousness when extending cervical spine — May be asymptomatic	— Avoid any passive or heavy resistive neck ROM — Needs immobilization if unstable (many symptoms) — Surgery if neurological signs progress — Gentle active ROM/low repetitions, no extension
Cervical or lumbar spine	Nerve compression secondary to OA	— Gradual, recurrent pain or pain after activity — Numbness, tingling, or pain in the extremities, sometimes only with certain movements	Avoid activity that results in numbness, tingling, or pain in the extremities

OA = osteoarthritis; RA = rheumatoid arthritis; UE = upper extremity; ROM = range of motion.

TABLE 24.4
Arthritis Stages, General Signs and Symptoms, and Exercise-Related Considerations

Stage	Signs and symptoms	Exercise considerations
Acute	Fatigue Joint pain Reduced joint tissue tensile strength attributable to inflammation Reduced joint nutrition	Teach energy conservation Perform non-weight-bearing and limited and slow ROM exercise Avoid stairs, carrying loads >10% body weight, fast walking, isotonic resistance training (40, 83) Perform isometrics at <40-50% of maximum voluntary contraction to limit blood flow reduction
Chronic	Permanent joint damage Pain at end of normal ROM Stiffness after rest Poor posture and ROM Joint deformities Pain with weight bearing Abnormal gait Weakness Contractures or adhesions Reduced aerobic endurance	Initiate walking and perform in water if necessary to reduce pain Perform low back flexion and abdominal strengthening exercise Perform lower body strengthening and ROM exercises (83) Avoid trunk extension (especially those with spinal stenosis) Use weights during resistance training that don't cause joint pain Maintain neutral spine position Avoid oral corticosteroids to reduce risk of osteoporosis and ligament laxity
Chronic with acute exacerbation	Inflammation and joint size greater than normal Joint tenderness, warmth, swelling Joint pain at rest and with motion Stiffness Functional limitations Hips and spine affected	Rest Normalize gait Same recommendations as for acute pahse

Note. ROM = range of motion.

Individuals may have laxity in the structures that support a joint because of the rheumatic process or because of the use of corticosteroids. Under these circumstances, the joint should be protected when the person performs exercise or normal activities. Protection is afforded by cautiously stretching to avoid extending beyond the functional range of motion. Vigorous stretching or manipulative techniques are contraindicated (60). In many cases, it may be necessary to provide external support to a joint. To protect smaller joints during activities and exercises, larger muscles and joints should be used.

Fatigue   Fatigue is common in those with rheumatic disease and can profoundly affect quality of life. Fatigue beginning in the afternoon and lasting until evening, and morning stiffness lasting for 3 or 4 hr, are common symptoms in individuals with RA, leaving only a few hours during midday when stiffness or fatigue are not a problem (123). Fatigue is a complex phenomenon related to exertion, deconditioning, depression, or a combination of these factors (11). An individual who is fatigued shortens the length of time of exercise, decreases exercise intensity, cannot tolerate progression of exercise volume, tends to be less motivated, and may become frustrated. Appropriate exercise training may decrease fatigue levels without worsening arthritis (91). As little as 15 min of exercise three times a week is sufficient to improve aerobic capacity in individuals with RA and fatigue (46).

Previous Joint Replacement   Total joint replacement surgery is common in individuals with arthritis.

Exercise Prescription Based on Disease Staging

Treatment goals	Recommendations
ACUTE STAGE OF ARTHRITIS	
Relieve pain and promote relaxation	Use relaxation techniques Medical supervision required
Minimize muscle atrophy	Mode: gentle isometrics performed at functional joint angles Intensity: <50% of maximum voluntary contraction Duration: 6 s while exhaling Frequency: 5-10 repetitions daily May increase blood pressure, intra-articular pressure, and joint contact forces
Prevent deformity and protect the joint structures	Use of supportive and assistive equipment for all pathologically active joints May benefit from immobilizing splints
CHRONIC STAGE OF ARTHRITIS	
Improve cardiorespiratory endurance	Mode: activities that use large muscle groups with repetitive motion (e.g., walking, swimming, dancing) Intensity: 60-80% of age-predicted MHR; 50-70% $\dot{V}O_2$max; RPE 12-16 on Borg scale (6-20) or 3-6 on Borg scale (1-10); Talk test: able to converse or sing a song comfortably Time: 20-60 min Frequency: 3 to 4 days per week, more frequently for lower intensity activity
Decrease contractures/adhesions	Perform flexibility exercises on shortened muscles; gradual progression as tolerated Even with unilaterally distributed LE joint dysfunction, perform active ROM within the normal range for all LE joints
Improve muscle strength, promote synovial nutrition, promote bone and cartilage integrity	Mode: isotonic or isometric muscle strengthening Perform in pain-free range Use functional movement patterns Intensity: perform 8-12 repetitions against gravity, then progress to resistance up to 70% of 1-repetition maximum as tolerated, 15 on Borg scale (6-20) Duration: progress to 8-10 exercises, 8-12 repetitions Frequency: 2 to 3 times per week with rest days between sessions Perform strengthening exercises for all areas of both LEs, even with unilaterally distributed LE joint dysfunction
Improve function	Practice functional activities such as stairs, sit-to-stand
CHRONIC STAGE OF ARTHRITIS WITH LOCAL EXACERBATION OF SYMPTOMS	
Improve cardiorespiratory endurance	Use uninvolved joints only Follow recommendations for chronic stage
Minimize muscle atrophy	For exacerbated joints, follow recommendations for acute stage For nonexacerbated joints, follow recommendations for chronic stage
GENERAL GOALS AND RECOMMENDATIONS FOR ALL STAGES OF ARTHRITIS	
Minimize joint stiffness and maintain available motion	For involved joints, including exacerbated joints, daily passive or active-assisted ROM[a] within limits of pain two or more times per day with gradual progression as tolerated
Protect the joint structures	Use supportive and assistive equipment for joints that are painful or cause gait deviations A rest day is crucial except for ROM and low-intensity cardiovascular exercise Avoid activities that stress joints, especially joints of hand and wrist

Note. LE = lower extremity; ROM = range of motion; MHR = maximum heart rate; RPE = rating of perceived exertion. [a]Passive ROM involves no muscle work to move the joint through the entire available range. Active assisted ROM involves limited muscle work as the joint is moved through the entire range. A portion of the weight of the limb is supported by a mechanical device or another limb, or another person carries a fraction of the weight of the limb.

Decreased joint range of motion occurs after surgery but does not typically affect function. Individuals with lower extremity joint replacement should avoid all high-impact activity. Those having had hip replacement should not flex their hip past 90° or adduct or internally rotate the hip past neutral. Many surgeons terminate these precautions after 2 to 6 months. Without physician confirmation, a conservative recommendation is to maintain these precautions with all exercise activity.

Time of Day   Morning stiffness is a problem for individuals with arthritis. It is important to be sensitive to the daily variability of symptoms, the difficulty of arising and performing activities of daily living, and, as a result, the difficulty of early morning activity. Moreover, a change in ability to perform exercise during periods of inclement weather is frequently reported by individuals with rheumatologic conditions. A drop in barometric pressure along with an increase in humidity can increase pain (67). For those with inflammatory arthritis characterized by prolonged morning stiffness, it is wise to prescribe exercise for the late morning or early afternoon (40).

Fibromyalgia   Fibromyalgia is a soft tissue pain disorder associated with inadequate, nonrestorative sleep, which often coexists with rheumatologic conditions. Fibromyalgia decreases the pain threshold and exercise tolerance and increases fatigue. In contrast to other rheumatologic conditions, an increase in pain with fibromyalgia does not signify an increase in the inflammatory process. Therefore, strength and cardiovascular training are not absolute contraindications with an exacerbation of fibromyalgia pain. An individual's ability to differentiate the pain of fibromyalgia from arthritic joint pain facilitates safe progression of the exercise program. An Arthritis Foundation aquatics program for fibromyalgia known as FIT (fibromyalgia interval training) recommends non-weight-bearing exercise in a pool. This helps some individuals with severe pain to persist with an exercise program.

Special Considerations Regarding Aquatic Therapy   RA is commonly associated with Raynaud's phenomenon and Sjogren's syndrome. Raynaud's phenomenon is a **vasospastic** problem presenting as blanching or cyanosis of the hands and feet when exposed to cold or emotional stress. It can result in pitting scars or gangrene. Individuals with Raynaud's phenomenon should avoid cool air and water and wear protective clothing including noncotton gloves, shirt, pants, and shoes. The choice of exercise modality should be dictated by the symptoms

attributable to arthritis and Raynaud's phenomenon. Sjogren's syndrome is an autoimmune condition characterized by dry mouth and eyes. It is caused by the infiltration of salivary and lacrimal glands with lymphocytes. Individuals with this condition may find chlorinated water and the air surrounding pools especially irritating to the eyes and are advised to wear goggles when in a pool.

Footwear   Use of appropriate footwear can reduce the risk of injury related to poor lower extremity mechanics and repeated shock. Commercial, lightweight athletic shoes that include hind foot control, a supportive midsole of shock-absorbing materials, a continuous sole, and forefoot flexibility can improve shock attenuation and biomechanics. Individuals with OA, RA, and AS with biomechanical faults in the lower extremity may need custom-made rigid or semirigid orthotics from a podiatrist, orthotist, or physical therapist.

Pulmonary Manifestations of Rheumatic Disease   Pulmonary disease can be associated with arthritic conditions. Rheumatoid arthritis is associated with interstitial lung disease (2). Those with AS often have restrictive lung disease caused by impairment of chest expansion. In most cases, pulmonary manifestations of rheumatologic disease are not absolute contraindications to exercise. A vital capacity of ≤ 1 L should be a relative contraindication to participating in pool therapy because of the restrictive effects of water against the chest wall (39). Chapters 19 and 20 review exercise specifics for those with pulmonary limitations.

Ankylosing Spondylitis   With AS, the bony fusion that occurs in the spine and sacroiliac joint cannot be prevented, but rehabilitative strategies can improve compromised range of motion and function (39). For this reason, therapy should focus on strengthening and stretching exercises to improve joint range of motion and on improving static and dynamic posture. Because individuals with AS tend to be younger and more active at diagnosis, exercise for AS can be performed at a higher intensity than for RA and OA (37, 103). When peripheral joints are involved, disease pathology is similar to RA and thus exercise recommendations specific for RA should apply. A phenomenon that can occur is called the "last joint" syndrome. With this bridging, ossification between vertebral bodies occurs at every level except one (47). This sole mobile segment is exposed to considerable stresses during exercise and present with localized pain and discitis. Rest, bracing, or surgical fusion may be necessary.

Corticosteroids   Systemic corticosteroids are a common treatment for RA and can lead to bone fragility and muscle atrophy. Muscle wasting from disuse attributable to pain and steroid-induced myopathy contribute to reductions in muscle strength (25). For those individuals with RA and using systemic steroids, exercise training should be modified so as not to adversely affect disease activity (70).

Body Composition   Obesity is a modifiable factor that negatively affects various arthritides. In particular, obesity is a strong risk factor for OA incidence, progression, and disability (79). A high body mass index (BMI) is associated with OA of the knee, hands, and feet combined (33). The mechanism is related to an increase in mechanical stress on weight-bearing joints (33). A 5-kg weight loss decreases the risk by 50% of developing knee OA within 10 years (36).

Rheumatoid arthritis is often associated with cachexia and a low BMI. However, in some cases, individuals with RA have high body mass indexes along with obesity-related health risks. Weight loss is difficult in this population because loss of muscle mass must be avoided. A study of obese subjects with RA found that a program of moderate physical training, reduced dietary energy intake, and a high-protein/low-energy supplement was successful in achieving a significant weight loss without loss of lean tissue (31).

Exercise Recommendations

The sequencing of exercises for individuals with arthritis is similar to that of the general population, beginning with a warm-up and ending with a cool-down. A warm-up should be performed to increase the tissue temperature throughout the body. As greater range and decreased stiffness evolve, the individual should be taught to judge whether it is safe to increase the range through which he or she is exercising (109). Superficial heat has a symbiotic effect and may be used immediately before warm-up. Skeletal muscle strengthening and cardiovascular conditioning exercises should be performed following warm-up and then should be followed by a cool-down period. Flexibility exercises should be performed during the cool-down.

Mode

Isotonic exercise is preferred over isometric exercise for dynamic strength training during the chronic stage of arthritis (96). Isotonic exercise is advantageous because it closely corresponds to everyday activities and therefore promotes improved daily function.

Low-intensity isometric exercise is preferred for muscle strengthening during the acute arthritic stage because it produces low articular pressures. Instructions should be to perform isometric contraction at no more than one half of the individual's maximum contraction for 6 s, while exhaling. Isometric exercise should be targeted at one muscle group at a time.

Aerobic training is ideally achieved by using a mode that minimizes the magnitude and rate of joint loading (117). The best types include walking, cycling, and pool exercise (32). Free-speed walking produces less hip joint contractile pressure than do isometric or standing dynamic hip exercises (117). However, faster walking speeds increase stress on knee joints (89). Poor lower extremity biomechanics, joint instability, or poor proprioception may contribute to undesirable joint forces when walking speed is increased (83). Without therapy to improve these impairments, increasing gait speed may have deleterious effects.

If walking is uncomfortable, if pain lasts more than 2 hr after walking, or if the individual has complicated biomechanics of the lower extremity, an alternative cardiovascular exercise should be used such as cycle ergometry, recumbent stair-stepping, upper body ergometry (especially in the absence of **osteoporosis**), and water-walking (21, 64, 85, 120). Cycle ergometry should be conducted with the seat height and crank length adjusted to limit knee flexion and minimize pedal load, which decreases knee joint stress (72).

Water is a good medium for cardiovascular work, range of motion exercise, and low-level strengthening. The buoyant quality of water can help patients perform passive and active joint range of motion. Strengthening exercise may be performed in the water. Water can offer resistance to motion taking place against buoyancy, when turbulence of the water is increased, by the use of a float attached to an extremity. An increase in speed increases the resistance. Many individuals with arthritis are able to tolerate longer and more vigorous workouts in the water than on land. Compliance with aquatic exercise decreases with water temperatures colder than 84° F, and cardiovascular stress increases with temperatures above 98° F (17). Contraindications to hydrotherapy include a history of uncontrolled seizures, incontinence of bowel or bladder, pressure sores or contagious skin rashes, and cognitive impairments that would jeopardize the patient's safety (58). If a great deal of assistance is needed with dressing, or fatigue or joint pain is caused by changing clothes, then pool therapy should not be used.

The mode and sequencing of exercise are important considerations for individuals with AS. An exercise program for an individual with AS should begin as soon after diagnosis of the condition as possible, beginning with exercises to improve spinal and peripheral joint motion before a strengthening program is initiated (39). Achieving functional range of motion of the hip joints should be emphasized, because a lack of such capability can be very disabling (19, 39, 95). The goal of muscle strengthening in this population is to maintain or approximate a **neutral spine** over the long term. A program to strengthen the back and hip extensors as well as general strengthening can be performed on land or in water. During strength training, the individual should be initially supervised with a goal of independently maintaining proper posture during spinal extension. High-impact activities should be avoided, because they are stressful to the spinal and sacroiliac joints. Swimmers with limited spinal and neck motion should use a mask, snorkel, and fins to avoid trunk and neck rotation. Sports that encourage extension are preferred over activities that require flexion (39). Contact sports are contraindicated for those with cervical spine (39) or peripheral joint involvement.

Frequency

The key to exercise therapy for arthritic conditions is to manipulate the intensity and duration for aerobic exercise, or the number of repetitions for resistance exercise, to achieve a training effect without causing joint discomfort. It is recommended that patients begin conservatively and increase the frequency before the intensity of exercise. People with arthritis should begin at about 3 days per week with a goal of 5 to 7 days per week.

Intensity

Cardiovascular exercise intensity should be guided by heart rate or rating of perceived exertion (38, 76). Standard training heart-rate range development can

be used (see chapter 7). A cardiovascular exercise intensity of 12 to 16 on the 15-point Borg scale or 3 to 6 on the 10-point Borg scale is recommended (see practical application 24.4).

Duration

Exercise duration and intensity are inversely related. Exercise intensity or the time/repetition variable of a specific exercise should be progressed when the exercise is not challenging and symptoms do not increase for two to three consecutive sessions. Conservative increments are recommended. In general, after 1 week of consistent exercise without an increase in symptoms, the duration of aerobic and resistive exercise or repetitions during resistive exercise should be increased toward the maximum recommended. Exercises can be progressed per session by increasing the duration of exercise or the number of exercises performed or by increasing total time spent exercising by decreasing any rest periods.

An increase in symptoms may require decreasing the intensity or duration/repetition of exercise for the affected joint. The "2-hr pain rule" is a helpful maxim for regulating exercise intensity. A localized increase in pain after an activity that lasts more than 2 hr suggests the need to decrease the exercise intensity or duration/repetition of the next exercise session.

Conclusion

The available data indicate that properly performed exercise is safe and effective for individuals with OA, RA, and AS (126). In the short term, exercise improves strength, enhances cardiovascular endurance, decreases stiffness, increases range of motion, decreases and prevents impairments, improves function, and prevents disability (4, 119). These benefits are in addition to the well-accepted benefits of exercise to the general population. However, precautions must be taken to ensure that exercise is safe and comfortable for people with arthritis.

Case Study 24

Medical History

Mrs. MZ is a 69-year-old African-American woman with a 15-year history of OA of the hands, cervical spine, right knee, and feet. She presented with a rotator cuff tear secondary to a fall the previous year, multiple sites of joint pain, up to 2 hr of morning stiffness, and evening fatigue. She

continued

occasionally uses a cane for ambulation. Her chief complaints are pain in her knees and hips (right worse than left), feet, hands, right shoulder, and lower back.

Diagnosis

A rheumatologic evaluation revealed mild fibromyalgia superimposed on OA. She had no known symptoms or risk factors of cardiovascular disease.

Exercise Test Results

An exercise test using cycle ergometry was notable for a resting heart rate of 60 beats · min⁻¹ and a peak metabolic equivalent (MET) of 7.7. She ambulated 1985 feet during a 12-min walk test with a perceived exertion of 15 on the Borg scale. Mrs. MZ underwent a 3-month supervised exercise program. Pain, stiffness, and difficulty with activities of daily living were dramatically improved by the end of the supervised exercise program. When she was ascending stairs, walking on a flat surface, shopping, getting in and out of a bath, and performing heavy domestic duties, pain with weight-bearing activity and morning stiffness improved from a preprogram difficulty of "very severe" to a postprogram difficulty of "mild."

After completing the program, she stopped exercising for 3 months both at home and at the Duke Center for Living because of an illness experienced by her husband. She resumed exercise 2 to 3 days per week and underwent follow-up exercise testing 10 months after the initial test. The results were 68 beats · min⁻¹ resting heart rate and a peak MET level achieved of 5.4. Mrs. MZ ambulated 1488 ft in a 12-min walk test, a decrease of 25%. Gait changes were noted during the walk, and she complained of left foot pain on termination of the test. The discussion of these findings with her served as a motivation to resume regular attendance with concentration on aerobic conditioning.

Development of the Exercise Prescription

Exercise commenced with the AFYAP I pool class for full body range of motion exercises and low-level strengthening, lower extremity muscle stretching, lower extremity strengthening, and cardiovascular conditioning. The exercise program also included physical therapy for her right shoulder. The first week of exercise consisted of lower extremity muscle flexibility exercises and two strengthening exercises. Lower extremity muscle flexibility exercises included seated hamstring stretch, calf stretch, and standing quadriceps muscle stretch. Isotonic leg presses and step-ups onto a 3-in. rise were added but then discontinued because of knee pain. After 1 week, Mrs. MZ tried walking to improve cardiovascular conditioning. She had an increase in foot and knee pain during and for approximately 24 hr after walking. Cycle ergometry did not increase her joint symptoms and was performed at each session thereafter. Mrs. MZ was instructed to walk in the pool for cardiovascular conditioning for 15 to 20 min after the AFYAP I pool class. Lower extremity strengthening via functional movements and isotonic machines was added each week. These included standing hip abduction, isotonic seated hip abduction and adduction, and seated leg curl. Strengthening exercises were initiated at 25% of the maximum voluntary contraction. Mrs. MZ was directed to report symptoms that occurred between sessions to the clinical exercise physiologist at the following session. Strengthening exercises were progressed slowly because of an increase in knee pain with high repetitions and resistance. Mrs. MZ reported an increase in shoulder, knee, lateral thigh, medial

continued

Case Study 24   *(continued)*

elbow, lateral elbow, and foot pain intermittently throughout the program. Exercises associated with an increase in joint or muscle symptoms were revised or discontinued. Fibromyalgia pain may have played a role in elbow symptoms, because the pain was associated with the muscle–tendon junction and not the joint.

Case Study 24   Discussion Questions

1. How could Mrs. MZ's program have been changed to improve her cardiovascular status?
2. Which cardiovascular equipment might be best for Mrs. MZ?
3. Could Mrs. MZ's postprogram compliance have been improved? If so, how?
4. What initial exercise intensity should be prescribed when fibromyalgia accompanies arthritis?

Glossary

aneural—Absence of nerve fibers.

articular—Relating to a joint.

asymmetric—Denoting a lack of symmetry between two or more like parts.

autoimmune—An immune response of the body against one of its own tissues.

avascular—Absence of blood vessels.

bursitis—Inflammation of one of the fluid-filled sacs located at sites of friction surrounding the joint.

chronicity—The state of being chronic.

crepitus—Crackling from the joint palpated on examination.

cysts—Abnormal sacs containing gas, fluid, or a semisolid material, with a membranous lining.

entheses—Sites where ligaments, tendons, or joint capsules are attached to bone.

enthesopathy—Inflammation at entheses.

effusion—Excess synovial fluid within a joint.

gel phenomenon—The sensation of difficulty moving a joint after a period of joint rest or immobility.

incidence—The number of new cases of a disease during a specific time interval.

intraosseous—Within bone.

isotonic—Denoting the condition when a contracting muscle shortens against a constant load, as when lifting a weight.

joint effusion—Increased fluid in synovial cavity of a joint.

neutral spine—The position in which the trunk and neck, and therefore the joints of the spine, are neither in flexion nor extension.

osteophyte—Bone spur.

osteoporosis—Decreased bone density, usually measured in the vertebral bodies of the lumbar spine or the femoral head of the hip.

prevalence—The total number of cases of a specific disease at a specific time.

spondyloarthropathy—A type of inflammatory arthritis involving ligament or tendon insertion sites (enthuses), leading to spinal and peripheral joint arthritis, usually in human lymphocyte antigen-B27-positive individuals.

subchondral sclerosis—Thickening of the bone beneath the cartilage layer of an arthritic joint.

symmetric—Equality or correspondence in the form of parts on the opposite sides of any body.

synovial joint—A joint in which the opposing bony surfaces are covered with a layer of hyaline cartilage or fibrocartilage and is nourished and lubricated by synovial tissue.

synovitis—Swelling within a joint attributable to inflammation of the synovial lining.

valgus—Bowlegged deformity.

varus—Knock-kneed deformity.

vasospastic—Contraction or spasm of the muscular coats of the blood vessels.

References

1. Anderson JJ, Felson DT. Factors associated with osteoarthritis of the knee in the first national Health and Nutrition Examination Survey (HANES I). Evidence for an association with overweight, race, and physical demands of work. Am J Epidemiol 1988; 128(1): 179-189.

2. Anderson R. Rheumatoid arthritis. In Primer on Rheumatic Diseases, Schumacher Jr. HR (Ed.), 1993; Arthritis Foundation, Atlanta, pp. 90-96.

3. Anonymous. Prevalence of leisure-time physical activity among persons with arthritis and other rheumatic conditions—United States, 1990-1991. Morb Mortal Wkly Rep 1997; 46(18): 389-393.

4. Baar M, Assendelft W, Dekker J, Oostendorp R, Bijlsma J. Effectiveness of exercise therapy in patients with osteo-

arthritis of the hip or knee: A systematic review of randomized clinical trials. Arthritis Rheum 1999; 42(7): 1361-1369.

5. Bacon P, Townend J. Nails in the coffin: Increasing evidence for the role of rheumatic disease in the cardiovascular mortality of rheumatoid arthritis. Arthritis Rheum 2001; 33(12): 2707.

6. Badley EM. The impact of disabling arthritis. Arthritis Care Res 1995; 8(4): 221-228.

7. Baker K, McAlindon T. Exercise for knee osteoarthritis. Curr Opin Rheumatol 2000; 12: 456-463.

8. Barry H. Activity for older persons and mature athletes. In Manual of Sports Medicine, Safran M, McKeag D, Van Camp S (Eds.), 1998; Lippincott-Raven, New York, pp. 184-189.

9. Barth WF. Office evaluation of the patient with musculoskeletal complaints. Am J Med 1997; 102(1A): 3S-10S.

10. Bellamy N, Buchanon W, Goldsmith C, Campbell J, Stitt L. Validation study of WOMAC: A health status instrument for measuring clinically important patient relevant outcome to antirheumatic drug therapy in patients with osteoarthritis of the hip or knee. J Rheum 1988; 15: 1833-1840.

11. Belza B. The impact of fatigue on exercise performance. Arthritis Care Res 1994; 7(4): 176-180.

12. Berman A, Studenski S. Musculoskeletal rehabilitation. Clin Geriatr Med 1998; 14(3): 641-659.

13. Berry MJ, Brubaker PH, O'Toole ML, Rejeski WJ, Soberman J, Ribisl PM, Miller HS, Afable RF, Applegate W, Ettinger WH. Estimation of VO2 in older individuals with osteoarthritis of the knee and cardiovascular disease. Med Sci Sports Exerc 1996; 28(7): 808-814.

14. Brandt K, Heilman D, Slemenda C, Mazzuca S, Braunstein E. Knee pain in elderly community subjects. Differences in lower extremity muscles strength, body weight, and depression scores among those with and without radiographic evidence of osteoarthritis. Tran Orthop Res 1999; 24: 222.

15. Bryant C, Mahler D, Froelicher V, Miller N. ACSM's Guidelines for Exercise Testing and Prescription. Williams & Wilkens, Baltimore, pp. 86-109, 1995.

16. Bullough P, Cawston T. Pathology and biochemistry of osteoarthritis. In Osteoarthritis, Doherty M (Ed.), 1994; Wolfe, London, pp. 29-58.

17. Bunning RD, Materson RS. A rational program of exercise for patients with osteoarthritis. Semin Arthritis Rheum 1991; 21(3 Suppl 2): 33-43.

18. Byers PH. Effect of exercise on morning stiffness and mobility in patients with rheumatoid arthritis. Res Nurs Health 1985; 8(3): 275-281.

19. Calin A, Elswood J, Rigg S, Skevington SM. Ankylosing spondylitis—An analytical review of 1500 patients: The changing pattern of disease. J Rheumatol 1988; 15(8): 1234-1238.

20. Cleland LG, Hill CL, James MJ. Diet and arthritis. Baillieres Clin Rheumatol 1995; 9(4): 771-785.

21. Coleman EA, Buchner DM, Cress ME, Chan BK, de Lateur BJ. The relationship of joint symptoms with exercise performance in older adults. J Am Geriatr Soc 1996; 44(1): 14-21.

22. Creamer P. Osteoarthritis pain and its treatment. Curr Opin Rheumatol 2000; 12: 450-455.

23. Daltroy LH, Robb-Nicholson C, Iversen MD, Wright EA, Liang MH. Effectiveness of minimally supervised home aerobic training in patients with systemic rheumatic disease. Br J Rheumatol 1995; 34(11): 1064-1069.

24. Daly MP, Berman BM. Rehabilitation of the elderly patient with arthritis. Clin Geriatr Med 1993; 9(4): 783-801.

25. Danneskiold-Samsoe B, Grimby G. The relationship between the leg muscle strength and physical capacity in patients with rheumatoid arthritis, with reference to the influence of corticosteroids. Clin Rheumatol 1986; 5(4): 468-474.

26. Davis M, Ettinger W, Neuhaus J. Obesity and osteoarthritis of the knee: Evidence from the National Health and Nutrition Examination Survey (NHANES I). Arthritis Rheum 1990; 20(3, Suppl 1): 34-41.

27. del Rincon I, Williams K, Stern M, Freeman G, Escalante A. High incidence of cardiovascular events in a rheumatoid arthritis cohort not explained by traditional cardiac risk factors. Arthritis Rheum 2001; 44(12): 2737-2745.

28. Dexter PA. Joint exercises in elderly persons with symptomatic osteoarthritis of the hip or knee. Performance patterns, medical support patterns, and the relationship between exercising and medical care. Arthritis Care Res 1992; 5(1): 36-41.

29. Dougados M, Gueguen A, Nguyen M, Thiesce A, Listrat V, Jacob L, Nakache J, Gabriel K, Lequesne M, Amor B. Longitudinal radiologic evaluation of osteoarthritis of the knee. J Rheumatol 1992; 19: 378-384.

30. Ekdahl C. Muscle function in rheumatoid arthritis. Assessment and training. Scand J Rheumatol Suppl 1990; 86: 9-61.

31. Engelhart M, Kondrup J, Hoie LH, Andersen V, Kristensen JH, Heitmann BL. Weight reduction in obese patients with rheumatoid arthritis, with preservation of body cell mass and improvement of physical fitness. Clin Exp Rheumatol 1996; 14(3): 289-293.

32. Ettinger WH, Jr., Burns R, Messier SP, Applegate W, Rejeski WJ, Morgan T, Shumaker S, Berry MJ, O'Toole M, Monu J, Craven T. A randomized trial comparing aerobic exercise and resistance exercise with a health education program in older adults with knee osteoarthritis. The Fitness Arthritis and Seniors Trial (FAST). JAMA 1997; 277(1): 25-31.

33. Felson D, Anderson J, Naimark A, Walker A, Meenan R. Obesity and knee osteoarthritis: The Framingham study. Ann Intern Med 1988; 109: 18-24.

34. Felson DT. The epidemiology of knee osteoarthritis: Results from the Framingham Osteoarthritis Study. Semin Arthritis Rheum 1990; 20(3 Suppl 1): 42-50.

35. Felson DT. Weight and osteoarthritis. Am J Clin Nutr 1996; 63(3 Suppl): 430S-432S.

36. Felson DT, Anderson JJ, Naimark A, Walker AM, Meenan RF. Obesity and knee osteoarthritis. The Framingham Study. Ann Intern Med 1988; 109(1): 18-24.

37. Fisher LR, Cawley MI, Holgate ST. Relation between chest expansion, pulmonary function, and exercise tolerance in patients with ankylosing spondylitis. Ann Rheum Dis 1990; 49(11): 921-925.

38. Frangolias DD, Rhodes EC. Metabolic responses and mechanisms during water immersion running and exercise. Sports Med 1996; 22(1): 38-53.

39. Gall V. Exercise in the spondyloarthropathies. Arthritis Care Res 1994; 7(4): 215-220.

40. Galloway MT, Jokl P. The role of exercise in the treatment of inflammatory arthritis. Bull Rheum Dis 1993; 42(1): 1-4.

41. Gecht M, Connell K, Sinacore J, Prohaska T. A survey of exercise beliefs and exercise habits among people with arthritis. Arthritis Care Res 1996; 9(2): 82-88.

42. Gussoni M, Margonato V, Ventura R, Veicsteinas A. Energy cost of walking with hip joint impairment. Phys Ther 1990; 70(5): 295-301.

43. Hakkinen A, Hakkinen K, Hannonen P. Effects of strength training on neuromuscular function and disease activity in patients with recent-onset inflammatory arthritis. Scand J Rheumatol 1994; 23(5): 237-242.

44. Hakkinen A, Malkia E, Hakkinen K, Jappinen I, Laitinen L, Hannonen P. Effects of detraining subsequent to strength training on neuromuscular function in patients with inflammatory arthritis. Br J Rheumatol 1997; 36(10): 1075-1081.

45. Hall J, Skevington SM, Maddison PJ, Chapman K. A randomized and controlled trial of hydrotherapy in rheumatoid arthritis. Arthritis Care Res 1996; 9(3): 206-215.

46. Harkcom TM, Lampman RM, Banwell BF, Castor CW. Therapeutic value of graded aerobic exercise training in rheumatoid arthritis. Arthritis Rheum 1985; 28(1): 32-39.

47. Haslock I. Ankylosing spondylitis: Management. In Rheumatology, Klippel J, Dieppe P (Eds.), 1994; Mosby, St. Louis, pp. 21-23.

48. Hasselbacher P. Joint Physiology. In Rheumatology, Klippel JH and Dieppe PA (Eds.), 1994; Mosby, London, pp. 1.3.1-1.3.6.

49. Hazes JM, van den Ende CH. How vigorously should we exercise our rheumatoid arthritis patients? Ann Rheum Dis 1996; 55(12): 861-862.

50. Helmick CG, Lawrence RC, Pollard RA, Lloyd E, Heyse SP. Arthritis and other rheumatic conditions: Who is affected now, who will be affected later? National Arthritis Data Workgroup. Arthritis Care Res 1995; 8(4): 203-211.

51. Hicks JE. Exercise in patients with inflammatory arthritis and connective tissue disease. Rheum Dis Clin North Am 1990; 16(4): 845-870.

52. Hoffman DF. Arthritis and exercise. Prim Care 1993; 20(4): 895-910.

53. Hubscher O. Pattern recognition in arthritis. In Rheumatology, Klippel J, Dieppe P (Eds.), 1998; Mosby, Philadelphia, pp. 231-236.

54. Hurley M. The role of muscle weakness in the pathogenesis of osteoarthritis. Rheum Dis Clin North Am 1999; 25(2): 283-298.

55. Hurt-Camejo E, Paredes S, Masana L, Camejo G, Sartipy P, Rosengren B, Pedreno J, Vallve J, Benito P, Wiklund O. Elevated levels of small, low-density lipoprotein with high affinity for arterial matrix components in patients with rheumatoid arthritis: Possible contribution of phospholipase A2 to this atherogenic profile. Arthritis Rheum 2001; 44(12): 2707-2710.

56. James M, Cleland L, Gaffney R, Proudman S, Chatterton B. Effect of exercise on ^{99m}Tc-DTPA clearance from knees with effusions. J Rheumatol 1994; 21: 501-504.

57. Jokl P. Prevention of disuse muscle atrophy in chronic arthritides. Rheum Dis Clin North Am 1990; 16(4): 837-844.

58. Kenney W. ACSM's Guidelines for Exercise Testing and Prescription. 1995, William & Wilkens, Baltimore, pp. 220-240.

59. Kirsteins AE, Dietz F, Hwang SM. Evaluating the safety and potential use of a weight-bearing exercise, Tai-Chi Chuan, for rheumatoid arthritis patients. Am J Phys Med Rehabil 1991; 70(3): 136-141.

60. Kisner C, Colby L. Therapeutic exercise: Foundations and techniques. 1990, FA Davis Co, Philadelphia.

61. Klepper S. Effects of an eight-week physical conditioning program on disease signs and symptoms in children with chronic arthritis. Arthritis Care Res 1999; 12(1): 52-60.

62. Klippel J. Primer on the Rheumatic Diseases. 1997, Arthritis Foundation, Atlanta, pp. 453-465.

63. Kovar P, Allegrante J, MacKenzie C, Peterson M, Gutin B, Charlson M. Supervised fitness walking in patients with osteoarthritis of the knee. Ann Intern Med 1992; 116: 529-534.

64. Kovar PA, Allegrante JP, MacKenzie CR, Peterson MG, Gutin B, Charlson ME. Supervised fitness walking in patients with osteoarthritis of the knee. A randomized, controlled trial. Ann Intern Med 1992; 116(7): 529-534.

65. Lane NE. Exercise: A cause of osteoarthritis. J Rheumatol Suppl 1995; 43: 3-6.

66. Lane NE, Buckwalter JA. Exercise: A cause of osteoarthritis? Rheum Dis Clin North Am 1993; 19(3): 617-633.

67. Lawrence J. The influence of climate on rheumatoic complaints. In Rheumatism in Populations, Lawrence J (Ed.), 1977; William Heinemann Medical Books Ltd, London, pp. 505-517.

68. Lawrence RC, Helmick CG, Arnett FC, Deyo RA, Felson DT, Giannini EH, Heyse SP, Hirsch R, Hochberg MC, Hunder GG, Liang MH, Pillemer SR, Steen VD, Wolfe

F. Estimates of the prevalence of arthritis and selected musculoskeletal disorders in the United States. Arthritis Rheum 1998; 41(5): 778-799.

69. Loeser RF, Jr. Evaluation of musculoskeletal complaints in the older adult. Clin Geriatr Med 1998; 14(3): 401-415.

70. Lyngberg KK, Harreby M, Bentzen H, Frost B, Danneskiold-Samsoe B. Elderly rheumatoid arthritis patients on steroid treatment tolerate physical training without an increase in disease activity. Arch Phys Med Rehabil 1994; 75(11): 1189-1195.

71. Machover S, Sapecky AJ. Effect of isometric exercise on the quadriceps muscle in patients with rheumatoid arthritis. Arch Phys Med Rehabil 1966; 47(11): 737-741.

72. Mangione KK, McCully K, Gloviak A, Lefebvre I, Hofmann M, Craik R. The effects of high-intensity and low-intensity cycle ergometry in older adults with knee osteoarthritis. J Gerontol A Biol Sci Med Sci 1999; 54(4): M184-M190.

73. March LM, Bachmeier CJ. Economics of osteoarthritis: A global perspective. Baillieres Clin Rheumatol 1997; 11(4): 817-834.

74. McAlindon TE, Felson DT, Zhang Y, Hannan MT, Aliabadi P, Weissman B, Rush D, Wilson PW, Jacques P. Relation of dietary intake and serum levels of vitamin D to progression of osteoarthritis of the knee among participants in the Framingham Study. Ann Intern Med 1996; 125(5): 353-359.

75. McAlindon TE, Jacques P, Zhang Y, Hannan MT, Aliabadi P, Weissman B, Rush D, Levy D, Felson DT. Do antioxidant micronutrients protect against the development and progression of knee osteoarthritis? Arthritis Rheum 1996; 39(4): 648-656.

76. McCardle W, Katch F, Katch V. Exercise Physiology: Energy, Nutrition, and Human Performance. 1986, Lea & Feibiger, Philadelphia.

77. McCurdy D. Genetic susceptibility to the connective tissue diseases. Curr Opin Rheum 1999; 11: 399-407.

78. Melton-Rogers S, Hunter G, Walter J, Harrison P. Cardiorespiratory responses of patients with rheumatoid arthritis during bicycle riding and running in water. Phys Ther 1996; 76(10): 1058-1065.

79. Messier S, Loeser R, Mitchell M, Valle G, Morgan T, Rejeski W, Ettinger W. Exercise and weight loss in obese older adults with knee osteoarthritis: A preliminary study. J Am Geriatr Soc 2000; 48: 1062-1072.

80. Miltner O, Schneider U, Graf J, Niethard F. Influence of isokinetic and ergometric exercises on oxygen partial pressure measurement in the human knee joint. Adv Exp Med Biol 1997; 411: 183-189.

81. Minor M. Rest and exercise. In Clinical Care in the Rheumatic Diseases, Wegener S, Belza B, Gall E (Eds.), 1997; American College of Rheumatism, Atlanta, pp. 73-78.

82. Minor M, Hewett J, Webel R, Dreisinger T, Kay D. Exercise tolerance and disease related measures in patients with rheumatoid arthritis and osteoarthritis. J Rheumatol 1988; 15: 905-911.

83. Minor MA. Exercise in the treatment of osteoarthritis. Rheum Dis Clin North Am 1999; 25(2): 397-415, viii.

84. Minor MA, Brown JD. Exercise maintenance of persons with arthritis after participation in a class experience. Health Educ Q 1993; 20(1): 83-95.

85. Minor MA, Hewett JE, Webel RR, Anderson SK, Kay DR. Efficacy of physical conditioning exercise in patients with rheumatoid arthritis and osteoarthritis. Arthritis Rheum 1989; 32(11): 1396-1405.

86. Minor MA, Hewett JE, Webel RR, Dreisinger TE, Kay DR. Exercise tolerance and disease related measures in patients with rheumatoid arthritis and osteoarthritis. J Rheumatol 1988; 15(6): 905-911.

87. Minor MA, Johnson JC. Reliability and validity of a submaximal treadmill test to estimate aerobic capacity in women with rheumatic disease. J Rheumatol 1996; 23(9): 1517-1523.

88. Minor MA, Lane NE. Recreational exercise in arthritis. Rheum Dis Clin North Am 1996; 22(3): 563-577.

89. Minor MA, Sanford MK. The role of physical therapy and physical modalities in pain management. Rheum Dis Clin North Am 1999; 25(1): 233-248, viii.

90. Neuberger GB, Kasal S, Smith KV, Hassanein R, DeViney S. Determinants of exercise and aerobic fitness in outpatients with arthritis. Nurs Res 1994; 43(1): 11-17.

91. Neuberger GB, Press AN, Lindsley HB, Hinton R, Cagle PE, Carlson K, Scott S, Dahl J, Kramer B. Effects of exercise on fatigue, aerobic fitness, and disease activity measures in persons with rheumatoid arthritis. Res Nurs Health 1997; 20(3): 195-204.

92. Nordemar R. Physical training in rheumatoid arthritis: A controlled long-term study. II. Functional capacity and general attitudes. Scand J Rheumatol 1981; 10(1): 25-30.

93. Nordemar R, Ekblom B, Zachrisson L, Lundqvist K. Physical training in rheumatoid arthritis: A controlled long-term study. I. Scand J Rheumatol 1981; 10(1): 17-23.

94. Noreau L, Moffet H, Drolet M, Parent E. Dance-based exercise program in rheumatoid arthritis. Feasibility in individuals with American College of Rheumatology functional class III disease. Am J Phys Med Rehabil 1997; 76(2): 109-113.

95. O'Driscoll SL, Jayson MI, Baddeley H. Neck movements in ankylosing spondylitis and their responses to physiotherapy. Ann Rheum Dis 1978; 37(1): 64-66.

96. O'Grady M, Fletcher J, Ortiz S. Therapeutic and physical fitness exercise prescription for older adults with joint disease: An evidence-based approach. Rheum Dis Clin North Am 2000; 26(3): 617-646.

97. O'Reilly S, Jones A, Doherty M. Muscle weakness in osteoarthritis. Curr Opin Rheumatol 1997; 9(3): 259-262.

98. Otterness I, Zimmerer R, Swindell A, Poole A, Saxne T, Heinegard D, Ionescu M, Weiner E. An examination of some molecular markers in blood and urine for discriminating patients with osteoarthritis from healthy

individuals. Acta Orthop Scan 1995; 66(Suppl 266): 142-164.

99. Philbin EF, Ries MD, French TS. Feasibility of maximal cardiopulmonary exercise testing in patients with end-stage arthritis of the hip and knee prior to total joint arthroplasty. Chest 1995; 108(1): 174-181.

100. Poole A, Rizkalla G, Ionescu M, Reiner A, Brooks E, Rorabeck C, Bourne R, Bogoch E. Osteoarthritis in the human knee: A dynamic process of cartilage matrix degradation, synthesis and reorganization. Agents & Actions Suppl 1993; 39: 3-13.

101. Praemer A, Furner S, Rice D. Musculoskeletal Conditions in the United States, 1992, American Academy of Orthopaedic Surgeons, Rosemont, IL, pp. 121-138.

102. Rall L, Roubenoff R. Body composition, metabolism, and resistance exercise in patients with rheumatoid arthritis. Arthritis Care Res 1996; 9(2): 151-156.

103. Rassmussen JO, Hansen TM. Physical training for patients with ankylosing spondylitis. Arthritis Care Res 1989; 2(1): 25-27.

104. Ries MD, Philbin EF, Groff GD. Relationship between severity of gonarthrosis and cardiovascular fitness. Clin Orthop 1995; 313: 169-176.

105. Santos H, Brophy S, Calin A. Exercise in ankylosing spondylitis: How much is optimum? J Rheumatol 1998; 25(11): 2156-2160.

106. Saxne T. Differential release of molecular markers in joint disease. Acta Orthop Scand 1995; (Suppl 266): 80-83.

107. Schilke JM, Johnson GO, Housh TJ, O'Dell JR. Effects of muscle-strength training on the functional status of patients with osteoarthritis of the knee joint. Nurs Res 1996; 45(2): 68-72.

108. Schweitzer M, Resnick D. Enthesopathy. In Rheumatology, Klippel J, Dieppe P (Eds.), 1994; Mosby, St. Louis, pp. 21-23.

109. Scully R, Barnes M. Physical therapy. In Physical Therapy, Scully RM, Barnes MR (Eds.), 1989; Lippincott, Philadelphia.

110. Semble E, Loeser R, Wise C. Therapeutic exercise for rheumatoid arthritis and osteoarthritis. Semin Arthritis Rheum 1990; 20(1): 32-40.

111. Shephard R, Shek P. Autoimmune disorders, physical activity, and training, with particular reference to rheumatoid arthritis. Exerc Immunol Rev 1997; 3: 53-67.

112. Spector T, Hart D, Nandra D, Doyle D, MacKillop N, Gallimore J, Pepys M. Low-level increases in serum C-reactive protein are present in early osteoarthritis of the

knee and predict progressive disease. Arthritis Rheum 1997; 40: 723-727.

113. Stenstrom CH. Home exercise in rheumatoid arthritis functional class II: Goal setting versus pain attention. J Rheumatol 1994; 21(4): 627-634.

114. Stenstrom CH. Therapeutic exercise in rheumatoid arthritis. Arthritis Care Res 1994; 7(4): 190-197.

115. Steultjens M, Dekker J, vanBaar M, Oostendorp R, Bijlsma J. Range of joint motion and disability in patients with osteoarthritis of the knee or hip. Rheumatology 2000; 39: 955-961.

116. Suomi R, Lindauer S. Effectiveness of Arthritis Foundation Aquatic Program on strength and range of motion in women with arthritis. J Aging Phys Activity 1997; 5: 341-351.

117. Tackson SJ, Krebs DE, Harris BA. Acetabular pressures during hip arthritis exercises. Arthritis Care Res 1997; 10(5): 308-319.

118. van Baar ME, Assendelft WJ, Dekker J, Oostendorp RA, Bijlsma JW. Effectiveness of exercise therapy in patients with osteoarthritis of the hip or knee: A systematic review of randomized clinical trials. Arthritis Rheum 1999; 42(7): 1361-1369.

119. van den Ende CH, Hazes JM, le Cessie S, Mulder WJ, Belfor DG, Breedveld FC, Dijkmans BA. Comparison of high and low intensity training in well controlled rheumatoid arthritis. Results of a randomised clinical trial. Ann Rheum Dis 1996; 55(11): 798-805.

120. van den Ende CH, Vliet Vlieland TP, Munneke M, Hazes JM. Dynamic exercise therapy in rheumatoid arthritis: A systematic review. Br J Rheumatol 1998; 37(6): 677-687.

121. Van Deusen J, Harlowe D. The efficacy of the ROM Dance Program for adults with rheumatoid arthritis. Am J Occup Ther 1987; 41(2): 90-95.

122. Verbrugge L, Patrick D. Seven chronic conditions: Their impact on US adults' activity levels and use of medical services. Am J Public Health 1994; 85: 173-182.

123. Weiss TE, Gum OB Biundo JJ, Jr. Rheumatic diseases. 1. Differential diagnosis. Postgrad Med 1976; 60(6): 141-150.

124. Yelin E, Callahan L. Economic cost and social and psychological impact of musculoskeletal conditions. Arthritis Rheum 1995; 38(10): 1351-1362.

125. Ytterberg S, Mahowald M, Krug H. Exercise for arthritis. Baillieres Clin Rheumatol 1994; 8(1): 161-189.

126. Ytterberg SR, Mahowald ML, Krug HE. Exercise for arthritis. Baillieres Clin Rheumatol 1994; 8(1): 161-189.

David L. Nichols, PhD
Assistant Research Professor
Institute for Women's Health
Texas Woman's University
Denton, TX

Osteoporosis

As middle age begins (approximately 35 years), small amounts of bone are lost every year. In women, this bone loss is accelerated during a 3- to 5-year period after menopause. Because women typically have lower bone mass than men, and because of the accelerated decline following menopause, they tend to be more prone to **osteoporosis.** Osteoporosis has classically been defined, for older women and men, as a pathological condition associated with an increased loss of bone mass, known as **osteopenia,** caused by increased bone resorption. As **bone mineral density (BMD)** declines, individuals are at risk for skeletal fractures. A second definition has also emerged to deal with the osteoporotic condition found in young female athletes as part of the female athlete triad (107). In this case, the definition of osteoporosis also includes the fact that these young women are dealing with a premature loss of bone mass. Osteoporosis is a major public health problem resulting in significant morbidity, mortality, and economic burden. Fractures result in impairments that increase the risk of pneumonias and can accelerate other underlying diseases, such as coronary artery disease.

Scope

The risk of fragility fracture approximately doubles for each standard deviation decrease in the measured

BMD. The level of BMD that is necessary to qualify as osteoporotic is still debated, primarily because of disagreements regarding what values should be used to identify the young adult mean BMD (the reference point where BMD has reached its peak before age-related bone mineral loss begins—it occurs at the end of puberty) (18-20). However, the World Health Organization (WHO) has defined osteoporosis as a bone mineral density measurement more than 2.5 standard deviations below the young adult mean (41).

The consequences of poor bone health have continued to escalate in recent years despite heightened awareness, a better understanding of prevention strategies, advances in technology for screening and diagnosis, and expanded treatment options. The major impact of osteoporosis is on older, postmenopausal, Caucasian women. Although osteoporosis is thought of primarily as a disease of women, an estimated 2 million men in the United States have osteoporosis and 3 million more may be at risk (64). In addition, low BMD is often found in young amenorrheic women. Osteoporosis now affects almost one out of every two women at some point in their lives. Medical costs are close to $10 billion per year. There are approximately 50,000 deaths each year as a result of complications from hip fracture (15), and although men are at a lower risk for sustaining a hip fracture than women, the mortality rate associated with hip fracture is higher in elderly men compared with women (67).

Physiology and Pathophysiology

Changes in bone are the result of the continual process of bone resorption and **bone formation**—known as bone remodeling. Bone remodeling does the following:

— Maintains the architecture and strength of the bone
— Regulates calcium levels
— Prevents fatigue damage

Remodeling is also important during periods of growth, puberty, and adolescence, when the majority of adult bone mass is laid down (13). Bone modeling and remodeling often occur simultaneously, and distinctions between them are not always apparent, but in general, **bone modeling** refers to alterations in the shape of the bone such as changes in length. Bone modeling usually ceases around the age of 18 to 20, when the skeleton stops growing, whereas

bone remodeling occurs throughout the lifespan. **Bone remodeling** is a complex process that has been more thoroughly described elsewhere (13, 65). In brief, bone remodeling is performed by individual basic multicellular units (sometimes referred to as bone multicellular units (22) or bone modeling units (103)) consisting of osteoclasts and osteoblasts (29, 66). These basic multicellular units have two roles: osteoclasts erode old bone (resorption), whereas osteoblasts form new bone (formation). Once growth has ceased, bone formation generally equals bone resorption; that is, the density of the bone remains unchanged (13). However, bone cells age, just like all other cells in the body. As a result, the amount of bone remodeling that occurs begins to decrease with increasing age. Osteoblasts seem to be more greatly affected by aging than osteoclasts, and thus bone formation decreases more than bone resorption, resulting in an age-related bone loss (22) that can lead to osteoporosis. The decrease in bone remodeling seen with aging may not be inevitable, however, and, as with other things, may be a result of the decreased physical activity and poorer nutrition of older people. Figure 25.1 illustrates the cycle of bone modeling and remodeling.

On the other hand, the loss of bone mass seen at menopause, or in young amenorrheic women, is a direct result of the decrease in endogenous estro-

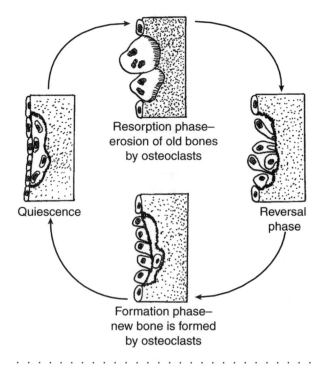

FIGURE 25.1   Bone modeling and remodeling.

gen. This decrease in estrogen levels does not alter bone formation activity but does increase in **bone resorption,** because bone resorption is inhibited by estrogen (82, 99). It is thought that over her lifetime, a woman loses as much as 50% of her peak **trabecular bone** mass (78), and the majority of that loss is attributable to an estrogen deficiency (76). Although this loss occurs mostly after menopause, bone loss has been seen in the perimenopausal years (30, 68) and is probably a result of decreased ovarian function (75).

The two most important factors in the development of osteoporosis are the amount of peak bone mass attained and the rate of bone loss (77). **Peak bone density** (or peak bone mass), for most purposes, is defined as the highest amount of bone mass attained during life. Although this definition is accurate, the term *peak bone mass* is often confused with the term **maximal bone mass.** Maximal bone mass should be defined as the highest bone mass a person could possibly achieve. Maximal bone mass would, theoretically, be controlled solely by genetic factors. Peak bone mass is also influenced to a certain extent by genetics, but the difference between maximal bone mass and peak bone mass is controlled by other factors such as physical activity, diet, and hormonal balance. It is unlikely that anyone ever reaches maximal bone mass.

Clinical Considerations

Osteoporosis may go undetected in many individuals until a fall or injury exposes the seriousness of the disease. A risk factor assessment can be conducted to determine the likelihood of the disease. Various sophisticated measurement techniques have evolved for assessing bone density that may be useful for disease diagnosis.

Signs and Symptoms

A variety of risk factors exist for osteoporosis (see table 25.1). A thorough medical history can help identify these risk factors. Unfortunately, these risk factors alone are not sufficient to distinguish between persons with low bone density or normal bone density (74, 101). However, for a postmenopausal (or amenorrheic), Caucasian woman with low body weight, who is not on hormone replacement therapy, the presence of any other risk factor would strongly suggest that this woman has low bone mass (74, 101).

Risk of osteoporotic fracture is related, for the most part, to a person's BMD, with fall frequency

---

TABLE 25.1

Risk Factors for Osteoporosis or Decreased Bone Mineral Density

Inherited factors	Environmental factors
Caucasian or Asian	Below normal weight
Female	Loss of menstrual function
Osteoporotic fracture in first-degree relative	Low calcium intake
Height <67 in.	Inactivity
Weight <127 lb	Prolonged corticosteroid use
	Smoking
	Excessive alcohol intake
	Caffeine

---

and bone geometry (the angle the bone is in when it encounters a force) also playing a role. But because fragility fractures are linked to BMD, and BMD later in life is determined to a great extent by peak bone mass, striving to make peak bone mass equal maximal bone mass could be the best protection against osteoporosis. Unfortunately, a fracture is often the first sign that a person has osteoporosis. Loss of bone mass occurs without outward symptoms, and by the time a fracture occurs, a patient may have lost as much as 30% or more of peak bone density. Even the best available treatments can increase bone density by only about 8% (104). So once a person reaches the level of bone density that is considered osteoporotic, he or she is unlikely to ever return to a level of normal BMD. This is true even in young amenorrheic women, who, after having been amenorrheic for extended periods of time, never regain sufficient bone mass to return to normal, even with the resumption of regular menstrual periods (21, 39).

Diagnostic and Laboratory Evaluations

The primary means of assessing bone health is to measure bone mineral density. Over the past 20 years, rapid advances have been made in techniques used for measuring bone density (93). These advances have resulted in increased precision, less radiation exposure, and the ability to measure fracture-prone sites (31, 93). BMD is most often measured in the spine, hip (femoral neck), and wrist, because these are the most common sites for fracture in osteoporosis. The accuracy of predicting BMD and fracture risk at one site based on measurement of BMD at another site is low (≤50%). Thus, a measurement at all three sites is preferable, especially the spine and hip, because the consequences of fracture are the highest at these two sites.

Measurements of **bone mineral content** are usually expressed as the amount of bone mineral (primarily hydroxyapatite) per unit of area, which gives BMD. BMD is generally expressed as grams of bone mineral content per square centimeter of bone (g · cm^{-2}) because most of the available technology can only measure area, not volume, of bone. **Quantitative computed tomography** is the only method currently available that provides an actual measurement of volumetric bone density. Other methods are available:

— **Single photon absorptiometry**
— **Dual photon absorptiometry**
— Single energy X-ray absorptiometry
— **Dual energy X-ray absorptiometry (DEXA)**

However, all of these methods express bone density as an area measurement (31). The most widely used densitometry technique is DEXA. Because quantitative computed tomography has the advantage of providing a three-dimensional image, thus allowing a separation of trabecular and **cortical bone,** it would seem to be the measurement of choice. However, it is a much more costly procedure

and has a higher radiation exposure than DEXA, and thus it is used less in clinical practice.

DEXA uses X-rays emitted at two different energy levels to distinguish bone tissue from the surrounding **soft tissue.** The information provided with a bone density measurement is not only the absolute value in terms of grams of bone mineral but also a comparison of that value to established normal values. A DEXA image of a femoral neck scan, along with its accompanying printout, is presented in figure 25.2. The subject's BMD is compared with reference standards.

Another new emerging technology for assessing bone health is **quantitative ultrasound.** Ultrasound does not measure bone density but rather measures two parameters called speed of sound and broadband ultrasound attenuation, which are related to the structural properties of bone. The advantage of ultrasound devices is that they are small, are portable, and use no ionizing radiation; thus, they may provide an attractive alternative to radiation-based densitometry.

Biochemical markers of bone metabolism are also sometimes used to monitor therapy or progression of bone disease. Although these markers generally show poor correlation with actual measures of bone density ($r = ≤.3$), they can be useful for assessing relative rates of bone formation or resorption. Another advantage of biochemical markers is that they can detect alterations in bone formation or resorption in a matter of weeks, whereas detecting changes in BMD may take several months (1). Although bio-

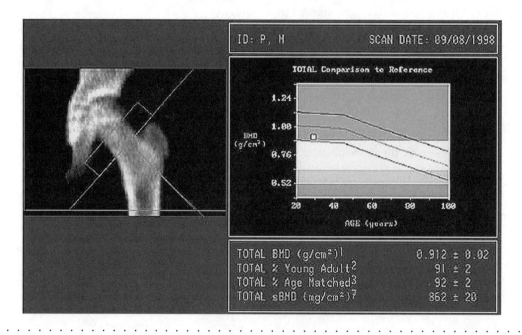

FIGURE 25.2   Femoral neck scan. Image of a proximal femur scan from a dual-energy x-ray absorptiometer.

chemical markers appear worthwhile for detecting changes in bone turnover in research studies, the clinical usefulness of any of these markers in monitoring individual patients has been questioned because of their wide day-to-day variability and their inability to predict bone loss.

With the use of DEXA technology and the aforementioned WHO guidelines on the level of bone density necessary to be considered osteoporotic, the diagnoses of osteoporosis might seem a simple matter. However, a multitude of factors can affect not only the individual's bone density but also the accuracy of the bone density measurement (10). Several advances in developing tools to evaluate bone health have been emerging, and a thorough review of the diagnostic and evaluation procedures can be found elsewhere (12). The clinician should be familiar with the critical anatomical locations where fractures typically occur and should have a fundamental understanding of the modalities used to evaluate BMD. Once osteoporosis has been confirmed via low bone mineral density and frequent fractures, the appropriate treatment should follow.

Treatment

Exercise can be useful to help increase, or at least maintain, bone mass in patients with low BMD. A number of studies in postmenopausal women have shown that exercise can increase bone density or prevent further bone loss compared with nonexercising controls. However, these studies have also pointed out that exercise without concomitant **estrogen replacement therapy** will usually result in further bone loss. The same situation can be seen in young amenorrheic athletes who continue to lose bone mass, despite their exercise training, as long as they continue without their menstrual cycles (21). Exercise should still be one of the first choices in the treatment of osteoporosis for both men and women, because it has the potential not only to increase bone mass but also to increase muscle strength and balance (which may help decrease falls).

Several nonpharmacological and pharmacological agents are available to increase, or slow the loss of, bone mass:

— Calcium supplementation
— Vitamin D supplementation
— Estrogen (or hormone) replacement therapy
— Selective estrogen receptor modulators (SERMs)
— Bisphosphonates
— **Calcitonin**

The role of calcium intake as a risk factor for osteoporosis, and as a therapeutic intervention, is controversial (40). The uncertainty arises from conflicting data in the literature. The controversy may stem from the fact that, although calcium is crucial to skeletal integrity, its role in the development of osteoporosis is more permissive than causal (36, 40). The evidence suggests that calcium is necessary for bone structure but its role is more passive depending on adequate hormonal regulation. The effect of calcium intake in postmenopausal women may depend on their stage in menopause. Calcium supplementation alone does not appear to protect against the accelerated bone loss during early menopause (18, 79), although it may be beneficial in women who are years beyond the onset of menopause (18). Vitamin D along with calcium supplementation does appear to increase BMD in individuals (19, 84). Whether the increase in BMD is attributable to the calcium or vitamin D, or the combination, is not really known. However, although increases in BMD are seen with calcium and vitamin D supplementation, no good evidence exists to suggest that they decrease fracture risk (40, 50, 97). Women are encouraged to take calcium supplements to prevent osteoporosis for several reasons:

1. Calcium is necessary for skeletal integrity.
2. Nutrition studies suggest that there are insufficient amounts of calcium in the average diet.
3. Calcium is not harmful.

Nevertheless, ensuring adequate dietary calcium intake is prudent because dietary consumption is low.

All of the current drugs with Food and Drug Administration (FDA) approval for osteoporosis are considered **antiresorptive therapy.** They halt the loss of bone or even increase bone mass by inhibiting bone resorption, while having no effect on bone formation. The majority of the drugs available are presented in table 25.2. It must be pointed out however, that all drug therapies approved by the FDA for treatment or prevention of osteoporosis are approved for use in postmenopausal women only.

Estrogen replacement therapy (ERT) has been used for many years in the treatment and prevention of osteoporosis in postmenopausal women. Studies have shown that estrogen therapy can halt the loss of, and often increase, bone mass and can reduce fracture risk in the postmenopausal woman (23, 38, 47, 54, 70). Some studies have pointed out, however, that not all women respond to therapy (94). More recent studies have explored the effectiveness

TABLE 25.2
Medical Therapies for the Treatment or Prevention of Osteoporosis

Drug class	Name of drug	Brand name
Estrogens[a]	Estrone sulfate	Ogen®
	Conjugated estrogen	Premarin®
	Transdermal estrogen	Estraderm®
	Estropipate	Ortho-Est®
	Esterified estrogen	Estratab®
	Conjugated estrogen +	Premphase®[b]
	Medroxyprogesterone acetate[b]	PremPro®[b]
SERMs	Raloxifene[d]	Evista®
	Tamoxifene[e]	Nolvadex®
Bisphosphonates	Alendronate[d]	Fosamax®
	Etidronate[e]	Didronel®
	Risedronate[d]	Actonel®
	Pamidronate[e]	Aredia®
Calcitonin[c]	Synthetic salmon calcitonin	MiaCalcin®
		Calcimar®
Others	Calcitriol[f]	
	Sodium fluoride[g]	
	Parathyroid hormone[h]	

Note. SERMs = selective estrogen receptor modulators.

[a]All estrogens have Food and Drug Administration (FDA) approval for prevention of osteoporosis, but only Premarin is approved for treatment. [b]Premphase and PremPro are estrogen and progesterone taken in combination, and both are FDA approved for treatment of osteoporosis. [c]Both calcitonins are approved for prevention, but only MiaCalcin is approved for treatment of osteoporosis. [d]FDA approved for both prevention and treatment of osteoporosis. [e]FDA approved but not with an osteoporosis indication. [f]Calcitriol is a vitamin D metabolite with FDA approval but not for osteoporosis. [g]Approval pending for an osteoporosis indication. [h]In clinical trials for treatment or prevention of osteoporosis.

of oral contraceptives or hormone replacement therapy on BMD in the amenorrheic athlete. Although these studies have generally shown positive results on bone mass, there are too few studies to draw definite conclusions (16, 37). Thus, ERT potentially can be used to treat low bone mass in the amenorrheic or postmenopausal female athlete. However, its use is contraindicated in women who are pregnant, have any history or suspected history

of breast cancer, or have thromboembolic disorders.

SERMs are antiresorptive agents under investigation for use in osteoporosis. Raloxifene has FDA approval for use in prevention of osteoporosis, but there are very few data on SERMs with regard to their therapeutic effects on existing low BMD. In comparison to ERT, SERMs offer several advantages and disadvantages. Advantages include the following (46):

— Favorable effects on bone and lipids, much like estrogen

— No stimulatory effect on breast or endometrial tissue, as seen with estrogen

Disadvantages include these (20, 56, 69):

— The increases in BMD seen with raloxifene or other SERMS are generally less than seen with estrogen.

— Unlike ERT, SERMs appear to have no beneficial effect on HDL cholesterol.

All things considered, SERMs are a good alternative to ERT for the woman with a history of breast cancer.

Calcitonin is a hormone used for calcium regulation by the body. Salmon calcitonin, in either an injectable or a nasal spray preparation, has proved effective in both increasing low bone mass and decreasing fracture risk in postmenopausal women (14, 32, 44, 72). The usefulness of calcitonin in the amenorrheic woman has not been evaluated, and recommendations for its use in this population are not available.

Bisphosphonates are one of the new, powerful, antiresorptive drugs being used to prevent and treat osteoporosis. However, only Alendronate and Risidronate are approved by the FDA for treatment or prevention of osteoporosis. Others have not been FDA approved for osteoporosis. Bisphosphonates increase bone mass by drastically reducing bone resorption (81), similar to the actions of estrogen or calcitonin. Bisphosphonates have been shown effective in reducing bone loss and decreasing fracture risk in postmenopausal women and men, as well as increasing bone strength (27, 59, 61, 63, 85, 98, 104). Bisphosphonates may also be useful in preventing the loss of bone that may occur from long-term corticosteroid treatment (73, 83). Oral bisphosphonates are poorly absorbed, and absorption can decrease if the recommended mode of administration is not followed. There are several side effects (28):

— Gastrointestinal problems

— Esophageal stricture

— Difficulty swallowing

— Active esophageal ulcer

— Esophagitis

— Renal insufficiency or failure

Patients who begin bisphosphonate therapy should be monitored for any of these side effects.

None of the drug therapies listed in table 25.2, or those under current investigation, have been studied in the premenopausal woman with low bone mass and do not have FDA approval for use in this population. For the premenopausal woman with low bone mass, treatment should always focus on the underlying cause, which is most likely an estrogen deficiency. This person should be counseled on ways to regain her menstrual cycles (such as increasing energy intake or decreasing energy expenditure). If this fails, or she refuses to comply, then some form of drug therapy may need to be considered to offset the inevitable bone loss seen with **amenorrhea.** The most reasonable choice would be some type of estrogen replacement, because it is the only osteoporosis therapy available with any data regarding treatment or safety efficacy in premenopausal women. There are still too few data, however, to make any specific treatment recommendations regarding dosing regimens in at-risk premenopausal women.

Graded Exercise Testing

The primary purposes of exercise testing are to aid in the diagnoses of coronary artery disease and to determine appropriate levels of exercise training. The value of exercise testing in a person with established osteoporosis should be carefully evaluated to make certain any potential benefits outweigh the risks. The majority of people with osteoporosis will be women (41); the accuracy of using exercise testing for diagnoses is lower in women than in men (52). Exercise testing can be of significant benefit in generating an exercise prescription. In addition, the vast majority of patients at risk for osteoporosis (postmenopausal women and elderly men) will be at a greater risk for heart disease, which could potentially be diagnosed with a stress test.

Medical Evaluation

A thorough understanding of the bone status of each osteoporosis patient cannot be emphasized enough. History of fractures, age, and amount of BMD can help determine the level of caution that must be taken to minimize inappropriate mechanical stress on high-risk joints.

Contraindications

In addition, because fractures in osteoporotic patients often occur with little trauma, the impact associated with exercise testing could lead to fractures, even when walking or bicycle protocols are used. On the other hand, the American College of Sports Medicine doesn't specifically state that osteoporosis is an absolute contraindication to exercise testing (3).

Recommendations and Anticipated Responses

If an exercise stress test is to be used, one that uses a bicycle protocol is probably the best choice, because that would involve the least trauma and impact on the bones. However, caution must still be taken when a bike protocol is used. An upright posture should be maintained by the patient at all times, because any sort of spinal flexion is contraindicated in people with osteoporosis. Treadmill protocols can be used if necessary, but a walking protocol should be used and care should be taken to ensure that the patient does not trip or fall. There are no specific studies regarding acute physiological responses to exercise testing in an osteoporotic population, but for those patients who can tolerate the exercise, there is no reason to believe their responses will be different from those individuals without osteoporosis. However, osteoporosis can sometimes mask the presence of coronary artery disease because it can prevent an individual from achieving the adequate heart rate and blood pressure necessary for accurate diagnoses. Pharmacological tests are available that can diagnose coronary artery disease without the use of exercise, and these types of tests might be more advisable in patients whose exercise capacity is clearly limited because of osteoporosis.

Exercise Prescription

Although studies have shown that several forms of exercise training have the potential to increase BMD, the optimal training program for improving or maintaining skeletal integrity has yet to be defined. Current experimental knowledge indicates that an **osteogenic** exercise regime should include the following factors (43, 90):

— Have load-bearing activities at high magnitude with few repetitions

— Create variable strain distributions throughout the bone structure (load the bone in directions to which it is unaccustomed)

— Be long term and progressive in nature

Resistance training (weightlifting) probably offers the best opportunity to meet these criteria on an individual basis, because it requires the least skill and has the added advantage of being highly adaptable to changes in both magnitude and strain distribution. In addition, strength and muscle size increases have been demonstrated following resistance training, even in the elderly. Practical application 25.1 reviews the literature regarding swimming and bone health.

PRACTICAL APPLICATION 25.1

Literature Review

Cross-sectional studies with athletes have indicated that swimming has little if any significant effect on BMD. Indeed, these studies have consistently found that swimmers had bone mineral densities that were significantly lower than other athletes and equal to, or in some cases lower than, than nonactive controls (25, 96). Cross-sectional studies are subject to bias because of their design, but, unfortunately, longitudinal studies looking at the effects of swimming on bone density in humans have not been done, although longitudinal studies in rats have indicated that swim training can actually increase BMD (92, 95). However, a recent abstract presented at the American Society of Bone and Mineral Research meeting presented limited evidence that swimming may benefit skeletal integrity (106). The investigators examined the effects of 2 years of swim training on BMD in postmenopausal women. Participants were divided into a swimming group (n = 22) and a control group (n = 19). Age, height, weight, calcium intake, and daily activity were similar between the groups. The swim program consisted of 1-hr sessions, and the participants attended an average of 1.5 sessions per week. BMD of the lumbar spine and the proximal femur (femoral neck, Ward's area, and trochanter) was measured by DEXA at baseline and at 1 and 2 years. Leg extensor strength was also measured at the same time points. At the end of 2 years, both groups showed a decrease in BMD at the lumbar spine with the rate of decline not different between the groups. However, at the sites on the proximal femur, the swim group had increases in BMD of 4.4%, 3.4%, and 5.7%, respectively, that were significantly higher than those of the control group (–0.2%, 1.4%, and 1.0%). Leg extensor strength was also significantly increased in the swim group. Although only preliminary, these results suggest that swim training may have a role in exercise prescription for osteoporosis not only to help increase muscle strength, which would aid in fall prevention, but also to help prevent bone loss or even increase bone density.

Special Exercise Considerations

Some special issues should be considered when one is developing an exercise program for a client with osteoporosis. Osteoporotic patients can fracture with little or no trauma, and thus high-impact activities such as running, jumping, or high-impact aerobics should be avoided in this population. Another activity that absolutely must not be done by people with osteoporosis is spinal flexion, which drastically increases the forces on the spine and increases the likelihood of a fracture; this includes exercises such as sit-ups or toe touches. Other activities that should be avoided are those that may increase the chance of falling, such as trampolines, step aerobics, or exercising on slippery floors (see table 25.3).

Exercise Recommendations

Resistance training combined with some sort of cardiovascular training (bicycling or walking) is probably the best exercise program for a patient with osteoporosis. Not only will such a program increase overall fitness and BMD, it will greatly reduce the risk of falling, which is one of the primary causes of fracture in osteoporosis.

Although the physiological responses to exercise in an osteoporotic population have not been specifically investigated, they should not be substantially different than those of an age-matched individual without osteoporosis. Similarly, the goals of an exercise program for someone with or without osteoporosis should also be the same. For a person just beginning an exercise program, those goals should include an increase in cardiovascular fitness, increased muscular strength, and an increase (or at least no decrease) in BMD. Heart disease remains the number one killer of men and women by a wide margin. So the goal of both women and men should be to increase their physical activity to reduce the risk of heart disease. The 1996 Surgeon General's report on health and physical activity (100) recommends approximately 30 min of moderate physical activity on most if not all days of the week. This would be a worthwhile goal for all people including those with osteoporosis. However, if the person with osteoporosis is just beginning an exercise program, the duration of exercise might need to be shortened initially to allow time for adjustment to the exercise to reduce the likelihood of musculoskeletal injury. As the person's fitness level increases, the amount of time exercising can be increased. For an osteoporotic individual, or for that matter any elderly person just beginning an exercise program, a simple walking program should provide the needed benefits along with safety. Practical application 25.2 contains exercise prescription guidelines for people with osteoporosis.

TABLE 25.3
The Osteoporotic Patient and Exercise

Beneficial exercises	Exercises to avoid
1. Modified sit-ups where only the head and shoulders come off the ground	1. Any type of abdominal crunches that involve spinal flexion
2. Lying leg lifts: the person lies flat on a firm surface with legs straight out in front and lifts legs 6 in. off the floor	2. Any movement involving spinal flexion such as toe touches or rowing machines
3. Back extension exercises	3. Jogging or running
4. Walking to help increase cardiovascular fitness and increase bone mineral density	4. High-impact aerobics or other activities that may jar the spine or hip
5. Strength-training exercises using either dumbbells or resistance-exercise machines; pay special attention to hips, thighs, backs, and arms	5. Leg adduction or abduction or squat exercises with any significant resistance
6. Exercises to improve balance or agility such as standing on one foot (with assistance if needed) for ≥15 s	6. Exercises or activities that increase the chance of falling, such as using trampolines or exercising on slippery floors

Exercise Prescription Guidelines

In designing an exercise program for the osteoporotic patient, the clinical exercise physiologist should bear in mind two primary goals:

1. Increase overall fitness (cardiovascular, muscular strength, balance and flexibility)
2. Increase or at least maintain BMD

However, these goals must be weighed against the need to create an exercise program that is safe for the person with osteoporosis (see table 25.3 for exercises to avoid). The vast majority of people with osteoporosis are postmenopausal women or elderly men in whom the risk of cardiovascular disease is significant. Thus, a walking program designed to increase cardiovascular fitness is surely the best choice, because this has the potential of not only improving cardiovascular fitness but also improving bone density (bicycle training is unlikely to improve BMD because it is not weight bearing). If the patient has no good recent history of exercise, the walking program should start slow and gradually progress over time. The ultimate goal would be to walk five times a week for 30-45 min at a speed of at least 3 mph or faster if possible (but never reaching a jogging speed). To reach that goal, the person may need to start at a slower pace and less distance, but the number of times per week can start at five. Just be sure that the patient can comfortably handle the speed and distance used for the first 2 weeks, and then the speed and distance can be increased gradually. The best choice is to increase only speed on a couple of days a week and increase the distance walked on the other days of the week.

A resistance-training program can help increase both bone density and muscular strength in the elderly. Although increases in bone density are generally only modest, strength gains can be much greater and will help reduce the risk of falling, which is a major cause of hip fracture. The resistance training program should begin slowly, starting out at 2 days per week and moving up to 3 or 4. One-repetition maximum testing (1RM) is not appropriate in this population. A 10RM test could be used but is not really necessary. Initially, simply set the resistance low enough (even if it is the minimum weight) that the person can easily complete 15 repetitions of an exercise without undue strain. Exercises that target all the major muscle groups should be chosen, especially for the legs and back. However, the client must avoid any exercise that involves spinal flexion (again, consult table 25.3). After the first 2 weeks in which one set of 15 repetitions is done for each exercise, a second set for each exercise can be done. Following a couple of weeks of 2 sets for each exercise, resistance can be added to increase the intensity of the program. Increase the resistance so that no more than 10 to 12 repetitions can be completed on the second set. After that, increase the weight gradually as needed to maintain the 10- to 12-repetition limit on the second set. Make certain that all exercises are always performed with slow, controlled movements.

Cardiovascular Training

No known studies have specifically examined cardiovascular adaptations in osteoporotic patients, but older adults can increase their fitness levels 10% to 30% with prolonged endurance training (2, 49). Since endurance training can decrease cardiovascular disease risk factors (i.e., high blood pressure and cholesterol), it should be incorporated as part of the exercise regimen of osteoporotic women and men. Moreover, a good majority of older women with osteoporosis will not be on estrogen replacement therapy and thus will be receiving none of the cardiovascular benefits associated with ERT (4). Weight-bearing endurance activities (e.g., walking) may be most beneficial for retaining bone density.

Resistance Training

Resistance training offers the most benefits for increases in muscular strength and bone density. Current recommendations suggest a single set of 15 repetitions of 8 to 10 exercises performed at least 2 days per week. This is a worthwhile goal for the person with osteoporosis, but a less strenuous program may be needed initially. Care should be taken to avoid the exercises previously mentioned that are dangerous for people with osteoporosis.

Balance and Agility Training

Certain exercises are quite beneficial for the osteoporotic patient. These include exercises designed to help with balance and agility in order to reduce falls. For instance, exercises that strengthen the quadriceps (i.e., knee extension) are helpful because poor strength in this muscle group has been linked to risk of falling. However, squats with free weights should be avoided because of the excess load that might be applied to the spine as well as the potential for spinal flexion during the squat lift. A specific exercise that helps build hip and low back strength as well as improve balance is standing on one foot for 5 to 15 s (see figure 25.3). Initially, patients should place their hands on a counter for support until they develop the strength and balance needed to perform the exercise without danger of falling. The osteoporotic patient should also

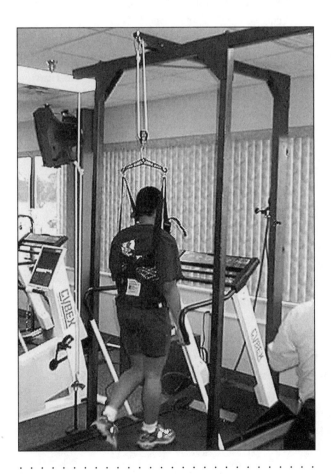

FIGURE 25.3   Balance exercise.

be encouraged to do spine extension (but not spinal flexion) exercises. Spine extension exercises can be performed in a chair and can strengthen the back muscles, which should help reduce the development of a dowager's hump and possibly reduce the risk of vertebral fracture. However, these and all exercises done by patients with osteoporosis should be performed with slow and controlled movements; jerky, rapid movement should be avoided. More complete information on these and other exercises for the osteoporotic patient can be found elsewhere.

Exercise may be useful both for increasing bone density to help prevent osteoporosis and as a therapeutic modality for patients in whom osteoporosis is already present. As a means of preventing osteoporosis, individuals need to engage in weight-bearing activities on a regular basis. This may be most beneficial among young adults, when bone modeling is at its highest. Nevertheless, it should also be remembered that excess exercise can sometimes be detrimental to bone health, especially if it leads to amenorrhea, as is often seen in female athletes. When a clinician is caring for an osteoporosis patient, however, caution must be observed in the type of exercise program to be used and the specific exercises done. Those patients with severe osteoporosis who are just beginning an exercise program should probably be supervised until it is determined that they can properly perform the exercises without danger to themselves.

Conclusion

Osteoporosis is an increasing health problem and should be a concern of everyone, even though the disease affects primarily women. As the population ages, the cost and problems associated with osteoporosis will continue to increase. However, osteoporosis should not be considered an inevitable consequence of aging. It is a preventable disease, but prevention must begin at an early age, perhaps even before puberty. Osteoporosis or osteopenia can be diagnosed with the use of DEXA technology, and bone density measurements should be seriously considered in anyone with existing risk factors for osteoporosis. Drug and exercise therapies are available to treat osteoporosis and its related problems, but once bone density is low enough to be considered osteoporotic, most of that lost density cannot be regained. Therefore, prevention should be the primary focus.

Medical History

Ms. RF, a 60-year-old Caucasian woman, is referred for the development of an exercise program. She is 9 years past menopause and has never been on hormone replacement therapy. She smoked for 35 years but quit 1 year ago after having been diagnosed with atherosclerosis. She has recently developed high blood pressure and would like to begin an exercise program to help in that regard. She is 5 feet 2 in. tall and of normal weight for her height, but she has never exercised regularly. She has recently had an exercise stress test which was normal, but no other physiological testing was done other than regular blood chemistries.

Development of Exercise Prescriptions

Based on the medical history, some sort of cardiovascular training program should be developed for this woman. Even though there is no confirmation at this time, low bone density should be suspected in this woman, and jogging or high-impact or step aerobics should not be considered. A walking program is surely the best choice, because this has the potential of not only improving her cardiovascular fitness but also improving bone density (bicycle training is unlikely to improve BMD because it is not weight bearing). Even though her stress test was normal, given her age and lack of any exercise history, the walking program should start slow and gradually progress over time. The ultimate goal would be for her to walk five times a week for 30-45 min at a speed of at least 3 mph or faster if possible (but never reaching a jogging speed). To reach that goal, Ms. RF may need to start at a slower pace and less distance, but the number of times per week can start at five. After it is determined that the patient can comfortably handle the speed and distance used for the first 2 weeks, then the speed and distance can gradually be increased. The best choice is to increase only speed on a couple of days a week and increase the distance walked on the other days of the week.

A resistance-training program should also be implemented for Ms. RF. Again, it should begin slowly, starting out at 2 days per week and moving up to 3 or 4. Initially, resistance should be low enough so that she can easily complete 15 repetitions of an exercise without undue strain. Exercises that target all the major muscle groups should be chosen but especially for the legs and back. However, any exercise that involves spinal flexion should not be done. Leg lifts or a modified sit-up, in which only the head is lifted off the floor, can be done to strengthen the abdominal muscles. After the first 2 weeks in which one set of 15 repetitions is done for each exercise, a second set for each exercise can be done. Following a couple of weeks of two sets for each exercise, resistance can be added to increase the intensity of the program. Increase the resistance so that no more than 10 to 12 repetitions can be completed on the second set. All exercises should be performed with slow, controlled movements.

Based on Ms. RF's case history, such an exercise program should be well tolerated and provide optimal health benefits. However, individual cases vary, and for the woman who has suffered fragility fractures as a result of osteoporosis, even the minimal exercise described here may not be tolerated initially. For such a woman, a more gradual, less intense program will be needed.

Case Study 25 Discussion Questions

1. What other information should be obtained from Ms. RF?

continued

2. What other physiological tests might be recommended?

3. Given the case history for Ms. RF presented here, and the fact that resistance training will offer little in the way of cardiovascular benefits beyond what will be achieved with the walking program, what will be the value of the resistance training program? Will the risks outweigh the benefits?

Glossary

amenorrhea—Absence of normal menses; for most studies, a woman is considered amenorrheic if she has less than three menses per year.

antiresorptive therapy—A term used to describe the various drug therapies currently available for treatment of osteoporosis. The term originates from the fact that the therapies halt the loss of bone by inhibiting bone resorption.

bone formation (modeling)—The process by which new bone is formed and deposited within the existing bone matrix. It is accomplished primarily by bone cells called osteoblasts.

bone mineral content—A measurement of the total amount of hydroxyapatite (calcium phosphate crystal) of bone and expressed as $g \cdot cm^{-2}$. It is synonymous with bone mass.

bone mineral density—Relative amount of bone mineral per measured bone width with values expressed as $g \cdot cm^{-2}$.

bone modeling—Alterations in the shape of the bone such as changes in length.

bone remodeling—A constant state of formation and resorption.

bone resorption—The process of eroding old bone from the existing bone matrix so that new bone can be formed in its place. Resorption is accomplished primarily by bone cells called osteoclasts. Bone resorption is greatly increased in estrogen-deplete women.

calcitonin—Hormone that is responsible for calcium regulation and inhibits bone resorption.

cortical bone—One of the two main types of bone tissue. It is hard, compact bone and is found mainly in the shafts of long bones. The other type is trabecular.

dual energy X-ray absorptiometry (DEXA)—A method for measuring bone mineral density and bone mineral content. It is based on the amount of radiation absorption, or attenuation, of the different body tissues. When bone mass is measured, the higher the attenuation of radiation by the bone, the greater the mass. Radiation exposure is minimal (<5 mR) compared with a chest X-ray (100 mR) or lumbar X-ray (600 mR).

dual photon absorptiometry—A method similar to DEXA for measuring bone density but one that relies on a radionuclide source as opposed to X-ray. The photon intensity is not as great as with DEXA, and precision is therefore reduced.

estrogen replacement therapy—Therapy that is useful for protecting bone loss in postmenopausal women.

maximal bone mass—The highest bone mass a person could possibly achieve.

osteogenic—Increasing bone mass.

osteopenia—Reduced bone mineral density, defined as being between 1 and 2.5 standard deviations below the young adult mean.

osteoporosis—A pathological condition associated with an increased susceptibility to fracture and decreased bone mineral density more than 2.5 standard deviations below the young adult mean.

peak bone density—The highest amount of bone mass achieved by an individual during his or her lifetime. Peak bone mass is assumed to be achieved in the second or third decade of life. The age at which peak bone mass is achieved will also vary based on which bone site is being measured.

quantitative computed tomography—The only method currently available that provides an actual measurement of volumetric bone density.

quantitative ultrasound—A device that measures structural properties of bone with sound waves. It uses no ionizing radiation like densitometry devices.

selective estrogen receptor modulators (SERMs)—Antiresorptive agents that have fewer side effects than estrogen replacement therapy and may be a good alternative to ERT for the woman with a history of breast cancer.

single photon absorptiometry—A method for determining bone mineral content by measuring the absorption by bone of a monoenergetic photon beam.

soft tissue—The total amount of tissue in the body minus bone mass as determined by DEXA.

trabecular bone—One of the two main types of bone tissue, also known as cancellous or spongy bone. It is made up of interlacing plates of bone tissue and is found mainly at the ends of long bones and within the vertebrae.

References

1. Akesson K. Biochemical markers of bone turnover. Acta Orthop Scand 1995; 66: 376-386.

2. American College of Sports Medicine. American College of Sports Medicine Position Stand. Exercise and physical

478 *PART VI Disorders of Bones and the Joints*

activity for older adults. Med Sci Sports Exerc 1998; 30: 992-1008.

3. American College of Sports Medicine. ACSM's Guidelines for Exercise Testing and Prescription. 6th ed. Philadelphia: Lippincott, Williams & Wilkins, 2000: 1-373.

4. Aygen EM, Karakucuk EI, Basbug M. Comparison of the effects of conjugated estrogen treatment on blood lipid and lipoprotein levels when initiated in the first or fifth postmenopausal year. Gynecol Endocrinol 1999; 13: 118-122.

5. Bailey DA, Faulkner RA, McKay HA. Growth, physical activity, and bone mineral acquisition. Exerc Sport Sci Rev 1996; 24: 233-266.

6. Bassey EJ, Rothwell MC, Littlewood JJ, Pye DW. Pre- and postmenopausal women have different bone mineral density responses to the same high-impact exercise. J Bone Miner Res 1998; 13: 1805-1813.

7. Bauer DC, Gluer CC, Cauley JA, et al. Broadband ultrasound attenuation predicts fractures strongly and independently of densitometry in older women. A prospective study. Study of Osteoporotic Fractures Research Group. Arch Intern Med 1997; 157: 629-634.

8. Bauer DC, Sklarin PM, Stone KL, et al. Biochemical markers of bone turnover and prediction of hip bone loss in older women: The study of osteoporotic fractures. J Bone Miner Res 1999; 14: 1404-1410.

9. Bonnick SL. The Osteoporosis Handbook. Dallas: Taylor, 1997: 1-180.

10. Bonnick SL. Bone Densitometry in Clinical Practice. Application and Interpretation. Totowa, NJ: Humana Press, 1998: 1-257.

11. Bonnick SL, Nichols DL, Sanborn CF, et al. Dissimilar spine and femoral z-scores in premenopausal women. Calcif Tiss Int 1997; 61: 263-265.

12. Bouxsein ML, Myers ER, Hayes WC. Biomechanics of age-related fractures. In: Marcus R, Feldman D, Kelsey JL, eds. Osteoporosis. San Diego: Academic Press, 1996: 373-393.

13. Buckwalter JA, Glimcher MJ, Cooper RR, Recker R. Bone biology. Part II: Formation, form, modeling, remodeling, and regulation of cell function. J Bone Joint Surg 1995; 77-A: 1276-1289.

14. Chesnut CH, Silverman S, Andriano K, et al. A randomized trial of nasal spray salmon calcitonin in postmenopausal women with established osteoporosis: The prevent recurrence of osteoporotic fractures study. Am J Med 2000; 109: 267-276.

15. Chrischilles EA, Shireman T, Wallace R. Costs and health effects of osteoporotic fractures. Bone 1994; 15: 377-385.

16. Cumming DC. Exercise-associated amenorrhea, low bone density, and estrogen replacement therapy. Arch Intern Med 1996; 156: 2193-2195.

17. Dalsky GP, Stocke KS, Ehsani AA, Slatopolsky E, Lee WC, Birge SJ. Weight-bearing exercise training and lumbar bone mineral content in postmenopausal women. Ann Intern Med 1988; 108: 824-828.

18. Dawson-Hughes B, Dallai GE, Krall EA, Sadowski L, Sahyoun N, Tannenbaum S. A controlled trial of the effect of calcium supplementation on bone density in postmenopausal women. N Engl J Med 1990; 323: 878-883.

19. Dawson-Hughes B, Harris SS, Krall EA, Dallal GE. Effect of calcium and vitamin D supplementation on bone density in men and women 65 years of age or older. N Engl J Med 1997; 337: 670-676.

20. Delmas PD, Bjarnason NH, Mitlak BH, et al. Effects of raloxifene on bone mineral density, serum cholesterol concentrations, and uterine endometrium in postmenopausal women. N Engl J Med 1997; 337: 1641-1647.

21. Drinkwater BL, Nilson K, Ott S, Chesnut CH. Bone mineral density after resumption of menses in amenorrheic athletes. JAMA 1986; 256: 380-382.

22. Eriksen EF, Langdahl BL. The pathogenesis of osteoporosis. Horm Res 1997; 48: 78-82.

23. Ettinger BF, Genant HK, Cann CE. Long-term estrogen replacement therapy prevents bone loss and fractures. Ann Intern Med 1985; 102: 319-324.

24. Evans WJ. Exercise training guidelines for the elderly. Med Sci Sports Exerc 1999; 31: 12-17.

25. Fehling PC, Alekel L, Clasey J, Rector A, Stillman RJ. A comparison of bone mineral densities among female athletes in impact loading and active loading sports. Bone 1995; 17: 205-210.

26. Feigenbaum MS, Pollock ML. Prescription of resistance training for health and disease. Med Sci Sports Exerc 1999; 31: 38-45.

27. Fogelman I, Ribot C, Smith R, Ethgen D, Sod E, Reginster JY. Risedronate reverses bone loss in postmenopausal women with low bone mass: Results from a multinational, double-blind, placebo-controlled trial. BMD-MN Study Group. J Clin Endocrinol Metab 2000; 85: 1895-1900.

28. Francis RM. Bisphosphonates in the treatment of osteoporosis in 1997: A review. Curr Ther Res 1997; 58: 656-678.

29. Frost HM. Perspectives: A proposed general model of the "mechanostat" (suggestions from a new skeletal-biologic paradigm). Anat Rec 1996; 244: 139-147.

30. Gambacciani M, Spinetti A, Taponeco F, et al. Bone loss in perimenopausal women: A longitudinal study. Maturitas 1994; 18: 191-197.

31. Genant HK, Engelke K, Fuerst T, et al. Noninvasive assessment of bone mineral and structure: State of the art. J Bone Miner Res 1996; 11: 707-730.

32. Grigoriou O, Papoulias I, Vitoratos N, et al. Effects of nasal administration of calcitonin in oophorectomized women: 2-year controlled double-blind study. Maturitas 1997; 28: 147-151.

33. Hans D, Dargent-Molina P, Schott AM, et al. Ultrasonographic heel measurements to predict hip fracture in elderly women: The EPIDOS prospective study. Lancet 1996; 348: 511-514.

34. Hans D, Srivastav SK, Singal C, et al. Does combining the results from multiple bone sites measured by a new quantitative ultrasound device improve discrimination of hip fracture? J Bone Miner Res 1999; 14: 644-651.

35. Harridge SD, Kryger A, Stensgaard A. Knee extensor strength, activation, and size in very elderly people following strength training. Muscle Nerve 1999; 22: 831-839.

36. Heaney RP. The role of nutrition in prevention and management of osteoporosis. Clin Obstet Gynecol 1987; 50: 833-846.

37. Hergenroeder AC, Smith EO, Shypailo R, Jones LA, Klish WJ, Ellis K. Bone mineral changes in young women with hypothalamic amenorrhea treated with oral contraceptives, medroxyprogesterone, or placebo over 12 months. Am J Obstet Gynecol 1997; 176: 1017-1025.

38. Hillard TC, Whitcroft SJ, Marsh MS, et al. Long-term effects of transdermal and oral hormone replacement therapy on postmenopausal bone loss. Osteoporos Int 1994; 4: 341-348.

39. Jonnavithula S, Warren MP, Fox RP, Lazaro MI. Bone density is compromised in amenorrheic women despite return of menses: A 2-year study. Obstet Gynecol 1993; 81: 669-674.

40. Kanis JA. The use of calcium in the management of osteoporosis. Bone 1999; 24: 279-290.

41. Kanis JA, Melton LJ, Christiansen C, Johnston CC, Khaltaev N. The diagnosis of osteoporosis. J Bone Miner Res 1994; 9: 1137-1141.

42. Kannus P, Haapasalo H, Sankelo M, et al. Effect of starting age of physical activity on bone mass in the dominant arm of tennis and squash players. Ann Intern Med 1995; 123: 27-31.

43. Kannus P, Sievanen H, Vuori I. Physical loading, exercise, and bone. Bone 1996; 18: 1S-3S.

44. Kapetanos G, Symeonides PP, Dimitriou C, Karakatsanis K, Potoupnis M. A double blind study of intranasal calcitonin for established postmenopausal osteoporosis. Acta Orthop Scand 1997; 275 (Suppl): 108-111.

45. Kerr D, Morton A, Dick I, Prince R. Exercise effects on bone mass in postmenopausal women are site-specific and load-dependent. J Bone Miner Res 1996; 11: 218-225.

46. Khovidhunkit W, Shoback DM. Clinical effects of raloxifene hydrochloride in women. Ann Intern Med 1999; 130: 431-439.

47. Kohrt WM, Birge SJJ. Differential effects of estrogen treatment on bone mineral density of the spine, hip, wrist and total body in late postmenopausal women. Osteoporos Int 1995; 5: 150-155.

48. Kohrt WM, Ehsani AA, Birge SJJ. Effects of exercise involving predominantly either joint-reaction or ground-reaction forces on bone mineral density in older women. J Bone Miner Res 1997; 12: 1253-1261.

49. Kohrt WM, Snead DB, Slatopolsky E, Birge S. Additive effects of weight-bearing exercise and estrogen on bone mineral density in older women. J Bone Miner Res 1995; 10: 1303-1311.

50. Komulainen MH, Kroger H, Tuppurainen MT, et al. HRT and Vit D in prevention of non-vertebral fractures in postmenopausal women: A 5 year randomized trial. Maturitas 1998; 31: 45-54.

51. Kujala UM, Kaprio J, Kannus P, Sarna S, Koskenvuo M. Physical activity and osteoporotic hip fracture risk in men. Arch Intern Med 2000; 160: 705-708.

52. Kwok Y, Kim C, Grady D, Segal M, Redberg R. Meta-analysis of exercise testing to detect coronary artery disease in women. Am J Cardiol 1999; 83: 660-666.

53. Lanyon LE. Using functional loading to influence bone mass and architecture: Objectives, mechanisms, and relationship with estrogen of the mechanically adaptive process in bone. Bone 1996; 18: 37S-43S.

54. Lees B, Pugh M, Siddle N, Stevenson JC. Changes in bone density in women starting hormone replacement therapy compared with those in women already established on hormone replacement therapy. Osteoporos Int 1995; 5: 344-348.

55. Looker AC, Bauer DC, Chesnut CH, et al. Clinical use of biochemical markers of bone remodeling: Current status and future directions. Osteoporos Int 2000; 11: 467-480.

56. Lufkin EG, Whitaker MD, Nickelsen T, et al. Treatment of established postmenopausal osteoporosis with raloxifene: A randomized trial. J Bone Miner Res 1998; 13: 1747-1754.

57. Maddalozzo GF, Snow CM. High intensity resistance training: Effects on bone in older men and women. Calcif Tissue Int 2000; 66: 399-404.

58. Matkovic V, Jelic T, Wardlaw GM, et al. Timing of peak bone mass in Caucasian females and its implication for the prevention of osteoporosis. Inference from a cross-sectional model. J Clin Invest 1994; 93: 799-808.

59. McClung M, Clemmesen B, Daifotis A, et al. Alendronate prevents postmenopausal bone loss in women without osteoporosis. A double-blind, randomized, controlled trial. Ann Intern Med 1998; 128: 253-261.

60. McCulloch RG, Bailey DA, Houston CS, Dodd BL. Effects of physical activity, dietary calcium intake and selected lifestyle factors on bone density in young women. CMAJ 1990; 142: 221-227.

61. Mortensen L, Charles P, Bekker PJ, Digennaro J, Johnston CCJ. Risedronate increases bone mass in an early postmenopausal population: Two years of treatment plus one year of follow-up. J Clin Endocrinol Metab 1998; 83: 396-402.

62. Notelovitz M, Martin D, Tesar R, et al. Estrogen therapy and variable-resistance weight training increase bone mineral in surgically menopausal women. J Bone Miner Res 1991; 6: 583-590.

63. Orwoll E, Ettinger M, Weiss S, et al. Alendronate for the treatment of osteoporosis in men. N Engl J Med 2000; 343: 604-610.

64. Orwoll ES. Osteoporosis in men. Endocrinol Metab Clin North Am 1998; 27: 349-367.

65. Parfitt AM. The cellular basis of bone remodeling: The quantum concept reexamined in light of recent advances in the cell biology of bone. Calcif Tissue Int 1984; 36 (Suppl 1): S37-S45.

66. Parfitt AM, Mundy GR, Roodman GD, Hughes DE, Boyce BF. A new model for the regulation of bone resorption, with particular reference to the effects of bisphosphonates. J Bone Miner Res 1996; 11: 150-159.

67. Poor G, Atkinson EJ, O'Fallon WM, Melton LJ. Determinants of reduced survival following hip fractures in men. Clin Orthop 1995; 319: 260-265.

68. Pouilles JM, Tremollieres F, Ribot C. [Vertebral bone loss in perimenopause. Results of a 7-year longitudinal study]. Presse Med 1996; 25: 277-280.

69. Prestwood KM, Gunness M, Muchmore DB, Lu Y, Wong M, Raisz LG. A comparison of the effects of raloxifene and estrogen on bone in postmenopausal women. J Clin Endocrinol Metab 2000; 85: 2197-2202.

70. Prestwood KM, Pilbeam CC, Burleson JA, et al. The short term effects of conjugated estrogen on bone turnover in older women. J Clin Endocrinol Metab 1994; 79: 366-371.

71. Pruitt LA, Jackson RD, Bartels RL. Weight-training effects on bone mineral density in early postmenopausal women. J Bone Miner Res 1992; 7: 179-185.

72. Reginster JY, Deroisy R, Lecart MP, et al. A double-blind, placebo-controlled, dose-finding trial of intermittent nasal salmon calcitonin for prevention of postmenopausal lumbar spine bone loss. Am J Med 1995; 98: 452-458.

73. Reid DM, Hughes RA, Laan RF, et al. Efficacy and safety of daily risedronate in the treatment of corticosteroid-induced osteoporosis in men and women: A randomized trial. European Corticosteroid-Induced Osteoporosis Treatment Study. J Bone Miner Res 2000; 15: 1006-1013.

74. Ribot C, Pouilles JM, Bonneu M, Tremollieres F. Assessment of the risk of post-menopausal osteoporosis using clinical factors. Clin Endocrinol (Oxford) 1992; 36: 225-228.

75. Richardson SJ, Nelson JF. Follicular depletion during the menopausal transition. Ann N Y Acad Sci 1990; 592: 13-20.

76. Richelson LS, Wahner HW, Melton LJ, Riggs BL. Relative contributions of aging and estrogen deficiency to postmenopausal bone loss. N Engl J Med 1984; 311: 1273-1275.

77. Riggs BL, Melton LJ. Involutional osteoporosis. N Engl J Med 1986; 311: 1676-1686.

78. Riggs BL, Wahner HW, Dunn WL, Mazess RB, Offord KP, Melton LJ. Differential changes in bone mineral density of the appendicular and axial skeleton with aging: Relationship to spinal osteoporosis. J Clin Invest 1981; 67: 328-335.

79. Riis B, Thomsen K, Christiansen C. Does calcium supplementation prevent postmenopausal bone loss: A double blind, controlled clinical trial. N Engl J Med 1987; 316: 173-177.

80. Rockwell JC, Sorenson AM, Baker S, et al. Weight training decreases vertebral bone density in premenopausal women: A prospective study. J Clin Endocrinol Metab 1990; 71: 988-993.

81. Rodan GA. Bone mass homeostasis and bisphosphonate action. Bone 1997; 20: 1-4.

82. Rodan GA. Control of bone formation and resorption: Biological and clinical perspective. J Cell Biochem Suppl 1998; 30-31: 55-61.

83. Saag KG, Emkey R, Schnitzer TJ, et al. Alendronate for the prevention and treatment of glucocorticoid-induced osteoporosis. Glucocorticoid-Induced Osteoporosis Intervention Study Group. N Engl J Med 1998; 339: 292-299.

84. Sairanen S, Karkkainen M, Tahtela R, et al. Bone mass and markers of bone and calcium metabolism in postmenopausal women treated with 1,25-dihydroxyvitamin D (Calcitriol) for four years. Calcif Tissue Int 2000; 67: 122-127.

85. Schneider PF, Fischer M, Allolio B, et al. Alendronate increases bone density and bone strength at the distal radius in postmenopausal women. J Bone Miner Res 1999; 14: 1387-1393.

86. Schott AM, Weill-Engerer S, Hans D, Duboeuf F, Delmas PD, Meunier PJ. Ultrasound discriminates patients with hip fracture equally well as dual energy X-ray absorptiometry and independently of bone mineral density. J Bone Miner Res 1995; 10: 243-249.

87. Sinaki M. Postmenopausal spinal osteoporosis: Physical therapy and rehabilitation principles. Mayo Clin Proc 1982; 57: 699-703.

88. Sinaki M, Mikkelsen BA. Postmenopausal spinal osteoporosis: Flexion versus extension exercises. Arch Phys Med Rehabil 1984; 65: 593-596.

89. Sinaki M, Wollan PC, Scott RW, Gelczer RK. Can strong back extensors prevent vertebral fractures in women with osteoporosis? Mayo Clin Proc 1996; 71: 951-956.

90. Snow CM. Exercise and bone mass in young premenopausal women. Bone 1996; 18: 51S-55S.

91. Snow CM, Shaw JM, Winters KM, Witzke KA. Long-term exercise using weighted vests prevents hip bone loss in postmenopausal women. J Gerontol A Biol Sci Med Sci 2000; 55: M489-M491.

92. Snyder A, Zierath JR, Hawley JA, Sleeper MD, Craig BW. The effects of exercise mode, swimming vs. running, upon bone growth in the rapidly growing female rat. Mech Ageing Dev 1992; 66: 59-69.

93. Sowers MR, Galuska DA. Epidemiology of bone mass in premenopausal women. Epidemiol Rev 1993; 15: 374-398.

94. Stevenson JC, Hillard TC, Lees B, Whitcroft SI, Ellerington MC, Whitehead MI. Postmenopausal bone loss:

Does HRT always work? Int J Fertil Menopausal Stud 1993; 38(Suppl 2): 88-91.

95. Swissa-Sivan A, Azoury R, Statter M, et al. The effect of swimming on bone modeling and composition in young adult rats. Calcif Tissue Int 1990; 47: 173-177.

96. Taaffe DR, Snow-Harter C, Connolly DA, Robinson TL, Brown MD, Marcus R. Differential effects of swimming versus weight-bearing activity on bone mineral status of eumenorrheic athletes. J Bone Miner Res 1995; 10: 586-593.

97. Tanizawa T, Imura K, Ishii Y, et al. Treatment with active vitamin D metabolites and concurrent treatments in the prevention of hip fractures: A retrospective study. Osteoporos Int 1999; 9: 163-170.

98. Thiebaud D, Burckhardt P, Kriegbaum H, et al. Three monthly intravenous injections of ibandronate in the treatment of postmenopausal osteoporosis. Am J Med 1997; 103: 298-307.

99. Turner RT, Riggs BL, Spelsberg TC. Skeletal effects of estrogen. Endocr Rev 1994; 15: 275-300.

100. U.S. Department of Health, and Human Services. Physical Activity and Health. A report of the Surgeon General. Atlanta: U.S. Department of Health and Human Services, Centers for Disease Control and Prevention, National Center for Chronic Disease Prevention and Health Promotion, 1996.

101. Weinstein L, Ullery B, Bourguignon C. A simple system to determine who needs osteoporosis screening. Obstet Gynecol 1999; 93: 757-760.

102. Welten DC, Kemper HCG, Post GB, et al. Weight-bearing activity during youth is a more important factor for peak bone mass than calcium intake. J Bone Miner Res 1994; 9: 1089-1095.

103. Weryha G, Leclere J. Paracrine regulation of bone remodeling. Horm Res 1995; 43: 69-75.

104. Wimalawansa SJ. A four-year randomized controlled trial of hormone replacement and bisphosphonate, alone or in combination, in women with postmenopausal osteoporosis. Am J Med 1998; 104: 219-226.

105. World Health Organization. Assessment of fracture risk and its application to screening for postmenopausal osteoporosis. Geneva: World Health Organization, 1994: 1-129.

106. Wu J, Ishimi Y, Yamakawa J, Tabata I, Higuchi M, Fukashiro S. The effects of swimming exercise on bone mineral density in postmenopausal women. J Bone Miner Res 1999; 14(Suppl 1): S385.

107. Yeager KK, Agostini R, Nattiv A, Drinkwater BL. The female athlete triad: Disordered eating, amenorrhea, osteoporosis. Med Sci Sports Exerc 1993; 25: 775-777.

Jan Perkins, PT, MSc
School of Rehabilitation and Medical Sciences
Central Michigan University
Mt. Pleasant, MI

J. Tim Zipple, MS, PT, OCS, FAAOMPT
School of Rehabilitation and Medical Sciences
Central Michigan University
Mt. Pleasant, MI

Nonspecific Low Back Pain

Most people in modern society experience nonspecific low back pain (NSLBP). It is a complicated and poorly understood phenomenon involving the interaction of a wide range of physical, social, and psychological factors. For an individual, the impact may be devastating; for society, the costs are enormous. Although exercise is widely used in the management of NSLBP, experts do not agree on the optimum type or dose of exercise. Furthermore, exercise prescription is complicated by the differences between acute and chronic NSLBP and by the great variability in other management strategies.

NSLBP is an umbrella term that includes pain in the lumbosacral area caused by a variety of somatic (musculoskeletal) dysfunctions. A specific origin of pathology, such as herniated disk, vertebral fracture, congenital malformation, or general medical condition, has not been identified. There appear to be as many different theories about causation as there are structures in the back. Indeed, "the actual origin of the pain is more of a philosophic statement of the training of the practitioner than hard, scientific fact"(3, p. 63). Also, most often "low back pain is not a symptom of a disease or degenerative process... it is a syndrome whose major and sometimes only symptom is pain" (3, p. 68). The best definition of NSLBP may be simply pain experienced in the lumbosacral region in the absence of major identifiable pathology.

This chapter focuses on management of subacute, recurrent, or chronic NSLBP. NSLBP caused by serious pathologies or requiring urgent medical or surgical intervention is not covered, and postsurgical rehabilitation routines are also excluded. For information on typical postsurgical rehabilitation and recovery paths, the reader is referred to other resources (e.g., see reference 19), which should be used in consultation with the healthcare workers caring for individual clients.

Scope

Most people, in all societal groups, will experience at least one episode of disabling NSLBP during their life. Back problems are "the most expensive musculoskeletal affliction and industrial injury and the most common cause of disability for Americans under the age of 45" (7, p. 192). On any given day, up to 28.7% of people interviewed will report some NSLBP (11). One-year **prevalence** rates vary from 3.9% to 63% (36).

Up to a point, NSLBP increases with age. A cohort study that examined individuals at age 14 and used survey follow-up at age 38 found that prevalence of NSLBP increased as the participants aged (with lifetime prevalence at 70% by age 38) and that rates for men and women were similar (23). Many others have noted an increase in NSLBP with age, with a peak between age 45 and 60, followed by a decline in reported pain (17, 25, 36). Perhaps this is attributable to the inherent stability of the spine associated with the loss of elasticity and increased stiffness seen with aging. It may also be that older individuals are more cautious with lifting. The relatively high rates of NSLBP noted in young adults and adolescents in cohort studies and the association with sports in boys (23) matches our clinical experience in which many adults describe a first episode of NSLBP in high school, frequently precipitated by athletic competition, heavy lifting, or trauma.

This cost of managing the minority of NSLBP patients whose acute episode becomes chronic is enormous. NSLBP costs may exceed $50 billion per year in the United States, with about 75% of this arising from expenses related to the minority who become disabled (21, 60, 65).

The average adult can expect between one and three episodes of NSLBP in a year. In most cases, these will resolve spontaneously without producing major disability. In a few cases, pain will continue and lead to chronic NSLBP.

Physiology and Pathophysiology

NSLBP is a symptom with many causes and consequently many pathologies. Here, the focus is on NSLBP for which precise descriptions are impossible. Opinions as to the structure causing pain in the majority of cases have varied over time and may indeed be more reflective of fashions in healthcare and contemporary technologies than actual pathologies.

There is an impressive list of structures with pain receptors that could cause NSLBP, including the anterior and posterior ligaments, interspinous ligament, yellow ligament, posterior annular fibers of the disk, intervertebral joint capsules, vertebral fascia, blood vessel walls, and paravertebral muscles (68, 71). Many experts consider the intervertebral disk to be the most frequently implicated structure in nontraumatic NSLBP, and popular treatments are often based on theories relating to disk pathologies (3). Others disagree and consider the importance of the disk as the source of NSLBP to have been "overrated" (68). It makes intuitive sense to suppose that NSLBP may be caused by multiple anatomic structures that may all respond similarly to the appropriate regimen. An approach that focuses on risk factors and **prevention** rather than pathologies is best.

Clinical Considerations

NSLBP is common in our society. If an individual's medical history includes NSLBP, it is important to determine if it is possibly associated with a more serious pathology that would require further physician evaluation prior to exercise testing and training. A number of risk factors have been shown to be associated with NSLBP; however, the strongest predictor of recurring NSLBP is the length of time between episodes of NSLBP. A number of different diagnostic tests can be used to assess NSLBP; however, these tests generally are not beneficial, unless it is speculated that the NSLBP is associated with a serious pathology. In respect to treatment for NSLBP, noninvasive treatment rather than surgery is strongly suggested in most cases, unless a patient is diagnosed with a disk herniation, where surgical treatment has a high rate of success. As a patient goes through treatment for acute NSLBP, a strong emphasis should be given to secondary prevention to decrease recurrence rates. In addition, because of the high **incidence** of NSLBP in our society, primary prevention is strongly encouraged, especially in occupational, household, and recreational pursuits with a high incidence of NSLBP.

Symptoms and Risk Factors

The individual presenting with NSLBP may complain of localized or generalized lumbosacral region pain of variable intensity, duration, and frequency. Radiating pain with a specific distribution of sensory changes, numbness, or lower extremity weakness can be associated with more serious pathology and can indicate specific tissue involvement. The client may have NSLBP with weight-bearing activities and complain of increasing pain with certain lumbar motions and postures. NSLBP may cause nocturnal discomfort that awakens the client when changing positions in bed. Symptoms are usually decreased with rest and anti-inflammatory medication. Any client presenting with symptoms such as those described in figure 26.1, or with new undiagnosed symptoms, chest pain, heart palpitations, fever, shortness of breath, unexplained weight loss, hernia, or unremitting spinal pain that is not relieved by rest, should consult with a physician prior to initiating an exercise program.

There are a number of established risk factors for NSLBP and disability. NSLBP is lower in those with greater back extensor muscle endurance, and people who have had previous episodes of NSLBP are more likely to experience additional back pain. This risk increases as the interval since the last episode shortens. People with recurrent or persistent back pain have decreased flexibility of hamstring and back extensor muscles and lower trunk muscle strength. The best predictor of an episode of back pain (and the only one sufficiently discriminative to be valuable in job selection) is a history of previous episodes, with risk increasing as time since the previous episode decreases (25, 58, 59). Other factors with back pain include obesity (29), smoking (27, 30), and motor vehicle driving (27). Sedentary occupations are also a possible risk factor (68). Heavy lifting, and lifting with twisting regardless of the weight, have continued to be considered risk factors for back pain (28, 47, 56). A relatively recent addition to hypothesized risk factors is the suggestion that there may be a genetic predisposition to degenerative disk disease (41).

It is important to note that the associations do not necessarily imply causation. For example, although there is a modest correlation between back pain and obesity, some experts believe that the association is unlikely to be causal (29, 31). Instead, obesity may play a part in simple NSLBP becoming

- Nocturnal back pain
- Back pain with constitutional symptoms: extreme fatigue, nausea, vomiting, diarrhea, fever, nonexertional sweating
- Unexplained weight loss
- Back pain with extreme weakness in one or both legs, numbness in the groin or rectum, bowel or bladder incontinence, difficulty initiating or stopping, or increased/decreased frequency of urination.
- Sacral pain in the absence of traumatic history
- Constant, intense back pain that does not vary with exertion or activity
- Back pain not relieved by rest or change in position
- Back pain that is relieved by leaning forward in a seated position (pancreas)
- Back pain that is associated with multiple sites and joints (rheumatoid arthritis)
- Severe, unremitting, localized back pain of 2 weeks duration
- Back pain associated with eating meals
- Rapid or weak pulse accompanied by decrease in blood pressure
- Back pain accompanied by a pulsing sensation or palpable abdominal pulse
- History of current gynecological disorders
- Nonorganic (psychogenic) signs such as overreaction to superficial palpation, minor axial spinal loading, glovelike/stockinglike sensory disturbances, or pain with "log-rolling" which minimally rotates the spine
- Any of a variety of sexual dysfunctions

FIGURE 26.1 "Red flag" indicators in nonspecific low back pain evaluations.

Adapted, by permission from C.C. Goodman and T.E.K. Snyder, 2000, *Differential diagnosis in physical therapy*, 3rd ed. (Philadelphia: W.B. Saunders Company), 454-456.

chronic or recurrent. The association with obesity is also substantial only in those with a **body mass index** greater than 29 (15).

Psychosocial and work environment factors are more predictive of back pain and disability than physical examination findings or mechanical stress at work (20, 53, 60, 65). Psychological distress, dissatisfaction with employment, low levels of physical activity, poor self-rated health, and smoking status are, along with poor spinal movement, associated with persistent back pain after an acute episode (60). For many years, there have been calls for a shift from treatment of pain to management of disability (65). This may be the key to cost control, because disability costs seem to be more linked to psychosocial determinants rather than physical attributes. Failure to address these issues will lead to inadequate management of individual cases and continued high costs to society (21).

The natural history of acute NSLBP is fairly well defined. For any given acute episode, resolution is anticipated in about 50% of patients within a month and for about 90% within a year (56). Because of this fairly positive outlook and its independence of treatment approach, expensive intervention is frequently reserved for the minority of people who do not improve in 2 months. Recently, some experts have reevaluated this optimistic outlook. The risk of additional episodes following a brief acute episode of back pain is 60% to 75% within a year (56, 62).

Diagnostic and Laboratory Evaluations

Diagnosis of specific low back pathology is fraught with difficulty. Despite the use of highly sophisticated tests, experts are often unable to give definitive explanations. Radiographic imaging will identify loss of disk space, malalignment (e.g., **scoliosis, spondylolisthesis**), and osteoarthritic changes such as **osteophytes** or **stenosis.** Often these do not correlate with the individual's signs and symptoms. **Magnetic resonance imaging (MRI)** is better at soft tissue examination than plain X-ray, but neither technology can identify the pain receptors responsible for the reported pain.

Most people who see physicians for back pain do not have serious pathology. In primary care, approximately 1% of patients will have back pain from systemic causes. Careful screening for "red flag" symptoms indicating serious pathology is an essential part of primary healthcare management (6, 7, 9) (see figure 26.1). For the vast majority of cases where there is not an underlying systemic disease or clearly definable injury, diagnosis may be less important than management.

The high incidence of "abnormal" findings makes routine use of MRI problematic (6, 7). By midlife, a high percentage of asymptomatic individuals will demonstrate MRI findings, with percentages increasing rapidly with age (7, 56). This is also true for other forms of diagnostic imaging such as X-ray, **computed axial tomography (CAT scan),** and **myelography** (figure 26.2). Although some radiographic abnormalities are strongly associated with back pain, many others have only a questionable relationship with pain or are very unlikely to be the source of a patient's pain (49, 63, 68). Two patients with similarly abnormal radiographic images of the lumbar spine may have vastly different clinical presentations and illness experiences.

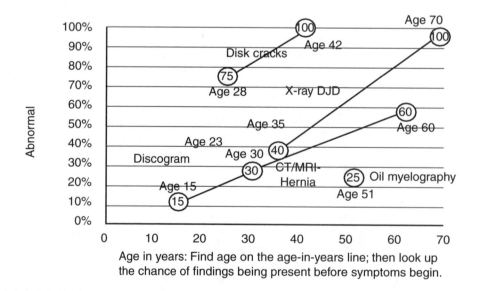

FIGURE 26.2   Disk findings in normal subjects. DJD = degenerative disk disease; CT = computed tomography; MRI = magnetic resonance imaging.

Adapted from Bigos, S.J. and Davis, G.E., "Scientific application of sports medicine principles for acute low back problems." *Journal of Orthopaedic Sports Physical Therapy* 24: 192-207, with permission of the Orthopaedic and Sports Sections of the American Physical Therapy Association.

Treatment

There are many medical management strategies for NSLBP. Common medical management includes a wide range of medications; prescription of exercise or passive modalities such as heat, massage, or **spinal traction;** facet joint injections; and surgeries such as **spinal discectomy, spinal decompression,** and **spinal fusion.** Current emphasis in primary care management of NSLBP is early return to activity, avoidance of needless surgery or use of unnecessary diagnostic tests, and ultimately cost-effective medical management of back pain.

Surgical rates and types of surgery vary widely across geographic areas. Outcomes show no corresponding variation, and there is a lack of agreement on indications for surgery or on successful outcomes. There has been a dramatic increase in surgical rates but no decrease in disability (14). There is little evidence from acceptable clinical trials that the expensive option of surgery offers any benefit over nonsurgical management in the majority of cases of back pain (48, 50). Unequivocal disk **herniation** is an important exception to this general statement. When strict criteria are adhered to in selecting surgical patients, the success rate can be high (67). Unfortunately, the wide variation in surgical rates nationally and globally attests to the lack of strict application of criteria by all practitioners.

The management of an isolated episode of acute back pain is less controversial than that of subacute, recurrent, or chronic pain. Current medical practice advocates conservative care for acute back pain in the absence of any red flag findings. Bed rest, if used at all, is limited to 24 to 48 hr; passive treatment modalities (such as hot packs, **transcutaneous electrical nerve stimulation, ultrasound,** and traction) are used sparingly; and early resumption of normal activities with or without additional exercise training is advocated (6, 7, 16, 67).

The limited studies looking at primary and secondary prevention (discussed subsequently) make it possible to suggest that some intervention, preferably in a community or worksite setting, may be beneficial in avoiding recurrence of back pain after an initial nonspecific injury. Low-stress aerobic activity is usually considered safe within 2 weeks, and "conditioning," especially of back extensors, is begun thereafter (6). Well-planned general exercise programs, continued as long as there is evidence that the patient is making functional progress, are suggested as part of rehabilitation (18). Few patients will have insurance coverage for this type of program in a rehabilitation center, because the present healthcare system focus is on short-term secondary prevention and tertiary prevention. Instead, individuals may need to continue exercise programs under the guidance of fitness professionals at health clubs and fitness or wellness centers. Many community wellness centers will have an important role in both secondary and primary prevention of nonspecific back pain.

Primary prevention is avoidance of a condition for which a person is at risk. If we can intervene before an injury or disease develops and stop it from appearing, this would be the optimal form of preventive health care. For traumatic back pain, frequently caused by vehicle accidents or falls, primary prevention includes standard safety precautions in household, recreational, and occupational situations (3). Safety equipment such as seatbelts should be used in all appropriate situations, and "watchful caution" should be practiced during activities that place individuals at risk of falling or encountering moving vehicles.

To summarize, primary prevention strategies for back pain include the following:

— Use safety equipment in work and leisure activities.

— Address risk factors of smoking, poor general fitness, obesity, stress, and poor seating.

— Perform balanced exercise programs that include both spinal flexion and extension.

— Avoid long period in one position—take breaks to move the spine out of fixed positions and balance postures.

— Avoid lifting with twisting.

The more common nontraumatic low back injuries are thought to be the result of a series of cumulative events in an individual with risk factors such that a relatively minor incident will precipitate symptoms (3). The best options for primary prevention of nontraumatic back pain come from addressing risk factors such as smoking, obesity, psychosocial stress, and poor seating for those exposed to occupational risk factors such as truck driving (3). Any early signs of dysfunction such as joint stiffness, minor aches and pains, or difficulties straightening the spine can minimize progression. Physical activity that promotes both cardiovascular and musculoskeletal fitness is beneficial, as is a balance of spinal flexion and extension activities and avoidance of prolonged loading of the spine in either flexion or extension. The spine is not meant for static positioning but for movement, and sustained postures or repetitive movements in any direction should be balanced with movement in the opposite direction.

In secondary prevention, treatment is used early in the course of a condition to cure that condition or prevent or slow its progression. In back pain, the aim is to prevent recurrences and to prevent acute and subacute back pain from becoming chronic. Secondary prevention strategies for back pain include the following:

— Catch problems early before an injury becomes disabling.
— Avoid or minimize bed rest.
— Encourage early return to activity, even in the presence of some pain.
— Keep use of passive modalities to a minimum, or use them to assist with a more active program.
— Be willing to adapt program to individual needs.
— Use behavioral strategies to encourage participation.

Because the majority of acute back pain episodes will resolve to a greater or lesser degree within a few weeks, emphasis is best placed on preventing progression to chronic pain and preventing recurrences. Emphasizing early return to activity even with some continuing pain; weaning from pain medications; avoiding dependence on passive modalities such as heat, rest, transcutaneous electrical nerve stimulation, or massage; and strongly emphasizing self-management are the preferred management strategies (3, 6). After recovery, a program that incorporates the measures suggested for primary prevention with extra consideration of the individual's particular at-risk activities will enhance outcomes.

Graded Exercise Testing

NSLBP is a diagnostic category that does not in itself indicate a need for a graded exercise test (GXT). Maximal or submaximal testing, although useful for prescription, is not required in general clinical practice unless history indicates possible coronary artery disease (52) or other medical conditions that would normally require a GXT before formulating an exercise prescription. American College of Sports Medicine (ACSM) guidelines suggest that older adults (men ≥45 or women ≥55) should have medical clearance including exercise testing and follow-up if vigorous exercise is planned (1). ACSM suggests similar precautions for those at increased risk for cardiovascular events and for those with known pulmonary, cardiac, or metabolic disease (1). Individuals with NSLBP should be screened routinely to identify those in any of these categories. Practical application 26.1 describes client–clinician interaction for patients with NSLBP.

Medical Evaluation

Medical evaluation should have cleared the individual referred with NSLBP for major pathologies, but a clinical exercise physiologist should be alert to the possibilities that serious pathologies have been missed. Figure 26.1 lists key "red flag" findings that indicate when medical evaluation may be required.

PRACTICAL APPLICATION 26.1

Client–Clinician Interaction

Information for Individuals With NSLBP

— Radiographic evidence of pathology doesn't correlate with level of spinal pain or disability.
— Studies provide evidence that adherence to a specific, progressive exercise program reduces the incidence and frequency of episodic NSLBP.
— Ergonomic adaptations and postural awareness reduce the incidence of episodic NSLBP.
— Minor lifestyle changes will permit compliance with self-management strategies for episodic NSLBP.
— Improvement in pain will not be instantaneous. Muscle strength can take 4 to 6 weeks to improve, and loss of pain may not be noted until the spine is stable and has time to adjust to muscle and ligament changes.
— After the initial rehabilitation stage, which may include progressive resistance exercises, daily exercise may entail simple aerobic exercises such as walking or light stretching.

Contraindications

A brief period of rest is commonly used for the first few days of an acute flare-up of back pain. During this period, GXT is inadvisable. There are also practical concerns relating to the patient's pain pattern and presentation that can affect GXT administration and evaluation. Because both upper extremity and lower extremity ergometers and treadmills require coordinated trunk mobility and stability, pain may interfere with the individual's ability to perform the test or reach maximal exercise levels. Selection of the means of testing should be tailored to the individual. For some, the seated position could be most painful and ambulation preferred, whereas for others a stationary bicycle would provoke less pain than treadmill testing. In any case, pain may prevent the person from reaching maximal exercise levels, and aggressive testing protocols could produce considerable posttesting soreness. A submaximal testing protocol using the exercise modality least likely to cause a flare-up in that person's pain may be most appropriate for individuals who report pain that is easily aggravated by activity.

Recommendations and Anticipated Responses

There are no specific GXT protocols for individuals with NSLBP. However, as discussed previously, if an individual is currently having an episode of NSLBP, GXT is not recommended because the person is unable to give an appropriate effort. When one's medical history reveals previous NSLBP, the mode of exercise testing should be carefully determined to avoid further exacerbation.

Exercise Prescription

Most programs of exercise for individuals with back pain include a combination of several forms of exercise and educational advice regarding lifestyle factors and general back care. With a single episode of back pain, it is not clear whether exercise will have any effect on the anticipated natural history of that episode. Exercise intervention after back pain has occurred is aimed at reducing risk factors and minimizing recurrences. With chronic and recurrent back pain, the case for more aggressive intervention is stronger. Increasing function and decreasing severity and frequency of back pain episodes should be the goal.

The clinical exercise physiologist may need to emphasize endurance over strength in the early training, adjusting repetitions and resistance to reflect this emphasis. Exercise will usually start with low levels of exercise done frequently and will progress through a system of exercising to quota. The focus should be on function rather than pain. A **periodization** schedule based on estimated maximums from low-intensity testing may be appropriate. In the absence of specific restrictions imposed by pathology, the exercise program should be designed to correct specific impairments found in initial comprehensive evaluation (55). Practical application 26.2 reviews the literature about exercise and low back pain.

Adherence to exercise is a problem for many people with recurrent or chronic NSLBP. Although function can improve quickly, it may take more than

continued

PRACTICAL APPLICATION 26.2

Literature Review

Exercises have been used to manage back pain for more than 100 years, yet there is a lack of well-controlled trials on the benefits of exercise. "Some form of exercise is probably the most commonly prescribed therapy for patients recovering from low back pain" (67, p. 363). Types of exercise favored have varied widely (38). Lumbar flexion exercises, usually based on the Williams flexion routine (69, 70), have had their vogue, as have hyperextension exercises focusing on increasing paravertebral muscle strength and endurance. At present, a balanced approach involving strengthening most spinal musculature and lower extremity muscles is arguably the most popular strategy, usually in combination with stretching exercises and recommendations for some form of aerobic conditioning.

When prescribed based on an individual evaluation, exercises are hypothesized to have multiple beneficial effects, including reduction of disk herniation, improved joint mobility, strengthening of weakened muscles, and stretching of adaptively shortened ligaments, capsules, and muscles. Increasingly, behavioral approaches to management are also included to address other aspects of the back pain syndrome than the simple physical limitation.

Popular approaches in rehabilitation have included variations on the Swedish Back School approach (72), which mainly uses education on back care; functional restoration using individually prescribed exercise combined with aggressive disability management (52); and a wide range of manual therapy approaches. Although these are based on widely different theories of causation, many use surprisingly similar positions and techniques (13). (See table below for selected examples of manual therapy philosophies frequently encountered.) The McKenzie approach, which uses treatments based on symptom response to movement and a theory of disk pathology (44), has become one of the more common manual therapy approaches in rehabilitation. Healthcare and wellness practitioners will also encounter individuals who have used a wide range of alternative approaches to therapeutic exercise and body-work. Selected approaches that are frequently encountered are listed in *Therapeutic Exercise Approaches* table on next page.

Manual therapy philosophies	Founder/history	Treatment philosophies
Chiropractic approach	Bone setters began treatment of malalignments in the late 1800s.	Treatment of subluxations of the spine by manipulations. Many chiropractors also use a variety of gentle techniques and joint oscillations to realign spinal segments.
Osteopathic approach	Founder: A.A. Still, a medical doctor in the United States in 1874.	Treatment of somatic dysfunctions with a variety of gentle muscle contractions and joint mobilizations, as well as high-velocity, low-amplitude thrust techniques.
Australian approach	Accredited to Geoffrey Maitland, a physiotherapist from Australia.	Treatment emphasis on subjective complaints and quality of pain, as well as behavior of pain and response to positions and mobilization techniques. Advocates use of comparable signs (pre–post treatment assessment) but avoids "diagnostic titles."
New Zealand approach	Accredited to Robin McKenzie and Stanley Paris, two physiotherapists from New Zealand.	Approach emphasizes self-management strategies. Three predisposing factors to low back pain: sitting postures, limited lumbar extension, and predominance of flexion activities in society. Divides back problems into postural, dysfunction, and derangement syndromes.
Norwegian approach	Accredited to Oddvar Holten, a physiotherapist in Oslo, Norway. Introduced in the United States by Freddy Kaltenborn and Olaf Eventh.	Uses a biomechanical approach to assess mobility of spinal and extremity joints. Uses Maitland and Kaltenborn mobilization grading system to normalize passive joint mobility and improve function.
Craniosacral therapy	Accredited to John Upledger, an osteopathic physician in the United States.	Treatment involves manipulation of the cranial sutures and sacrum to adjust the flow of cerebral spinal fluid, which oscillates caudally and cranially (craniosacral rhythm). Disruptions in the rhythm lead to a variety of autonomic and musculoskeletal dysfunction.
Variety of soft tissue mobilization approaches	Myofascial release: Fred Mitchell, Sr, DO Rolfing: Ida P Rolf, PhD Hellerwork and Soma: Joseph Heller and Bill Williams, PhD Bindegewemassage: Elisabeth Dicke Swedish remedial techniques	Soft tissue approaches run the gamut of gentle myofascial stretching to vigorous mobilization of deep tissues and internal cavities. Realignment and restructuring of tissues lead to normalization of posture and muscle function.

continued

Therapeutic exercise approaches	Founder/history	Treatment philosophies
Pilates	Joseph Pilates, developed while working as a nurse in World War I.	Use of specialized equipment to promote balance, strength, proper posture, and agility.
T'ai chi	Passed on from many Chinese generations including Wu style, Yang style, Ch'en style, and Chuan style.	Use of slow, balanced movements of extremities around stable trunk through a series of 108 postures. Uses concept of power centering during standing postures, balancing ying and yang. Slow form for strengthening and meditation, fast form used as defense.
Feldenkrais	Developed by Moshe Feldenkrais, a physicist, judo expert, and athlete.	Observed his own unnatural movement patterns and created a system of movement awareness exercises. Recognition of mental and emotional activity that perturb all aspects of human performance. Uses visual, auditory, and kinesthetic cues to alter movement.
Alexander technique	Frederick M. Alexander, developed during late 1800s.	Use of specific movements of the body to promote bodily awareness and health. Self-awareness of harmful movement patterns allows subject to correct and move in harmony with breathing pattern.
Trager approach	Milton Trager, MD, who was a former boxer and acrobat.	Concentration, repetition, and refinement of movement, targeted through unconscious mind. Use of active effortless movements called "Mentastics."
Yoga	Origins in Egypt and India 5000 years ago.	Uses a series of stretching positions that liberate the natural flow of energy in the body, with slow meditative assumptions of the poses and strong emphasis on deep relaxation. Improves flexibility, posture, and body awareness. Improves circulation.

One approach to using exercise in a clinical setting focuses on initially establishing normal movement patterns, then moving on to stretching tight structures, strengthening with early progression through increasing repetitions, and adding weighted training exercises (18). Great emphasis is placed on form and maintenance of good muscle control throughout the entire range of movement. This approach is compatible with the functional restoration philosophy of treatment and, in our opinion, is a logical dynamic approach to back care.

There have been some excellent attempts to validate exercise selection and link experimental research and clinical practice. One interesting technique combines electromyography and computer modeling to estimate tissue loads during activities (26, 42, 43). In an article specifically written for clinicians, McGill (42) synthesized several studies and made specific exercise suggestions. While emphasizing that any program must be tailored to the specifics of the individual patient, he suggested that the "ideal" early exercise program for a patient with back pain would avoid loading the spine through range of motion yet would provide sufficient challenge to allow a muscle conditioning effect. Whereas a healthy athlete is able to load the spine through range, a different approach is suggested for the less trained individual. Other treatments that are currently in vogue actually unload the spine during early rehabilitation, with progressive loading and resistance training then added as recovery occurs. Examples include some programs incorporating aquatic therapy and use of unloading harnesses for treadmill ambulation (57).

2 months of training for individuals with chronic back pain to experience significant pain relief (38, 39). Discontinuing exercise after the completion of a program is as common among individuals with back pain (18) as it is in the general population. Support, encouragement, and easy availability of follow-up programs in the community may be especially important until an exercise habit is well established and whenever a program has to be restarted after a lapse for any reason.

Supervision is valuable, particularly in the early stages, to ensure use of correct form and encourage exercise adherence. In one study, chronic back pain patients were randomly assigned to either supervised or unsupervised training for flexibility, muscle strength, and aerobic conditioning according to a periodization schedule (55). After 6 months, there were marked differences in the experimental groups. The group given independent exercise averaged only 31.95 of a planned 96 sessions. The other group, assigned a certified strength and conditioning specialist who supervised them, completed an average of 90.75 sessions. Because long-term follow-up was not done, it is not known whether benefits and exercise participation were continued, but this work supports the commonsense idea that supervision enhances adherence.

Special Exercise Considerations

Once a person has been cleared by a medical professional, there are few contraindications to an exercise program. However, many patients with recurrent or chronic pain may be so deconditioned and fearful of exercise or movement that progression should be slow and initial exercise levels low.

Special exercise considerations for patients with NSLBP include the following:

— The patient should obtain medical clearance for exercise.

— The exercise specialist should monitor for red flag findings.

— Deconditioning may be greater than in sedentary healthy individuals.

— The exercise specialist should use caution if loading through the spine and should consider unloading in some cases for pain management.

— Progression should be slow and initial exercise levels low.

— Smokers may need a slower progression.

One caution that should apply is to avoid overstressing or overtraining. With overtraining, large muscle groups are allowed to compensate or substitute for smaller, deconditioned muscle groups. An emphasis on form and evaluation of postexercise response is important in determining exercise level. The goal is to provide adequate stress for a training effect but to avoid stressing any tissue beyond tolerable levels and to do this without using abnormal mechanics (24). The clinical exercise specialist must remember that a person with NSLBP may have been very inactive for a considerable period of time and so will need to start with a lower intensity program than a sedentary but healthy individual. Persons with chronic NSLBP have been shown to have selective atrophy of type 2 muscle fibers in the back muscles. This has consequences for fatigue resistance and may help explain the poor back muscle endurance seen in individuals with back pain (51).

In general, it is best to avoid exercises that provide compression loading of the spine in injured individuals until late in rehabilitation. Early exercise focus should be on safe performance of rhythmic exercises with low resistance and an emphasis on correct form that avoids substitution patterns. Initially, exercises such as squats or calf raises may be done with less than full body weight, and then they can progress to full weight-bearing and finally to additional resistance. For most individuals, the use of hand-held weights may be more appropriate for adding resistance than a bar behind the neck and across the shoulders for exercises like lunges, squats, or calf raises. Similarly, leg presses that rely on adding weight through the shoulders and spine should usually be avoided in early treatment.

Unless they can be performed with supervision, dead lifts should be avoided or used with extreme caution with patients with a history of back pain. However, some clinicians do use these in their rehabilitation programs for patients with back pain (18). The argument in favor of dead lifts states that they are an important means of strengthening erector spinae. Correct training in technique can help avoid future exacerbations and permit retraining for stressful activities. This argument has validity, but the exercise remains controversial because of the high potential for reinjury. Unless an exercise advisor is confident of his or her ability to instruct individuals in this lift and is certain that correct form will continue to be used by the client, it is best left out of an exercise program.

Given the association between smoking and back pain, caution should be used in exercise prescription with smokers. Although most published work takes a judgmental approach to the issue of smoking or merely documents the association, it might be better to acknowledge the fact that some individuals will not be able to quit and to use a slower exercise progression for smokers than nonsmokers (18).

Exercise Recommendations

Most individuals with NSLBP should finish rehabilitation with a general exercise prescription that matches ASCM guidelines for healthy adults (1). Adoption of these is appropriate in the early stages of exercise training to allow for common problems such as greatly decreased exercise tolerance, lack of flexibility, poor neuromuscular coordination, and pain with movement or loading through the spine. Practical application 26.3 includes general guidelines for prescribing exercise for the population with NSLBP.

Cardiovascular Training

Aerobic exercise is often used in rehabilitation of NSLBP, even though evidence is not conclusive (4, 10). Despite this, most experts still recommend aerobic exercise with subacute and chronic back pain because of its known benefits in other areas of health, hypothesized psychological benefits, or the theoretical rationale with back pain. General fitness is considered desirable for many reasons and has several hypothesized benefits in general pain management.

Given known benefits of aerobic exercise for general health and mental well-being, it is reasonable to include a graduated exercise program with an individualized exercise prescription. Maximal or submaximal testing, although useful for prescription, is not required in general clinical practice un-

less history indicates possible coronary artery disease (52). Any aerobic activity that interests the client may be used, but in general, high-impact activities such as jogging or exercise requiring sustained spinal flexion (as in some bicycling) are considered poor choices. However, the mode of activity that will aggravate NSLBP differs for each individual. As individuals with NSLBP start an aerobic exercise program, they should start slowly and be sensitive to activities that precipitate NSLBP. Otherwise, aerobic exercise training should follow the ACSM guidelines for promoting health and fitness. These recommendations are meant as a starting point for clients with NSLBP that would allow progression in resistance for strengthening exercises and cardiovascular fitness. People with NSLBP may need to initiate shorter duration training sessions at lower intensities until low-level exercise tolerance is achieved. Evidence of progressive low back pain during and after exercise, residual postexercise muscle soreness lasting greater than 48 hr, impairment in tolerance for activities of daily living, or evidence of neuropathy requires reevaluation by a qualified healthcare practitioner before exercise is resumed. Modifications in the parameters of resistance or fitness training will be prescribed by the appropriate healthcare provider.

Resistance Training

Available evidence suggests that the more important component of resistance training, at least for the typical NSLBP and for prevention, is endurance (5, 37, 43). **Dynamic endurance** may be more critical than **static endurance** (45). A well-designed program will consider both strength and endurance,

---

PRACTICAL APPLICATION 26.3

Exercise Prescription Summary

— Mode: Aerobic conditioning of large muscle groups may include walking, hiking, cycling, bicycling, cross-country skiing, aerobic dance, rowing, stair climbing, swimming, or in-line skating. Strength-training resistance may come from standard resistance machines, pulley resistance machines, rubber tubing or bands, dumbbells, or the use of gravity for resistance.

— Frequency: For aerobic conditioning, 3 to 5 days per week submaximal training is recommended. For strength training, submaximal and maximal resistance training should occur two to three times per week.

— Intensity: Aerobic training intensity should be targeted at 55% to 90% of maximal heart rate or 40% to 85% of maximal oxygen uptake reserve. Early stage strength-training intensity should start with submaximal resistance loads using higher repetitions of 20 to 30. Late stage strength training should include maximal resistance for 8 to 12 repetitions.

— Duration: Aerobic conditioning should continue for 20 to 30 min per session, while strength training performed at proper speeds may require 30 to 60 min of training time.

with literature and clinical expertise supporting a somewhat heavier emphasis on endurance. Many individuals with back pain have difficulty tolerating positions for more than a brief period of time. Poor endurance can predispose them to injuries at relatively low loads. Particularly if sustained positions are required for vocational or avocational activities, the clinical exercise physiologist should consider training for these specific positions (18).

Most programs encourage strengthening of back and abdominal muscles. To adequately strengthen all muscles involved, a variety of exercises are required. The following specific suggestions are from the series of articles McGill and colleagues (26, 42, 43) that focus on safety through minimizing spinal loading while providing adequate stimulus for muscle training. Based on **electromyography**

studies for iliopsoas muscle activity and intervertebral disk pressure, there is no evidence to support the use of bent-knee sit-ups over straight-leg sit-ups. Instead of sit-ups, various types of "curl-ups" or "crunches" are suggested to strengthen rectus abdominis muscles, including curl-ups with one leg straight and hands used to maintain a neutral lumbar curve (figure 26.3*a*). Others believe that the individual performing crunches should independently stabilize the pelvis to promote lumbopelvic neuromuscular control. This neutral curve may be most important in early recovery.

For strengthening the lateral oblique muscles, the horizontal side support exercise shown in figure 26.3, *c* and *d*, is suggested (26, 42, 43). It activates these muscles without causing high lumbar compressive loading. It has the additional advantage of

FIGURE 26.3   Strengthening exercises commonly used in low back rehabilitation.

training quadratus lumborum, which is considered important in lumbar stabilization.

Back extensions are suggested for strengthening the erector spinae muscles (18). Hyperextensions (of back with fixed legs or legs with fixed spine) do not produced any additional benefit over extension to neutral in a postsurgical population but can cause transient back pain in one third of patients (40). Evaluating individual patient responses to end-range exercise may help minimize problems and increase responsiveness to treatment. It is import to avoid loading the spine through range of motion in patients beginning a program (42, 43).

Extension of both back and legs simultaneously in prone lying (figure 26.4) is often suggested (7), but high lumbar spine compression is produced (42, 49, 68). Instead, back extensors can be strengthened in other ways. Alternatives include resisted back extension with fixed pelvis and initially less demanding exercises that involve extension of one arm, one leg, or both in kneeling (figure 26.3, *e* and *f*). Equipment that isolates the lumbar spine from pelvis and legs may produce greater strength improvement than more traditional Roman chair exercises (46, 54). However, most programs have been done without such sophisticated equipment, and lack of access to equipment should not be a reason for avoiding lumbar extension training. A simple exercise to start a program that does not need special equipment and that could be used to teach form is the "good morning exercise" done with the pelvis blocked by a bench or table (figure 26.3*g*).

Upper extremity and lower extremity muscle strengthening should be part of any general prevention program. In the upper extremities, latissimus dorsi exercises are important because this muscle is involved in back protection and some movement initiation (18). There are anatomical connections linking lumbodorsal fascia with gluteals, hamstrings, and latissimus dorsi (8), linking them to overall back function. Exercises such as the lunge and squat (with hand-held weights providing resistance) provide training for trunk stabilization as well as desirable lower extremity strengthening (18).

We concur with the suggestions from the literature given previously but would also consider spe-

cific rotator muscle strengthening. As suggested previously, a potential source of back pain may be the selective loss of strength and endurance of the smaller muscles of the low back such as the rotators and multifidi. Specific exercises for these muscles may be helpful. Figure 26.3*f* shows an exercise that will encourage activation of the thoracolumbar rotators, but this exercise may allow substitution of the larger muscle groups. A simple seated exercise (later progressed to standing to promote lumbo-pelvic coordination) would use a pulley, rubber band, or tubing as resistance (figure 26.5).

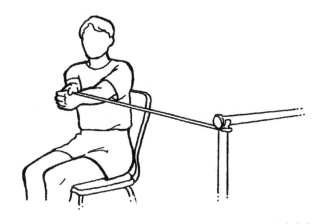

FIGURE 26.5 Seated resisted trunk rotation exercise.

Range of Motion

General recommendations for flexibility exercises are impossible to make (42, 43). Instead, a stretching program must be developed based on the patient's history and examination findings. The limited data available do not support the view that greater flexibility of the spine prevents injury. It is important to maintain adequate flexibility at hips and knees for lifting, and if stretching of the spine is done, it is probably best done in an unloaded position (42). If there are deficits in lower extremity flexibility, these should be addressed individually. Spinal flexibility should not be emphasized until the person has the muscular strength and endurance to control a mobile spine (43). We recommend starting with low-resistance, high-repetition strengthening to re-educate neuromuscular control of increased spinal movement that occurs with stretching.

Individuals with a history of back pain have reduced back and hamstring flexibility (5). This finding is associated with recurrence or persistence of NSLBP. On the other hand, those with very mobile spines are at increased risk of experiencing NSLBP during the following year (5). Apart from this, there

FIGURE 26.4 Lumbar hyperextension exercise.

is no demonstrated link between spinal flexibility and back pain.

Most programs incorporate flexibility exercises into their exercise routines. In the absence of clear evidence, it seems reasonable to continue to do so unless hypermobility or instability is a particular problem. When in doubt, it is best to have the person evaluated for abnormal flexibility before advocating spinal stretching exercises. Nonspinal stretches, usually calf muscles and hamstrings stretches (18, 61), are part of most back programs. Other areas and muscles that are frequently tight include the iliotibial band, hip adductors, hip flexors, and quadriceps. Specific muscle length tests should be admin-

istered by qualified individuals to establish an individualized stretching program.

A reasonable approach would be to evaluate for tightness and design an exercise program that includes stretches to correct any impairments noted. A few commonly used stretches for the low back and lower extremities are shown in figure 26.6. Many variations are equally suitable. Any exercise that involves repeated flexion should be balanced with stretches into extension. Caution should be used with individuals who have known disk pathology, because repeated flexion exercises can shift disk nuclear material posteriorly (46). Passive hyperextension stretches (such as pushing up from prone

FIGURE 26.6   Stretches commonly used in low back rehabilitation.

by using the arms with the hips staying in the floor) may be a better option (65). A balance in stretching movements and inclusion of extension as well as flexion stretches are important.

Frequency

Only continued adherence to an exercise program is likely to offer lasting benefit (40). In a study with intensive training of people with chronic back pain, all subgroups of studied patients improved with the 3-month training program, but a year later only those who continued training at least once a week had remained significantly improved (40). Training at least twice a week is usually suggested (40, 52). Given the population targeted in this chapter, two to three sessions a week is most appropriate for a maintenance program. In early rehabilitation, daily exercise is often suggested with frequency decreasing as intensity increases and exercise tolerance approaches population norms.

Intensity

There are few guidelines specific to the management of NSLBP regarding intensity selection. Those available usually suggest that exercise needs to be more intense than that normally provided for patients. Self-selected intensity or exercise to pain tolerance often leads to inadequate exercise levels. Although pain may not improve for several months in many people, an intensive program will result in greater functional and psychological benefit than a less aggressive approach. Others also advocate a quota approach to prescribing exercise intensity to prevent underexercising and suggest using operant-conditioning behavioral tactics (32-35, 55).

In setting exercise intensity, we suggest a trial and error approach, such as the DeLorme method of determining maximal resistance, to select resistance weights that allow 20 to 30 repetitions with good neuromuscular control in a pain-free or minimal range of motion. This initial program will promote endurance and control of movement. As the person progresses, resistance should increase while the number of repetitions decreases to 8 to 12, compatible with ACSM strength training guidelines (1, 2).

Duration

Although the ACSM recommends 20 to 30 min of aerobic exercise per session, weight training may require 30 to 60 min per session to complete a prescribed program. The total resistance training program will include the intermittent rest periods between intensive sets to recuperate.

Conclusion

NSLBP is one of society's most common problems. Details of suitable exercise programs are limited, but research shows that most people with nonspecific back pain can use exercise to restore function and decrease disability. Programs should be initiated with modifications to accommodate individual impairments and should progress to maintenance programs that are as close as possible to those suggested for general fitness with healthy populations.

Although individuals with back pain may need extra encouragement to begin and continue a general exercise program, those who are able to do so can hope for considerable improvement in function and quality of life. Reassurance, education, and encouragement with self-management are important components of care. Individuals are too rarely given this information, and open communication with individuals who have back pain can be one of the key aspects of management. For the vast majority of individuals, their back pain is nonspecific and can be best managed with conservative yet active treatment strategies.

Case Study 26

Medical History

Mrs. AB is a 28-year-old Caucasian bank worker. She has had recurrent back pain since the age of 16. At that time she had an awkward fall while playing softball. She was taken to her local emergency department, where she was given the diagnosis of "muscle strain" and treated with painkillers and muscle relaxants. Although her back improved very quickly, she feels that the pain never completely resolved. As she continued through high school she noticed that while she had continuing periods of pain they seemed to be less severe when she was physically active.

continued

At age 21, she experienced another acute episode of back pain that began suddenly when she sat down on a couch. This time she did not seek medical help but treated herself with over-the-counter pain medication. The pain slowly resolved over several months.

Diagnosis

Two years ago on July 4th weekend, she was camping with her husband and daughter. As she helped her husband pick up the tent she had a sudden onset of severe back pain. She saw both her physician and chiropractor for treatment. Her X-ray results were "normal," but she was told her MRI showed a bulging disk between the fourth and fifth lumbar vertebra. Her pain was in her back and right buttock. It slowly improved but continued to bother her for the next several months.

Exercise Prescription and Program

In October, her chiropractor suggested she start fitness exercise. The previous January she had joined her town's wellness center but her attendance had been sporadic. She had completely stopped working out when she injured her back, and the manager of the facility had put her membership on hold while she was receiving treatment for her back pain.

Under the guidance of the fitness center manager, she began a program of treadmill walking to tolerance and resumed a modified version of the program of resistance exercise she had started when joining the facility. The center is adjacent to the bank where she works and she now goes there after work several days a week. The bank she works for is one of the major corporate sponsors of the wellness center and she is able to take advantage of a reduced membership rate as a result.

Currently, she experiences back pain at least weekly. She does not take time off work with this but may modify her activities slightly. With the severe episodes described previously, she returned to work despite considerable pain within 3 to 5 days of the episode. Her supervisor insisted on purchasing an ergonomic chair for her use at work and she finds this very helpful. Her job allows her to change position frequently and she is never required to either sit or stand for long periods of time. She has noticed that any prolonged posture will aggravate her back pain for several days.

Similarly she has found that beginning any new sporting activity will also increase pain. Last summer she coached her daughter's softball team and found that the frequent squatting activities flared her back up considerably. Depending on the activity, it may take days to weeks to settle back down to the usual pain level. She continues to see her chiropractor once a month.

She exercises 3 to 4 nights a week at the wellness center and has found that her episodes of back pain worsen and the frequent low levels of pain increase in severity whenever she fails to do this. She feels that her back is still improving but very, very slowly. Her goal continues to be complete pain elimination.

After the last major episode of back pain, she was started on a modified version of her earlier resistance program by the fitness facility manager. She does three sets of 12 to 15 repetitions per set of the following exercises: chest press (two varieties), seated rowing, leg curls, triceps pull-down, and knee extension. All are done on weight equipment. She does two sets of 30 abdominal crunches. She also uses a back extension machine designed to isolate back extensors where she does three sets of 20 reps with a resistance of 80 lb. When she started this after her injury, she was only able to lift 30 lb in a very limited range. The flexibility of her spine has improved but she is cautious of allowing extension of her back much beyond a straightened position with this or any other exercise.

continued

Prior to the episode she had been using a leg press machine that loaded through the shoulders and a lat pull-down. Both were extremely painful after the injury and were discontinued from her program. Although she works out regularly, the weights have been the same on the machines for about a year. She uses the center treadmill for a walking program and does light stretching of arms and legs prior to her workout. She has tight hamstrings and reduced active and passive back extension.

Case Study 26   Discussion Questions

1. How well does Mrs. AB fit the profile of the typical individual with nonspecific back pain?
2. What characteristics in her work and recreational situation have helped her deal with her pain?
3. Without a more detailed examination of her specific muscle weaknesses, to what extent will a generalized strength training program produce the desired outcome (pain-free spinal mobility and physical tolerance of all work-related tasks)?
4. Do you believe, based on the recurrent nature of her symptoms, that present management strategies are adequate?
5. What specific recommendations could you make for her based on the provided information?

Glossary

body mass index—A classification system used to determine whether an individual meets ideal height–weight proportions.

computed axial tomography (CAT or CT scan)—Tomography (moving X-ray tube and film) where a pinpoint radiographic beam sweeps transverse planes of tissue and a computerized analysis of the variance in absorption produces a precise image of that area.

dynamic endurance—Classification of exercise where concentric-eccentric shifting occurs until muscular fatigue is induced. An example of dynamic endurance is biceps curls until fatigue occurs and the subject is unable to continue full motion against resistance.

electromyography—A diagnostic neurological test to study the potential (electrically measured activity) of a muscle at rest, the reaction of muscle to contraction, and the response to muscle insertion of a needle. The test is an aid in ascertaining whether a patient's illness is directly affecting the spinal cord, muscles, or peripheral nerves.

herniation—Development of an abnormal protrusion or projection of an intervertebral disk.

incidence—The frequency of occurrence of any event or condition over a period of time and in relation to the population in which it occurs.

magnetic resonance imaging (MRI)—The diagnostic test that uses principles of magnetism to generate an electromagnetic field around the body causing certain atoms in the nucleus of the body cells to "line up." Then, by sending and receiving radio signals, which are fed into a computer, the device records the position of those atoms, providing a distinct picture of the tissues being investigated. The patient lies inside a large, tunnel-like tube for 30-60 minutes while the images are being formulated by the computer. This diagnostic imaging technique has been found to have certain advantages over radiographs and CT scans in the diagnosis of spinal disorders.

myelography—Radiographic inspection of the spinal cord and nerve roots by use of a radiopaque contrast medium (a substance that causes the absorbing tissues to appear darker or lighter on a radiograph) injected into the intrathecal space. Air or oil dye may be used as contrasting agents.

osteophyte—A bony excrescence or outgrowth, usually branched in shape.

periodization—A system of fractioning larger periods of muscle training into smaller phases or cycles. Intensity, frequency, sets, repetitions, and rest periods are altered to reduce the risk of overtraining and minimize uncomfortable responses.

prevalence—The number of cases of a disease present in a specified population at a given time. This may be given at one identified time (point prevalence), or in a specified time period, such as 2 weeks or a year (period prevalence).

prevention—Three general categories of intervention strategies to limit the impact of potential or established disease in the population.

primary—Intervention geared toward removing or reducing the risk factors of disease.

secondary—Intervention that promotes early detection and treatment of disease with the goals of preventing disease recurrences or progression, promoting recovery, and avoiding complications.

scoliosis—Abnormal lateral curvature (side-bending and rotational components) of the spine. Usually the curvature consists of two curves, the original abnormal curve and a second compensatory curve in the opposite direction (also referred to as an S-curve).

spinal decompression—Surgical intervention to excise bony or soft tissue structures that exert pressure on neural tissues in the spine.

spinal discectomy—Surgical intervention to excise the portion of the herniated disk that is causing compression on neural tissue. The extent of the tissues removed is based on the extent of the intervertebral disk herniation.

spinal fusion—Surgical intervention to fixate unstable hypermobile vertebral segments through the use of metal plates, screws, wires, and autologous bony transplants.

spinal traction—Use of specialized harness systems and electronic winch or manually applied distractive forces on the spine in a variety of spinal and bodily positions. The purpose of this modality is to separate vertebra and stretch the associated soft tissues, decompressing nerve roots and relieving symptoms.

spondylolisthesis—Forward subluxation (malalignment) of superior lumbar vertebra on an inferior vertebra, leading to traction or compression of nerve roots and intervertebral supportive soft tissues and causing associated irritation and nociceptive input. This condition may be benign, depending on the amount of the slippage.

static endurance—Classification of exercise where isometric contractions of muscle groups lead to anaerobic exhaustion. An example of static endurance is a prolonged trunk extension position until fatigue occurs and the subject is unable to hold the position.

stenosis—Constriction or narrowing of a passage or orifice. In spinal stenosis, there is a congenital or degenerative narrowing of the intervertebral or vertebral foramen (opening) leading to compressive forces on the nerve roots that travel through these openings.

transcutaneous electrical nerve stimulation—Use of small battery-operated or plugged-in devices for delivery of electrical current across the skin to provide patients with pain relief, artificial contraction of muscles, fatigue of spastic muscles, and pulsations to decrease swelling in a joint. The stimulation is given through electrode pads placed directly on the skin over the muscles selected to be stimulated or inhibited from nociceptive input or in areas determined by nerve supply or to acupuncture points. The underlying theories are based on the gate theory of pain control, where sensory stimulation inhibits pain transmission at the spinal cord level, or stimulation of Aδ and C fibers, to cause the release of endogenous opiates.

ultrasound—Use of sonic waves, delivered by a vibrating quartz crystal, to deliver heat or medication to healing musculoskeletal structures. A variety of machines deliver the ultrasonic waves via a transducer rubbed directly over the skin using gel or water as transmitting medium.

References

1. American College of Sports Medicine. 2000. ACSM's Guidelines for Exercise Testing and Prescription. 6th ed. Philadelphia: Lippincott, Williams & Wilkins.

2. American College of Sports Medicine. 1998. ACSM position stand on the recommended quantity and quality of exercise for developing and maintaining cardiorespiratory and muscular fitness and flexibility in healthy adults. *Medicine and Science in Sports and Exercise* 30: 975-991.

3. Barbis, J.M. 1992. Prevention and management of low back pain. In *Prevention practice: Strategies for physical therapy and occupational therapy,* ed. J. Rothman and R.E. Levine, 63-72. Philadelphia: Saunders.

4. Battié, M.C., S.J. Bigos, L.D. Fisher, T.H. Hansson, A.L. Nachemson, D.M. Spengler, M.D. Wortley, and J. Zeh. 1989. A prospective study of the role of cardiovascular risk factors and fitness in industrial back complaints. *Spine* 14: 141-147.

5. Biering-Sörensen, F. 1984. Physical measurements as risk indicators for low-back trouble over a one year period. *Spine* 9: 106-119.

6. Bigos, S., O. Bowyer, G. Braen, K. Brown, R. Deyo, S. Haldeman, J. Hart, E. Johnson, R. Keller, D. Kiddo, M. Liang, R. Nelson, M. Nordin, B. Owen, M. Pope, R. Schwartz, D. Stewart, J. Susman, J. Triano, L. Tripp, D. Turk, C. Watts, and J. Weinstein. 1994. *Acute low back problems in adults. Clinical Practice Guideline No. 14.* AHCPR Publication No. 95-0642. Rockville, MD: Agency for Health Care Policy and Research, Public Health Service, U.S. Department of Health and Human Services.

7. Bigos, S.J., and G.E. Davis. 1996. Scientific application of sports medicine principles for acute low back pain. *Journal of Orthopedic and Sports Physical Therapy* 24: 192-207.

8. Bogduck, N., and L.T. Twomey. 1991. *Clinical anatomy of the lumbar spine,* 2nd ed. Melbourne, Australia: Churchill Livingstone.

9. Borenstein, D.G., R.A. Deyo, and N.J. Marcus (article consultants). 1998. A low-tech approach to low-back pain. *Patient Care* April 30: 84-86, 91-92, 94, 100, 103.

10. Cady, L.D., D.P. Bischoff, E.R. O'Connell, P.C. Thomas, and J.H. Allan. 1979. Strength and fitness and subsequent back injuries in firefighters. *Journal of Occupational Medicine* 21: 269-272.

11. Cassidy, J.D., L.J. Carroll, and P. Côte. 1998. The Saskatchewan Health and Back Pain Survey: The prevalence of low back pain and related disability in Saskatchewan adults. *Spine* 23: 1860-1867.

12. Cedraschi, C., M. Nordin, A.L. Nachemson, and T.L Vischer. 1998. Health care providers should use a common language in relation to low back pain patients. *Baillieres Clinical Rheumatology* 12: 1-5.

13. Darling, D. 1993. In search of the perfect treatment. In *Back pain rehabilitation,* ed. B. D'Orazio, 3-31. Stoneham, MA: Butterworth-Heinemann.

14. Deyo, R.A. 1996. Low back pain. A primary care challenge. *Spine* 21: 2826-2832.

15. Deyo, R.A., and J.E. Bass. 1989. Lifestyle and low-back pain. The influence of smoking and obesity. *Spine* 14: 501-506.

16. Deyo, R.A., A.K. Diehl, and M. Rosenthal. 1986. How many days of bed rest for acute low back pain? A randomized clinical trial. *New England Journal of Medicine* 315: 1064-1070.

17. Deyo, R.A., and Y-J. Tsui-Wu. 1987. Descriptive epidemiology of low back pain and its related medical care in the United States. *Spine* 12: 264-268.

18. D'Orazio, B.P. (ed) 1999. *Low back pain handbook.* Boston: Butterworth-Heinemann.

19. D'Orazio, B.P., C. Tritsch, S.A. Vath, and M.A. Rennie. 1999. Postoperative lumbar rehabilitation. In *Low back pain handbook,* ed. B.P. D'Orazio, 277-308. Boston: Butterworth-Heinemann.

20. Frymoyer, J.W. 1992. Predicting disability from low-back pain. *Clinical Orthopedics and Related Research* 279: 101-109.

21. Frymoyer, J.W., and W.L. Cats-Baril. 1991. An overview of the incidences and costs of low back pain. *Orthopedic Clinics of North America* 22: 263-271.

22. Hadler, N.M., and T.S. Carey. 1998. Low back pain: An intermittent and remittent predicament of life. *Annals of the Rheumatic Diseases* 57:1-2.

23. Harreby, M., Kjer, J., Hesselsøe, G., and Neergaard, K. 1996. Epidemiological aspects and risk factors for low back pain in 38-year-old men and women: A 25-year prospective cohort study of 640 school children. *European Journal of the Spine* 5: 312-318.

24. Jackson, R. 1999. Postural dynamics: Functional causes of low back pain. In *Low back pain handbook,* ed. B.P. D'Orazio, 159-191. Boston: Butterworth-Heinemann.

25. Jayson, M.I. 1996. ABC of work related disorders: Back pain. *BMJ* 313: 355-358.

26. Juker D., S.M. McGill, P. Kropf, and T. Steffen. 1998. Quantitative intramuscular myoelectric activity of lumbar portions of psoas and the abdominal wall during a wide variety of tasks. *Medicine and Science in Sports and Exercise.* 30: 301-310.

27. Kelsey, J.L., P.B. Githens, T. O'Conner, U. Weil, J.A. Calogero, T.R. Holford, A.A. White III, S.D. Walter, A.M. Ostfeld, and W.O. Southwick. 1984. Acute prolapsed lumbar intervertebral disc. An epidemiological study with special reference to driving automobiles and cigarette smoking. *Spine* 9: 608-613.

28. Kelsey, J.L., P.B. Githens, A.A. White III, T.R. Holford, S.D. Walter, T. O'Conner, A.M. Ostfeld, U. Weil, W.O. Southwick, and J.A. Calogero. 1984. An epidemiological study of lifting and twisting on the job and risk for acute prolapsed lumbar intervertebral disc. *Journal of Orthopedic Research* 2: 261-266.

29. Leboeuf-Yde C. 2000. Body weight and low back pain. A systematic literature review of 56 journal articles reporting on 65 epidemiological studies. *Spine* 25: 226-237.

30. Leboeuf-Yde C. 1999. Smoking and low back pain. A systematic literature review of 41 journal articles reporting 47 epidemiological studies. *Spine* 24: 1463-1470.

31. Leboeuf-Yde, C., K.O. Kyvik, and N.H. Bruun. 1999. Low back pain and lifestyle. Part II: Obesity. Information from a population-based sample of 29,424 twin subjects. *Spine* 24: 779-784.

32. Lindström, I., C. Ohlund, C. Eek, L. Wallin, L-E. Peterson, W.E. Fordyce, and A.L. Nachemson. 1992. The effect of graded activity on patients with subacute low back pain: A randomized prospective clinical study with on operant-conditioning behavioral approach. *Physical Therapy* 72: 279-293.

33. Linton, S.J. 1994. Chronic back pain: Activities training and physical therapy. *Behavioral Medicine* 20: 105-111.

34. Linton, S.J. 1994. Chronic back pain: Integrating psychological and physical therapy. *Behavioral Medicine* 20: 101-104.

35. Linton, S.J., L.A. Bradley, I. Jenson, E. Spangfort, and L. Sundell. 1989. The secondary prevention of low back pain: A controlled study with follow-up. *Pain* 36: 197-207.

36. Loney, P.L., and P.W. Stratford. 1999. The prevalence of low back pain in adults: A methodological review of the literature. *Physical Therapy* 79: 384-396.

37. Luoto, S., M. Heliovaara, H. Hurri, and M. Alaranta. 1995. Static back endurance and the risk of low back pain. *Clinical Biomechanics* 10: 323-324.

38. Manniche, C. 1996. Clinical benefit of intensive dynamic exercises for low back pain. *Scandinavian Journal of Medicine and Science in Sports* 6: 82-87.

39. Manniche, C., K. Asmussen, B. Lauritsen, H. Vinterberg, H. Karbo, S. Abildstrup, K. Fischer-Nielsen, R. Krebs, and K. Ibsen. 1993. Intensive dynamic back exercises with or without hyperextension in chronic back pain after surgery for lumbar disc protrusion. *Spine* 18: 560-567.

40. Manniche, C., E. Lundberg, I. Christensen, L. Bentzen, and G. Hesselsøe. 1991. Intensive dynamic back exercises for chronic low back pain: A clinical trial. *Pain* 47: 53-63.

41. Matsui, H., M. Kanamori, H. Ishihara, K. Yudoh, Y. Naruse, and H. Tsuji. 1998. Familial predisposition for lumbar degenerative disc disease. A case-control study. *Spine* 23: 1029-1034.

42. McGill, S.M. 1998. Low back exercises: Evidence for improving exercise regimens. *Physical Therapy* 78: 754-765.

43. McGill, S.M. 1998. Low back exercises: Prescription for the healthy back and when recovering from injury. In *American College of Sports Medicine's resource manual for guidelines for exercise testing and prescription,* 3rd ed, ed. J.L. Roitman, 116-126. Baltimore: Williams & Wilkins.

44. McKenzie, R.A. 1989. *The lumbar spine: Mechanical diagnosis and therapy.* Minneapolis: Orthopedic Physical Therapy Products.

45. Moffroid, M.T. 1997. Endurance of trunk muscles in persons with chronic low back pain: Assessment,

performance, training. *Journal of Rehabilitation Research and Development* 34: 440-447.

46. Mooney, V., and S. Leggett. 1996. Back pain: Which exercises can help, which can harm? *Consultant* 36:2543-2546, 2548, 2553-2554.

47. Mundt, D.J., J.L. Kelsey, A.L. Golden, H. Pastides, A.T. Berg, J. Sklar, T. Hosea, M.M. Panjabi, and the Northeast Collaborative group on Low Back Pain. 1993. An epidemiological study of non-occupational lifting as a risk factor for herniated lumbar intervertebral disc. *Spine* 18: 595-602.

48. Nachemson, A.L. 1994. Chronic pain—The end of the welfare state? *Quality of Life Research* 3(Suppl. 1): S11-S17.

49. Nachemson, A.L. 1976. The lumbar spine. An orthopedic challenge. *Spine* 1: 59-71.

50. Nachemson, A.L. 1992. Newest knowledge of low back pain. A critical look. *Clinical Orthopedics and Related Research* 279: 8-20.

51. Ng, J.K-F., C.A. Richardson, V. Kippers, and M. Parnianpour. 1998. Relationship between muscle fiber composition and functional capacity of back muscles in healthy subjects and patients with back pain. *Journal of Orthopedic and Sports Physical Therapy* 26: 389-402.

52. Oldridge, N.B., and J.E. Stoll. 1997. Spinal disorders and low back pain. In *Exercise testing and exercise prescription for special cases,* 2nd ed, ed. J.S. Skinner, 139-152. Philadelphia: Lea & Febiger.

53. Papageorgiou, A.C., G.J. Macfarlane, E. Thomas, P.R. Croft, M.I. Jayson, and A.J. Silman. 1997. Psychosocial factors in the workplace—Do they predict new episodes of low back pain? *Spine* 22: 1137-1142.

54. Pollock, M.L., S.H. Leggett, J.E. Graves, A. Jones, M. Fulton, and J. Cirulli. 1989. Effective resistance training of lumbar extensor strength. *American Journal of Sports Medicine* 17: 624-629.

55. Rainville, J., J. Sobel, C. Hartigan, G. Monlux, and J. Bean. 1997. Decreasing disability in chronic back pain through aggressive spine rehabilitation. *Journal of Rehabilitation Research and Development* 34: 383-393.

56. Riihimäki, H. 1991. Low back pain, its origin and risk indicators. *Scandinavian Journal of Work, Environment, and Health* 17: 81-90.

57. Ritz S., T. Lorren, S. Simpson, T. Mondry, and M. Comer. 1993. Rehabilitation of degenerative disease of the spine. In *Rehabilitation of the spine. Science and practice,* ed. S.H. Hochschuler, H.B. Cotler, and R.D. Guyer, 457-477. St. Louis: Mosby.

58. Smedley, J., P. Egger, C. Cooper, and D. Coggon. 1997. Prospective cohort study of predictors of incident low back pain in nurses. *BMJ* 314: 1225-1228.

59. Smedley, J., H. Inskip, C. Cooper, and D. Coggon. 1998. Natural history of low back pain. A longitudinal study in nurses. *Spine* 1998 23: 2422-2426.

60. Thomas, E., A.J. Silman, P.R. Croft, A.C. Papageorgiou, M.I.V. Jayson, and G.J. Macfarlane. 1999. Predicting who develops chronic low back pain in primary care: A prospective study. *BMJ* 318: 1662-1667.

61. Tollison C.D., and M.L. Kriegel. 1988. Physical exercise in the treatment of low back pain. Part II: A practical regimen of stretching exercise. *Orthopedic Review* 17: 913-923.

62. Van den Hoogen, H.J.M., B.W. Koes, J.T.M. van Eijk, L.M. Bouter, and W. Devillé. 1998. On the course of low back pain in general practice: A one year follow up study. *Annals of the Rheumatic Diseases* 57: 13-19.

63. Van Tulder, M.W., W.J. Assendelft, B.W. Koes, and L.M. Bouter. 1997. Spinal radiographic findings and non-specific low back pain. A systematic review of observational studies. *Spine* 22: 427-434.

64. Violinn, E., J. Mayer, P. Diehr, D. Van Koevering, F.A. Connell, and J.D. Loeser. 1992. Small area analysis of surgery for low back pain. *Spine* 17: 575-579.

65. Waddell, G. 1987. A new model for the treatment of low back pain. *Spine* 12: 632-644.

66. Walker, W.C., R.A. Chiappini, and R.M. Buschbacher. 1993. Rational prescription of exercise in managing low back pain. *Critical Reviews in Physical and Rehabilitation Medicine* 5: 219-226.

67. Weisel, S.W., H.L. Feffer, and R.H. Rothman. 1990. A lumbar spine algorithm. In *The lumbar spine.* ed. J.N. Weinstein and S.W. Wiesel, 358-368. Philadelphia: Saunders.

68. White A.A. III, and M.K. Panjabi. 1990. The clinical biomechanics of spine pain. In *Clinical biomechanics of the spine,* 2nd ed, 379-474. Philadelphia: Lippincott.

69. Williams, P.C. 1937. Lesions of the lumbosacral spine, part I. *Journal of Bone and Joint Surgery* 19: 343-363.

70. Williams, P.C. 1937. Lesions of the lumbosacral spine, part II. *Journal of Bone and Joint Surgery* 19: 690-703.

71. Wyke, B. 1970. The neurological basis of thoracic spine pain. *Rheumatology and Physical Medicine* 10: 356-367.

72. Zachrisson-Forssell, M. 1980. The Swedish back school. *Physiotherapy* 66: 112-114.

David R. Gater, Jr., MD, PhD
Medical Director, University of Michigan Model Spinal Cord Injury Care System
Research Associate, Ann Arbor VAMC
Assistant Professor, Physical Medicine and Rehabilitation
University of Michigan
Ann Arbor, MI

Spinal Cord Injury

The spinal cord serves as the major conduit for motor, sensory, and autonomic neural information transmission between the brain and the body. A **spinal cord injury (SCI)** affects conduction of neural signals across the site of the injury or lesion. An SCI is classified by the lowest segment of the spinal cord with normal sensory and motor function on both sides of the body and may be defined as complete (without sensory in the lowest sacral segment) or incomplete (partial preservation of sensory or motor function below the neurological level, including the lowest sacral segment). **Tetraplegia** (the preferred term over quadriplegia), refers to "the impairment or loss of motor and/or sensory function in the cervical segments of the spinal cord due to damage of neural elements within the spinal canal" (57). It is distinguished from **paraplegia** in that it includes dysfunction of the arms, whereas both tetraplegia and paraplegia involve impairment of function of the trunk, legs, and pelvic organs. Table 27.1 lists the American Spinal Injury Association's definitions of five degrees (completeness) of SCI (57). Table 27.2 lists specific clinical syndromes as they relate to injuries to specific spinal cord locations.

Scope

Approximately 250,000 to 400,000 U.S. citizens have an SCI (66). The incidence of SCI approaches 52 new

---

TABLE 27.1
· · · · · ·
American Spinal Injury Association Impairment Scale

Degree of impairment	Conditions of impairment
A	Complete: No motor or sensory function is preserved in the sacral segments S4-S5.
B	Incomplete: Sensory but not motor function is preserved below the neurological level and includes the sacral segments S4-S5.
C	Incomplete: Motor function is preserved below the neurological level, and more than half of key muscles below the neurological level have a muscle grade <3.
D	Incomplete: Motor function is preserved below the neurological level, and at least half of key muscles below the neurological level have a muscle grade of ≥3.
E	Normal: Motor and sensory function are normal.

Adapted, by permission, from R.J. Marino et al., 2000, *International standards for neurological classification of spinal cord injury* (Chicago, IL: American Spinal Injury Association), 18-19.

TABLE 27.2
· · · · · ·
Clinical Syndrome Resulting in Spinal Cord Injury

Syndrome	Cause	Physiological deficits
Central cord syndrome	Incomplete cervical spinal cord injury with cord damage	Weakness, sensory deficits in upper extremities (less in lower)
Brown-Sequard syndrome	Unilateral cord lesion	Ipsilateral proprioceptive and motor deficit, contralateral pain impairment and temperature deficit below level of injury
Anterior cord syndrome	Anterior cord ischemia (T10-L2)	Below level of injury, pain impairment and temperature deficit
Conus medullaris syndrome	Upper and lower motor neuron damage	Bowel, bladder, lower extremity areflexia, and flaccidity; preserved or facilitated reflexes
Cauda equina syndrome	Lumbosacral nerve root injury	Bowel, bladder, and lower extremity areflexia and severe dysesthetic pain

---

injuries per million per year. Almost 40% die before reaching a hospital. Hence, 8000 to 10,000 Americans will survive a new SCI each year, with more than 90% returning to a private residence following rehabilitation. Men are affected four times as frequently as women, and 77% of SCI injuries occur in those between 16 and 45 years old (54% for 16- to 30-year-olds). The majority of these injuries occur in young men involved in motor vehicle collisions. Acts of vio-lence, falls (mostly elderly), and sports injuries (mostly diving) are other primary causes of SCI. Pediatric SCI is often congenital, presenting at birth with incomplete formation and closure of neural and skeletal bony elements referred to as **spina bifida** or **myelomeningocele.**

Recent advances in acute care have significantly reduced the number of complete SCIs, particularly in cervical injuries. Since 1994, 30% of SCIs have resulted in **incomplete tetraplegia,** whereas only 23% are considered **complete tetraplegia** at the time of discharge from a rehabilitation facility. In the same time period, 20% of SCIs resulted in **incomplete paraplegia,** whereas 26% still had **complete paraplegia** at the time of discharge. A small percentage of SCIs (<1%) completely resolve by the time of hospital discharge, without neurological deficits.

The medical costs associated with SCI are staggering. Average lifetime medical costs exceed $1.1 million per individual, and average lifetime foregone earnings/fringe benefits per case may exceed $2.1 million (38). Even with the advent of managed care, the initial hospitalization and rehabilitation length of stay will exceed 60 days and $80,000 for tetraplegia and 40 days and $50,000 for thoracic paraplegia. As SCI care and technology have improved, life expectancy has likewise improved. A 20-year-old man with newly acquired tetraplegia now has an additional life expectancy of 33 years or more, whereas the 20 year-old with paraplegia may expect to live an additional 44 years or more (23).

Physiology and Pathophysiology

The spinal cord, a portion of the central nervous system, links the conscious and subconscious functions of the brain with the peripheral and autonomic nervous systems. It extends through and is protected by the spinal column, a flexible segment of interdigitating bones and disks arranged to maximize mobility and reduce risk of injury. There are 33 total vertebrae (7 cervical, 12 thoracic, 5 lumbar, 5 sacral, and 3-5 coccygeal). A pair of nerves arises from each vertebral segment. Figure 27.1 depicts these segments and nerves. The spinal cord is about 25% shorter than the spinal column, and thus subsequent spinal nerves exit from the cord as the **cauda equina.** The vascular supply to the cord comes from an anterior and two posterior spinal arteries at each vertebral level, supplying the anterior two thirds and posterior one third of the cord, respectively.

Neural tracts of the central nervous system and cell bodies of the **peripheral nervous system** are susceptible to primary and secondary injury. Primary injury can damage neural tracts, cell bodies, and vascular structures supplying the cord. Secondary injury occurs because of hemorrhage and local edema within the cord, which compromise vascular supply, resulting in local ischemia. Infarction of the gray matter will occur within 4 to 8 hr after injury if blood flow cessation remains. Inevitably, necrosis,

or cell death, occurs and can enlarge over the ensuing one or two vertebral levels above and below the area of trauma. **Gliosis, astrocyte,** and **syringomyelia** may form during the next several months. Formation of fibrous and glial scarring is the final phase of the injury process.

Injury to the spinal cord obstructs the transmission of neural messages through the cord and results in the loss of somatic and autonomic control over trunk, limbs, and viscera below the site of the lesion. Systemic responses to exercise seen in nondisabled individuals are blunted in persons with SCI. This diminishes their ability to perform physical activity and exercise, both at the conscious (somatic) and subconscious (autonomic) level.

Somatic Nervous System Disruption

The **somatic nervous system** consists of motor (efferent) and sensory (afferent) neural pathways connecting the brain to the body via the spinal cord. Complete SCI interrupts transmission of these signals, and voluntary movement and sensory perception are absent below the lesion. Neurological classification of SCI is standardized as seen in figure 27.2.

Autonomic Nervous System Disruption

The **autonomic nervous system** coordinates automatic life-sustaining processes and organizes visceral responses to somatic reactions. It is composed of sympathetic and parasympathetic divisions, which regulate the action of smooth muscle and glands. Essential functions of the autonomic nervous system during exercise include modulating heart rate, stroke volume, blood pressure, blood flow, ventilation, thermoregulation, and metabolism. The **sympathetic nervous system** and **parasympathetic nervous system** make up the autonomic nervous system. Function affected by level of SCI is provided in table 27.3.

Systemic Adaptation

The following sections review the systemic effects and adaptations to SCI.

— Cardiovascular: In response to acute cervical and upper thoracic SCI, bradycardia is common and attributable to loss of sympathetic nervous system influences, with no effect on the parasympathetic nervous system (49). This usually resolves in 2 to 6 weeks (24, 64). Reduced sympathetic influence on peripheral and splanchnic vascular beds reduces

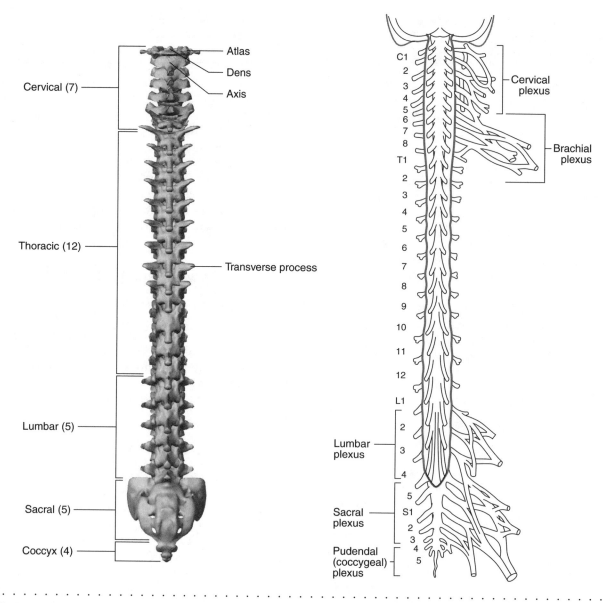

FIGURE 27.1　Anatomy of the spinal cord.

Adapted, by permission, from R.S. Behnke, 2001, *Kinetic Anatomy* (Champaign, IL: Human Kinetics), 130, 171.

peripheral vascular resistance, which enhances venous pooling and **orthostatic hypotension.**

— **Autonomic dysreflexia:** For persons with SCI above T6, an uncontrolled outflow of sympathetic activity in response to stimuli below the SCI (e.g., distended bowel or bladder, lacerations, fractures, pressure sores, sunburn) may occur, resulting in life-threatening paroxysmal hypertension (11, 18, 62). Reflex bradycardia and vasodilation, manifested as flushing, headache, hyperhydrosis, piloerection, pupillary dilation, and blurred vision, may occur (11). Such symptoms suggest an autonomic crisis that can lead to intracerebral hemorrhages, seizures, arrhythmias, and death if not immediately and appropriately treated.

— Pulmonary: Ventilation is impaired in most SCI patients because of paralysis of rib cage and abdominal musculature, reduced pulmonary compliance, and reduced diaphragmatic excursion, and thus some patients may require ventilator assistance. Tetraplegia below C5 typically spares voluntary control of the diaphragm, although inspiration remains impaired. Ventilation will worsen as the level of disability increases (16). However, this may be improved with aggressive spirometry and exercise training (17).

— Bowel and bladder function: Cervical, thoracic, and lumbar spinal cord lesions increase the risk of gastric and duodenal ulcers, increase bowel motility and hyperreflexia, and eliminate voluntary control

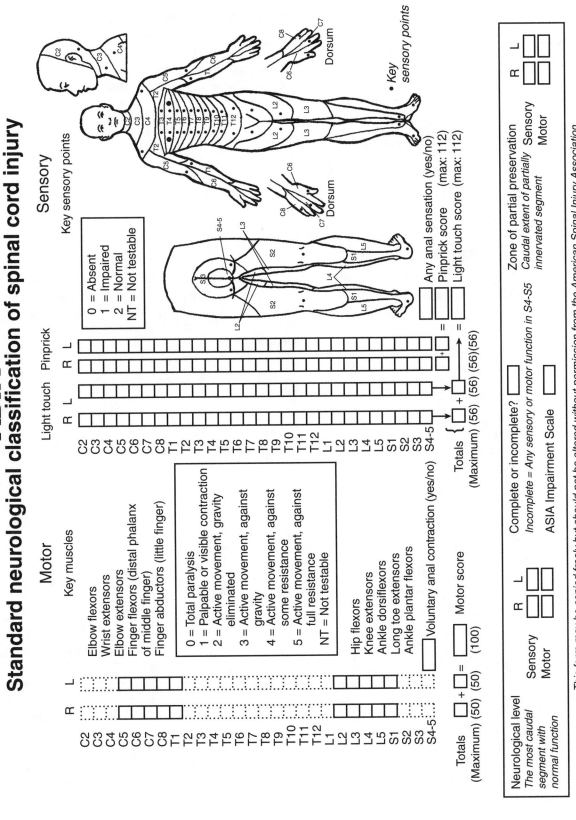

FIGURE 27.2 Tool used to classify the neurological extent of a spinal cord injury.

American Spinal Injury Association: International Standards for Neurological Classification of Spinal Cord Injury, revised 2002; Chicago, IL, American Spinal Injury Association; 2002.

507

TABLE 27.3
. . . . .
Effects of Level of Injury on Somatic and Autonomic Function

Level of injury	Somatic: movements preserved include all above each level affected	Autonomic: functions preserved include all above each level affected
C1-3	Chin	Cranial level parasympathetic nerves are not affected by SCI.
C4	Shoulders shrug and head turn	
C5	Shoulder movement and elbow flexion	
C6	Wrist extension	
C7	Elbow, wrist, finger extension	
C8	Lowest level of tetraplegia; finger flexion	
T1-10	Begins level of paraplegia; variable sensory deficits; intercostals control	Sympathetic nerves are from T1-L5; SCI above this level affects smooth muscles, organs and glandular sympathetic mediated effects (e.g., heart rate, stroke volume, ventilation, sweating, splanchnic vasoconstriction, and skeletal muscle vasodilation).
T6-12	Trunk stabilizers	
T1-S1	Paraspinal muscles	
L2	Hip flexion	
L3	Knee extension	
L4	Ankle dorsiflexion	
L5	Toe extension; hip abduction	
S1	Ankle plantar flexors, hip extensors	Affects lower digestive structures, gall bladder, and bladder function.
S2-S5	Bowel, bladder sphincter control	

Note. SCI = spinal cord injury.

of defecation (84). Hyperreflexia of the bladder wall and sphincter muscles results in greater risk of **vesiocoureteral reflux, hydronephrosis,** and acute renal failure if not appropriately managed. Bladder spasms and loss of voluntary sphincter control may lead to urinary incontinence. Urinary tract infections, as well as renal/bladder stones, are more frequent in SCI than in the nondisabled population.

— Hyperreflexia: Spasticity is seen in up to 60% of individuals with SCI (58) and is attributable to central inhibition of spinal reflex arcs (89). Some individuals with SCI are able to use their spasticity to assist them in performing mobility and **activities of daily living (ADL)** tasks, but many find that the spasticity is painful, disrupts sleep, interferes with function, causes muscle contractures, and can lead to shear- or pressure-induced skin breakdown.

— Thermoregulation: The interruption of autonomic pathways in tetraplegia results in partially **poikilothermic** responses to thermal stress and a reduced ability to dissipate environmental and internally generated heat by sweating (75). This is because the majority of the sweat response is limited to regions above the level of SCI.

— Endocrine: Autonomic dysfunction and somatic paralysis in SCI may also affect metabolic and hormonal function. This includes the sympathomedullary response and the adrenocortical-pituitary axis, resulting in flattened circadian rhythms and poorly regulated corticosteroid responses (87). Glucose intolerance often occurs and is often accompanied by hyperinsulinemia (1). Although thyroid function may be acutely altered in SCI, thyroid function tests are generally normal in healthy SCI adults (8). Conversely, testosterone and free testosterone levels in men with SCI are often reduced (83), whereas growth hormone release is blunted and chronically depressed (7). Following the acute phase of SCI, ovulatory menstrual cycles are fairly well preserved (70).

— **Osteopenia:** Neurogenic osteopenia in complete tetraplegia results from the withdrawal of stress and strain upon bone (88). Bone mineral loss is rapid during the first 4 months of SCI (32). Homeostasis at 67% of original bone mass is achieved at about 16 months postinjury. Thus, the risk of fractures is increased.

Clinical Considerations

The management of SCI is clinically complex and beyond the scope of the clinical exercise professional working with these patients on their long-term fitness management. Issues that must be addressed in the acute phases of SCI are summarized in figure 27.3. Co-morbidities that may affect the long-term rehabilitation of these patients include brain injury, thoracic contusions and fractures, intra-abdominal trauma, upper and lower extremity fractures, plexopathies, and peripheral neuropathies. Typically, medical management issues take precedence over fitness concerns during the first 3 months postinjury, and the altered physiology must constantly be considered as the person with SCI is reintegrated into community and recreational pursuits.

Signs and Symptoms

The common signs and symptoms of chronic SCI are listed in table 27.4. Each of these poses potential issues during the daily life of the patient with SCI. The clinical exercise professional working with these patients should be aware of these and able to assist when needed.

Diagnostic and Laboratory Evaluations

The diagnosis of SCI is largely based on the physical examination according to the American Spinal Injury Association's criteria listed in table 27.2. Motor function sensory levels (pin prick and light

---

TABLE 27.4

Signs and Symptoms of Spinal Cord Injury

Signs	Symptoms
Motor paralysis (BLOI)	Impaired or absent voluntary motor function
Sensory loss (BLOI)	Impaired or absent sensation
Hyperreflexia (UMN lesion)	Brisk DTRs, spasticity, spasms, clonus
Flaccidity (LMN lesion)	Flaccid paralysis with absent DTRs
Hypotension	Dizziness or loss of consciousness
Pulmonary dysfunction	Require accessory muscles of respiration
Neurogenic bladder	Urinary incontinence, urinary tract infection
Neurogenic bowel	Fecal incontinence, constipation

Note. BLOI = below the level of injury; UMN = upper motor neuron; DTRs = deep tendon reflexes; LMN = lower motor neuron.

Adapted from A.H. Ropper, 1994, Trauma of the head and spine. In *Harrison's Principles of Internal Medicine*, edited by K.J. Isselbacher, E. Braunwald, J.D. Wilson, J.B. Martin, A.S. Fauci, and D.L.Kasper (St. Louis: McGraw Hill).

Acute hospitalization management issues

- Spine management: imaging of cervical, thoracic, lumbar, sacral spine
- Surgical or orthotic stabilization of unstable spinal column injuries and spinal cord decompression
- Range of motion limitations to allow complete bony and soft tissue healing of spinal elements

Respiratory management issues

- Assisted ventilation often required for high cervical injuries
- Secretion management essential because of impaired cough and increased parasympathetic nervous system influence on pulmonary secretions
- Assisted cough required for SCI above T6 attributable to intercostals and abdominal muscle paralysis

Cardiovascular management issues

- Relative bradycardia attributable to impaired sympathetic nervous system in SCI above T6, occasionally requires pacemaker placement
- Hypotension attributable to systemic vasodilation resulting from impaired sympathetic drive
- Venous stasis can result in deep venous thromboses and/or pulmonary emboli

Functional mobility issues

- Upper extremity range of motion, strengthening, and endurance within limitations of orthotics and medical management
- Bed mobility (including side to side, supine to prone to supine, supine to sit)
- W/C mobility (including forward/backward propulsion, turning, uneven terrain, curbs, ramps, hills)
- Transfers (including bed to W/C to bed, W/C to toilet to W/C, W/C to bath to W/C, W/C to floor to W/C, W/C to car to W/C)
- Activities of daily living including feeding, grooming, dressing, bathing, toileting
- Bladder management training, typically with intermittent catheterization or alternative
- Bowel management training, typically with digital stimulation and manual evacuation or alternative
- Skin management training with monitoring and pressure relief techniques
- Equipment evaluation for personal care, mobility, and public accessibility
- Home and vehicle evaluation for accessibility
- Psychological and social adjustment to spinal cord injury
- Introduction to vocational and recreational opportunities for persons with SCI

Note. SCI = spinal cord injury; W/C = wheelchair.

FIGURE 27.3   Management issues during SCI acute hospitalization and rehabilitation.
Note. SCI = spinal cord injury; W/C = wheelchair.

touch) are assessed and the injury is listed as complete if no motor or sensory function is spared at the S4 to S5 level; any preservation of function spared at these sacral levels denotes an incomplete SCI. The diagnosis of SCI is assisted with the use of electrodiagnostic studies, most notably **somatosensory evoked potentials** (13).

When one is assessing the exercise literature, it is important to note the motor and impairment levels for SCI subjects, because the level and completeness of the injury significantly affect the degree to which the somatic and autonomic nervous systems contribute to exercise responses. The International Stoke Mandeville Wheelchair Sports Federation classification system may help the clinical exercise professional to determine the degree of impairment as it relates to exercise (table 27.5).

The spine is considered unstable when two or more of the spinal columns are damaged (involving soft tissue and/or bony elements) and surgical stabilization is warranted (figure 27.4) (22). It is important to note, however, that spinal cord damage may occur in the presence of apparent spine stability, and that neurological dysfunction does not always occur in the presence of an unstable spine. Several examples of spinous fractures and dislocations are provided in table 27.6.

Treatment

The care and management of the person with SCI have significantly advanced, extending longevity with improved quality of life but also unmasking the problems associated with physical inactivity, blunted autonomic and hormonal responses, and upper extremity overuse syndromes common to chronic survivors of SCI.

Clinical practice guidelines developed by the Consortium for Spinal Cord Medicine and published by the Paralyzed Veterans of America (www.pva.org) describe expected outcomes for SCI and discuss the

---

TABLE 27.5

The International Stoke Mandeville Wheelchair Sports Federation Classification System for Spinal Cord Injury Competition in Sport

Level	Criteria
IA	All participants with cervical lesions with complete or incomplete quadriplegia who have involvement of both hands, weakness of triceps (up to and including grade 3 on manual muscle testing scale[a]), and severe weakness of the trunk and lower extremities interfering with trunk balance and the ability to walk.
IB	All participants with cervical lesions with complete or incomplete quadriplegia who have involvement of upper extremities but less than IA with preservation of normal or good triceps (4 or 5 on testing scale) and with a generalized weakness of the trunk and lower extremities interfering significantly with trunk balance and the ability to walk.
IC	All participants with cervical lesions with complete or incomplete quadriplegia who have involvement of upper extremities but less than IB with preservation of normal or good triceps (4 or 5 on testing scale) and normal or good finger flexion and extension (grasp or release) but without intrinsic hand function and with a generalized weakness of the trunk and lower extremities interfering significantly with trunk balance and the ability to walk.
IIA	Participants with complete or incomplete paraplegia below T1 down to and including T5 or comparable disability with total abdominal paralysis or poor abdominal muscle strength (0-2 on testing scale) and no useful trunk sitting balance.
IIB	Participants with complete or incomplete paraplegia or comparable disability below T5 down to and including T10 with upper abdominal and spinal extensor musculature sufficient to provide some element of trunk sitting balance but not normal.
III	Participants with complete or incomplete paraplegia or comparable disability below T10 down to and including L2 without quadriceps or very weak quadriceps with a value up to and including 2 on the testing scale and gluteal paralysis.
IV	Participants with complete or incomplete paraplegia or comparable disability below L2 with quadriceps in grades 3-5.
V	Participants with minimal muscle deficit

[a]Manual muscle testing scale: (1) muscle flicker without joint movement, (2) full joint range of motion with gravity eliminated, (3) full joint range of motion against gravity, (4) greater than antigravity but less than normal motor strength, (5) age- and sex-matched normal strength.

Reprinted, by permission, from the International Stoke Mandeville Wheelchair Sports Federation, 2002.

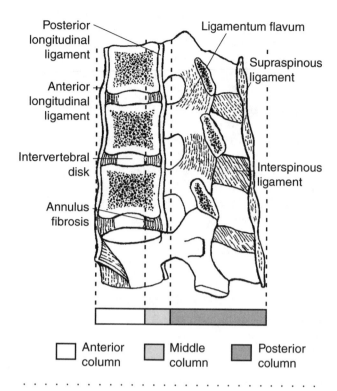

Posterior longitudinal ligament
Anterior longitudinal ligament
Intervertebral disk
Annulus fibrosis
Ligamentum flavum
Supraspinous ligament
Interspinous ligament

☐ Anterior column   ☐ Middle column   ■ Posterior column

FIGURE 27.4   Denis three-column model.

Adapted, by permission, from F. Denis, 1983, "The three column spine and its significance in the classification of acute thoracolumbar spinal injuries," *Spine* 8(8): 817-831.

prevention and management of autonomic dysreflexia, thromboembolism, bowel management, depression, and pressure ulcers in SCI. Future clinical practice guidelines must address the application and prescription of exercise.

Graded Exercise Testing

Graded exercise testing can be used to assess aerobic fitness or training effects in asymptomatic or athletic populations, screen individuals at risk for heart disease, determine progress in rehabilitation, demonstrate maximal strength and power capacities, and assist with the exercise prescription (77). Additional benefits to persons with SCI include the opportunity to establish a relationship between fitness and posttraumatic return to gainful employment and to determine how the fitness level of a person with SCI changes over time (21).

Medical Evaluation

Comorbid medical issues common among patients with SCI that should be considered for exercise testing and training include respiratory complications, coronary artery disease, peripheral arterial disease, **circulatory hypokinesis** leading to hypotensive responses, obesity, type 2 diabetes mellitus, pressure sores, **joint contractures,** and osteopenia. There is also a high incidence of smoking (~35%) and upper extremity overuse problems. Finally, medications

TABLE 27.6
Types and Causes or Results of Vertebral Fractures and Dislocations

Type of injury	Common cause/result of fracture
Atlantoaxial dislocation	Rheumatoid arthritis; can result in respiratory failure
Atlanto-occipital dislocation	Most common in children; typically causes death
Jefferson's fracture	Burst ring of atlas from descending force on vertex of scull (e.g., diving accident)
Hangman's fracture	Hyperextension and longitudinal distraction of upper cervical spine (e.g., chin striking steering wheel in automobile accident)
Teardrop fracture	Vertebral body compression with anterior bony fragment
Compression or Burst fracture	Retropulsed bony fragment into spinal cord
Hyperextension of cervical spine	With cervical stenosis, central cord syndrome
Thoracolumbar fracture	High-impact spinal cord injury associated with rib fractures

Adapted from A.H. Ropper, 1994, Trauma of the head and spine. In *Harrison's Principles of Internal Medicine*, edited by K.J. Isselbacher, E. Braunwald, J.D. Wilson, J.B. Martin, A.S. Fauci, and D.L.Kasper (St. Louis: McGraw Hill).

PRACTICAL APPLICATION 27.1

Client–Clinician Interaction

SCI patients must overcome tremendous physical, emotional, spiritual, and intellectual obstacles to succeed in life. They must wake up early to bathe, dress, groom, and perform bowel and bladder care. Seated pressure relief must be performed every 15 to 20 min. Bodily functions must also be managed according to a timed regimen. Obstacles are everyday occurrences. Often the person with SCI who consults an exercise physiologist is one of the most disciplined, motivated, and enthusiastic clients the exercise clinician will encounter. Prepare to be surprised!

It is essential to understand the goals and obstacles of the person with SCI. This requires direct interaction with the patient. Whenever possible, speak at eye level with the client. Gain an understanding of the client's daily routine, environmental and transportation barriers that must be overcome, and concerns the person may have about an exercise routine. Provide empathetic listening. Be prepared to discuss the application of exercise benefits to the client's functional abilities, and become familiar with accessible facilities in the community.

The exercise clinician should consider spending a 24-hr day without standing. During this time, only perform **community mobility** and seated activities of daily living (including car, tub, and toilet transfers) from a wheelchair, to more fully appreciate your SCI client's perspective and needs. This experience can be eye opening for clinical exercise professionals, allowing them a glimpse into the SCI patient's world. This may be helpful when considering the daily life of these patients and how they might incorporate an exercise or physical activity regimen.

should be listed and assessed for their effect on exercise tolerance. Commonly prescribed medications used in SCI, their indications, and common side effects are provided in table 27.7.

Heart disease is the second leading cause of death in patients with SCI, accounting for approximately 22% of all deaths (38). Unfortunately, silent myocardial ischemia caused by disrupted visceral afferent fibers in higher levels of SCI may prevent an individual from recognizing symptoms such as angina (6). Amputations for peripheral arterial disease are sometimes required because of diminished healing and the development of ulcers. Obesity is a problem attributable to reduced (12-54%) basal energy expenditure. As a result, the risk for glucose intolerance and type 2 diabetes is increased (34).

Pressure ulcers occur as the result of shear, friction, and unrelieved pressure, usually over a bony prominence, which damages underlying tissue. This problem is exacerbated in persons with tetraplegia because they are unable to feel the sensations of tingling, discomfort, and pain. Appropriately prescribed wheelchair-seating systems with scheduled pressure relief every 15 to 30 min should reduce the risk of pressure ulcers in the exercising individual with SCI (26). Phantom or neuropathic pain (e.g., burning, tingling, electrical sensations) may occur from a region below the SCI and can adversely affect exercise ability (12).

Upper extremity overuse injuries in wheelchair-reliant individuals occur most frequently at the shoulder (19, 81). Hip, knee, and plantar flexion contractures may result from unbalanced muscle forces in wheelchair-reliant individuals over a period of time, but these do not usually affect the individual's function unless he or she desires to stand or walk (incomplete SCI). Heterotopic ossification (bony overgrowth within the joint space) can occasionally limit range of motion to such an extent that the affected individual loses the ability to transfer and perform certain ADLs. Management is with gentle range of motion exercises, nonsteroidal anti-inflammatory agents, bisphosphonates, and occasionally surgical resection (80).

Spasticity can affect exercise ability. Treatment includes the removal of any stimuli inducing increased tone, daily prolonged stretch of affected muscle groups, and pharmacological or surgical management (47).

Understanding the level of SCI and associated medical problems can help the clinical exercise professional understand the physical limitations of each individual patient. Table 27.8 lists ADLs and functional ability by level of SCI.

Contraindications

It is recommended that men 40 or more years of age and women 50 or more years of age who have SCI be considered at moderate risk for untoward events during exercise, independent of traditional CAD risk. Persons with SCI should obtain medical clearance from a physician knowledgeable in SCI care,

	TABLE 27.7		
	Medications Commonly Used in Spinal Cord Injury		
Medication (generic)	**Brand name**	**Indication**	**Common side effects**
Amitriptyline	Elavil	Neuropathic pain	Dry mouth, constipation, arrhythmias, hypotension
Carbamazepine	Tegretol	Neuropathic pain	Drowsiness, dizziness, aplastic anemia, agranulocytosis
Diazepam	Valium	Spasticity	Drowsiness, dry mouth, dizziness, hypotension
Enoxaparen	Lovenox	DVT or PE	Bleeding, bruising
Gabbapentin	Neurontin	Neuropathic pain	Drowsiness, dry mouth, dizziness, hypotension
Imipramine	Tofranil	Bladder incontinence	Dry mouth, constipation, arrhythmias, hypotension
Lioresol	Baclofen	Spasticity	Drowsiness; hallucinations and seizures if abruptly discontinued
Nortriptyline	Pamelar	Neuropathic pain	Dry mouth, constipation, arrhythmias, hypotension
Oxybutynin	Ditropan	Bladder spasms	Drowsiness, dry mouth, dizziness, hypotension
Phenoxybenzamine	Dibenzyline	Autonomic dysreflexia	Dizziness, hypotension
Prazosin	Minipress	Autonomic dysreflexia	Dizziness, hypotension
Terazosin	Hytrin	Autonomic dysreflexia	Dizziness, hypotension
Tizanidine	Zanaflex	Spasticity	Drowsiness, dry mouth, dizziness, hypotension
Tolteridine	Detrol	Bladder spasms	Drowsiness, dry mouth, dizziness, hypotension
Warfarin	Coumadin	DVT or PE	Bleeding, bruising

Note. DVT = deep vein thrombosis; PE = pulmonary embolus.

including a 12-lead electrocardiogram and risk profile assessment, prior to performing an exercise test.

Standard contraindications for exercise testing noted in chapter 6 should be applied to the SCI population. The following are common contraindications in patients with SCI. Without adequate assessment and treatment, these should be considered absolute reasons not to perform an exercise test.

— Autonomic dysreflexia resulting from recent fracture, may precipitate spasms, or increase the risk of fatty emboli, hypertensive crisis, or cerebrovascular events

— Orthostatic hypotension, with the risk of syncope

— Recent deep vein thrombosis or pulmonary embolism

— Pressure ulcers, which increase the risk of autonomic dysreflexia during exercise

Common relative contraindications include the following:

— Active tendinitis (e.g., rotator cuff, elbow flexors, wrist flexors/extensors)

— Chronic heterotopic ossification

— Peripheral neuropathy

— Pressure ulcers of grade 2 or less

— Spasticity

TABLE 27.8
Functional Ability by Level of Spinal Cord Injury

Activity	C3-C4	C5	C6	C7	C8-T1	T2-T6	T7-T10	T11-L2	L3-S3
Feeding	D	Modified I	I	I	I	I	I	I	I
Grooming	D	Modified I	I	I	I	I	I	I	I
Upper extremity dressing	D	Min A	Modified I	Modified I	I	I	I	I	I
Lower extremity dressing	D	Modified I	Min A	Modified I	I	I	I	I	I
Bathing	D	D	Modified I	Modified I	I	I	I	I	I
Pulmonary hygiene	D	Assisted cough	Assisted cough	Assisted cough	Assisted cough	Assisted cough	I	I	I
Bowel management	D	D	Modified I	Modified I	I	I	I	I	I
Bladder management	D	D	Mod A	Modified I	I	I	I	I	I
Bed mobility	D	Mod A	Modified I	I	I	I	I	I	I
Wheelchair propulsion	D	Modified I	Modified I	I	I	I	I	I	I
Wheelchair transfers	D	Max A	Modified I	I	I	I	I	I	I
Pressure relief	D	D	I	I	I	I	I	I	I
Driving	D	Modified I	Modified I	Modified I	Modified I	Modified I	Modified I	Modified I	Modified I
Ambulation	N/A	N/A	N/A	N/A	Exercise only	KAFO + Loft	KAFO + Loft	KAFO + Loft	AFO

Note. D = totally dependent; Min A = minimal (25%) assistance required from another person; Mod A = moderate (50%) assistance required from another person; Max A = maximal (75%) assistance required from another person; Modified I = independent with modified equipment; I = independent; N/A = not applicable; KAFO = knee–ankle–foot orthosis (bracing) required; AFO = ankle–foot orthosis (bracing) required; Loft = Lofstrand (forearm) crutches required.

Bladder and bowel evacuation should be implemented immediately before the graded exercise test to minimize the risk of exertional incontinence or autonomic dysreflexia. Manual stimulation or **disimpaction** is required to regularly maintain fecal continence, and, in some cases, colostomy is warranted for bowel management. Bladder management is most often performed by intermittent catheterization, although indwelling catheters and bladder diversion techniques are sometimes warranted. Environmental consideration must also be made because of the SCI patient's difficulty with body temperature regulation.

Recommendations and Anticipated Responses

Field testing is the easiest and least expensive, as well as mobility-specific, method of evaluation in selected wheelchair users (31). However, recent reports have failed to significantly correlate field testing with actual $\dot{V}O_2$peak, possibly because of variability in terrain, wind speed, temperature, and humidity (86). Arm crank ergometry is the most often used test mode with SCI (73). Wheelchair ergometry is mobility specific for most. Several systems have been developed and tested, including wheelchairs mounted on a motorized treadmill (85), low-friction rollers (56), and specialized devices to simulate overground propulsion (15, 52). When compared with arm crank ergometry, wheelchair ergometry results in similar or greater $\dot{V}O_2$peak responses with lower peak power output (37), indicating reduced mechanical efficiency.

Several devices are available to assess all-extremity oxygen consumption, which may be appropriate for persons with incomplete SCI and for monitoring aerobic fitness of those using combined upper extremity and lower extremity **functional electrical stimulation (FES).** An improved venous return, via inclusion of the lower body muscle pump, increases stroke volume and cardiac output (55, 61, 68).

Protocols should employ incremental graded advances in resistance or power output requirement with periodic discontinuance for blood pressure, heart rate, and electrocardiogram determination (36, 52). A typical wheelchair ergometry protocol employs an initial resistance of 25 W, with 5- to 10-W increases every 2 to 3 min to symptom-limited fatigue. It is common for persons with SCI to only achieve 40 to 100 W at peak exercise. Population-specific prediction equations should be used when estimating $\dot{V}O_2$peak (43). Standard test termination rationale should be used (see chapter 6).

For individuals undergoing testing to rule out ischemic heart disease, postexercise echocardiography (51) or nuclear imaging studies (6) may improve the sensitivity of the exercise stress test. In a study using standard exercise testing, only 5 of 13 subjects with known myocardial ischemia had ST-segment changes indicative of ischemia (6).

The arm crank or wheelchair ergometer should be adjusted appropriately to allow optimal efficiency and reduce musculoskeletal injuries at the shoulder, elbow, and wrist. Straps applied to the torso improve trunk stability. Wheelchair gloves or flexion mitts with Velcro® straps can prevent blisters, lacerations, and abrasions, especially for those tetraplegics whose hands and fingers are insensate or unable to sufficiently grasp. Velcro straps and cuffed weights are commonly used for resistance training equipment modifications. Abdominal binders and leg wraps may improve pulmonary dynamics and venous return, which reduce the risk of hypotension.

Body composition analysis is important in this population because BMI is not representative, and obesity is a common problem. Although recently touted as the gold standard for determining body composition in SCI (79), dual energy X-ray absorptiometry (DEXA) introduces significant error, because it does not account for hydration status. Four-compartment body composition modeling should be used to determine percent body fat in persons with SCI until regression-based equations specific to DEXA in SCI are established (40). Additional testing appropriate for the person with SCI prior to exercise training may include pulmonary function testing, quantified strength and flexibility measures, DEXA to determine bone mineral density, radiographs of paralyzed extremities to exclude asymptomatic fractures, lipid profiles, and HbA1c to rule out glucose intolerance and diabetes.

Because of sympathetic impairment, peak heart rate rarely exceeds 120 beats · min⁻¹ in those with complete tetraplegia and T1-T3 paraplegia. Although variable responses occur in T4-T6 paraplegia, most persons with SCI below T7 are able to reach their age-adjusted peak heart rate. Similar trends are reported for blood pressure responses. In general, $\dot{V}O_2$peak and peak power output are significantly diminished in patients with SCI (4, 9, 45, 54, 74, 76). However, the lower the injury, the less the impairment. $\dot{V}O_2$peak values range from 12 ml · kg⁻¹ · min⁻¹ for tetraplegic patients to more than 30 ml · kg⁻¹ · min⁻¹ in low level injury paraplegic patients. In these same groups, peak power output ranges from less than 40 to more than 100 W, respectively.

Persons with SCI and a resting systolic blood pressure below 100 mmHg should be closely monitored during exercise. As they approach peak exercise, the risk of a hypotensive response increases despite the use of leg wraps and abdominal binders. Should symptomatic hypotension occur, exercise testing should be halted and the person should be tilted back in his or her wheelchair to elevate the lower extremities above the level of the heart, promoting venous return.

Exercise Prescription

The exercise prescription for a person with SCI should focus on the typical parameters (see chapter 7). However, because of potential complexities, the prescription should ideally be developed by a team composed of an exercise physiologist (physiological responses), a physical therapist (orthopedic limitations), and a physician (medical concerns and oversight). It is initially optimal for the person with SCI to begin an exercise training program under the supervision of either a physical therapist or clinical exercise physiologist. Practical application 27.2 discusses functional responses to exercise training in people with SCI.

Special Exercise Considerations

Because of their reduced body temperature regulatory ability, persons with tetraplegia should only exercise in a mildly temperate climate (outside during

PRACTICAL APPLICATION 27.2

Literature Review

Tasks such as feeding, grooming, hygiene, dressing, bathing, transfers, and toileting are referred to as **activities of daily living (ADLs),** whereas tasks of **community mobility** include traversing sidewalks, stairs or ramps, paths, and environmental barriers (e.g., curbs, speed bumps). Patients with chronic (>20 years) SCI will require greater physical assistance as they age. The percentage of maximal heart rate while performing ADLs and community mobility tasks is higher in people with tetraplegia than paraplegia, and an inverse relationship exists between physical capacity and physical strain (20, 44). In one study, only 29% of SCI subjects with $\dot{V}O_2$peak less than 15 ml $\cdot$ kg^{-1} $\cdot$ min^{-1} were able to perform independent ADLs (41).

Upper Extremity Aerobic Training

Critical review of upper extremity aerobic conditioning studies demonstrates variability in exercise prescription and results, partially attributable to the level of SCI. For instance, Gass et al. (33) trained seven subjects (four with C5-6 lesions, three with T1-T4 lesions) five times weekly to exhaustion on a graded exercise test protocol by using a wheelchair on a treadmill ergometer. After 7 weeks of training with this ergometry system, the mean $\dot{V}O_2$peak had increased from 9.5 to 12.7 ml $\cdot$ kg^{-1} $\cdot$ min^{-1} and endurance time had increased by 4.4 min, suggesting considerable change in functional capacity (33). Knutsson et al. (50) evaluated 20 SCI patients with complete and incomplete SCI between C5 and L1. For the 10 persons assigned to the training group, a 40% increase in peak work rate (40-57 W) was reported, although no significant change was noted in the three subjects with SCI above T6. Taylor et al. (82) assessed the effects of ACE performed 30 min per day, five days per week for 8 weeks in individuals with paraplegia. The trained group significantly improved $\dot{V}O_2$peak from 22.8 to 26.3 ml $\cdot$ kg^{-1} $\cdot$ min^{-1} without significant changes in maximal heart, postexercise lactate, or body fat.

More recently, McLean and Skinner (59) matched 14 tetraplegic subjects, by peak power output, to either a supine or seated exercise training regimen to assess for changes in postural position on stroke volume, cardiac output, and exercise capacity. Their subjects performed arm crank ergometry exercise in either a sitting or a supine position at 60% of their $\dot{V}O_2$peak three times a week for 10 weeks with progressive increments in either duration or resistance; no significant differences were found in stroke volume or cardiac output, although absolute $\dot{V}O_2$peak increased from 720 to 780 ml $\cdot$ min^{-1}, suggesting peripheral adaptations.

Resistance Training

It seems inherent from the large number of shoulder and upper extremity musculoskeletal problems encountered by persons with SCI that a prophylactic, structured, and progressively resistive strengthening program focusing on scapular, rotator cuff, and pectoral muscles would increase strength and reduce the risk for overuse injury, likely improving the ability of these people to perform functional tasks in the community. Little information is available, however, to accurately determine the effects of resistance training on strength, power, muscle mass, or functional abilities in SCI. Nilsson et al. (65) reported increases in dynamic strength (16%) and endurance (80%) when comparing bench press before and after a 7-week combined arm crank ergometry/resistance training program in paraplegic adults. Chawla et al. (14) reported that a resistance training program

continued

including bench press, incline press, lateral raises, incline curls, lat pulls, and triceps stretch improved ADL function in 10 patients with SCI. Unfortunately, specific ADLs and quantitative measures of strength were not reported. From these few studies, it appears promising that specific strength training will benefit patients with SCI.

Functional Electrical Stimulation

In 1987, medical guidelines were developed for patient participation in FES rehabilitation, including medical criteria for inclusion and exclusion. Computerized FES is a neuromuscular aid used to restore purposeful movement of limbs paralyzed by upper motor neuron lesions. Numerous reviews are devoted to FES and its potential to stimulate beneficial exercise adaptations in patients with SCI (35, 67, 68). Briefly, FES of the lower extremities can be used to do the following:

— Stimulate skeletal muscle strength (25, 69, 71) and endurance (25, 42, 60)
— Improve energy expenditure and increase stroke volume (29, 42)
— Increase total body peak power, $\dot{V}O_2$peak, and ventilatory rate (3, 42, 60)
— Reverse myocardial disuse atrophy (63)
— Increase high-density lipoprotein levels and improve body composition (5)
— Improve self-perception (78)
— Increase lower extremity bone mineral density (10, 39)

Despite these encouraging findings, functional gains in upper extremity strength, aerobic capacity, community mobility, and ADLs were not demonstrated in response to FES lower extremity training. After 8 weeks of inactivity that followed a 12-week training program, the 45% improvement in $\dot{V}O_2$peak attained by FES plus leg ergometry training was reduced by approximately 50%, whereas power output returned to pretraining levels. Interestingly, submaximal heart rate, peak heart rate, and peak ventilatory volume did not change at any time point.

A logical and intuitive progression in the development of FES lower extremity exercise training has been the combined use of concurrent arm crank ergometry and FES leg cycle ergometry (LCE), termed **hybrid** exercise. As expected from the combination of upper and lower extremity exercise peak power, $\dot{V}O_2$peak, stroke volume, and cardiac output (Q) significantly increase during hybrid exercise bouts with SCI subjects (27, 61) and during combined upper extremity rowing plus lower extremity FES (53).

appropriate weather or inside in a climate-controlled area) to avoid the risk of hypo- and hyperthermia. Appropriate seating and positioning are necessary to reduce the risk of pressure sores, autonomic dysreflexia, spasticity, and musculoskeletal trauma.

Few commercial fitness centers are able to accommodate the needs of the person with SCI, including wheelchair access. A recent survey of physical fitness facilities in a major metropolitan city demonstrated that none of the 34 facilities reviewed met all of the 1990 Americans with Disabilities Act (ADA) requirements for accessibility (28). Other potential barriers include individual constraints attributable to transportation availability and required assistance. Finances may also be a concern for many patients. These issues have a direct impact on compliance.

Adapted or adaptable equipment, some of which was outlined in the exercise testing recommenda-

tions section, is required for proper and safe exercise. Other recommendations include the use of upper arm bands or Co-Ban® tape to prevent abrasions at the medial upper arm with wheelchair propulsion. Abdominal binders and leg wraps may be used to facilitate improved pulmonary dynamics and greater venous return. FES-LCE and hybrid systems are now commercially available but remain expensive to purchase and maintain relative to arm crank ergometry. Velcro straps and cuffed weights are commonly used for resistance training to improve or create a grip.

Exercise Recommendations

Individuals with complete lesions at or above C4 are limited to using FES-LCE or hybrid exercise equipment because of problems with arm function. For

Exercise Prescription

		Cardiovascular	Resistance	Flexibility
Mode		Arm crank ergometry Wheelchair ergometry Arm crank cycling Community wheeling Seated aerobics Aquatics Wheelchair recreation FES + leg cycle ergometry ACE and FES + leg ergometry	Therabands Wrist weights Body weight Dumbbells Free weights W/C accessible machines FES-isokinetic	Active assisted (anterior shoulder, pectoral muscles, rotator cuff) Passive assisted (hip flexors, knee flexors, plantar flexors)
Frequency		3-7 days per week	1-3 sets, 2-3 days per week	7 days per week
Intensity		RPE 11-14 50-85% $\dot{V}O_2$peak or peak power output 30-80% HRreserve 60-90% HRpeak "Talk test"	8-12 reps at 60-75% 1-RM	As tolerated
Duration		20-60 min; continuous or interval	30-60 min	5-15 min
Progression		Slow (<5% per week)	Increase resistance when 12 reps achieved	Work to increase ROM as tolerated
Precautions		Avoid exertional hypotension. May initially require multiple sets of 5-10 min duration. Monitor for autonomic dysreflexia. Avoid thermal stress. Include warm-up and cool-down.	Avoid Valsalva maneuver. Provide spotter. Use seat belt and chest strap for balance. Use adaptive grip and mitts. Don't exceed stress limits of wheelchair.	Stretch to strain, not pain. Avoid Valsalva maneuver. Don't overstress insensate joints. Provide midshaft support for long osteopenic limbs.

Note. FES = functional electrical stimulation; ACE = arm crank ergometry; RPE = rating of perceived exertion; RM = repetition maximum; HR = heart rate; ROM = range of motion.

instance, the person with C5 tetraplegia will require exercises that do not require active wrist extension, elbow extension, or grasp, whereas those with C6 lesions have active wrist extension but little or no elbow extension or grasp. Table 27.3 can be used to determine the effects of the level of SCI on the ability to exercise. Additionally, persons with incomplete SCI who have relative sparing of sensation in the lower extremities may not be able to tolerate FES-LCE or hybrid exercise modes because of pain.

The initial stages of an aerobic exercise program should focus more on developing the habit of exercise, rather than the intensity and duration of exercise, because exercise adherence is poor in this population. Succinct, precise, and quantifiable goals will optimize chances of a successful outcome. The patient should make the informed choice of exercise mode, because this will partially affect patient compliance. Also, individuals should be told to expect delayed-onset muscle soreness (DOMS). The onset and duration of DOMS are similar to those for nondisabled individuals.

Cardiovascular Training

The most appropriate method to monitor and prescribe intensity of aerobic exercise for the individual with SCI is controversial. Heart rate responses in tetraplegia and high-level paraplegia are lower than in nondisabled individuals because of a poor sympathetic response (59). Although variable, it appears that 30% to 80% of heart rate reserve corresponds to 50% to 85% of $\dot{V}O_2$peak in those with high-level paraplegia and tetraplegia (46). Using a percentage of peak power output is also recommended (59). However, this is cumbersome because it requires continual reevaluation to maintain optimal training intensity since peak power output increases with training. Rating of perceived exertion (RPE) may also be used to guide exercise training intensity. This method is used successfully in cardiac transplant patients who, like patients with SCI, have reduced peak heart rates secondary to cardiac denervation (48). Although not investigated in patients with SCI, RPE values of 11 to 14 may be used during exercise training because this corresponds to approximately 60% to 90% of $\dot{V}O_2$peak in nondisabled individuals. Another possible method is the "talk test." Nondisabled individuals able to speak during exercise without feeling short of breath are typically at appropriate exercise intensity. Once again, however, this method has not been validated in the SCI population.

The duration of a single aerobic exercise bout will vary depending on fitness level, but a goal is to follow the recommendations for the general population (2). Exercise progression should occur as tolerated to a maximum of 60 min duration at the prescribed intensity range.

Aerobic exercise should be performed no fewer than 3 days a week to maintain fitness but may occur up to 5 days a week for optimal gains without negative consequences, particularly when performed at relatively low intensity levels (30). Persons with SCI may require increased exercise frequency to optimize caloric expenditure for weight management but should be judiciously monitored to reduce the incidence of upper extremity overuse syndromes. Persons with SCI who have $\dot{V}O_2$peak less than 15.5 ml · kg^{-1}· min^{-1} (<3 metabolic equivalents) may require multiple bouts of exercise daily, each lasting 5 to 15 min, until they are able to tolerate 20- to 30-min sessions.

Resistance Training

Minimally, scapular stabilization and rotator cuff resistance exercises should be used in all SCI patients capable of voluntary control of these muscles. Initial intervention should include two sets of 10 repetitions, with 6-s isometric contractions for shoulder protractors, retractors, elevators, and depressors, as well as for internal and external shoulder rota-

tors. As the patient tolerates, progression to dynamic exercise should occur.

Theraband type exercises are useful initially, but a plateau in gains can be expected because of the limitation in resistance of this device. Although dumbbells and free weights may be used under close supervision, paralyzed lower extremities and truncal musculature significantly reduce a person's ability to balance even small objects when lying supine or when seated without significant truncal support. When the person is using free weights or isotonic or isokinetic machines, wheelchair brakes should be set prior to lifting, and care should be taken not to exceed the wheelchair's weight and stress limitations as provided by the manufacturer. Standard recommendations for intensity and progression should be followed as discussed in chapter 7.

Range of Motion

Range of motion exercise should be performed daily and should focus on all major joints, especially those with contracture and spasticity. Both active and passive assisted methods of static stretching can be used. During passive stretching, it is important to carefully work through the range of motion in joints lacking sensation, because the individual cannot determine when the maximal range of the joint has been reached. It may also be important in select individuals to support the midshaft of long bones, if the patient has osteopenia, to reduce the chance of fracture.

Conclusion

The SCI patient presents with unique obstacles and considerations with which the clinical exercise professional must be familiar to provide safe and optimal exercise testing and training oversight. These individuals tend to be sedentary and are excellent candidates for regular exercise training. Because of the many potential medical and possible exercise-related problems, a team approach to evaluation, exercise prescription development, and exercise training guidance is recommended.

Case Study 27

Medical History

Ms. BF is a 28-year-old white female with thoracic paraplegia since age 19 caused by a motor vehicle accident. Up to that time she was athletic and played volleyball in high school. Since her rehabilita-

continued

tion, she has been wheelchair reliant for community mobility but is otherwise independent for activities of daily living, including bowel/bladder management and driving with hand controls. She received an MBA this year, has a full-time job as an accountant for a law firm, but is otherwise quite sedentary. Her body weight has remained stable at 150 lb for the past 4 years, and her body mass index (BMI) = 31. Other exam findings include blood pressure (BP) = 100/60; heart rate (HR) = 85; electrocardiogram (ECG) with normal sinus rhythm; normal heart, lung, and bowel sounds; intact skin, mildly reduced range of motion at the shoulder (extension, internal and external rotation); 5° hip, knee, and plantar flexion contractures bilaterally; 5/5 upper extremity with 0/5 lower extremity motor strength; absent pin prick and light touch sensation everywhere below the xyphoid process; and 3+ (brisk) reflexes at both the knees and ankles. She reports poor sleep quality and excessive fatigue and intermittently notes pounding headaches and sweating associated with an overfull bladder. Her medications include Detrol for bladder spasms and Lioresol for lower extremity spasms and spasticity.

Diagnosis

Ms. BF has T6 American Spinal Injury Association A paraplegia with neurogenic bowel and bladder, spasticity, and occasional autonomic dysreflexia. She likely has impaired pulmonary function attributable to lower intercostal, paraspinal, and abdominal muscle paralysis. Although she is fully functional with her upper extremities, she will have poor to fair sitting balance because of paraspinal, abdominal, and lower extremity paralysis. One suspects that her elevated BMI is the result of her sedentary lifestyle. The fact that she has completed a graduate degree and holds a full-time job suggests she has adjusted well to her SCI.

Exercise Test Results

Ms. BF will most likely want to know how exercise can affect her current lifestyle and health, and whether it is worth investing time already committed to a busy and productive life. Encouraging her to share short- and long-term life goals will provide an opportunity to educate her about the acute and chronic benefits of exercise, particularly as they can improve her community mobility and modify her elevated risks for chronic diseases, including coronary artery disease, obesity, diabetes, osteoporosis, and upper extremity overuse syndromes. As an initial step to increasing Ms. BF's physical activity level, an exercise evaluation is suggested.

Ms. BF's exercise evaluation is performed on an upper body ergometer by using a discontinuous protocol with an initial workload of 24 W at a constant 50 rev · min^{-1} and 6-W increments every 3 min. Her peak exercise performance yielded the following results: peak HR = 172, BP = 130/50, $\dot{V}O_2$peak = 16.4 ml · kg^{-1} · min^{-1}, peak power output = 42 W; ECG revealed isolated premature ventricular contractions (PVCs) with no evidence of ST-segment depression. Ms. BF stopped the test herself because of shoulder fatigue and dizziness, associated with a 10 mm Hg decrease in systolic blood pressure. She experienced some exertional dyspnea but no other symptoms.

Other Procedures

In the past, Ms. BF's physician has asked her to try FES. She was initially evaluated and thought to be a good candidate but never seriously considered FES.

continued

Case Study 27 *(continued)*

Development of Exercise Prescription

Ms. BF is interested in improved community mobility, losing weight, and maintaining good health. She is also motivated to begin a regular physical activity program. She will begin under the guidance of an outpatient physical therapist, with a goal of performing her routine at a medical fitness center under the guidance of a clinical exercise physiologist. Her maximal heart rate response on the exercise test will be used to set a target heart rate zone based on her heart rate reserve, and she will be introduced to a wheelchair ergometer as well as the upper body ergometer. Her goal is to perform between 20 and 60 min of aerobic exercise 3 to 5 days each week, although her initial workouts may be divided into two or three 10-min sessions, with 10-15 min recovery between sessions to reduce her risk of hypotension. Ms. BF will also begin a resistance-training program emphasizing shoulder stabilization with Therabands 2 to 3 days/week, progressing to machines or free weights as accessible. Range of motion exercises will be performed daily to improve sitting posture and shoulder biomechanics as well as to prepare for an eventual trial with FES exercise.

Case Study 27 Discussion Questions

1. What is your interpretation of the medical history?

2. What might you discuss with this patient regarding her examination findings and lifestyle, and how would you do this?

3. Discuss the exercise test with respect to mode, protocol, and her physiological responses. Considering these results, what are your recommendations for exercise?

4. Do you think that FES might benefit Ms. BF? Why or why not?

5. At what heart rate should she perform aerobic exercise?

6. What specific precautions should be considered for individuals with SCI who perform exercise?

7. What types of strength and range of motion exercises should she perform?

8. How might you keep Ms. BF motivated?

Glossary

activities of daily living (ADLs)—Bathing, dressing, grooming, toileting, feeding, and transferring.

astrocyte—A star-shaped neural cell that provides nutrients, support, and insulation for neurons of the central nervous system.

autonomic dysreflexia—Sudden, exaggerated reflex increase in blood pressure in persons with SCI above T6, sometimes accompanied by bradycardia, in response to a noxious stimulus originating below the level of SCI.

autonomic nervous system—Components of the nervous system responsible for coordinating life-sustaining processes and organization of visceral responses to somatic reactions.

cauda equina—Lumbosacral spinal nerve roots forming a cluster at the terminal region of the spinal cord that resembles a horse's tail.

circulatory hypokinesis—Insufficient vascular tone resulting in hypotension, as increased metabolic demands of upper extremity exertion are not matched by appropriate hemodynamic responses.

community mobility—Locomotion and transportation of an individual through his or her community.

complete paraplegia—Motor and sensory dysfunction of the trunk, legs, and pelvic organs resulting from SCI, without motor or sensory sparing below the level of the injury.

complete tetraplegia—Motor and sensory dysfunction of the arms, trunk, legs, and pelvic organs resulting from SCI, without motor or sensory function spared below the level of the injury.

disimpaction—Manual removal of fecal material from rectal vault.

functional electrical stimulation—Externally applied electrical stimulation of neuromuscular elements to activate paralyzed muscles in precise sequence and intensity to restore muscular function.

gliosis—Excess of astroglia in damaged areas of central nervous system.

hybrid—Combined use of concurrent upper extremity exercise and lower extremity functional electrical stimulation ergometry.

hydronephrosis—Kidney with dilated renal pelvis and collecting system attributable to ureteral obstruction or backflow (reflux) from bladder.

incomplete paraplegia—Incomplete motor and sensory dysfunction of the trunk, legs, and pelvic organs resulting from SCI.

incomplete tetraplegia—Incomplete motor and sensory dysfunction of the arms, trunk, legs, and pelvic organs resulting from SCI.

insensate—Lacking sensation.

joint contractures—Reduced passive range of motion at a joint caused by shortened tendons, typically associated with unbalanced spasticity.

myelomeningocele—Congenital open neural tube defect with disruption of skin, bone, and neural elements; usually involves spinal cord dysfunction despite surgical closure.

orthostatic hypotension—Decrease of at least 20 mm Hg in systolic blood pressure when an individual moves from a supine position to a standing position.

osteopenia—Bone density between 1 and 2.5 standard deviations below average for young people.

paraplegia—Motor and sensory dysfunction of the trunk, legs, and pelvic organs resulting from SCI.

parasympathetic nervous system—Craniosacral portion of the central nervous system that promotes anabolic activity and energy conservation.

peripheral nervous system—Sensory and motor components of the nervous system that have extensions outside of the brain and spinal cord.

poikilothermic—Having body temperature that varies with the environment.

somatic nervous system—Neural elements over which one has conscious awareness and control.

somatosensory evoked potentials—Physiological or electrical stimulation of afferent peripheral nerve fibers with subsequent monitoring of (evoked) electrical activity in the central nervous system, usually at the levels of the spinal cord, midbrain, and cortex.

spasticity—Velocity-dependent tone, typically associated with hyperreflexia.

spina bifida—Congenital neural tube defect with varying degrees of skin, bone, and neural element involvement.

spinal cord injury (SCI)—Damage involving the spinal cord.

sympathetic nervous system—Lumbosacral portion of the central nervous system that promotes the classic "fight or flight" response to a given stimuli.

syringomyelia—Chronic syndrome characterized pathologically by cavitation and gliosis of the spinal cord (usually cervical or thoracic), medulla, or both.

tetraplegia—Motor and sensory dysfunction of the arms, trunk, legs, and pelvic organs resulting from spinal cord injury.

vesicoureteral reflux—Backflow of urine from the bladder into the upper urinary tracts.

References

1. Aksnes AK, Hjeltnes N, Wahlstrom EO, Katz A, Zierath JR, Wallberg-Henriksson H. Intact glucose transport in morphologically altered denervated skeletal muscle from quadriplegic patients. American Journal of Physiology 271(3 Pt 1): E593-600, 1996.
2. American College of Sports Medicine. Position stand: The recommended quantity and quality of exercise for developing and maintaining cardiorespiratory and muscular fitness in healthy adults. Medicine and Science in Sports and Exercise 30(6): 975-991, 1998.
3. Barstow TJ, Scremin AM, Mutton DL, Kunkel CF, Cagle TG, Whipp BJ. Gas exchange kinetics during functional electrical stimulation in subjects with spinal cord injury. Medicine and Science in Sports and Exercise 27(9): 1284-1291, 1995.
4. Barstow TJ, Scremin AM, Mutton DL, Kunkel CF, Cagle TG, Whipp BJ. Peak and kinetic cardiorespiratory responses during arm and leg exercise in patients with spinal cord injury. Spinal Cord 38(6): 340-345, 2000.
5. Bauman WA, Alexander LR, Zhong Y-G, Spungen AM. Stimulated leg ergometry training improves body composition and HDL-cholesterol values. Journal of the American Paraplegia Society 17(4): 201, 1994.
6. Bauman WA, Raza M, Chayes Z, Machac J. Tomographic thallium-201 myocardial perfusion imaging after intravenous dipyridamole in asymptomatic subjects with quadriplegia. Archives of Physical Medicine and Rehabilitation 74: 740-744, 1993.
7. Bauman WA, Spungen AM. Disorders of carbohydrate and lipid metabolism in veterans with paraplegia or quadriplegia: A model of premature aging. Metabolism 43(6): 749-756, 1994.
8. Bauman WA, Spungen AM. Metabolic changes in persons after spinal cord injury. Physical Medicine and Rehabilitation Clinics of North America 11(1): 109-140, 2000.
9. Bernard PL, Mercier J, Varray A, Prefaut C. Influence of lesion level on the cardioventilatory adaptations in paraplegic wheelchair athletes during muscular exercise. Spinal Cord 38(1): 16-25, 2000.
10. Bloomfield SA, Mysiw WJ, Jackson RD. Bone mass and endocrine adaptations to training in spinal cord injured individuals. Bone 19(1): 61-68, 1996.
11. Braddom RL, Rocco JF. Autonomic dysreflexia: A survey of current treatment. American Journal of Physical Medicine and Rehabilitation 70: 234-241, 1991.
12. Bryce TN, Ragnarsson KT. Pain after spinal cord injury. Physical Medicine and Rehabilitation Clinics of North America 11(1): 157-168, 2000.

13. Bursell JP, Little JW, Stiens SA. Electrodiagnosis in spinal cord injured persons with new weakness or sensory loss: Central and peripheral etiologies. Archives of Physical Medicine and Rehabilitation 80(8): 904-909, 1999.

14. Chawla JC, Bar C, Creber I, Price J, Andrews B. Techniques for improving the strength and fitness of spinal cord injured patients. Paraplegia 17: 185-189, 1979-80.

15. Cooper RA. A force/energy optimization model for wheelchair athletics. IEEE Transactions on Systems, Man and Cybernetics 20(2): 444-449, 1990.

16. Cooper RA, Baldini FD, Langbein WE, Robertson RN, Bennett P, Monical S. Prediction of pulmonary function in wheelchair users. Paraplegia 31(9): 560-570, 1993.

17. Crane L, Klerk K, Ruhl A, Warner P, Ruhl C, Roach KE. The effect of exercise on pulmonary function in persons with quadriplegia. Paraplegia 32: 435-441, 1994.

18. Curt A, Nitsche B, Rodic B, Schurch B, Dietz V. Assessment of autonomic dysreflexia in patients with spinal cord injury. Journal of Neurological and Neurosurgical Psychiatry 62(5): 473-477, 1997.

19. Curtis KA, Black K. Shoulder pain in female wheelchair basketball players. Journal of Orthopaedic and Sports Physical Therapy 29(4): 225-231, 1999.

20. Dallmeijer AJ, Hopman MTE, van AS HHJ, van der Woude LHV. Physical capacity and physical strain in persons with tetraplegia, the role of sport activity. Spinal Cord 34: 729-735, 1996.

21. Davis GM. Exercise capacity of individuals with paraplegia. Medicine and Science in Sports and Exercise 25(4): 423-432, 1993.

22. Denis F. The three column spine and its significance in the classification of acute thoracolumbar spinal injuries. Spine 8(8): 817-831, 1983.

23. DeVivo MJ, Krause JS, Lammertse DP. Recent trends in mortality and causes of death among persons with spinal cord injury. Archives of Physical Medicine and Rehabilitation 80(11): 1411-1419, 1999.

24. Dixit S. Bradycardia associated with high cervical spinal cord injury. Surgical Neurology 43: 514, 1995.

25. Faghri PD, Glaser RM, Figoni SF. Functional electrical stimulation leg cycle ergometer exercise: Training effects on cardiorespiratory responses of spinal cord injured subjects at rest and during submaximal exercise. Archives of Physical Medicine and Rehabilitation 73(11): 1085-1093, 1992.

26. Ferguson-Pell MW. Technical considerations: Seat cushion selection. In: Choosing a Wheelchair System: Journal of Rehabilitation Research and Development Clinical Supplement #2, Todd SP (Ed). Department of Veterans Affairs, Veterans Health Administration, Rehabilitation Research & Development Service, Scientific and Technical Publications Section, Baltimore, pp. 47-73, 1992.

27. Figoni SF, Glaser RM, Collins SR. Peak physiologic responses of trained quadriplegics during arm, leg and hybrid exercise in two postures (Abstract). Medicine and Science in Sports and Exercise 27(5): S83, 1995.

28. Figoni FS, McLain L, Bell AA, et al. Accessibility of physical fitness facilities in the Kansas City metropolitan area. Topics in Spinal Cord Injury and Rehabilitation 3(3): 66-78, 1998.

29. Figoni SF, Rodgers MM, Glaser RM, Hooker SP, Feghri PD, Ezenwa BN, Mathews T, Suryaprasad AG, Gupta SC. Physiologic responses of paraplegics and quadriplegics to passive and active leg cycle ergometry. Journal of the American Paraplegia Society 13(3): 33-39, 1990.

30. Franklin BA (Ed). ACSM's Guidelines for Exercise Testing and Prescription, 6th edition. American College of Sports Medicine. Philadelphia, Lippincott Williams & Wilkins, 2000.

31. Franklin BA, Swnatek KI, Grais SL, Johnstone KS, Gordon S, Timmis GC. Field test estimation of maximal oxygen consumption in wheelchair users. Archives of Physical Medicine and Rehabilitation 71: 574-578, 1990.

32. Garland DE, Stewart CA, Adkins RH, Hu SS, Rosen C, Liotta FJ, Weinstein DA. Osteoporosis after spinal cord injury. Journal of Orthopedic Research 10(3): 371-378, 1992.

33. Gass GC, Watson J, Camp EM, Court HJ, McPherson LM, Redhead P. The effects of physical training on high level spinal lesion patients. Scandinavian Journal of Rehabilitation Medicine 12: 61-65, 1980.

34. Gater DR, Yates JW, Clasey JL. Relationship between glucose intolerance and body composition in spinal cord injury. Medicine & Science in Sports and Exercise 32(5): S148, 2000.

35. Glaser RM. Functional neuromuscular stimulation. Exercise conditioning of spinal cord injured patients. International Journal of Sports Medicine 15(3): 142-148, 1994.

36. Glaser RM, Janssen TWJ, Suryaprasad AG, Gupta SC, Mathews T. The physiology of exercise. In: Physical Fitness: A Guide for Individuals With Spinal Cord Injury, Apple DF (Ed). Department of Veterans Affairs, Veterans Health Administration, Rehabilitation Research & Development Service, Scientific and Technical Publications Section, Baltimore, pp. 1-24, 1996.

37. Glaser RM, Sawka MN, Brune MF, Wilde SW. Physiological responses to maximal effort wheelchair ergometry and arm crank ergometry. Journal of Applied Physiology 48: 1060-1064, 1980.

38. Go BK, DeVivo MJ, Richards JS. The epidemiology of spinal cord injury. In: Spinal Cord Injury: Clinical Outcomes from the Model Systems, Stover SL, DeLisa JA, Whiteneck GG (Eds). Rockville, MD, Aspen, pp. 21-51, 1995.

39. Hangartner TN, Rodgers MM, Glaser RM, Barre PS. Tibial bone density loss in spinal cord injured patients: Effects of FES exercise. Journal of Rehabilitation Research and Development 31(1): 50-61, 1994.

40. Heymsfeld SB, Lichtman S, Baumgartner RN, et al. Body composition of humans: Comparison of two improved

four-compartment models that differ in expense, technical complexity, and radiation exposure. American Journal of Clinical Nutrition 52(1): 52-58, 1990.

41. Hjeltnes N, Jansen T. Physical endurance capacity, functional status and medical complications in spinal cord injured subjects with long-standing lesions. Paraplegia 28: 428-432, 1990.

42. Hooker SP, Figoni SF, Rodgers MM, Glaser RM, Mathews T, Suryaprasad AG, Gupta SC. Physiologic effects of electrical stimulation leg cycle exercise training in spinal cord injured persons. Archives of Physical Medicine and Rehabilitation 73(5): 470-476, 1992.

43. Hooker SP, Greenwood JD, Hatae DT, et al. Oxygen uptake and heart rate relationship in persons with spinal cord injury. Medicine and Science in Sports and Exercise 25(10): 1115-1119, 1993.

44. Janssen TWJ, van Oers CAJM, Rozendaal EP, Willemsen EM, Hollander AP, van der Woude LHV. Changes in physical strain and physical capacity in men with spinal cord injuries. Medicine and Science in Sports and Exercise 28(5): 551-559, 1996.

45. Janssen TW, van Oers CA, van Kamp GJ, TenVoorde BJ, van der Woude LH, Hollander AP. Coronary heart disease risk indicators, aerobic power, and physical activity in men with spinal cord injuries. Archives of Physical Medicine and Rehabilitation 78(7): 697-705, 1997.

46. Janssen TWJ, van Oers CAJM, van der Woude LHV, Hollander AP. Physical strain in daily life of wheelchair users with spinal cord injuries. Medicine and Science in Sports and Exercise 26(6): 661-670, 1994.

47. Katz RT. Management of spasticity. American Journal of Physical Medicine and Rehabilitation 67: 108-116, 1988.

48. Kavanagh T. Physical training in heart transplant recipients. Journal of Cardiovascular Risk 3(2): 154-159, 1996.

49. Kawamoto M, Sakimura S, Takasaki M. Transient increase of parasympathetic tone in patients with cervical spinal cord trauma. Anaesthesia and Intensive Care 21: 218-221, 1993.

50. Knutsson E, Lewenhaupt-Olsson E, Thorsen M. Physical work capacity and physical conditioning in paraplegic patients. Paraplegia 11(3): 205-216, 1973.

51. Langbein WE, Edwards SC, Louie EK, Hwang MH, Nemchausky BA. Wheelchair exercise and digital echocardiography for the detection of heart disease. Rehabilitation Research and Development Reports 34: 324-325, 1996.

52. Langbein WE, Maki KC, Edwards LC, Hwang MH, Sibley P, Fehr L. Initial clinical evaluation of a wheelchair ergometer for diagnostic exercise testing: A technical note. Journal of Rehabilitation Research and Development 31(4): 317-325, 1994.

53. Laskin JJ, Ashley EA, Olenik LM, Burnham R, Cumming DC, Steadward RD, Wheeler GD. Electrical stimulation-assisted rowing exercise in spinal cord injured people. A pilot study. Paraplegia 31(8): 534-541, 1993.

54. Lassau-Wray ER, Ward GR. Varying physiological response to arm-crank exercise in specific spinal injuries. Journal of Physiological Anthropology and Applied Human Science 19(1): 5-12, 2000.

55. Loudon JK, Cagle PE, Figoni SF, Nau KL, Klein RM. A submaximal all-extremity exercise test to predict maximal oxygen consumption. Medicine and Science in Sports and Exercise 30(8): 1299-1303, 1998.

56. Lundberg A. Wheelchair driving: Evaluation of a new training outfit. Scandinavian Journal of Rehabilitation Medicine 12: 67-72, 1987.

57. Marino RJ, Ditunno JF, Donovan WH, et al. International Standards for Neurological and Functional Classification of Spinal Cord Injury, Revised 2000. American Spinal Injury Association, Chicago, 2000.

58. Maynard FM, Karunas RS, Adkins RH, Richards JS, Waring III WP. Management of the neuromusculoskeletal systems. In: Spinal Cord Injury: Clinical Outcomes From the Model Systems, Stover SL, DeLisa JA, Whiteneck GG (Eds), Bethesda, MD, Aspen, pp 145-169, 1995.

59. McLean KP, Skinner JS. Effect of body training position on outcomes of an aerobic training study on individuals with quadriplegia. Archives of Physical Medicine and Rehabilitation 76:139-150, 1995.

60. Mohr T, Podenphant J, Biering-Sorensen F, Galbo H, Thamsborg G, Kjaer M. Increased bone mineral density after prolonged electrically induced cycle training of paralyzed limbs in spinal cord injured man. Calcified Tissue International 61(1): 22-25, 1997.

61. Mutton DL, Scremin AME, Barstow TJ, Scott MD, Kunkel CF, Cagle TG. Physiologic responses during functional electrical stimulation leg cycling and hybrid exercise in spinal cord injured subjects. Archives of Physical Medicine and Rehabilitation 78: 712-718, 1997.

62. Naftchi NE. Mechanism of autonomic dysreflexia. Contributions of catecholamine and peptide neurotransmitters. Annals of the New York Academy of Science 579: 133-148, 1990.

63. Nash MS, Bilsker MS, Kearney HM, Ramirez JN, Applegate B, Green BA. Effects of electrically-stimulated exercise and passive motion on echocardiographically-derived wall motion and cardiodynamic function in tetraplegic persons. Paraplegia 33(2): 80-89, 1995.

64. Nash MS, Bilsker S, Marcillo AE, Isaac SM, Botelho LA, Klose J, Green BA, Rountree MT, Shea JD. Reversal of adaptive left ventricular atrophy following electrically-stimulated exercise training in human tetraplegics. Paraplegia 29: 590-599, 1991.

65. Nilsson S, Staff PH, Pruett ED. Physical work capacity and the effect of training on subjects with long-standing paraplegia. Scandinavian Journal of Rehabilitation Medicine 7(2): 51-56, 1975.

66. National Spinal Cord Injury Association. Spinal cord injury statistics. Available at www.spinalcord.org/resource/Factsheets/factsheet2.html 1999.

67. Petrofsky JS. Thermoregulatory stress during rest and exercise in heat in patients with a spinal cord injury. European Journal of Applied Physiology 64(6): 503-507, 1992.

68. Phillips W, Burkett LN, Munro R, Davis M, Pomeroy K. Relative changes in blood flow with functional electrical stimulation during exercise of the paralyzed lower limbs. Paraplegia 33(2): 90-93, 1995.

69. Ragnarsson KT, Pollack S, O'Daniel W, Edgar R, Petrofsky J, Nash MS. Clinical evaluation of computerized functional electrical stimulation after spinal cord injury: A multicenter pilot study. Archives of Physical Medicine and Rehabilitation 69: 672-677, 1988.

70. Reame NE. A prospective study of the menstrual cycle and spinal cord injury. American Journal of Physical Medicine and Rehabilitation 71: 15-21, 1992.

71. Rodgers MM, Glaser RM, Figoni SF, Hooker SP, Ezenwa BN, Collins SR, Mathews T, Suryaprasad AG, Gupta SC. Musculoskeletal responses of spinal cord injured individuals to functional neuromuscular stimulation-induced knee extension exercise training. Journal of Rehabilitation Research and Development 28(4): 19-26, 1991.

72. Ropper AH. Trauma of the head and spine. In: Harrison's Principles of Internal Medicine (13th ed.), Isselbacher KJ, Braunwald E, Wilson JD, Martin JB, Fauci AS, Kasper DL (Eds.). St. Louis, McGraw Hill, 1994.

73. Sawka MN, Foley ME, Pimental NA, Toner MM, Pandolf KB. Determination of maximal aerobic power during upper-body exercise. Journal of Applied Physiology: Respiratory, Environmental and Exercise Physiology 54: 113-117, 1983.

74. Schmid A, Huonker M, Barturen JM, Stahl F, Schmidt-Trucksass A, Konig D, Grathwohl D, Lehmann M, Keul J. Catecholamines, heart rate, and oxygen uptake during exercise in persons with spinal cord injury. Journal of Applied Physiology 85(2): 635-641, 1998.

75. Schmidt KD, Chan CW. Thermoregulation and fever in normal persons and in those with spinal cord injuries. Mayo Clinic Proceedings 67(5): 469-475, 1992.

76. Schneider DA, Sedlock DA, Gass E, Gass G. VO_2 peak and the gas-exchange anaerobic threshold during incremental arm cranking in able-bodied and paraplegic men. European Journal of Applied Physiology and Occupational Physiology 80(4): 292-297, 1999.

77. Sharkey BJ, Graetzer DG. Specificity of exercise, training, and testing. In: ACSM's Resource Manual for Guidelines for Exercise Testing and Prescription, Second Edition, Durstine JL, King AC, Painter PL, Roitman JL,

Zwiren LD (Eds.). Philadelphia, Lea & Febiger, pp. 82-92, 1993.

78. Sipski ML, Delisa JA, Schweer SA. Functional electrical stimulation bicycle ergometry: Patient perceptions. American Journal of Physical Medicine and Rehabilitation 68(3): 147-149, 1989.

79. Spungen AM, Bauman WA, Wang J, Pierson RN. Measurement of body fat in individuals with tetraplegia: A comparison of eight clinical methods. Paraplegia 33(7): 402-408, 1995.

80. Subarrao JV, Garrison SJ. Heterotopic ossification: Diagnosis and management, current concepts and controversies. Journal of Spinal Cord Medicine 22(4): 273-283, 1999.

81. Subarrao JV, Klopfstein J, Turpin R. Prevalence and impact of wrist and shoulder pain in patients with spinal cord injury. Journal of Spinal Cord Medicine 18(1): 9-13, 1995.

82. Taylor AW, McDonell E, Brassard L. The effects of an arm ergometer training programme on wheelchair subjects. Paraplegia 24: 105-114, 1986.

83. Tsitouras PD, Zhong YG, Spungen AM, Bauman WA. Serum testosterone and growth hormone/insulin-like growth factor-I in adults with spinal cord injury. Hormone and Metabolic Research 27(6): 287-292, 1995.

84. Ugalde V, Litwiller SE, Gater DR. Physiatric anatomic principles, bladder and bowel anatomy for the physiatrist. Physical Medicine & Rehabilitation: State of the Art Reviews 10: 547-568, 1996.

85. Van Der Woude LHV, Veeger HEJ, Rozendal RH, Van Ingen Schenau GJ, Rooth F, Van Nierop P. Wheelchair racing: Effects of rim diameter and speed on physiology and technique. Medicine and Science in Sports and Exercise 20(5): 492-500, 1988.

86. Vinet A, Bernard PL, Poulain M, Varray A, Le Gallais D, Micallef JP. Validation of an incremental field test for the direct assessment of peak oxygen uptake in wheelchair-dependent athletes. Spinal Cord 34(5): 288-293, 1996.

87. Wang YH, Huang TS. Impaired adrenal reserve in men with spinal cord injury: Results of low- and high-dose adrenocorticotropin stimulation tests. Archives in Physical Medicine and Rehabilitation 80(8): 863-866, 1999.

88. Wolff J. Das Gesetz der Transformation der Knochen. Berlin, Ahirshwald, 1892.

89. Young RR. Spasticity: A review. Neurology 44(Suppl 9): S12-S20, 1994.

Charles P. Lambert, PhD
Nutrition, Metabolism, and Exercise Laboratory
Department of Geriatrics
University of Arkansas for Medical Sciences
Little Rock, AR

Multiple Sclerosis

Multiple sclerosis (MS) is characterized by random or sporadic patches of inflammation of the central nervous system that result in **demyelination.** The demyelination, in turn, causes **plaques,** which can become permanent scars (45). **Sclerosis** refers to the condition of demyelination and has many causes, and the term multiple sclerosis refers to the multiple areas of demyelinated tissue of the central nervous system (12).

Scope

In the United States, MS is most common in Caucasians of Northern European descent (45). As of 1996, there were approximately 265,000 cases of MS in the United States, with 64% afflicting females (14). Approximately 20% of people with MS have a close relative with the disease, and 5% have a sibling with MS (45). The gene DR2 appears to be involved in the response of the immune system and is common in individuals from Northern Europe. However, only a small percentage of people who have this gene develop MS, suggesting an environmental component (45).

The initial diagnosis of MS is most common in individuals between the ages of 15 and 50. In the United States, most affected individuals live in the north (i.e., above 40° latitude) (10). The prevalence rate in 1996 was 58.3 cases per 100,000 people. The incidence rate in 1996 was 3.2 cases occurring per 100,000 live births in the United States (14).

MS can be devastating, as demonstrated by the fact that within 10 years of diagnosis, more than 50% of individuals with MS will become unemployable (13). In an investigation that correlated healthcare costs to the level of neurological dysfunction in U.S. military veterans, the average total healthcare cost was $35,000 per year, and these costs were related to the severity of neurological dysfunction (4). This relationship has also been reported in Canadian citizens (5).

Physiology and Pathophysiology

MS can follow at least seven courses of clinical progression (table 28.1) (23). Early indicators of MS include **asymmetry of reflexes,** exceedingly rapid reflexes, or **pathological reflexes** (27). Muscle spasticity is usually not present until the disease reaches the advanced stages (27). Some individuals do not demonstrate clinical signs of MS during life, and MS is only diagnosed during autopsy. Many who have one or two attacks frequently completely recover and never incur any permanent disability (45). Some individuals with MS have frequent attacks, and al-

though they do not completely recover they may never become "disabled." This latter scenario occurs in approximately 25% of all cases of MS (45). For many patients, MS progression is slow, and after 10 to 25 years, walking without assistance is often difficult or impossible.

Clinical Considerations

The common problem for those with MS is an increasing inability to ambulate. This results from the multitude of problems associated with the disease. Proper treatment can help to delay or diminish the debilitation of the patient. The following section reviews clinical indicators of the disease and state-of-the-art treatment.

Signs and Symptoms

The initial symptoms of MS are associated with the areas of the central nervous system that are demyelinated to the greatest degree. The most common initial symptom is weakness in one or more limbs, **optic neuritis,** and **paresthesia** (8). When MS is present for 8 to 14 years, general weakness is observed in about 80% of patients. Advanced MS often presents with weakness of both upper and lower limbs, with one side typically weaker than the other. Spasticity markedly affects voluntary movement.

TABLE 28. 1
Clinical Courses of Multiple Sclerosis

Type	Characteristic
Relapsing-remitting	Characterized by disease relapses with either a full recovery or a deficit after recovery; no progression of disease symptoms in recovery stage
Primary-progressive	Disease progression from onset with infrequent plateaus and temporary small improvements; clinical status is continuously worsening with no distinctive relapses
Secondary-progressive	Begins as relapsing–remitting but continually progresses either with or without infrequent relapses, plateaus, and remissions
Relapsing-progressive	Combination of relapse with subsequent disease progression; no consensus clinical definition
Progressive-relapsing	Progressive from onset with short definite relapses with or without full recovery
Benign	Neurologic system remains fully functional 15 years after the initial onset
Malignant	Progressive and rapid resulting in death or multiple neurologic system disability shortly after onset

Systemic fatigue is common and typically is noted after routine work such as house cleaning. Cerebellar symptoms occur but are difficult to distinguish from spasticity, weakness, **vertigo,** and sensory loss. **Nystagmus** is also a common sign of MS (8). Table 28.2 provides a list of common signs and symptoms of MS.

TABLE 28.2
Common Signs and Symptoms of Multiple Sclerosis

Symptoms	Signs
Weakness	Optic neuritis
Spasticity	Paresthesia
Vertigo	Nystagmus
Systemic fatigue	
Sensory loss	
Memory impairment	

Psychological effects such as an impaired cognitive ability and memory loss can occur early after disease onset. A strong correlation exists between the level of physical disability and defect of cognition (26). Depression is also associated with MS and appears to be affected by living circumstances (e.g., nonambulatory, house-bound). However, the potential for the disease process directly causing depression also exists.

Urinary urgency, **incontinence,** and sexual dysfunction in both men and women are often present. Impaired thermoregulation is also common with MS. It was reported that about 40% of MS patients have an abnormal sweating response (6, 32). Abnormal cardiovascular function during exercise (e.g., low peak heart rate) can be caused by autonomic cardiovascular reflex dysfunction and may be seen in up to 50% of patients with MS (34, 43, 47).

Diagnostic and Laboratory Evaluations

From 1965 through the 1980s, the diagnostic criteria for MS were limited (27, 42). When five of the six criteria were met, and included number 6 of figure 28.1, the clinical diagnosis of MS was satisfied.

1. Examination of objective neurological indicators.
2. Onset of symptoms between the ages of 10 and 50 years.
3. Neurological signs and symptoms of central nervous system white matter disease.
4. Dissemination of disease activity in time: two or more attacks that lasted at least 24 hr and were separated by 1 month or more; or 6 months of progression of signs and symptoms.
5. Two or more nonconnected anatomical areas affected.
6. No alternative clinical explanation.

FIGURE 28.1   Criteria for diagnosing multiple sclerosis from 1965 to the mid-1980s.

In the 1970s and 1980s, advances in imaging (e.g., **computed tomography,** or **CT**; **magnetic resonance imaging,** or **MRI**) and the development of **evoked response testing** allowed for the visualization and evaluation of sclerotic lesions. In addition, obtaining cerebrospinal fluid samples to detect specific immune factors has also advanced diagnostic ability. In 1983, new diagnostic criteria that incorporated these diagnostic tools and information obtained from immunological analyses were developed (38). However, even today diagnosis initially focuses on the medical history and neurological examination, and then the decision to perform an MRI, a CT scan, or other testing should be made. Figure 28.2 shows an MRI of the brain of a MS patient. Practical application 28.1 provides more information about imaging techniques.

Treatment

Common skeletal muscle problems associated with MS include **spasticity, tremors,** and pain. Spasticity reduces one's ability to coordinate muscle contractions for movements such as walking. Tremors can also affect the ability to coordinate skeletal muscle contraction during physical activity. The primary treatment for these conditions in patients with MS is drug therapy. Several categories of medications are used to treat MS, including immunosuppressants, corticosteroids, immunomodulators, and antispastics. Musculoskeletal pain can be treated with acetaminophen or aspirin (45). If pain

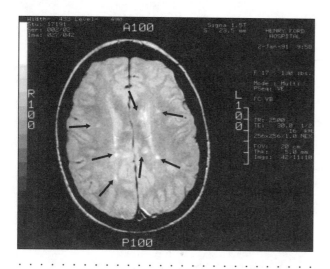

FIGURE 28.2 Magnetic resonance image depicting multiple (arrows) areas of demyelination within the white matter of the brain.

Reprinted, by permission, from C.A. Brawner and J.R. Schairer, 2000, "Multiple sclerosis; Case report from Henry Ford Hospital," *Clinical Exercise Physiology* 2(1): 17.

is attributed to plaques within the nervous system, tricyclic antidepressants are commonly used and are effective (45).

Fatigue is common and has two primary mechanisms in patients with MS. One type is systemic and results in persistent tiredness; the other involves fatigue following physical activity and typically affects the legs. Amantadine and Pemoline (Cylert) reduce systemic long-term fatigue. Pemoline improves symptoms in up to 60% of patients with MS (45). Depression is also common in patients with MS. The use of tricyclic antidepressants or monoamine oxidase inhibitors is useful (41, 45).

An increase in urinary frequency may be the result of an irritable bladder wall. This appears to be a manifestation of the demyelination associated with the disease but may also be secondary to bladder infection caused by an inability to control bladder function. Imipramine (Tofranil), which is an antidepressant, and isopropamide (Darbid), oxybutynin (Ditropan), propantheline (Pro Banthine), and

PRACTICAL APPLICATION 28.1

Imaging Techniques

Computed tomography (CT) is a method of scanning capable of producing high-resolution, high-contrast images (48). A complete CT picture is a composite of several sliced images (9). During this procedure, an X-ray beam and detector move through a 360° arc around the patient, who is on a motorized couch. A highly focused X-ray beam irradiates the subject. The image produced corresponds to the amount of X-ray beam that passes through each section and is inversely proportional to the density of the tissue (9). The scan takes only a few minutes to complete.

The detector system converts the X-rays to an electron stream. The stream is then digitized and quantified into a system of measurement known as the Hounsfield unit. These units are used to differentiate between tissue densities. Dense structures, such as bone, absorb a large quantity of X-ray and appear white. Air shows up as very dark or black. Water is typically the standard to which other tissues are compared and is given a Hounsfield unit of zero (48). CT scans of the brain may demonstrate ventricular enlargement, low-density periventricular abnormalities, or focal regions of enhancement attributable to contrast agent injection in patients with definite MS. In addition to its use in patients with MS, CT is also useful for detecting calcification of the coronary arteries.

Magnetic resonance imaging (MRI) uses magnets and radiofrequency waves to produce high-quality images of anatomy in several planes (9). MRI does not require the use of ionizing radiation (2). Instead, radiofrequency pulses pass through a patient who is placed in a large magnetic field. The computer assembles the image from 1- to 10-mm scanned slices or sections. MRI takes approximately 30 to 45 min.

Tissues with high hydrogen proton numbers, such as fat, will emit a strong signal. A computer program converts the signal to an intense or bright image. When less hydrogen is present, such as in bone, a signal with a low intensity will be produced, showing up as dark or black. The resulting image has excellent contrast resolution.

More than 90% of patients with MS will have a positive MRI or CT scan with increased brightness or whiteness at the affected areas of the brain. Because of its excellent resolution, MRI is the most commonly used imaging technique available for MS-affected tissue.

dicyclomine (Bentyl), which are all anticholinergics, can be used to reduce the irritative effects.

Immunosuppressive drugs are used in an attempt to suppress autoimmune effects on myelin. These medications may be best for patients with MS who have very slow disease progression (45). Immunomodulators are agents that alter the function of the immune system with a goal of delaying MS progression and short-term **exacerbation** (40, 45). These are effective for relapsing-remitting MS, and the reduction in the rate of **relapse** may be as much as 29% (15, 22, 40). Additionally, immunomodulator treatment reduces the number of new or enlarged lesions measured by MRI (8, 25, 33, 46).

The use of corticosteroids does not appear to affect MS progression or exacerbation, but these drugs are effective at reducing the recovery time from an exacerbative event. However, side effects are associated with long-term use that are important for a clinical exercise professional to understand, including bone softening and high blood pressure. These drugs may also result in adrenal insufficiency. Short-term side effects include acne, psychosis, infection, and ulcers (45).

A variety of interventions have been used in the treatment of MS (figure 28.3). Of these, only **thalamotomy** and thalamic stimulation have demonstrated potential benefits. However, the evidence is equivocal and these therapies should be used with caution, because they are costly, currently are not approved by the Food and Drug Administration, and are considered investigational (45).

Natural killer and nonspecific suppressor cell function is impaired in MS patients (1, 3). This could result in an inability to suppress an acute autoimmune response leading to disease exacerbation. Exercise improves natural killer and suppressor cell function in individuals free of disease (21, 31). Exercise may also reduce some secondary effects of MS. These include impaired bladder and bowel function (35). Finally, exercise may positively affect psychological health and quality of life in these patients (35).

Graded Exercise Testing

Exercise testing is useful in the MS patient to determine the safety and effectiveness of exercise training. Optimally, each patient should be evaluated prior to testing. Resulting information can be used to develop a specific exercise prescription. Practical application 28.2 discusses client–clinician interaction for patients with MS.

- Acupuncture
- Dorsal column stimulation
- Hyperbaric oxygen
- Transcutaneous nerve stimulation
- Thalamotomy
- Thalamic stimulation
- Sympathectomy
- Ganglionectomy
- Surgical spinal cord relaxation
- Vertebral artery surgery
- Ultrasound
- Magnetotherapy
- Dental occlusal therapy
- Replacement of mercury amalgam fillings
- Hysterectomy
- Surgical implantation of pig brain into the abdominal wall (i.e., cellular therapy)

FIGURE 28.3 Interventional treatments for multiple sclerosis.

Adapted from W.A. Sibley, 1996, *Therapeutic Claims in Multiple Sclerosis: A Guide to Treatments*, (New York: Demos Vermande).

Medical Evaluation

Although MS is a disease with a multitude of signs and symptoms, rarely does an individual exhibit a majority of these (table 28.2). Several are important with respect to exercise safety. Problems with vision are exhibited by some. Issues related to sight during exercise should be addressed in these cases. Those with cognitive deficits may have an impaired ability to comprehend the purpose of an exercise test, provide informed consent, provide feedback, and follow an exercise prescription. In this case, someone responsible for the individual should accompany him or her. Loss of strength or numbness in a leg or arm or poor flexibility can affect the ability to perform certain types of exercise and must be considered. Classification of the level of MS-related disability based on the medical evaluation can be made using the Kurtzke Functional Systems (table 28.3) and the Kurtzke Expanded Disability Status Scale (table 28.4). These tables are used together to objectively rate a patient's level of functional disability.

Review of the medical history should be used to determine any prescribed or over-the-counter medications an individual is using. Medications to treat muscle spasticity often result in fatigue, which may

PRACTICAL APPLICATION 28.2

Client–Clinician Interaction

The systemic fatigue noted by individuals with MS varies from day to day. The clinical exercise professional must anticipate the need to reduce exercise training volume (i.e., intensity, duration) if an individual is fatigued. This requires daily assessment of an individual for indicators of fatigue. This is best assessed by verbal communication. Appropriate questions include, "How do you feel today?" and "Are you tired today?"

The clinical exercise professional must also be prepared to recognize and discuss indicators of depression. However, it is unlikely that most patients will divulge this information, and many times the clinician must ask questions to discern the degree of depression. Indicators of stress may be related to depression. These include poor sleep habits, noncompliance with lifestyle change, and elevated scores on standardized questionnaires. The exercise clinician should share the positive effects of exercise training on various psychological variables (e.g., mood, depression, anxiety). If necessary, patients with excessive depression should be referred to a mental health professional.

Individuals with MS may have cognitive problems. Because of this, the exercise clinician may have to repeat directions or clarify explanations. Many individuals with MS die of heart disease and stroke, and this may occur at a younger age if the disease has forced them into a sedentary lifestyle at an early age. The exercise clinician should counsel these patients about the risk factors associated with coronary artery and cardiovascular disease in those with MS.

It is important to provide constant motivation to individuals with MS to exercise regularly. Common reasons for nonadherence to exercise training are fatigue and muscle soreness. The clinical exercise professional must stress that although exercise training may be fatiguing and cause soreness in the short term, the soreness will likely cease and fatigue will reduce over the long term. Additionally, these patients should understand that resistance exercise will strengthen the skeletal muscles and allow them to perform activities of daily living with less fatigue.

TABLE 28.3
Kurtzke Functional Systems

Category	Rating scale range
Pyramidal	0-6
Bowel and bladder	0-6
Cerebellar	0-5
Visual (optic)	0-6
Brainstem	0-5
Cerebral (mental)	0-5
Sensory	0-6
Other[a]	0-3

Note. Ratings range from normal function (0), to signs only without disability (1), to increasing levels of disability (2-6).

[a]Any other neurological findings attributed to multiple sclerosis (i.e., fatigue).

Adapted from Ponichtera-Mulcare et al. (37).

limit maximal exercise capacity. It is common for skeletal muscle fatigue to occur prior to cardiorespiratory fatigue. During standard exercise testing, this may limit the ability to diagnose coronary artery disease (i.e., a reduced test sensitivity attributable to a false-negative test result). If a diagnostic test is desired, it may be best to use an imaging type test (see chapter 6).

Contraindications

Standard graded exercise testing contraindications are reviewed in chapter 6 and should be followed for patients with MS (16). There are no specific contraindications common among patients with MS.

Recommendations and Anticipated Responses

The exercise professional should carefully consider the mode of exercise used, because each has specific advantages. Use of a cycle ergometer does not require much balance compared with a treadmill. For individuals who have **ataxia** or who walk using an assistive device, a cycle ergometer should be best. Additionally, the resistance of a cycle ergometer can be adjusted low enough to be within most patients'

TABLE 28.4

Kurtzke Expanded Disability Status Scale

Rating	Disability	Functional limitations	Comparison to Kurtze FS rating
0	Normal neurological exam	None	0 in all FS categories
1.0	None	None	Grade 1 in one FS category
1.5	None	None	Grade 1 in more than one FS category
2.0	Minimal	Affects one FS	Grade 2 in one FS category
2.5	Minimal	Affects two FS	Grade 2 in two FS categories
3.0	Moderate	Affects three to four FS	Grade 2 in three to four FS categories
3.5	Moderate–fully ambulatory	Affects three to four FS	Grade 3 in one FS category and 1-2 in two categories
4.0	Mildly severe disability: fully ambulatory without aid, self-sufficient, up 12 hours per day	Affects one or more FS	Grade 4 in one FS category with 0-1 in others
4.5	Moderately severe disability: same as 4.0 with some limitation of ADLs or needs minimal assistance	Affects one or more FS	Grade 4 in one FS category with 0-1 in others
5.0	Severe: walk only 200 m without rest or aid; impaired ADLs	Affects one or more FS	Grade 5 in one FS category with 0-1 in others
5.5	Severe: walk only 100 m without rest or aid; impaired ADLs	Affects one or more FS	Grade 5 in one FS category with 0-1 in others
6.0	Severe: intermittent or unilateral aid to walk 100 m	Affects two or more FS	Grade 3+ in two or more FS categories
6.5	Very severe: constant aid (cane, crutches, braces) to ambulate 20 m	Affects two or more FS	Grade 3+ in two or more FS categories
7.0	Extremely severe: unable to walk 5 m with aid; wheelchair restricted; can wheel self; sits up 12+ hr per day; can transfer self from chair	Affects more than one FS	Grade 4+ in more than 1 FS category
7.5	Extremely severe: only takes a few steps; cannot wheel self; full day in wheelchair	Affects more than one FS	Grade 4 + in more than 1 FS category
8.0	No ambulation; restricted to bed or chair; retains self-care and can use arms	Affects many FS	Grade 4+ in several FS categories
8.5	Bed ridden; minimal arm use; some self-care	Affects many FS	Grade 4+ in many FS categories
9.0	Bed ridden; no self care; can talk and eat	Affects all FS	Mostly grade 4+ for all FS categories
9.5	Bed ridden; cannot communicate, eat, or swallow	Affects all FS	Grade 4+ for all FS categories
10	Death attributable to multiple sclerosis		

Note. FS = functional systems; ADL = activities of daily living.

capabilities. Some individuals may be better suited to using a treadmill because walking is the mode that ambulatory individuals use daily. The risk of falling should be considered and can be quickly evaluated during the walk from the waiting room to the testing room.

A low-level protocol beginning with a 1- to 2-min warm-up period may be best for most patients (28). Each 3-min stage should increase by no more than 2 metabolic equivalents (METs) for the treadmill, 12 to 25 W for leg ergometry, and 8 to 12 W for arm ergometry. Some younger patients with little disability may tolerate a high-level protocol. Table 28.5 reviews the physiological responses to exercise of the patient with MS.

Individuals with MS may have an attenuated systolic blood pressure response during exercise (35, 43). This may be related to autonomic dysfunction resulting from sclerotic plaques in paraventricular and cardiovascular autonomic nuclei (i.e., cardiovascular control center) region of the brain (47). Other evidence suggests that an attenuated pressor response may be attributable to a reduced skeletal muscle metabolic response affecting the skeletal muscle chemoreflex (29).

Individuals with MS may also have abnormal temperature regulation during exercise. Approximately 40% of these patients have an abnormal sweating response that is related to disease severity (6, 32). This increases the risk of hyperthermia. Use of electric fans to improve evaporative and convective cooling, attention to fluid replacement, and climate control (i.e., room temperature 72-76°, low humidity) are recommended to reduce the risk of heat-related illness.

Skeletal muscle fatigue may occur before patients reach peak cardiovascular levels. This may be caused by an inability to recruit additional skeletal muscle as other skeletal muscle fatigues or by impaired skeletal muscle metabolism (18, 19).

Exercise Prescription

When deciding about appropriateness of exercise and level of supervision, the clinical exercise physiologist can use the functional disability system presented in tables 28.3 and 28.4. Common disease-associated problems include fatigue, imbalance, incoordination, numbness or loss of sensation, and loss of strength. Supervision should be considered in these individuals, with the level of supervision dependent on the level of disability. Those with known impairment of heart rate or blood pressure responses during exercise may require monitoring. Practical application 28.3 reviews the literature about MS and exercise.

Special Exercise Considerations

Because individuals with MS have problems with thermoregulation, the previous recommendations for cooling with an electric fan and controlling room temperature should be heeded. In addition, it is recommended that fluid be replaced during exercise at the rate at which fluid is lost from the body (determined by weighing the patient before and after exercise). Because of the potential for heat-related illness, patients with MS should be educated about this potential problem and the steps they need to take to avoid problems during exercise training.

TABLE 28.5
Physiological Responses During Exercise of Individuals With Multiple Sclerosis (MS) Relative to Normal Healthy Individuals

Physiological response	Response of MS patients relative to the response of normal healthy individuals
Submaximal oxygen consumption during treadmill walking	Increased (50)
Submaximal and maximal arterial blood pressure	Decreased (34)
Temperature regulation	Decreased (6, 32)
Skeletal muscle fatigue	Earlier (18, 19)

Literature Review

Differences in physiological variables in individuals with MS compared with non-MS subjects provide a basis for using exercise training in this population. These negatively affected variables include the following:

— Muscle strength (2, 20, 36)
— Muscle fatigue resistance (19, 44)
— Motor unit firing rates (39)
— Muscle activation (39)
— Cardiorespiratory fitness (37)
— Muscle oxidative capacity (17)
— Walking speed (17)
— Habitual physical activity (30)

The following paragraphs provide a brief review of the effects of these physiological abnormalities and of the exercise training studies performed to date.

Skeletal muscle fatigue is more pronounced in MS than in non-MS subjects (44). Kent-Braun et al. (18) reported that the half-time of the skeletal muscle phosphocreatine (PCr) resynthesis rate after depletion was 48% slower in individuals with MS compared with non-MS subjects. This impaired PCr resynthesis rate appears to be the result of a reduced concentration of oxidative enzymes in people with MS (17). Kent-Braun et al. (17) also reported that the skeletal muscle oxidative capacity of individuals with MS is reduced compared with healthy subjects and is similar to that observed in individuals with complete spinal cord transection. It is hypothesized that the reduced skeletal muscle oxidative capacity in patients with MS is consistent with reduced physical activity levels (30) and this may be associated with the increased levels of fatigue described by these patients.

In the skeletal muscle of those with MS, cross-sectional area is reduced for type 1, 2a, and 2b muscle fibers (17). This reduced fiber size may be partially related to disuse and reduced skeletal muscle strength. In healthy subjects, resistance training increases the size of all skeletal muscle fiber types but may increase type 2 fibers to a greater extent than type 1 fibers (24). To date, there are no published reports of the effects of resistance training on muscle fiber size in individuals with MS.

Variable	Resistance training	Aerobic training
Muscle strength	Increased	Increased (35)
Muscle fatigue	Decreased (49)	Decreased (11)
Systemic fatigue	Unknown	Unknown
Ability to complete tasks of daily living	Increased	Unknown
Cardiorespiratory function	No change	Increased (35)
Skeletal muscle oxidative capacity	Increased	Increased
Disease progression	Unknown	Unknown
Immune function	Unknown	Unknown
Psychological variables	Unknown	Improved (35)

continued

Few studies have evaluated the efficacy of exercise training in individuals with MS. Petajan et al. (35) examined the influence of 15 weeks of aerobic exercise training on $\dot{V}O_2$peak, body composition, isometric strength, blood lipids, mood, fatigue, and disease status in patients with MS. $\dot{V}O_2$peak increased by 22%, and this change was not related to the level of neurological impairment. This suggests that patients with varying degrees of MS severity can improve $\dot{V}O_2$peak by aerobic exercise training. Additionally, this training program increased the isometric strength of the knee and shoulder extensors and the shoulder and elbow flexors (35). Although not significant ($p = .07$), there was a trend toward a reduction of body fat percent. Serum triglyceride concentration decreased by 17% without a change in the total, high-density lipoprotein, or low-density lipoprotein cholesterol levels. The exercise training group subjects reported improvements in quality of life scores. Changes in mood status were equivocal, with early improvement and later regression.

Svensson et al. (49) reported five case studies in individuals with MS who performed knee flexion for 4 to 8 weeks using free weights, a pulley apparatus, and an isokinetic dynamometer, at 40% to 50% of the directly determined 1-repetition max. Patients performed 50 repeated maximum contractions of the limb with the most pronounced symptoms using an isokinetic dynamometer. Following this training, an increase in peak repetition torque was noted. Gehlsen et al. (11) demonstrated improvement in strength, total work, and fatigue following a 10-week aquatics exercise program of freestyle swimming and shallow water calisthenics. Peak torque for knee extension was improved for all velocities (i.e., 60, 120, 180, 240, and 300° · s⁻¹) except 0° s⁻¹ (i.e., isometric contraction). Total work for the 50 knee extensions at 180° · s⁻¹ increased by 192% at 5 weeks and 330% at 10 weeks. Leg fatigue, measured as the percentage decline in peak torque, was 12.8% less from pretrial to midtrial and 14.1% less from pretrial to posttrial. These changes suggest that muscle endurance training is beneficial for patients with MS.

No randomized controlled trial has assessed the effects of range of motion training in the MS patient population.

Exercise Recommendations

It appears that the same type of improvement (e.g., increased $\dot{V}O_2$max, improved psychological variables, decreased muscle fatigue, and increased skeletal muscle strength and oxidative capacity) occurs at a similar rate in patients with MS compared with healthy individuals (35). There appears to be little risk of disease exacerbations. Practical application 28.4 discusses the exercise prescription for patients with MS.

Cardiovascular Training

To improve cardiovascular function, individuals with MS should perform standard aerobic exercise at 50% to 70% of $\dot{V}O_2$peak, three sessions per week, for 30 min per session. Some patients, especially those with severe fatigue, disease exacerbations, or poor fitness, may benefit from a reduced initial exercise intensity (e.g., 40-50% of $\dot{V}O_2$peak). For those patients whose autonomic nervous system is affected and have an attenuated exercise heart rate response, guiding exercise intensity using either the rating of perceived exertion scale, percent peak

METs, or a work rate equivalent to the desired percent $\dot{V}O_2$peak is appropriate. For those with ambulatory limitations, cycling or swimming may be a better choice than walking. Progression should be based on a patient's ability to tolerate exercise, focusing primarily on avoiding excessive fatigue and disease exacerbation.

Resistance Training

Strength training should follow standard training practices for clinical populations including recommendations for intensity, sets, repetitions, and frequency of exercise (see chapter 7). The training routine should focus on the major muscle groups using standard types of resistive movements. If a patient is quite weakened, low resistances should be used. Avoidance of free weights will eliminate problems with balance and coordination. For patients with MS, it may be best to perform strength training on noncardiovascular conditioning days to avoid excessive fatigue. However, patients who are stable and have only minor side effects from the disease may be able to perform resistance training following an aerobic training session. Modifications, such as

Exercise Prescription

	Cardiovascular	Muscle strength	Muscle endurance	Flexibility
Frequency	3 times per week	3 times per week	3 times per week	5-7 times per week
Duration	~30 min	~30 min	~10 min	~15 min
Intensity	50-85% of $\dot{V}O_2$peak; 60-75% of HR peak; 50-85% of HR reserve	80% 1RM 1-2 sets 8-15 repetitions	40-50% of 1RM 1-3 sets 15+ repetitions	NA
Mode	Aerobic equipment or exercises	Machine weights, dumbbells, elastic cords	Machine weights, dumbbells, elastic cords	Static stretching, may be assisted by a partner or using a device such as a towel
Considerations	Balance on some equipment; temperature regulation; excessive fatigue; attenuated HR or BP response	Unilateral weakness; balance; muscle spasticity; excessive fatigue	See muscle strength considerations	Balance; avoid excessive joint range and muscle spasticity aggravation
Recommendations	Consider seated modes or handrail use on treadmill; maintain hydration, maintain a thermoneutral environment, and use a fan for evaporative and convective cooling	Work to correct imbalances; rest at least 1 day between sessions; avoid free weights unless "spotter" is used	See muscle strength considerations	Perform while seated or on the floor; perform either after warm-up or cool-down

Note. HR = heart rate; 1RM = 1-repetition maximum; NA = not applicable; BP = blood pressure.

training muscle groups unilaterally because of differences in strength between the limbs or because of range of motion limitation, may be required. These need to be considered on a patient-by-patient basis.

Muscle Endurance Training

Muscle endurance training should be performed after strength training and should consist of one set to muscular failure. Intensity should be approximately 40% to 50% of the 1-repetition maximum weight.

Range of Motion

Because patients with MS may have a reduced range of motion of specific affected joints, a general flexibility program, designed to improve range of motion, is recommended. This should be performed 5 to 7 days per week and should follow a strength or cardiovascular training session. A static stretching routine focusing on all the major muscle groups and joints, with special emphasis on areas with spastic "flare-ups," is recommended. Patients should also be encouraged to stretch even if other exercise training is not performed on that day. Chapter 7 provides general information about range of motion training.

Conclusion

Because patients with MS can have progressive deterioration of their physical ability, they should following an exercise training routine. This may limit the amount of physical disability experienced by the individual and if, performed prudently, is of little risk to the patient. Specific recommendations for this patient population are lacking, but standard, low-level exercise training is recommended and appears to be safe and effective for most individuals.

Medical History

Mrs. NB, a 29-year-old Caucasian female, was diagnosed with MS 6 years ago. At that time she had problems with ataxia and **diplopia.** She has had one to two exacerbations of MS per year since that time. An increase in disease stability was noted after she started recombinant interferon β-1b (Betaseron) 5 years ago. She stopped the interferon therapy briefly during a pregnancy 4 years ago, and after the pregnancy she had three exacerbations during the following year despite being back on Betaseron. With each exacerbation, her symptoms of ataxia, vertigo, and diplopia worsened. These symptoms involve primarily the left side of her body. These extremities also "cramp" on occasion. Overall, she functions reasonably well, taking care of her 4-year-old daughter. She rests daily while her daughter goes to preschool. She had to quit working within a year after her diagnosis because of problems with ataxia and fatigue.

An MRI of her brain demonstrated multiple white matter lesions consistent with MS. A spinal fluid examination demonstrated elevated immunoglobulin G synthesis rate and oligoclonal bands consistent with the diagnosis of MS.

She is taking no other medication on a regular basis and has not been involved in a regular exercise program. She occasionally gets a urinary tract or upper respiratory infection that necessitates antibiotics. Cramping in her extremities has not been bad enough to warrant a muscle relaxant on a regular basis. She notes that when she walks more than a couple of blocks, she feels weakness in her left leg and often needs to rest for a few minutes before walking further. She also feels at times that her left leg may "give out."

On examination there is mild incoordination and diffuse hyperreflexia, which is more pronounced in the left side extremities. She also has a few beats of nystagmus.

Diagnosis

Her diagnosis is relapsing–remitting multiple sclerosis, Grade 4.0 on the Kurtzke Expanded Disability Status Scale (tables 28.3 and 28.4). Grade 4.0 means that the individual is fully ambulatory without aid, is self-sufficient, is up and about some 12 hr a day despite relatively severe disability, and is able to walk without aid or rest for at least 500 m, and that the disease affects one or more functional systems.

Exercise Evaluation

An exercise test was performed to assess functional ability and to rule out cardiac origin of her arm weakness while walking. The patient performed a bicycle protocol of 3-min stages at 25, 50, and 75 W. There was no sign of electrocardiographic abnormalities. Exercise was discontinued because of volitional fatigue.

Resting values:

Heart rate = 92

Blood pressure = 100/60

Electrocardiogram: normal sinus rhythm and within normal limits

continued

Peak exercise values:

Heart rate = 188

Blood pressure = 150/50

Rating of perceived exertion = 19

Peak $\dot{V}O_2$ = 17.5 ml $\cdot$ kg^{-1} $\cdot$ min^{-1}

Peak METs = 5.0

Exercise Prescription

Because the patient is just beginning an exercise program, and because of the ataxia and fatigue she experiences, she should start out at a low intensity and duration for both aerobic and strength training. Limb strength is reduced on left side because of increased MS-related symptoms.

— Cardiovascular training: She should begin aerobic training on a cycle ergometer at 60% of her heart rate reserve (~150 beats $\cdot$ min^{-1}) for 30 min three times per week. The intensity should be increased gradually, as tolerated, over a few weeks to a heart rate that is 85% of heart rate reserve (~172 beats $\cdot$ min^{-1}).

— Strength training: She should perform resistance training two times per week at 40% of 1-repetition maximum (1RM) on her left side and 60% of 1RM on her right side. This will require using hand weights, dumbbells, or resistance training machines. Initially, only one set of eight repetitions should be performed and should focus on the major muscle groups. Exercises such as leg extensions, leg curls, dumbbell chest press, and seated rowing are appropriate. The volume should be increased gradually over several weeks to two sets of eight repetitions and expand to other skeletal muscle groups.

— Muscular endurance training: An endurance skeletal muscle training program can be administered on the days that resistance exercise is performed. A suggestion is two times per week, at 40% to 50% of 1RM, performing one set to muscular failure for the gastrocnemius, quadriceps, hamstring, biceps and triceps brachii, chest, and abdominal muscles.

— Flexibility training: She should also perform stretching exercise before or after each exercise session focusing on the ankles, knees, hips, lower back, shoulders, wrists, and neck. Static stretches should be held for 10 to 20 s and performed at least twice for each muscle group.

Case Study 28   Discussion Questions

1. What can you say about the ability of the graded exercise test to detect myocardial ischemia in this patient versus others who might be more limited by MS?

2. How should the prescribed exercise be modified for a person with a greater level of disability?

3. How should the prescribed exercise be modified for a person with less disability?

4. What mode of aerobic and resistance exercise should this individual use?

5. What is an advantage of using non-weight-bearing exercise in an individual with MS with a high degree of disability?

6. What physiological variables need to be monitored very closely in this individual? Why?

7. Do you think individuals with MS should exercise during an exacerbation of disease activity? Why?

8. What are the potential benefits in terms of functional capacity of resistance training for individuals with MS?

Glossary

asymmetry of reflexes—The reflexes are not the same on both sides of the body.

ataxia—An inability to coordinate muscle activity during voluntary movement; most often caused by disorders of the cerebellum or the posterior columns of the spinal cord; may involve the limbs, head, or trunk.

computed tomography (CT)—Imaging technique that involves an X-ray which gives a film with a detailed cross-section of the structure of the tissue.

demyelination—The loss of the myelin covering that insulates the nerve tissue.

diplopia—Double vision that is caused by a disorder of the nerves that innervate the extraocular muscles or by impaired function of the muscles themselves.

evoked response testing—Test in which brain electrical signals are recorded as they are elicited by specific stimuli of the somatosensory, auditory, and visual pathways.

exacerbation—An increase in signs and symptoms of MS.

incontinence—Lack of control of urination or defecation.

magnetic resonance imaging (MRI)—Medical imaging that uses nuclear magnetic resonance as its energy source and allows for a higher resolution and no radiation compared with computed tomography. A detailed cross-section of the structure of the tissue can be obtained using MRI.

multiple sclerosis—A debilitating disease characterized by multiple areas of scar tissue replacing myelin around axons in the central nervous system.

nystagmus—Rhythmic movements of the eyes that are involuntary.

optic neuritis—Inflammation of the optic nerve.

paresthesia—A subjective feeling such as numbness, "pins and needles," or tingling.

pathological reflexes—Reflexes in which the time course or magnitude of the response is not normal.

plaques—Scarring of axons in the central nervous system attributable to demyelination.

relapse—Reversion back to an active disease process in MS after a remission.

remission—Recovery period from the active disease process.

sclerosis—Tissue hardening occurring because scar tissue replaces lost myelin around axons in the central nervous system.

spasticity—One type of increase in muscle tone at rest; characterized by increased resistance to passive stretch, velocity dependence and asymmetry about joints (i.e., greater in the flexor muscles at the elbow and the extensor muscles at the knee). Exaggerated deep tendon reflexes and clonus are additional manifestations.

thalamotomy—Destruction of a selected portion of the thalamus for the relief of pain and involuntary movements.

tremor—Repetitive, often regular, oscillatory movements caused by alternate, or synchronous, but irregular contraction of opposing muscle groups.

vertigo—A sensation of spinning or whirling motion.

References

1. Antel, J., M. Bania, A. Noronha, and S. Neely. Defective suppressor cell function mediated by T8+ cell lines from patients with progressive multiple sclerosis. *J Immunol* 137: 3436-3439, 1986.

2. Armstrong, L.E., D.M. Winant, P.R. Swasey, M.E. Seidle, A.L.Carter, and G. Gehlsen. Using isokinetic dynamometry to test ambulatory patients with multiple sclerosis. *Phys Ther* 63: 1274-1279, 1983.

3. Benczur, M., G.G. Petranyl, G. Palffy, M. Varga, M. Talas, B. Kotsy, I. Foldes, and S.R. Hollan. Dysfunction of natural killer cells in multiple sclerosis: A possible pathogenic factor. *Clin Exp Immunol* 39: 657-662, 1980.

4. Bourdette, D.N., A.V. Prochazka, W. Mitchell, P. Licari, and J. Burks. Health care costs of veterans with multiple sclerosis: Implications for rehabilitation of MS. VA Multiple Sclerosis Rehabilitation Study Group. *Arch Phys Med Rehabil* 74: 26-31, 1993.

5. Canadian Burden of Illness Study Group. Burden of illness of multiple sclerosis: Part I: Cost of Illness. *Can J Neurol Sci* 25: 23-30, 1998.

6. Cartlidge, N.E. Autonomic function in multiple sclerosis. *Brain* 95: 661-664, 1972.

7. Chen, N.Y.M., T.L. Pope, and D.J. Ott. Basic Radiology. New York: McGraw-Hill, 1996.

8. Compston, A. McAlpine's Multiple Sclerosis (3rd ed.). London: Churchill Livingstone, 1998.

9. Erkonen, W.E. Radiology 101: The Basics and Fundamentals of Imaging. Philadelphia: Lippincott-Raven, 1998.

10. Frankel, D.I. Multiple sclerosis. In: Neurological Rehabilitation (2nd ed.). D.A. Umpherd (Ed.). St. Louis: Mosby, 1990, pp. 531-550.

11. Gehlsen, G.M., S.A. Grigsby, and D.M. Winant. Effects of an aquatic fitness program on the muscular strength and endurance of patients with multiple sclerosis. *Phys Ther* 64: 653-657, 1984.

12. Glanze, W.D., K.N. Anderson, and L.E. Anderson. Mosby's Medical, Nursing, and Allied Health Dictionary (3rd ed.). St. Louis: C.V. Mosby Company, 1990.

13. Inman, R.P. Disability indices, the economic costs of illness, and social insurance: The case of multiple sclerosis. *Acta Neurol Scand* 70: 46-55, 1984.

14. Jacobson, D.L., S.J. Gange, N.R. Rose, and N.M. Graham. Epidemiology and estimated population burden of selected autoimmune diseases in the United States. *Clin Immunol Immunopathol* 84: 223-243, 1997.

15. Johnson, K.P., B.R. Brooks, J.A. Cohen, C.C. Ford, J. Goldstein, R.P. Lisak, L.W. Myers, H.S. Panitch, J.W. Rose, R.B. Schiffer, T. Vollmer, L.P. Weiner, and J.S. Wolinsky. Extended use of glatiramer acetate (Copaxone) is well tolerated and maintains its clinical effect on multiple sclerosis relapse rate and degree of disability. Copolymer 1 Multiple Sclerosis Study Group. *Neurology* 50: 701-708, 1998.

16. Kenney, W.L., R.H. Humphrey, and C.X. Bryant. ACSM's Guidelines for Exercise Testing and Prescription. Baltimore: Williams & Wilkins, 1995.

17. Kent-Braun, J.A., A.V. Ng, M. Castro, M.W. Weiner, D. Gelinas, G.A. Dudley, and R.G. Miller. Strength, skeletal muscle composition, and enzyme activity in multiple sclerosis. *J Appl Physiol* 83(6): 1998-2004, 1997.

18. Kent-Braun, J.A., K.R. Sharma, R.G. Miller, and M.W. Weiner. Postexercise phosphocreatine resynthesis is slowed in multiple sclerosis. *Muscle Nerve* 17: 835-841, 1994.

19. Kent-Braun, J.A., K.R. Sharma, M.W. Weiner, and R.G. Miller. Effects of exercise on muscle activation and metabolism in multiple sclerosis. *Muscle Nerve* 17: 1162-1169, 1994.

20. Lambert, C.P., R.L. Archer, and W.J. Evans. Muscle strength and fatigue during isokinetic exercise in individuals with multiple sclerosis. *Med Sci Sports Exerc* 33: 1613-1619, 2001.

21. Lambert, C.P., M.G. Flynn, W.A. Braun, and E. Mylona. Influence of acute submaximal exercise on T-lymphocyte suppressor cell function in healthy young men. *Eur J Appl Physiol* 82: 151-154, 2000.

22. Liu, C., and L.D. Blumhardt. Randomised, double blind, placebo-controlled study of subcutaneous interferon beta-1a in relapsing-remitting multiple sclerosis: a categorical disability trend analysis. *Mult Scler* 8: 10-14, 2002.

23. Lublin, F.D., and S.C. Reingold. Defining the clinical course of multiple sclerosis: Results of an international survey. National Multiple Sclerosis Society (USA) Advisory Committee on Clinical Trials of New Agents in Multiple Sclerosis. *Neurology* 46: 907-911, 1996.

24. MacDougall, J.D. Morphological changes in human skeletal muscle following strength training and immobilization. In: *Human Muscle Power*, N.L. Jones, N. McCartney, and A.J. McComas (Eds.). Champaign, IL: Human Kinetics, 1986, pp. 269-288.

25. Mancardi, G.L., F. Sardanelli, R.C. Parodi, E. Melani, E. Capello, M. Inglese, A. Ferrari, M.P. Sormani, C. Ottonello, F. Levrero, A. Uccelli, and P. Bruzzi. Effect of copolymer-1 on serial gadolinium-enhanced MRI in relapsing remitting multiple sclerosis. *Neurology* 50: 1127-1133, 1998.

26. McIntosh-Michaelis, S.A., M.H. Roberts, S.M. Wilkinson, I.D. Diamond, D.L. McLellan, J.P. Martin, and J.J. Spackman. The prevalence of cognitive impairment in a community survey of multiple sclerosis. *Br J Clin Psychol* 30 (Pt 4): 333-348, 1991.

27. Miller, A. Diagnosis of multiple sclerosis. *Semin Neurol* 18: 309-316, 1998.

28. Mulcare, J.A., Multiple sclerosis. In: *Exercise Management for Persons With Chronic Diseases and Disabilities* (1st ed.). J.L. Durstine (Ed.), Champaign, IL: Human Kinetics, 1997, pp. 189-193.

29. Ng, A.V., H.T. Dao, R.G. Miller, D.F. Gelinas, and J.A. Kent-Braun. Blunted pressor and intramuscular metabolic responses to voluntary isometric exercise in multiple sclerosis. *J Appl Physiol* 88: 871-880, 2000.

30. Ng, A.V., and J.A. Kent-Braun. Quantitation of lower physical activity in persons with multiple sclerosis. *Med Sci Sports Exerc* 29: 517-523, 1997.

31. Nieman, D.C., D.A. Henson, G. Gusewitch, B.J. Warren, R.C. Dotson, D.E. Butterworth, and S.L. Nehlsen-Cannarella. Physical activity and immune function in elderly women. *Med Sci Sports Exerc* 25: 823-831, 1993.

32. Noronha, M.J., C.J. Vas, and H. Aziz. Autonomic dysfunction (sweating responses) in multiple sclerosis. *J Neurol Neurosurg Psychiatry* 31:19-22, 1968.

33. Paty, D.W., and D.K. Li. Interferon beta-1b is effective in relapsing-remitting multiple sclerosis. II. MRI analysis results of a multicenter, randomized, double-blind, placebo-controlled trial. UBC MS/MRI Study Group and the IFNB Multiple Sclerosis Study Group. *Neurology* 43: 662-667, 1993.

34. Pepin, E.B., R.W. Hicks, M.K. Spencer, Z.V. Tran, and C.G. Jackson. Pressor responses to isometric exercise in multiple sclerosis. *Med Sci Sports Exerc* 28:656-660, 1996.

35. Petajan, J.H., E. Gappmaier, A.T. White, M.K. Spencer, L. Mino, and R.W. Hicks. Impact of aerobic training on fitness and quality of life in multiple sclerosis. *Ann Neurol* 39:432-441, 1996.

36. Ponichtera, J.A., M.A. Rodgers, and R.M. Glaser. Concentric and eccentric isokinetic lower extremity strength in multiple sclerosis and able bodied. *J Orthop Sports Phys Ther* 16: 114-122, 1992.

37. Ponichtera-Mulcare, J.A., T. Mathews, R.M. Glaser, and S.C. Gupta. Maximal aerobic exercise in ambulatory and semiambulatory patients with multiple sclerosis. *Med Sci Sports Exerc* 26: S29, 1994.

38. Poser, C.M., D.W. Paty, L. Scheinberg, W.I. McDonald, F.A. Davis, G.C. Ebers, K.P. Johnson, W.A. Sibley, D.H. Silberberg, and W.W. Tourtellotte. New diagnostic criteria for multiple sclerosis: Guidelines for research protocols. *Ann Neurol* 13: 227-231, 1983.

39. Rice, C.L., T.L. Vollmer, and B. Bigland-Ritchie. Neuromuscular responses of patients with multiple sclerosis. *Muscle Nerve* 15: 1123-1132, 1992.

40. Rudick, R.A., D.E. Goodkin, L.D. Jacobs, D.L. Cookfair, R.M. Herndon, J.R. Richer, A.M. Salazar, J.S. Fischer, C.V. Granger, J.H. Simon, J.J. Alam, N.A. Simonian, M.K. Campion, D.M. Bartozzak, D.N. Bourdette, J. Braiman, C.M. Brownscheidle, M.E. Coats, S.L. Cohan, S.S. Dougherty, R.P. Kinkel, M.K. Mass, F.E. Munschauer, R.L. Priore, R.H. Whitham, et al. Impact of interferon beta-1a on neurologic disability in relapsing multiple sclerosis. The Multiple Sclerosis Collaborative Research Group. *Neurology* 49: 358-363, 1997.

41. Schiffer, R.B. and N.M. Wineman. Antidepressant pharmacotherapy of depression associated with multiple sclerosis. *Am J Psychiatry* 147: 1493-1497, 1990.

42. Schumacher, G.A., G.W. Beebe, R.F. Kibler, L.T. Kurland, J.F. Kurtzke, F. McDowell, B. Nagler, W.A. Sibley, W.W.

Tourtellotte, and T.L. Willmon. Problems of experimental trials of therapy in multiple sclerosis. *Ann NY Acad Sci* 122:552-568, 1965.

43. Senaratne, M.P.J., D. Carroll, K.G. Warren, and T. Kappagoda. Evidence for cardiovascular autonomic nerve dysfunction in multiple sclerosis. *J Neurol Neurosurg Psychiatry* 4: 947-952, 1984.

44. Sharma, K.R., J. Kent-Braun, M.A. Mynhier, M.W. Weiner, and R.G. Miller. Evidence of an abnormal intramuscular component of fatigue in multiple sclerosis. *Muscle Nerve* 18: 1403-1411, 1995.

45. Sibley, W.A. *Therapeutic Claims in Multiple Sclerosis. A Guide to Treatments.* New York: Demos Vermande, 1996.

46. Simon, J.H., L.D. Jacobs, M. Campion, et al. Magnetic resonance studies of intramuscular interferon beta-1a for relapsing multiple sclerosis. The Multiple Sclerosis Collaborative Research Group. *Ann Neurol* 43: 79-87, 1998.

47. Sterman, A.B., P.K. Coyle, D.J. Panasci, and R. Grimson. Disseminated abnormalities of cardiovascular autonomic function in multiple sclerosis. *Neurology* 35: 1665-1668, 1985.

48. Stimac, G.K. *Introduction to Diagnostic Imaging.* Philadelphia: Saunders, 1992.

49. Svensson, B., B. Gerdle, and J. Elert. Endurance training in patients with multiple sclerosis: Five case studies. *Phys Ther* 74:1017-1026, 1994.

50. Oligata R., J.M. Burgunder, and M. Mumenthaler. Increased energy cost of walking in multiple sclerosis: effect of spasticity, ataxia, and weakness. *Arch Phys Med and Rehabil* 69: 846-849, 1988.

William Saltarelli, PhD
School of Health Sciences
Central Michigan University
Mt. Pleasant, MI

Children

There is general agreement among pediatric health specialists that the habitual physical activity behavior of children and adolescents is a major health issue (61). High levels of physical activity may play important roles in determining cardiorespiratory, skeletal, and psychological health of children (68, 72). The positive influence of habitual physical activity on various conditions such as coronary artery disease, hypertension, obesity, stroke, hypercholesterolemia, colon cancer, diabetes, osteoporosis, and low functional capacity in adults is well documented (14). However, the benefits of physical activity and its influence on these diseases and risk factors in children are not as clear. To address this issue in children, an advisory committee (International Consensus Conference on Physical Activity Guidelines for Adolescents) composed of scientists interested in the influence of physical activity and health in children was formed in 1992 to develop a consensus on the effects of physical activity on youth and adolescents. The results of this effort were published in the November 1994 issue of *Pediatric Exercise Science*. The following is a brief synopsis of the committee's findings and represents the most recent and comprehensive scientific review of the topic. Furthermore, the focus of this chapter is to review what is known about physical activity and children including basic exercise physiology, exercise testing protocols, and suggestions for prescribing exercise for this special population.

Scope

The most recent data on the physical activity patterns of children come from the Youth Risk Behavior Surveillance System (YRBSS) completed under the auspices of the Centers for Disease Control and Prevention in 1999. The YRBSS includes national, state, and local school-based surveys of children and adolescents age 12 to 21 years (24). These surveys were large-scale cross-sectional studies that consisted of self-reported 7-day recall questions. The best way to answer the question of whether today's children are physically active is to compare the YRBSS's data with *Healthy People 2000* objectives (61).

© Human Kinetics

Light to Moderate Physical Activity

The goal for *Healthy People 2000* with respect to moderate physical activity was to have children and adolescents (ages 6-17) engage in moderate physical activity for at least 30 min · day⁻¹, at least 5 days per week. The YRBSS data found that only 26.7% of students nationwide had walked or bicycled for at least 30 min on 5 or more days per week preceding the survey. African American and Hispanic students were more likely than white students to have participated in moderate physical activity (24). Also, females were less active than males, and activity decreased with age in both sexes.

Vigorous Physical Activity

The goal of *Healthy People 2000* was for at least 75% of children and adolescents (ages 6-17) to engage in vigorous physical activity that promotes cardiorespiratory fitness 3 or more days per week for at least 20 min. The YRBSS survey data revealed that 64.7% of children and adolescents engage in vigorous physical activity to the intensity of sweating/breathing hard for 20 min per session, 3 or more days a week. Males (72.3%) were more likely to engage in vigorous activity than females (57.1%). White children were more vigorously active than African Americans/Hispanics, and vigorous activity declined for all groups as they got older. In addition, for both boys and girls, the level of vigorous physical activity declined markedly from age 12 to 18. For example, 70% of 12-year-old boys reported that they were active, whereas 18-year-olds were 48% active.

Flexibility and Strength Development Activities

The goal of *Healthy People 2000* was to increase to at least 40% the proportion of children and adolescents who perform physical activities that enhance and maintain muscular strength, muscular endurance, and flexibility. This objective was addressed by two survey questions in the YRBSS. The first question asked if students participated in stretching exercises (e.g., toe touching, knee bending, and leg stretching) on 3 or more days preceding the survey. Results indicated that 51.3% of the students had participated in these activities. In addition, ninth grade girls (59.8%) were significantly more likely than older girls to do stretching exercise (41%). The second question focused on strengthening exercise (e.g., push-ups, sit-ups, and weightlifting) performed on 3 or more days of the 7 days preceding the survey. Overall, boys (58.1%) were more likely to participate in this activity than girls (43.2%). As with stretching, 9th grade girls (52%) were more likely to participate in strengthening exercises than older girls (34%). Overall, more than half of students nationwide (51.4%) had engaged in strengthening exercises on 3 of the 7 days preceding the survey.

Physiology and Pathophysiology

The following is a summary of the physical activity guidelines for adolescents as developed by the International Consensus Conference on Physical Activities Guidelines for Adolescents and published in the November 1994 issue of *Pediatric Exercise Science*. The following specific topics address the influence of physical activity on aerobic fitness, blood pressure, obesity, blood lipids, skeletal health, and psychological health.

Aerobic Fitness

In adults, it is generally accepted that habitual physical activity is related to aerobic capacity (maximal **oxygen uptake** or $\dot{V}O_2$max) and that fit individuals have lower mortality and morbidity rates (15). However, these relationships are not as clear in children. Lack of physical activity is known to be an independent risk factor for cardiovascular disease in adults, but again little is known about the influence of physical activity on these same risk factors in children. Recent evidence strongly links cardiovascular disease risk factors with atherosclerotic lesions in children as young as 2 years of age (79, 81, 83).

The correlation between physical activity and aerobic capacity in children is small to moderate, with correlation coefficients around $R = .16$ (54). Morrow and Freedson (54) also concluded from their review of the literature that neither walking nor daily physical activity provides the intensity to increase aerobic capacity in children. They suggested that the lack of relationship between physical activity and aerobic fitness could be attributable to one of the following:

— Available measurement techniques for physical activity are poor.

— Children have a generally acceptable level of fitness unrelated to physical activity.

— Aerobic systems are not trainable in children.

Analysis of data from the National Children and Youth Fitness Survey by Dotson and Ross (30) concluded that children who scored lowest in the mile run/walk (estimating $\dot{V}O_2$max) reported participating in fewer high-intensity cardiorespiratory activities. Other conclusions concerning activity and aerobic capacity from the review by Morrow and Freedson (54) included these:

1. Walking does not appear to provide a significant stimulus to increase aerobic power in children.

2. Daily physical activity probably has a weak association with aerobic capacity ($\dot{V}O_2$max) in children.

3. Specific intensity of activity necessary for aerobic benefit is not known precisely but may be at a higher intensity than for adults.

Blood Pressure

It is estimated that 2.8 million children and adolescents have hypertension (9). Hypertension in children is defined as an average systolic and/or diastolic blood pressure above the 95th percentile for age and sex (58) (see table 29.1 for specific normal and hypertensive values). Studies on adults and children have consistently shown that physical activity does not lower resting blood pressure in normotensive individuals. However, this is not the case with hypertensive subjects, as shown by numerous controlled studies (35, 81). A more extensive review of the subject can be found in an article by Alpert and Wilmore (3), who recommended that hypertensive children engage in regular physical activity to increase basic fitness and reduce levels of fatness.

Adiposity and Obesity

The third National Health and Nutrition Examination Survey (NHANES III) reported that approximately one child in five in the United States is overweight (42, 56). The NHANES definition of "overweight" is body mass index (BMI) greater than the 95th percentile. Analysis of past NHANES data revealed that over the past 30 years, the number of obese children in the United States has more than doubled. Tables 29.2 and 29.3 provide data on different body composition measurements including overweight and overfat definitions and normal values.

Recent studies have reviewed the cardiovascular risk of adolescent obesity (33, 55, 87). Overweight children were at least 2.4 times more likely to have elevated total cholesterol, low-density lipoprotein cholesterol, triglycerides, low high-density lipoprotein cholesterol, low fasting insulin, and elevated blood pressure. In addition, Must and colleagues (56) found that adults who were overweight as children have increased mortality and morbidity rates irrespective of adult weight. The causes of childhood obesity are complex and were reviewed in depth by Schlicker et al. (76). However, low physical activity (sedentary lifestyle) and unhealthy diet are important contributing factors.

Blood Lipids

The atherosclerotic process has been shown to begin in early childhood and this process is influenced

TABLE 29.1
Blood Pressure Values in Children of Various Ages

	Percentile	Age	Boys	Girls
Systolic blood pressure (mmHg)	90th, borderline hypertension	10	115	115
		12	119	119
		14	125	125
		16	130	130
	95th, hypertension	10	119	119
		12	125	123
		14	128	128
		16	134	134
Diastolic blood pressure (mmHg)	90th, borderline hypertension	10	75	75
		12	77	77
		14	78	78
		16	81	81
	95th, hypertension	10	80	80
		12	83	81
		14	83	82
		16	85	85

Note. Blood pressure values are reported in the literature by percentiles of height. This table reflects values at the 50th percentile for height. Normal blood pressure ≤90th percentile; borderline hypertensive ≥90th percentile but <95th percentile; hypertension >95th percentile.

Data from Bartosh & Aronson (12).

TABLE 29.2
Body Composition and Overfat

Age	Boys BMI (85%)	Boys BMI (95%)	Girls BMI (85%)	Girls BMI (95%)
10	19.5	22.5	20.0	23.0
11	20.2	23.2	20.5	23.5
12	20.2	23.2	21.5	25.5
13	23.0	25.2	22.5	26.2
14	22.5	26	26.2	27.0
15	23.5	26.5	24.0	28.0
16	24.0	27.5	24.5	29.0
17	25.0	28.0	25.0	29.5
18	25.5	29.0	25.5	30.5

Note. >85th percentile is considered at risk for overfat; >95th percentile is over fat. BMI = body mass index.

Data from National Center for Health Statistics, (57).

TABLE 29.3
Obesity Risk From Percent Fat and Sum of Skinfold Measurements

Risk of obesity	Sum of skinfolds (mm)		Body fat (%)	
	BOYS	**GIRLS**	**BOYS**	**GIRLS**
Optimal	12-25	16-34	10-20	15-27
Moderate risk	26-33	34-42	20-25	17–32
High	>34	>42	>25	>32

Note. For ages 10-16 years. Percent fat and sum of skinfolds from calf and triceps.

From T.G. Lohman, 1992, *Advances in body composition assessment* (Champaign, IL: Human Kinetics), 94. Reprinted by permission of T.G. Lohman.

by hyperlipidemia (13). In addition, it has been shown that abnormal blood lipids and lipoproteins do **track** from childhood to adulthood (43, 59).

The importance of hyperlipidemia in children can also be found in the conclusions of the Report of the Expert Panel on Blood and Cholesterol (5):

— Atherosclerosis or its precursors begin in young people.

— Elevated cholesterol levels early in life play a role in the development of adult atherosclerosis.

— Eating patterns and genetics affect blood cholesterol levels and the risk of coronary heart disease.

— Lowering blood lipid levels in children and adolescents may be beneficial.

Table 29.4 presents normal lipid levels for children 5 to 19 years of age.

Skeletal Health

The problem of bone loss, osteoporosis, and bone fracture risk in our society is well documented (23, 31, 52). Although this disease process is complex, the pathophysiology includes the following:

— Failure to attain a sufficiently high peak bone mass (bone mineral density) during the growing years

— Failure to maintain peak bone mass for a sufficient period of time during the adult years

— Accelerated bone loss in later years (10)

It is estimated that 90% of the total adult bone mass is deposited by the end of adolescence (50). It ap-

pears that bone mineral content and its relationship to physical activity is similar to what is observed in adults, in that high-impact weight-bearing activities increase bone mineral content.

The following recommendations were put forth by Bailey and Martin (10), from their review of physical activity and skeletal health in children.

1. A lifelong commitment to physical activity and exercise must be made.

2. Weight-bearing activities are better than weight-supported activities such as swimming and cycling.

3. Short, intense daily activity is better than prolonged activity done infrequently.

4. Activities that increase muscle strength should be promoted, as these will enhance bone density.

5. Activities should work all large muscle groups.

6. Immobilization and periods of immobility should be avoided; where this is not possible (as in bed rest during sickness), even brief daily weight-bearing movements can help to reduce bone loss.

Psychological, Social, and Emotional Health

Psychological and mental health is extremely important in children and adolescents and includes the following: depression, anxiety, stress, self-esteem, self-concept, hostility, anger, and intellectual function (22, 80). In a more recent review of the literature, Tortolero, Tayler and Murray (82) concluded the following:

TABLE 29.4
Healthy Blood Lipid Values for Children

HDL-C (MG · DL^{-1})^a

Age	Females	Males
6-11	>66	>69
12-14	>63	>63
15-19	>64	>57

TOTAL CHOLESTEROL (MG · DL^{-1})

Age	Females	Males
0-5	None	None
6-9	<173	<172
10-14	<174	<179
15-19	<175	<167

LDL-C (MG · DL^{-1})

Age	Females	Males
6-11	<114	<115
12-14	<114	<115
15-19	<118	<120

TRIGLYCERIDES (MG · DL^{-1})

Age	Females	Males
0-5	None	None
6-9	<62	<64
10-14	<77	<70
15-19	<72	<84

Note. Abnormal values set at 75% percentile. HDL-C = high-density lipoprotein cholesterol; LDL-C = low-density lipoprotein cholesterol. ^{a}The healthy values for HDL-C are those reported by the Mayo Clinic. However, there is inconsistency in the literature as to ideal values for children.

Mayo clinic data (51).

1. There exists a strong relationship between physical activity and improved self-efficacy, greater perceived physical competence, greater perceived health and well-being, and decreased depression and stress.

2. Moderate positive relationships have been found between physical activity and self-concept, self-esteem, and greater alcohol use (a negative outcome).

3. An inconclusive relationship exists between physical activity and body image, academic functioning, social skills, anxiety, hostility, aggression, suicide, sexual activity, and tobacco use (82).

Assessment of Physical Activity in Children

Physical activity is a complex behavior and therefore a challenge to assess, especially in children and adolescents. When one thinks of all the movements and activities of children and adolescents, it is easy to conclude that there is no "best" method of assessment. An excellent general review of the most popular methods of physical activity assessment was presented by Harro et al. (36). For large-scale field assessments such as in physical education classes, methods such as heart-rate monitors, activity monitors, direct observation, and self-report questionnaires seem most appropriate. Welk and Wood (85) provided an excellent review of the attributes of various assessment techniques.

Because of their simplicity, paper-and-pencil recall assessments are popular. The Seven-Day Physical Activity Recall developed by Sallis et al. (73), Physical Activity Questions for Children (37), and the Previous Day Physical Activity Recall (PDPAR) are examples of these easily administered assessments. The PDPAR is used in FITNESSGRAM, which provides computer feedback to children. Children enter their physical activity levels for each 30-min block of time during the day. FITNESSGRAM recommends that children record one weekend and two weekdays of physical activity. To help prompt the responses, the children are provided with a sample list of activities divided into categories based on the activity pyramid. The activity pyramid categories include lifestyle, aerobic activity, aerobic sports, muscular activity, flexibility, and rest. For each activity, students are asked to rate the intensity of the activity with categories of light, moderate, and vigorous activity (32). Practical application 29.1 describes client–clinician interaction between the exercise physiologist and children.

This section briefly reviews each area of fitness (aerobic, anaerobic, muscular strength and endurance, body composition, and flexibility) with respect to age- and sex-related differences. Refer to table

Client–Clinician Interaction

The pediatric exercise physiologist Oded Bar-Or often states in his presentations that "children are not small adults" when referring to responses and adaptations to exercise. This advice is very important to keep in mind when evaluating the exercise performance of children. The following is a brief review of special considerations that must be addressed when testing children.

Laboratory Environment

— The lab must be safe and well illuminated.

— Staff should be trained and have warm, friendly personalities to establish a positive relationship with children as soon as they enter the facility.

Unique Safety Issues

— Two testers are essential to ensure constant visual and verbal contact with the child.

— During treadmill testing, two spotters should be used with one at the subject's side and one behind the subject.

Pretest Protocols

1. Establish a relationship with the children as they enter the facility.
2. Completely explain the test to both parents and children. Be sure both know exactly what will take place and what the child will be asked to do.
3. Following the explanation of the procedure, parental consent and the child's assent documents should be completed. A child's assent document (for children 6 and older) should be written in age-appropriate language and be short and to the point.
4. Although offering an incentive to the child is controversial, it can be extremely helpful in eliciting a maximal effort.

Laboratory Equipment

When appropriate, testing equipment should be modified for the size and maturity of the subject.

29.5, for information on metabolic, anaerobic, and cardiorespiratory factors of children and their relationship to adult responses.

Aerobic Capacity

Laboratory evaluation of cardiorespiratory capacity (aerobic capacity) is indicated for children for a variety of reasons including diagnosis of medical conditions, assessment of exercise-induced symptoms, measurement of exercise response following surgery, or use in exercise prescription. With the interest in youth fitness, aerobic tests have taken on more importance. The American College of Cardiology and the American Heart Association have published selection criteria for identifying children to be tested for cardiorespiratory endurance in a clinical setting (33). Readers are directed to this comprehensive reference for clinical indicators and criteria.

The most commonly used indicator of cardiorespiratory capacity is maximal oxygen uptake, which is defined as the highest volume of oxygen that can be consumed during an exercise bout per unit of time. $\dot{V}O_2$max is commonly expressed in absolute terms (L/minute) and in relative terms (ml · kg^{-1} · min^{-1}). Prior to puberty, absolute $\dot{V}O_2$max is slightly higher in boys compared with girls and improves with age and growth in both sexes. Following puberty however, this relationship changes as boys increase $\dot{V}O_2$max with age and as their cardiovascular, pulmonary, and skeletal systems grow. $\dot{V}O_2$max (absolute) in girls in this same age group remains relatively constant (11). This sex difference is possibly attributable to the increase in non-metabolic weight (fat) in females, a decrease in habitual physical activity, or both. Many exercise physiologists prefer to express aerobic capacity in relative terms using mass in kilograms as the relative factor.

TABLE 29.5
Physiological Characteristics of the Exercising Child

FUNCTION	COMPARISON WITH ADULTS	IMPLICATIONS FOR EXERCISE PRESCRIPTION
METABOLIC:		
Aerobic		
VO_{2max} (L/min)	Lower function of body mass	
VO_{2max} (ml/kg/min)	Similar	Can perform endurance tasks reasonable well
Submaximal oxygen demand (economy)	Cycling: similar (18% to 30% mechanical efficiency); walking and running: higher metabolic cost	Greater fatigability in prolonged high-intensity tasks (running and walking); greater heat production in children at a given speed or running
Anaerobic		
Glycogen stores	Lower concentration and rate of utilization of muscle glycogen	
Phosphofructokinase (PFK)	Glycolysis limited because of low level of PFK	Ability of children to perform intense anaerobic tasks that last 10 to 90 seconds is distinctly lower than that of adults
LA max	Lower maximal blood lactate levels	
Phosphagen stores	Stores and breakdown of ATP and CrP are the same	Same ability to deal metabolically with very brief intense exercise
Oxygen transient	Faster reaching of steady state than adults Shorter half-time of oxygen increase in children	Children reach metabolic steady state faster/Children contract a lower oxygen deficit/Faster recovery/Children, therefore, are well suited to intermittent activities
LAmax	Lower at a given percent of VO_{2max}	May be reason why children perceive a given workload as easier
Heart rate at lactate threshold	Higher	
CARDIOVASCULAR:		
Maximal cardiac output (Qmax) Q at a given VO_2	Lower because of size difference Somewhat lower	Immature cardiovascular system means child is limited in bringing internal heat to surface for dissipation when exercising intensely in the heat
Maximal stroke volume (SVmax)	Lower because of size and heart volume difference	
Stroke volume at a given VO_2	Lower	
Maximal heart rate (HRmax)	Higher	Up to maturity HRmax is between 195 and 215 meats/min
Heart rate at submax work	At a given power output and at relative metabolic load, child has a higher heart rate	Higher heart rate compensates for lower stroke volume
Oxygen-carrying capacity	Blood volume, hemoglobin concentration, and total hemoglobin are lower in children	
Oxygen content in arterial and venous blood (CaO_2 – CvO_2)	Somewhat higher	Potential deficiency of peripheral blood supply during maximal exertion in hot climates
Blood flow to active muscle	Higher	
Systolic and diastolic pressures	Lower maximal and submaximal	No known beneficial or detrimental effects on working capacity of child

FUNCTION	COMPARISON WITH ADULTS	IMPLICATIONS FOR EXERCISE PRESCRIPTION
CARDIOPULMONARY RESPONSE:		
Maximal minute ventilation V_{Emax} (L/min)	Smaller	Early fatigability in tasks that require large respiratory minute volumes
V_{Emax} (ml/kg/min)	Same as adolescents and young adults	
Vesubmax; ventilatory equivalent	VE at any given VO_2 is higher in children	Less efficient ventilation would mean a greater oxygen cost of ventilation/May explain the relatively higher metabolic cost of submaximal exercise
Respiratory frequency and tidal volume	Marked by higher rate (tachypnea) and shallow breathing response	Children's physiologic dead space is smaller than that of adults; therefore; alveolar ventilation is still adequate for gas exchange
PERCEPTION (RPE)	Exercising at a given physiologic strain is perceived to be easier by children	Implications for initial phase of heat acclimatization
THERMOREGULATORY:		
Surface area	Per unit mass is approximately 36% greater in children (percentage is variable, depends on size of child, i.e., surface area per mass may be higher in younger children and lower in older ones)	Greater rate of heat exchange between skin and environment/In climatic extremes, children are at increased risk of stress
Sweating rate	Lower absolute and per unit of surface area. Greater increase in core temperature required to start sweating	Great risk of heat-related illness on hot, humid days because of reduced capacity to evaporate sweat/Lower tolerance time in extreme heat
Acclimatization to heat	Slower physiologically, faster subjectively	Children require longer and more gradual program of acclimatization; special attention during early stages of acclimatization
Body cooling in water	Faster because of higher surface per heat, producing unit mass; lower thickness of subcutaneous fat	Potential hypothermia
Body core heating during dehydration	Greater	Prolonged activity: hydrate well before and enforce fluid intake during activity

Maximum aerobic power in relative terms remains relatively constant across a wide range of ages (6-16) in boys, reflecting an increase in body mass and an equal increase in the ability to use oxygen. In girls, however, VO_2max in relative terms declines continuously especially at puberty, and the decline continues through life (11).

It is well established in adults that the best measure of aerobic capacity is a test of maximum oxygen uptake. When we discuss the concept of maximum oxygen uptake (VO_2max) in children, it is tempting to compare children's responses and values to adults. It must be kept in mind that children are not small adults. In adults, exercise physiologists usually equate high relative VO_2max values with superior endurance performance. This concept does not hold true for children, especially very young children. For example, compare the VO_2max of an average young healthy adult from a treadmill protocol of about 40 to 45 ml · kg⁻¹ · min⁻¹ to the average VO_2max of a 10-year-old child of about 50 to 55 ml · kg⁻¹ · min⁻¹. The average mile run time (performance) of these adults of will be about 7 to 8 min, whereas the average time of the children is much slower at about 10 to 12 min. Therefore, high $\dot{V}O_2$max in children does not necessarily produce good endurance performance times. Many other variables influence running performance including body composition, skill, running economy, practice, and pacing. In addition, studies comparing endurance performance with laboratory $\dot{V}O_2$max determination indicate lower correlations (R = .6-.7) (48) in children than the much stronger correlation of about R = .9 in adult studies (27). A final note to keep in mind is that $\dot{V}O_2$max (relative) remains relatively constant through the growing years in children while their endurance performance steadily improves.

Medical Evaluation

Before aerobic capacity is tested in children and adolescents, preliminary information should be gathered (as with adults) including a complete medical history, parental consent, child's assent, and resting physiological measures. It is extremely important to acquire child's assent and for the subject and parents to fully understand the test and what will be expected of them. Also, establishing a positive relationship with the subject and all laboratory personnel is critical to the success of the test. A positive and safe atmosphere is essential. Resting physiological data of heart rate (HR), blood pressure, and electrocardiogram should be measured to ensure that the child is healthy.

Contraindications

Contraindications to testing are similar to American College of Sports Medicine (ACSM) guidelines for adults and are reviewed for children by Rowland (69) and include:

— Acute inflammatory cardiac disease (e.g., pericarditis, myocarditis, acute rheumatic heart disease)

— Uncontrolled congestive heart failure

— Acute myocardial infarction

— Acute pulmonary disease (e.g., acute asthma, pneumonia)

— Severe systemic hypertension (e.g., blood pressure >240/120 mmHg)

— Acute renal disease (e.g., acute glomerulonephritis)

— Acute hepatitis (within 3 months after onset)

— Drug overdose affecting cardiorespiratory response to exercise (e.g., digitalis toxicity, quinidine toxicity) (69)

Careful observation of the subject for indications to terminate the test is critical to overall test safety. If an electrocardiograph is administered, adult termination criteria should be applied (ACSM guidelines). Other termination criteria for children include the following:

— Excessive increase in systolic blood pressure over 240 mmHg or diastolic pressure over 120 mmHg

— Progressive decrease in systolic blood pressure

— Pallor or clamminess of the skin

— Pain, headache, dyspnea, or nausea (8)

There is much confusion with respect to the definition of a $\dot{V}O_2$max test in children. It is extremely difficult to attain clear adult criteria of a plateau of oxygen uptake in children (11). This is probably attributable to children stopping because of fatigue more often than adults or their unique cardiovascular system. Many researchers prefer to use the term *peak* oxygen uptake rather than *maximal* oxygen uptake to measure endurance capacity. Rowland even suggested the term *exhaustive* to indicate the best effort the child is willing to provide (69). Our laboratory uses a three criteria format. Children must attain two of the first three criteria subsequently for a peak test to be achieved. Also listed are other criteria found in the literature.

1. Attainment of 95% of age-predicted maximum HR

2. Respiratory exchange ratio (RER) greater than 1.05

3. Rating of perceived exertion (Borg RPE) of 18 or greater (68)

Additional criteria:

4. Subjective signs of exhaustion

5. Child stops because of fatigue

6. Lactate levels of about 6 to 7 mmol $\cdot$ L^{-1} for 7- to 8-year-olds or 4 to 5 mmol $\cdot$ L^{-1} for 5- to 6-year-olds (64)

Success in eliciting a maximal effort takes planning and specific strategies. Children must be sure they understand that a maximal effort is expected, and this concept must be reinforced throughout the test by constant encouragement. Statements such as "You're looking great," "Keep going," and "You can go more stages" really help. Although controversial on ethical grounds, rewards or incentives for each stage have been used to motivate children.

The physiological differences between cycle and treadmill protocols must also be taken into account when interpreting maximal testing data. Their differences follow those seen in adults and include the following:

1. Peak heart rate and $\dot{V}O_2$max are greater during treadmill protocols.
2. RER at maximal effort is usually higher in cycle ergometry.

Recommendations and Anticipated Responses

Most laboratory testing is performed with progressive workloads applied on cycle ergometers or motorized treadmills. Rowland (67) presented a complete treatment of this topic including mode and protocol selection. As with adults, cycle tests hold some advantage over those performed on a treadmill. Cycle ergometers are safer and more appropriate, especially with obese subjects, and are more conducive to data collection (blood pressure, HR, RPE, and lactate). However, there are some concerns with cycle tests:

— Some children may end the test prematurely because of leg fatigue.

— Some children have trouble keeping to a standard cadence or pedal speed.

— Cycle data have been shown to be less reliable than treadmill data.

— RER and blood lactate have been found to be higher on cycles.

One advantage of motorized treadmills is that children are familiar with walking and running. However, the use of this mode requires some practice, with 3 to 5 min needed for children to become accustomed to and feel comfortable with the machine. It is recommended that two spotters be used, one in back and one to the side, especially when metabolic data are being collected.

— Treadmill ergometry: Most protocols used to assess the aerobic capacity of children, whether maximal or submaximal in nature, are modifications of adult protocols. In clinical settings, the Balke protocol (see table 29.6) has been modified for children by altering stage duration and accommodating different fitness levels.

When $\dot{V}O_2$max is assessed in healthy children it is advisable to initially establish a comfortable running speed for each child, usually between 3 and 5 mph. This initial trial also lets children become accustomed to the equipment and laboratory personnel. Speed is kept constant at this predetermined rate throughout the test. Subsequent stages are 2 min in duration with the grade increased 2% each stage until volitional fatigue or a maximum criterion is achieved.

— Cycle ergometry: Similar to treadmill protocols, cycle ergometry protocols for children are usually modifications of those used with adults. Table 29.7 presents protocols that have been modified for children. Most cycle protocols require a pedal cadence of 50 to 60 rpm. The Adams Submaximal Progression Continuous Cycle protocol, also called a PWC170, can be used to evaluate relative aerobic power in children (1). This protocol consists of three stages that are usually submaximal in intensity. Heart rate is monitored during each of the 6-min stages. Performance is evaluated according to the mechanical power that the child produces at the HR of 170 beats · min^{-1} (PWC170). The resistance of the initial stage is dependent on the child's weight in kilograms and is presented in table 29.8. Modification of the PWC170 has included shorter (3-min)

TABLE 29.6

Modified Balke Treadmill Protocol

Subject	Speed (mph)	Initial grade (%)	Increment (%)	Stage duration (min)
Poorly fit	3.0	6	2	2
Sedentary	3.25	6	2	2
Active	5.00	0	2-1/2	2
Athlete	5.25	0	2-1/2	2

Reprinted, by permission, T.W. Rowland, 1993, *Pediatric laboratory exercise testing: Clinical guidelines* (Champaign, IL: Human Kinetics), 36.

TABLE 29.7
Cycle Ergometer Protocols for Pediatric Populations

Author	Rate (rev · min^{-1})	Body measure (height or surface area)	Initial load	Increment	Stage duration (min)
McMaster	50	Height (cm)	(W)	(W)	
		<120	12.5	12.5	2
		120-140	12.5	25	2
		140-160	25	25	2
		>160	25	50 (male)	2
				25 (female)	2
James[a]	60-70	Surface area (m^2)	(kg · m · min^{-1})	(kg · m · min^{-1} × 2)	
		<1.0	200	100	3
		1.0-1.2	200	200	3
		>1.2	200	300	3
Godfrey	60	Height (cm)	(W)	(W)	
		<120	10	10	1
		120-150	15	15	1
		>150	20	20	1

[a]If more than three levels of exercise are necessary, add 100-200 kg · m · min^{-1} until exhaustion.

Reprinted, by permission, T.W. Rowland, 1993, *Pediatric laboratory exercise testing: Clinical guidelines* (Champaign, IL: Human Kinetics), 23.

TABLE 29.8
The Adams Submaximal Progressive Continuous Cycling Protocol, Initial Power Settings by Body Weight Groups

Body weight (kg)	1st stage power (W)	2nd stage power (W)	3rd stage power (W)
<30	16.5	33	50
30-39.9	16.5	50	83
40-59.9	16.5	50	100
≥60	16.5	83	133

Adapted from Adams (1).

stages and a higher heart rate of 190 (PWC190). (74). Table 29.9 provides average $\dot{V}O_2$max values for boys and girls using treadmill and cycle protocols.

— Field tests of aerobic capacity: Many different field tests have been used to estimate aerobic capacity in children (40, 70). The 1-mile run/walk is one of the cardiorespiratory endurance measures included in both the Physical Best (4, 5) and FITNESS-GRAM (32) health fitness test batteries. Numerous

studies have addressed the reliability and validity of these tests to estimate $\dot{V}O_2$max (20, 39). Regression equations for both the 1-mile run/walk and PACER tests (multi-stage shuttle run) have been developed to estimate $\dot{V}O_2$max and are presented in figure 29.1.

— The multistage shuttle run test can be used to estimate $\dot{V}O_2$max according to the procedure described by Leger et al. (44). Each subject runs back and forth on a 20-m course starting at a speed of

8.5 km · hr⁻¹ (2.36 m · s⁻¹). The running speed is increased by 0.5 km · hr⁻¹ (0.14 m · s⁻¹) every minute. The running pace is regulated by a prerecorded audio tape, which signals when the subject needs to be at one or the other end of the 20-m course. Subjects keep completing subsequent stages until they cannot keep up the progression of speed. Their speed at the last stage completed is used to calculate $\dot{V}O_2$max. Table 29.10 gives healthy zone standards for the mile run/walk, pacer laps, and $\dot{V}O_2$max.

Anaerobic Capacity

It has been shown that children produce less anaerobic power/capacity than adults and that anaerobic power increases with age (11). The lower anaerobic

TABLE 29.9
Aerobic Capacity in Children

Protocols	Boys	Girls
	Average $\dot{V}O_2$max (ml · kg⁻¹ · min⁻¹)	
General normal values	42-52	35-47
Treadmill	48.0 ± 5	39.0 ± 5
Cycle	42.0 ± 5	36.0 ± 5

Data from Aerobics Institute (35) and Armstrong et al. (9).

Mile run/walk

$\dot{V}O_2$max (ml · kg⁻¹ · min⁻¹)
= .21 (age × sex) − 0.84 (BMI)
= 8.41 (MRW) + 0.34 (MRW²) + 108.94

Where: age = age in years; sex = 0 for females and 1 for males; BMI = body mass index (kg · m⁻²); MRW = run walk time in minutes.

Pacer: multistage shuttle

$\dot{V}O_2$max (ml · kg⁻¹ · min⁻¹)
= [3.238 · (max speed)] − [3.248 · (age)] + [0.1536 · (max speed · age)] + 31.025

Where: max speed (km · hr⁻¹) = 8.5 + 0.5 (# of stages completed + age in years)

FIGURE 29.1 Oxygen uptake equations for run/walk and pacer tests.
Adapted from Cureton, K.J. et al. (29) and FITNESSGRAM Cooper Institute (32).

TABLE 29.10
Healthy Zone Standards for Aerobic Capacity

Age	Boys	Girls	Boys	Girls	Boys	Girls
	MILE RUN/WALK (MIN)		**PACER (LAPS)**		**$\dot{V}O_2$MAX (ML · KG⁻¹ · MIN⁻¹)**	
10	9:00-9:30	9:30-12:30	17-55	35-37	42-52	39-47
11	8:30-11:00	9:00-12:00	23-61	37-49	42-52	38-46
12	8:00-10:30	9:00-12:00	29-68	13-40	42-52	37-45
13	7:30-10:00	9:00-11:30	35-74	15-42	42-52	36-44
14	7:00-9:30	8:30-11:00	41-80	18-44	42-52	35-43
15	7:00-9:00	8:00-10:30	46-85	23-50	42-52	35-43
16	7:00-8:30	8:00-10:30	52-90	28-56	42-52	35-43
17	7:00-8:30	8:00-10:00	57-64	34-61	42-52	35-43
18+	7:00-8:30	8:00-10:00	57-94	34-61	42-52	35-43

Data from FITNESSGRAM Cooper Institute (32).

power/capacity of children has been attributed to lower lactate production, possibly attributable to lower levels of glycolytic enzymes such as phosphofructokinase, lower sympathetic activity, lower glycogen storage capabilities, or higher ventilatory thresholds (expressed as a percentage of $\dot{V}O_2$max) (11). In addition, recovery $\dot{V}O_2$ (oxygen debt) has been found to be less in children as well as maximal exercise blood pH. All or some of these factors may contribute to the lower anaerobic capability in children. When males and females are compared with respect to anaerobic performance, males generally outperform females at all ages, and the rate of improvement with age is greater in males. Medical evaluation concerns and contraindications for anaerobic capacity testing are similar to those described for aerobic capacity testing.

The evaluation of anaerobic capacity in pediatric populations has been used to follow the progress of neuromuscular diseases (e.g., cerebral palsy and muscular dystrophy; Bar-Or, 1983). In healthy children, the Wingate test is used along with various field tests to evaluate athletic performance.

— Wingate anaerobic cycle power test: The Wingate anaerobic power test was developed to assess both short-term and moderate-term anaerobic capacities in adults. This test lasts 30 s and requires a subject to pedal a cycle ergometer as fast as possible. The number of revolutions is counted in each 5-s period for a total of 30 s. From the number of revolutions completed, peak power (most revolutions in any 5-s interval) and 30-s anaerobic capacity can be determined. The test can be performed by the legs or by arms. Bar-Or (11) stated that the Wingate anaerobic test is feasible for use with healthy and disabled children as young as 6 years. Practical application 29.2 provides the specific protocol.

— Field tests of anaerobic capacity: A simple test of anaerobic performance is the timed 50-yd run. This test is included in many fitness testing batteries and provides information on children's anaerobic capacity. It should be noted, however, that performance on a 50-yd run is influenced by running experience, genetics, and motor performance, especially in young children.

— Standing vertical jump: Sargent (75) developed the vertical jump test to measure leg power in adults. The application of this test in children is the same as in adults, with subjects required to jump vertically as high as they can. Subjects' ability to exert leg power is determined from the height of the jump compared with the height reached by fingers when the children are standing erect with the arms extended. This field test is easily learned and perfected

by children and adolescents. Two numerical values can be used to represent results of the vertical jump test. First, the simplest number is the difference between the initial reach and jump performance. However, peak power in watts can also be calculated by the following formula:

$$\text{Peak power (watts)} = [78.5 \times \text{VJ (cm)}] + [60.6 \times \text{mass (kg)}] - [15.3 \times \text{height (cm)}] + 431$$

where VJ = vertical jump (jump height minus reach height), mass = body mass in kg, and height = body height in cm.

An easier method that uses the preceding equation is the Lewis nomogram, which uses jump height and weight (2).

Muscular Strength

As recently as the late 1970s, pediatric exercise experts and medical doctors believed that prepubescent children could not benefit from strength or resistance training because this developmental group lacked the prerequisite circulatory hormones (84). Furthermore, many believed that the stress imposed by this training was not safe and could injure bones, especially at the growth plates. Since that time, numerous controlled studies have provided compelling evidence that strength or resistance training produces strength gains in both prepubescent girls and boys (60, 77, 86). As with adults, the effectiveness of training appears to depend on intensity, volume, and duration. However, these specific factors have not been established with certainty in children. An excellent reference is a document sponsored by American Orthopedic Society for Sports Medicine (21), which concludes that strength training for the prepubescent

— improves muscular strength and endurance,

— improves motor skills,

— protects against injury (sports),

— has positive psychological benefits, and

— provides a forum for the introduction of safe and proper training.

Medical Evaluation

One of the more serious injury concerns is the potential for strength or resistance training to cause skeletal damage to the epiphysis or growth plates of children. Although there are some reports of epiphyses fractures during late puberty, there is only one reported case of such injury being caused by weight training. Most injuries caused by training are of the muscle strain nature, which result from improper

Wingate Protocol

Equipment

MONARCH mechanically braked cycle or arm ergometer.

Protocol

1. The subject performs a 3- to 4-min warm-up.
2. The subject observes a 3-min rest period.
3. Standard braking force should be set based on subject's weight (in kilograms) from *Wingate Protocol* table.
4. The subject begins pedaling at 0 kp resistance, and then the proper load calculated from *Wingate Protocol* table.
5. Count the maximum revolutions in 30 s as well as for each 5-s segment. Be sure to give encouragement.
6. Provide a 3-min cool-down at low resistance. Cool-down is very important to help the cardiovascular system recover.
7. Peak power is calculated from the number of revolutions in the best 5-s interval.

   Peak anaerobic power (kgm/5 s) = revolution in 5 s $\times$ force (kp) $\times$ flywheel circumference

   Conversion to watts = peak = 2 $\times$ kgm/5 s
8. Anaerobic capacity is calculated from the total revolutions per 30 s.

   Anaerobic capacity in 30 s (kgm/30 s) = revolution/30 s $\times$ flywheel circumference $\times$ force (kp)

   Conversion to watts = anaerobic capacity = kgm/30 s

BODY WEIGHT (KG)	GIRLS		BOYS	
	Leg (kp)	Arm (kp)	Leg (kp)	Arm (kp)
20-24.9	1.3-1.7	0.8-1.0	1.4-1.8	0.8-1.1
25-29.9	1.7-2.0	1.0-1.2	1.8-2.0	1.1-1.3
30-34.9	2.0-2.3	1.2-1.4	2.1-2.5	1.3-1.5
35-39.9	2.3-2.7	1.4-1.6	2.5-2.7	1.5-1.6
40-44.9	2.7-3.0	1.6-1.8	2.8-3.2	1.7-1.9
45-49.9	3.0-3.3	1.8-2.0	3.2-3.5	1.9-2.1
50-54.9	3.3-3.7	2.0-2.2	3.5-3.9	2.1-2.3
55-59.9	3.7-4.0	2.2-2.4	3.9-4.2	2.3-2.5
60-64.9	4.0-4.3	2.4-2.6	4.2-4.6	2.5-2.8
65-69.9	4.3-4.7	2.6-2.8	4.6-4.9	2.8-3.0

Reprinted, by permission, T.W. Rowland, 1993, *Pediatric laboratory exercise testing: Clinical guidelines* (Champaign, IL: Human Kinetics), 169.

lifting techniques or attempts at maximal lifts. There is no evidence indicating that weight training is more risky with respect to musculoskeletal injury than other youth sports (16, 17, 62, 63).

Contraindications

It is important to be sure that children are emotionally mature enough to begin a strength program. Adults usually suggest resistance training to children; however, the child must be enthusiastic about participation. Otherwise, there are no contraindications, unless the child has preexisting musculoskeletal problems.

The following are specific guidelines for resistance training for the preadolescent put forth by Blimke (17), including safety and program information.

— Encourage resistance training as only one of a variety of normal recreational and sport activities.

— Encourage using a variety of different training modalities, such as free weights, springs, machines, and body weight.

— Discourage interindividual competition, and stress the importance of personal improvement.

— Discourage extremely high-intensity (loading) efforts, such as maximal or near-maximal lifts with free weights or weight machines.

— Avoid isolated eccentric training.

— Encourage a circuit system approach to capitalize on possible cardiorespiratory benefits.

— If using weight training machines, select either those that have been designed for children or those for which the loads and levers can be easily adjusted to accommodate the reduced strength capacity and size of children.

— Provide experienced supervision, preferably by an adult, when free weights or training machines are used in training.

— Preclude physical and medical contraindications.

— Provide instruction in proper technique, and demand that children use this technique.

— Have children warm up with calisthenics and stretches.

— Begin with exercises that use body weight as resistance before progressing to free weights or weight training machines.

— Individualize training loads when using free weights and training machines.

— Train all major muscle groups, and both flexors and extensors.

— Exercise muscles through their entire range of motion.

— Alternate days of training with rest days, and do not allow children to train more than three times per week.

— When children use free weights or machines, they should progress gradually from light loads, high repetitions (>15), and few sets (2-3) to heavier loads, fewer repetitions (6-8), and moderate numbers of sets (3-4).

— Instruct children to cool down after training with stretching exercises for major joints and muscle groups.

— When selecting equipment, check for durability, stability, sturdiness, and safety.

— Instruct children to heed sharp or persistent pain as a warning and seek medical advice.

Recommendations and Anticipated Responses

The assessment of muscular strength and endurance in children is needed for normal evaluation and for proper training prescriptions. Although many types of machines (isokinetic, variable resistance, Nautilus) are used, this chapter presents assessment with free weights and the protocols of popular fitness test batteries that all feature low-cost and simple equipment. In children, **repetition maximum** (RM) can easily be determined. Kramer and Fleck (41) suggested that RM for six or fewer repetitions is sufficient to measure strength (see figure 29.2).

Many programs have as their primary goal to increase muscular endurance. Kramer and Fleck (45) stated that having the child do as many repetitions as possible at a specified percentage of his or her body weight or his or her 6RM resistance is sufficient to evaluate muscular endurance. They also suggested using 60% to 80% of 6RM to test relative local muscle endurance. Tests are terminated when repetitions lack proper technique or safety is a factor (41).

— Field tests of muscular strength and endurance: For evaluating the strength and endurance of abdominal muscles, the most popular field tests include the sit-up and curl-up. FITNESSGRAM uses the curl-up, as described in detail in the *FITNESSGRAM Test Administration Manual* (32). The advantage over the traditional sit-up is that the curl-up minimizes the use of the hip flexors used in a sit-up and is safer with respect to low back/spine compression. The student being tested lies on a mat, knees bent at 140°, arms straight and parallel to the trunk

1. The child warms up with 5 to 10 repetitions at 50% of the estimated 6RM.

2. After 1 min of rest and some stretching, the child performs six repetitions at 70% of the estimated 6RM.

3. The child repeats step 2 at 90% of the estimated 6RM.

4. After about 2 min of rest, depending on the effort needed to perform the 90% set, the child performs six repetitions with 100% or 105% of the estimated 6RM.

5. If the child successfully completes six repetitions in step 4, add 2.5% to 5% of the resistance used in step 4 and have the child attempt six repetitions after 2 min of rest. If the child does not complete six repetitions in step 4, subtract 2.5% to 5% of the resistance used in step 4 and have the child attempt six repetitions after 2 min of rest.

6. If the first part of step 5 is successful (the child lifts 2.5-5% more resistance than used in step 4), retest the child starting with higher resistances after at least 24 hr of rest, because performance of more sets will be greatly affected by fatigue. If the second part of step 5 is successful (the child lifts 2.5-5% less than the resistance used in step 4), this is the child's 6RM. If the second part of step 5 is not successful (the child does not lift 2.5-5% less resistance used in step 4), retest the child after at least 24 hr of rest, starting with less resistance.

. . . .

FIGURE 29.2   Estimating children's repetition maximum (RM).

Adapted from Kramer and Fleck (41).

with palms resting on mat. Curl-ups are completed in a slow and controlled manner at a 20 per minute cadence (one curl every 3 s). Students are stopped after a maximum of 75 curl-ups are completed or they lose proper form and/or cadence. However, used more often is the traditional sit-up (5). Once a signal is given, the student sits up (forearms touch the thigh) as many times as possible in 1 min.

— Upper arm and shoulder girdle strength/endurance: Traditionally, the pull-up or chin-up has been used to assess arm and shoulder strength. A major problem with this test is that the majority of children cannot complete at least one repetition.

These zero scores are not helpful in assessment, and they can be demoralizing to children. Many believe that the pull-up is not appropriate for children who cannot accomplish as least one repetition. With this problem in mind, FITNESSGRAM includes the 90° push-up, modified pull-up, and flexed arm hang.

The 90° push-up is completed when the student being tested assumes a prone position on the mat with hands placed under the shoulders; fingers stretched out; legs straight, parallel, and slightly apart; and toes tucked under. The student pushes up off mat with the arms until the arms are straight. The back should be kept in a straight line during the movement. A push-up is then recorded when the student lowers the body using the arms until the elbows bend to 90° angle and the upper arm is parallel to the floor. Push-ups continue at a rhythm of 20 per minute or 1 push-up every 3 s. Students are stopped when their form or cadence falters (32).

Body Composition

Body composition assessment has recently become important in the field of pediatric exercise and medicine. It has been shown that body composition has an important influence on both field and laboratory performance in children (67). There are no specific medical issues to address before assessing body composition. In assessing body composition, the tester needs to conduct the test in a place that will respect the child's privacy, and the test must be sensitive to the child's self-image. Also, the way in which individual results are presented to the child should require parental involvement.

Methods commonly available for measuring body composition in children include the following:

— BMI
— Skinfold measurement
— Hydrostatic weighing
— Bioelectrical impedance
— Dual energy X-ray absorptiometry (DEXA)

Choosing a method of body composition assessment for children is difficult because of many developmental factors that influence the accuracy of various prediction equations. Hydrostatic weighing is considered the "gold standard" for the estimation of percent body fat in adults. As outlined by Lohman (46), this method is problematic for use in children not only because the process is difficult for children (i.e., being submerged and blowing air out of lungs) but also because children undergo changes in their chemical composition of fat-free mass (FFM) especially during puberty (46). For example, protein and mineral content increase about 5% and the water content decreases about 9% from birth to adulthood (67). In fact, the chemical maturity of FFM is not reached until late adolescence (48). Therefore, the use of adult equations for relating body density to percent fat in children has been found to overestimate fat between 7% and 13% (18). When performing hydrostatic weighing on children, Lohman proposed using the Siri equation for estimating body fat from density. This method incorporates age-specific values for FFM.

— Skinfold measurements: A common method to estimate body fatness is the anthropometric measurement of skinfold thickness. Although this method has its drawbacks, it is a simple and relatively inexpensive approach that estimates body fat with an accuracy of 3% to 4%. It is important that testers practice and become accurate before reporting values to children. Slaughter et al. (78) provided accurate equations using two skinfold sites (either the calf and triceps or calf and subscapular). These sites are relatively easy to locate and measure in children. Notice that Slaughter's regression equations

account for racial and developmental differences (figure 29.3).

The simple sum of skinfolds (calf and triceps or triceps and subscapular) can be very useful not only when normal values are compared but when weight loss (or fat loss) is considered. When the sum of skinfolds is used as a criterion measure, the problem of accurate regression equation is eliminated.

— BMI: BMI ($kg \cdot m^{-2}$) is the easiest and least invasive body composition measurement. Many large epidemiological studies use this method because research has shown that the risk of chronic disease increases with BMI. However, a weakness of BMI is the inability to distinguish between muscle tissue and fat; therefore, some subjects may have a high BMI attributable to extensive muscle mass. This method should be considered when skinfold measurements are not feasible. Table 29.2 presents healthy BMI values.

Flexibility

Flexibility is defined as the ability to move joints through a full range of motion. It is a widely accepted concept that children are extremely flexible and

---

Triceps and calf skinfolds

% Fat = 0.735 ΣSF + 1.0 males, all ages
% Fat = 0.610 ΣSF + 5.0 females, all ages

Triceps and subscapular skinfolds (>35 mm)

% Fat = 0.783 ΣSF + I males
% Fat = 0.546 ΣSF + 9.7 females

Triceps and subscapular skinfold (<35 mm)

% Fat = 1.21 (ΣSF) – 0.008 (ΣSF)^2 + I males
% Fat = 1.33 (ΣSF) – 0.013 (ΣSF)^2 + 2.5 females (2.0 blacks, 3.0 whites)
I = Intercept which varies with maturation level and racial group for males as follows:

Age	Black	White
Prepubescent	-3.5	-1.7
Pubescent	-5.2	-3.4
Postpubescent	-6.8	-5.5
Adult	-6.8	-5.5

Note. SF = skinfold. Calculations were derived by using the equation in Slaughter et al. (78).

....

FIGURE 29.3   Prediction equations of percent fat from triceps and calf or from triceps and subscapular skinfolds in children.

Reprinted, by permission, from T.G. Lohman, 1992, *Advances in Body Composition Assessment*, (Champaign, IL: Human Kinetics), 94.

therefore flexibility should not be a priority in activities or training. Exceptions seem to be children in sports such as gymnastics and dancing, where the importance of flexibility is required and appreciated. However, flexibility training is recommended at all ages to ensure safe activity. Most research studies show a decline in flexibility as children get older. Clark (26) concluded that boys tend to lose flexibility after the age of 10 and girls after age 12. In fact, Milne et al. (53) found that flexibility in both boys and girls declined between kindergarten and second grade. Girls as a group are usually more flexible than boys, which may reflect the activities in which girls participate. In addition, the YRBSS data showed that girls were more apt to engage in flexibility training than boys (25). There are no contraindications or specific medical concerns, unless the child has pre-existing musculoskeletal problems.

— Trunk extension test (trunk lift): This test is included in the FITNESSGRAM test battery (32). The objective is to lift the upper body off the floor using the large muscles of the back and hold this position to allow for measurement. The subject lies on a mat in a prone position, toes pointed, and hands placed under the thighs. The subject lifts the upper body off the floor in a very slow and controlled manner to a maximum height of 12 in. This position is held long enough for the tester to measure the distance from the floor to the chin.

— Traditional sit-and-reach flexibility test: The Physical Best health-related test battery includes the traditional sit-and-reach test as a measure of low back and hamstring flexibility. This test is completed with a sit-and-reach box. With shoes removed and knees straight, students are asked to bend forward as far as possible. Although the sit-and-reach test is a component of many health-related fitness test batteries, it has been criticized because it may only be an acceptable test of hamstring flexibility and may not be adequate to test the low back (26).

— Back-saver sit-and-reach: Another version of the sit-and-reach is an optional test included in FITNESSGRAM called the back-saver sit-and-reach. This test is administered similar to the traditional sit-and-reach except each leg is tested separately. This test is reported to cause less pressure on the anterior portion of the lumbar vertebra. Stretching one hamstring at a time avoids excessive flexing of the lumbar spine and hyperextension of both knees (32).

PRACTICAL APPLICATION 29.3

Literature Review

Aerobic Fitness

The **plasticity** of aerobic fitness in children is currently being debated and studied. Plasticity refers to the extent that normal growth-related changes (improvements) in maximal aerobic capacity can be altered by changes in physical activity. More succinctly stated, plasticity refers to the question, Can increases in the level of physical activity improve aerobic fitness ($\dot{V}O_2$max)? Plasticity, therefore, can be thought of as trainability. Numerous studies have established that improvement in $\dot{V}O_2$max after a period of aerobic training appears to be less in children than adults. Rowland (67) reviewed 13 studies that attempted to correlate habitual activity in children with their level of aerobic capacity ($\dot{V}O_2$max). He found only five of the studies concluded that a significant correlation exists between levels of physical activity and aerobic capacity (67). All of these studies measured habitual physical activity of children and were not training studies.

A possible explanation for the lack of support for a significant training effect for habitual physical activity of children can be found in the literature. For example, studies assessing the intensity of physical activity in children have consistently shown that only a small percentage of children meet the guidelines that call for at least 20 min of sustained activity eliciting between 60% and 90% of maximal heart rate. Specifically, Armstrong et al. (9) found that only 13% of boys and 6.5% of girls attained heart rates over 160 beats · min^{-1} for a 20-min period during a 3-day assessment period.

If physical activity is not a major contributor to $\dot{V}O_2$max, then what about formal endurance training? Adult training using ACSM guidelines for intensity, duration, mode, and frequency (large muscle groups, rhythmic activities, 20-60 min, 3-5 days per week equivalent to 65-90% maximum heart rate) usually produces between a 5% and 35% improvement in 12 weeks (8). Mahon (47)

continued

reviewed three controlled training studies on children less than 8 years of age. In these studies, the experimental group showed a 12.5% increase in $\dot{V}O_2$max while the control group increased only 7.5%. Studies with children 8 to 13 years old showed an average 13.8% increase in $\dot{V}O_2$ while the controls increased only 0.7% (54). For adolescents 13 years of age or older, an increase of 6.8% in $\dot{V}O_2$ was found while the control group had no change in $\dot{V}O_2$. From these data it is clear that children and adolescents can adapt to training by increasing $\dot{V}O_2$max but at a much lower rate than adults. When one is considering $\dot{V}O_2$ changes in children, it should be pointed out that initial fitness and genetic endowment can also influence responses.

In the previously mentioned training studies, the average frequency was 3 days per week, the average duration was 30 min, and the intensities were greater than 160 beats $\cdot$ min^{-1} and as high as 85% HRmax. The modes were continuous running, weight training, aerobics, and jumping rope. It is therefore clear that children as young as 8 years old can increase $\dot{V}O_2$max with training, but their increase is not as great as that of adults.

It is generally agreed that most children are not active enough to improve aerobic capacity or $\dot{V}O_2$max. Most studies that have shown improvement in $\dot{V}O_2$max are training studies using very high intensity guidelines. Therefore, for children not in formal training programs, little correlation has been seen between physical activity and aerobic capacity. The goals, therefore, of promoting physical activity in children should be as follows:

1. To promote physical activity habits to be carried to adulthood
2. To modify cardiovascular disease factors including, blood pressure, blood lipids, and body composition

Anaerobic Fitness

Very little information is available concerning exercise prescription and trainability of anaerobic systems in children by employing short-burst activities. Rowland (67) indicated that the major reason for this lack of information is that an accurate, noninvasive method of assessing anaerobic metabolism similar to $\dot{V}O_2$max in aerobic systems does not exist (67). Consequently, short-burst activities, which are very common in the habitual activity of children, are poorly understood. The limited available research indicates that children can increase their anaerobic power by training. For example, studies by Grodjinovski et al. (34), Rotstein et al. (66), and Sargent et al. (75) have shown children to improve anaerobic performance from 4% to 14% following interval-type training. Grodjinovski's 6-week training study consisted of one group riding a cycle ergometer for five 10-s all-out bouts followed by three 30-s all-out bouts (34). The other group ran three all-out 40-m runs followed by three all-out 150-m runs during each training session. Both groups trained 3 days per week. Improvement in anaerobic performance was about 4% in each group. Although this study shows that anaerobic improvement can be improved by training, it also gives some information on duration, intensity, and frequency of training. This study indicates a training frequency of 3 weeks of short-duration and high-intensity efforts.

Exercise Prescription

As stated in the most recent edition of ACSM's *Guidelines for Exercise Testing and Prescription* (19), the art of exercise prescription is the successful integration of exercise science with behavioral techniques that result in long-term program compliance and attainment of the individual's goals. This is an extremely challenging task with adults but even more difficult with children. Compounding the problem is the lack of specific definitive information on the appropriate duration and intensity of exercise for children.

The International Consensus Conference on Physical Activity Guidelines for Adolescents concluded that all youth should be physically active daily as part of play, games, sports, work, transportation, recreation, physical education, or planned exercise. These guidelines also stated that additional aerobic benefits may be achieved by engaging in moderate to vigorous physical activities using large muscle groups such as running, cycling, and swimming. The general intensity, frequency, and duration of these activities should be a minimum of 3 days per week for a minimum of 30 min at an intensity of 75% heart rate reserve (72). For younger children, The National Association for Sports and Physical Education (28) issued physical activity guidelines for elementary school aged children and recommended the following:

— Elementary school aged children should accumulate at least 30 to 60 min of age-appropriate and developmentally appropriate physical activity from a variety of activities on all, or most, days of the week.

— An accumulation of more than 60 min, and up to several hours per day, of age-appropriate and developmentally appropriate activity is encouraged.

— Some of the child's activity each day should be in periods lasting 10 to 15 min or more and should include moderate to vigorous activity. This activity will typically be intermittent in nature, involving alternating moderate to vigorous activity with brief periods of rest and recovery.

— Children should not have extended periods of inactivity.

Cardiovascular Training

The ACSM (7) set forth general recommendations in 1988 for aerobic exercise prescription in children and adolescents:

— Although children are generally quite active, children generally choose to participate in activities that consist of short-burst, high-energy exercise. Children should be encouraged to participate in sustained activities that use large muscle groups.

— The type, intensity, and duration of exercise activities need to be based on the maturity of the child, medical status, and previous experiences with exercise.

— Regardless of age, the exercise intensity should start out low and progress gradually.

— Because of the difficulty in monitoring heart rates with children, the use of a modified Borg scale is a more practical method of monitoring exercise intensity in children.

— Children are involved in a variety of activities throughout the day. Because of this, a specific time should be dedicated to sustained aerobic activities.

— The duration of the exercise session will vary depending on the age of the children, their previous exercise experience, and the intensity of the exercise session.

— Because it is often quite difficult to get children to respond to sustained periods of exercise, the session periods need to be creatively designed.

An important omission in these ACSM guidelines is proper intensity. Rowland (67) and others have suggested the use of adult guidelines to increase aerobic fitness, and studies have shown higher intensity to be effective. However, these guidelines seem proper for trained athletes. Sallis and Patrick (72) suggested a more moderate approach:

— Guideline 1: All adolescents should be physically active daily, or nearly every day, as part of play, games, sports, work, transportation, recreation, physical education, or planned exercise, in the context of family, school, and community activities (consistent with objective 1.3 from *Healthy People 2000*) (61).

Adolescents should do a variety of physical activities as part of their daily lives. These activities should be enjoyable, involve a variety of muscle groups, and include some weight-bearing activities. The intensity or duration of the activities is probably less important than the fact that energy is expended and a habit of daily activity is established. Adolescents are encouraged to incorporate physical activity into their lifestyles by doing such things as walking up stairs, walking or riding a bicycle for errands, having conversations while walking with friends, parking at the far end of parking lots, and doing household chores.

— Guideline 2: Adolescents should engage in three or more sessions per week of activities that last 20 min or more and that require moderate to vigorous levels of exertion (consistent with objective 1.4 from *Healthy People 2000*) (61). Moderate to vigorous activities are those that require at least as much effort as brisk or fast walking. A diversity of activities that use large muscle groups are recommended as part of sports, recreation, chores, transportation, work, school physical education, or planned exercise. Examples include brisk walking, jogging, stair climbing, basketball, racket sports, soccer, dance, swimming laps, skating, strength (resistance) training, lawn mowing, strenuous homework, cross-country skiing, and cycling.

Note: Vigorous physical activities are rhythmic, repetitive activities that require the use of large muscle groups to elicit a heart response of 60% or more of a subject's maximum heart rate adjusted for age. An exercise heart rate of 60% of maximum heart rate for age is sufficient for cardiorespiratory conditioning. Maximum heart rate equals roughly 220 beats · min⁻¹ minus age.

Specific aerobic fitness prescription guidelines for children and adolescents are reviewed in the *Physical Best Teachers Guide* (6). Practical application 29.4 contains these guidelines, which are divided into three objectives: basic health-related fitness, intermediate health-related fitness, and athletic performance. This approach allows for differences in abilities, interests, and fitness objectives of children.

Resistance Training

Training prescriptions for children show great variability but seem to generally follow adult prescriptions with the exception of lower resistance and high repetitions. Practical application 29.5 reviews training principals for children with respect to improving muscular strength and endurance for basic health and athletic performance. Supervision to ensure proper technique is most important when prescribing resistance training to children (71).

Range of Motion

Flexibility or adequate range of motion has been identified as one of the components of health-related fitness. However, many questions exist concerning flexibility training:

— Should low-intensity aerobic activity be performed before stretching?
— Which general method of stretching—static or ballistic—is better or safer?
— How much time should be dedicated to flexibility training in children and adolescents?

A review of the literature provides no definitive answers to these questions. However, general principals of flexibility training can be stated. The *Physical Best Teachers Guide* states that children should engage in at least 5 min of low-level aerobic activity before stretching. With respect to the static versus ballistic question, the *Physical Best Teachers Guide* suggests **static flexibility training** for the child whose objective is basic or intermediate health-related fitness, and a more **ballistic approach** when the objective is increasing athletic performance. The final

PRACTICAL APPLICATION 29.4

Cardiovascular Training Recommendations

	Basic health-related fitness	Intermediate health-related fitness	Athletic performance fitness
Frequency	3 times per week	3-5 times per week	5-6 times per week
Intensity	50-60% HRmax	60-75% HRmax	65-90% HRmax
Time	30 min total, accumulated	40-60 min total, accumulated	60-120 min total, accumulated
Type	Walking, jogging, dancing, games, and activities that require minimal equipment demands.	Jogging, running, fitness-based games and activities, intramural and local league sports.	Training programs, running, aerobics, interscholastic, and community sports programs.
Overload	Not necessary to bring child to overload during base level.	Be creative with activity to increase tempo or decrease rest period; 1-3 times per week.	Program design should stress variable intensities and durations to bring student into overload; 2-3 times per week.
Progression and specificity	Let student "get the idea" of movement. Progression is minimal.	Introduce program design and incorporate variation.	Specific sets, repetitions, and exercises to meet desired outcomes.

Note. HR = heart rate.

Reprinted, by permission, from AAHPERD, 1999, *Physicaleducation for lifelong fitness: The physical best teacher's guide* (Champaign, IL: Human Kinetics), 88.

Training Principles Applied to Muscular Strength and Muscular Endurance, Based on Fitness Goals

	Basic health-related fitness	Intermediate health-related fitness	Athletic performance fitness
Frequency	2-3 times per week; allow for minimum 1-day rest between training sessions.	3-4 times per week; alternating upper and lower body segments will allow for consecutive training days.	4-5 times per week; training activities are specific to sport participation.
Intensity	Very light, <40% of a "projected" maximal effort.[a]	Light to moderate, 50-70% of "projected" maximal effort.[a]	Specific load adaptations required for sport participation.
Time	1-2 sets of 6-12 repetitions.	1-3 sets of 6-15 repetitions.	3-5 sets of 5-20 repetitions.
Type	Body weight, single and and multijoint activities involving major muscle groups.[a]	Resistance exercises such as leg press, bench press, pull-ups, and additional presses and pulls.[a]	Advanced sport-specific, multijoint lifts (clean pulls, power presses, Olympic-style lifts).
Overload	Not necessary to bring child to overload during base level.	Introduce one of the components of overload; 1-2 times per week.	Program design should stress variable intensities and durations to bring student into overload; 2-3 times per week.
Progression and specificity	Let student get the idea of correct movement. Progression is minimal.	Introduce program design and incorporate variation.	Specific sets, repetitions, and exercises to meet desired outcomes.

[a]Projected maximal effort (1-repetition max, or 1RM) can be calculated from submaximal testing. For example, if a child bench presses 125 lb, 10 times his calculated 4RM = (number of reps × 0.03) + 1 × weight lifted or (10 × 0.3) + 1 × 125 = 0.3 + 1 × 125 = 162.5 1RM.

Reprinted, by permission, from AAHPERD, 1999, *Physicaleducation for lifelong fitness: The physical best teacher's guide* (Champaign, IL: Human Kinetics), 100.

question concerning time and emphasis on flexibility training depends on the specific objective and amount of time available. Although no clear recommendations can be found in the literature, a suggestion of 5% of the available time seems reasonable. Practical application 29.6 summarizes the FITT principle for flexibility training.

Conclusion

This chapter focuses on physical activity and the basic exercise physiology of children and adolescents. Of primary interest is the question, "Are the physical activity patterns of children adequate for optimal health?" The answer to this question is extremely difficult; however, recent survey data indicate that only 64% of children and adolescents engage in vigor-

ous physical activity and only 26% of the same group engaged in light to moderate physical activity within 7 days of the survey. Physical activity seems to have a positive effect on obesity, blood lipids, skeletal health, and psychological factors. More study is needed to quantify and qualify the amount of activity that will maximally influence these factors.

Another largely unanswered question in this age group involves proper exercise prescription guidelines. Most published guidelines for the enhancement of aerobic capacity, anaerobic capacity, muscular strength, and flexibility are simply modified from adult research. This chapter summarizes the most recent research with respect to the proper intensity, duration, and frequency of activity to enhance these physiological attributes. Guidelines are also presented to design safe and effective training programs for children and adolescents.

The FITT Principle Applied to Flexibility Training, Based on Fitness Goals

	Basic health-related fitness	Intermediate health-related fitness	Athletic performance fitness
Frequency	Before and after each activity/exercise session (minimum of 3 times per week)	Before and after each activity/exercise session (daily)	Before and after each training session
Intensity	To mild tension, or slight muscular discomfort	To mild tension, or slight muscular discomfort	To mild tension, or slight muscular discomfort, at a level appropriate for sport participation
Time	10-15 s; 2 times per stretch	10-15 s; 3 times per stretch	Dependent on static, dynamic, or ballistic (usually conducted by qualified trainer/coach)
Type	Static; major muscle groups	Static; major muscle groups, introduction of dynamic stretching	Usually dynamic and/or ballistic; major muscle groups and sport-specific stretches
Overload	Not necessary at base level	Ask student to identify level of stretch intensity; if appropriate for activity, have student stretch slightly farther than previous same stretch	Because dynamic and ballistic stretches dominate advanced level, overload is not appropriate to ballistic stretching
Progression and specificity	Start very easy into stretch; slow movements with minimal applied resistance to muscle involved	Stretch major core muscles first, then move to extremities; introduce dynamic flexibility.	Start with easy multijoint dynamic movements, progressing to more resistive dynamic movements, followed by moderate static and/or proprioceptive neuromuscular facilitation stretching

Reprinted, by permission, from AAHPERD, 1999, *Physicaleducation for lifelong fitness: The physical best teacher's guide* (Champaign, IL: Human Kinetics), 115.

Case Study 29

Mr. JT is a normal 10-year-old white male. He is sedentary most of the time and would rather play computer games than play outdoors or participate in sports. Mr. JT's mother is concerned about his long-term health. She asked a staff member at a university human performance laboratory to assess his health and suggest healthy lifestyle changes. The tests and results are as follows:

Body Composition Results
Height: 157.5 cm

continued

Weight: 70.5 kg

Skinfold thickness: calf 23 mm and triceps 22 mm

Cardiorespiratory Endurance

$\dot{V}O_2$max (treadmill) = 41 ml $\cdot$ kg^{-1} $\cdot$ min^{-1} and 2900 L $\cdot$ min^{-1}

Mile run = 9 min 30 sec

Pacer test = 19 laps

Flexibility

Sit-and-reach = 22 cm

Muscular strength

Sit-ups per minute = 20

Pull-ups = 0

Blood lipids/glucose

Total cholesterol = 200 mg $\cdot$ dl^{-1}

High-density lipoprotein = 32 mg $\cdot$ dl^{-1}

Triglycerides = 190 mg $\cdot$ dl^{-1}

Low-density lipoprotein = 220 mg $\cdot$ dl^{-1}

Glucose = 119 mg $\cdot$ dl^{-1}

Blood pressure = 130/86 mmHg

Case Study 29   Discussion Questions

1. Is Mr. JT at risk with respect to body composition, body mass index, percent fat, or sum of skinfolds?

2. Is Mr. JT at risk with respect to cardiorespiratory endurance in treadmill testing, mile run, or pacer results?

3. Is Mr. JT at risk with respect to blood lipids, glucose, or blood pressure?

4. As you sit down with Mr. JT and his mom following testing, how would you approach the subject with the results?

5. What suggestions would you give Mr. JT to improve his results? If you suggest increasing physical activity, how would you justify this to Mr. JT? Explain why physical activity is or is not beneficial for Mr. JT.

6. Write an exercise prescription for Mr. JT to improve the following:

   Aerobic endurance

   Muscular strength and endurance

   Flexibility

   Blood pressure

   Blood test results (high-density lipoprotein-cholesterol)

Glossary

absolute oxygen uptake—Oxygen uptake expressed in liters per minute (L · min⁻¹).

ballistic flexibility (stretching)—Stretching using active muscle movement with a bouncing-type action.

dynamic flexibility (stretching)—Slow and constant stretch held for a period of time.

oxygen uptake (consumption)—A measure of a person's ability to take in and use oxygen.

plasticity—The extent to which normal maturation of maximal aerobic power can be altered by changes in physical activity.

relative oxygen uptake—Oxygen uptake expressed in ml of oxygen per kilogram of body weight per minute (ml · kg⁻¹ · min⁻¹).

repetition maximum (RM)—Maximum number of repetitions possible with a given resistance.

tracking—The concept that risk factors or other conditions which are expressed in childhood will persist and also be expressed in adulthood.

References

1. Adams, F.H. The physical working capacity of normal children. *Pediatrics,* 28: 55, 1961.

2. Adams, G.M. *Exercise Physiology—Laboratory Manual.* Boston: McGraw-Hill, 1988.

3. Alpert, B.S., and J.H. Wilmore. Physical activity and blood pressure in adolescents. *Pediatric Exercise Science,* 6(4): 361-380, 1994.

4. American Alliance for Health, Physical Education, Recreation and Dance. *AAHPERD Health Related Physical Fitness Test Manual.* Washington DC: American Alliance for Health, Physical Education, Recreation and Dance, 1992.

5. American Alliance for Health, Physical Education, Recreation and Dance. *Physical Best.* Reston, VA: American Alliance for Health, Physical Education, Recreation and Dance, 1988.

6. American Alliance for Health, Physical Education, Recreation and Dance. *Physical Best Teachers Guide.* Champaign, IL: Human Kinetics, 1999.

7. American College of Sports Medicine. Opinion statement on physical fitness in children and youth. *Medicine and Science Sport Exercise,* 204: 422-423, 1988.

8. American Heart Association. *1999 Heart and Stroke Facts Statistical Update.* Dallas: American Heart Association, 1998.

9. Armstrong, N., L. Williams, J. Balding, P. Gentle, and B. Kirby. Cardiopulmonary fitness, physical activity patterns, and selected coronary risk factor variables in 11 to 16 year olds. *Pediatric Exercise Science,* 3: 219-228, 1991.

10. Bailey, D.A., and A.D. Martin. Physical activity and skeletal health in adolescents. *Pediatric Exercise Science,* 6: 330-347, 1994.

11. Bar-Or, O. *Pediatric Sports Medicine for the Practitioner.* New York: Springer-Verlag, 1983.

12. Bartosh, S.M., and A.J. Aronson. Childhood hypertension—An update on etiology, diagnosis and treatment. *Pediatric Cardiology,* 46(2): 235-250, 1999.

13. Berenson, G.S., S.R. Sprinivasn, B. Weihang, W.P. Newman, R.E. Tracy, and W.A. Wattigney. Association between multiple cardiovascular risk factors and atherosclerosis in children and young adults. *New England Journal of Medicine,* 338: 1650-1656, 1998.

14. Blair, S.N. Changes in physical fitness and all-cause mortality. *Journal of the American Medicine Association,* 273: 1093-1098, 1995.

15. Blair, S.N., H.W. Kohl III, R.S. Paffenbarger, D.G. Clark, K.H. Cooper, and L.W. Gibbons. Physical fitness and all-cause mortality: A prospective study of healthy men and women. *Journal of the American Medical Association,* 262: 2395-2401, 1989.

16. Blimkie, C.J.R., and D.G. Sale. Strength development and trainability during childhood. In E.V. Praash (Ed.), *Pediatric Anaerobic Performance.* Champaign, IL: Human Kinetics, 1986, p. 196.

17. Blimkie C. Resistance training during preadolescence. *Sports Medicine,* 15(6): 389-407, 1993.

18. Boileau, R.A., T.G. Lohman, M.H. Slaughter, C.A. Horswill, and R.J. Stillman. Problems associated with determining body composition in maturing youngsters. In E.W. Brown and C.F. Branta (Eds.), *Competitive Sports for Children and Youth.* Champaign, IL: Human Kinetics, 1998, pp. 3-16.

19. American College of Sports Medicine. *ACSM's Guidelines for Exercise Testing and Prescription,* 6th ed. Philadelphia: Lippincott, Williams and Wilkins, 2000.

20. Bono, M.J., J.J. Roby, F.G. Micale, J.F. Sallis, and W.E. Shepard. Validity and reliability of predicting maximum oxygen uptake via field tests in children and adolescents. *Pediatric Exercise Science,* 3: 250-255, 1991.

21. Cahill, B.R. (Ed.). Proceedings of the conference on strength training and the prepubescent. *American Orthopedic Society for Sports Medicine,* 14, 1988.

22. Calfas, K.J., and W.C. Taylor. Effects of physical activity on psychological variables in adolescents. *Pediatric Exercise Science,* 6: 406-425, 1994.

23. Cassell, C.S., M. Benedict, G. Uetrect, M. Ranz, and B. Specker. Bone mineral density in young gymnasts and swimmers. *Medicine and Science in Sports and Exercise,* 25(Suppl.): S49, 1993.

24. Centers for Disease Control. Vigorous physical activity among high school students. *Morbidity and Mortality Weekly Reports,* 41: 33-35, 1990.

25. Chestnut, C. Theoretical overview: Bone development, peak bone mass, bone loss, and fracture risk. *American Journal of Medicine,* 91(5B): 25-45, 1991.

26. Clark, H.H. Joint and body range of movement. *Physical Fitness Research Digest*, 5: 16-18, 1975.

27. Cooper, K.H. A means of assessing maximal O_2 uptake. *Journal of the American Medical Association*, 203: 201-204, 1968.

28. Corbin, C.B., and R.P. Pangrazi. *Physical Activity for Children: A Statement of Guidelines*. Reston, VA: National Association for Sport and Physical Education, 1999.

29. Cureton, K.J., M.A. Sloniger, J.P. O'Bannon, D.M. Block, and W.P. McCormick. A generalized equation for prediction of VO_2 peak from one-mile run/walk performance. *Medicine and Science in Sports and Exercise*, 27(3): 445-451, 1995.

30. Dotson, C.O., and J.G. Ross. Relationships between activity patterns and fitness. *Journal of Physical Education, Recreation and Dance*, 56(1): 86-90, 1985.

31. Drinkwater, B.L. Physical exercise and bone health. *Journal of the American Medical Women's Association*, 45(3): 91-97, 1990.

32. *FITNESSGRAM Test Administrator's Manual*, 2nd ed. Champaign, IL: Human Kinetics, 2001.

33. Gibbons, R.J., G.J. Balady, and W.J. Beasley. ACC/AHA guidelines for exercise testing. A report of the American College of Cardiology/American Heart Association Task Force on Practice Guidelines (Committee on Exercise Testing). *Journal of the American College of Cardiology*, 30: 260-315, 1997.

34. Grodjinovsky, A., O. Inbar, R. Dotan, and O. Bar-Or. Training effect on the anaerobic performance of children as measured by the Wingate Anaerobic test. In K. Berg and B.O. Eriksson (Eds.), *Children and Exercise IX*. Baltimore: University Park Press, 1979, pp. 139-145.

35. Hansen, H.S., K. Froberg, N. Hyldebrandt, and J.R. Nielsen. A controlled study of eight months of physical training and reduction of blood pressure in children: The Odense schoolchild study. *British Medical Journal*, 303: 682-685, 1991.

36. Harro, M., C. Riddoch, N. Armstrong, and W. Van Mechelen. Physical activity. In N. Armstrong and W. Van Mechehen (Eds.), *Pediatric Exercise Science and Medicine*. Oxford, UK: Oxford University, 2000, p. 472.

37. Kowalski, K.C., P.R.E. Crocker, and R.A. Faulkner. Validation of the Physical Activity Questionnaire for older children. *Pediatric Exercise Science*, 9: 174-186, 1997.

38. Krahenbuhl, G.S., J.S. Skinner, and N.M. Kohrt. Development of aerobic power in children. *Exercise and Sport Sciences Reviews*, 13: 503-538, 1985.

39. Krahenbuhl, G.S., R.P. Pangrazi, G.W. Petersen, L.N. Burkett, and M.J. Schneider. Field testing of cardiorespiratory fitness in primary school children. *Medicine and Science in Sports and Exercise*, 10: 208-213, 1978.

40. Krahenbuhl, G.S., R.P. Pangrazi, L.N. Burkett, M.J. Schneider, and G.W. Petersen. Field estimation of $\dot{V}O_2$max in children eight years of age. *Medicine and Science in Sports*, 9: 37-40, 1977.

41. Kramer, W.J., and S.J. Fleck. *Strength Training for Young Athletes*. Champaign, IL: Human Kinetics, 1993.

42. Kuczmarski, R.J., et al. Varying body mass index cutoff points to describe overweight prevalence among U.S. adults: NHANES III (1988-1994). *Obesity Research*, 5: 542, 1997.

43. Lauer, R.M., J. Lee, and W.R. Clarke. Factors affecting the relationship between childhood and adult cholesterol levels: The Muscatine study. *Pediatrics*, 82: 309-318, 1988.

44. Leger, L.A., D. Mercier, C. Gadoury, and J. Lambert. The multistage 20 meter shuttle run test for aerobic fitness. *Journal of Sport Sciences*, 6: 93-101, 1988.

45. Lohman, T.G. Assessment of body composition in children. *Pediatric Exercise Science*, 1(1): 19-30, 1989.

46. Lohman, T.G. The use of skinfold to estimate body fatness in children and youth. *Journal of Physical Education, Recreation and Dance*, 58: 98-102, 1987.

47. Mahon, A.D. Exercise Training. In N. Armstrong and W. Van Mechelen (Eds.), *Pediatric Exercise Science and Medicine*. Oxford, UK: Oxford University Press, 2000, pp. 201-219.

48. Malina, R.M., and C. Bouchard. *Growth, Maturation, and Physical Activity*. Champaign, IL: Human Kinetics, 1991, pp. 87-150.

49. Massicotte, D.R., R. Gauther, and R.P. Markon. Prediction of VO_2max from running performance in children aged 10-17 years. *Journal of Sports Medicine and Physical Fitness*, 25: 10-17, 1985.

50. Matkovic V., D. Fonatana, C. Tominac, P. Goel, and C. Chestnut. Factors which influence peak bone mass formation: A study of calcium balance and the inheritance of bone mass in adolescent females. *American Journal of Clinical Nutrition*, 52: 878-888, 1990.

51. Mayo Medical Laboratories. *Test Catalog*. Rochester, MN: Mayo Press, 2001.

52. Melton, L.J. Osteoporosis. In R. Berg and J. Cassels (Eds.), *The Second Fifty Years: Promoting Health and Preventing Disability*. Washington, DC: National Academy Press, 1990, p. 76.

53. Milne C., V. Seefeldt, and P. Reuschlein. Relationship between grade, sex, race, and motor development in young children. *Research Quarterly*, 47: 726, 1976.

54. Morrow, J.R., and P.S. Freedson. Relationship between habitual physical activity and aerobic fitness in adolescents. *Pediatric Exercise Science*, 6(4): 315-329, 1994.

55. Must A., and R.S. Strauss. Risks and consequences of childhood and adolescent obesity. *International Journal of Obesity and Related Metabolic Disorders*, 23(Suppl. 2): 52-11, 1999.

56. Must, A., P.F. Jacques, G.E. Dallal, C.J. Bajema, and W.H. Dietz. Long-term morbidity and mortality of overweight adolescents. *New England Journal of Medicine*, 327: 1350-1355, 1992.

57. National Center for Health Statistics. Prevalence of overweight among children and adolescents. Accessed 1/3/99 at http://www.cdc.gov/nchs/products/pubs/pubd/hestats/over99fig1.htm.

58. National Institutes of Health. Report of the Second Task Force on Blood Pressure Control in Children. *Pediatrics,* 103: 1175-1182, 1999.

59. Orchard, T.J., R.P. Donahue, L.H. Kuller, P.N. Hodges, and A.L. Drash. Cholesterol screening in childhood: Does it predict adult hypercholesterolemia? The Beaver County experience. *Journal of Pediatrics,* 103: 687-691, 1983.

60. Pfeiffer, R.D., and R.S. Francis. Effects of strength training on muscle development in pre-pubescent, pubescent, and postpubescent males. *The Physician and Sportsmedicine,* 14: 134-143, 1986.

61. Public Health Service. *Healthy People 2000: National Health Promotion and Disease Prevention Objectives* (DHHS Publication No. PHS 91-50212). Washington, DC: U.S. Department of Health and Human Services, 1990.

62. Risser W.L. Musculoskeletal injuries caused by weight training. *Clinical Pediatrics,* 29(6): 305-310, 1990.

63. Risser W.L. Weight training injuries in children and adolescents. *American Family Physician,* 44: 2104-2108, 1991.

64. Rivera-Brown, A., M.A. Rivera, and U.R. Fontera. Applicability of criteria for $\dot{V}O_2$max in active adolescents. *Pediatric Exercise Science,* 4: 331-339, 1992.

65. Roberts, S. Exercise prescription recommendations for children. *ACSM Certified News,* 7(1): 3, 1997.

66. Rotstein, A.R., R. Dotan, O. Bar-or, and G. Tenenbaum. Effects of training on anaerobic threshold, maximal aerobic power and anaerobic performance of preadolescent boys. *International Journal of Sports Medicine,* 7: 281-286, 1986.

67. Rowland, T.W. *Developmental Exercise Physiology.* Champaign, IL: Human Kinetics, 1996.

68. Rowland, T. W. *Exercise and Children's Health.* Champaign, IL: Human Kinetics, pp. 27-83, 1990.

69. Rowland, T.W. (Ed.). *Pediatric Laboratory Testing—Clinical Guidelines.* Champaign IL: Human Kinetics, 1993.

70. Safrit, M.J. *Complete Guide to Youth Fitness Testing.* Champaign, IL: Human Kinetics, 1995.

71. Sale, D.G. Strength training in children. In C.V. Gisolfi and D.R. Lamb (Eds.), *Perspectives in Exercise Science and Sports Medicine* (Vol. 2). Indianapolis: Benchmark Press, 1989, pp. 165-222.

72. Sallis J.F., and K. Patrick. Physical activity guidelines for adolescents: Consensus statement. *Pediatric Exercise Science,* 6: 302-314, 1994.

73. Sallis, J.F., M.J. Buono, J.J. Roby, F.G. Micate, and J.A. Nelson. Seven-day recall and other physical activity self-reports in children and adolescents. *Medicine & Science in Sports and Exercise,* 25: 99-108, 1993.

74. Saltarelli, W. *The effects of pace training on children's performance time and heart rate response during a one-mile run.* Doctoral dissertation. University of Toledo, 1989.

75. Sargeant, A. J., P. Dolan, and A. Thorne. Effects of supplementary physical activity on body composition, aerobic, and anaerobic power in 13-year-old boys. In R.A. Binkborst, H.C.G. Kemper, and W.H. Saris (Eds.), *Children and Exercise XI.* Champaign, IL: Human Kinetics, 1985, pp. 140-150.

76. Schlicker, S.A., S.T. Borra, and C. Regan. *Nutrition Reviews* 52: 11-20, 1996.

77. Sewell, L., and L.J. Micheli. Strength training for children. *Journal of Pediatric Orthopedics* 6: 143-146, 1986.

78. Slaughter, M.H., T.G. Lohman, R.A. Boileau, C.A. Horswill, R.H. Stillman, M.D. Van Loan, and D.A. Bemben. Skinfold equations for estimation of body fatness in children and youth. *Ham. Ber* 56: 681-689, 1984.

79. Strong J.P., G.T. Malcom, C.A. McMahan, R.E. Tracy, W.P. Newman, E.E. Herderick, and J.F. Cornhill. Prevalence and extent of atherosclerosis in adolescents and young adults. *Journal of the American Medical Association,* 281: 727-735, 1999.

80. Taylor, C.B., J.F. Sallis, and R. Needle. The relation of physical activity and exercise to mental health. *Public Health Reports,* 100: 195-202, 1985.

81. Tipton, C.M. Exercise training and hypertension: An update. *Exercise and Sport Sciences Reviews,* 19: 447-505, 1991.

82. Tortolero, S.R., T.W. Taylor, and N.G. Murry. Physical activity, physical fitness and social, psychological and emotional health. In N. Armstrong and W. Van Mechehen (Eds.), *Pediatric Exercise Science and Medicine.* Oxford University Press, 2000, pp. 273-291.

83. Twisk, J.W., W. Mechelen, H.C.G. vanKemper, and G.B. Post. The relation between "long term exposure" to lifestyle during youth and young adulthood and risk factors for cardiovascular disease. *Journal of Adolescent Health,* 20:309, 1997.

84. Vrijens, J. Muscle strength development in pre and post pubescent age. *Med Sport* 11: 152-158, 1978.

85. Welk, G.J., and K. Wood. Physical activity assessments in physical education—A practical review of instruments and their use in the curriculum. *Journal of Physical Education, Recreation and Dance,* 71: 30-40, 2000.

86. Weltman A., C. Janney, C.B. Rians, K. Strand, B. Berg, S. Tippett, J. Wise, B.R. Cahill, and F.I. Katch. The effects of hydraulic resistance strength training in pre-pubescent males. *Medicine and Science in Sports and Exercise,* 18: 629-638, 1986.

87. Weston, A.T., R. Petosa, and R.R. Pate. Validation of an instrument for measurement of physical activity in youth. *Medicine and Science in Sports and Exercise,* 29: 138-143.

Nicole Y.J.M. Leenders, PhD
The Ohio State University
General Clinical Research Center
Columbus, OH

The Elderly

Geriatrics is a branch of medicine that deals with the problems and diseases associated with elderly people (>65 years) and the aging process. As people age, they go through physiological processes that are natural and likely enhanced by certain conditions (e.g., inactivity, environmental factors). For the purpose of description, the elderly are divided into the **old age** (65-74 years of age), the **very old age** (75-84 years of age), and the **oldest old** (older than 85 years of age) (61, 94). Although these groups chronologically categorize the elderly, they are a heterogeneous group of individuals in terms of physiological status and there is overlap between groups.

Scope

Between 1965 and 1995, the over 65-year-old population in the United States grew by 82%, to 13% of the population. This represents a quadrupling since 1900. It is expected that by 2025 there will be approximately 62 million people over the age of 65 (20% of the population) (1).

A child born in 1997 in the United States has a life expectancy of 76.5 years, compared with only 47.5 years in 1900. Women born in 1997 will outlive males, on average, by 6.5 years, and thus the female to male ratio increases with age. Minority populations will represent 25% of the elderly population in 2030, up from 15% in 1997. Currently, on average,

© Digital Vision

whites outlive blacks by 6 years (46). The prevalence of chronic disease and functional impairment increases with age. In 1995-1996, greater than 50% of the elderly in the United States had at least one **disability,** and multiple chronic diseases and impairments are common (1).

The high prevalence of these chronic diseases (figure 30.1) causes increased health and medical care costs and reliance on medical services and assisted living. Money spent in 1995 for care of the elderly was 35% of total healthcare expenditure (26). Falls and **dementia** account for much of this cost (9, 26). Falls are a leading cause of injury, disability, and death of the elderly (73, 102, 106).

A decline in general health status and reduced mobility negatively affects an older individual's ability to carry out daily living activities. For instance, many have difficulty with bathing, dressing, and preparing food (15, 53), and about one-third perceive their health as fair to poor. Not surprisingly, the rate of physical inactivity increases with age (figure 30.2). Only 25% of the elderly report regular participation in sustained **physical activity** five or more times per week for 30 min or more per occasion (109). Blacks and males are less active than whites and females, respectively, across every age group (49).

Physiology and Pathophysiology

Anatomical and physiological changes of several organ systems can lead to functional disability and increased risk of premature death (100). Table 30.1 reviews the effects of aging on the organ systems. Sudden decline of organ function is commonly at-

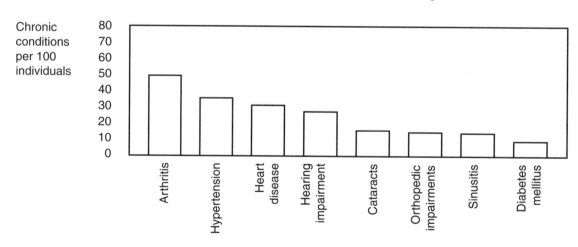

FIGURE 30.1   Prevalence of chronic diseases per 100 individuals in adults 65 years of age or older.

Adapted from Administration on Aging: Profile of Older Americans: 1998.

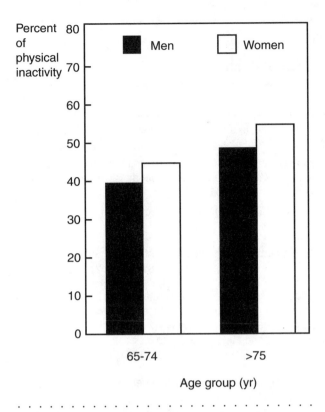

FIGURE 30.2   Percentage of older individuals self-reporting physical inactivity.

Adapted from Blackman, D.K. et al., 1999.

tributable to chronic disease, whereas aging effects result in a gradual impairment (87). There are two theories of aging:

1. The genetic theory: proposes aging is programmed by cellular signals from birth to death
2. The damage theory: proposes aging is caused by progressive translation and transcription errors in cells.

Changes in organ system function are influenced by diet, environmental factors, and genetics. Each of these may affect gene expression. The aging process may be attenuated by modifying factors that enhance the aging process (5, 86, 100). Table 30.2 reviews the major organ systems and specific effects associated with aging. Many of these changes are augmented by physical inactivity and are attenuated when exercise volume is maintained as aging occurs.

Clinical Considerations

The most common chronic diseases present in the elderly are coronary artery disease, arthritis, hyper-

tension, diabetes mellitus, and obesity. Each of these is covered in detail in other chapters of this textbook. Of these, arthritis is the most prevalent, coronary artery disease is the leading cause of death (26), and Alzheimer's disease is the ninth leading cause of death (46).

Diagnostic and Laboratory Evaluations

Many older individuals have a history of falling, are frail (i.e., weak, vulnerable), and are limited in their ability to perform activities of daily living (ADLs). This does not necessarily preclude an individual from exercise participation but does direct the focus of a preexercise examination. Agility, coordination, gait, and balance should be assessed to identify individuals at risk of falling. For example, difficulty standing up from a chair, difficulty performing a slow walking speed, and poor ability to maintain balance during a push on the shoulder are predictors of falling (24, 73). A goniometer can be used to obtain measurements of range of motion (ROM) in joints such as the hip, ankle, knee, shoulder, and elbow. Limited ROM in these joints is related to limitations in performing ADLs. Table 30.3 reviews preexercise evaluation tests. Evaluation of mental (depression) and intellectual impairment (dementia, Alzheimer's disease) should also be made, because these may affect the ability to perform appropriate exercise and exercise compliance.

Graded Exercise Testing

Low- to moderate-risk elderly individuals can participate in **moderate intensity exercise,** defined as an intensity of 3 to 6 **metabolic equivalents (METs),** or 40% to 60% $\dot{V}O_2$max, without an exercise test (3). As a screening, these individuals should complete a questionnaire that can identify any preexisting conditions (table 30.4). Based on responses, consultation with a physician or an exercise test may be recommended before the person begins an exercise program (33). If the individual has one or more risk factors for or any signs or symptoms of coronary artery disease, or if he or she wishes to begin a high-intensity exercise program, an exercise evaluation (physical exam and exercise test) should be performed (3). Practical application 30.1 discusses client–clinician interaction for elderly clients.

Medical Evaluation

When necessary (i.e., for high-risk individuals), a thorough medical history and physical examination should be performed by a qualified physician. The

TABLE 30.1
· · · · · · ·
Aging Effects on Organ Systems

Organ system	Effects
Skeletal muscle	Sarcopenia: loss of mass, strength, contractile speed, and power. Loss of mass at rate of 1.2 kg per decade from the 5th to 9th decade attributable to hypoplasia and atrophy (33, 40, 42, 62, 83). Hypertrophy attributable to compensatory to loss of motor units (4). Motor unit remodeling: reduced axonal sprouting rate, nerve fiber contacts per muscle fiber, and type 2 fiber density (attributable to cross-reinnervation) (28, 109); increase in motor unit size (41). Myosin phenotype: increased slow isoform (54, 85). Increased risk of osteoporosis, frailty, fractures, and arthritis. Decreased flexibility (95).
Body composition	Decreased lean mass and total water; increased fat; reduced bone mass starting in 3rd decade and accelerating in the 5th, increasing osteoporosis risk (14, 38, 40, 57, 87, 94, 102).
Cardiovascular	Arterial stiffening and reduced dilatory capacity; increased peripheral resistance and blood pressure; ventricular thickening; increased left ventricular end-diastolic volume; reduced baroreceptor sensitivity and orthostatic tolerance; 5-15% per decade decline in peak $\dot{V}O_2$ attributable to lower peak cardiac output (heart rate and stroke volume) and augmented by an increase in body fat; increased cardiovascular disease risk (25, 37, 38, 58, 62, 76, 87, 97, 103, 105).
Metabolic	Insulin insensitivity and glucose intolerance, and elevated plasma insulin levels leading to increased type 2 diabetes risk; increased obesity prevalence (109).
Respiratory	Increased chest wall stiffening and elastic recoil; pulmonary artery stiffening and increased pressure; reduced inspiratory and expiratory capacity; reduced pulmonary function (VC, FEV_1); increased dead space volume; reduced peak ventilation by up to 35% (25, 100).
Nervous	Reduced stimulus for visceral arterial constriction for blood redistribution during exercise; reduced cardiac β-adrenergic stimulation; central nervous deficits leading to decreased cognition, memory, learning ability, reaction time, and sleep; altered gait and balance; increased risk of dementia and Alzheimer's; impaired hearing and sight (9, 10, 19, 75, 94).
Energy expenditure and energy intake	Decrease in daily energy expenditure attributable to decrease in resting metabolic rate and physical activity; decrease in resting metabolic rate related to a decrease in fat-free mass and increase in fat mass; low caloric and protein intake (79, 80, 81, 91, 109).
Thermoregulation	Decreased ability to regulate body temperature when homeostasis is challenged; decreased amount of sweat per active sweat gland; reduced response to increased blood flow during exercise attributable to structure and response of cutaneous blood vessels; inadequate ability to reduce splanchnic blood flow during exercise (50, 51, 90).

Note. VC = vital capacity; FEV_1 = forced expiratory volume in 1 s.

American College of Sports Medicine recommends a physical exam for all men older than 45 years and women older than 55 years age prior to beginning a program of **vigorous intensity physical activity** (i.e., >70% $\dot{V}O_2$max) (3). The general methods of a medical evaluation are presented in chapter 4. For the elderly, emphasis should be placed on assessing specific areas of risk. This includes the cardiovascular system and the musculoskeletal system (3, 5, 44). For an office evaluation of functional capac-

ity for ADLs, questions about bathing, dressing, and getting in and out of bed should be asked. A standardized questionnaire, such as the Yale Physical Activity Scale, which is specifically designed for the elderly population, may be used to assess activity level (27).

Contraindications

Chapter 6 reviews absolute and relative contraindications to exercise testing. Common contraindications

TABLE 30.2
Physiological Changes Associated With Aging

System	Change
CARDIOVASCULAR	
Rest	
— Heart rate	Decreases
— Stroke volume	Increases
— Systolic and diastolic blood pressure	Increases
Maximum exercise	
— Heart rate	Decreases
— Oxygen consumption	Decreases
— Cardiac and vascular responses to β-adrenergic stimulation	Decrease
RESPIRATORY	
Maximum exercise	
— Maximum ventilation	Decreases
— Tidal volume	Decreases
— Breathing frequency	Increases
— Vital capacity	Decreases
— Residual volume	Increases
MUSCULOSKELETAL	
Muscle mass and strength	Decrease
Elasticity in connective tissue	Decreases
Balance	Decreases
Coordination	Decreases
Bone density	Decreases
METABOLIC	
Glucose tolerance	Decreases
Insulin action	Decreases
Metabolic rate	Decreases
THERMOREGULATION	
Thirst sensation	Decreases
Skin blood flow	Decreases
Sweat production per sweat gland	Decreases

Adapted from (5, 7, 50, 58, 100).

TABLE 30.3
Preexercise Training Evaluations

Test	Measurement	Outcome	Risk
Chair stand	Stand from a chair of standard height, unaided and without using arms	Ability Time required	Unable ≥ 2.0 s
Step-ups	Step-ups onto a single 23-cm step in 10 s	Ability Number of times	Unable <3 in 10 s
Walking speed	6-m walk	Time Number of steps RPE Heart rate Blood pressure Gait abnormalities such as asymmetry	<0.6 m $\cdot$ s^{-1}
Tandem walk	Walking along a 2-m line, 5 cm wide	Number of errors (off line, touching examiner or another object)	≥ 8 errors
One-leg stand	Stand on one leg	Ability Time	<2 s
Functional reach	Maximal distance an individual can reach forward beyond arm's length, while maintaining a fixed base of support in the standing position	Inches	
Timed "up and go"(pods)	Stand up from standard chair, walk distance of 3 m, turn, walk back to chair, and sit down again.	Time	>10 s
Range of motion	Using a goniometer, assess the following: Shoulder abduction (SA), flexion (SF), extension (SE) Elbow flexion (EF), extension (EE) Hip flexion (HF), extension (HE) Knee flexion (KF), extension (KE) Ankle dorsiflexion (DF), plantar flexion (PF)	Degrees	$<90°$ (SA); $<150°$ (SF); $<20°$ (SE) $<140°$ (EF); $<20°$ (EE) $<90°$ (HF), within $10°$ (HE) $<90°$ (KF); not within $<10°$ full KE Unable to perform DF and PF

Note. RPE = rating of perceived exertion.
Adapted from (7, 17, 55, 73, 82).

---

in the elderly include elevated resting blood pressure (diastolic >115 mmHg or systolic >200 mmHg), moderate valvular heart disease, electrolyte abnormalities, complex ventricular ectopy, ventricular aneurysm, uncontrolled metabolic diseases such as diabetes, and neuromuscular, musculoskeletal, and rheumatoid disorders (3).

Recommendations and Anticipated Responses

The elderly are at increased risk of obesity, osteoporosis, hypertension, low aerobic fitness level, poor balance, ambulatory instability, neuromuscular incoordination, and vision and hearing impairment. The exercise professional must consider these when selecting an exercise testing protocol (96, 97). Walk-

<table>
<tr><td colspan="3">TABLE 30.4

Exercise Questionnaire</td></tr>
</table>

Questions	Yes	No
A.	Do you get chest pains while at rest and/or during exertion?		
B.	If the answer to question A is yes, is it true that you have not had a physician diagnose these pains yet?		
C.	Have you ever had a heart attack?		
D.	If the answer to question C is yes, was your heart attack within the last year?		
E.	Do you have high blood pressure?		
F.	If you don't know the answer to question E, answer this: Was your blood pressure reading more than 150/100?		
G.	Are you short of breath after extremely mild exertion and sometimes even at rest or at night in bed?		
H.	Do you have any ulcerated wounds or cuts on your feet that do not seem to heal?		
I.	Have you lost 10 lb or more in the past 6 months?		
J.	Do you get pain in your buttocks or the back of your legs—thighs and calves—when you walk?		
K.	While at rest, do you frequently experience fast irregular heartbeats or, at the other extreme, very slow beats? (Although a low heart rate can be a sign of an efficient and well-conditioned heart, a very low rate can also indicate a complete heart block.)		
L.	Are you currently being treated for any heart or circulatory condition such as vascular disease, stroke, angina, hypertension, congestive heart failure, poor circulation in the legs, valvular heart disease, blood clots, or pulmonary disease?		
M.	As an adult, have you ever had a fracture of the hip, spine, or wrist?		
N.	Did you fall more than twice in the past year (no matter what the reason)?		
O.	Do you have diabetes?		

Key: If the answer is yes to one of these questions, an individual should be advised to undergo an evaluation by a physician.

Reprinted, by permission, from W.J. Evans, 1999, "Exercise training guidelines for the elderly," *Medicine and Science in Sports and Exercise* 31:12-17.

ing is appropriate for most. Handrail support is important when elderly people use a treadmill to provide balance assistance. Also, cycling should be considered for those with balance or gait problems. Performing an exercise test during the morning hours may be best, because many individuals are less fatigued than later in the day (88).

The maximal attainable work rate during a graded exercise test for an older individual is likely less than 7 METs. A low-intensity testing protocol with small increments in work rate (1.0-2.0 METs per stage) is generally recommended. Popular protocols that fit these parameters include the Naughton and Balke, as well as cycling increases of 25 to 30 W per stage. Ramping-type protocols can also be considered. Chapter 6 provides details about exercise testing protocols. Also, because it takes longer for an elderly person to reach steady state in $\dot{V}O_2$, minute ventilation, and heart rate (HR), a warm-up period (e.g., >5 min) at a low level of intensity may be useful.

Peak heart rate, stroke volume, cardiac output, and $\dot{V}O_2$ are lower in the elderly than for younger

Client–Clinician Interaction

Many older individuals do not have a spouse, close children, or friends to rely on for socialization, assistance, and support (32). Although with age, social relationships may change from family to more formalized organizations or nonfamily members, many elderly live in social isolation and are very lonely. This is important because epidemiological studies have demonstrated a relationship between social support and both mental and physical health (20, 39, 108). Furthermore, several studies demonstrate lack of social support to be a major risk factor for depression, morbidity, and mortality (12, 45, 77, 78, 92, 101, 104).

Participation in an organized physical activity program provides an excellent opportunity for interaction and forming social networks, which in turn can contribute to feelings of well-being and improve quality of life and physical performance (74, 83). Furthermore, a favorable attitude on the part of significant others (i.e., spouse, friends) toward an exercise program is frequently associated with better compliance with exercise programs. The clinical exercise professional can help by identifying persons who require social support. In a structured exercise setting, the exercise professional should also be prepared to provide social support to these individuals by making conversation.

In an effort to increase social interaction for elderly persons, one can consider exercise programs that offer either supervised or community-based sessions (44, 53). Examples of methods to improve social interaction for the elderly participating in an exercise program include a "buddy" exercise system, where individuals are matched up with those of similar ability to perform their exercise together; social occasions including picnics, holiday parties, and birthday celebrations; and placement of exercise equipment that allows individuals to face each other. Each of these opportunities can be facilitated by an exercise professional responsible for programming in the structured exercise setting (e.g., cardiac rehabilitation programs and senior community centers).

individuals. Potential mechanisms include cardiac **β-receptor** down-regulation (76), physical inactivity, and skeletal muscle morphological changes. And because maximal **power** output is lower in the elderly, at any absolute submaximal work rate an elderly individual is closer to anaerobic threshold compared with someone who is younger (84, 94). Cardiac output at maximal exercise is generally 20% to 30% lower compared with younger individuals (76).

Mental deficiencies such as dementia or Alzheimer's disease may limit the ability of an individual to comprehend directions during a graded exercise test. These individuals should be tested in a comfortable environment, preferably with a familiar individual present. Similar recommendations can be made for individuals who have hearing or vision impairments. For those individuals, a clear description by video or audio is recommended (2, 3).

With regard to strength and ROM testing, general testing procedures, such as the 1-repetition maximum (1RM) test, can be used. However, because a 1RM test may increase the risk of injury, one may administer the modified 1RM evaluation (see chapter 7) or simply provide general recommendations for low-resistance strength training without prior strength assessment in most elderly individuals. Using weight machines is best, because they reduce

balance requirements and the risk of injury. The focus of evaluation should include those muscles used for typical ADLs, such as the hip extensors, flexors, and abductors; knee extensors and flexors; ankle dorsiflexors and plantar flexors; and shoulder extensors, flexors, and abductors (55). Chapter 7 provides basic methodology for testing strength and ROM.

Exercise Prescription

For most elderly individuals, the 1995 joint recommendation of the American College of Sports Medicine and the Centers for Disease Prevention is appropriate (79, 109). This is presented in chapter 7. The intensity of activity should be light to moderate and can be achieved by walking for 30 min each day. Any other aerobic type activity will also suffice. For those with a low fitness level, health benefits can be obtained when short bouts of moderate intensity physical activity, lasting no less than 10 min, are performed (52, 109). Standard heart rate or work rate guided exercise training may also be employed in the elderly. Chapter 7 also provides details of general exercise prescription development. Practical application 30.2 reviews the literature concerning exercise for the elderly.

Literature Review

Cardiovascular Training

In healthy, sedentary elderly men and women, $\dot{V}O_2$peak can increase 10% to 15% if moderate-intensity (40-50% $\dot{V}O_2$peak) training is performed. Greater improvements of up to 30% are observed when exercise training is performed 3 or more days per week, for 30 to 45 min duration at an intensity greater than 50% of $\dot{V}O_2$peak (29, 56, 98, 99).

In elderly men, increases in $\dot{V}O_2$peak resulting from endurance training are attributable to an increased cardiac output and a widening of the arteriovenous oxygen difference. Increases in cardiac output are attributable entirely to an increased stroke volume. Maximal heart rate does not change, or decreases slightly, with endurance training (97, 103). Increases in stroke volume are mediated by left ventricular volume overload **hypertrophy** (29) and increased blood volume, which enhances the Frank-Starling response. The increase in $\dot{V}O_2$peak in older women observed after a 9 month training program was attributable to a widened arteriovenous oxygen difference without hemodynamic improvement (99).

Aerobic training also improves insulin sensitivity and glucose homeostasis in the elderly (21, 22, 48). Insulin action and skeletal muscle **GLUT 4** concentrations improved in elderly individuals who used a cycle ergometer at 70% of $\dot{V}O_2$peak for 60 min on 7 consecutive days (22). These changes may reduce the risk of type 2 diabetes or help to better control blood glucose levels.

Although many studies demonstrate improved cardiorespiratory function following moderate-to high-intensity exercise training, this type of training may not always be the best choice. For instance, when an elderly group participated in a structured physical activity program, spontaneous physical activity during the remainder of the day was reduced (43, 66).

Combining strength and endurance training is also beneficial for the elderly individual. One study showed that after 6 months of combined resistance and endurance training, older healthy individuals increased their $\dot{V}O_2$peak (11%) and their upper and lower body strength (23). The ability to carry out normal daily tasks such as carrying groceries, transferring laundry, vacuuming, making a bed, climbing stairs, and floor sweeping improved and translated to carrying 14% more weight and moving 10% faster. The amount of improvement was dependent on the initial conditioning level, with the least fit demonstrating the most improvement (23).

Resistance Training

Elderly individuals, including the oldest old and very frail elderly, demonstrate physiological adaptations to strength training (35, 36, 40, 47, 107). How much adaptation occurs depends on the frequency, volume, mode, type of training, and initial training state (33). Resistance training programs lasting from 8 weeks to 1 year can increase muscle strength and mass in the elderly, regardless of age and sex.

Increases in strength in the elderly are the result of both muscle hypertrophy and neuromuscular adaptation (107). However, several studies have reported large increases in skeletal muscle strength in the elderly that are proportionally greater than increases in cross-sectional area (35, 36, 38).

In addition to strength gains, regular resistance training improves gait, balance, and overall functional capacity (13, 16, 35, 47). Strength training also increases bone mineral density and content (72), increases metabolic rate (18), assists with the maintenance of body weight by decreasing fat mass and increasing lean mass, and improves insulin action and plasma levels (67). The gains observed during resistance training can be expected to persist for several weeks after training is ceased (60).

Range of Motion

Several studies have demonstrated that regular ROM training improves the flexibility of the spine, hips, ankles, knees, and shoulders in older individuals (63, 64, 68-71, 86) whereas few have shown no effect (30). Improvements in flexibility can be expected to increase the effective range of strength gains and improve ambulatory ability.

Special Exercise Considerations

Because many older individuals have at least one chronic disease and are sedentary or only minimally active, beginning an exercise program with intermittent bouts of moderate-intensity activity may be best. This will allow most people to achieve early success with an exercise regimen and may improve compliance by reducing pain, fatigue, and injuries. A goal should be to work toward 30 min of continuous exercise. Additionally, persons beginning an exercise program should be instructed on proper technique and training load increases to limit their susceptibility to injury. The clinical exercise professional should also stress that regular physical activity can enhance quality of life.

It may be difficult for the elderly person to perform physical activity or exercise at home because of lack of space or equipment. Alternatives include senior citizen and community centers, health clubs and church facilities, shopping malls, and community swimming pools. These locations may also provide fellowship that is important for the single person. Social activities that may be of interest to the elderly include an organized walk or bicycle tour or competitive events such as the Senior Olympics or Master's swimming (2, 7). Whatever facility is used or event is performed, it must be safe and free from barriers. Additionally, many older individuals, especially those with a low fitness level, have a lower tolerance for heat and cold. Thus, environmental conditions, fluid intake, and clothing are very important considerations (5, 51, 94).

Exercise Recommendations

In general, an active lifestyle preserves and enhances skeletal muscle strength and endurance, flexibility, cardiorespiratory fitness, and body composition, which otherwise would diminish and increase chronic disease risk with advancing age. It is therefore important to develop optimal exercise interventions for the elderly that increase daily physical activity levels (i.e., energy expenditure) and thereby delay or counteract the effects of risk factors on chronic disease. Practical application 30.3 provides exercise recommendations for the elderly.

Cardiovascular Training

Low cardiorespiratory fitness is a risk factor for cardiovascular disease and all-cause mortality (11). Low $\dot{V}O_2$peak is associated with reduced ability to perform ADLs including climbing stairs and brisk walking (8). A small improvement in cardiovascular fitness is associated with a lower risk of death (31).

Healthy, sedentary older men and women can increase their cardiorespiratory fitness by performing aerobic exercise training (29, 34, 56, 65, 76, 85, 98, 99, 103).

Common physical activities that the elderly population should engage in are walking (outdoors, indoors, or treadmill), cycling, gardening, swimming (laps or water aerobics), and golfing while walking with a pull cart (109). Individuals who are unable to perform ambulatory activities may be candidates to perform seated chair activities, stationary cycling, and water activities. T'ai chi may also be beneficial to improve strength and balance (59, 89). High-intensity activities such as running, rowing, aerobic/gravity riders, and stair-steppers may not be appropriate.

Low- to moderate-intensity exercise programs can be performed daily. Higher intensity exercise sessions (>70% heart rate reserve) should only be performed 3 to 5 days per week (3). This allows for recovery days, which is important for the elderly, who recover more slowly than younger persons. Older individuals with a low exercise capacity may benefit from multiple daily sessions of short duration, whereas the more capable individual can benefit from three sessions per week with exercise bouts performed once per day (3, 5). The appropriate intensity can be prescribed using the heart rate reserve, heart rate percent, or rating of perceived exertion methods (see chapter 7).

For the healthy older individual, it is recommended that exercise be performed minimally for 30 min but not beyond 60 min in duration. If an individual beginning an exercise program is predominantly sedentary, has severe chronic disease, or has a very low fitness level, a minimum of 30 min of continuous exercise may not be possible. Sessions of as little as 10 min two or three times a day are appropriate in this situation. General health benefits are still obtained with these approaches (3).

During the initial stages of an exercise program, the intensity should be low (40-60% heart rate or $\dot{V}O_2$ reserve). This is an important period for promoting physical activity adherence. Intensity and duration can be increased every 2 to 3 weeks, as tolerated, until the desired total exercise volume is reached. The exercise professional working with elderly clients must recognize indicators of musculoskeletal or orthopedic injury, boredom, and decreased exercise tolerance, which may lead to cessation of exercise training (5, 44, 52).

Resistance Training

Strength training has the potential to improve functional capacity and quality of life of the elderly person (35). In general, most elderly individuals can

Exercise Recommendations for the Elderly

Mode	Frequency	Intensity	Duration	Special considerations
AEROBIC TRAINING More often: walking, cycling, pool activity, seated aerobics, ADLs. Less often: jogging, swimming laps, rowing, aerobic dance.	Daily at low to moderate intensity or 3-5 times per week for moderate to high intensity.	ADLs at comfortable pace. Low to moderate: 40% or 50-70% $\dot{V}O_2$peak or HRR. Moderate to high: >70% $\dot{V}O_2$peak or HRR.	Goal is 30 min continuous and up to 60 min. Intervals may be as short as 8-10 min.	May need to start with short bouts initially and build to 30 min continuously. Higher intensity may be difficult, especially for those who are sedentary. Consider beginning at low to moderate intensity. Progress duration and frequency initially, as this may improve compliance. Comorbidities such as arthritis, osteoporosis, and heart disease need to be considered. See these chapters for further special considerations.
RESISTANCE TRAINING More often: multi-station machine type (e.g., Universal), elastic bands, hand weights. Less often: free weights.	2-3 times per week	60-80% of 1RM. Progress by reassessing 1RM or as tolerated by the individual.	8-20 repetitions. 20-30 min per session.	Free weights may be difficult for an individual to balance, and assistance should always be available. Focus on major muscle groups that are used to perform ADLs (shoulders, back, arms, abdominals, and legs). See chapters reviewing heart diseases, osteoporosis, and arthritis for further special considerations.
RANGE OF MOTION TRAINING Static stretching. See chapter 7 for examples.	Daily, but especially following an aerobic and/or resistance training session.	Subject should feel a mild stretch without inducing pain. Progress range of stretch based on lack of discomfort experienced.	5-30 min total, with two 30-s bouts on each muscle group. Involve all major muscle groups (neck, shoulders, arms, lower back, quadriceps, hamstrings, calves, ankles).	Avoid ballistic movements and the Valsalva maneuver during the stretching routine. A brief routine performed before or after aerobic or resistance exercise may only focus on the muscle groups used and last as little as 5 min. Consider using alternate methods such as seated movements and use of a towel or elastic band to assist those with difficulties performing standard flexibility training. An overall routine can last up to 20-30 min. See chapters reviewing osteoporosis and arthritis for further special considerations.

Note. ADL = activities of daily living; HRR = heart rate reserve; RM = repetition maximum.

participate in a resistance training program that is individually designed. Those with hypertension or arthritis or at risk of osteoporotic fracture need to be evaluated by a physician prior to initiating a resistance training program (13, 33).

Free weights can be used if individual supervision is provided. However, exercises with free weights have an inherent stability component that may make them difficult for some elderly persons to perform. If no weight equipment is available, milk jugs filled with sand, cans of food, or elastic bands can be used for resistance (33, 106). Proper technique is important to limit the risk of injury. Correct breathing patterns should be taught so that participants avoid the **Valsalva** maneuver (i.e., breath holding), which can reduce venous return and increase blood pressure. At least 48 hr of rest should occur between sessions for optimal recovery to reduce the risk of injury (33, 62, 83, 109).

A standard program (see chapter 7) should be followed: two to three sets of a selected weight that can be lifted 8 to 12 times (33, 62). Increasing the number of repetitions to 20 and decreasing the amount of resistance lifted may promote muscle endurance (33). The resistance training session should last between 20 and 30 min. Progress should be monitored and progression of intensity/resistance can occur every 2 to 3 weeks.

Range of Motion

The objective of ROM exercise is to increase or maintain joint ROM and decrease stiffness. This is important for the elderly, because flexibility is lost with age. In addition to traditional ROM exercises, activities such as yoga, dancing, and t'ai chi can be performed (17). ROM exercises should be performed at least three times per week but preferably every day.

Prior to performing an ROM routine, the participant should begin with a warm-up period of light intensity aerobic-type exercise. ROM exercises should then focus on all of the major joints of the body (i.e., hip, back, shoulder, knee, trunk, and neck region) with increases in the degree of stretch as an individual can tolerate. Chapter 7 reviews general recommendations.

Conclusion

More than half of elderly people have at least one disability or chronic condition. Participation in a regular physical activity/exercise program has many physiological health benefits including reducing the risk and lessening the impact of many chronic diseases. Physical activity of light to moderate intensity helps to improve health, whereas moderate- to high-intensity physical activity with an emphasis on aerobic endurance improves cardiorespiratory function ($\dot{V}O_2$) as well as health. Elderly individuals also demonstrate improvements during resistance training by increasing muscle mass and strength. This improves gait, balance, overall functional capacity, and bone health. There are also psychological benefits associated with regular physical activity and exercise. In general, the elderly person can improve physical and mental health by performing regular physical activity, and this should be encouraged by all medical and exercise professionals.

Case Study 30

Medical History

Mrs. KA is a 70-year-old white female who has a history of falling and osteoarthritis in her hands. She has been a housewife for most of her life and has never performed any type of regular physical activity. She lives by herself and has smoked half a pack of cigarettes a day for the past 40 years. Her body mass index is 30; her blood pressure is 140/76 mmHg.

Diagnosis

Mrs. KA went to see her doctor with complaints about her inability to complete normal daily living tasks such as carrying bags, walking around the block without worry of falling, and going up and down her basement stairs. She was unable to stand from a chair of standard height without using arms and unable to step up onto a 23-cm step. She also indicated that she feels lonely during most

continued

days of the week. A resting electrocardiogram showed no abnormalities. No significant diagnostic findings were made. The physician recommended she begin an exercise routine and ordered an exercise test.

Exercise Test Results

The patient underwent an exercise test. Her electrocardiogram demonstrated isolated premature ventricular contractions and normal ST segments at peak exercise. The following values were noted:

Resting heart rate = 84

Resting blood pressure = 138/80

Peak heart rate = 145

Peak blood pressure = 220/90

Estimated $\dot{V}O_2$peak = 13 ml $\cdot$ kg^{-1} $\cdot$ min^{-1}

Development of Exercise Prescription

Mrs. KA wishes to become stronger so she is able to perform general ADLs and remain independent. In discussion with her physician, she was referred to the local YMCA to join the elderly adult fitness program. She went somewhat reluctantly, and after consultation with the exercise professional on staff she set a goal of becoming more active on a regular basis to maintain her independent living status.

Case Study 30 Discussion Questions

1. What is an appropriate exercise activity and exercise intensity for Mrs. KA?

2. What are the major concerns for Mrs. KA during exercise activity?

3. Based on Mrs. KA's medical history and living situation, what are important barriers to her continued participation in the physical activity program?

4. What special precautions were likely taken during her exercise test?

5. What motivational strategies can be used to ensure her exercise adherence?

6. What are the major risks of Mrs. KA's participation in an exercise program?

7. What other tests might you perform on Mrs. KA (diagnostic or functional)?

8. What might be appropriate recommendations to progress Mrs. KA to walking three times a week for 30 min or on a daily basis?

Glossary

β-receptor—A cell receptor that is activated by a β-agonist such as epinephrine, norepinephrine, or dopamine.

dementia—A progressive decline in mental function, in memory, and in acquired intellectual skills.

disability—Loss of physical function.

geriatrics—A branch of medicine that deals with the problems and diseases associated with elderly people (>65 years) and the aging process.

GLUT 4—Insulin-regulated glucose transporter responsible for the removal of glucose from blood and delivery to the inner cell membrane.

hypertrophy—An increase in cell size.

metabolic equivalent (MET)—An expression of the rate of energy expenditure at rest. 1 MET = 1 kcal $\cdot$ kg^{-1} $\cdot$ hr^{-1} = 3.5 ml $\cdot$ kg^{-1} $\cdot$ min^{-1}.

moderate intensity physical activity—Activities that cause small increases in breathing and heart rate; ~50-70% $\dot{V}O_2$peak.

old age—Between 65 and 74 years of age.

oldest old—Older than 85 years of age.

power—Performance of work expressed per unit of time.

physical activity—Any bodily movement produced by skeletal muscles that results in energy expenditure.

Valsalva—An expiratory effort against a closed glottis that increases intrathoracic pressure and reduces venous return.

very old age—Between 75 and 84 years of age.

vigorous intensity physical activity—Activities that result in large increases in breathing and heart rate; >70% $\dot{V}O_2$peak.

References

1. American Association of Retired Persons. *A Profile of Older Americans.* 1990. Washington, DC: Author.

2. American College of Sports Medicine's Resource Manual for Guidelines for Exercise Testing and Prescription. 1998. 3rd ed. Baltimore: Williams & Wilkins.

3. American College of Sports Medicine. ACSM's Guidelines for Exercise Testing and Prescription. 2000. 6th ed. Baltimore: Lippincott Williams & Wilkins.

4. Anniansson A, Grimby G, Hedberg M. Compensatory muscle fiber hypertrophy in elderly men. J Appl Physiol 1992; 73: 812-816.

5. Barry HC, Eathorne SW. Exercise and aging. Issues for the practitioner. Med Clin North Am 1994; 78: 357-376.

6. Bayles C. Frailty. In American College of Sports Medicine's Exercise Management for Persons With Chronic Diseases and Disabilities. 1997. J.J. Durstine (Ed.), pp. 112-118. Champaign, IL: Human Kinetics.

7. Binder EF, Birge SJ, Spina R, Ehsani AA, Brown M, Sinacore DR, Kohrt WM. Peak aerobic power is an important component of physical performance in older women. J Gerontol 1999; 54A: M353-356.

8. Birdt TA. Alzheimer's disease and other primary dementia. In Harrison's Principles of Internal Medicine. 1998. A.S. Fauci et al. (Eds.), pp. 2348-2356. New York: McGraw-Hill.

9. Blackman DK, Kamimoto LA, Smith SM. Overview: Surveillance for selected public health indicators affecting older adults United States. MMWR 1999; 48(SS08): 1-6.

10. Blair SN, Kampert, JB, Kohl HW, et al. Influences of cardiorespiratory fitness and other precursors on cardiovascular disease and all-cause mortality in men and women. JAMA 1996; 276: 205-210.

11. Blazer DG. Social support and mortality in an elderly community population. Am J Epidemiol 1982; 115: 684-694.

12. Bloomfield SA. Changes in musculoskeletal structure and function with prolonged bed rest. Med Sci Sports Exerc 1997; 29: 197-206.

13. Bloomfield SA. Bone, ligament and tendon. In Perspectives in Exercise Science and Sports Medicine. 1995. C.V. Gisolfi, D.R. Lamb, E. Nadel (Eds.), pp. 175-227. Carmel, IN: Cooper.

14. Boult C, Kane RL, Loius TA, Boult L, McCaffrey D. Chronic conditions that lead to functional limitation in the elderly. J Gerontol 1994; 49: M28-M36.

15. Brown AB, McCartney N, Sale DG. Positive adaptations to weight-lifting training in the elderly. J Appl Physiol 1990; 69: 1725-1733.

16. Buchner DM, Cress ME, Wagner EH, de Lateur BJ, Price R, Abrass IB. The Seattle FICSIT/Move It study: The effect of exercise on gait and balance in older adults. J Am Geriatr Soc 1993; 41: 321-325.

17. Campbell WW, Crim MC, Dallal GE, Young VR, Evans WJ. Increased energy requirements and changes in body composition with resistance training in older adults. Am J Clin Nutr 1994; 60: 167-175.

18. Chodzko-Zajko WJ, Moore KA. Physical fitness and cognitive function in aging. Exerc Sport Sci Rev 1994; 22: 195-220.

19. Colantonio A, Kasl SV, Ostfeld AM, Berkman LF. Psychosocial predictors of stroke outcomes in an elderly population. J Geront Biol Sci 1993; 48: S261-S268.

20. Cononie CC, Goldberg AP, Rogus E, Hagberg JM. Seven consecutive days of exercise lowers plasma insulin responses to an oral glucose challenge in sedentary elderly. J Am Geriatr Soc 1994; 42: 394-398.

21. Cox JH, Cortright RN, Dohm GL, Houmard JA. Effect of aging on response to exercise training in humans: Skeletal muscle GLUT-4 and insulin sensitivity. J Appl Physiol 1999; 86: 2019-2025.

22. Cress EM, Buchner DM, Questad KA, Esselman PC, deLateur BJ, Schwartz RS. Exercise: Effects on physical functional performance in independent older adults. J Gerontol 1999; 54A: M242-M248.

23. Davis JW, Ross PD, Nevitt MC, Wasnich RD. Risk factors for falls and for serious injuries on falling among older Japanese women in Hawaii. J Am Geriatr Soc 1999; 47: 792-798.

24. Dempsey JA, Seals DR. Aging, exercise and cardiopulmonary function. In Perspectives in Exercise Science and Sports Medicine. 1995. C.V. Gisolfi, D.R. Lamb, E. Nadel (Eds.), pp. 237-298. Carmel, IN: Cooper.

25. Desai MM, Zhang P, Hennessy CH. Surveillance for morbidity and mortality among older adults—United States, 1995-1996. MMWR 1999; 48(SS08): 7-25.

26. DiPietro L, Caspersen CJ, Ostfeld AM. A survey for assessing physical activity among older adults. Med Sci Sports Exerc 1995; 25: 628-642.

27. Doherty TJ, Vandervoort AA, Taylor AW, Brown WF. Effects of motor units losses on strength in older men and women. J Appl Physiol 1993; 74: 868-874.

28. Ehsani AA, Ogawa T, Miller TR, Spina RJ, Jilka SM. Exercise training improves left ventricular systolic function in older men. Circulation 1991; 83: 96-103.

29. Engels H-J, Drouin J, Zhu W, Kazmierski JF. Effects of low-impact, moderate-intensity exercise training with and without wrist weights on functional capacities and mood status on older adults. Gerontology 1998; 44: 239-244.

30. Erikssen G, Liestol K, Bjornholt J, Thaulow E, Erikssen J. Changes in physical fitness and changes in mortality. Lancet 1998; 352: 759-762.

31. Ernst JM, Cacioppo JT. Lonely hearts: Psychological perspectives on loneliness. Appl Prev Psychol 1999; 8: 1-22.

32. Evans WJ. Exercise training guidelines for the elderly. Med Sci Sports Exerc 1999; 31: 12-17.

33. Ferketich AM, Kirby TE, Alway SE. Cardiovascular and muscular adaptations to combined endurance and strength training in elderly women. Acta Physiol Scand 1998; 164: 259-267.

34. Fiatarone MA, O'Neill EF, Ryan ND, Clements KM, Solares GR, Nelson ME, Roberts SB, Kehaias JJ, Lipsizt LA, Evans WJ. Exercise training and nutritional supplementation for physical frailty in very elderly people. New Engl J Med 1994; 330: 1769-1775.

35. Fiatarone MA, Marks EC, Ryan ND, Meredith CN, Lipsitz LA, Evans WJ. High-intensity strength training in nonagenarians. Effects on skeletal muscle. JAMA 1990; 263: 3029-3034.

36. Fleg JL, O'Connor F, Gerstenblith G, Becker LC, Clulow J, Schulman SP, Lakatta EG. Impact of age on the cardiovascular response to dynamic upright exercise in healthy men and women. J Appl Physiol 1995; 78: 890-900.

37. Fleg JL, Lakatta EG. Role of muscle loss in the age-associated reduction in VO_2max. J Appl Physiol 1988; 65: 1147-1151.

38. Ford SE, Ahluwalia IB, Galuska D. Social relationships and cardiovascular disease risk factors: Findings from the third national health and nutrition examination survey. Prev Med 2000; 30: 83-92.

39. Frontera WR, Meredith CN, O'Reilly KP, Knuttgen HG, Evans WJ. Strength conditioning in older men: Skeletal muscle hypertrophy and improved function. J Appl Physiol 1988; 64: 1038-1044.

40. Galganski ME, Fuglevand AJ, Enoka RM. Reduced control of motor output in a human hand muscle of elderly subjects during submaximal contractions. J Neurophysiol 1993; 69: 2108-2115.

41. Gallagher D, Ruts E, Visser M, Heshka S, Baumgartner RN, Wang J, Pierson RN, Pi-Sunyer FX, Heymsfield SB. Weight stability masks sarcopenia in elderly men and women. Am J Physiol Endocrinol Metab 2000; 279: E366-E375.

42. Goran MI, Poehlman ET. Total energy expenditure and energy requirements in healthy elderly persons. Metabolism 1992; 41: 744-753.

43. Heath G. Exercise programming for older adults. In American College of Sports Medicine's Resource Manual for Guidelines for Exercise Testing and Prescription. 1998, 3rd ed., pp. 516-520. Baltimore: Williams & Wilkins.

44. House JS, Landis KR, Umberson D. Social relationships and health. Science 1988; 241: 540-545.

45. Hoyert DL, Kochanek KD, Murphy SL. Deaths: Final Data for 1997. National Vital Statistics Reports. 1999. Washington, DC: U.S. Department of Health and Human Services, Centers for Disease Control and Prevention, National Center for Health Statistics, National Vital Statistics Systems.

46. Hunter GR, Treuth MS, Weinsier RL, Kekes-Szabo T, Kell SH, Roth DL, Nicholson C. The effects of strength conditioning on older women to perform daily tasks. J Am Geriatr Soc 1995; 43: 756-760.

47. Kahn SE, Larson VG, Beard JC, Cain KC, Fellingham GW, Schwarts RS, Veith RC, Stratton JR, Cerqueira MD, Abrass IB. Effect of exercise on insulin action, glucose tolerance and insulin secretion in aging. Am J Physiol 1990; 258: E937-E943.

48. Kamimoto LA, Easton AN, Maurice E, Husten CG, Macera CA.. Surveillance for five health risks among older adults—United States, 1993-1997. MMWR 1999; 48(SS08): 89-130.

49. Kenney WL. Thermoregulation at rest and during exercise in healthy older adults. Exerc Sport Sci Rev 1997; 25: 41-76.

50. Kenney WL. The older athlete: Exercise in hot environments. Sport Science Exchange 1993; 6: 44.

51. King AC, Martin JE. Physical activity promotion: Adoption and maintenance. In American College of Sports Medicine's Resource Manual for Guidelines for Exercise Testing and Prescription. 1998, 3rd ed., pp. 564-569. Baltimore: Williams & Wilkins.

52. King AC, Haskell WL, Taylor CB, Kraemer HC, DeBusk RF. Group vs home-based exercise training in healthy older men and women. A community-based clinical trial. JAMA 1991; 266: 1535-1542.

53. Klitgaard H, Mantoni M, Schiaffino S, Ausoni S, Gorza L, Laurent-Winter C, Schnohr P, Saltin B. Function, morphology and protein expression of aging skeletal muscle: A cross-sectional study of elderly men with different training backgrounds. Acta Physiol Scand 1990; 140: 41-54.

54. Koch M, Gottschalk M, Baker DI, Palumbo S, Tinetti ME. An impairment and disability assessment and treatment protocol for community-living elderly persons. Phys Ther 1994; 74: 286-294.

55. Kohrt WM, Malley MT, Coggan AR, Spina RJ, Agawa T, Ehsani AA, Bourey RE, Martin III WH, Holloszy JO. Effects of gender, age, and fitness level on response of VO_2max to training in 60-71 year olds. J Appl Physiol 1991; 71: 2004-2011.

56. Kuczmarski RJ, Flegal KM, Campbell SM, Johnson CL. Increasing prevalence of overweight among U.S. adults. JAMA 1994; 272: 205-211.

57. Lakatta EG. Cardiovascular aging research: The next horizons. J Am Geriatr Soc 1999; 47: 613-625.

58. Lan C, Lai J, Chen S, Wong M. 12-month t'ai chi training in the elderly: Its effect on health fitness. Med Sci Sports Exerc 1998; 30: 345-351.

59. Lemmer JT, Hurlbut DE, Martel GF, Tracy BL, Ivey FM, Metter EJ, Fozard JL, Fleg JL, Hurley BF. Age and gender responses to strength training and detraining. Med Sci Sports Exerc 2000; 32: 1505-1512.

60. Manton KG, Corder LS, Stallard E. Estimates of change in chronic disability and institutional incidence and prevalence rates in the U.S. elderly population from the 1982, 1984, and 1989 national long term care survey. J Gerontol 1993; 48: S153-S166.

61. Mazzeo RS, Cavanagh P, Evans WJ, Fiatarone M, Hagberg J, McAuley E, Starzell J. Exercise and physical activity for older adults. Position stand. Med Sci Sports Exerc 1998; 30: 992-1008.

62. McMurdo MET, Rennie L. A controlled trial of exercise by residents of older people's home. Age Ageing 1993; 22: 11-15.

63. McMurdo MET, Burnett L. Randomized controlled trial of exercise in the elderly. Gerontology 1992; 38: 292-298.

64. Meredith CN, Frontera WR, Fisher WC, Hughes VA, Herland JC, Edwards J, Evans WJ. Peripheral effects of endurance training in young and old subjects. J Appl Physiol 1989; 66: 2844-2849.

65. Meijer EP, Westerterp KR, Verstappen FTJ. Effect of exercise training on total daily physical activity in elderly humans. Eur J Appl Physiol 1999; 80: 16-21.

66. Miller JP, Pratley RE, Goldberg AP, Gordon P, Rubin M, Treuth MS, Ryan AS, Hurley BF. Strength training increases insulin action in healthy 50- to 65-yr-old men. J Appl Physiol 1994; 77: 1122-1127.

67. Mills EM. The effect of low-intensity aerobic exercise on muscle strength, flexibility, and balance among sedentary elderly persons. Nurs Res 1994; 43: 207-211.

68. Morey MC, Schenkman M, Studenski SA, Chandler JM, Crowley JM, Sullivan Jr. RJ, Pieper CF, Doyle ME, Higginbotham MB, Horner RD, MacAller H, Puglisi CM, Morris KG, Weinberger M. Spinal-flexibility-plus-aerobic versus aerobic-only training: Effects of a randomized clinical trial on function in at-risk older adults. J Gerontol 1999; 54A: M335-M342.

69. Morey MC, Cowper PA, Feussner JR, DiPasquale R, Crowley JM, Sullivan RJ. Two-year trends in physical performance following supervised exercise among community-dwelling older veterans. J Am Geriatr Soc 1991; 39: 549-554.

70. Morey MC, Cowper PA, Feussner JR, DiPasquale RC, Crowley JM, Kitzman DW, Sullivan RJ. Evaluation of a supervised exercise program in a geriatric population. J Am Geriatr Soc 1989; 37: 348-354.

71. Nelson ME, Fiatarone MA, Morganti CM, Trice I, Greenberg RA, Evans WJ. Effects of high-intensity strength training on multiple risk factors for osteoporotic fractures. JAMA 1994; 272: 1909-1914.

72. Nevitt MC, Cummings SR, Kidd S, Black D. Risk factors for recurrent non-syncopal falls. JAMA 1989; 261: 2663-2668.

73. Nichols JF, Omizo DK, Peterson KK, Nelson KP. Efficacy of heavy-resistance training for active women over sixty: Muscular strength, body composition, and program adherence. J Am Geriatr Soc 1993; 41: 205-210.

74. Nieman DC. Fitness and Sports Medicine. 1990. Palo Alto, CA: Bull.

75. Ogawa T, Spina RJ, Martin III WH, Kohrt WM, Schechtman K, Holloszy JO, Ehsani AA. Effects of aging, sex and physical training on cardiovascular responses to exercise. Circulation 1992; 86: 494-503.

76. Orth-Gomer K, Rosengren A, Wilhelmsen L. Lack of social support and incidence of coronary heart-disease in middle-aged Swedish men. Psychol Med 1993; 55: 37-43.

77. Orth-Gomer K, Johnson JV. Social network interaction and mortality. A six-year follow-up study of a random sample of Swedish population. J Chron Dis 1987; 40: 949-957.

78. Pate RR, Pratt M, Blair SN, Haskell WL, Macera CA, Bouchard C, Buchner D, Ettinger W, Heath GW, King AC, Kriska A, Leon AS, Marcus BH, Morris J, Paffenbarger Jr. RS, Patrick K, Pollock ML, Rippe JM, Sallis JF, Wilmore JH. Physical activity and public health. A recommendation from the Centers for Disease Control and Prevention and the American College of Sports Medicine. JAMA 1995; 273: 402-407.

79. Piers LS, Soares MJ, McCormack LM, O'Dea K. Is there evidence for an age-related reduction in metabolic rate? J Appl Physiol 1998; 85: 2196-2204.

80. Poehlman ET, Arciero PJ, Goran MI. 1994. Endurance exercise in aging humans: Effects on energy metabolism. Exerc Sport Sci Rev 1994; 22: 251-284.

81. Podsiadlo D, Richardson S. The timed "Up & Go": A test of basic functional mobility for frail elderly persons. J Am Geriatr Soc 1991; 39: 142-148.

82. Pollock ML, Franklin BA, Balady GJ, Chaitman BL, Fleg JL, Fletcher B, Limacher M, Pina IL, Stein RA, Williams M, Bazzarre T. Resistance exercise in individuals with and without cardiovascular disease. Benefits, rationale, safety and prescription. An advisory from the committee on exercise, rehabilitation and prevention, council on clinical cardiology, American Heart Association. Circulation 2000; 101: 828-833.

83. Prioux J, Ramonatxo M, Hayoit M, Mucci P, Prefaut C. Effect of aging on the ventilatory response and lactate kinetics during incremental exercise in man. Eur J Appl Physiol 2000; 81: 100-107.

84. Proctor DN, Balagopal P, Nair KS. Age-related sarcopenia in humans is associated with reduced synthetic rates of specific muscle proteins. J Nutr 1998; 128: 351S-355S.

85. Raab DM, Agre JC, McAdam M, Smith EL. Light resistance and stretching exercise in elderly women: Effect upon flexibility. Arch Phys Med Rehabil 1988; 69: 268-272.

86. Resnick NM. Geriatric medicine. In Harrison's Principles of Internal Medicine. 1998, A.S. Fauci et al. (Eds.). 14th ed., pp. 37-46. New York: McGraw-Hill.

87. Rimmer J. Alzheimer's disease. In American College of Sports Medicine's Exercise Management for Persons With Chronic Diseases and Disabilities. 1997. J.J. Durstine (Ed.), pp. 227-229. Champaign, IL: Human Kinetics.

88. Rohm Young D, Apple LJ, Jee S, Miller ER. The effects of aerobic exercise and t'ai chi on blood pressure in older people: Results of a randomized trial. J Am Geriatr Soc 1999; 47: 277-284.

89. Rowell LB. Human Cardiovascular Control. 1993. Oxford, UK: Oxford University Press.

90. Ryan AS, Craig LD, Finn SC. Nutrient intakes and dietary patterns of older Americans: A national study. J Gerontol 1992; 47: M145-M150.

91. Seeman TE, Kaplan GA, Knudsen L, Cohen R, Guralnik J. Social network ties and mortality among the elderly in the Alameda county study. Am J Epidemiol 1978; 126: 714-723.

92. Shephard RJ. Aging, Physical Activity and Health. 1997. Champaign, IL: Human Kinetics.

93. Shephard RJ. The scientific basis of exercise prescribing for the very old. J Am Geriatr Soc 1990; 38: 62-70.

94. Skinner JS. Importance of aging for exercise testing and exercise prescription. In Exercise Testing and Exercise Prescription for Special Cases. 1993. J.S. Skinner (Ed.), 2nd ed., pp. 75-86. Philadelphia: Lea & Febiger.

95. Spina RJ. Cardiovascular adaptations to endurance exercise training in older men and women. Exerc Sport Sci Rev 1999; 22: 317-332.

96. Spina RJ, Miller TR, Bogenhagen WH, Schechtman KB, Ehsani AA. Gender-related differences in left ventricular filling dynamics in older subjects after endurance exercise training. J Gerontol 1996; 51A: B232-B237.

97. Spina RJ, Ogawa T, Miller TR, Kohrt WM, Ehsani AA. Effect of exercise training on left ventricular performance in older women free of cardiopulmonary disease. Am J Cardiol 1993; 71: 99-104.

98. Spirduso WW. 1995. Physical Dimensions of Aging. Champaign, IL: Human Kinetics.

99. Steinbach U. Social networks, institutionalization, and mortality among elderly people in the United States. J Gerontol 1992; 47: S183-S190.

100. Stevens JA, Harbraouck LM, Durant TM, Dellinger AM, Batabyal PK, Crosby AE, Valluru BR, Kresnow M, Huerro JL. Surveillance for injuries and violence among older adults. MMWR 1999; 48(SS8): 27-50.

101. Stratton JR, Levy WC, Cerqueria MD, Schwartz RS, Abrass IB. Cardiovascular responses to exercise. Effects of aging and exercise training in healthy men. Circulation 1994; 89: 1648-1655.

102. Sugisawa H, Liang J, Liu X. Social networks, social support, and mortality among older people in Japan. J Gerontol 1994; 49: S3-S13.

103. Tanaka H, DeSouza CA, Jones PP, Stevenson ET, Davy KP, Seals DR. Greater rate of decline in maximal aerobic capacity with age in physically active vs. sedentary health women. J Appl Physiol 1997; 83: 1947-1953.

104. Tinetti ME, Baker DI, McAvay G, Claus EB, Garret P, Gottschalk M, Koch ML, Trainor K, Horwitz RI. A multifactorial intervention to reduce the risk of falling among elderly people living in the community. New Engl J Med 1994; 331: 821-827.

105. Tracy BL, Ivey FM, Hurlbut D, Martel GF, Lemmer JT, Siegel EL, Metter EJ, Fozard JL, Fleg JL, Hurley BF. Muscle quality II: Effects of strength training in 65- to 75-yr-old men and women. J Appl Physiol 1999; 86: 195-201.

106. Unchino BN, Cacioppo JT, Kiecolt-Glaser JK. The relationship between social support and physiological processes: A review with emphasis on underlying mechanisms and implications for health. Psychol Bull 1996; 119: 488-531.

107. U.S. Department of Health and Human Services. 1996. *Physical Activity and Health: A Report of the Surgeon General.* Atlanta: U.S. Department of Health and Human Services, Centers of Disease Control and Prevention, National Center for Chronic Disease Prevention and Health Promotion.

108. Withers RT, Smith DA, Tucker RC, Brinkman M, Clark DG. Energy metabolism in sedentary and active 49- to 70-yr-old women. J Appl Physiol 1998; 84: 1333-1340.

109. White TP. Skeletal muscle structure and function in older mammals. In Perspectives in Exercise Science and Sports Medicine. 1995. C.V. Gisolfi, D.R. Lamb, E. Nadel (Eds.), pp. 115-174. Carmel, IN: Cooper.

Farah A. Ramírez-Marrero, PhD
Associate Professor
Department of Physical Education and Recreation and HIV/AIDS
Research and Education Center
University of Puerto Rico, Río Piedras Campus
San Juan, Puerto Rico

Female-Specific Issues

Around the world, women's involvement in physical activity, regular exercise, and sports has increased during the past 25 years. An emphasis on the health benefits of physical activity and exercise is partly responsible for this change. Another reason is the influence of media attention to women's sports, especially in the United States with the passage of the Title IX legislation in 1972, which reduced societal prejudices and obstacles for females. Young girls now find sports-related female role models who are physically fit, strong, and competitive. Additionally, research confirms the multiple physiological and psychological benefits of increased physical activity, exercise, and sports participation among women of all ages.

As the number of women involved in vigorous exercise training and competition increases, so increases the incidence of menstrual cycle, reproductive function, and orthopedic disturbances among highly trained female athletes, compared with the female nonathletic or sedentary population. This chapter focuses on these physiological responses and reviews their relationship to exercise.

Menstrual cycle changes during intense training and competition were reported by elite female athletes as early as the 1950s and 1960s (20). Many female athletes happily welcomed the reduced frequency or cessation of their menses and justified it as a natural, harmless result of their intense exercise training. Health professionals believed that

© Human Kinetics

these changes were temporary and, therefore, were not concerned about the long-term health consequences. However, there is an increased risk of low **bone mineral density (BMD)** among young female athletes who experience the cessation of their menstrual cycle (10). In addition, many female athletes believe that a low body weight is a requirement for successful sports participation. This may be responsible for the large amount of disordered eating practices in females compared with males.

Disordered eating practices increase the risk of developing menstrual cycle disturbances and of bone demineralization. Reductions in BMD at a young age may never be regained and may result in premature **osteoporotic fractures**. In 1992, the American College of Sports Medicine (ACSM) Women's Task Force addressed the three interrelated medical disorders known as the female athlete triad. These medical disorders include **disordered eating** (e.g., anorexia nervosa, bulimia nervosa), **amenorrhea** (i.e., absence or suppression of menstruation), and **osteoporosis** (i.e., reduction in bone mass). The ACSM published a position stand in 1997 to establish that the female athlete triad is a serious syndrome that affects sports performance and causes

medical and psychological morbidity and mortality. The syndrome also affects some girls and women who participate in a wide variety of physical activities (1).

Amenorrhea

Physicians classify amenorrhea as primary or secondary, depending on the conditions in which it occurs. **Primary amenorrhea** is diagnosed when **menarche,** which is the initiation of menses, has not occurred by age 16, or no sexual development is present by age 14 (1, 45, 69). **Secondary amenorrhea** is generally diagnosed when menstrual bleeding has not occurred for at least three to six consecutive menstrual cycles in women who already had at least one previous menstruation (1, 69). Practical application 31.1 reviews the literature about amenorrhea and exercise.

Scope

All women who perform vigorous exercise training are at risk of developing menstrual cycle disturbance. Table 31.1 lists the various types of menstrual disturbances. Females participating in sports that require strenuous physical training, who also have a low body weight or lean body physique, are at an increased risk of developing menstrual cycle disturbances. The prevalence of amenorrhea in the general population is between 2% and 5%, whereas in women engaged in vigorous exercise training it is

TABLE 31.1	
Types of Menstrual Disturbances	
Type	**Definition**
Delayed menarche	Menstruation not started by age 16
Shortened luteal phase	Duration less than normal 10-16 days
Anovulatory cycles	Menstrual cycle without egg release
Oligomenorrhea	Irregular or inconsistent menstrual cycles
Amenorrhea	Complete cessation of menstrual cycle

Literature Review

In a series of elegant studies, Loucks and coworkers (38, 39, 41, 42) evaluated the independent and combined effect of exercise training and dietary restriction on the menstrual cycle. Low blood levels of T_3 (i.e., 3,5,3'-triiodothyronine) were used to identify energy deficiency. Although a low T_3 level is not associated with sickness, it accurately reflects thyroid function and is commonly used to detect hypothyroidism and starvation. Low T_3 levels are consistently observed among amenorrheic athletes compared with regularly menstruating athletes (37, 50).

In one study (38), the following six groups of women were compared: sedentary-energy balanced, sedentary-energy deprived, light exercise-energy balanced, light exercise-energy deprived, heavy exercise-energy balanced, and heavy exercise-energy deprived. A low T_3 level was found in each of the energy-deprived groups but not in the energy-balanced groups, regardless of exercise training status. It was suggested that dietary modifications to prevent energy deficiencies might prevent the low T_3 levels and also prevent athletic amenorrhea. These results suggest that it may not be necessary to modify the exercise training routine (i.e., reduce training volume) to restore a normal menstrual cycle.

A threshold of energy intake that abruptly induces T_3 suppression was identified to occur at energy intakes of 20 to 25 kcal $\cdot$ kg^{-1} of lean body mass per day (39). In a separate observation, when a low T_3 level was induced by exercise alone, without dietary restriction, the suppression of luteinizing hormone (LH) pulsatility was smaller than previous investigations instituting energy deficits by combining both methods. This suggests a possible protective effect of exercise against menstrual cycle dysfunction resulting from low energy availability (41). Aggressive 24-hr refeeding protocols in energy-deficient subjects increased T_3 levels without restoring LH pulsatility (42).

In summary, exercise by itself does not appear to trigger menstrual cycle disturbances, but it is part of the equation. Energy deficiency caused by increased exercise energy expenditure and reduced energy intake will compromise the energy available for reproductive functions and thyroid metabolism, will alter LH secretion, and likely will result in amenorrhea. Aggressive refeeding for a relatively long period of time (i.e., >24 hr) might be needed to completely reverse LH pulse suppression.

---

between 5% and 46% (26). Among elite runners and professional ballet dancers, the prevalence is 40% and 66%, respectively (45, 52).

The prevalence of menstrual cycle disturbance is likely underreported because abnormal luteal phases and **anovulatory** cycles are not always obvious, as these can occur even when menstrual bleeding is present. The following discussion focuses on the physiology and pathophysiology of amenorrhea.

Physiology and Pathophysiology

A normal menstrual cycle occurs when there is precise synchronization of hormonal events in the hypothalamus, anterior pituitary gland, and ovaries (practical application 31.2 and figure 31.1). The cascade of events regulating the menstrual cycle is initiated by the hypothalamus with the release of **gonadotropin releasing hormone** (GnRH). As the female matures, the hypothalamus begins consistent hourly bursts of GnRH secretion. At about age 10, the pituitary begins to progressively secrete more

follicle stimulating hormone and **luteinizing hormone,** which culminates with the initiation of monthly menstrual cycles beginning between ages 11 to 16 (i.e., menarche). The control and timing of these pulses determine the normal events of the menstrual cycle (25).

Athletic amenorrhea is diagnosed when the suppression of the menstrual cycle is attributed to very intense exercise training. This requires that pregnancy, abnormalities of the reproductive tract, failure of the ovaries, pituitary abnormalities, and tumors be ruled out as the cause (45). A proposed mechanism is an imbalance of the hypothalamic-pituitary-gonadal axis (36). Infertility is a consequence of menstrual cycle suppression. However, a more serious negative effect is reduced estrogen production, also known as **hypoestrogenism** (40). Low estrogen levels in premenopausal women are related to premature bone mineral loss and increased risk of stress fractures and osteoporosis (17).

Because the psychological and physiological profile of amenorrheic females is highly variable, it is

Normal Menstrual Cycle Events

Gland	Function
Hypothalamus	Starts secreting gonadotropin releasing hormone (GnRH) in a pulsatile fashion every 60-90 min
Pituitary gland	Secretes follicle stimulating hormone (FSH) and luteinizing hormone (LH) in a pulsatile fashion
Ovaries	Follicular phase — Approximately 14 days — Secretes estrogens (primarily estradiol) to stimulate the growth of the endometrial lining — A marked increase in estradiol at the end of the follicular phase may serve as a signal for the LH surge from the pituitary gland Ovulation: The LH surge from the pituitary gland stimulates the release of an egg (ovum) from the follicle Luteal phase — Approximately 10-16 days — The follicle transforms into a corpus luteum — Starts the secretion of progesterone to prepare the endometrium for the fertilized egg — If no egg is implanted, estrogen and progesterone levels drop and the endometrial lining degenerates — The drop in estrogen and progesterone levels may stimulate the hypothalamus to release GnRH and start the cycle again

Adapted from (6, 35, 53).

difficult to identify the precise causative factors of athletic amenorrhea. Possible causes are listed in table 31.2 (5, 21, 36, 40, 41, 45, 61, 73). The most popular and accepted of these explanations are low body fat (17), physical and psychological stress (40, 61), and low energy availability (41).

TABLE 31.2
Potential Causes of Athletic Amenorrhea

Potential cause	Effect
Low body weight or fat	Delayed menarche and amenorrhea
Intense training prior to menarche	Delayed menarche
Increased psychological stress	Increased endogenous opioids = depressed gonadotropin releasing hormone production
Low energy availability	Abnormal luteal phase

The low body fat hypothesis is based on early observations of amenorrheic runners who were more likely to have lower initial body weight for height, greater weight loss with onset of exercise training, and a lower percent body fat than regularly menstruating runners (15, 21, 62, 67, 72). It was suggested that those females with a body fat level below a critical value were at high risk of becoming amenorrheic (21).

The physical and psychological stress hypothesis suggests that the demands of sports training and competition stimulate the secretion of stress-related neurohormones including endogenous opioids (i.e., endorphins), dopamine, and cortisol. These neurohormones are believed to directly and indirectly alter GnRH production and secretion from the hypothalamus (5, 45, 63, 69), resulting in disruption of the menstrual cycle.

The low energy availability hypothesis suggests that a dietary energy deficit (i.e., consumption of fewer calories than required for a given basal metabolic rate and physical activity level) generates a signal that somehow disrupts the GnRH **pulsatile** secretion and results in an abnormal luteal phase of the menstrual cycle. If the energy deficit persists, the

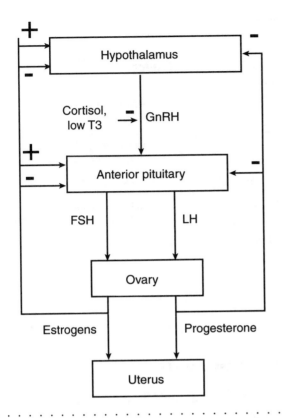

FIGURE 31.1   Hormonal feedback: Events that regulate the menstrual cycle. GnRH = gonadotropin releasing hormone; FSH = follicle stimulating hormone; LH = luteinizing hormone.

Reprinted, by permission, from A.J. Pearl, 1993, *Athletic Female* (Champaign, IL: Human Kinetics), 47.

body's natural reaction is to conserve essential physiological functions (e.g., cardiovascular, thermoregulation) and compromise the less critical functions (e.g., reproduction, growth) (31, 41, 42).

Disordered Eating

Losing body weight can be an ongoing, frustrating struggle for many women who constantly strive to achieve the "right" body shape for sport or for life. Abnormal eating behaviors are usually the result of attempts to reach an unrealistically low body weight or body fat and are defined as a subclass of multiple abnormal eating practices (69). These practices include fasting, diet pills, laxatives, diuretics, vomiting, and binge eating and are related to the disorders of anorexia nervosa and bulimia nervosa (1).

Anorexia nervosa is a psychological disorder characterized by an exaggerated pursuit of thinness or an obsessive fear of becoming obese. There is a constant energy deficit resulting in a considerable loss of body weight (i.e., at least 15% below ideal body weight) and fat and a refusal to maintain a minimally normal body weight for age and height (14, 31, 60). Females who are anorexic are at risk of developing amenorrhea. Amenorrhea is one of the diagnostic criteria cited by the American Psychiatric Association (DSM-IV) for anorexia nervosa (31).

Bulimia nervosa is characterized by recurrent episodes of binge eating, typically followed by purging, which may be self-induced vomiting, use of laxatives, or use of diuretics (14, 59, 65). People with bulimia experience a fear of becoming obese. However, unlike anorexia nervosa, bulimia may not be accompanied by weight loss. Even if purging does not occur, fasting or excessive exercise is usually present as a compensatory behavior (36, 67). Some experts believe that bulimia is simply a variant of anorexia nervosa.

Scope

The American College of Physicians has estimated that 10% to 15% of adolescent girls and young women have clinically evident anorexia or bulimia nervosa (14). Others have estimated that in adolescent and young adult women, the prevalence of anorexia nervosa and the prevalence of bulimia are approximately 1% and 2% to 4%, respectively (31).

Published reports suggest that female athletes have a similar or higher prevalence of disordered eating behaviors compared with the general female population, with a range of 15% to 62% for selected sports (1, 14, 41, 55). In contrast, an assessment of the prevalence of disordered eating among 1445 females competing in Division II of the National Collegiate Athletics Association revealed that no athlete met the established clinical criteria for anorexia nervosa and only 1.1% met the criteria for bulimia (30). However, this study suggested that female student-athletes have some clinical indicators associated with disordered eating behaviors (i.e., 9% bulimia, and 3% anorexia nervosa).

Some recognized sociocultural, biological, and psychological factors associated with the development of disordered eating behavior are presented in table 31.3. Ethnic factors may play a role, as Caucasians and Hispanics appear to be at a higher risk of disordered eating behaviors than African American adolescent females (55). If we consider the strong societal and cultural influence on human behavior, it is not difficult to understand how many young females connect thinness with success, beauty, and power. A successful and confident woman, as observed on television and in magazines, is represented by a tall and very thin, sometimes skeletal-looking

TABLE 31.3

Factors Associated With Disordered Eating Behavior

Sociocultural	Biological	Psychological	Sports related[a]
Thinness equals success	Abnormal neurochemistry controlling sense of hunger	Dysfunctional family	Pressure to reach a specific weight or low body fat to optimize performance
Thinness equals beauty		Physical/sexual abuse	Lack of nutritional knowledge
Thinness equals power		Low self-esteem	Emphasis on lean appearance that is evaluated by judges
		Lack of identity	
		Competitiveness[a]	
		Perfectionism[a]	Weight classification
		Compulsiveness[a]	

[a]Factors likely to prevail among female athletes.

female. Also, the need to be accepted and liked by family members and friends combined with low self-esteem may be a factor.

Female athletes with disordered eating behaviors are more likely to be highly competitive and have a perfectionist attitude. The particular sport in which they train and compete might also influence the development of disordered eating behaviors. For example, sports that require weight classifications, or sports in which the athlete's appearance is evaluated by a panel of judges, are more likely to include athletes at risk of developing eating disorders (31). High-risk sports for the development of disordered eating behavior include:

— Dance
— Figure skating
— Gymnastics
— Distance running
— Cycling
— Cross-country skiing
— Volleyball
— Swimming
— Diving
— Horse racing
— Martial arts
— Rowing

It is important to clarify that the sports are not arranged in any particular order (69).

Osteoporosis

Osteoporosis is defined as a decrease in bone density that enhances bone fragility and increases the risk of fracture (33, 69). Athletes who perform weight-bearing and weight-training activities have up to a 40% higher BMD compared with sedentary individuals (16). Female athletes with disordered eating and menstrual cycle disturbances are at a high risk of suffering serious consequences on bone health. Despite the mechanical loading effect of weight-bearing exercise training, female athletes who are amenorrheic tend to have a lower BMD, lose bone at an accelerated rate over time (e.g., 2-6% per year), and have a higher prevalence of stress fractures than both menstruating athletes and sedentary women (1, 69). The spinal BMD of young amenorrheic female athletes is similar to the spinal BMD of a 70- to 80-year-old postmenopausal women (19). Skeletal tissue responds to mechanical loading and hormone replacement therapy. Therefore, both regular physical activity and normal plasma estrogen levels are crucial for the maintenance of normal BMD (21). Further discussion and details on osteoporosis and exercise may be found in chapter 25.

There are many reports of low BMD in female athletes with menstrual cycle disturbances, but the actual prevalence of osteoporosis in young female athletes is not known (table 31.4). The reduced BMD in amenorrheic athletes is associated with an increased risk of stress fracture (19, 34, 72). An interesting case report of a nontraumatic femoral shaft fracture was reported in a 32-year-old white female

	TABLE 31.4		
Bone Mineral Density (BMD) in Female Athletes With Menstrual Cycle Disturbances			
Subjects	**Site**	**Observations**	**Authors**
38 amenorrheic athletes	Spinal bone	22-29% less bone mass than eumenorrheic athletes	Cann et al. (10)
14 amenorrheic athletes	Lower lumbar vertebrae	Lower BMD compared with 14 eumenorrheic athletes matched for age, height, weight, sport, dietary intake, and training program	Drinkwater et al. (17)
22 amenorrheic professional ballet dancers	Spine, wrist, metatarsals	BMD lower compared with eumenorrheic ballet dancers	Warren et al. (71)
29 amenorrheic athletes	Lumbar spine, femoral neck, trochanter, Ward triangle, intertrochanteric region, femoral shaft, tibia	BMD lower compared with 20 eumenorrheic athletes	Rencken et al. (54)

with **oligomenorrhea** during the 13th mile of a half marathon (19). This case report highlights the serious health implications of menstrual cycle disturbances for the competitive as well as the recreational female athlete.

The low BMD observed during an extended period of menstrual cycle disturbance may never return to normal, even with hormone replacement therapy or the resumption of a normal menstrual cycle. The use of exogenous estrogens and calcium supplementation in amenorrheic athletes may halt the loss of bone mass but has little effect on regaining lost bone density (18). Therefore, hormone replacement therapy is only useful for acute, short-term interventions because young female athletes are less likely to adhere to prolonged medical treatments. It is important to educate females who perform vigorous exercise training about the importance of regaining a normal menstrual cycle by increasing their caloric intake, reducing their training volume, or both. Formerly amenorrheic female athletes who decreased training volume by 10% to 20% and increased their body weight by 2% to 3% had a 6% increase of their BMD during the year immediately following menstrual cycle resumption (18).

The mechanical loading effect of exercise training on BMD is considered a protective factor. However, this effect is sport specific. It appears that the anatomical site receiving the direct impact caused by the sport-specific bodily movement receives most of the osteogenic benefit (56, 69, 66). Young female gymnasts, compared with young distance runners with similar menstrual irregularities, have a higher whole-body lumbar spine and femoral neck BMD (56). However, the interindividual variability (e.g., genetic predisposition for resisting or developing one or more of the female triad disturbances) and the identification of "high-risk sports" make it difficult to draw conclusions regarding the influence of genetics and exercise training on bone health status.

Clinical Considerations

Females with symptoms of one of the three components of the female athlete triad should be closely monitored for the other two because of their strong association. Once identified, the behavioral and neuroendocrine mechanisms of the female triad can be treated. Treatment should be preferably under the guidance of a multidisciplinary team that includes a physician, a dietitian, a psychologist, and an exercise professional. In cases involving athletes, the coaches and close family members should also be involved in the treatment plan. The team goal is to educate about harmful practices and their results, to help restore a normal energy-balanced diet and, if appropriate, to help resume a normal menstrual cycle. Female athletes are also confronted with other issues that might predispose them to different types of injuries and health consequences when no education is provided. Some of these issues are exercise training during pregnancy, control of body temperature, use of anabolic steroids, and orthopedic considerations.

Amenorrhea

All women who exercise vigorously and experience cessation of their menstrual cycle should have a clinical examination. The primary factors to be considered in the exam are presented in figure 31.2. It is important to understand that amenorrhea is likely the result of many factors that affect the normal hypothalamic function. Body weight and body fat reduction, as well as physical and emotional stress, are the most recognized factors of amenorrhea associated with increased physical activity levels.

Recommended laboratory tests for the diagnosis of amenorrhea are included in table 31.5 (45). It is important to remember that before amenorrhea can be diagnosed, one must rule out pregnancy and any abnormalities of the pituitary gland and reproductive tract.

If amenorrheic females, with appropriate counseling, are able to match total daily caloric intake with total daily energy expenditure, either by increasing caloric intake (e.g., 3-5%) or decreasing exercise training intensity (e.g., 10-20%), they usually regain a normal menstrual cycle within 2 or 3 months (18). If an active female is not willing to make these changes, then the recommended treatment should be (45, 52) as follows:

1. Hormone replacement therapy (e.g., 0.625 mg of conjugated estrogen on days 1-25 and progesterone on days 14-25) and increased calcium intake (e.g., 1200-1500 mg per day) to prevent osteoporosis and possibly coronary artery disease. If the woman is sexually active, the use of low-dose oral contraceptives to prevent bone loss is acceptable (e.g., <50 μg estrogen daily) (64).

2. **Bromocriptine** administration to reduce prolactin levels if hyperprolactinemia is diagnosed.

3. Nutritional and psychological counseling.

4. Reassessment every 3 to 6 months.

Menstrual cycle history

- Age at menarche
- Previous menstrual irregularities
- Use of birth control pills
- Pregnancy

Nutritional habits

- Vegetarian diet
- Low body weight
- Low body fat
- Eating disorders

Physical activity

- Sudden increases in frequency, duration, and intensity of exercise
- More than one exercise session a day

Social and emotional support

- Family relationships
- Relationships with friends
- Coping skills

Estrogen deficiency

- Vaginal dryness
- Dyspareunia
- Hot flashes

Androgen excess

- Polycystic ovarian syndrome
- Baldness
- Acne
- Facial, chest, abdominal hair

Others

- Galactorrhea
- Stress fractures
- Low bone density

FIGURE 31.2   Evaluation of athletic amenorrhea.

Adapted from L.A. Marshall, 1994, Clinical evaluation of amenorrhea. In *Medical and Orthopedic Issues of Active and Athletic Women*, edited by R. Agostini (Philadelphia: Hanley & Belfus), 152-163.

TABLE 31.5
. . . . .
Laboratory Tests for the Diagnosis of Amenorrhea

Test for	Normal values
Pregnancy	+ or –
Elevated thyroid-stimulating hormone	2-210 mU $\cdot$ L^{-1}
Reduced T$_3$ (3,5,3'-triiodothyronine)	1.2-1.5 nmol $\cdot$ L^{-1} or 110-230 ng $\cdot$ dl^{-1}
Elevated prolactin (mid-morning before breakfast)	3-19 ng $\cdot$ m^{-1}
Progestin challenge	Vaginal bleeding within 10 days of progesterone injection or oral medroxyprogesterone acetate
Follicle stimulating hormone (FSH)	>40 IU $\cdot$ L^{-1} or <5 IU $\cdot$ L^{-1} with no withdrawal bleeding after progestin challenge
Luteinizing hormone (LH)	6-30 ImU $\cdot$ ml^{-1}
Elevated LH/FSH	>2:1
Elevated testosterone	24-47 ng $\cdot$ dl^{-1}
Elevated dehydroepiandrosterone sulfate	50-400 ng $\cdot$ dl^{-1}
Bone mineral density	1.0-1.1 g $\cdot$ cm^{-2}

Disordered Eating

Female athletes with eating disorders are often in denial of their problem. They typically try to disguise their behavior by using baggy or loose clothing and avoid eating in front of other people. Some characteristic signs and symptoms of anorectics and bulimics are presented in table 31.6. These characteristics are divided into two categories: starvation in anorectics and purging in bulimics. However, women who suffer from anorexia or bulimia share many signs and symptoms, including criticism of their bodies, frequent mood swings, menstrual irregularities, and extreme preoccupation with food, calories, and body weight.

Serious medical complications can result as a consequence of a disordered eating behavior. Some of the most damaging health problems are caused by the abuse of laxatives and diuretics. These drugs will not help to reduce body fat but will cause an excessive loss of body water, leading to dehydration and electrolyte imbalance. A low potassium level (**hypokalemia**) can induce cardiac arrhythmias resulting in inadequate cardiac output and possibly death. Laxative abuse can also result in gastrointestinal complications like bleeding, loss of normal colonic peristalsis, and gastric dilation (53). Some recommended laboratory evaluations that aid in the diagnosis of disordered eating include these (31):

— Urinalysis to assess changes in pH (normal pH = 4.6-8.0)

— Complete blood count (e.g., screen for infection, malignancy, anemia, and any inflammatory process)

— Blood chemistry evaluation (e.g., electrolytes, liver, kidney, and thyroid functions)

— Electrocardiography if bradycardia (i.e., pulse <50 beats $\cdot$ min^{-1}) or electrolyte abnormalities are present. Look for the following electrocardiographic signs:

1. Low voltage (QRS amplitude <0.5 mV in all limb leads)

2. Low (<3 mm) or inverted T-waves (usually represents ischemia)

3. Prolonged QT intervals (normal = 0.34-0.42 s)

A family physician should be consulted to diagnose disordered eating behaviors and arrange for the appropriate treatment. Treatment for females with

TABLE 31.6

Disordered Eating in Female Athletes: Signs and Symptoms of Anorexia Nervosa (Starvation) and Bulimia (Purging)

Signs	Symptoms
ANOREXIA NERVOSA	
Bradycardia	Chronic fatigue
Hypotension	Lethargy
Hypothermia	Cold intolerance
Lanugo	Cold and discolored extremities
Dehydration	Dry hair
Electrolyte imbalance	Dry skin
Compulsive, excessive exercise	Constipation
Loss of fat and muscle	Abdominal pain
Extreme restriction of food intake	Diarrhea
Weight fluctuations over short periods	Lightheadedness
Anemia	Headaches
Frequent consumption of water and diet sodas	Lack of concentration
Amenorrhea	Stress fractures
Osteoporosis	Depression
BULIMIA	
Enlarged parotid glands	Chest or throat pain
Edema at face and extremities	Fatigue, headaches
Tooth enamel erosion	Abdominal pain, constipation, or diarrhea
Excessive use of laxatives	Bloating
Metacarpal calluses	Depression
Weight fluctuations	Hysterical behavior
Disappearing or making trips to the bathroom after eating large meals	

Adapted from Agostini et al. (3), and Pomeroy and Mitchell (53) disordered eating is intended to help normalize eating patterns and behavior, increase self-esteem, and improve communication about issues related to their condition. It is important to provide a comfortable, nonjudgmental environment where the female with disordered eating can get helpful information, direction, and support (14). If the female's eating patterns are not normal and she is energy deficient but otherwise healthy, nutritional counseling might be the only intervention needed. However, if the female is at risk of a serious eating, a stronger approach is desirable and both nutritional counseling and psychological counseling are recommended. Supervision of every meal might be initially required, and in extreme cases hospitalization might be the best option. Additional eating disorder information and questionnaires available over the Internet are presented in practical application 31.3.

Nutritional counseling should include evaluation of dietary intake (e.g., dietary recall questionnaire), assessment of exercise energy expenditure (e.g., physical activity questionnaire or accelerometry), and body composition assessment (e.g., skinfolds or circumferences) on a monthly or bimonthly basis. Female athletes or young girls at risk of eating disorders need assistance to determine a normal healthy body weight and percent body fat range and to perceive these as normal. One method commonly used is to have the individual sign a written contract to gain a goal amount of body weight per week. Intervention from a mental health professional is crucial to evaluate life stressors, strategies to cope with daily situations, and issues of self-esteem, all of which may be related to the underlying problem (31).

Exercise should be limited when electrolyte and electrocardiogram abnormalities are present. Physicians might consider hospitalization when the following signs are present: body weight 30% below normal, hypotension, dehydration, electrolyte abnormalities, severe psychological depression, and failure to respond to treatment. Approximately 20% of nonathletes treated for disordered eating behavior are unable to overcome their struggle with body image and weight control. A 19% mortality rate of anorexia nervosa patients is reported (53). The mortality rate of people with bulimia is not known. Suicide, cardiovascular failure, gastrointestinal perforations, or bloodstream infections are typically the immediate cause of death (31). However, the majority of patients with disordered eating behavior have the capacity to improve their condition with proper treatment and supervision (58).

Osteoporosis

Osteoporosis is a silent disease with little or no possibility of being detected in young women unless medical counseling is sought for related conditions such as menstrual cycle dysfunction, eating disorders, or stress fractures. The hypoestrogenic status of amenorrheic, energy-deficient females stimulates the process of bone demineralization. These weakened, porous bones are less resistant to the forces of intense and continuous skeletal muscle contractions. Therefore, energy-deficient and amenorrheic females are also at a higher risk of experiencing stress fractures, which is an indicator of premature osteoporosis.

PRACTICAL APPLICATION 31.3

Eating Disorders Information and Screening Available on the Internet

The questionnaires presented in the following Web sites are intended for educational purposes only. They do not constitute a professional evaluation. If a disordered eating behavior is suspected, please refer to a qualified healthcare professional for a thorough clinical evaluation.

— Eating Disorder Test (http://caringonline.com/eatdis/misc/edtest.htm)

— National Eating Disorder Screening Program (www.nmisp.org/eat.htm)

— Are You Dying to be Thin? A Questionnaire (www.medainc.org/questionnaire.htm)

— Eating Disorder Questionnaire from University of Iowa Hospitals (www. uihc.uiowa.edu/pubinfo/eatdis/quiz/htm)

— Eating Disorders Screening from Online Psych (www.allhealth.com/onlinepsych/menatlhealth/olpgen/0,6103,7123_127645,00.htm)

— Questionnaire by William Smedley, MD (http://caringonline.com/eatdis/misc/smedtest.htm)

Early detection of bone demineralization is difficult because not all young females with menstrual cycle disturbances will develop low bone mass density and subsequent signs and symptoms. Female athletes in particular should have their menstrual status and dietary habits evaluated regularly. They need to be educated about bone health status and to consult a healthcare professional when needed. If any signs or symptoms of a disordered eating behavior or menstrual irregularities are observed or suspected, bone density evaluations might be considered.

Measurements of BMD should be compared with same-age, -sex, and -race controls. Fractures are more likely to occur when BMD is two standard deviations below the normal value or below the 90th percentile of values from a like-group with osteoporotic fractures. Once diagnosed, proper treatment should be designed to recover normal menstrual cycles and estrogen levels. As previously mentioned, the female must increase dietary energy intake to balance her energy expenditure, including an increase in daily calcium intake (e.g., 1200-1500 mg · day^{-1}). Estrogen replacement therapy and oral contraceptives are also alternatives for those active females unwilling to change diet or exercise patterns. There are no guarantees that bone mass will recover back to normal levels. In fact, a more realistic goal would be to attenuate or stop the demineralization process. Follow-up evaluations are desirable to assess the effect of intervention in reversing osteoporosis in young female athletes.

Thermoregulation

Women's ability to control and regulate body temperature at rest and during exercise is influenced by many factors. Body surface area, body fat percentage, menstrual cycle status, physical fitness level, and thermoregulatory effectiveness (i.e., the ability to dissipate heat) are some of these factors. The average woman has more body fat, lower physical fitness levels, higher skin and core temperature, a higher heart rate, lower sweating rate, and later onset of sweating compared with the average man (6, 68). Many of these characteristics are beneficial during exercise in the cold (e.g., body fat, higher body temperature). However, during exercise in hot environments, these characteristics reduce the ability to regulate body temperature. However, a large body surface area to mass ratio does allow women to have a faster heat exchange between the skin and the environment (6).

The menstrual cycle is characterized by a follicular (preovulatory) and luteal phase (postovulatory).

Body core temperature during the luteal phase is 0.3 to 0.6° C higher compared with the follicular phase. Because of this, the threshold for the onset of sweating is set at a higher body temperature during the luteal phase. This difference is attributed to the higher level of progesterone secretion occurring after ovulation. Despite this, a woman's ability to tolerate physical activities in hot environments does not appear to be influenced by the phase of the menstrual cycle.

The physiological interrelationship between hormonal release during the menstrual cycle and the aforementioned sex-related thermoregulatory characteristics is not completely understood. However, it is clear that exposing women to exercise in hot environments improves thermoregulatory mechanisms. Both heat acclimation and physical training decrease the threshold of skin vasodilation and sweating response and increase sweat production in trained females (6). These improve the rate of heat dissipation. Physical training also improves exercise tolerance and results in lower core body temperature and heart rate at submaximal work rates.

Dehydration during exercise causes body temperature to increase more rapidly because of reduced sweat production. If sweat is not available at the skin surface, the cooling effect of sweat evaporation is absent, and the body's ability to dissipate heat is diminished. Dehydration also affects the cardiovascular system (i.e., reduced stroke volume and blood pressure, reduced maximal oxygen consumption), alters muscle metabolism, and reduces the intensity at which anaerobic threshold occurs (46, 48). Therefore, proper hydration before, during, and after exercise is extremely important (table 31.7).

Menstrual cycle disturbances do not seem to affect the thermoregulatory effectiveness of athletic females during exercise in hot environments. However, these women have a lower core body temperature at rest and delayed heat production during exercise in cold air compared with their regularly menstruating counterparts (3, 68). During exercise, amenorrheic women have a reduced heart rate response and a higher skin blood flow than eumenorrheic women, suggesting a reduction in heat storage by improved thermoregulatory responses (32). A lower progesterone level is postulated as the reason for the reduced cardiovascular strain and better heat dissipation observed in amenorrheic females at rest and during exercise. However, the mechanism for this is not understood.

Nonathletic females with disordered eating behaviors have reported symptoms of vasodilation and swelling of their hands and feet in warm temperatures. These symptoms are only temporary but tend

TABLE 31.7

· · · · · ·

Recommendations: Fluid and Other Nutrients

	Amount	Timing
Fluids: Should be cooler than ambient temperature (15-22° C). Flavored drinks might enhance palatability and promote voluntary drinking.	**BEFORE EXERCISE** 17 oz 8-16 oz	2 hr before 10-15 min before
	DURING EXERCISE 3-10 oz	Every 10-20 min
	AFTER EXERCISE At least a pint for every pound of body weight lost	Unrestricted
	Amount	**Notes**
Calcium intake	800-1200 mg · day^{-1} if normally menstruating 1200-1500 mg · day^{-1} if amenorrheic	Calcium carbonate is the recommended supplement
Iron intake	From animal and vegetable sources. Vitamin C helps absorption.	No supplements recommended Could become toxic

Note. Monitor diet to enhance adequate intake of vitamins, minerals, proteins, and calories.

Adapted from Murray (49)

---

to be uncomfortable for the individual. This is especially true when the person performs physical activity or exercise. These symptoms are not as common in athletic females with disordered eating behavior.

Pregnancy

Exercise training during pregnancy is a common practice. Scientific evidence suggests that physical activity of moderate intensity improves maternal and fetal well-being. Women who continue to exercise during pregnancy usually experience the following (2, 9, 13, 27):

— Improved cardiovascular function

— Limited weight gain and body fat retention

— Improved digestion

— Reduced constipation

— Reduced back pain

— Improved attitude and mental state

— Easier labor or reduction in possible complications during labor

— Reduced odds of cesarean delivery

— Faster recovery

— Better fitness level

Moreover, moderate exercise appears to reduce diastolic blood pressure in pregnant women at risk of hypertension (74). The offspring of exercising pregnant women may have a reduced fat cell growth rate without compromising other body cell growth, a high stress tolerance, and an advanced neurological developmental rate (8, 13).

Regular physical training and pregnancy often influence the same physiological variables, resulting in an additive effect. For example, both exercise training and pregnancy improve cardiovascular fitness level (i.e., increased maximal oxygen consumption), cardiac output, heart rate response to exercise, total blood volume, red blood cell volume, and ventilatory response to exercise. Many of these adaptations are associated with increased total body weight, amount of metabolically active tissue, and hormonal changes that occur during pregnancy.

The conservative point of view emphasizes that exercise during pregnancy is potentially harmful. Medical and safety concerns regarding pregnancy and exercise are presented in table 31.8. However, it is important to point out the lack of scientific evidence to support a cause-and-effect relationship between exercise and pregnancy-related problems such as **ectopic pregnancies,** spontaneous abortion, fetal or placenta abnormalities, premature membrane rupture, or premature labor. Moreover, the quality and quantity of breast milk are not altered by regular aerobic exercise during pregnancy or postpartum.

Light to moderate physical activity (i.e., intensity <60% $\dot{V}O_2$max, duration 20-30 min) is recommended for pregnant women who have not previously engaged in vigorous exercise training. There is no evidence that running, jumping, or increasing body temperature causes abnormal or difficult pregnancies. However, a general consensus is that pregnant women should avoid starting an intense exercise training program because of possible risk of injuries and compromised cardiovascular function. Exercise should also be modified or ceased if a pregnant woman experiences any of the following (2, 12):

— Calf pain or swelling of ankles, hands, or face
— Acute illness
— Decreased fetal movement
— Vaginal bleeding in early pregnancy
— Persistent nausea or vomiting, dizziness, and headaches
— Chest pain or excessive palpitations
— Sudden onset of abdominal or pelvic pain

Regular physical activity during pregnancy is safe when all necessary precautions are taken to ensure the safety of both the pregnant women and the fetus. It is important to continuously monitor the comfort level of the pregnant woman, rest periods, nutrition, body position, and hydration during exercise training. Both exercise in the supine position after the first semester and dehydration can reduce the woman's cardiac output, fetal blood flow, and adequate nutrients to the baby. Each of these has the potential to adversely affect the normal process of growth and development. Pregnant women should avoid any indicators of dehydration (e.g., thirst, weight loss 0.8-1.0 kg) and consume enough calories (e.g., approximately 300 kcal per day above normal) to support the metabolic demands of both pregnancy and exercise (12, 66, 72). To ensure proper

hydration, pregnant women should follow the recommendations presented in table 31.7.

Maximal exercise testing during pregnancy is usually avoided unless it is a medical necessity (2). A detailed exercise prescription is not necessary to provide a safety measure for most healthy, physically active, pregnant women (12). Despite this exercise prescription, recommendations are presented. For general health and cardiovascular fitness, three to four exercise sessions per week of 30 min or less in duration are considered safe for pregnant women (12, 72). The use of heart rates to monitor exercise intensity could present a problem because there is large individual variation in heart rate during pregnancy. Nonetheless, heart rates exceeding 150 beats · min^{-1} have been associated with reduced fetal blood flow (72). The use of rating of perceived exertion (e.g., Borg scale = 13-14) is considered more reliable than heart rate for monitoring exercise intensity during pregnancy. Pregnant women should never reach the point of fatigue or exhaustion during exercise, because the metabolic demand might shift blood away from the placenta, thus compromising fetal blood flow. Resistance training is usually considered safe as long as hydration, exercise intensity, and perceived exertion are carefully monitored. Range of motion or flexibility routines can be used as long as movements pressuring the abdominal area or placing pregnant women on their backs are avoided. These body positions may reduce fetal blood flow.

Anabolic Steroids

The use of **anabolic steroids** to improve strength, sports performance, and physical appearance has become popular among athletic females who have strong incentives to seek any competitive edge. Many young female athletes competing for athletic scholarships perceive the use of these drugs as an "investment" in their future (22). A prevalence of 0.5% to 2.0% of anabolic steroid use among female adolescents is reported (4). It appears that steroid use and abuse, particularly among the athletic population, has not declined in the United States despite many legal and educational efforts (75).

The general perception among many athletic females is that anabolic steroids will improve muscle strength and size, training intensity, and performance (44). The scientific evidence has shown that anabolic steroids potentially increase protein synthesis rate, muscle size, and muscle strength when used in combination with heavy resistance or strength training. Anabolic steroids can also benefit female athletes by reversing the catabolic effect of

TABLE 31.8

Exercise and Pregnancy: Medical and Safety Concerns for the Mother and Fetus

Mother		Fetus	
CONCERN	**SOLUTION**	**CONCERN**	**SOLUTION OR EFFECT**
Poor balance while running or jogging because of shifts in weight distribution and center of gravity.	Slow down, run cautiously, and never run alone.	Direct fetal trauma. Tissue and fluid surrounding fetus provide protection.	No scientific data.
Overheating and dehydration. Pregnancy elevates body core temperature by approximately 0.5° C, elevating resting metabolic rate by 15-20%. Excessive sweating might reduce blood volume.	Drink plenty of fluids before, during, and after exercise. Use appropriate exercise clothing and avoid exercise during extremely hot and humid weather.	Hyperthermia and reduced fetal blood flow. Might cause neural tube defects, growth retardation, reduced birth weight, or fetal abnormalities.	No scientific data.
Leg, hip, and abdominal pain. Reduced circulation to lower extremities during late pregnancy, extra weight to carry.	Stretch and warm up before any exercise session. Wear cushioned and comfortable shoes.	Reduced fetal blood flow.	No scientific data.
Nutrient availability. Pregnancy increases energy requirements by approximately 300 kcal · day^{-1}.	It is expected for pregnant women to gain 25-40 lb.	Substrate availability and hypoxia: reduced fetal glucose and oxygen availability. Might cause growth retardation, reduced birth weight, or fetal abnormalities.	No scientific data.
Reduced oxygen availability for aerobic exercise. Cardiovascular drift: added blood circulation to placenta.	Modify exercise intensity. Never exercise to the point of fatigue or exhaustion. Avoid intense and prolonged exercise. Monitor heart rate and rates of perceived exertion.	Reduced fetal blood flow. Intense exercise causes a redistribution of blood flow: more to muscles, less to other areas including the placenta.	Light to moderate physical activities are considered safe for mother and fetus.
Musculoskeletal injury. Ballistic movements and sudden postural changes can increase the risk of injury. However, the risk of injury for fit pregnant females should be lower.	Continuous/aerobic exercises are more acceptable than intermittent/anaerobic exercises.	Umbilical cord entanglement. Can reduce blood flow to important fetal organs.	No scientific data.

Note. A physician should be consulted before any exercise program is considered during pregnancy. Ask about contraindications to exercise and a list of high-risk sports to avoid during pregnancy.

stress-induced cortisol levels known to increase during intense exercise training and competition, particularly in those who are amenorrheic.

There are also many negative side effects reported among female anabolic steroid users. The continuous self-administration of multiple anabolic-androgens increases appetite, aggressiveness, irritability, sex drive, acne, body hair, and clitoral size among females (44). It also tends to deepen the voice. Other side effects include menstrual abnormalities, changes in lipoprotein profiles, and mental depression symptoms associated with anabolic steroid withdrawal (24, 44). Anabolic steroids induce menstrual cycle irregularities by interfering with the amplitude or daily total of GnRH pulses, leading to anovulatory cycles. These events can cause hypoestrogenism with the previously mentioned negative effects on bone mass and cardiovascular health. The abuse of anabolic steroids could predispose athletic women to premature **atherosclerosis** and other diseases.

Orthopedic Issues

The risk of sustaining a sports-related injury, such as a serious knee injury, is four to six times more likely in female, versus male, athletes participating in sports that require frequent quick stops, jumps, and cutting movements (28, 29, 51). For instance, in sports like handball, soccer, basketball, and alpine skiing, women experience proportionately more injuries to the anterior cruciate ligament compared with men (23, 51, 57). Most of these injuries occur in noncontact situations when the athlete lands from a jump, makes a lateral pivot while running, or perform a high-speed plant-and-cut movement (28, 51).

It is estimated that more than 30,000 knee injuries occur per year among intercollegiate and high school female athletes in the United States (28). The higher incidence of knee injuries observed in women's sports compared with men's is attributed to anatomical factors such as an increased pelvic width, poor alignment of lower extremity, knee instability, increased tibial rotation, and an increased **Q angle** (i.e., male = 13°, female = 18°) (11, 29). Other factors including weaker quadriceps musculature and hormonal changes may alter ligament compliance and bone strength (23, 29, 43).

Studies have looked at the relationship between the menstrual cycle and the risk of sports-related injuries among athletic females. Some have observed an increased risk of injury in female athletes who have irregular menstrual periods (7, 34). Others have found a higher risk of injury during the pre-

menstrual than the menstrual phase of the cycle (47, 51). The suggested mechanism is an increased ligamentous laxity and decreased neuromuscular performance associated with the fluctuation of female sex hormones throughout the menstrual cycle (28). Because oral contraceptives stabilize the hormonal levels during the menstrual cycle, they have been associated with a reduced risk of sports-related injury (28, 47). Moreover, the physiological and psychological symptoms associated with the menstrual cycle (e.g., discomfort, swelling, irritability) can affect fitness components such as balance and coordination and, therefore, increase the risk of injuries in the female sports participant.

Exercise Prescription

It is important to recognize that not all athletic women are underweight, underfat, and overstressed. Most athletic women do not experience abnormal reproductive function. However, it is true that some become obsessed with their weight and their body image. These women are most likely to become victims of one or more of the triad disorders, and their health will be at risk.

All female athletes should have a preparticipation physical examination that includes information about eating behavior, history of menstrual irregularities, history of stress fractures, mental depressive symptoms, pressure to lose weight, and history of exercise intensity, frequency, and duration. If there is any sign that the athlete is at risk of one or more of the triad disorders, a multidisciplinary team should be consulted (i.e., sport nutritionist, sport psychologist, team physician) for follow-up visits and more clinical evaluations prior to continued exercise training. Physical fitness testing is also appropriate to help the athletic female achieve specific goals and allow comparison with other normative data. Fitness testing should include measurement and evaluation of body composition, cardiorespiratory endurance, muscular strength and endurance, and flexibility. An exercise prescription for each of these areas should be developed for each individual. Chapter 7 reviews the general concepts of exercise prescription that apply to these women.

Exercise prescription for women with signs and symptoms of the female triad must include a determination of the proper amount of exercise training versus the amount of dietary intake so that energy balance is achieved. Athletic women should include resistance and flexibility training in their training program.

Conclusion

Participation in physical activities, exercise, and sports is important for all girls' and women's health and well-being. Those instances in which excessive exercise and abnormal eating patterns are used by females with the purpose of achieving an un-realistically low body weight and slim image need to be assessed and addressed. Eating disorders, amenorrhea, and osteoporosis are the components of the "athletic female triad" and represent a real threat for girls' and women's health. Sports training during pregnancy, thermoregulation, and the use of performance-enhancing drugs also represent poten-tial problems for female athletes. Clinical exercise professionals have the responsibility of educating, evaluating, and treating athletic girls and women so that success in sport and in life can be achieved.

Case Study 31

Medical History

Ms. VH is a 17-year-old Caucasian female athlete who was first evaluated at the Sports Health and Exercise Science Center located at the Olympic Center in Salinas, Puerto Rico. The team that handled this case consisted of a sport nutritionist, a family physician, a psychologist, and a psychiatrist. The athlete had never previously had a serious medical problem. Her height was 62 in. and weight 110 lb. She competed in the long jump and triple jump. Her typical training schedule consisted of two sessions of approximately 2 hr per day, 5 days per week of strength, aerobic, and specific event practice. She expended approximately 500 to 1000 kcal per day in training.

Diagnosis

The athlete initially was evaluated by her coach and then referred to the sports medicine clinic. A family physician, a psychologist, and a nutritionist met with the athlete. She complained of being consistently tired and often dizzy. She had been sleeping more than usual for about a year. This was also about the same time that she began to induce vomiting to control her weight but now vomiting comes "naturally." She also expressed feelings of depression. Her family apparently was not interested or involved in her life.

Test Results

A dietary recall analysis revealed that her caloric intake was about 750 kcal per day. Her blood tests showed high hematocrit and serum sodium levels.

Other

A clinical psychologist's evaluation indicated that this athlete had clinical depression.

Treatment

The athlete met with an exercise professional to educate her about how sport performance is impaired when not enough energy is available. A nutritionist counseled her and the athlete agreed to start a dietary plan of 1500 kcal per day for 2 weeks. She would then be reevaluated. She was also told about the increased risk of injury related to her current practices. Finally, Zoloft (an antide-pressant) was prescribed.

continued

Case Study 31   *(continued)*

Case Study 31   Discussion Questions

1. What other clinical problems is this athlete at risk for?
2. What other screening and clinical tests must be performed for a complete diagnosis and treatment plan?
3. What sports are most likely to lead females to develop this condition? Why?
4. What exercise/training recommendations should be given to this athlete?
5. What kind of specific nutritional counseling should be provided?

Glossary

amenorrhea—Absence or suppression of menstruation.

anabolic steroids—Testosterone derivatives, or steroid hormones resembling testosterone, which stimulate the building up of body tissue.

anorexia nervosa—Loss of appetite associated with intense fear of becoming obese.

anovulatory—Not accompanied with the discharge of an ovum.

atherosclerosis—The most common form of arteriosclerosis.

bone mineral density (BMD)—Measurement of skeletal mineral content by volume of bone in a region of interest.

bromocriptine—Ergot derivative that suppresses secretion of prolactin. Used to stimulate ovulation in the galactorrhea-amenorrhea syndrome.

bulimia—Disorder that includes recurrent episodes of binge eating, self-induced vomiting and diarrhea, excessive exercise, strict diet, and exaggerated concern about body shape.

disordered eating—Inappropriate eating behaviors leading to insufficient energy intake.

dysmenorrhea—Pain in association with menstruation.

ectopic pregnancy—Implantation of the fertilized ovum outside of the uterine cavity.

follicle-stimulating hormone—Hormone produced by the anterior pituitary to stimulate the growth of the follicle in the ovary and spermatogenesis in the testes.

gonadotropin-releasing hormone—Releasing hormone produced in the hypothalamus that acts on the pituitary and causes the release of gonadotropic substances, luteinizing hormone, and follicle-stimulating hormone.

hypoestrogenic—Decreased plasma estrogen levels.

hypokalemia—Extreme potassium depletion in the circulating blood.

luteinizing hormone—Hormone secreted by the anterior lobe of the pituitary to stimulate the development of the corpus luteum.

menarche—The beginning of menstrual function.

oligomenorrhea—Scanty or infrequent menstrual flow.

osteoporosis—Disease process that results in reduction of bone mass per unit of volume.

osteoporotic fracture—Broken bone caused by a reduction in the mass of the bone per unit of volume.

primary amenorrhea—Delay of menarche beyond age 18.

pulsatile—Characterized by a rhythmical pulsation.

Q angle—Acute angle formed by a line from the anterior superior iliac spine of the pelvis through the center of the patella and a line from tibial tubercle through the patella.

secondary amenorrhea—Cessation of menses in a woman who has previously menstruated.

References

1. American College of Sports Medicine. Position Stand on the Female Athlete Triad. *Med. Sci. Sports Exerc.*, 29: i-ix, 1997.

2. American College of Sports Medicine. *ACSM's Guidelines for Exercise Testing and Prescription.* 6th ed. Philadelphia: Lippincott Williams & Wilkins, 2000.

3. Agostini, R., B.L. Drinkwater, and M.D. Johnson. *The Female Athlete Triad* [Video]. ACSM's Hot Topics and Fundamentals of Sports Medicine Series: A Physician's Guide, 1996 (LDN: 476). Available from ACSM, P.O. Box 1440, Indianapolis, IN 46206-1440.

4. Bahrke, M.S., C.E. Yesalis, and K.J. Brower. Anabolic-androgenic steroid abuse and performance-enhancing drugs among adolescents. *Child Adolesc. Psychiatr. Clin. N. Am.*, 7(4): 821-838, 1998.

5. Bale, P. Body composition and menstrual irregularities of female athletes: Are they precursors of anorexia? *Sports Med.*, 17: 347-352, 1994.

6. Bar-Or, O. Thermoregulations in females from a life span perspective. In: *Exercise and the Female: A Life Span Approach*, O. Bar-Or, D.R. Lamb, and P.M. Clarkson (Eds). Carmel, IN: Cooper, 1996, Vol. 9, pp. 249-288.

7. Beckvid-Henriksson, G., C. Schnell, and A. Linden-Hirschberg. Women endurance runners with menstrual dysfunction have prolonged interruption of training due to injury. *Gynecol. Obstet. Invest.*, 49(1): 41-46, 2000.

8. Bell, R., and S. Palma. Antenatal exercise and birthweight. *Aust. N.Z. J. Obstet. Gynaecol.*, 40(1): 70-73, 2000.

9. Bungum, T.J., D.L. Peaslee, A.W. Jackson, and M.A. Perez. Exercise during pregnancy and type of delivery in nulliparae. *J. Obstet. Gynecol. Neonatal Nurs.*, 29(3): 258-264, 2000.

10. Cann, C.E., M.C. Martin, H.K. Genant, and R.B. Jaffe. Decreased spinal mineral content in amenorrheic women. *JAMA*, 251: 623-626, 1984.

11. Ciullo, J.V. Lower extremity injuries. In: *The Athletic Female*, A.J. Pearl (Ed.). Champaign: Human Kinetics, 1993, pp. 267-298.

12. Clapp, J.F. Exercise during pregnancy. In: *Exercise and the Female: A Life Span Approach*, O. Bar-Or, D.R. Lamb, and P.M. Clarkson (Eds.). Carmel, IN: Cooper, 1996, Vol. 9, pp. 413-451.

13. Clapp, J.F. Exercise during pregnancy. A clinical update. *Clin. Sports Med.*, 19(2): 273-286, 2000.

14. Clark, N. Eating disorders among athletic females. In: *The Athletic Female*, A.J. Pearl (Ed.). Champaign: Human Kinetics, 1993, pp. 141-157.

15. Dale, E., D.H. Gerlach, and A.L. Wilhite. Menstrual dysfunction in distance runners. *Obstet. Gynecol.*, 54: 47-53, 1979.

16. Dalsky, G.P. Effect of exercise on bone: Permissive influence of estrogen and calcium. *Med. Sci. Sports Exerc.*, 22: 281-285, 1990.

17. Drinkwater, B.L., K. Nilson, C.H. Chesnut, J. Bremner, S. Shainholtz, and M.B. Sothworth. Bone mineral content of amenorrheic and eumenorrheic athletes. *N. Engl. J. Med.*, 311: 277-281, 1984.

18. Drinkwater, B.L. Amenorrhea, body weight, and osteoporosis. In: *Eating, Body Weight and Performance in Athletes*, K.D. Brownell, J. Rodin, and J.H. Wilmore (Eds.). Philadelphia: Lea & Febiger, 1992, pp. 235-247.

19. Drinkwater, B.L. The female athlete triad. Presented at the 13th Annual Conference on Exercise Sciences and Sports Medicine. San Juan, Puerto Rico, March 1999.

20. Erdelyi, G.J. Gynecological survey of female athletes. *J. Sports Med. Phys. Fitness*, 2:174-179, 1962.

21. Frisch, R.E., and J.W. McArthur. Menstrual cycles: Fatness as a determinant of minimum weight for height necessary for their maintenance or onset. *Science*, 185: 949-951, 1974.

22. Gorman, C. Girls on steroids. Among young athletes, these dangerous drugs are all the rage. Is your daughter using them? *Time Magazine*, 152(6): 93, 1998.

23. Gray, J., J.E. Taunton, D.C. McKenzie, D.B. Clement, J.P. McConkey, and R.G. Davidson. A survey of injuries to the anterior cruciate ligament of the knee in female basketball players. *Int. J. Sports Med.*, 6(6): 314-316, 1985.

24. Gruber, A.J., and H.G. Pope. Psychiatric and medical effects of anabolic-androgenic steroid use in women. *Psychother. Psychosom.*, 69(1): 19-26, 2000.

25. Guyton, A.C. *Textbook of Medical Physiology.* 8th ed. Philadelphia: Saunders, 1991, pp. 899-914.

26. Harber, V.J. Menstrual dysfunction in athletes: An energetic challenge. *Exerc. Sport Sci. Rev.*, 28(1): 19-23, 2000.

27. Heffernan, A.E. Exercise and pregnancy in primary care. *Nurse Pract.*, 25(3): 42, 49, 53-56, 2000.

28. Hewett, T.E. Neuromuscular and hormonal factors associated with knee injuries in female athletes. Strategies for intervention. *Sports Med.*, 29(5): 313-327, 2000.

29. Hutchinson, M.R., and M.L. Ireland. Knee injuries in female athletes. *Sports Med.*, 19(4): 288-302, 1995.

30. Johnson, C., P.S. Powers, and R. Dick. Athletes and eating disorders: The National Collegiate Athletic Association study. *Int. J. Eat. Disord.*, 26(2): 179-188, 1999.

31. Johnson, M.D. Disordered eating. In: *Medical and Orthopedic Issues of Active and Athletic Women*, R. Agostini (Ed.). Philadelphia: Hanley & Belfus, 1994, pp. 141-151.

32. Kolka, M.A., and L.A. Stephenson. Effect of luteal phase elevation on core temperature on forearm blood low during exercise. *J. Appl. Physiol.*, 82: 1079-1083, 1997.

33. Lemcke, D.P. Osteoporosis and menopause. In: *Medical and Orthopedic Issues of Active and Athletic Women*, R. Agostini (Ed.). Philadelphia: Hanley & Belfus, 1994, pp. 175-182.

34. Lloyd, T., S.J. Triantafyllou, E.R. Baker, P.S. Houts, J.A. Whiteside, A. Kalenak, and P.G. Stumpf. Women athletes with menstrual irregularity have increased musculoskeletal injuries. *Med. Sci. Sports Exerc.*, 18(4): 374-379, 1986.

35. Loucks, A.B., and S.M. Horvath. Exercise-induced stress responses of amenorrheic and eumenorrheic runners. *J. Clin. Endocrinol. Metab.*, 59: 1109-1120, 1984.

36. Loucks, A.B. Effects of exercise training on the menstrual cycle: Existence and mechanisms. *Med. Sci. Sports Exerc.*, 22: 275-280, 1990.

37. Loucks, A.B., G.A. Laughlin, J.F. Mortola, L. Girton, J.C. Nelson, and S.S.C. Yen. Hypothalamic-pituitary-thyroid function in eumenorrheic and amenorrheic athletes. *J. Clin. Endocrinol. Metab.*, 75: 514-518, 1992.

38. Loucks, A.B., and R. Callister. Induction and prevention of low-T_3 syndrome in exercising women. *Am. J. Physiol.*, 264: R924-R930, 1993.

39. Loucks, A.B., and E.M. Heath. Induction of low-T_3 syndrome in exercising women occurs at a threshold of energy availability. *Am. J. Physiol.*, 266: R817-R823, 1994.

40. Loucks, A.B. The reproductive system. In: *Exercise and the Female: A Life Span Approach*, O. Bar-Or, D.R. Lamb, and P.M. Clarkson (Eds.). Carmel, IN: Cooper, 1996, Vol. 9, pp. 41-70.

41. Loucks, A.B., M. Verdun, and E.M. Heath. Low energy availability, not stress of exercise, alters LH pulsatility in exercising women. *J. Appl. Physiol.*, 84: 37-46, 1998.

42. Loucks, A.B., and M. Verdun. Slow restoration of LH pulsatility by refeeding in energetically disrupted women. *Am. J. Physiol.*, 275: R1218-R1226, 1998.

43. Lund-Hanssen, H., J. Gannon, L. Engebretsen, K. Holen, and S. Hammer. Isokinetic muscle performance in

healthy female handball players and players with a unilateral anterior cruciate ligament reconstruction. *Scand. J. Med. Sci. Sports*, 6(3): 172-175, 1996.

44. Malarkey, W.B., R.H. Strauss, D.J. Leizman, M. Liggett, and L.M. Demers. Endocrine effects in female weight lifters who self-administer testosterone and anabolic steroids. *Am. J. Obstet. Gynecol.*, 165: 1385-1390, 1991.

45. Marshall, L.A. Clinical evaluation of amenorrhea. In: *Medical and Orthopedic Issues of Active and Athletic Women*, R. Agostini (Ed.). Philadelphia: Hanley & Belfus, 1994, pp. 152-163.

46. Millard-Stafford, M., P.B. Sparling, L.B. Rosskopf, T.K. Snow, L.J. DiCarlo, and B.T. Hinson. Fluid intake in male and female runners during a 40-km field run in the heat. *J. Sports Sci.*, 13: 257-263, 1995.

47. Möller-Nielsen, J., and M. Hammar. Women's soccer injuries in relation to the menstrual cycle and oral contraceptive use. *Med. Sci. Sports Exerc.*, 21(2): 126-129, 1989.

48. Moquin, A., and R.S. Mazzeo. Effect of mild dehydration on the lactate threshold in women. *Med. Sci. Sports Exerc.*, 32(2): 396-402, 2000.

49. Murray, B. Fluid replacement: The American College of Sports Medicine Position Stand. *GSSI*, 9(4): i-vii, 1996.

50. Myerson, M., B. Gutin, M.P. Warren, M.T. May, I. Contento, M. Lee, F.X. Pi-Sunyer, R.N. Pierson, and J. Brooks-Gunn. Resting metabolic rate and energy balance in amenorrheic and eumenorrheic runners. *Med. Sci. Sports Exerc.*, 23: 15-22, 1991.

51. Myklebust, G., S. Maehlum, I. Holm, and R. Bahr. A prospective cohort study of anterior cruciate ligament injuries in elite Norwegian team handball. *Scand. J. Med. Sci. Sports* 8(3): 149-153, 1998.

52. Otis, C.L. Exercise-associated amenorrhea. *Clin. Sports Med.*, 11(2): 351-362, 1992.

53. Pomeroy, C., and J.E. Mitchell. Medical issues in the eating disorders. In: *Eating, Body Weight and Performance in Athletes*, K.D. Brownell, J. Rodin, and J.H. Wilmore (Eds.). Philadelphia: Lea & Febiger, 1992, pp. 202-221.

54. Rencken, M.L., C.H. Chesnut, and B.L. Drinkwater. Bone density at multiple skeletal sites in amenorrheic athletes. *JAMA*, 276: 238-240, 1996.

55. Rhea, D.J. Eating disorder behavior of ethnically diverse urban female adolescent athletes and non-athletes. *J. Adolesc.*, 22(3): 379-388.

56. Robinson, T.L., C. Snow-Harter, D.R. Taaffe, D. Gillis, J. Shaw, and R. Marcus. Gymnasts exhibit higher bone mass than runners despite similar prevalence of amenorrhea and oligomenorrhea. *J. Bone Miner. Res.*, 10: 26-35, 1995.

57. Roos, H., M. Ornell, P. Gardsell, L.S. Lohmander, and A. Lindstrand. Soccer after anterior cruciate ligament injury—an incompatible combination? A national survey of incidence and risk factors and a 7-year follow-up of 310 players. *Acta Orthop. Scand.*, 66(2): 107-112, 1995.

58. Root, M.P.P. Recovery and relapse in former bulimics. *Psychotherapy*, 27: 397-403, 1990.

59. Round table. Eating disorders in young athletes. *Phys. Sportsmed.*, 13: 89-106, 1985.

60. Sanborn, C.F., M. Horea, B.J. Siemers, and K.I. Dieringer. Disordered eating and the female athlete triad. *Clin. Sports Med.*, 19(2): 199-213, 2000.

61. Sanborn, C.F., and W.W. Wagner. The female athlete and menstrual irregularity. In: *Sport Science Perspectives for Women*, J. Puhl, C.H. Brown, and R.O. Voy (Eds.). Champaign, IL: Human Kinetics, 1988, pp. 111-130.

62. Schwartz, B., D.C. Cumming, E. Riordan, M. Selye, S.S. Yen, and R.W. Rebar. Exercise-associated amenorrhea: A distinct entity? *Am. J. Obstet. Gynecol.*, 141: 662-670, 1981.

63. Shangold, M.M. Sports and menstrual function. *Phys. Sportsmed.*, 8: 66-69, 1980.

64. Shangold, M.M., R.W. Rebar, A.C. Wentz, and I. Schiff. Evaluation and management of menstrual dysfunction in athletes. *JAMA*, 263:1665-1669, 1990.

65. Slavin, J.L. Eating disorders in women athletes. In: *Sport Science Perspectives for Women*, J. Puhl, C.H. Brown, and R.O. Voy (Eds.). Champaign, IL: Human Kinetics, 1988, pp. 189-197.

66. Slavin, J.L., J.M. Lutter, S. Cushman, and V. Lee. Pregnancy and exercise. In: *Sport Science Perspectives for Women*, J. Puhl, C.H. Brown, and R.O. Voy (Eds.). Champaign, IL: Human Kinetics, 1988, pp 151-160.

67. Speroff, L., and D.B. Redwine. Exercise and menstrual function. *Phys. Sportsmed.*, 8: 42-52, 1980.

68. Stephenson, L.A., and M.A. Kolka. Thermoregulation in women. *Exerc. Sport Sci. Rev.*, 21: 231-262, 1993.

69. Vereeke-West, R. The female athlete: The triad of disordered eating, amenorrhea and osteoporosis. *Sports Med.*, 26: 63-71, 1998.

70. Warren, M.P., J. Brooks-Gunn, L.H. Hamilton, W.L. Fiske, and W.G. Hamilton. Scoliosis and fractures in young ballet dancers. *N. Engl. J. Med.*, 314: 1348-1353, 1986.

71. Warren, M.P., J. Brooks-Gunn, R.P. Fox, C. Lancelot, D. Newman, and W.G. Hamilton. Lack of bone accretion and amenorrhea: Evidence for a relative osteopenia in weight-bearing bones. *J. Clin. Endocrinol. Metab.*, 72: 847-853, 1991.

72. Warren, M.P., and M.M. Shangold. *Sports Gynecology: Problems and Care of the Athletic Female*. Cambridge, UK: Blackwell Science, 1997, pp. 26-30, 113-136.

73. Wentz, A. Body weight and amenorrhea. *Obstet. Gynecol.*, 56: 482-487, 1980.

74. Yeo, S., N.M. Steele, M.C. Chang, S.M. Leclaire, D.L. Ronis, and R. Hayashi. Effect of exercise on blood pressure in pregnant women with a high risk of gestational hypertensive disorders. *J. Reprod. Med.*, 45(4): 293-298, 2000.

75. Yesalis, C.E., C.K. Barsukiewicz, A.N. Kopstein, and M.S. Bahrke. Trends in anabolic-androgenic steroid use among adolescents. *Arch. Pediatr. Adolesc. Med.*, 151(12): 1197-1206, 1997.

INDEX

Note: Page numbers followed by italicized f or t refer to the figure or table on that page, respectively.

ABOUT THE EDITORS

Jonathan Ehrman, PhD, is associate director of preventive cardiology at Henry Ford Hospital in Detroit. He has a 17-year background in clinical exercise physiology. He is certified as an ACSM exercise specialist and program director.

Dr. Ehrman has written numerous manuscripts and abstracts, and he edits a section in the journal *Clinical Exercise Physiology*. He is an American College of Sports Medicine fellow and a member of its certification committee. Dr. Ehrman earned his PhD in clinical exercise physiology from Ohio State University.

Paul Gordon, PhD, is an ACSM-certified exercise specialist who teaches graduate clinical exercise physiology courses in the School of Medicine at West Virginia University. He has directed several cardiopulmonary rehabilitation programs and served as an examiner and coordinator for the ACSM exercise specialist certification.

Dr. Gordon is an American College of Sports Medicine fellow. He earned his PhD in exercise physiology from the University of Pittsburgh.

Paul Visich, PhD, has been a clinical exercise professor for 10 years and currently teaches at Central Michigan University. He worked eight years in a clinical setting that included cardiac and pulmonary rehabilitation and primary disease prevention.

Dr. Visich is chair of the Professional Education Committee of the American College of Sports Medicine. He earned a PhD in exercise physiology and an MPH in epidemiology from the University of Pittsburgh.

Steven Keteyian, PhD, has more than 20 years of experience working as a clinical exercise physiologist. He is program director of preventive cardiology at the Henry Ford Heart and Vascular Institute.

Dr. Keteyian is an American College of Sports Medicine fellow and author of two college textbooks. He earned his PhD from Wayne State University.